Critical Properties of ϕ^4-Theories

Critical Properties of ϕ^4-Theories

Hagen Kleinert
Verena Schulte-Frohlinde

Freie Universität Berlin, Germany

World Scientific
New Jersey • London • Singapore • Hong Kong

Published by

World Scientific Publishing Co. Pte. Ltd.

P O Box 128, Farrer Road, Singapore 912805

USA office: Suite 1B, 1060 Main Street, River Edge, NJ 07661

UK office: 57 Shelton Street, Covent Garden, London WC2H 9HE

British Library Cataloguing-in-Publication Data
A catalogue record for this book is available from the British Library.

CRITICAL PROPERTIES OF ϕ^4-THEORIES

ISBN 981-02-4658-7
ISBN 981-02-4659-5 (pbk)

Printed in Singapore.

Preface

During the past 25 years, field theory has given us much understanding of critical phenomena. Development in this area was extremely rapid and has reached a certain degree of maturity. Perturbative calculations of the critical exponents in $4 - \varepsilon$ dimensions have been carried out to the level of five loops, in 3 dimensions to seven loops, with great effort. The resulting power series diverge and their evaluation requires resummation methods, which have been developed at the same time to a considerable degree of accuracy.

The present monograph started life as lecture notes for a course on quantum field theory delivered regularly by the first author (H.K.) at the Freie Universität Berlin since the early seventies. In 1989, two of his students, J. Neu and the second author (V. S.-F.) attended this course while working on their Master's thesis. Ambitiously, they undertook the arduous task of recalculating the five-loop renormalization constants of $O(N)$-symmetric ϕ^4-theories published by K.G. Chetyrkin, S.G. Gorishny, S.A. Larin, and S.V. Tkachov, and discovered several errors.[1] They traveled to Moscow to discuss their findings with the Russian authors who confirmed them after repeated checks. The correct results were subsequently published in a joint paper.[2] In her Ph.D. thesis, V. S.-F. extended the five-loop calculations to a mixture of interactions with $O(N)$ and cubic symmetry.[3] The complete five-loop results are contained in this book.

At present it would be extremely difficult to increase the number of loops in the exact calculations any further without injection of new ideas. We therefore believe it is time to put together the available field-theoretic techniques in this monograph, so that future workers on this subject may profit from it.

We are grateful to J. Neu for many discussions at an early stage in the preparation of this book until 1991. He wrote the computer program for enumerating the Feynman diagrams with the associated weight factors in Chapter 14. More recently, Dr. E. Babaev, Dr. A. Pelster, Dr. Pai-Yi Hsiao, Dr. C. Bevillier contributed with comments.

We would further like to thank Dr. J.A. Gracey for several useful communications on the large-n limits of the critical exponents, and to Drs. Butera and Comi for permission to use their high-temperature expansions in Chapter 20 which we made available as files on the internet pages of this book (http://www.physik.fu-berlin.de/~kleinert/re.html#b8).

Most importantly, we are indebted to Dr. B. Kastening for his intensive reading of the book. His corrections and useful suggestions greatly helped improve the final draft. Many printing and stylistic errors were pointed out by Dr. Annemarie Kleinert and by Jeremiah Kwok, our editor of World Scientific Publishing Company.

None of the above persons can, of course, be blamed for the errors introduced in the subsequent correction process.

---

[1] J. Neu, FU-Berlin MS Thesis, 1991, V. Schulte-Frohlinde, FU-Berlin MS Thesis, 1991.

[2] H. Kleinert, J. Neu, V. Schulte-Frohlinde, K.G. Chetyrkin, S.A. Larin, Phys. Lett. B *272*, 39 (1991) (hep-th/9503230); Erratum ibid. *319*, 545 (1993).

[3] V. Schulte-Frohlinde, FU-Berlin PhD Thesis, 1996. Results are published in H. Kleinert and V. Schulte-Frohlinde, Phys. Lett. B *342*, 284 (1995) (cond-mat/9503038).

All quoted papers published by our research group in Berlin can be downloaded from the internet, the more recent ones from the Los Alamos server (http://xxx.lanl.gov/find), the older ones from our local server (http://www.physik.fu-berlin.de/~kleinert/re0.html).

Finally, H.K. thanks his secretary Ms. S. Endrias for her invaluable help in finishing the book.

Hagen Kleinert and *Verena Schulte-Frohlinde*
Berlin, May 2001

Contents

List of Figures

List of Tables

1

Introduction

Systems containing a large number of particles exhibit a great variety of phase transitions. Most common are *first-* and *second-order transitions*. A transition is said to be first-order if the internal energy changes discontinuously at a certain temperature. Such a transition is accompanied by the release or absorption of *latent heat*. Important examples are melting and evaporation processes. Second-order transitions involve no latent heat, and the internal energy changes continuously with temperature. The derivative of the internal energy with respect to the temperature diverges at the transition temperature T_c, which is also called the *critical temperature*. The most important examples for materials undergoing second-order transitions are ferromagnets, superfluids, and superconductors.

There also exist phase transitions of higher order in which the first appearance of a divergence occurs in some higher derivative of the internal energy with respect to the temperature. A famous extreme example is the *Kosterlitz-Thouless transition* [1] of a Coulomb gas in two space dimensions. The same type of transition is also found in thin films of ^{4}He at temperatures of a few degrees Kelvin where the films become superfluid. In this transition, the internal energy may be differentiated any number of times with respect to the temperature and does not show any divergence. Instead, the temperature behavior exhibits an essential singularity of the form $e^{\text{const} \times (T-T_c)^{-1/2}}$.

The present text is devoted to a field-theoretic description of second-order transitions. Transitions of the first and higher than second order will not be considered.

1.1 Second-Order Phase Transitions

An important property of second-order phase transitions is the divergence, at the critical temperature T_c, of the length scale, over which the system behaves coherently. This is accompanied by a divergence of the size of thermal fluctuations. As a consequence, many physical observables show, near T_c, a power behavior in the temperature difference $|T - T_c|$ from the critical point, i.e., they behave like $|T - T_c|^p$. The power p is called a *critical exponent* of the associated observable.

We shall focus our attention on those physical systems whose relevant thermal fluctuations near the transition temperature can be described by some N-component *order field* $\phi(\mathbf{x}) = (\phi_1(\mathbf{x}) \cdots \phi_N(\mathbf{x}))$. An order field is the space-dependent generalization of Landau's famous *order parameter*, which characterizes all second-order transitions in a molecular field approximation. The energy of a field configuration is described by some functional of the order field $E = E[\phi(\mathbf{x})]$. To limit the number of possible interaction terms, certain symmetry properties will be assumed for the energy in the N-dimensional field space.

The thermal expectation value of the order field will be denoted by

$$\Phi(\mathbf{x}) \equiv \langle \phi(\mathbf{x}) \rangle. \tag{1.1}$$

This expectation value is nonzero below, and zero above the transition temperature [2]. The high-temperature state with zero expectation value is referred to as the *normal phase*, the low-temperature phase with nonzero expectation value is the *ordered phase*. In most systems, the nonzero expectation value $\Phi(\mathbf{x})$ is a constant, for example in ferromagnetic systems where the ordered phase shows a uniform magnetization. Only systems in the normal phase will be described in detail in this book. Thus we shall ignore a great number of interesting physical systems, for example cholesteric and smectic liquid crystals, which have a periodic $\Phi(\mathbf{x})$ and are of great industrial importance.

The ferromagnetic transition is the archetype for a second-order phase transition, in which a paramagnetic normal phase goes over into a ferromagnetic ordered phase when temperature is lowered below T_c, which is here called the *Curie temperature*. The microscopic origin lies in the lattice constituents. In ferromagnetic materials like Fe, Co, and Ni, these possess permanent magnetic moments which tend to line up parallel to one another to minimize the exchange energies of the electrons. At nonzero temperatures, they are prevented from perfect alignment by thermal fluctuations which increase the entropy of the many possible directional configurations. Disorder is also favored by the fact that alignment creates magnetic field energy, which causes a breakup into Weiss domains. This mechanism will be ignored. The local magnetic moments in a continuum approximation of these systems are described by a three-component order field $\phi(\mathbf{x}) = (\phi_1(\mathbf{x}), \phi_2(\mathbf{x}), \phi_3(\mathbf{x}))$.

At zero temperature, the alignment forces give the system a constant global order. The thermal expectation value of the field is nonzero, exhibiting a *spontaneous magnetization* $\Phi \equiv \langle \phi(\mathbf{x}) \rangle$ which serves as an order parameter, and which we shall alternatively denote by $\mathbf{M}$ when dealing with magnetic systems. The vector $\mathbf{M}$ may point in any fixed spatial direction.

If the temperature is raised, the size of $\mathbf{M}$ decreases due to the disordering effect of thermal fluctuations. When approaching the Curie temperature T_c, the spontaneous magnetization tends to zero according to a power law

$$|\mathbf{M}| \sim |T - T_c|^\beta. \tag{1.2}$$

Above the Curie temperature, it is identically zero, and the system is in the normal phase.

From the symmetry point of view, the existence of a spontaneous magnetization below T_c implies a *spontaneous breakdown of rotational symmetry*. The energy functional of the system is rotationally invariant in the space of the order field which happens to coincide here with the configuration space; whereas below T_c there exists a preferred direction defined by the spontaneous magnetization vector $\mathbf{M}$. The ordered state has a reduced symmetry, being invariant only under rotations around this direction.

In general one speaks of a spontaneous symmetry breakdown if the symmetry group in field space reduces to a subgroup when passing from the disordered high-temperature phase to the ordered low-temperature phase.

Another important system undergoing a second-order phase transition is liquid ^4He. At a temperature $T_\lambda = 2.18\text{K}$, called the λ-point, it becomes superfluid. At the transition, the phase factor $e^{i\theta(\mathbf{x})}$ of its many-body wave function which fluctuates violently in the normal state, becomes almost constant with only small long-wavelength fluctuations. Since the phase factor $e^{i\theta(\mathbf{x})}$ may be viewed as a two-dimensional vector $(\cos\theta(\mathbf{x}), \sin\theta(\mathbf{x}))$, this transition may also be described by a two-component order field $\phi(\mathbf{x})$. For experimental studies of its phase transition, liquid ^4He is especially well suited since its thermal conductivity is very high in the superfluid state, which permits an extremely uniform temperature to be established in a sizable sample, with variations of less than 10^{-8} K, as we shall see below in Fig. 1.1.

1.2 Critical Exponents

It is usually possible to design scattering experiments which are sensitive to the thermal fluctuations of the order field $\boldsymbol{\phi}(\mathbf{x})$. They are capable of measuring the full correlation function

$$G_{ij}(\mathbf{x} - \mathbf{y}) = \langle \phi_i(\mathbf{x})\phi_j(\mathbf{y})\rangle \qquad (1.3)$$

which in the normal phase is proportional to δ_{ij}:

$$G_{ij}(\mathbf{x} - \mathbf{y}) = \delta_{ij}G(\mathbf{x} - \mathbf{y}). \qquad (1.4)$$

1.2.1 Correlation Functions

Take, for example, an order field $\boldsymbol{\phi}(\mathbf{x})$ with only a single component $\phi(\mathbf{x})$ describing the local density $\rho(\mathbf{x})$ of a liquid near its critical point. In this case, the *correlation function*

$$G(\mathbf{x} - \mathbf{y}) = \langle \phi(\mathbf{x})\phi(\mathbf{y})\rangle \qquad (1.5)$$

can be measured by neutron scattering. The information on $G(\mathbf{x} - \mathbf{y})$ is contained in the so-called *structure factor* $S(\mathbf{x})$, whose experimental measurement is sketched in Appendix 1A.

In a phase with broken symmetry where the order field $\boldsymbol{\phi}(\mathbf{x})$ has a nonzero thermal expectation value $\boldsymbol{\Phi} = \langle \boldsymbol{\phi}(\mathbf{x})\rangle$, the physically interesting quantity is the *connected correlation function*, which describes the fluctuations of the deviations of the order field $\boldsymbol{\phi}(\mathbf{x})$ from its expectation value $\boldsymbol{\Phi}$, to be denoted by $\delta\boldsymbol{\phi}(\mathbf{x}) \equiv \boldsymbol{\phi}(\mathbf{x}) - \boldsymbol{\Phi}$:

$$G_{c\,ij}(\mathbf{x} - \mathbf{y}) = \langle \delta\phi_i(\mathbf{x})\delta\phi_j(\mathbf{y})\rangle = \langle \phi_i(\mathbf{x})\phi_j(\mathbf{y})\rangle - \Phi_i\Phi_j. \qquad (1.6)$$

In a ferromagnet, $\boldsymbol{\phi}(\mathbf{x})$ describes the local magnetization, and $\delta\boldsymbol{\phi}(\mathbf{x})$ its deviations from the spontaneous magnetization $\boldsymbol{\Phi} = \mathbf{M}$. The connected correlation function $G_{c\,ij}(\mathbf{x} - \mathbf{y})$ is an anisotropic tensor, which may be decomposed into a longitudinal and a transverse part with respect to the direction of $\boldsymbol{\Phi}$:

$$G_{c\,ij}(\mathbf{x}) = \frac{\Phi_i\Phi_j}{\Phi^2}G_{c\,L}(\mathbf{x}) + \left(\delta_{ij} - \frac{\Phi_i\Phi_j}{\Phi^2}\right)G_{c\,T}(\mathbf{x}). \qquad (1.7)$$

For temperatures closely above T_c, the correlation functions have a universal scaling behavior. To characterize this behavior, we identify first a microscopic length scale a below which microscopic properties of the material begin to become relevant. The size of a is usually defined by the spacing of a lattice or the size of molecules. Only for distances much larger than a can a field-theoretic description of the system in terms of an order field be meaningful. One then observes correlation functions whose behavior can be explained by field theory. For $T \gtrsim T_c$, they have a typical scaling form

$$G(\mathbf{x}) \sim \frac{1}{r^{D-2+\eta}}\, g(\mathbf{r}/\xi), \qquad r \equiv |\mathbf{x}| \gg a\,, \qquad (1.8)$$

where the function g falls off exponentially for large distances $r \to \infty$:

$$g(\mathbf{x}/\xi) \sim \exp(-r/\xi). \qquad (1.9)$$

For $T \lesssim T_c$, the same type of behavior is found for $G_{c\,L}(\mathbf{x})$, but with different functions $g(\mathbf{r}/\xi)$ and parameter ξ. The characteristic length scale ξ over which this falloff occurs is called the *correlation* or *coherence length* ξ. For $r \ll \xi$, but still much larger than a, the function $g(\mathbf{r}/\xi)$ becomes independent of $\mathbf{r}$, so that the correlation function behaves like a pure power $1/r^{D-2+\eta}$.

For $T \lesssim T_c$, the same type of behavior is found for the longitudinal connected correlation function $G_{cL}(\mathbf{x})$, but with different functions $g(\mathbf{r}/\xi)$ and coherence length ξ.

Near the critical temperature, the correlation length ξ diverges according to a power law

$$\xi \sim |T - T_c|^{-\nu}, \tag{1.10}$$

which defines the critical exponent ν. In principle, the exponent could be different when T_c is approached from above or below. Experimentally, however, they seem to be equal. In the field theory of critical phenomena, this equality will emerge as a prediction [in Eqs. (1.87) and (1.90) at the mean-field level, and in Section 10.10 in general]. The proportionality constant in (1.10), however, will be different for the two approaches from above and below. The ratio of these proportionality constants is called the *amplitude ratio*.

At $T = T_c$, the correlation length is infinite and the correlation function shows a pure power behavior

$$G_c(\mathbf{x}) \sim \frac{1}{r^{D+\eta-2}}, \qquad r \equiv |\mathbf{x}| \quad \text{at} \quad T = T_c, \tag{1.11}$$

which defines the critical exponent η. Higher correlation functions have a similar power behavior, all characterized by the single exponent η.

As power laws do not contain any length scale, this implies that at $T = T_c$ the large-scale properties of the system are free of any length scale. Such properties are generically called *critical*. The critical properties are independent of the microscopic nature of the system. They are *universal* properties, independent of lattice structures and atomic composition. The same is true for the behavior in the immediate neighborhood of T_c, where a system approaches the critical point and the correlation function has a behavior of the form (1.8).

One therefore distinguishes systems at or very close to a phase transition by the *universality classes* of their critical exponents. Each class depends only on the symmetry properties of the energy as a functional of the order parameter, apart from the space dimension of the many-body system.

For temperatures closely below T_c, the longitudinal correlation function $G_{cL}(\mathbf{x})$ has a similar behavior as (1.8) for $T \lesssim T_c$ only with a slightly different correlation length which, however, diverges with the same critical exponent ν for $T \to T_c$. At T_c it is characterized by the same critical exponent η as in (1.11).

The tranverse correlation functions have a special property: if the field energy is symmetric under $O(N)$ symmetry transformations of the N field components, spatially constant fluctuations transverse to $\mathbf{\Phi}$ are symmetry transformations and do not expend any energy. For this reason, transverse fluctuations have an infinite correlation length and behave like (1.11) for *all* T, not just at T_c. This is a manifestation of the *Nambu-Goldstone theorem*, which states that the spontaneous breakdown of a continuous symmetry in a phase transition gives rise to long-range modes associated with each generator of the symmetry.

The exponents η and ν of the correlation function $G(\mathbf{x})$ above T_c, and of $G_{cL}(\mathbf{x})$, $G_{cT}(\mathbf{x})$ below T_c will be calculated in Chapter 10 for a great variety of physical systems.

Experimentally one can easily measure the susceptibility tensor $\chi_{ij}(\mathbf{k})$ of a system at a finite wave vector $\mathbf{k}$, denoted by $\chi_{ij}(\mathbf{k})$. It is proportional to the Fourier transforms of the correlation function $G_{cij}(\mathbf{x} - \mathbf{y})$, and has the same invariant decompositions (1.4) and (1.7) as the correlation function above and below T_c, respectively. Thus we decompose, above T_c,

$$\chi_{ij}(\mathbf{x} - \mathbf{y}) = \delta_{ij}\chi(\mathbf{x} - \mathbf{y}), \tag{1.12}$$

with

$$\chi(\mathbf{x} - \mathbf{y}) \propto G(\mathbf{x} - \mathbf{y}), \tag{1.13}$$

and below T_c:

$$\chi_{c\,ij}(\mathbf{x}) = \frac{\Phi_i \Phi_j}{\Phi^2} \chi_{c\,L}(\mathbf{x}) + \left(\delta_{ij} - \frac{\Phi_i \Phi_j}{\Phi^2} \right) \chi_{c\,T}(\mathbf{x}), \tag{1.14}$$

with longitudinal and transverse parts

$$\chi_{L,T}(\mathbf{k}) \propto G_{c\,L,T}(\mathbf{k}) = \int d^D x \, e^{i\mathbf{k}\mathbf{x}} \, G_{c\,L,T}(\mathbf{x}). \tag{1.15}$$

1.2.2 Other Critical Exponents

There are other critical exponents which can be measured in global thermodynamic experiments, where one observes certain thermodynamic potentials or their derivatives with respect to temperature or external fields. Important examples are the specific heat C, the susceptibility χ_{ij} at zero external field, the magnetization $\mathbf{M} = \mathbf{\Phi}$ at zero external field below the critical temperature, or the magnetic equation of state $\mathbf{M}(\mathbf{B})$ at $T = T_c$. They have the following characteristic critical power behaviors:

$$
\begin{aligned}
C &\sim |T - T_c|^{-\alpha}, & \tag{1.16} \\
\chi_L(\mathbf{0}) &\sim |T - T_c|^{-\gamma}, & \tag{1.17} \\
\mathbf{k}^2 \chi_T(\mathbf{k}) &\underset{\mathbf{k}=0}{\sim} |T - T_c|^{\eta\nu}, & \tag{1.18} \\
|\mathbf{M}| &\sim |T - T_c|^{\beta}, & \text{for } T < T_c, \tag{1.19} \\
|\mathbf{M}| &\sim |\mathbf{B}|^{1/\delta}, & \text{for } T = T_c. \tag{1.20}
\end{aligned}
$$

These will all be derived in Chapter 10. An experimental critical behavior is shown for the

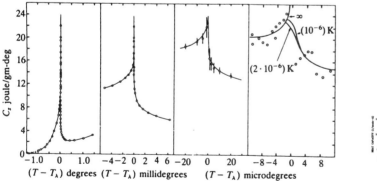

FIGURE 1.1 Specific heat near superfluid transition at $T_\lambda \approx 2.18$ K measured with increasing temperature resolutions. The curve has the typical λ-shape which is the reason for calling it λ-transition. Note that at higher resolutions, the left shoulder of the peak lies above the right shoulder. The data are from Ref. [3]. The forth plot is broadened by the pressure difference between top and bottom of the sample. This is removed by the microgravity experiment in the space shuttle yielding the last plot (open circles are irrelevant here) [4]. They show no pressure broadening even in the nK regime around the critical temperature.

specific heat of the superfluid transition of ^{4}He in Fig. 1.1. The curve for the specific heat has the form of a Greek letter λ, and for this reason one speaks here of a λ-transition, and denotes the critical temperature T_c in this case by T_λ. Just like the critical exponent ν, the

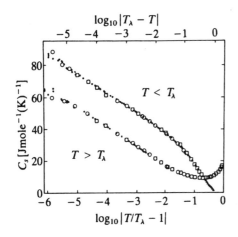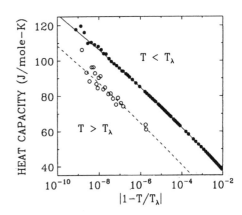

FIGURE 1.2 Specific heat of ^{4}He near superfluid transition plotted against $\log_{10}|T/T_\lambda - 1|$. The early data on the left-hand side by G. Ahlers [5] yield $\alpha \approx -0.026 \pm 0.004$ (for these an upper scale shows $\log_{10}|T_\lambda - T|$ with T measured in units of K). The right-hand side shows the space shuttle data of J.A. Lipa et al. in Refs. [4, 6], which yield $\alpha = -0.01056 \pm 0.0004$.

exponents α and γ could also in principle be different for the approaches of T_c from above and from below. Experiments, however, suggest their equality. Only the proportionality constants in (1.16)–(1.18) will be different for the two approaches, with well-defined amplitude ratios. This is illustrated by the plot of the experimental specific heat of the superfluid ^{4}He transition in Fig. 1.2 against the logarithm of $|T - T_\lambda|$. The data points lie approximately on straight lines corresponding to a logarithmic behavior

$$C \approx (A_\pm/\alpha)\left(1 - \alpha \log_{10}|T/T_\lambda - 1|\right) + \text{const}, \tag{1.21}$$

for the approaches from above and below, respectively. From his early data, G. Ahlers [5] determined the parameters α and $A_\pm$ by the best fits shown in Fig. 1.2, which gave $\alpha = -0.026 \pm 0.004$, $(A_- + A_+)/2 = 1.3 \pm 0.02$, and $A_+/A_- = 1.112 \pm 0.022$. The observed behavior (1.21) is compatible with a leading power $C \propto A_\pm |T - T_\lambda|^{-\alpha}$ of Eq. (1.16), owing to the smallness of the critical exponent α. For a finite α, a double-logarithmic plot would have been appropriate (as below in Fig. 1.3 determining the exponent ν).

The accuracy of Ahler's experiments was limited by the pressure difference within the sample caused by the earth's gravitational field. The small pressure dependence of the critical temperature smears out the singularity at T_λ to a narrow round peak, as seen on the right-hand plot of Fig. 1.1. The smearing was diminished by two orders of magnitude when the measurement was repeated at zero gravity in the space shuttle [4]. This permitted a much closer approach to the critical point than on earth, yielding the data on the right-hand plot of Fig. 1.2. They are fitted very well by a function

$$C = (A_\pm/\alpha)|t|^{-\alpha}\left(1 + D|t|^\Delta + E|t|\right) + B, \quad t \equiv T/T_\lambda - 1, \tag{1.22}$$

with [6]

$$\alpha = -0.01056 \pm 0.0004, \quad \Delta = 0.5, \qquad A_+/A_- = 1.0442 \pm 0.001, \tag{1.23}$$

$$A_-/\alpha = -525.03, \qquad\qquad D = -0.00687, \qquad E = 0.2152, \quad B = 538.55 \ (\text{J/mol K}). \tag{1.24}$$

The exponent Δ of the subleading singularity in t governs the approach to scaling and defines an associated critical exponent ω by the relation $\Delta = \omega\nu$, as will be derived in Eq. (10.152).

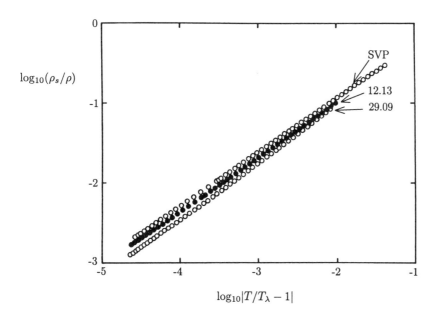

FIGURE 1.3 Doubly logarithmic plots of superfluid density ρ_s divided by the total density as a function of temperature. The slope is 2/3 [7]. The labels on the curves indicate the pressures in bar. The label SVP refers to the saturated vapor pressure.

The equality of the critical exponents will also be explained by the Ginzburg-Landau theory in Eqs. (1.43) and (1.50), and in general in Section 10.10.

For superfluid ^{4}He below T_c, an important experimental quantity is the *superfluid density* ρ_s which determines the stiffness of the fluctuations of the phase angle $\theta(\mathbf{x})$ of the ground state wave function (recall the discussion at the end of Section 1.1). It will be defined below in Eq. (1.122), being proportional to the $\mathbf{k} \to 0$ limit of $\mathbf{\Phi}^2/\mathbf{k}^2\chi_T(\mathbf{k})$. From Eqs. (1.18) and (1.19) we see that its critical behavior is $2\beta - \eta\nu = (D-2)\nu$. Its measurement therefore supplies us directly with information on the exponent ν which determines the rate of divergence of the coherence length in Eq. (1.10):

$$\rho_s \propto |T_c - T|^{(D-2)\nu}. \tag{1.25}$$

This will be proved in Subsection 10.10.5. The experimental data in Fig. 1.3 show precisely this behavior with $\nu \approx 2/3$.

1.2.3 Scaling Relations

Several relations between the critical exponents were suggested on general theoretical grounds and found to hold experimentally. Different exponents in thermodynamic potentials were related to each other by Widom [8], who assumed these potentials near T_c to be generalized homogeneous functions of their variables [9]. For the free energy of a system with a magnetization M, he conjectured the general functional form

$$F(t, M) = |t|^{2-\alpha}\,\psi(|t|/M^{1/\beta}), \tag{1.26}$$

where t is the reduced temperature characterizing the relative temperature distance from the critical point,

$$t \equiv T/T_c - 1, \tag{1.27}$$

and ψ some smooth function of its arguments. The scaling property (1.26) was first proposed by Kadanoff [10], who argued semiphenomenologically that not only thermodynamic potentials but also the correlation functions should be generalized homogeneous functions of two variables. For these he deduced the general form for $t \gtrsim 0$

$$G(\mathbf{x}) = \frac{f(\mathbf{x}/|t|^{-\nu})}{r^{D-2+\eta}}, \qquad r \equiv |\mathbf{x}|, \tag{1.28}$$

thus combining the properties (1.8), (1.10), and (1.11) in a single expression.

Such considerations showed that only two of the thermodynamic critical exponents (1.16)–(1.20) are independent of each other. There are two *scaling relations* between them:

$$\alpha + \beta(1 + \delta) = 2, \qquad \text{(Griffith)} \tag{1.29}$$
$$\gamma + \beta(1 - \delta) = 0. \qquad \text{(Widom)} \tag{1.30}$$

A combination of these is

$$\alpha + 2\beta + \gamma = 2. \qquad \text{(Rushbrook)} \tag{1.31}$$

Moreover, the two independent thermodynamic critical exponents are directly related to the two critical exponents η and ν which parametrized the power behavior of the correlation function in (1.28). The relations are:

$$\alpha = 2 - \nu D, \tag{1.32}$$
$$\beta = \frac{\nu}{2}(D - 2 + \eta), \tag{1.33}$$
$$\gamma = \nu(2 - \eta), \tag{1.34}$$
$$\delta = \frac{D + 2 - \eta}{D - 2 + \eta}. \tag{1.35}$$

The scaling relations (1.32), (1.33), and (1.35) involving the dimension D are referred to as *hyperscaling relations.*

These relations are satisfied in all exactly solvable models (for instance, the two-dimensional Ising model, and the spherical model for $D = 1, 2, 3$, both to be described in more detail in the next section) and by computer simulations of nonsolvable models. They are also confirmed by experimental observations of critical properties in many physical systems described by these models.

It is instructive to derive one of the scaling relations (1.32)–(1.35) between thermodynamic critical exponents and those of the correlation functions, for example (1.34). The susceptibility tensor at zero wave vector and external field in the normal phase is obtained from the correlation function (1.4) by the spatial integral

$$\chi_{ij} \propto \int d^D x \, \langle \phi_i(\mathbf{x}) \phi_j(\mathbf{y}) \rangle = \delta_{ij} \chi = \delta_{ij} \int d^D x \, G_c(\mathbf{x} - \mathbf{y}). \tag{1.36}$$

Inserting the general scaling form (1.8) with (1.9), we obtain

$$\chi \propto \int d^D x \, \frac{1}{r^{D-2+\eta}} e^{-r/\xi} \propto \xi^{2-\eta} \propto |T - T_c|^{-\nu(2-\eta)}. \tag{1.37}$$

Comparison with (1.17) yields the scaling relation (1.34). Similar derivation holds for $T < T_c$ and the longitudinal susceptibility χ_L defined in (1.15).

1.3 Models for Critical Behavior

In order to describe correctly a second-order phase transition, a model must have the following properties. At T_c, fluctuations must occur on arbitrarily large length scales. In the neighborhood of T_c, a correlation length which diverges for $T \to T_c$ should be the only relevant length scale of the system, making the correlation functions of the model at the critical point scale independent. The experimentally observed scaling forms (1.26) and (1.28) of the thermodynamic functions and the correlation functions should be reproduced.

1.3.1 Landau Theory

The construction of field-theoretic models is guided by the existence of successful self-consistent *molecular-field theories* for a great variety of many-body systems. In these theories, the interacting system is approximated by a noninteracting system subject to the influence of a self-consistent external field, whose temperature dependence is determined by an averaging procedure over molecular properties of the system. The prototype for a mean-field approximation is the Weiss theory for ferromagnetism. Another well-known example is the Van der Waals theory of the fluid-gas transitions.

Existing mean-field theories of second-order phase transitions are unified in *Landau's theory of phase transitions* [11]. The Gibbs free energy $\Gamma(T, \mu, V, \mathbf{M})$ is assumed to depend on the temperature T, the chemical potential μ, the volume V, and some order parameter $\mathbf{M}$, which has in general several components. The energy is invariant under some group of linear transformations among these components, depending on the symmetry of the system, which for simplicity will be assumed to be isotropic. The symmetry restricts the form of the different powers of the multi-component order parameter. For a vector $\mathbf{M}$, for instance, rotational symmetry does not permit a cubic term $\mathbf{M}^3$. The power series expansion is truncated after the fourth power in $\mathbf{M}$, whose coefficient is assumed to be positive, to guarantee stability.

It is then easy to show that if symmetry forbids the existence of a cubic term, the transition can only be of second order. A cubic term would change the transition to first order. In a magnetic system with rotational symmetry, where the order parameter is the magnetization vector $\mathbf{M}$, the Landau expansion of the Gibbs free energy has the general form

$$\Gamma(T, \mu, V, \mathbf{M}) = V \left[\frac{A_2}{2} \mathbf{M}^2 + \frac{A_4}{4!} \left(\mathbf{M}^2 \right)^2 \right], \tag{1.38}$$

where V is the volume of the system. The coefficients A_2, A_4 depend only on T and μ. Minimizing $\Gamma(T, \mu, V, \mathbf{M})$ with respect to $\mathbf{M}$ gives two possible minima, in which the vector $\mathbf{M}$ has an arbitrary direction and a size $M = |\mathbf{M}|$ given by

$$M = 0, \qquad \text{for} \quad A_2 \geq 0, \quad A_4 > 0, \tag{1.39}$$

$$M = \sqrt{-6A_2/A_4}, \quad \text{for} \quad A_2 < 0, \quad A_4 > 0. \tag{1.40}$$

A phase transition occurs if A_2 changes its sign. We identify this place with the critical point by setting near the zero

$$A_2 \equiv a_2 \left(\frac{T}{T_c^{\mathrm{MF}}} - 1 \right) \equiv a_2 \tau, \tag{1.41}$$

and approximating a_2 and A_4 by a constant near T_c^{MF}. The relative temperature distance from the mean-field critical temperature T_c^{MF} is denoted by τ, in contrast to the similar quantity with the true critical temperature T_c defined in Eq. (1.27). From Eqs. (1.40) and (1.41) we obtain the temperature behavior

$$M \propto \tau^{1/2} \quad \text{for} \quad T < T_c^{\mathrm{MF}}, \tag{1.42}$$

corresponding to a critical exponent $\beta = 1/2$ in Eq. (1.19) within the Landau theory.

At the minimum, the Gibbs free energy (1.38) is

$$\Gamma_{\min} = \begin{cases} 0 & \text{for} \quad T \ge T_c^{\mathrm{MF}}, \\ -V\dfrac{3a_2^2}{2A_4}\tau^2 & \text{for} \quad T < T_c^{\mathrm{MF}}. \end{cases} \tag{1.43}$$

The energy below T_c^{MF} is the so-called *condensation energy* F_s. From this we deduce that the specific heat C looks like a step function $\Theta(t)$ in the immediate neighborhood of T_c^{MF}. This determines the critical exponent α in Eq. (1.16) as being zero in this *Landau approximation*, for the approach to T_c^{MF} from above as well as from below. It also shows that the proportionality constants in Eq. (1.16) are different in these two approaches, with a well-defined amplitude ratio.

The magnetic susceptibility tensor χ_{ij} of the system with zero wave vector $\mathbf{k}$ is found from the Gibbs free energy as follows. First we add to the energy (1.38) an interaction term with an external magnetic field $\mathbf{B}$:

$$\Gamma_{\mathrm{ext}} = -V\mathbf{M} \cdot \mathbf{B}. \tag{1.44}$$

The total Gibbs free energy

$$\Gamma + \Gamma_{\mathrm{ext}} = V\left[\frac{A_2}{2}\mathbf{M}^2 + \frac{A_4}{4!}\left(\mathbf{M}^2\right)^2 - \mathbf{M} \cdot \mathbf{B}\right] \tag{1.45}$$

must now be minimized in the presence of $\mathbf{B}$ to obtain the magnetization $\mathbf{M}$ produced by the magnetic field. Its direction is parallel to $\mathbf{B}$ and its magnitude $M(B)$ is given by the solution of the cubic equation

$$A_2 M + \frac{A_4}{6}M^3 - B = 0, \tag{1.46}$$

where $B = |\mathbf{B}|$. At the critical point where $A_2 = 0$, we have $M^3 \propto B$, so that the critical exponent of the relation (1.20) is determined to be $\delta = 3$.

The susceptibility tensor $\chi_{ij}(\mathbf{M})$ at zero wave vector and finite magnetization is given by the derivative matrix $\partial M_i / \partial B_j$, which is the inverse of the second-derivative matrix of the Gibbs free energy

$$\chi_{ij}^{-1}(\mathbf{M}) = \frac{\delta B_i}{\delta M_j} = \frac{\delta^2 \Gamma[\mathbf{M}]}{\delta M_i \delta M_j}. \tag{1.47}$$

In the ordered phase, the susceptibility is anisotropic, and has the same decomposition into longitudinal and transverse parts with respect to the direction of $\mathbf{M}$ as $G_{c\,ij}$ in Eq. (1.7):

$$\chi_{ij}(\mathbf{M}) = \chi_L(M)\frac{M_i M_j}{M^2} + \chi_T(M)\left(\delta_{ij} - \frac{M_i M_j}{M^2}\right). \tag{1.48}$$

The longitudinal part is found by differentiating (1.46) with respect to B yielding

$$\left(A_2 + \frac{A_4}{2}M^2\right)\chi_L(B) = 1. \tag{1.49}$$

Setting $B = 0$ and inserting (1.39), (1.40), and (1.41), we obtain the longitudinal susceptibility at zero field

$$\chi_L(0) = \begin{cases} \dfrac{1}{a_2}\tau^{-1} & \text{for } T \geq T_c^{\text{MF}}, \\[2mm] \dfrac{1}{2a_2}|\tau|^{-1} & \text{for } T < T_c^{\text{MF}}. \end{cases} \tag{1.50}$$

The critical exponent γ defined in (1.17) is thus predicted by Landau's theory to be equal to unity. This holds for the approach to T_c^{MF} from above as well as from below. Moreover, the proportionality constants in Eq. (1.17) are different in these two approaches with a well-defined amplitude ratio.

The transverse susceptibility $\chi_T(0)$ is for $T \geq T_c^{\text{MF}}$ the same as the longitudinal one, since the system is isotropic at zero magnetic field. Below the critical temperature, however, the nonzero spontaneous magnetization (1.40) breaks the rotational symmetry of the Gibbs free energy (1.38) making $\chi_T(0)$ different from $\chi_L(0)$. In fact, to the lowest order, the magnetization changes by a large amount upon a small transverse variation of the magnetic field, making the transverse susceptibility infinite. This is a manifestation of the general *Nambu-Goldstone theorem* which states that the spontaneous breakdown of a continuous symmetry in a phase transition causes a divergent susceptibility for variations associated with the remaining symmetry.

The critical exponents of the Landau theory

$$\alpha = 0, \quad \beta = 1/2, \quad \gamma = 1, \quad \delta = 3 \tag{1.51}$$

can be shown to satisfy the scaling relations (1.29)–(1.31).

Landau's theory makes no predictions for the exponents η and ν of the correlation functions since the order parameter is assumed to take the same value everywhere. In order to calculate these, the order parameter has to be generalized to a space-dependent order field, as was first done by Ginzburg and Landau. This extension is discussed in Section 1.4.

The Landau values for the critical exponents give only a rough estimate of the size of the critical exponents. They differ qualitatively and quantitatively from experimental values. In particular, they do not distinguish between systems of different symmetry and different spontaneous symmetry breakdown, whereas the experimental critical exponents depend sensitively on these. As we shall see below, this is caused by the neglect of local fluctuations, which become extremely important near the transition temperature.

1.3.2 Classical Heisenberg Model

A famous model which serves to demonstrate the importance of fluctuations is the *classical Heisenberg model* of ferromagnetic systems. The energy is specified on a lattice as a sum over nearest-neighbor interactions between local spin vectors $\mathbf{S}_i$, which are conventionally normalized to unit length,

$$E = -\frac{J}{2}\sum_{\{i,j\}} \mathbf{S}_i \cdot \mathbf{S}_j \quad \text{with } J > 0. \tag{1.52}$$

The index pairs i, j run over all nearest neighbor pairs. The spin vectors $\mathbf{S}_i$ fluctuate and represent the order field. The expectation $\mathbf{M} \equiv \langle \mathbf{S}_i \rangle$ is the magnetization of the system. It is nonzero only at low temperatures. Above a critical temperature T_c, the system is normal with $\mathbf{M} = 0$.

A special case of this model is the famous *Ising model* where the direction of the vector $\mathbf{S}_i$ is restricted to a single axis, pointing parallel or antiparallel to it. The phase transition of magnetic systems with strong anisotropy can be described by this model. Then $\mathbf{S}$ can be replaced by a scalar with positive and negative signs. The order field of this model $\phi(\mathbf{x})$ has only one component $S(\mathbf{x})$. The symmetry which is spontaneously broken in the low-temperature phase is the reflection symmetry $S(\mathbf{x}) \rightarrow -S(\mathbf{x})$.

The energy (1.52) can be generalized from three-vectors $\mathbf{S}_i$ to N-dimensional vectors to describe a spontaneous breakdown of $O(N)$ symmetry.

The partition function associated with the generalized Heisenberg model is

$$Z = \sum_{\text{spin configurations}} e^{-E/k_B T}, \tag{1.53}$$

where the sum runs over all possible spin configurations. In order to specify this sum mathematically, let us rescale the spin vectors to unit length by replacing $\mathbf{S} \rightarrow |\mathbf{S}|\mathbf{n}$, and denote each lattice point by a vector $\mathbf{x}$ on a simple cubic lattice, i.e., we let $\mathbf{x}$ take any value

$$\mathbf{x}_{(m_1, m_1, \ldots, m_D)} \equiv \Sigma_{i=1}^{D} m_i\, \mathbf{i}, \tag{1.54}$$

where $\mathbf{i}$ are the basis vectors of a D-dimensional hypercubic lattice, and m_i are integer numbers. The sum over products $\sum_{\{i,j\}} \mathbf{n}_i \mathbf{n}_j$ can then be rewritten as

$$\sum_{\{i,j\}} \mathbf{n}_i \mathbf{n}_j = \sum_{\{i,j\}} [\mathbf{n}_i(\mathbf{n}_j - \mathbf{n}_i) + 1] = 2\sum_{\mathbf{x}}\sum_{\mathbf{i}} \{\mathbf{n}(\mathbf{x})[\mathbf{n}(\mathbf{x}+\mathbf{i}) - \mathbf{n}(\mathbf{x})] + 1\}$$

$$= \sum_{\mathbf{x}} \left\{ -\sum_{\mathbf{i}}[\mathbf{n}(\mathbf{x}+\mathbf{i}) - \mathbf{n}(\mathbf{x})]^2 + 2D \right\}. \tag{1.55}$$

We then introduce the lattice gradients

$$\nabla_{\mathbf{i}}\, \mathbf{n}(\mathbf{x}) \equiv \frac{1}{a}[\mathbf{n}(\mathbf{x}+\mathbf{i}) - \mathbf{n}(\mathbf{x})], \tag{1.56}$$

$$\overline{\nabla}_{\mathbf{i}}\, \mathbf{n}(\mathbf{x}) \equiv \frac{1}{a}[\mathbf{n}(\mathbf{x}) - \mathbf{n}(\mathbf{x}-\mathbf{i})], \tag{1.57}$$

where a is the lattice spacing, and rewrite (1.55) as

$$\sum_{\{i,j\}} \mathbf{n}_i \mathbf{n}_j = \sum_{\mathbf{x}} \left\{ -a^2 \sum_{\mathbf{i}}[\nabla_{\mathbf{i}}\, \mathbf{n}(\mathbf{x})]^2 + 2D \right\}. \tag{1.58}$$

We can now use the lattice version of partial integration [12]

$$\sum_{\mathbf{x}}\sum_{\mathbf{i}} \nabla_{\mathbf{i}} f(\mathbf{x})g(\mathbf{x}) = -\sum_{\mathbf{x}}\sum_{\mathbf{i}} f(\mathbf{x})\overline{\nabla}_{\mathbf{i}} g(\mathbf{x}), \tag{1.59}$$

which is valid for all lattice functions with periodic boundary conditions, to express (1.58) in terms of a lattice version of the Laplace operator $\overline{\nabla}_{\mathbf{i}}\nabla_{\mathbf{i}}$, where repeated lattice unit vectors are summed. Then

$$\sum_{\{i,j\}} \mathbf{n}_i \mathbf{n}_j = \sum_{\mathbf{x}} \left\{ a^2\, \mathbf{n}(\mathbf{x})\overline{\nabla}_{\mathbf{i}}\nabla_{\mathbf{i}}\, \mathbf{n}(\mathbf{x}) + 2D \right\}. \tag{1.60}$$

Ignoring the irrelevant constant term $\Sigma_{\mathbf{x}} 2D$, we can write the sum over all spin configurations in the partition function (1.53) as a product of integrals over a unit sphere at each lattice point

$$Z = \prod_{\mathbf{x}} \left[\int \mathbf{n}(\mathbf{x}) \right] \exp \left\{ \frac{1}{2} \sigma a^2 \sum_{\mathbf{x}} \mathbf{n}(\mathbf{x}) \overline{\nabla}_{\mathbf{i}} \nabla_{\mathbf{i}} \mathbf{n}(\mathbf{x}) \right\} \tag{1.61}$$

where we have introduced the dimensionless parameter

$$\sigma \equiv \frac{J}{k_B T}. \tag{1.62}$$

This parameter characterizes how easily the direction of the unit vectors can be turned in the opposite direction at a certain temperature. It will be called the *stiffness* of the directional fluctuations.

It is easy to estimate the temperature at which this partition function has a phase transition. For this we liberate the vectors $\mathbf{n}(\mathbf{x})$ from lying on a unit sphere, allowing them to run through the entire N-dimensional space, calling them $\boldsymbol{\phi}(\mathbf{x})$. To keep the system unchanged, we enforce the unit length with the help of Lagrange multipliers $\lambda(\mathbf{x})$. Thus we rewrite the partition function (1.61) as

$$Z = \prod_{\mathbf{x}} \left[\int \boldsymbol{\phi}(\mathbf{x}) \int \lambda(\mathbf{x}) \right] \exp \left(-\frac{1}{2} \sigma a^2 \sum_{\mathbf{x}} \left\{ -\boldsymbol{\phi}(\mathbf{x}) \overline{\nabla}_{\mathbf{i}} \nabla_{\mathbf{i}} \boldsymbol{\phi}(\mathbf{x}) + \lambda(\mathbf{x}) [\boldsymbol{\phi}^2(\mathbf{x}) - 1] \right\} \right). \tag{1.63}$$

A continuous field formulation of this generalized Heisenberg model with the same critical properties has the field energy

$$E[\boldsymbol{\phi}, \lambda] = \frac{1}{2} \int d^D x \left\{ \partial_i \boldsymbol{\phi}(\mathbf{x}) \partial_i \boldsymbol{\phi}(\mathbf{x}) + a^2 \lambda(\mathbf{x}) \left[\boldsymbol{\phi}^2(\mathbf{x}) - 1 \right] \right\}. \tag{1.64}$$

The integrals in the lattice partition function (1.63) are now Gaussian and can be performed using the generalization of the integral formula

$$\int d\xi e^{-a\xi^2/2} = \sqrt{\frac{2\pi}{a}} \tag{1.65}$$

for positive c-numbers a to symmetric positive $L \times L$ matrices $A_{\alpha\beta}$:

$$\prod_{\alpha} \left[\int d\xi_{\alpha} \right] e^{-\xi_{\alpha} A_{\alpha\beta} \xi_{\beta}/2} = \frac{\sqrt{2\pi}^L}{\sqrt{\det A}}, \tag{1.66}$$

where $\det A$ is the determinant of the matrix A. With the help of the well-known matrix formula

$$\det A = e^{\operatorname{tr} \log A}, \tag{1.67}$$

where tr denotes the trace of the matrix A, formula (1.66) can be rewritten as

$$\prod_{\alpha} \left[\int d\xi_{\alpha} \right] e^{-\xi_{\alpha} A_{\alpha\beta} \xi_{\beta}/2} = \sqrt{2\pi}^L e^{-\operatorname{tr} \log A/2}. \tag{1.68}$$

Since the lattice Laplacian $\overline{\nabla}_{\mathbf{i}} \nabla_{\mathbf{i}}$ is a symmetric positive matrix, we can express (1.63) in the form

$$Z = \operatorname{const} \times \prod_{\mathbf{x}} \left[\int \lambda(\mathbf{x}) \right] \exp \left[-\frac{N}{2} \operatorname{tr} \log \left[-\overline{\nabla}_{\mathbf{i}} \nabla_{\mathbf{i}} + \lambda(\mathbf{x}) \right] + \frac{\sigma a^2}{2} \sum_{\mathbf{x}} \lambda(\mathbf{x}) \right]. \tag{1.69}$$

In the limit of large N, the integrals over $\lambda(\mathbf{x})$ in this partition function can be performed exactly. The generalized Heisenberg model is then referred to as the *spherical model*.

The integration is possible since for large N, the integrals can be treated by the saddle point approximation. For any function $f(\xi)$ with a single smooth extremum at ξ_m, this approximation is

$$\int d\xi\, e^{-Nf(\xi)} \approx \sqrt{\frac{2\pi}{Nf''(\xi_m)}}\, e^{-Nf(\xi_m)}. \tag{1.70}$$

For the functional in the exponent of (1.69), the extremum is given by a constant $\lambda(\mathbf{x}) \equiv \lambda$ satisfying the so-called gap equation [13]

$$Nv_{a^2\lambda}^D(\mathbf{0}) = \sigma, \tag{1.71}$$

where

$$v_{a^2\lambda}^D(\mathbf{x}) \equiv a^{-2}\big[(-\overline{\nabla}_i\nabla_i) + a^2\lambda(\mathbf{x})\big]^{-1}(\mathbf{x}) \tag{1.72}$$

is the dimensionless lattice version of the Yukawa potential in D dimensions

$$V_{m^2}^D(\mathbf{x}) = \int \frac{d^Dk}{(2\pi)^D} \frac{e^{i\mathbf{k}\mathbf{x}}}{\mathbf{k}^2 + m^2}. \tag{1.73}$$

The lattice potential has the Fourier representation

$$v_{a^2\lambda}^D(\mathbf{x}) = \prod_i\left[\int_{-\pi/a}^{\pi/a} \frac{d(ak_i)}{2\pi}\right] \frac{e^{i\mathbf{k}\mathbf{x}}}{\sum_{i=1}^D(2 - 2\cos ak_i) + a^2\lambda}. \tag{1.74}$$

The denominator can be rewritten as $\int_0^\infty ds\, e^{-s[\sum_{i=1}^D(2-2\cos ak_i)+a^2\lambda]}$, leading to the multiple integral

$$v_{a^2\lambda}^D(\mathbf{0}) = \int_0^\infty ds\, e^{-s(a^2\lambda+2D)} \prod_{i=1}^D\left[\int_{-\pi}^\pi \frac{d\kappa_i}{2\pi} e^{2s\cos\kappa_i}\right]. \tag{1.75}$$

The integrations over k_i can now easily be performed, and we obtain the integral representation

$$v_{a^2\lambda}^D(\mathbf{0}) = \int_0^\infty ds\, e^{-s(a^2\lambda+2D)}[I_0(2s)]^D, \tag{1.76}$$

where $I_0(2s)$ is the modified Bessel function. Integrating this numerically, we find for $D = 3, 4, \ldots, \ldots$ the values shown in Table 1.1 [14]. A power series expansion of the Dth power of

TABLE 1.1   Values of lattice Yukawa potential $v_{l^2}^D(\mathbf{0})$ of reduced mass l^2 at origin for different dimensions and l^2. The lower entries show the approximate values from the hopping expansion (1.78).

D	$v_0^D(\mathbf{0})$	$v_1^D(\mathbf{0})$	$v_2^D(\mathbf{0})$	$v_3^D(\mathbf{0})$	$v_4^D(\mathbf{0})$
3	0.2527	0.1710	0.1410	0.1214	0.1071
	0.2171	0.1691	0.1407	0.1214	0.1071
4	0.1549	0.1271	0.1105	0.0983	0.0888
	0.1496	0.1265	0.1104	0.0983	0.0888

the modified Bessel function in (1.76),

$$[I_0(2s)]^D = 1 + Ds^2 + D(2D - 1)\frac{s^4}{4} + D(6D^2 - 9D + 4)\frac{s^3}{36} + \ldots, \tag{1.77}$$

leads to the so-called *hopping expansion* for $v^D_{a^2\lambda}(\mathbf{0})$:

$$v^D_{a^2\lambda}(\mathbf{0}) = \frac{1}{2D + a^2\lambda} + \frac{2D}{(2D + a^2\lambda)^3} + \frac{6D(2D - 1)}{(2D + a^2\lambda)^5} + \frac{20D(6D^2 - 9\,D + 4)}{(2D + a^2\lambda)^7} + \mathcal{O}(D^{-9}), \quad (1.78)$$

which converges rapidly for large D, and yields for $D = 3, 4$ to the approximate values shown in the lower entries of Table 1.1 . They lie quite close to the exact values in the upper part.

The lattice potential at the origin $v^D_{a^2\lambda}(\mathbf{0})$ in the gap equation (1.71) is always smaller than the massless potential $v^D_0(\mathbf{0})$. A nonzero value for λ can therefore only be found for sufficiently small values of the stiffness σ, i.e., for sufficiently high temperatures T [recall (1.62)]. The temperature T_c at which the gap equation (1.71) has the solution $\lambda = 0$ determines the Curie point. Thus we have

$$T_c = \frac{J}{\sigma_c k_B}, \quad (1.79)$$

where σ_c is the critical stiffness

$$\sigma_c = N v^D_0(\mathbf{0}). \quad (1.80)$$

This result, derived for large N, turns out to be amazingly accurate even for rather small N. As an important example, take $N = 2$ where the model consists of planar spins and is referred to as XY-model. For $D = 3$, it describes accurately the critical behavior of the superfluid transition in helium near the λ-transition. From the approximation (1.80) and the value $v^D_0(\mathbf{0}) \approx 0.2527$ in Table 1.1 we estimate

$$\sigma_c \approx 0.5054. \quad (1.81)$$

In Monte-Carlo simulation of this model one obtains, on the other hand [15],

$$\sigma_c \approx 0.45. \quad (1.82)$$

Thus, in three dimensions, we can use the large-N result (1.80) practically for all $N \geq 2$.

Apart from the spherical model in all dimensions, there also exist exact results also for the two-dimensional Ising model. In general, only approximate results are available. They are obtained from analyzing series expansions in powers of the temperature or the inverse temperature (low- and high-temperature expansions, respectively). Another tool to obtain approximate results are computer simulations using so-called Monte Carlo techniques.

Critical exponents of the different models are listed in Table 1.2 . The differences show that the critical exponents depend sensitively on the internal symmetry of the order parameter, apart from the space dimension. By studying these models on different types of lattices, one finds that the critical exponents are *not* influenced by the symmetry group of the lattice nor by any microscopic property of the system.

1.4   Fluctuating Fields

The long-range character of critical fluctuations and the universality of the critical properties suggest that it is possible to calculate these exponents by a phenomenological field theory rather than a microscopic model. One may neglect the lattice completely and describe the partition function of the system near T_c^{MF} by a functional integral over a continuous local order field $\phi(\mathbf{x})$. The order field may be considered as an average of the localized magnetic moments of some lattice model over a few lattice spacings. The energy functional is assumed to have a Taylor expansion in the order field $\phi(\mathbf{x})$ and in its gradients. The lowest gradient energy accounts for the nearest-neighbor interactions. It has the effect of suppressing short-wavelength fluctuations.

Exponent	Landau	Ising $D = 2$	Ising $D = 3$	Helium II $D = 3$	Heisenberg $D = 3$	Spherical $D = 3$
α	0 (disc)	0 (log)	0.1097 (.0012)	-0.011 (.004)	-0.122 (.009)	-1
β	1/2	1/8	0.3258(.0014)	0.3470(.0014)	0.3662(.0025)	1/2
γ	1	7/4	1.2378(.0006)	1.3178(.0010)	1.3926(.0010)	2
δ	3	15	4.8055(.0140)	4.7950(.0140)	4.7943(.0140)	5
η	0	1/4	0.0355(.0009)	0.0377(.0006)	0.0374(.0004)	0
ν	1/2	1	0.6301(.0005)	0.6715(.0007)	0.7096(.0008)	1

TABLE 1.2 Critical exponents. The numbers in the Landau column of the last two rows are derived in Section 1.4 after introducing gradient terms into the Gibbs free energy as proposed by Ginzburg and Landau. For the remaining columns see Chapters 17, 19, 20, and 21, in particular Table 20.2 and Eq. (20.97). The numbers in parentheses are error estimates.

1.4.1 Ginzburg-Landau Energy Functional

The symmetry group of the high-temperature phase limits the possible interaction terms in the energy functional $E[\phi]$. In a ferromagnet, the order parameter is a real scalar three-component field. Reflection symmetry $\phi \rightarrow -\phi$ eliminates all terms with odd powers. The restriction to the vicinity of the critical point eliminates all but three terms:

$$E[\phi] = \int d^D x \left\{ \frac{A_1}{2} \partial_i \phi(\mathbf{x}) \partial_i \phi(\mathbf{x}) + \frac{A_2}{2} \phi^2(\mathbf{x}) + \frac{A_4}{4!} [\phi^2(\mathbf{x})]^2 \right\}. \tag{1.83}$$

This is the famous *Ginzburg-Landau energy functional*. It differs from Landau's expansion of the Gibbs free energy (1.38) by the first gradient term (apart from the more general notation). If the system is symmetric only under a subgroup of the group $O(N)$, there may be more terms. An important case where the symmetry is only hypercubic (cubic for $N = 3$) will be considered in Chapter 18.

The parameters A_1 and A_4 in the expansion (1.83) depend weakly on the temperature, whereas A_2 vanishes at the critical temperature T_c^{MF} according to (1.41), with a weakly temperature-dependent coefficient a_2, as in the expansion (1.38). An approximate energy functional of the form (1.83) can be derived by a few simple manipulations from the functional integral representation of the partition function of the classical Heisenberg model (1.52) [16].

The Ginzburg-Landau energy functional (1.83) allows us to derive immediately an approximate critical exponent η governing the power behavior of the correlation function at the critical point [see (1.11)]. There A_2 vanishes according to (1.41). Neglecting the interaction, the connected correlation function (1.6) is given by the inverse of the differential operator between the ϕ-fields, as we shall see later in Eq. (2.34):

$$G_{ij}(\mathbf{x}) = \langle \phi_i(\mathbf{x}) \phi_j(\mathbf{0}) \rangle = \delta_{ij} G(\mathbf{x}) = \delta_{ij} \frac{k_B T}{A_1} \int \frac{d^D k}{(2\pi)^D} \frac{e^{i \mathbf{k} \mathbf{x}}}{\mathbf{k}^2} = \delta_{ij} \frac{k_B T}{A_1} \frac{1}{r^{D-2}}. \tag{1.84}$$

Comparison with (1.11) yields the approximation $\eta = 0$ listed in Table 1.2 .

The exponent ν governing the temperature behavior of the correlation length [see (1.10)] is obtained by neglecting the fourth-order terms in the fields in (1.83), considering only the quadratic part

$$E[\phi] = \int d^D x \left[\frac{A_1}{2} \partial_i \phi(\mathbf{x}) \partial_i \phi(\mathbf{x}) + \frac{A_2}{2} \phi^2(\mathbf{x}) \right]. \tag{1.85}$$

For $T \geq T_c^{\mathrm{MF}}$, this can be rewritten as

$$E[\phi] = \frac{A_2}{2} \int d^D x \left[\xi^2 \partial_i \phi(\mathbf{x}) \partial_i \phi(\mathbf{x}) + \phi^2(\mathbf{x})\right], \tag{1.86}$$

where

$$\xi^2 \equiv \frac{A_1}{A_2} = \xi_0^2 \tau^{-1} \quad \text{with} \quad \xi_0^2 \equiv \frac{A_1}{a_2} \tag{1.87}$$

defines the characteristic length scale of the fluctuations. The correlation function is now given by

$$G_{ij}(\mathbf{x}) = \langle \phi_i(\mathbf{x}) \phi_j(\mathbf{0}) \rangle = \delta_{ij} \frac{k_B T}{A_1} \int \frac{d^D k}{(2\pi)^D} \frac{e^{i\mathbf{k}\mathbf{r}}}{\mathbf{k}^2 + 1/\xi^2} = \mathrm{const} \times \delta_{ij} \frac{e^{-r/\xi}}{r^{D-2}}. \tag{1.88}$$

Thus ξ measures directly the coherence length defined in Eq. (1.9).

From Eq. (1.87) we derive the mean-field approximation for the critical exponent ν as T approaches T_c^{MF} from above. Comparison with (1.10) shows that $\nu = 1/2$, as listed in Table 1.2 .

For $T < T_c^{\mathrm{MF}}$, the quadratic term is found by expanding the expectation value $\Phi = \langle \phi(\mathbf{x}) \rangle$, yielding

$$E[\phi] = |A_2| \int d^D x \left\{\xi^2 \partial_i \delta\phi(\mathbf{x}) \partial_i \delta\phi(\mathbf{x}) + [\delta\phi(\mathbf{x})]^2\right\}, \tag{1.89}$$

where

$$\xi^2 = -\frac{A_1}{2A_2} = \frac{A_1}{2a_2}|\tau|^{-1} = \frac{1}{2}\xi_0^2 |\tau|^{-1}, \quad \text{with} \quad \xi_0^2 \equiv \frac{A_1}{a_2}. \tag{1.90}$$

The *zero-temperature coherence length* in the Ginzburg-Landau theory is

$$\xi(0) = \xi_0/\sqrt{2}. \tag{1.91}$$

Comparing the temperature behaviors of (1.90) and (1.87) we see that the critical exponents ν are equal for the approach to T_c^{MF} from below and from above. We also see that the proportionality constants in Eq. (1.10) are different in these two approaches.

Below T_c^{MF}, the connected correlation function $G_{c\,ij}(\mathbf{x}) = \langle \delta\phi_i(\mathbf{x}) \delta\phi_j(\mathbf{0}) \rangle$ has the decomposition (1.7) into a longitudinal part and a transverse part with respect to Φ. These have the spatial behavior

$$G_{c\,L}(\mathbf{x}) = \frac{k_B T}{A_1} \int \frac{d^D k}{(2\pi)^D} \frac{e^{i\mathbf{k}\mathbf{r}}}{\mathbf{k}^2 + 1/\xi^2} = \mathrm{const} \times \frac{e^{-r/\xi}}{r^{D-2}}. \tag{1.92}$$

$$G_{c\,T}(\mathbf{x}) = \frac{k_B T}{A_1} \int \frac{d^D k}{(2\pi)^D} \frac{e^{i\mathbf{k}\mathbf{r}}}{\mathbf{k}^2} = \mathrm{const} \times \frac{1}{r^{D-2}}. \tag{1.93}$$

While the longitudinal part of the connected correlation function below T_c^{MF} has the same r-dependence as the full correlation function above T_c^{MF} in (1.88), except for the different correlation length (1.90) by a factor $\sqrt{2}$, the transverse part has an infinite range. At zero momentum, the longitudinal susceptibility $\chi_L(\mathbf{k})$ defined in (1.15) is proportional to $\xi^2 \propto |t|^{-1}$, so that we find once more the Landau approximation to the critical exponent γ in Eq. (1.17) to have the value $\gamma = 1$, as in (1.51).

Note that the hyperscaling relations (1.32), (1.33), and (1.35) involving the dimension D are fulfilled by the critical exponents of the Ginzburg-Landau theory only for $D = 4$. The special role of this dimension in the ϕ^4-theory will be fully appreciated in Chapter 7.

1.4.2   Ginzburg Criterion

A rough understanding of this special role is already possible at the level of the Ginzburg-Landau theory, as first pointed out by Ginzburg in 1960 [17]. Let us estimate the fluctuations of the order field. For this we separate the fields $\phi_i(x)$ and their expectation values Φ_i into size and direction parts by writing

$$\phi_i(x) = \phi(x)\, n_i(x), \qquad \langle \phi_i \rangle \equiv \Phi_i \equiv \Phi n_i. \tag{1.94}$$

The unit direction vector $\mathbf{n}$ breaks spontaneously the $O(N)$ symmetry, and the decomposition (1.7) of the connected correlation function (1.6) becomes

$$G_{c\,ij}(\mathbf{x} - \mathbf{y}) = \langle \delta\phi_i(\mathbf{x})\delta\phi_j(\mathbf{y}) \rangle = \langle \phi_i(\mathbf{x})\phi_j(\mathbf{y}) \rangle - \Phi_i \Phi_j = \frac{n_i n_j}{n^2} G_{c\,L}(\mathbf{x}) + \left(\delta_{ij} - \frac{n_i n_j}{n^2} \right) G_{c\,T}(\mathbf{x}). \tag{1.95}$$

Using Eq. (1.92) we now see that the mean square deviation of the field from its expectation value is

$$\langle [\phi(\mathbf{x}) - \Phi]^2 \rangle = \frac{k_B T}{A_1} \int \frac{d^D k}{(2\pi)^D} \frac{1}{\mathbf{k}^2 + 1/\xi^2} = \frac{k_B T}{A_1} V_{1/\xi^2}^D(\mathbf{0}) \tag{1.96}$$

where $V_{m^2}^D(\mathbf{0})$ is the momentum integral

$$V_{m^2}^D(\mathbf{0}) = \int \frac{d^D k}{(2\pi)^D} \frac{1}{\mathbf{k}^2 + m^2} \tag{1.97}$$

which diverges at large k. The integral is equal to the Yukawa potential (1.73) for a particle of mass m, evaluated at the origin.

In order to derive the desired fluctuation information we imagine decomposing the field system into a lattice of cubic patches with an edge length $\xi_l = l\xi$ of the order of the coherence length ξ. One usually assumes size parameters l of the order of unity which ensures an approximate independence of the patches. The fluctuation width (1.96) within such a patch is calculated with the momenta in the integral (1.97) limited by $k_{\max} \approx \pi/\xi_l$. Denoting the surface of a unit sphere in D dimensions by

$$S_D \equiv \frac{2\pi^{D/2}}{\Gamma(D/2)}, \tag{1.98}$$

which will be derived later in Appendix 8A, we obtain the patch version of (1.97):

$$V_{1/\xi^2}^D(\mathbf{0}) \approx \xi_l^{2-D} w_{l^2}^D(\mathbf{0}), \tag{1.99}$$

with the reduced patch version of the Yukawa potential at the origin

$$w_{l^2}^D(\mathbf{0}) \approx \frac{S_D}{(2\pi)^D} \int_0^\pi dq\, q^{D-1} \frac{1}{q^2 + l^2}. \tag{1.100}$$

For $D = 3, 4$, this has the values shown in Table 1.3 . Ginzburg decomposed in three dimensions $1/(k^2 + l^2) = 1/k^2 - l^2/k^2(k^2 + l^2)$, and rewrote

$$w_{l^2}^3(\mathbf{0}) \approx \frac{S_3}{(2\pi)^3} \frac{1}{l} \left[\pi - \int_0^\pi dq\, \frac{l^2}{q^2 + l^2} \right]. \tag{1.101}$$

The second integral is now convergent and, assuming the lattice spacing parameter l to be equal to unity, Ginzburg approximated it as $\int_0^\pi dk\, 1/(k^2 + 1) \approx \pi/2$, thus estimating $w_1^3(\mathbf{0}) \approx 1/(2\pi^2) \times \pi/2 \approx 0.0795$, which lies reasonably close to the value 0.0952 in Table 1.3 .

TABLE 1.3   Values of reduced patch version $w_{l^2}^D(\mathbf{0})$ of Yukawa potential of mass l^2 at origin for different dimensions and l^2.

D	$w_0^D(\mathbf{0})$	$w_1^D(\mathbf{0})$	$w_2^D(\mathbf{0})$	$w_3^D(\mathbf{0})$	$w_4^D(\mathbf{0})$
3	0.1591	0.0952	0.0863	0.0809	0.0769
4	0.0625	0.0474	0.0439	0.0416	0.0400

Alternatively, we can approximate the patch version of the Yukawa potential (1.97) by a Yukawa potential on a simple cubic lattice of spacing ξ_l, or equivalently, replace the reduced version $w_{l^2}^D(\mathbf{0})$ of Eq. (1.100) by the reduced lattice potential [compare (1.74)]

$$v_{l^2}^D(\mathbf{0}) = \int_{-\pi}^{\pi} \frac{d^D\kappa}{(2\pi)^D} \frac{1}{\sum_{i=1}^{D}(2 - 2\cos\kappa_i) + l^2}. \tag{1.102}$$

In terms of $v_{l^2}^D(\mathbf{0})$, the fluctuation width (1.96) can be written as

$$\langle[\phi(x) - \Phi]^2\rangle = k_B T\, l^{2-D} (2|\tau|)^{D/2-1} \frac{1}{a_2}\left(\frac{a_2}{A_1}\right)^{D/2} v_{l^2}^D(\mathbf{0}). \tag{1.103}$$

Fluctuations will become important if this width is of the order of $\Phi^2 = 6|\tau|a_2/A_4$. This gives the condition

$$|\tau| \le \tau_G \equiv [l^{2-D} K v_{l^2}^D(\mathbf{0})]^{2/(4-D)} \tag{1.104}$$

where K is the parameter

$$K \equiv 2^{D/2-1} k_B T_c^{\mathrm{MF}} \frac{A_4}{6a_2^2}\left(\frac{a_2}{A_1}\right)^{D/2}. \tag{1.105}$$

The right-hand side of Eq. (1.104) defines the *Ginzburg temperature*

$$T_G = T_c^{\mathrm{MF}}\left(1 - \tau_G\right), \tag{1.106}$$

above which the mean-field approximation becomes unreliable. The size fluctuations shift the critical temperature below the mean-field critical temperature (*Ginzburg's criterion*).

It is interesting to compare the energy δE_G in a fluctuation pocket of coherence volume ξ^D near the Ginzburg temperature with the thermal energy $k_B T_c^{\mathrm{MF}}$. From the condensation energy density (1.43) and the coherence length (1.90), we find

$$\delta E_G = \xi^D \frac{3a_2^2}{2A_4}\tau_G^2 = \frac{1}{2^{2+D/2}} \frac{6a_2^2}{A_4}\left(\frac{A_1}{a_2}\right)^{D/2} \tau_G^{2-D/2}. \tag{1.107}$$

Using (1.105) and (1.104), this yields an energy ratio

$$\frac{\delta E_G}{k_B T_c^{\mathrm{MF}}} = \frac{1}{8K}\tau_G^{2-D/2} = \frac{1}{8} v_{l^2}^D(\mathbf{0}) l^{2-D}. \tag{1.108}$$

This ratio is considerably smaller than unity. For instance in three dimensions with $l = 1$, the right-hand side is roughly equal to 0.02. Thus, at the Ginzburg temperature, the Boltzmann factor for the fluctuation pockets is close to unity $e^{-\delta E_G/k_B T} \approx e^{-0.02}$. This implies that an alternative estimate of the Ginzburg temperature from the condition $\delta E_G \approx k_B T_c^{\mathrm{MF}}$, which is sometimes found in the literature [18], grossly overestimates the size of τ_G.

We are now in a position to understand the underlying reason for the special role of $D = 4$ dimensions. From Eq. (1.103), and also (1.107), we see that as T approaches T_c^{MF} from below in less than four dimensions, the fluctuations *increase*. In more than four dimensions, on the other hand, they decrease and become irrelevant. For this reason one always observes mean-field behavior in more than four dimensions. This will be seen in more detail in Chapter 10. The dimension $D_u = 4$ is called the *upper critical dimension*.

The length parameter l is a free parameter in this criterium. The above and Landau's estimates are both based on the assumption that $l \approx 1$ which corresponds to the physical situation that the order field $\Phi(\mathbf{x})$ is properly defined only up to a length scale of the order of the coherence length. In an ordinary superconductor, this is indeed the case. The order field describes Cooper pairs of electrons whose wave functions extend over a coherence length, requiring a cutoff related to this size. In modern superconductors, where the phase transition occurs at higher critical temperature of the order of 100 K, however, there is the theoretical possibility that Cooper pairs could have a much smaller diameter than the coherence length [19]. In this case, l would be much smaller than unity. If the Cooper pairs are bound extremely strongly, another effect becomes important: quantum fluctuations begin driving the phase transition. In this limit, the Cooper pairs form an almost free gas of almost point-like bosons which undergo Bose-Einstein condensation [20]. The relevant length scale is then the De Broglie wavelength of thermal motion

$$\lambda = \frac{2\pi\hbar}{\sqrt{2Mk_BT}}, \tag{1.109}$$

where M is the mass of the Cooper pairs.

1.4.3 Kleinert Criterion

The Ginzburg temperature does not tell us the full story about the onset of fluctuations. In an $O(N)$-symmetric theory, the order field performs not only size fluctuations, but also directional fluctuations. These are of long-range and therefore will be more violent than the size fluctuations. They destroy the order at a much lower temperature than Ginzburg's T_G, as shown by Kleinert [21]. To find the relevant temperature where these become important we ignore size fluctuations and introduce a normalized order field $\mathbf{n} \equiv \ /\sqrt{\phi^2}$. From Eq. (1.83) we see that this field has a pure gradient energy

$$E_{\mathbf{n}} = \frac{A_1}{2}\ ^2 \int d^D x \, [\partial_i \mathbf{n}(\mathbf{x})]^2. \tag{1.110}$$

We now proceed in analogy with Ginzburg's analysis and imagine that the field system consists of a simple-cubic lattice of patches of size $\xi_l = l\xi$. Then we can associate with (1.110) the energy of a classical $O(N)$-symmetric Heisenberg model on a simple cubic lattice with unit spacing

$$E_{\mathbf{n}} = \frac{A_1}{2}\ ^2 \xi_l^{D-2} \sum_{\{i,j\}} \mathbf{n}_i(\mathbf{n}_j - \mathbf{n}_i) = \frac{A_1}{2}\ ^2 \xi_l^{D-2} \sum_{\{i,j\}} \mathbf{n}_i\mathbf{n}_j - DA_1\ ^2\xi_l^{D-2}, \tag{1.111}$$

which is completely equivalent to the energy (1.52) with a parameter J given by

$$J \equiv A_1\ ^2 \xi_l^{D-2}. \tag{1.112}$$

The ratio J/k_BT is the stiffness σ of the directional fluctuations introduced in Eq. (1.62). Inserting here the mean-field temperature behavior of $|\ | = |\mathbf{M}|$ from Eq. (1.40), and the coherence length $\xi = \xi_0/\sqrt{2|\tau|} = \sqrt{A_1/2a_2|\tau|}$ from (1.90), we see that J depends on the temperature as follows:

$$J = \frac{l^{D-2}}{2^{D/2-1}} \frac{6a_2^2}{A_4} \left(\frac{A_1}{a_2}\right)^{D/2} |\tau|^{2-D/2} = k_B T_c^{\mathrm{MF}} \frac{l^{D-2}}{K} |\tau|^{2-D/2}. \tag{1.113}$$

Now, as discussed above in Subsection 1.3.2, the classical Heisenberg model has a phase transition where directional fluctuations disorder the ordered state if J is of the order of $k_B T$. More specifically, the transition for an O(N)-symmetric energy occurs for $D > 3$ at a critical value J_c which is roughly given by the critical stiffness σ_c in Eq. (1.80).

$$\sigma_c \approx \frac{J_c}{k_B T_c^{MF}} \approx N v_0^D(\mathbf{0}). \tag{1.114}$$

In $D = 3$ dimensions, the exact value of $v_0^D(0)$ gives an estimate for the critical stiffness $\sigma_c \approx 0.2527N$. For the XY-model with $N = 2$, this is equal to 0.5054. Monte-Carlo simulations for $N = 2$, on the other hand, yield $\sigma_c \approx 0.45$ [22].

Thus we see that the directional fluctuations destroy the order in a reduced temperature interval

$$\tau_K \equiv [l^{2-D} K N v_0^D(\mathbf{0})]^{2/(4-D)}. \tag{1.115}$$

This is the *Kleinert criterion*, the directional analog of the Ginzburg criterion in Eq. (1.104). Due to directional fluctuations, the transition occurs roughly at a temperature

$$T_K = T_c^{MF} (1 - \tau_K) \tag{1.116}$$

smaller than T_G (and, of course, much smaller than T_c^{MF}).

In the limit of large N, directional fluctuations are always responsible for the destruction of the ordered state. This explains why the critical exponents of the ϕ^4-theory to be derived in this text and those of the Heisenberg model have the same series expansions in powers of $1/N$ in any dimension $D > 2$.

For finite N, a small coherence length $\xi(0)$ requires a very small coupling strength in the mean-field energy to make directional fluctuations important. Note that in the symmetry-broken phase, a small $A_4 > 0$ implies a deep degenerate minimum with O(N) symmetry. The coupling must be larger than zero to have a potential minimum at $\tau < 0$.

What is a simple experimental signal for the dominance of directional fluctuations? In magnetic systems one measures, in the neighborhood of the critical point, the spontaneous magnetization Φ, the coherence length ξ, and further the longitudinal and transverse parts of the susceptibility defined in Eq. (1.15), all as a function of temperature. From the longitudinal and transverse parts of the susceptibility one finds the correlation functions $G_L(\mathbf{k})$ and $G_T(\mathbf{k})$. From these one may determine the mean-field transition temperature T_c^{MF} by plotting ξ^{-2} or $1/G_L(\mathbf{0})$ versus temperature. In the mean-field regime these are straight lines which intercept the temperature axis at $T = T_c^{\mathrm{MF}}$. We may extract the mean-field parameter (1.105) by plotting any of the combinations

$$K_L(t) \equiv |t|^{2-D/2} \frac{1}{\xi^D} \frac{G_L(\mathbf{0})}{k_B T_c \Phi^2}, \tag{1.117}$$

$$K_T(t) \equiv |t|^{2-D/2} \frac{\mathbf{q}^2}{\xi^{D-2}} \frac{G_T(\mathbf{q})}{k_B T_c \Phi^2}\bigg|_{\mathbf{q}\to 0}, \tag{1.118}$$

$$\bar{K}_L(t) \equiv |t|^{2-D/2} \frac{1}{\xi^{D-2}} \frac{[dG_L^{-1}(\mathbf{q})/d\mathbf{q}^2]^{-1}}{k_B T_c \Phi^2}\bigg|_{\mathbf{q}\to 0}, \tag{1.119}$$

versus $t = T/T_c - 1$. In the mean-field regime, all three combinations determine experimentally the constant K in (1.105), as can easily be verified using Eqs. (1.93), (1.92), (1.90), and (1.40). From these K-values we calculate the reduced temperatures τ_G and τ_K using (1.104) and (1.115). The ratio of the two is

$$\frac{\tau_K}{\tau_G} = \left(\frac{N v_0^D(\mathbf{0})}{v_{l^2}^D(\mathbf{0})} \right)^{2/(4-D)}, \tag{1.120}$$

which in three dimensions with $l = 1$ is roughly $2.26 \, N^2$. Thus the temperature T_K of Eq. (1.116), where directional fluctuations destroy the order, always lies *far below* the Ginzburg temperature T_K of Eq. (1.106).

In superfluid helium, we may plot, by analogy with (1.118),

$$K_T(t) \equiv |t|^{2-D/2} \frac{M^2 k_B T_c}{\xi^{D-2} \hbar^2 \rho_s}, \tag{1.121}$$

where M is the atomic mass and ρ_s is the superfluid mass density, which at the mean-field level is defined by writing the gradient energy (1.110) as

$$E_{\mathbf{n}} = \frac{\rho_s}{2 k_B T} \frac{\hbar^2}{M^2} \int d^3x \, [\partial \mathbf{n}(\mathbf{x})]^2. \tag{1.122}$$

In the critical regime, the three quantities in (1.118) and (1.121) go universally to zero like $|t|^{2-D/2}$, since $\xi \propto |t|^{-\nu}$, $\chi_L(\mathbf{0}) \approx |t|^{(\eta-2)\nu}$, $q^2 \chi_T(\mathbf{q})|_{\mathbf{q} \to 0} \approx |t|^{\eta\nu}$, $\Phi^2 \approx |t|^{\nu(D-2+\eta)}$, $\rho_s \approx |t|^{(D-2)\nu}$, where $t \equiv T/T_c - 1$, and ν and η are the critical exponents defined above. This dependence follows directly from (1.10), (1.17), (1.18), (1.19), (1.33), and (1.25), and will be derived in Chapter 10.

Experimentally, the superfluid density of bulk helium cannot be fitted by a mean-field approximation [23]. If we nevertheless try to fit roughly a mean-field curve to the superfluid density, we obtain $\rho_s/\rho \approx 2(1 - T/T_c^{\mathrm{MF}})$, where $\rho = M/a^3$ is the total mass density, with $a \approx 3.59 \, \text{Å}$ [24], and T_c^{MF} differs from T_c by only about 5%, such that it may be neglected for the present estimate. The factor $k_B T_c$ at $T_c = 2.18 \, \text{K}$ can be expressed as $k_B T_c \approx 2.35 \hbar^2/M a^3$. With $\xi(0) \approx 2 \, \text{Å}$, Eq. (1.121) yields an estimate for K of about 2. Inserting this into Eqs. (1.104) and (1.115), we find

$$\tau_K \approx 1, \qquad \tau_G \approx 0.12. \tag{1.123}$$

The large size of $\tau_K \approx 1$ reflects the bad quality of a mean-field approximation. The numbers in the estimates (1.123) depend quite sensitively on the choice of the lattice spacing parameter. An increase of l by a factor 2 decreases τ_K by a factor 4, which is an appropriate estimate for τ_K. In any case, we may conclude that the superfluid transition in helium is initiated by size fluctuations, not by directional fluctuations. Still, the immediate neighborhood of the critical point contains both types of fluctuations in a universal way. The above statement concerns only the onset of the critical behavior and the type of fluctuations which drive the system into the critical behavior.

We may define an experimental quantity $X_K(T)$ which exhibits directly the onset of directional fluctuations:

$$X_K(T) = k_B T_c |t|^{D/2-1} \xi^{D-2} \Phi^2 \left. \frac{G_T^{-1}(\mathbf{q})}{q^2} \right|_{\mathbf{q} \to 0} = k_B T_c |t|^{D/2-1} \xi^{D-2} \Phi^2 \left. \frac{dG_L^{-1}(\mathbf{q})}{d\mathbf{q}^2} \right|_{\mathbf{q} \to 0}, \tag{1.124}$$

or in superfluid helium

$$X_K(T) = |t|^{D/2-1}\xi^{D-2}\frac{\hbar^2\rho_s}{M^2 k_B T_c}. \tag{1.125}$$

In the mean-field regime, this falls off linearly towards the mean-field critical temperature T_c^{MF}. Near the critical value σ_c, however, the linear falloff changes into the critical power behavior $|t|^{D/2-1}$.

As in the case of the Ginzburg criterion, the parameter l may differ considerably from unity in extreme cases, such as in some models of high-T_c superconductors. The discussion at the end of the previous subsection applies here as well. The directional fluctuations play a crucial role in pion physics [25].

1.5   General Remarks

In order to properly define a field theory, it is necessary to regularize the short-distance behavior of the system. This may be done with the help of the original lattice of the system or by an auxiliary mathematical one. Another possibility is to work in a continuous spacetime, but assuming all momenta to be confined to a sphere of a large radius Λ, called a *momentum space cutoff*. Such a cutoff supplies a smallest length scale over which the order parameter can vary. Since very short distances become visible only under ultraviolet light, Λ is also called an *ultraviolet cutoff* or *UV cutoff*. A continuous field theory with an ultraviolet cutoff Λ has no ultraviolet divergences.

In quantum field theory, a system described by the above Ginzburg-Landau functional (1.83) is called a ϕ^4-theory. Equipped with different symmetries, these ϕ^4-theories have become the most appealing theoretical tool to study the critical phenomena in a great variety of statistical systems. Their relevance for understanding these phenomena was emphasized by Wilson, and Fisher [26]. Using field-theoretic techniques, it has been possible to understand quite satisfactorily all second-order phase transitions of magnetic systems in Table 1.2 and their generalizations. These explain other important phase transitions. An O(2)-symmetric ϕ^4-theory, for example, has the same critical properties as superfluid ^{4}He at the transition to the normal phase. If the critical exponents of O(N) models are continued analytically to $N = 0$, one obtains the values observed in diluted solutions of *polymers* [27].

The calculation of observable consequences of such a quantum field theory proceeds via perturbation theory. Certain correlation functions are obtained as power series in the coupling strength A_4, which is usually denoted by g. The expansion coefficients receive contributions from a rapidly growing number of *Feynman integrals* which are multiple integrals in energy-momentum space.

Initially, quantum field theory is formulated in a Minkowski space with one time and $D-1$ space dimensions. The associated Feynman integrals have singularities in the energy subintegrals, which can be removed by a rotation of the integration contour in the complex energy plane by 90 degrees. This operation, called *Wick rotation*, makes the integrations run along the imaginary energy axis. In quantum field theory, the singularities in the integral correspond to states in a Hilbert space representing particle-like excitations. The Wick rotation removing these singularities changes the quantum field theory into a theory of statistically fluctuating fields. After the Wick rotation, the Planck constant $\hbar$, which in the quantum field theory controls the size of quantum fluctuations, plays the role of the temperature T governing the size of thermal fluctuations. In applications to statistical physics, $\hbar$ may directly be replaced by $k_B T/E_0$, where E_0 is some energy scale. The squared mass m^2 of the particle-like excitations in the original quantum field is proportional to the parameter A_2 in the Ginzburg-Landau functional (1.83). The critical point where thermal fluctuations become violent is characterized by the vanishing of the mass m.

The different types of Feynman integrals are organized most efficiently by means of *Feynman diagrams*. The coefficients of the different powers g^p are associated with Feynman diagrams containing, for a ϕ^4 interaction, a number of closed lines called *loops*. The number of diagrams grows exponentially fast with the power p, like $(p-1)!! \equiv 1 \cdot 3 \cdots (p-1)$.

At the critical point, i.e. for zero mass, a system of fluctuating fields with a quartic self-interaction has the important scaling properties described above. In $D > 4$ dimensions, all exponents have their mean-field values.. The dimension $D = 4$ separating the two types of behavior is called the *critical dimension* of the theory, more precisely the above-defined upper critical dimension $D_u = 4$. In this text, we shall only be concerned with the upper critical dimension. There exists also a *lower critical dimension* $D_l = 2$ which appears in field-theoretic descriptions of the same systems by means of vector fields of unit length with an energy functional (1.69). Expansions around the lower critical dimension are not the subject of this book, although they will be related to the expansions around four dimensions in Section 19.8. We shall use the generic term critical dimension mainly for D_u.

The critical exponents governing the power laws of correlation functions at the critical dimension $D_u = 4$ are different from those observed in nature. It was an important discovery of Wilson and Fisher [26] that the differences could be explained by the difference between the physical space dimension D and the upper critical dimension $D_u = 4$. They found a way to continue the correlation functions analytically from their four-dimensional form to the physical three-dimensional one. During this continuation, they maintained their pure power form of the correlation functions and changed only the numerical values of the critical exponents. The important mathematical tool for this continuation was the calculation of all Feynman integrals in an arbitrary continuous number of dimensions $D = 4 - \varepsilon$, via a power series expansion in ε. The physical exponents are obtained for $\varepsilon = 1$.

Critical phenomena are determined by the long-wavelength fluctuations of a system. As such, they are independent of the short-distance properties of the system. They are indistinguishable for a wide variety of microscopically quite different physical systems. This is the universality property discussed before in Section 1.2.

The initial idea to study critical phenomena of a field theory by Kadanoff [28], employed a repeated application of a so-called *blocking transformation*. It is based on integrating out the fluctuations with short wavelengths while rescaling in a specific way the parameters mass, coupling constant, and field normalization which govern the remaining long-range fluctuations. After a few iterations, the changes stabilize in a fixed point. At this point, the correlation length becomes infinite, corresponding to the limit $m \to 0$, and all correlation functions show a power behavior typical for critical phenomena. Ultimately, this method turned out to be completely equivalent to an application of the so-called *renormalization group* in quantum field theory.

Field-theoretic calculations of critical phenomena in a spacetime continuum have two obstacles. First, all perturbative terms consist of divergent integrals and require a regularization to control the divergences. This obstacle is overcome by the fact that if we limit the interaction to a quartic term in the fields, the theory is renormalizable. This implies that these divergences can be absorbed into a few physical parameters characterizing the theory. They differ from the initial parameters by factors called *renormalization constants*. This removes the first obstacle.

Note that the renormalizability is not of physical relevance for any real many-body system which always possess an intrinsic short-distance scale such as a lattice spacing. The renormalizability is merely of technical advantage enabling us to apply field-theoretic techniques developed for theories in continuous spacetime to the critical phenomena in many-body systems.

The effort in isolating the infinities is useful since the renormalization constants turn out to contain all necessary information on the critical exponents. This comes about as follows. The critical theory near $D = 4$ dimensions can be renormalized only after introducing some arbitrary fixed mass parameter μ. This appears in all correlation functions in conjunction with the physical coupling constant g. The renormalizability of the theory has the consequence that theories with different μ and different g are not completely independent from one another. There exist families of indistinguishable correlation functions characterized by a set of parameters $\mu, g(\mu), m(\mu)$. These families are found by solving the so-called *renormalization group equations*, which are first-order differential equations in the mass parameter μ. These differential equations contain so-called *renormalization group functions* whose power series expansions in ε govern the critical exponents. They are completely determined by the divergences of the Feynman integrals.

The second obstacle in calculating the critical exponents is more serious: Due to the exponentially-fast growing number of Feynman diagrams, the expansion coefficients of the powers g^p increase in size like $(p-1)!!$. This implies that the power series diverge even for very small values of the coupling strength g. Their radius of convergence is zero, and the same thing is true for the series expansion in powers of $\varepsilon = 4 - D$ for the critical exponents.

In order to extract useful information from such expansions, elaborate *resummation methods* have been developed. The progress in the field theory of critical phenomena was possible only by simultaneous progress in resummation theory.

The purpose of the present book is to give an introduction to the techniques of calculating the power series for the critical exponents up to the order ε^5 and resumming them. These series will first be obtained for single quartic self-interaction term with $O(N)$ symmetry, then for a combination of such terms with $O(N)$ and cubic symmetry.

As explained above, perturbation theory requires the generation of a great number of diagrams and their weight factors. For higher orders, this can be done reliably only with the help of computer-algebraic calculations. The counting displayed in this text relies on programs developed first in collaboration with J. Neu [29]. These revealed a counting error of the five-loop diagrams in the standard literature. Helpful for the counting process was a unique representation of the diagrams by a matrix with integer-valued elements, and a simple method for extracting the multiplicity of the diagrams from this matrix (see Chapter 14).

The calculation of Feynman integrals corresponding to the various diagrams is performed in dimensional regularization according to the rules of 't Hooft and Veltman [30]. The divergences of the theory are removed by counterterms defined in a so-called *minimal subtraction scheme.* (MS-scheme). The renormalization group functions obtained in this scheme are of maximal simplicity.

The calculation of the renormalization constants requires a recursive subtraction of the divergences of subdiagrams. This is done with a so-called *R-operation* invented by Bogoliubov [31] which works diagram-wise.

In the minimal subtraction scheme, the renormalization constants are independent of mass and external momenta. This allows their calculation via massless integrals with only one external momentum. The external vertices where this momentum enters a diagram may be chosen arbitrarily as long as the infrared behavior of the integral remains unchanged (*IR-rearrangement*). A Russian group, Chetyrkin, Kataev, and Tkachov [32] developed algorithms for the reduction of such massless integrals to nested one-loop integrals, which turn out to be expressible in terms of Gamma functions. These algorithms are applicable to only a few generic two- and three-loop diagrams, since they are based on a subdiagram of triangle form. They have so far not been generalized to other forms, such as square diagrams.

A further class of diagrams becomes calculable when transforming the massless, dimensionally regularized integrals into a dual form by Fourier transformation. These dual integrals are solvable or may be reduced to solvable integrals by applying certain reduction algorithms developed in **x**-space by Kazakov [33]. He called this algorithm *method of uniqueness*. We prefer instead the name *method of ideal index constellation*.

Most of the diagrams can be brought to one of the calculable forms by IR-rearrangement. For many diagrams, only one arrangement of the external vertices enables us to calculate the corresponding integral. In some cases, IR-rearrangement is successful only if IR-divergences are taken into account. In dimensional regularization, the IR-divergences manifest themselves in the same way as ultraviolet ones as poles in ε. They can, therefore, be subtracted by a procedure analogous to the R-operation, called the R^*-*operation*.

In this way, all integrals up to five loops can be calculated algebraically, with only six exceptions. These require special individual methods combining partial integrations, clever differentiations and applications of the R-operation. For the calculation of all integrals up to five loops, a computer-algebraic program was developed. The methods were available in the literature, but required several corrections.

At the six-loop level, new generic types of diagrams are encountered requiring new methods for their calculation. Subdiagrams of the square type cannot be avoided, and IR-rearrangement is of no help. This is the reason why calculations have not yet been extended to six loops [34].

In the case of fields with several indices and tensorial interactions, each Feynman diagram describes not only a Feynman integral in momentum space, but also a corresponding index sum. This sum leads to certain symmetry factors for each diagram. In this work, a combination of $O(N)$ and cubic symmetry was considered as an interaction, thus allowing the description of many universality classes of critical exponents. For $N = 2, 3$, the $O(N)$ and the mixed $O(N)$-cubic symmetry cover all possible symmetries, assuming only one length scale. For $N = 4$, there are, in principle, many more symmetries. However, two-loop calculations by Toledano, Michel, and Brézin [35] found only two other universality classes of critical behaviors, besides the isotropic and the cubic one. The five-loop calculation of the integrals presented in this book permits extension of these results to the five-loop level.

The results of the integrals together with the symmetry factors yield the critical exponents of the systems as an expansion in ε up to ε^5. For the combination of $O(N)$ and cubic symmetry, these expansions extend former calculations in $D = 4 - \varepsilon$ by two orders, and former calculations by Mayer, Sokolov, and Shalayev [36] in fixed dimension $D = 3$ by one order.

For a comparison with experiments, the series for the critical exponents have to be evaluated at $\varepsilon = 1$. As pointed out before, the series are divergent requiring special techniques for their resummation. The most simple technique, the *Padé approximation*, is the easiest to apply, since it does not require additional information on the series. A more powerful method, however, uses the knowledge of the behavior of all power series at high orders g^k. The series are re-expanded into functions with the same behavior at higher orders. This method can be applied to both, the series in g and the series in ε. It has recently been employed by the present authors [37] for the resummation of series which contain an additional interaction in cubic symmetry, thereby confirming the result from Padé approximations.

The most efficient method for evaluating the perturbation expansions for the critical exponents, however, is variational perturbation theory which has been developed only recently by one of the authors [38]. This method has led to theoretical values for the critical exponent α whose accuracy matches that of the satellite experiments described above. This method will be described in detail and applied in Chapters 19 and 20.

Amplitude ratios will not be discussed in this book. There are two reasons for this: First, they are much harder to calculate than critical exponents, such that their perturbation expansions have only been carried only up to the third order. Second, experiments do not yet provide us with reliable data which can be compared with the theoretical results. We therefore refer the reader to the literature on this subject [39].

Appendix 1A   Correlations and Structure Factor

In Born approximation, the differential cross section of neutron scattering by a liquid whose molecules form an ensemble of scattering potentials $V(\mathbf{x}_j)$, reads

$$\frac{d\sigma}{d\Omega d\omega} = \frac{m^2}{4\pi^2\hbar^4 p}\frac{p'}{p}V(\mathbf{q})S(\mathbf{q},\omega), \qquad \omega = (\mathbf{p}^2 - \mathbf{p}'^2)/2M,$$

where

$$S(\mathbf{q},\omega) \equiv \sum_n e^{-E_n/k_B T} \sum_{n'} \left| \left[\sum_{j=1}^N e^{i\mathbf{q}\mathbf{x}_j/\hbar} \right]_{n'n} \right|^2 \delta\left(\omega + E_{n'} - E_n\right), \qquad p \equiv |\mathbf{p}|,\ p' \equiv |\mathbf{p}'|$$

is the *dynamic structure factor* of the liquid, and $\mathbf{p}, \mathbf{p}'$ are the initial and final neutron momenta. The labels n, n' refer to initial and final wave functions of the liquid, $d\Omega$ the solid angle, $d\omega$ the energy interval of the outgoing neutrons, and $V(\mathbf{q}) \equiv \int d^3x e^{i\mathbf{q}\mathbf{x}/\hbar}V(\mathbf{x})$ is the Fourier transformed interaction potential. In $\mathbf{x}$-space, the structure factor is

$$S(\mathbf{x},t) = \int \frac{d^3q}{(2\pi\hbar)^3} \int \frac{d\omega}{2\pi\hbar} e^{i(\mathbf{q}\mathbf{x}-\omega t)/\hbar} S(\mathbf{q},\omega) = \langle \rho(\mathbf{x},t)\rho(\mathbf{0},0) \rangle.$$

Integrating this over all times yields the (static) structure factor

$$S(\mathbf{x}) \equiv \int \frac{d^3q}{(2\pi\hbar)^3} e^{i\mathbf{q}\mathbf{x}/\hbar} S(\mathbf{q},0). \tag{1A.1}$$

It can be extracted from elastic scattering data and yields direct information on the correlation function $G(\mathbf{x} - \mathbf{y})$.

Notes and References

The classic reference on critical phenomena is
H.E. Stanley, *Introduction to Phase Transitions and Critical Phenomena*, Oxford Science Publications, Oxford, 1971.

Detailed introductions in the quantum field theory of critical phenomena are given in the textbooks by
S. Ma, *Modern Theory of Critical Phenomena*, Benjamin/Cummings, New York, 1976;
D.J. Amit, *Field Theory, the Renormalization Group, and Critical Phenomena*, McGraw-Hill,

New York, 1978;

P. Ramond, *Field Theory, A Modern Primer*, Benjamin/Cummings, New York, 1981;

G. Parisi, *Statistical Field Theory*, Addison Wesley, New York, 1988;

C. Itzykson and J.-M. Drouffe, *Statistical Field Theory*, Cambridge University Press, Cambridge, 1989;

M. Le Bellac, *Quantum and Statistical Field Theory*, Vol I: *Foundations*, Vol II: *Modern Applications*, Oxford Science Publications, Oxford, 1991;

J. Zinn-Justin, *Quantum Field Theory and Critical Phenomena*, Clarendon, Oxford, 1989;

J.J. Binney, N.J. Dowrick, A.J. Fisher, and M.E.J. Newman, *The Theory of Critical Phenomena. An Introduction to the Renormalization Group*, Clarendon Press, Oxford, 1993.

Reviews on critical phenomena are found in
M.E. Fisher, Rep. Prog. Phys. **30**, 731 (1967);
L.P. Kadanoff *et al.*, Rev. Mod. Phys. **39**, 395 (1967);
and on the renormalization group by
K.G. Wilson and J. Kogut, Phys. Rep. **12**, 75 (1974);
E. Brézin, J.C. Le Guillou, and J. Zinn-Justin, in *Phase Transitions and Critical Phenomena*, Vol. 6, ed. C. Domb and M.S. Green, Academic Press, New York, 1976.

For more details on correlation functions and structure factors, see the original paper by
L. Van Hove, Phys. Ref. **95**, 249 (1954) or the textbooks
S.W. Lovesey, *Theory of Neutron Scattering from Condensed Matter*, Vols I and II, in *International Series of Monographs on Physics*, Clarendon, Oxford, 1984;
E. Balcar and S.W. Lovesey, *Theory of Magnetic Neutron and Photon Scattering*, Clarendon, Oxford, 1989.

Table 1.2 is an updated version of a table in
D.J. Amit, *Field Theory, the Renormalization Group, and Critical Phenomena*, McGraw-Hill, 1978.
The table is based on resummation results in $D=3$ dimensions from Chapter 20

The individual citations in the text refer to:

[1] J. Kosterlitz and D. Thouless, J. Phys. C **7**, 1046 (1973).

[2] For disorder fields, the opposite is true. The theory of such fields is developed in
   H. Kleinert, *Gauge Fields in Condensed Matter*, Vol. I *Superflow and Vortex Lines, Disorder Fields and Phase Transitions*, World Scientific, Singapore, 1989 (www.physik.fu-berlin.de/~kleinert/re.html#b1).

[3] The data are from
   W.M. Fairbank, M.J. Buckingham, and C.F. Keller, in *Proceedings 1965 Washington Conference on Critical Phenomena*, ed. M.S. Green and J.V. Sengers, Nat'l. Bur. Stand. Misc. Publ. **273**, 71 (1966).

[4] J.A. Lipa, D.R. Swanson, J. Nissen, T.C.P. Chui, and U.E. Israelson, Phys. Rev. Lett. **76**, 944 (1996).
   See also related data in
   D.R. Swanson, T.C.P. Chui, and J.A. Lipa, Phys. Rev B **46**, 9043 (1992);
   D. Marek, J.A. Lipa, and D. Philips, Phys. Rev B **38**, 4465 (1988);

L.S. Goldner and G. Ahlers, Phys. Rev. B **45**, 13129 (1992);
L.S. Goldner, N. Mulders, and G. Ahlers, J. Low Temp. Phys. **93**, 131 (1992).

[5] The data in Fig. 1.2 are from
G. Ahlers, Phys. Rev. A **3**, 696 (1971); and his lecture at *1978 Erice Summer School on Low Temperature Physics*, ed. J. Ruvalds and T. Regge, North Holland, Amsterdam, 1978.
See also
T.H. McCoy, L.H. Graf, Phys. Lett. A **58**, 287 (1972).

[6] The latest best fit is given in Ref. [15] of J.A. Lipa, D.R. Swanson, J. Nissen, Z.K. Geng, P.R. Williamson, D.A. Stricker, T.C.P. Chui, U.E. Israelson, and M. Larson, Phys. Rev. Lett. **84**, 4894 (2000).

[7] The data are from
D.S. Greywall and G. Ahlers, Phys. Rev. A **7**, 2145 (1973)].

[8] B. Widom, J. Chem. Phys. **43**, 3892, 3898 (1965).

[9] A derivation of this property in field theory will be given in Chapter 7.

[10] L.P. Kadanoff, Physics **2**, 263 (1966).

[11] L.D. Landau, J.E.T.P. **7**, 627 (1937).

[12] For the handling of lattice gradients see Section 2.2 of the textbook by
H. Kleinert, *Path Integrals in Quantum Mechanics, Statistics and Polymer Physics,* World Scientific Publishing Co., Singapore 1995.

[13] For a more detailed derivation see Eq. (5.166) on page 462 of the textbook in Ref. [2].

[14] See Eq. (6.125) on p.168, and Eq. (6A.41) on page 239, and the tables on pages 178 and 241 of the textbook in Ref. [2].

[15] See pages 390 and 391 of the textbook in Ref. [2].

[16] For details see, for example, the textbook in Ref. [2].

[17] V.L. Ginzburg, Fiz. Twerd. Tela **2**, 2031 (1960) [Sov. Phys. Solid State **2**, 1824 (1961)].
See also the detailed discussion in Chapter 13 of the textbook
L.D. Landau and E.M. Lifshitz, *Statistical Physics*, 3rd edition, Pergamon Press, London, 1968.

[18] D.S. Fisher, M.P.A. Fisher, and D.A. Huse, Phys. Rev. B **43**, 130 (1991). See their Eq. (4.1).

[19] See the model discussed in E. Babaev and H. Kleinert, Phys. Rev. B **59**, 12083 (1999) (cond-mat/9907138), and references therein.

[20] See, for example, the textbook of Landau and Lifshitz cited in Ref. [17]. There is also a many-particle path integral description of this phenomena in Chapter 7 of the textbook in Ref. [12].

[21] H. Kleinert, Phys. Rev. Lett. **84**, 286 (2000) (cond-mat/9908239).

[22] See pages 390 and 391 of the textbook in Ref. [2].

[23] See Fig. 5.3 on page 428 of the textbook in Ref. [2].

[24] See pp. 256–257 of the textbook in Ref. [2].

[25] See H. Kleinert and B. Van den Bossche, Phys. Lett. B **474**, 336 (2000) (hep-ph/9907274);
H. Kleinert and E. Babaev, Phys. Lett. B **438**, 311 (1998) (hep-th/9809112).

[26] K.G. Wilson, Phys. Rev. B **4**, 3174, 3184 (1971); K.G. Wilson and M.E. Fisher, Phys.
Rev. Lett. **28**, 240 (1972);
K.G. Wilson, Phys. Rev. Lett. **28**, 548 (1972);
and references therein like, for example,
A.A. Migdal, Sov. Phys. JETP **32**, 552 (1971).

[27] P.G. de Gennes, Phys. Lett. A **38**, 339 (1972).

[28] L.P. Kadanoff *et al.*, Rev. Mod. Phys. **39**, 395 (1967).

[29] J. Neu, M.S. thesis, FU-Berlin (1990).
See also the recent development in
H. Kleinert, A. Pelster, B. Kastening, and M. Bachmann, Phys. Rev. E **62**, 1537 (2000)
(hep-th/9907168).

[30] G. 't Hooft and M. Veltman, Nucl. Phys. B **44**, 189 (1972).

[31] N.N. Bogoliubov and O.S. Parasiuk, Acta Math. **97**, 227 (1957).

[32] K.G. Chetyrkin, A.L. Kataev, and F.V. Tkachov, Nucl. Phys. B **174**, 345 (1980).

[33] D.I. Kazakov, Phys. Lett. B **133**, 406-410 (1983); Theor. Math. Phys. **61**, 84 (1985).

[34] Calculations of more than 5 loops in ε-expansion may be done for certain classes of
diagrams with the help of knot theory:
D.J. Broadhurst, Z. Phys. C **32**, 249 (1986);
D. Kreimer, Phys. Lett. B **273**, 177 (1991);
D.J. Broadhurst and D. Kreimer, Int. J. Mod. Physics C **6**, 519 (1995); UTAS-HHYS-96-
44 (hep-th/9609128).

[35] J.-C. Toledano, L. Michel, P. Toledano, and E. Brézin, Phys. Rev. B **31**, 7171 (1985).

[36] I.O. Mayer, A.I. Sokolov, and B.N. Shalayev, Ferroelectrics, **95**, 93 (1989).

[37] H. Kleinert and V. Schulte-Frohlinde, Phys. Lett. B **342**, 284 (1995) (hep-th/9503230);
See also
H. Kleinert, S. Thoms, and V. Schulte-Frohlinde, Phys. Rev. A **56**, 14428 (1997) (cond-mat/9611050).

[38] The final and most powerful version of variational perturbation theory used in this book
was developed in the textbook in Ref. [12] and
H. Kleinert, Phys. Rev. D **57**, 2264 (1998) (www.physik.fu-berlin.de/˜kleinert/257); addendum Phys. Rev. D **58**, 107702 (1998) (cond-mat/9803268); Phys. Rev. D **60**, 085001
(1999) (hep-th/9812197); Phys. Lett. A **277**, 205 (2000) (cond-mat/9906107); Phys. Lett.

B **434**, 74 (1998) (cond-mat/9801167); Phys. Lett. B **463**, 69 (1999) (cond-mat/9906359); Phys. Lett. A. **264**, 357 (2000) (hep-th/9808145). Prior works are cited in Chapters 19 and 20.

[39] P.C. Hohenberg, A. Aharony, B.I. Halperin, and E.D. Siggia, Phys. Rev. B **13**, 2986 (1976);
C. Bervillier, Phys. Rev. B **14**, 4964 (1976);
F.J. Wegner, Phys. Rev. B **5**, 4529 (1972);
M.-C. Chang and A. Houghton, Phys. Rev. B **21**, 1881 (1980);
C. Bagnuls and C. Bervillier, Phys. Rev. B **24**, 1226 (1981);
J.F. Nicoll and P.C. Albright, Phys. Rev. B **31**, 4576 (1985);
G.M. Avdeeva and A.A. Migdal, JETP Lett. **16**, 178 (1972);
E. Brézin, D.J. Wallace, and K.G. Wilson, Phys. Rev. Lett. **29**, 591 (1972); Phys. Rev. B **7**, 232 (1972);
D.J. Wallace and R.K.P. Zia, Phys. Lett. A **46**, 261 (1973); J. Phys. C **7**, 3480 (1974);
A.R. Rajantie, Nucl. Phys. B **480**, 729 (1996); Addendum **513**, 761 (1998);
V. Dohm, Z. Phys. B **60**, 61 (1985);
R. Schloms and V. Dohm, Nucl. Phys. B **328**, 639 (1989); Phys. Rev. B **42**, 6142 (1990); Addendum **46**, 5883 (1992);
F.J. Halfkann and V. Dohm, Z. Phys. B **89**, 79 (1992);
S.A. Larin, M. Mönnigmann, M. Strösser, and V. Dohm, Phys. Rev. B **58**, 3394 (1998);
M. Strösser, L.A. Larin, and V. Dohm, Nucl. Phys. B **540**, 654 (1999);
S.S.C. Burnett, M. Strösser, and V. Dohm, Nucl. Phys. B **504**, 665 (1997) ; Addendum **509**, 729 (1998);
H. Kleinert and B. Van den Bossche, cond-mat/0011329.

2

Definition of ϕ^4-Theory

A thermally fluctuating field theory is defined by means of a functional integral representation of the partition function, in which the fields are coupled linearly to external sources. This constitutes a generating functional, from which all thermodynamic quantities and correlation functions of the system can be obtained by functional differentiation, as we shall now see.

2.1 Partition Function and Generating Functional

All objects of study in this book are described in terms of an N-component fluctuating field $\phi = (\phi_1(\mathbf{x}) \cdots \phi_N(\mathbf{x}))$ in D euclidean space dimensions. The field components interact with each other via a term of fourth order in the fields, $\lambda T_{\alpha\beta\gamma\delta}\phi_\alpha\phi_\beta\phi_\gamma\phi_\delta$, $(\alpha, \beta, \gamma, \delta = 1, \ldots, N)$, where the parameter $\lambda > 0$ characterizes the interaction strength and is called the *coupling constant* of the theory. The quantity $T_{\alpha\beta\gamma\delta}$ is a *coupling tensor*. Many structural properties of the theory do not depend on the number N of field components. When discussing such properties, the subscripts $\alpha, \beta, \gamma, \delta$ and the tensor $T_{\alpha\beta\gamma\delta}$ will be omitted everywhere, except in Chapters 6, 17, and 18, which derive particular consequences of the tensor structure.

The field $\phi(\mathbf{x})$ performs thermal fluctuations. We shall investigate the properties of the system mainly in the normal phase where the expectation value of the field ϕ is zero, which happens in the temperature regime in which the symmetry of the system is unbroken [see the definition following Eq. (1.1)]. Then the fluctuations take place around zero. Their size is controlled by a local energy functional consisting of two terms:

$$E[\phi] = E_0[\phi] + E_{\text{int}}[\phi]. \tag{2.1}$$

The first term

$$E_0[\phi] = \int d^D x \, \frac{1}{2} \Big\{ [\partial_{\mathbf{x}}\phi(\mathbf{x})]^2 + m^2\phi^2(\mathbf{x}) \Big\} \tag{2.2}$$

is quadratic in the field, and is called the *free-field energy*. The second term

$$E_{\text{int}}[\phi] = \int d^D x \, \frac{\lambda}{4!} \, \phi^4(\mathbf{x}) \tag{2.3}$$

is of fourth-order in the field and is called the *interaction energy*. The parameter m is called the *mass* of the field. This name derives from the fact that if the theory were to be continued analytically in one coordinate, say x_D, to a time-like variable $t = ix_D$, this would change the squared length of a vector from $\mathbf{x}^2 = x_1^2 + \ldots + x_{D-1}^2 + x_D^2$ to $x^2 = x_1^2 + \ldots + x_{D-1}^2 - t^2$, and the free-field energy would turn into an action describing the propagation of relativistic particles of mass m in a Minkowski spacetime. Instead of thermal fluctuations, the field $\phi(\mathbf{x})$ would then perform quantum fluctuations, which can equivalently be described by a *quantum field operator* $\hat{\phi}(x)$, which is capable of creating and annihilating particles of mass m in a multiparticle Hilbert space. No such Hilbert space exists in the present euclidean formulation, but the nomenclature of calling m a mass has nevertheless become customary. Moreover, because of the intimate relationship of the theories by analytic continuation in the time variable $x_D \to -it$, the entire fluctuating field theory is often referred to as a quantum field theory, in contrast to a nonfluctuating field theory, which is similarly referred to as a classical field theory, by analogy.

All thermodynamic properties of the system are described by the partition function of the fluctuating field which is given by the functional integral

$$Z^{\text{phys}} = \int \mathcal{D}\phi(\mathbf{x})\, e^{-E[\phi]/k_B T}, \tag{2.4}$$

where k_B is Boltzmann's constant and T the temperature. The measure of functional integration is defined by the product of integrals at each space point $\mathbf{x}$, multiplied by some irrelevant normalization factor $\mathcal{N}$:

$$\int \mathcal{D}\phi(\mathbf{x}) \equiv \mathcal{N} \prod_{\mathbf{x}} \int d\phi(\mathbf{x}). \tag{2.5}$$

The space points are initially assumed to lie on a narrow spatial lattice, whose spacing is assumed to approach zero.

We shall abbreviate the notation by renormalizing the field, the mass, and the coupling constant λ in such a way that $k_B T = 1$. Thus we shall write the partition function as

$$Z^{\text{phys}} = \int \mathcal{D}\phi(\mathbf{x})\, e^{-E[\phi]}. \tag{2.6}$$

For zero coupling constant λ, this reduces to the free-field partition function

$$Z_0^{\text{phys}} = \int \mathcal{D}\phi(\mathbf{x})\, e^{-E_0[\phi]}. \tag{2.7}$$

The functional integral is now Gaussian, and can be evaluated exactly.

For the upcoming derivation of perturbation expansions it will be convenient to renormalize the partition function by this value, and define the reduced partition function

$$Z \equiv \frac{Z^{\text{phys}}}{Z_0^{\text{phys}}}, \tag{2.8}$$

whose free part Z_0 is equal to unity:

$$Z_0 = 1. \tag{2.9}$$

While the partition function (2.4) describes all thermodynamic properties of the system, it does not give any information on the local properties of the system which are observed in scattering experiments. This information is carried by the *correlation functions* of the field $\phi(\mathbf{x})$ (see Appendix 1A for details):

$$\begin{aligned} G^{(n)}(\mathbf{x}_1, \dots, \mathbf{x}_n) &\equiv \langle \phi(\mathbf{x}_1) \cdots \phi(\mathbf{x}_n) \rangle \\ &= Z^{-1} \int \mathcal{D}\phi(\mathbf{x})\, \phi(\mathbf{x}_1) \cdots \phi(\mathbf{x}_n)\, e^{-E[\phi]}. \end{aligned} \tag{2.10}$$

They are also called *n-point functions* or *Green functions*. There is a compact way of describing all correlation functions of the system with the help of one functional object. One introduces an auxiliary external field $j(\mathbf{x})$ called a *current*, and adds to the energy functional (2.1) a linear interaction energy of this current with the field $\phi(\mathbf{x})$,

$$E_{\text{source}}[\phi, j] = -\int d^D x\, \phi(\mathbf{x}) j(\mathbf{x}), \tag{2.11}$$

thus extending it to a total energy

$$E[\phi, j] = E[\phi] + E_{\text{source}}[\phi, j]. \tag{2.12}$$

The partition function formed with this energy

$$Z[j] = (Z_0^{\text{phys}})^{-1} \int \mathcal{D}\phi(\mathbf{x}) \, e^{-E[\phi, j]} \tag{2.13}$$

is a functional of $j(\mathbf{x})$ whose value at $j(\mathbf{x}) \equiv 0$ is equal to the normalized partition function (2.8).

The functional derivatives of $Z[j]$ with respect to $j(\mathbf{x})$ evaluated at $j \equiv 0$ yield obviously the correlation functions of the system:

$$G^{(n)}(\mathbf{x}_1, \ldots, \mathbf{x}_n) = Z^{-1} \left[\frac{\delta}{\delta j(\mathbf{x}_1)} \cdots \frac{\delta}{\delta j(\mathbf{x}_n)} Z[j] \right]_{j \equiv 0}. \tag{2.14}$$

For this reason, $Z[j]$ is referred to as the *generating functional* of the theory.

2.2   Free-Field Theory

The properties of the theory defined by the generating functional (2.13) will be investigated with the help of perturbative expansions. This is an expansion of $Z[j]$ around the generating functional $Z_0[j]$ of the free-field theory in powers of the coupling constant λ. For this we rewrite the free-field energy functional (2.2) after a partial integration of the gradient term as

$$\begin{aligned}
E_0[\phi] &= \frac{1}{2} \int d^D x \, \phi(\mathbf{x}) \left(-\partial_{\mathbf{x}}^2 + m^2 \right) \phi(\mathbf{x}) \\
&= \frac{1}{2} \int d^D x_1 d^D x_2 \, \phi(\mathbf{x}_1) \, D(\mathbf{x}_1, \mathbf{x}_2) \, \phi(\mathbf{x}_2). \tag{2.15}
\end{aligned}$$

In the second line we have expressed the differential operator $-\partial_{\mathbf{x}}^2 + m^2$ as a symmetric *functional matrix* $D(\mathbf{x}_1, \mathbf{x}_2) = D(\mathbf{x}_2, \mathbf{x}_1)$. This is a matrix with continuous indices $\mathbf{x}_1$ and $\mathbf{x}_2$. Its explicit form is

$$D(\mathbf{x}_1, \mathbf{x}_2) = \delta^{(D)}(\mathbf{x}_1 - \mathbf{x}_2) \left(-\partial_{\mathbf{x}_2}^2 + m^2 \right). \tag{2.16}$$

The difference between (2.15) and (2.2) is a surface term which vanishes if the ϕ-field satisfies periodic boundary conditions. This will always be assumed.

Given such a quadratic energy, we can immediately calculate the free-field functional integral (2.7). Remembering the definition (2.5), we see that

$$Z_0^{\text{phys}} = \int \mathcal{D}\phi(\mathbf{x}) \, e^{-E_0[\phi]} = \mathcal{N} \left[\prod_{\mathbf{x}} \int d\phi(\mathbf{x}) \right] e^{-E_0[\phi]} = \frac{1}{\sqrt{\text{Det} D}}, \tag{2.17}$$

where $\text{Det} D$ is the functional determinant of $D(\mathbf{x}_1, \mathbf{x}_2)$. This is defined by a straightforward generalization of a well-known identity for matrices:

$$\text{Det} D = e^{\text{Tr} \log D} = e^{\text{Tr} \log[1 + (D-1)]}. \tag{2.18}$$

Hence

$$Z_0^{\text{phys}} = e^{-\frac{1}{2} \text{Tr} \log D}. \tag{2.19}$$

The proof of this formula is simple. A single Gaussian integral yields

$$\int \frac{d\phi}{\sqrt{2\pi}} e^{-a\phi^2/2} = \frac{1}{\sqrt{a}}. \tag{2.20}$$

For a diagonal matrix M with matrix elements $M_{ii} = a_i$, we obtain similarly

$$\prod_i \int \frac{d\phi_i}{\sqrt{2\pi}} e^{-a_i\phi_i^2/2} = \frac{1}{\sqrt{\Pi_i a_i}}. \tag{2.21}$$

An arbitrary symmetric matrix M can be diagonalized by a rotation of the integration variables ϕ_i which does not change the volume of integration, so that we find

$$\prod_i \int \frac{d\phi_i}{\sqrt{2\pi}} e^{-\phi_i M_{ij}\phi_j/2} = \frac{1}{\sqrt{\Pi_i a_i}} = \frac{1}{\sqrt{\det M}}. \tag{2.22}$$

Since the measure of functional integration in (2.17) is defined by the product of integrals at each space point $\mathbf{x}$, this formula applies with the indices i replaced by $\mathbf{x}$, and the determinant on the right-hand side of (2.22) replaced by the functional determinant on the right-hand side of (2.17). The normalization factor $\mathcal{N}$ picks up all infinite factors arising in the limit of zero lattice spacing.

With the explicit result (2.17) we can write the reduced partition function (2.8) as

$$Z = e^{\frac{1}{2}\mathrm{Tr}\,\log D} \int \mathcal{D}\phi(\mathbf{x})\, e^{-E[\phi]}. \tag{2.23}$$

It will be useful to absorb the prefactor into the measure of integration and define

$$\int \mathcal{D}'\phi(\mathbf{x}) \equiv e^{\frac{1}{2}\mathrm{Tr}\,\log D} \int \mathcal{D}\phi(\mathbf{x}), \tag{2.24}$$

so that we write the reduced partition function as

$$Z = \int \mathcal{D}'\phi(\mathbf{x})\, e^{-E[\phi]}, \tag{2.25}$$

and the generating functional (2.13) as

$$Z[j] = \int \mathcal{D}'\phi(\mathbf{x})\, e^{-E[\phi,j]}. \tag{2.26}$$

Setting here $\lambda = 0$, we obtain the generating functional of the free-field theory:

$$Z_0[j] = \int \mathcal{D}'\phi(\mathbf{x})\, e^{-\{E_0[\phi]+E_{\mathrm{source}}[\phi,j]\}}. \tag{2.27}$$

This can be integrated as easily as for zero sources. The field still occurs at most quadratically in the exponent, and this permits us to absorb the current term into the field $\phi(\mathbf{x})$ by a shift

$$\phi(\mathbf{x}) \to \phi'(\mathbf{x}) = \phi(\mathbf{x}) - \int d^D x'\, D^{-1}(\mathbf{x}, \mathbf{x}') j(\mathbf{x}'), \tag{2.28}$$

where the symbol $D^{-1}(\mathbf{x}, \mathbf{x}')$ denotes the inverse functional matrix of $D(\mathbf{x}, \mathbf{x}')$, defined by

$$\int d^D x'\, D(\mathbf{x}, \mathbf{x}') D^{-1}(\mathbf{x}', \mathbf{x}'') = \delta^{(D)}(\mathbf{x} - \mathbf{x}''). \tag{2.29}$$

With $D(\mathbf{x}, \mathbf{x}')$, the inverse functional matrix $D^{-1}(\mathbf{x}, \mathbf{x}')$ is also symmetric in its arguments. The shift (2.28) transforms the generating functional (2.27) into

$$Z_0[j] = \int \mathcal{D}'\phi'(\mathbf{x}) \, e^{-E_0[\phi']} e^{\frac{1}{2}\int d^Dx d^Dx' \, j(\mathbf{x}) D^{-1}(\mathbf{x},\mathbf{x}')j(\mathbf{x}')}. \tag{2.30}$$

The functional integral over the shifted field $\phi'(\mathbf{x})$ yields the same constant as in (2.19), leading to

$$Z_0[j] = e^{\frac{1}{2}\int d^Dx d^Dx' \, j(\mathbf{x}) D^{-1}(\mathbf{x},\mathbf{x}')j(\mathbf{x}')}. \tag{2.31}$$

For $j \equiv 0$, this coincides with the reduced free-field partition function (2.7) with unit normalization.

The correlation functions of the free-field theory are now obtained from the functional derivatives of $Z_0[j]$ at $j(\mathbf{x}) \equiv 0$:

$$
\begin{aligned}
G_0^{(n)}(\mathbf{x}_1, \ldots, \mathbf{x}_n) &= \left[\frac{\delta}{\delta j(\mathbf{x}_1)} \cdots \frac{\delta}{\delta j(\mathbf{x}_n)} Z_0[j] \right]_{j\equiv 0} \\
&= \left[\frac{\delta}{\delta j(\mathbf{x}_1)} \cdots \frac{\delta}{\delta j(\mathbf{x}_n)} e^{\frac{1}{2}\int d^Dx d^Dx' \, j(\mathbf{x}) D^{-1}(\mathbf{x},\mathbf{x}')j(\mathbf{x}')} \right]_{j\equiv 0}.
\end{aligned} \tag{2.32}
$$

Only correlation functions with an even number of fields are nonzero. They are given by the functional derivatives

$$G_0^{(n)}(\mathbf{x}_1, \ldots, \mathbf{x}_n) = \frac{1}{2^{n/2} \, (n/2)!} \left\{ \frac{\delta}{\delta j(\mathbf{x}_1)} \cdots \frac{\delta}{\delta j(\mathbf{x}_n)} \left[\int d^Dx d^Dx' \, j(\mathbf{x}) D^{-1}(\mathbf{x}, \mathbf{x}')j(\mathbf{x}') \right]^{n/2} \right\}_{j\equiv 0}. \tag{2.33}$$

For $n = 2$, this equation yields the *free two-point function* or the *free propagator* of the field

$$
\begin{aligned}
G_0^{(2)}(\mathbf{x}_1, \mathbf{x}_2) &= \frac{1}{2} \left[\frac{\delta}{\delta j(\mathbf{x}_1)} \frac{\delta}{\delta j(\mathbf{x}_2)} \int d^Dx d^Dx' \, j(\mathbf{x}) D^{-1}(\mathbf{x}, \mathbf{x}')j(\mathbf{x}') \right]_{j\equiv 0} \\
&= \frac{1}{2} \left[D^{-1}(\mathbf{x}_1, \mathbf{x}_2) + D^{-1}(\mathbf{x}_2, \mathbf{x}_1) \right] \\
&= D^{-1}(\mathbf{x}_1, \mathbf{x}_2).
\end{aligned} \tag{2.34}
$$

Thus we may replace $D^{-1}(\mathbf{x}, \mathbf{x}')$ in Eqs. (2.31)–(2.33) by the free two-point function $G_0^{(2)}(\mathbf{x}, \mathbf{x}')$. This function will occur so often in this text that it is convenient to drop the superscript and write $G_0(\mathbf{x}, \mathbf{x}')$ for $G_0^{(2)}(\mathbf{x}, \mathbf{x}')$.

Equation (2.33) implies that all n-point functions of the free-field theory can be expressed as sums of products of two-point functions. The product rule of differentiation leads to a sum of $n!$ terms. Since $G_0^{(2)}(\mathbf{x}, \mathbf{x}')$ is symmetric in its arguments, and multiplication is commutative, the $n!$ terms decompose into groups of $2^{n/2} \times (n/2)!$ identical terms. This number is canceled out by the denominator in front of the right-hand side of (2.33) coming from the Taylor expansion of the exponential in (2.32). Hence we remain with a sum of $n!/[2^{n/2}(n/2)!] = (n-1)!!$ different products of $n/2$ propagators $G_0^{(2)}$ with no numerical prefactor:

$$
\begin{aligned}
G_0^{(n)}(\mathbf{x}_1, \ldots, \mathbf{x}_n) &= \sum_{i=1}^{(n-1)!!} G_0^{(2)}(\mathbf{x}_{\pi_i(1)}, \mathbf{x}_{\pi_i(2)}) \cdots G_0^{(2)}(\mathbf{x}_{\pi_i(n-1)}, \mathbf{x}_{\pi_i(n)}) \\
&= \sum_{i=1}^{(n-1)!!} \prod_{j=1}^{n/2} G_0^{(2)}(\mathbf{x}_{\pi_i(2j-1)}, \mathbf{x}_{\pi_i(2j)}).
\end{aligned} \tag{2.35}
$$

The indices $\pi_i(j)$ $(1 \leq i \leq (n-1)!!,\ 1 \leq j \leq n)$ label the position arguments for all independent set of pairs: $\{\ [\pi_i(1), \pi_i(2)], \ldots, [\pi_i(n-1), \pi_i(n)]\ \}$. The right-hand side of (2.35) is known as *Wick's expansion*.

The easiest way to enumerate the $(n-1)!!$ different pair combinations in Wick's expansion is by lining up the n indices

$$1\ 2\ 3\ 4\ \ldots\ n-2\ n-1\ n, \tag{2.36}$$

and introducing a *pair contraction* marked by two common dots, for instance:

$$\dot{1}\ \dot{2}\ 3\ 4\ \ldots\ n-2\ n-1\ n. \tag{2.37}$$

One then forms the $n-1$ single contractions starting out from the index 1:

$$\begin{aligned}
& \dot{1}\ \dot{2}\ 3\ 4\ \ldots\ n-2\ n-1\ n \quad + \quad \dot{1}\ 2\ \dot{3}\ 4\ \ldots\ n-2\ n-1\ n \\
+\ & \dot{1}\ 2\ 3\ \dot{4}\ \ldots\ n-2\ n-1\ n \quad + \quad \ldots \quad + \quad \dot{1}\ 2\ 3\ 4\ \ldots\ n\ \dot{-}\ 2\ n-1\ n \\
+\ & \dot{1}\ 2\ 3\ 4\ \ldots\ n-2\ n\ \dot{-}\ 1\ n \quad + \quad \dot{1}\ 2\ 3\ 4\ \ldots\ n-2\ n-1\ \dot{n}.
\end{aligned} \tag{2.38}$$

The uncontracted $n-2$ indices in each term are now treated once more in the same way, etc., until all indices are contracted. The $(n-1)!!$ terms obtained in this way are precisely the indices of the position arguments of the free propagators on the right-hand side of Eq. (2.35).

Note that the functional matrix (2.16) is invariant under arbitrary translations $\mathbf{a}$ of the spatial arguments of the field $\phi(\mathbf{x})$

$$\mathbf{x} \to \mathbf{x} + \mathbf{a}. \tag{2.39}$$

This property goes over to its inverse, the propagator, which therefore depends only on the difference between its arguments. It is then useful to introduce a propagator function with only a single argument:

$$G_0^{(2)}(\mathbf{x}, \mathbf{x}') = G_0(\mathbf{x}, \mathbf{x}') \equiv G_0(\mathbf{x} - \mathbf{x}'). \tag{2.40}$$

For the free n-point functions $G_0^{(n)}(\mathbf{x}_1, \ldots, \mathbf{x}_n)$, translational invariance has the consequence

$$G_0^{(n)}(\mathbf{x}_1, \ldots, \mathbf{x}_n) = G_0^{(n)}(\mathbf{x}_1 - \mathbf{x}_n, \ldots, \mathbf{x}_{n-1} - \mathbf{x}_n, 0), \tag{2.41}$$

which can be directly verified using the Wick expansion (2.35) and the translational invariance (2.40) of the two-point function.

2.3 Perturbation Expansion

An expansion of the generating functional $Z[j]$ in (2.13) in a power series of the coupling constant λ yields the so-called *perturbation expansion*

$$Z[j]\ =\ Z_0[j]\ +\ \sum_{p=1}^{\infty} Z_p[j], \tag{2.42}$$

which reads more explicitly

$$\begin{aligned}
Z[j]\ =\ & Z_0[j] + \sum_{p=1}^{\infty} \frac{1}{p!} \lambda^p \left[\frac{d^p}{d\lambda^p} Z[j] \right]_{\lambda=0} \\
=\ & Z_0[j] + \sum_{p=1}^{\infty} \frac{1}{p!} \left(\frac{-\lambda}{4!} \right)^p \int \mathcal{D}'\phi(\mathbf{x}) \int d^D z_1 \cdots d^D z_p \\
& \times \phi^4(\mathbf{z}_1) \cdots \phi^4(\mathbf{z}_p)\, e^{-\{E_0[\phi] + E_{\text{source}}[\phi, j]\}}.
\end{aligned} \tag{2.43}$$

For zero external current, the terms on the right-hand side contain integrals over correlation functions of $4p$ fields. Thus we may write the perturbation series of the partition function as

$$Z \equiv Z[j \equiv 0] = 1 + \sum_{p=1}^{\infty} Z_p, \tag{2.44}$$

where the pth-order contribution is given by

$$Z_p = \frac{1}{p!} \left(\frac{-\lambda}{4!} \right)^p \int d^D z_1 \cdots d^D z_p \, G_0^{(4p)}(\mathbf{z}_1, \mathbf{z}_1, \mathbf{z}_1, \mathbf{z}_1, \ldots, \mathbf{z}_p, \mathbf{z}_p, \mathbf{z}_p, \mathbf{z}_p), \tag{2.45}$$

and where $G_0^{(4p)}$ are free $4p$-point functions with the Wick expansion (2.35).

Applying the derivatives in (2.14) to (2.43), we find the perturbation series for n-point functions

$$G^{(n)}(\mathbf{x}_1, \ldots, \mathbf{x}_n) = Z^{-1} \left[\frac{\delta}{\delta j(\mathbf{x}_1)} \cdots \frac{\delta}{\delta j(\mathbf{x}_n)} Z[j] \right]_{j \equiv 0}$$

$$= Z^{-1} \left[G_0^{(n)}(\mathbf{x}_1, \ldots, \mathbf{x}_n) + \sum_{p=1}^{\infty} \frac{1}{p!} \left(\frac{-\lambda}{4!} \right)^p \int d^D z_1 \cdots d^D z_p \right. \tag{2.46}$$

$$\left. \times \int \mathcal{D}'\phi(\mathbf{x}) \, \phi(\mathbf{x}_1) \cdots \phi(\mathbf{x}_n) \phi^4(\mathbf{z}_1) \cdots \phi^4(\mathbf{z}_p) \, e^{-\{E_0[\phi] + E_{\text{source}}[\phi, j]\}} \right]_{j=0}.$$

The functional integral on the right-hand side may be expressed in terms of free n-point functions $G_0^{(n)}(\mathbf{x}_1, \ldots, \mathbf{x}_n)$ as

$$G^{(n)}(\mathbf{x}_1, \ldots, \mathbf{x}_n) = Z^{-1} \left[G_0^{(n)}(\mathbf{x}_1, \ldots, \mathbf{x}_n) + \sum_{p=1}^{\infty} G_p^{(n)}(\mathbf{x}_1, \mathbf{x}_2, \ldots, \mathbf{x}_n), \right] \tag{2.47}$$

with the contributions of pth order with $p \geq 1$ given by the integrals

$$G_p^{(n)}(\mathbf{x}_1, \ldots, \mathbf{x}_n) = \frac{1}{p!} \left(\frac{-\lambda}{4!} \right)^p \int d^D z_1 \cdots d^D z_p$$

$$\times \; G_0^{(n+4p)}(\mathbf{z}_1, \mathbf{z}_1, \mathbf{z}_1, \mathbf{z}_1, \ldots, \mathbf{z}_p, \mathbf{z}_p, \mathbf{z}_p, \mathbf{z}_p, \mathbf{x}_1, \ldots, \mathbf{x}_n). \tag{2.48}$$

The free-field correlation functions $G_0^{(n+4p)}$ in the sum may now be Wick-expanded as in Eq. (2.35) into sums over products of propagators G_0. Due to the proliferating number of terms with increasing order p, the evaluation of these series rapidly becomes complicated. Even their organization is involved, although it may be economized with the help of Feynman diagrams to be introduced in the next chapter.

An important property should be observed at this place. By (2.41), the free n-point functions $G_0^{(n)}(\mathbf{x}_1, \ldots, \mathbf{x}_n)$ are invariant under the spatial translations (2.39):

$$G_0^{(n)}(\mathbf{x}_1, \ldots, \mathbf{x}_n) = G_0^{(n)}(\mathbf{x}_1 + \mathbf{a}, \ldots, \mathbf{x}_n + \mathbf{a}). \tag{2.49}$$

The perturbation expansion (2.48) shows that this property goes over to the interacting n-point functions:

$$G^{(n)}(\mathbf{x}_1, \ldots, \mathbf{x}_n) = G^{(n)}(\mathbf{x}_1 + \mathbf{a}, \ldots, \mathbf{x}_n + \mathbf{a}). \tag{2.50}$$

This is, of course, a general consequence of the fact that the full interacting field energy (2.1) is invariant under arbitrary translations of the spatial argument of the field:

$$\phi(\mathbf{x}) \rightarrow \phi(\mathbf{x} + \mathbf{a}). \tag{2.51}$$

For the interacting two-point function, translational invariance implies that it depends only on the difference of its arguments:

$$G^{(2)}(\mathbf{x}, \mathbf{x}') \equiv G(\mathbf{x}, \mathbf{x}') \equiv G(\mathbf{x} - \mathbf{x}'), \tag{2.52}$$

where we have introduced a bilocal short notation $G(\mathbf{x}, \mathbf{x}')$ for the interacting propagator, just as for the free propagator in (2.40).

2.4 Composite Fields

Later, in Section 7.3.1, we shall see a special role played by a certain class of correlation functions with coinciding spatial arguments of some of its fields. For these we introduce the notation

$$G^{(l,n)}(\mathbf{y}_1, \dots, \mathbf{y}_l, \mathbf{x}_1, \dots, \mathbf{x}_n) \equiv \frac{1}{2^l} \langle \phi^2(\mathbf{y}_1) \cdots \phi^2(\mathbf{y}_l) \phi(\mathbf{x}_1) \cdots \phi(\mathbf{x}_n) \rangle. \tag{2.53}$$

They can, of course, all be generated by appropriate functional derivatives of $Z[j]$. More conveniently, however, we introduce a new generating functional extended by a source term coupled to the square of the field $\phi(\mathbf{x})$:

$$Z[j, K] = \int \mathcal{D}'\phi(\mathbf{x}) \, e^{-E_0[\phi] + \int d^D x [j(\mathbf{x})\phi(\mathbf{x}) + \frac{1}{2} K(\mathbf{x})\phi^2(\mathbf{x})]}. \tag{2.54}$$

From this, the correlation functions (2.53) are obtained by the functional derivatives

$$G^{(l,n)}(\mathbf{y}_1, \dots, \mathbf{y}_l, \mathbf{x}_1, \dots, \mathbf{x}_n) \tag{2.55}$$

$$= Z^{-1} \left[\frac{\delta}{\delta K(\mathbf{y}_1)} \cdots \frac{\delta}{\delta K(\mathbf{y}_l)} \frac{\delta}{\delta j(\mathbf{x}_1)} \cdots \frac{\delta}{\delta j(\mathbf{x}_n)} Z[j] \right]_{j \equiv K \equiv 0}.$$

For $K(\mathbf{x}) = \text{constant} = k$, the source term can be combined with the mass term in the energy functional (2.2). As a consequence, the correlation functions for a mass $m^2 - k$ can be written as a power series in k:

$$G^{(n)}(\mathbf{x}_1, \dots, \mathbf{x}_n) \Big|_{m^2 \to m^2 - k} = \sum_{l=0}^{\infty} \frac{k^l}{l!} \int d^D y_1 \cdots d^D y_l \, G^{(l,n)}(\mathbf{y}_1, \dots, \mathbf{y}_l, \mathbf{x}_1, \dots, \mathbf{x}_n). \tag{2.56}$$

This equation shows that $G^{(1,n)}$ can be obtained by differentiating $G^{(n)}(\mathbf{x}_1, \dots, \mathbf{x}_n)$ with respect to the mass. If we emphasize the mass dependence of $G^{(n)}(\mathbf{x}_1, \dots, \mathbf{x}_n)$ by an extra argument m, we have

$$-\frac{\partial}{\partial m^2} G^{(n)}(\mathbf{x}_1, \dots, \mathbf{x}_n, m) = \int d^D y \, G^{(1,n)}(\mathbf{y}, \mathbf{x}_1, \dots, \mathbf{x}_n), \tag{2.57}$$

or, expressed in terms of field expectation values:

$$-\frac{\partial}{\partial m^2} \langle \phi(\mathbf{x}_1) \cdots \phi(\mathbf{x}_n) \rangle = \frac{1}{2} \int d^D y \, \langle \phi^2(\mathbf{y}) \phi(\mathbf{x}_1) \cdots \phi(\mathbf{x}_n) \rangle. \tag{2.58}$$

The differentiation $-m^2 \partial_{m^2}$ is called a *mass insertion*, since it inserts a mass term $m^2 \phi^2(\mathbf{x})/2$ into a correlation function.

Notes and References

Textbooks on quantum field theory are listed at the end of Chapter 1. The notation in the present book follows
H. Kleinert, *Gauge Fields in Condensed Matter*, Vol. I, *Superflow and Vortex Lines, Disorder Fields and Phase Transitions*, World Scientific, Singapore, 1989 (www.physik.fu-berlin.de/~kleinert/re.html#b1).

3

Feynman Diagrams

In the previous chapter we have derived perturbative expansion formulas for the partition function and the n-point correlation functions. The expansion coefficients are sums of multiple integrals whose number grows rapidly with increasing order. Their organization is simplified by means of diagrammatic techniques due to Feynman. All expansions will be performed in the normal phase of the system which possesses the symmetry of the energy functional [recall the remarks following Eq. (1.1)].

3.1 Diagrammatic Expansion of Correlation Functions

According to Eq. (2.47), the n-point correlation functions $G^{(n)}$ possesses an expansion in powers of the coupling constant λ as:

$$G^{(n)}(\mathbf{x}_1, \ldots, \mathbf{x}_n) = Z^{-1} \sum_{p=0}^{\infty} G_p^{(n)}(\mathbf{x}_1, \ldots, \mathbf{x}_n), \tag{3.1}$$

where the expansion terms $G_p^{(n)}(\mathbf{x}_1, \ldots, \mathbf{x}_n)$ of order λ^p are given by the integrals (2.48). Using the generating formula (2.32), these can be obtained from the functional derivatives

$$G_p^{(n)}(\mathbf{x}_1, \ldots, \mathbf{x}_n) = \left(\frac{-\lambda}{4!} \right)^p \frac{1}{p!} \frac{1}{(2p+n/2)! \, 2^{2p+n/2}} \tag{3.2}$$

$$\times \frac{\delta}{\delta j(\mathbf{x}_1)} \cdots \frac{\delta}{\delta j(\mathbf{x}_n)} \left[\prod_{i=1}^{p} \int d^D z_i \left(\frac{\delta}{\delta j(\mathbf{z}_i)} \right)^4 \right] \left[\int d^D x \, d^D y \, j(\mathbf{x}) G_0(\mathbf{x}, \mathbf{y}) j(\mathbf{y}) \right]^{2p+n/2}.$$

The $4p + n$ functional derivatives on the right-hand side produce a sum of $(4p + n)!$ terms consisting of products of free two-point correlation functions $G_0(\mathbf{x}, \mathbf{x}')$. Among these, $(2p + n/2)! \, 2^{2p+n/2}$ are identical because of the symmetry in the arguments and the commutativity of the associated propagators. This multiplicity factor cancels out the factorials in the denominators of (3.2) which had their origin in the Taylor expansion of the exponential (2.32). One is left with a sum of $(4p + n - 1)!!$ terms, each a product of $(4p + n)/2$ free propagators.

The counting is clearest by writing each term in such a way that coinciding $\mathbf{z}$-arguments of the free propagators are initially distinguished. Their coincidence is enforced in an extended integral by means of δ-functions. In this way we obtain the expression

$$G_p^{(n)}(\mathbf{x}_1, \ldots, \mathbf{x}_n) \equiv \frac{1}{p!} \left(\frac{-\lambda}{4!} \right)^p \int d^D z_1 \cdots d^D z_p \int d^D y_1 \cdots d^D y_{4p+n} \prod_{l=1}^{n} \left[\delta^{(D)}(\mathbf{y}_{4p+l} - \mathbf{x}_l) \right]$$

$$\times \prod_{k=1}^{p} \left[\delta^{(D)}(\mathbf{y}_{4k-3} - \mathbf{z}_k) \delta^{(D)}(\mathbf{y}_{4k-2} - \mathbf{z}_k) \delta^{(D)}(\mathbf{y}_{4k-1} - \mathbf{z}_k) \delta^{(D)}(\mathbf{y}_{4k} - \mathbf{z}_k) \right]$$

$$\times \sum_{i=1}^{(4p+n-1)!!} \prod_{j=1}^{\frac{4p+n}{2}} G_0\big(\mathbf{y}_{\pi_i^{(4p+n)}(2j-1)}, \mathbf{y}_{\pi_i^{(4p+n)}(2j)}\big). \tag{3.3}$$

We have given the pair indices $\pi_i(2j-1), \pi_i(2j)$ a superscript which indicates the total number of indices from which the pairs have been selected. The sum includes all possible permutations of the $\mathbf{y}_i$ variables, except for those which correspond only to an interchange of the spatial arguments within a propagator, or to an interchange of identical propagators as a whole. Those permutations have already been accounted for by factors $(2p+l)! \, 2^{(2p+n/2)}$, which are subsequently canceled by the denominators of the Taylor expansion of the exponential (2.32). The remaining permutations in the sum (3.3) are called *relevant permutations*. This restriction of the sum has to be kept in mind when carrying out the $\mathbf{y}$- and $\mathbf{z}$-integrations.

Each product in the sum (3.3) is pictured by a *Feynman diagram* or *Feynman graph*. We will use $G_p^{(n)}$ to denote both, the integral representation and the diagram representation. The position variables $\mathbf{x}_l$ and $\mathbf{z}_k$ are represented by points, and the free propagators $G_0(\mathbf{y}, \mathbf{y}')$ by lines

$$\underset{\mathbf{y} \qquad \mathbf{y}'}{\bullet\!\!-\!\!-\!\!-\!\!-\!\!\bullet} \;\; \stackrel{\wedge}{=} \;\; G_0(\mathbf{y}, \mathbf{y}') \qquad\qquad (3.4)$$

connecting the points $\mathbf{y}$ and $\mathbf{y}'$. The points $\mathbf{y}_{4p+1} = \mathbf{x}_1, \ldots, \mathbf{y}_{4p+n} = \mathbf{x}_n$ are endpoints of a line. They are called *external points* and the corresponding lines are called *external lines*. The points $\mathbf{y}_1, \ldots, \mathbf{y}_{4p}$ correspond to integration variables. They are called *internal points*. They carry δ-functions in (3.3) that enforce the coincidence of groups of four internal points. Such coinciding points with four emerging lines are called *vertices*. In this introductory discussion, they are represented by a small circle marked by the variable of integration. Later we shall always include the coupling constant into the definition of the vertex and omit specifying the associated integration variable. This will be indicated by a dot:

$$\times \;\; \stackrel{\wedge}{=} \;\; -\lambda \int d^D z \, . \qquad\qquad (3.5)$$

The diagrams associated with an n-point function are called *n-point diagrams*. The sum in (3.3) runs over all diagrams which can be drawn for p vertices and n external points. Each diagram appears repeatedly with permuted line and vertex labels.

After carrying out the $\mathbf{y}$-integrations, many terms in the sum (3.3) coincide. As an example, consider the diagrammatic representation of the expansion coefficients $G_1^{(2)}$. According to Wick's rule, there are $(4 \times 1 + 2 - 1)!! = 5 \times 3 \times 1 = 15$ contractions, and an equal number of diagrams. Four of them are pictured in Fig. 3.1. All other diagrams in the sum are either

(a)   (b)   (c)   (d)

FIGURE 3.1 Diagrams contributing to expansion of $G_1^{(2)}$.

of type (a) and (b), or of type (c) and (d), but with the indices of the $\mathbf{y}$-variables permuted. Omitting the labels $\mathbf{y}_1, \ldots, \mathbf{y}_4$ of the four endpoints of lines coinciding with the point $\mathbf{z}_1$, the diagrams of type (a) and (b), as well as those of type (c) and (d), become indistinguishable.

The number of times with which each Feynman integral occurs among the functional derivatives in (3.2), which is equal to the number of Wick contractions leading to the same Feynman integral, is called the *multiplicity M_G* of a Feynman integral.

The multiplicity M_G is usually combined with the factor $1/4!^p p!$ accompanying the pth-order expansion term of the exponential function in (2.46). The result is the so-called *weight factor* W_G of a diagram G:

$$W_G = \frac{M_G}{4!^p p!}. \tag{3.6}$$

It is this factor which eventually accompanies each Feynman diagram in the perturbation expansion.

Initially, each vertex possesses four emerging lines and there are 4! ways of labeling them. Thus there are in general 4! diagrams of each type. In many cases, however, this number is reduced because the sum in (3.3) does not cover all possible index permutations. This can be seen analytically upon performing the integrations over $\mathbf{y}_1, \ldots, \mathbf{y}_4$. For a vertex at $\mathbf{z}$, the variables $\mathbf{y}_1, \ldots, \mathbf{y}_4$ may appear all in different propagators,

$$\int d^D y_1 \ldots d^D y_4 \delta^{(D)}(\mathbf{y}_1 - \mathbf{z}) \delta^{(D)}(\mathbf{y}_2 - \mathbf{z}) \delta^{(D)}(\mathbf{y}_3 - \mathbf{z}) \delta^{(D)}(\mathbf{y}_4 - \mathbf{z})$$
$$\times\, G_0(\mathbf{y}_1, \bar{\mathbf{y}}_1)\, G_0(\mathbf{y}_2, \bar{\mathbf{y}}_2)\, G_0(\mathbf{y}_3, \bar{\mathbf{y}}_3)\, G_0(\mathbf{y}_4, \bar{\mathbf{y}}_4)$$
$$= G_0(\mathbf{z}, \bar{\mathbf{y}}_1)\, G_0(\mathbf{z}, \bar{\mathbf{y}}_2)\, G_0(\mathbf{z}, \bar{\mathbf{y}}_3)\, G_0(\mathbf{z}, \bar{\mathbf{y}}_4), \tag{3.7}$$

where $\bar{\mathbf{y}}_1, \ldots, \bar{\mathbf{y}}_4$ are an arbitrary set of variables. As there are 4! ways to arrange $\mathbf{y}_1, \ldots, \mathbf{y}_4$ in the integrand, there are 4! terms which coincide after the integration, giving an overall factor 4!. However, instead of (3.7), we could also encounter a situation in which two $\mathbf{y}_i$-variables appear in the same propagator.

$$\int d^D y_1 \ldots d^D y_4 \delta^{(D)}(\mathbf{y}_1 - \mathbf{z}) \delta^{(D)}(\mathbf{y}_2 - \mathbf{z}) \delta^{(D)}(\mathbf{y}_3 - \mathbf{z}) \delta^{(D)}(\mathbf{y}_4 - \mathbf{z})$$
$$\times\, G_0(\mathbf{y}_1, \mathbf{y}_2)\, G_0(\mathbf{y}_3, \bar{\mathbf{y}}_3)\, G_0(\mathbf{y}_4, \bar{\mathbf{y}}_4)$$
$$= G_0(\mathbf{z}, \mathbf{z})\, G_0(\mathbf{z}, \bar{\mathbf{y}}_3)\, G_0(\mathbf{z}, \bar{\mathbf{y}}_4). \tag{3.8}$$

Here, the permutation of $\mathbf{y}_1$ and $\mathbf{y}_2$ amounts to an interchange of the arguments of the propagator. Such permutations are not of the relevant type and are thus excluded from the sum over the permutations in (3.3). For this reason, the $\mathbf{y}$-integration yields now only $4!/2$ identical terms, thus resulting in a factor $4!/2$. Diagrammatically, $G_0(\mathbf{z}, \mathbf{z})$ produces a self-connection of a vertex. Examples for this case are the diagrams (a) or (b) in Fig. 3.1. Instead of 4! different ways to label the lines of the vertex, we find only four possibilities to choose the line connecting the vertex to $\mathbf{x}_1$, and three possibilities to choose the line connecting the vertex to $\mathbf{x}_2$. Altogether, the diagram contributes with a factor $3 \times 4 = 12$.

Other irrelevant permutations occur if more than one propagator has the same combination of the arguments $\bar{\mathbf{y}}_i$

$$\int d^D y_1 \ldots d^D y_4 \delta^{(D)}(\mathbf{y}_1 - \mathbf{z}) \delta^{(D)}(\mathbf{y}_2 - \mathbf{z}) \delta^{(D)}(\mathbf{y}_3 - \mathbf{z}) \delta^{(D)}(\mathbf{y}_4 - \mathbf{z})$$
$$\times\, G_0(\mathbf{y}_1, \bar{\mathbf{y}}_1)\, G_0(\mathbf{y}_2, \bar{\mathbf{y}}_1)\, G_0(\mathbf{y}_3, \bar{\mathbf{y}}_1)\, G_0(\mathbf{y}_4, \bar{\mathbf{y}}_4)$$
$$= G_0(\mathbf{z}, \bar{\mathbf{y}}_1)\, G_0(\mathbf{z}, \bar{\mathbf{y}}_1)\, G_0(\mathbf{z}, \bar{\mathbf{y}}_1)\, G_0(\mathbf{z}, \bar{\mathbf{y}}_4). \tag{3.9}$$

In this case, permutations of the indices $1 \leftrightarrow 2 \leftrightarrow 3$ are irrelevant and the sum contains only $4!/3! = 4$ identical terms. In the diagrammatic illustration, this case corresponds to a triple connection of the vertices $\bar{\mathbf{y}}_1$ and $\mathbf{z}$. For double and fourfold connections we find a factor $4!/2! = 12$ and $4!/4! = 1$, respectively.

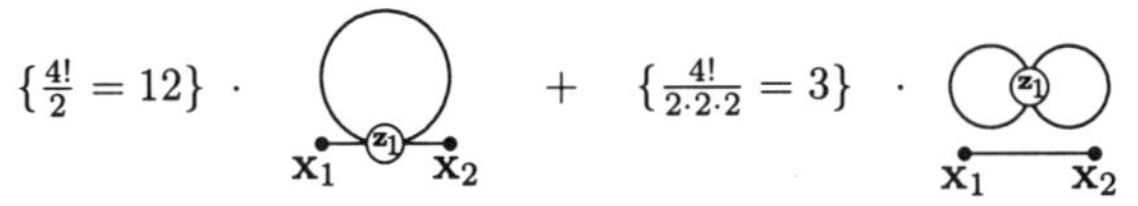

$$\{\tfrac{4!}{2} = 12\} \cdot \quad \underset{\mathbf{x}_1 \quad \mathbf{x}_2}{\bigcirc} \quad + \quad \{\tfrac{4!}{2\cdot2\cdot2} = 3\} \cdot \quad \underset{\mathbf{x}_1 \quad \mathbf{x}_2}{\infty}$$

FIGURE 3.2 Diagrams in expansion of $G_1^{(2)}$. The multiplicities add up to $(4p + n - 1)!!$ with $n = 2$ and $p = 1$: $12 + 3 = (4 \times 1 + 2 - 1)!!$.

It will be convenient to use the one-to-one relation between Feynman integrals and diagrams to write *diagrammatic equations* for any correlation function to be calculated. The first order two-point function whose diagrams we just calculated has then the diagrammatic equation:

$$G_1^{(2)}(\mathbf{x}_1, \mathbf{x}_2) \equiv \frac{1}{2} \underset{\mathbf{x}_1 \quad \mathbf{x}_2}{\bigcirc} + \frac{1}{8} \underset{\mathbf{x}_1 \quad \mathbf{x}_2}{\bullet\!-\!\!-\!\!\bullet} \; \infty \; . \tag{3.10}$$

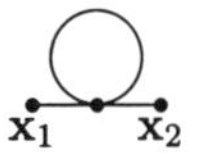

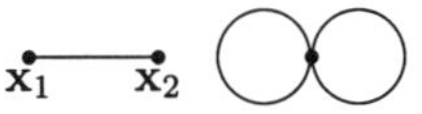

The diagrammatic contributions to the correlation function $G_1^{(2)}$ are summarized in Fig. 3.2. The first diagram has a self-connection giving rise to a factor $1/2$. The second diagram contains two self-connections and one double connection leading to the factor $1/(2\cdot2\cdot2)$. To check the resulting multiplicative factor 3 we note that there are obviously only three different ways of choosing two out of four points if the order is irrelevant. So only the following combinations appear in the sum: $G_0(\mathbf{y}_1, \mathbf{y}_2)G_0(\mathbf{y}_3, \mathbf{y}_4)$, $G_0(\mathbf{y}_1, \mathbf{y}_3)G_0(\mathbf{y}_2, \mathbf{y}_4)$, $G_0(\mathbf{y}_1, \mathbf{y}_4)G_0(\mathbf{y}_2, \mathbf{y}_3)$. The second-order contributions $G_2^{(2)}$ to the two-point function are shown in Fig. 3.3.

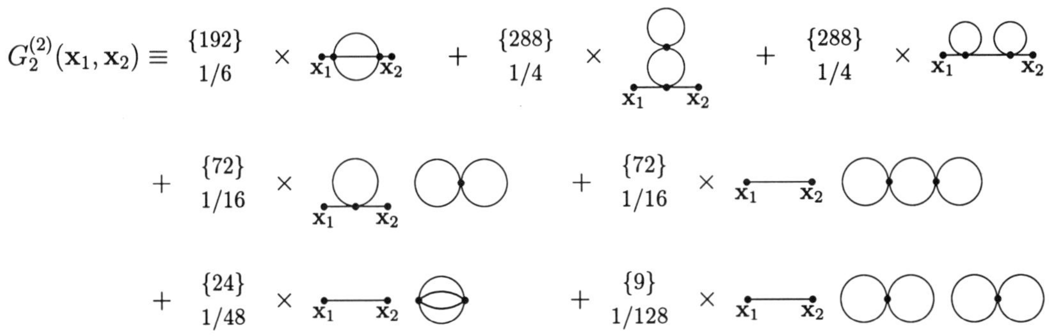

$$G_2^{(2)}(\mathbf{x}_1, \mathbf{x}_2) \equiv \begin{matrix} \{192\} \\ 1/6 \end{matrix} \times \underset{\mathbf{x}_1 \quad \mathbf{x}_2}{\ominus} + \begin{matrix} \{288\} \\ 1/4 \end{matrix} \times \underset{\mathbf{x}_1 \quad \mathbf{x}_2}{\,} + \begin{matrix} \{288\} \\ 1/4 \end{matrix} \times \underset{\mathbf{x}_1 \quad \mathbf{x}_2}{\,}$$

$$+ \begin{matrix} \{72\} \\ 1/16 \end{matrix} \times \underset{\mathbf{x}_1 \quad \mathbf{x}_2}{\,} + \begin{matrix} \{72\} \\ 1/16 \end{matrix} \times \underset{\mathbf{x}_1 \quad \mathbf{x}_2}{\,}$$

$$+ \begin{matrix} \{24\} \\ 1/48 \end{matrix} \times \underset{\mathbf{x}_1 \quad \mathbf{x}_2}{\,} + \begin{matrix} \{9\} \\ 1/128 \end{matrix} \times \underset{\mathbf{x}_1 \quad \mathbf{x}_2}{\,}$$

FIGURE 3.3 Diagrams in expansion of $G_2^{(2)}$. The curly brackets indicate the multiplicities, the numbers underneath them show the weight factors. The vertices are indicated with the short notation (3.5).

After performing all the **y**-integration in (3.3), the function $G_p^{(n)}(\mathbf{x}_1, \ldots, \mathbf{x}_n)$ is a sum of terms which differ in the arrangement of their $\mathbf{x}_i$ and $\mathbf{z}_i$ arguments in the propagators. The terms with permuted $\mathbf{z}_i$-arguments differ only in the labeling of the vertices, and this is irrelevant to the $\mathbf{z}_i$-integrations. Thus the $\mathbf{z}_i$-integration produces only an integer factor. In pth order perturbation theory, we have p vertices allowing initially for $p!$ permutations. This number is reduced to the relevant permutations in the sum (3.3). The irrelevant one are the $\mathbf{z}_i$-permutations, which are equivalent to a mere rearrangement of some propagators $G_0(\mathbf{z}_i, \mathbf{z}_j)$. They are called *identical vertex permutations* (IVP), since in the diagrammatic representation the reordering of the arguments of the propagators is equivalent to interchanging the corresponding vertices, performed with the vertices attached to the same lines. The number of identical vertex permutations will be denoted by N_{IVP}.

(a)   (b)

FIGURE 3.4 (a) Fourth-order diagram in $G_4^{(2)}$. (b) Same diagram with assignment of integration variables to internal points. The vertices $\mathbf{z}_3$ and $\mathbf{z}_4$ can be interchanged without cutting any line. This is an identical vertex permutation.

An example is shown in Fig. 3.4. The analytic expression for the right-hand diagram (b) is

$$\int d^D z_1 d^D z_2 d^D z_3 d^D z_4 G_0(\mathbf{x}_1, \mathbf{z}_1) G_0(\mathbf{z}_1, \mathbf{z}_2) G_0(\mathbf{z}_1, \mathbf{z}_4) G_0(\mathbf{z}_1, \mathbf{z}_3)$$
$$\times [G_0(\mathbf{z}_3, \mathbf{z}_4)]^2 G_0(\mathbf{z}_3, \mathbf{z}_2) G_0(\mathbf{z}_4, \mathbf{z}_2) G_0(\mathbf{z}_2, \mathbf{x}_1). \tag{3.11}$$

The interchange $\mathbf{z}_3 \leftrightarrow \mathbf{z}_4$ is irrelevant. The number of identical vertex permutations is therefore $N_{\mathrm{IVP}} = 2$. In contrast, the permutations involving $\mathbf{z}_1$ or $\mathbf{z}_2$ are relevant. The corresponding vertices cannot be interchanged in the diagram without cutting the lines to the external vertices.

Let us count the multiplicity M_G of the diagram. Each vertex can be connected to four lines in 4! ways, and since there are 4 vertices, there are initially $4! \cdot 4! = 7\,962\,624$ ways of drawing the diagram. There is one double-connection between vertices $\mathbf{z}_3$ and $\mathbf{z}_4$ so that this number has to be divided by 2. These two vertices can, moreover, be interchanged by an identical vertex permutation, such that we arrive at a multiplicity

$$M_G = \frac{4!^4 \cdot 4!}{2 \cdot 2} = 1\,990\,656. \tag{3.12}$$

This number will be found in the complete list of all multiplicities of two-point diagrams in Table 14.3 on page 264, where it appears with the label 4-3 or No. 3 in the first and last columns, respectively. The number 3 refers to the running number of this diagram in Appendix A.2 on page 451.

The associated weight of the diagram is, according to the definition (3.6),

$$W_G = \frac{1\,990\,656}{7\,962\,624} = \frac{1}{4}. \tag{3.13}$$

This number will be found in Appendix B.2 on page 464 (in the third column of the third four-loop entry).

The reader may proceed in this way to check the weight factors of the diagrammatic expansion of the second-order correction to the two-point function $G_2^{(2)}(\mathbf{x}_1, \mathbf{x}_2)$ in Fig. 3.3. There are altogether $(2 + 4 \times 2 - 1)!! = 945$ expansion terms.

Collecting the factors we find the general formula for the multiplicity M_G:

$$M_G = \frac{4!^P\, p!}{2!^{S+D} 3!^T 4!^F N_{\mathrm{IVP}}},$$

S = number of self-connections,
D = number of double connections,
T = number of triple connections,
F = number of fourfold connections,
N_{IVP} = number of identical vertex permutations,   (3.14)

and for the weight W_G

$$W_G = \frac{1}{2!^{S+D}3!^T4!^F N_{\mathrm{IVP}}}, \tag{3.15}$$

Formula (3.15) allows us to determine the multiplicity of simple diagrams by inspection, although the application to higher-order diagrams has the difficulty mentioned above of finding the number of identical vertex permutations. One way to solve this problem will be indicated in Chapter 14.

The above formula (3.15) determines the weight of a diagram with a fixed configuration of positions $\mathbf{x}_1, \ldots, \mathbf{x}_n$ at the ends of the external legs. These appear, however, in various permutations. To find the number of these permutations, we connect the endpoints of the external lines to an extra fictitious vertex labeled by $\mathbf{x}_0$. There are initially $n!$ ways of doing this. But because of the identity of the lines, the number of different connections is smaller by a factor $2!^{D'}3!^{T'}4!^{F'}$, where D', T', F' count the number of double, triple, and fourfold connections from the vertices of G to the extra vertex, respectively. Actually, fourfold connections do not appear. Also, the vertices where external lines enter can show additional symmetries if the labeling of the external vertices is neglected. Such a symmetry is accounted for by a factor $N_{\mathrm{IVP}}^{\mathrm{ext}}$ in the denominator. Thus there are

$$N_{\mathrm{perm}} = \frac{n!}{2!^{D'}3!^{T'}4!^{F'}N_{\mathrm{IVP}}^{\mathrm{ext}}} \tag{3.16}$$

different configurations of position labels of the external lines $\mathbf{x}_1, \ldots, \mathbf{x}_n$ for each diagram.

The weight of a diagram G irrespective of the external positions is then given by:

$$W_G^{\mathrm{unlabeled}} = W_G \times N_{\mathrm{perm}} = \frac{1}{2!^{S+D}3!^T4!^F N_{\mathrm{IVP}}} \times \frac{n!}{2!^{D'}3!^{T'}4!^{F'}N_{\mathrm{IVP}}^{\mathrm{ext}}}. \tag{3.17}$$

The value for $W_G^{\mathrm{unlabeled}}$ equals the value W_G of the corresponding amputated diagram which will be introduced in Chapter 14, in Eq. (14.1).

In the example of the diagram in Fig. 3.4, there are two external lines. These give rise to a factor $n! = 2$ in the numerator of formula (3.17), and thus of (3.13). However, when neglecting the labeling of the vertices $\mathbf{x}_1$ and $\mathbf{x}_2$ the vertices at $\mathbf{z}_1$ and $\mathbf{z}_2$ can also be permuted without changing the diagram giving rise to a factor $N_{\mathrm{IVP}}^{\mathrm{ext}} = 2$. Hence the number of irreducible vertex permutations in the denominator in (3.13) is increased by a factor 2 as well, thus yielding the same multiplicity as before.

Another example with a nontrivial configuration of external points is discussed at the end of Section 3.4.

3.2   Diagrammatic Expansion of the Partition Function

Let us also find a diagrammatic representation for the perturbation series of the partition function in Eq. (2.44). The expansion terms (2.45) are quite similar to those for $G_p^{(n)}$ in (2.48), except that all spatial variables are integrated out. Expanding the free correlation functions in (2.45) à la Wick, we obtain an expression analogous to (3.3):

$$Z_p \equiv \frac{1}{p!}\left(\frac{-\lambda}{4!}\right)^p \int d^D z_1 \cdots d^D z_p \int d^D y_1 \cdots d^D y_{4p}$$

$$\times \prod_{k=1}^{p}\left[\delta^{(D)}(\mathbf{y}_{4k-3} - \mathbf{z}_k)\delta^{(D)}(\mathbf{y}_{4k-2} - \mathbf{z}_k)\delta^{(D)}(\mathbf{y}_{4k-1} - \mathbf{z}_k)\delta^{(D)}(\mathbf{y}_{4k} - \mathbf{z}_k)\right]$$

$$\times \sum_{i=1}^{(4p-1)!!} \prod_{j=1}^{2p} G_0\big(\mathbf{y}_{\pi_i^{(4p)}(2j-1)}, \mathbf{y}_{\pi_i^{(4p)}(2j)}\big) \; . \tag{3.18}$$

The right-hand side contains a sum over products of free propagators whose lines all end at internal points. There are no external vertices or legs. The corresponding diagrams are called *vacuum diagrams*. The diagrammatic expansion of the partition function consists of the sum of all vacuum diagrams, as shown in Fig. 3.5 up to order λ^2, including their weight factors. These are the same vacuum diagrams which appeared before in disconnected pieces of the diagrams contributing to the expansion terms $G_p^{(2)}$ shown in Fig. 3.3. The simultaneous appearance of vacuum diagrams in these expansions has the pleasant consequence that, when dividing the sum of all diagrams in $G^{(2)}$ by those in Z and re-expanding the ratio in powers of λ, the contributions from the vacuum diagrams will cancel each other out. This will be seen in more detail below.

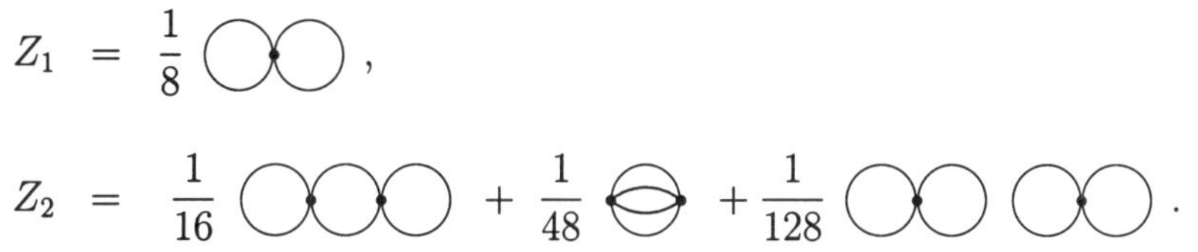

$$Z_1 \;=\; \frac{1}{8}\;\raisebox{-0.5em}{\includegraphics{}} \;,$$

$$Z_2 \;=\; \frac{1}{16}\;\raisebox{-0.5em}{\includegraphics{}} \;+\; \frac{1}{48}\;\raisebox{-0.5em}{\includegraphics{}} \;+\; \frac{1}{128}\;\raisebox{-0.5em}{\includegraphics{}} \;.$$

FIGURE 3.5 Vacuum diagrams in Z_1 and Z_2 with their respective weight factors.

3.3 Connected and Disconnected Diagrams

It is useful to distinguish *connected* and *disconnected diagrams*. For a disconnected diagram, the associated Feynman integral in $G_p^{(n)}$ or in Z_p factorizes. Examples are shown in Fig. 3.6. The connected diagrams contained in $G_p^{(n)}$ are denoted by $G_{pc}^{(n)}$.

$$G_0(\mathbf{x}_1,\mathbf{x}_2)\int d^D z\,[G_0(\mathbf{z},\mathbf{z})]^2$$

(a)

$$\int d^D z\, G_0(\mathbf{x}_1,\mathbf{z})G_0(\mathbf{x}_2,\mathbf{z})G_0(\mathbf{z},\mathbf{z})$$
$$\times \int d^D z'\, G_0(\mathbf{x}_3,\mathbf{z}')G_0(\mathbf{x}_4,\mathbf{z}')G_0(\mathbf{z}',\mathbf{z}')$$

(b)

FIGURE 3.6 Two disconnected diagrams and the associated integral expressions.

3.3.1 Multiplicities of Disconnected Diagrams

The multiplicity M_G of a diagram G consisting of n disconnected parts is calculated in the usual way. If a number n_{id} of the n disconnected parts are identical, the number of identical vertex permutations includes a factor $n_{\mathrm{id}}!$, since these parts can be permuted in $n_{\mathrm{id}}!$ ways. In general, there may be m sets of $n_{\mathrm{id}}^{(i)}$ ($i = 1, \ldots, m$) identical diagrams, with a number of permutations of identical diagrams $(n_{\mathrm{id}}^{(1)})! \cdots (n_{\mathrm{id}}^{(m)})!$. The weight-factor of the disconnected diagram is given by the product of the weight factors of the connected diagrams.

The composition rules of disconnected diagrams from their connected components are expressed most compactly in terms of the generating functional of all correlation functions. If we write

$$Z[j] = e^{W[j]}, \tag{3.19}$$

then $W[j]$ is the generating functional for the connected correlation functions. We shall prove this in Chapter 5. Here we only note that an exponential function of any diagram yields, with the correct weight factors, the sum of all disconnected diagrams composed of any number of copies of this diagram. For a disconnected diagram with two or more identical components, the inverse factorials in the expansion coefficients of the exponential produce reduction factors to the products of individual weights of the components, yielding a combined weight factor specified by Eq. (3.17). In the example $\infty\ \infty$, they produce a factor $1/2$ correcting the product of the individual weights $1/64$ to the correct total weight $1/128$.

For $j \equiv 0$, Eq. (3.19) relates the disconnected and connected diagrams for the vacuum to each other.

There exist simple relations between the numbers of diagrams of two- and four-point functions and those of the vacuum diagrams. These will be found in Chapters 5 and used further in Chapter 14.

3.3.2   Cancellation of Vacuum Diagrams

The diagrammatic expression for $G_p^{(n)}$ consists of all possible Feynman diagrams that can be drawn for n external points and p vertices. It contains connected and disconnected diagrams. The disconnected diagrams are composed of *external components* involving the external lines, and *vacuum components* containing only vacuum diagrams. If the number of vertices of the vacuum component is called p_V, the number of vertices of the external component is $p - p_V$. For $n = 2$, the external part contains only one connected lower-order diagram. For $n = 4$, the external part may be composed of two two-point diagrams, and so on.

As all possible contractions are contained in $G_p^{(n)}$, all possible compositions of external and vacuum diagrams occur. The weight factors for disconnected diagrams are the product of the weight factors of the constituents times the factor for the permutations of the identical diagrams. Thus $G_p^{(n)}$ can be written as:

$$G_p^{(n)}(\mathbf{x}_1, \ldots, \mathbf{x}_n) = \sum_{k=0}^{p} G_{p-k\,\mathrm{ext}}^{(n)}(\mathbf{x}_1, \ldots, \mathbf{x}_n) Z_k, \tag{3.20}$$

where G_{ext} denotes the diagrams of the corresponding order which have no vacuum component and contain only external parts. If Eq. (3.20) is summed over all orders, the partition sum can be factored out:

$$\begin{aligned}
\sum_{p=0}^{\infty} G_p^{(n)}(\mathbf{x}_1, \ldots, \mathbf{x}_n) &= \sum_{p=0}^{\infty} \sum_{k=0}^{p} G_{p-k\,\mathrm{ext}}^{(n)}(\mathbf{x}_1, \ldots, \mathbf{x}_n) Z_k \\
&= \sum_{p'=0}^{\infty} G_{p'\,\mathrm{ext}}^{(n)}(\mathbf{x}_1, \ldots, \mathbf{x}_n) \sum_{k=0}^{\infty} Z_k \\
&= \sum_{p'=0}^{\infty} G_{p'\,\mathrm{ext}}^{(n)}(\mathbf{x}_1, \ldots, \mathbf{x}_n) Z.
\end{aligned} \tag{3.21}$$

Insertion of this equation into Eq. (3.1) yields

$$G^{(n)}(\mathbf{x}_1, \ldots, \mathbf{x}_n) = \sum_{p=0}^{\infty} G_{p\,\text{ext}}^{(n)}(\mathbf{x}_1, \ldots, \mathbf{x}_n), \tag{3.22}$$

where the two-point function satisfies $G_{p\,\text{ext}} = G_{p\,\text{c}}$.

3.4 Connected Diagrams for Two- and Four-Point Functions

After dropping the vacuum components, the diagrams for two- and four-point functions can easily be written down. They are always connected for the two-point function (this would not be true for a ϕ^3-interaction). The lowest contributions are

$$G(\mathbf{x}_1, \mathbf{x}_2) = G_c(\mathbf{x}_1, \mathbf{x}_2) = \underset{\mathbf{x}_1 \quad \mathbf{x}_2}{\bullet\!\!-\!\!\bullet} + \frac{1}{2} \underset{\mathbf{x}_1 \quad \mathbf{x}_2}{\bigcirc}$$

$$+ \frac{1}{4} \underset{\mathbf{x}_1 \quad \mathbf{x}_2}{\bigcirc\!\!\bigcirc} + \frac{1}{6} \underset{\mathbf{x}_1 \quad \mathbf{x}_2}{\bullet\!\!\bigcirc\!\!\bullet} + \frac{1}{4} \underset{\mathbf{x}_1 \quad \mathbf{x}_2}{\bullet\!\bigcirc\bigcirc\!\bullet} + \mathcal{O}(g^3). \tag{3.23}$$

The diagrams for the 4-point function without vacuum components are either connected, or they decompose into a product of two connected diagrams which occur in the expansion of the two-point function. With the help of the product rule in the last section we can always separate out the disconnected contributions, writing

$$G^{(4)}(\mathbf{x}_1, \mathbf{x}_2, \mathbf{x}_3, \mathbf{x}_4) = G_c^{(4)}(\mathbf{x}_1, \mathbf{x}_2, \mathbf{x}_3, \mathbf{x}_4) + G_c(\mathbf{x}_1, \mathbf{x}_2)G_c(\mathbf{x}_3, \mathbf{x}_4) \tag{3.24}$$

$$+ \; G_c(\mathbf{x}_1, \mathbf{x}_3)G_c(\mathbf{x}_2, \mathbf{x}_4) + G_c(\mathbf{x}_1, \mathbf{x}_4)G_c(\mathbf{x}_2, \mathbf{x}_3).$$

The connected component has the expansion

$$G_c^{(4)}(\mathbf{x}_1, \mathbf{x}_2, \mathbf{x}_3, \mathbf{x}_4) = \underset{\mathbf{x}_1 \; \mathbf{x}_3}{\overset{\mathbf{x}_2 \; \mathbf{x}_4}{\times}} + \frac{1}{2}\left(\underset{\mathbf{x}_1 \quad \mathbf{x}_3}{\overset{\mathbf{x}_2 \quad \mathbf{x}_4}{\times\!\!-\!\!\times}} + 2 \text{ perm.}\right) + \frac{1}{2}\left(\underset{\mathbf{x}_1}{\overset{\mathbf{x}_3}{\times\!\!\bigcirc\!\!\times}}_{\mathbf{x}_4} + 3 \text{ perm.}\right)$$

$$+ \frac{1}{2}\left(\underset{\mathbf{x}_1 \quad \mathbf{x}_4}{\overset{\mathbf{x}_3 \; \mathbf{x}_2}{\bigtriangleup}} + 5 \text{ perm.}\right) + \frac{1}{4}\left(\underset{\mathbf{x}_1 \quad \mathbf{x}_3}{\overset{\mathbf{x}_2 \quad \mathbf{x}_4}{\times\!\!\bigcirc\!\!\bigcirc\!\!\times}} + 2 \text{ perm.}\right)$$

$$+ \frac{1}{2}\left(\underset{\mathbf{x}_1 \quad \mathbf{x}_4}{\overset{\mathbf{x}_2 \quad \mathbf{x}_3}{\times\!\!\bigcirc\!\!\times}} + 2 \text{ perm.}\right) + \frac{1}{4}\left(\underset{\mathbf{x}_1 \quad \mathbf{x}_3}{\overset{\mathbf{x}_2 \quad \mathbf{x}_4}{\times\!\!\bigcirc\!\!\times}} + 11 \text{ perm.}\right)$$

$$+ \frac{1}{4}\left(\underset{\mathbf{x}_1 \quad \mathbf{x}_3 \quad \mathbf{x}_4}{\overset{\mathbf{x}_2}{\bullet\!\!\bigcirc\!\!\bigcirc\!\!\bullet}} + 3 \text{ perm.}\right) + \frac{1}{4}\left(\underset{\mathbf{x}_1 \quad \mathbf{x}_3 \quad \mathbf{x}_4}{\overset{\mathbf{x}_2}{\bigcirc\!\!\bullet\!\!\bigcirc}} + 5 \text{ perm.}\right)$$

$$+ \frac{1}{4}\left(\underset{\mathbf{x}_1 \quad \mathbf{x}_2 \quad \mathbf{x}_4}{\overset{\mathbf{x}_3}{\bullet\!\!\bigcirc\!\!\bullet}} + 3 \text{ perm.}\right) + \frac{1}{6}\left(\underset{\mathbf{x}_1 \quad \mathbf{x}_3 \quad \mathbf{x}_4}{\overset{\mathbf{x}_2}{\bullet\!\!\bigcirc\!\!\bullet}} + 3 \text{ perm.}\right) + \mathcal{O}(g^4). \tag{3.25}$$

It is useful to illustrate the calculation of the weights for one of the diagrams with a nontrivial configuration of external points $\mathbf{x}_1, \ldots, \mathbf{x}_4$, for instance any one the diagrams in the second-last parentheses. Each has one self- and one double connection, and the number of identical vertex permutations is $N_{\text{IVP}} = 1$. From Eq. (3.15) we obtain its weight $W_G = 1/4$. If the ends are connected to an external vertex, we obtain the diagram

There is a triple connection to the extra fictitious vertex on the left, such that the number N_{perm} of configurations of Eq. (3.16) is $N_{\text{perm}} = 4!/3! = 4$, which explains the number of terms in parentheses. Multiplying this additional factor to the weight factor $W_G = 1/4$ of the labeled diagram according to formula (3.17), we find $W_G^{\text{unlabeled}} = 1$ which is the total weight of all diagrams in the second-last parentheses, irrespective of the configurations of the external points.

3.5   Diagrams for Composite Fields

In Eq. (2.53), we introduced correlation functions containing composite operators $\phi^2(\mathbf{x})$. For these we showed in Eqs. (2.57) and (2.58) that their spatial integrals within an expectation value of ordinary fields can be generated by forming derivatives of the n-point functions $G^{(n)}(\mathbf{x}_1, \ldots, \mathbf{x}_n)$ with respect to the mass. In this book, we shall deal in detail only with a single insertion of $\phi^2(\mathbf{x})$ into G (see Chapter 12):

$$G^{(1,2)}(\mathbf{y}, \mathbf{x}_1, \mathbf{x}_2) \equiv \frac{1}{2}\langle \phi^2(\mathbf{y})\phi(\mathbf{x}_1)\phi(\mathbf{x}_2)\rangle. \tag{3.26}$$

Its spatial integral is generated by the derivative with respect to the squared mass:

$$\frac{\partial}{\partial m^2} G(\mathbf{x}_1, \mathbf{x}_2) = -\int d^D y \, G^{(1,2)}(\mathbf{y}, \mathbf{x}_1, \mathbf{x}_2). \tag{3.27}$$

The perturbative calculation of the quantity on the right-hand side follows Wick's expansion rule. The resulting Feynman diagrams are quite simply related to those of G, whose differentiation with respect to the squared mass generates successively in every line a $-\phi^2/2$ -insertion. The latter is represented diagrammatically by a $-\phi^2$-vertex, which is a vertex with two legs carrying a factor -1 (since there are two ways of connecting two lines to it). In a Feynman diagram, it will be denoted by a dot on the line. The negative sign is implied by this point. For the free correlation function G we obtain the diagrammatic equation

$$\int d^D y \, G^{(1,2)}(\mathbf{y}, \mathbf{x}_1, \mathbf{x}_2) = \frac{\partial}{\partial m^2} \underset{\mathbf{x}_1 \qquad \mathbf{x}_2}{\bullet\!\!-\!\!\!-\!\!\bullet} = \underset{\mathbf{x}_1 \qquad \mathbf{x}_2}{\bullet\!\!-\!\!\bullet\!\!-\!\!\bullet}. \tag{3.28}$$

Since two legs can be connected with the $-\phi^2$-vertex in two ways without generating a different diagram, every diagram appears twice, thereby canceling the factor $1/2$ of the $-\phi^2/2$ -insertion in (3.26).

    Lines in a diagram are called *topologically equivalent* if they are part of a double, triple, or quadruple connection, or if they are transformed into one another by an identical vertex permutation. The differentiation of topologically equivalent lines leads to identical diagrams with a $-\phi^2$-vertex, causing a multiplicity factor. For example, the $-\phi^2/2$-insertion on a threefold connection leads to three equivalent diagrams.

The first terms of the diagrammatic expansion of $G^{(1,2)}$ are

$$\int d^D y \, G^{(1,2)}(\mathbf{y}, \mathbf{x}_1, \mathbf{x}_2) \;=\; \underset{\mathbf{x}_1 \quad \mathbf{x}_2}{\bullet\!\!-\!\!\bullet\!\!-\!\!\bullet} \;+\; \frac{1}{2} \underset{\mathbf{x}_1 \quad \mathbf{x}_2}{\bigcirc} \;+\; \frac{1}{2} \underset{\mathbf{x}_1 \quad \mathbf{x}_2}{\bigcirc} \;+\; \frac{1}{2} \underset{\mathbf{x}_1 \quad \mathbf{x}_2}{\bigcirc} \;+\; \mathcal{O}(g^2).$$

The number of terms grows rapidly with increasing order, so that we restrict the display here to the first order. The second-order diagrams will be shown explicitly in Eq. (4.43).

The correlation function $G^{(1,2)}$ is always connected, i.e., $G^{(1,2)} = G_c^{(1,2)}$, because of the restriction of our study to the normal phase of the system.

Notes and References

For more details see the textbooks on quantum field theory listed at the end of Chapter 1. We have followed the textbook
H. Kleinert, *Gauge Fields in Condensed Matter*, Vol. I, *Superflow and Vortex Lines, Disorder Fields and Phase Transitions*, World Scientific, Singapore, 1989 (www.physik.fu-berlin.de/~kleinert/re.html#b1).

4

Diagrams in Momentum Space

At the end of Section 2.3, we have observed that all correlation functions of the local field theory under consideration are invariant under spatial translations. This invariance has the consequence that Feynman integrals are greatly simplified by Fourier transformation. It is therefore useful to set up rules for composing Feynman integrals directly in momentum space.

4.1   Fourier Transformation

The Fourier transform of a function $F(\mathbf{x})$ is defined by

$$F(\mathbf{k}) \equiv \int d^D x \, e^{-i \mathbf{k} \cdot \mathbf{x}} \, F(\mathbf{x}). \tag{4.1}$$

The original function $F(\mathbf{x})$ is retrieved from $F(\mathbf{k})$ by the inverse Fourier transformation

$$F(\mathbf{x}) \equiv \int \frac{d^D k}{(2\pi)^D} \, e^{i \mathbf{k} \cdot \mathbf{x}} \, F(\mathbf{k}). \tag{4.2}$$

By applying a Fourier transformation to all arguments of the n-point functions $G^{(n)}(\mathbf{x}_1, \ldots, \mathbf{x}_n)$, we obtain the n-point functions in momentum space $G^{(n)}(\mathbf{k}_1, \ldots, \mathbf{k}_n)$.

4.1.1   Free Two-Point Function

The Fourier transform of the free two-point function $G_0(\mathbf{x}, \mathbf{x}') = G_0^{(2)}(\mathbf{x}, \mathbf{x}')$ reads

$$G_0(\mathbf{k}, \mathbf{k}') = \int d^D x \, d^D x' \, e^{-i (\mathbf{k} \cdot \mathbf{x} + \mathbf{k}' \cdot \mathbf{x}')} \, G_0(\mathbf{x}, \mathbf{x}'). \tag{4.3}$$

As observed after Eq. (2.34), the free two-point function depends only on the difference of its spatial variables, due to translational invariance. This simplifies the momentum space representation, which becomes

$$\begin{aligned}
G_0(\mathbf{k}, \mathbf{k}') &= \int d^D x' \, e^{-i (\mathbf{k}' + \mathbf{k}) \cdot \mathbf{x}'} \int d^D x \, e^{-i \mathbf{k} \cdot (\mathbf{x} - \mathbf{x}')} G_0(\mathbf{x} - \mathbf{x}') \\
&= (2\pi)^D \, \delta^{(D)}(\mathbf{k} + \mathbf{k}') \, G_0(\mathbf{k}).
\end{aligned} \tag{4.4}$$

Thus the Fourier components $G_0(\mathbf{k}, \mathbf{k}')$ depend only on one momentum variable $\mathbf{k}$, and the function $G_0(\mathbf{k})$ is simply the Fourier transform of the free $\mathbf{x}$-space Green function of Eq. (2.40) with a single argument $G_0(\mathbf{x} - \mathbf{x}')$, which has the Fourier representation

$$G_0(\mathbf{x}, \mathbf{x}') = G_0(\mathbf{x} - \mathbf{x}') = \int \frac{d^D k}{(2\pi)^D} \, e^{i \mathbf{k} \cdot (\mathbf{x} - \mathbf{x}')} \, G_0(\mathbf{k}). \tag{4.5}$$

In momentum space, the free propagator has a simple algebraic expression which follows directly from Eq. (2.34) and the definition of the functional inverse in (2.29). The free propagator satisfies the integral equation:

$$\int d^D x' \, D(\mathbf{x}, \mathbf{x}') G_0(\mathbf{x}', \mathbf{x}'') = \delta^{(D)}(\mathbf{x} - \mathbf{x}''). \tag{4.6}$$

Inserting the expression (2.16) for $D(\mathbf{x}, \mathbf{x}')$, and the Fourier representation (4.5) for $G_0(\mathbf{x}', \mathbf{x}'')$, we obtain

$$\int d^D x' \, \delta^{(D)}(\mathbf{x} - \mathbf{x}')(-\partial_{\mathbf{x}}^2 + m^2) \int \frac{d^D k}{(2\pi)^D} \, e^{i\,\mathbf{k}\cdot(\mathbf{x}'-\mathbf{x}'')} \, G_0(\mathbf{k})$$

$$= \int \frac{d^D k}{(2\pi)^D} \, (\mathbf{k}^2 + m^2) \, G_0(\mathbf{k}) \, e^{i\,\mathbf{k}\cdot(\mathbf{x}-\mathbf{x}'')} = \delta^{(D)}(\mathbf{x} - \mathbf{x}''). \tag{4.7}$$

By comparing this with the Fourier representation of the δ-distribution,

$$\int \frac{d^D k}{(2\pi)^D} \, e^{i\,\mathbf{k}\cdot(\mathbf{x}-\mathbf{x}'')} = \delta^{(D)}(\mathbf{x} - \mathbf{x}''), \tag{4.8}$$

we find the momentum space representation of the free propagator

$$G_0(\mathbf{k}) = \frac{1}{\mathbf{k}^2 + m^2}. \tag{4.9}$$

The Fourier transform of the n-point function is, of course, defined by

$$G^{(n)}(\mathbf{k}_1, \ldots, \mathbf{k}_n) = \int d^D x_1 \cdots d^D x_n \, e^{-i\,(\mathbf{k}_1\cdot\mathbf{x}_1 + \ldots + \mathbf{k}_n\cdot\mathbf{x}_n)} G^{(n)}(\mathbf{x}_1, \ldots, \mathbf{x}_n). \tag{4.10}$$

4.1.2 Connected *n*-Point Function

Since the correlation function associated with a disconnected Feynman diagram factorizes into those of the connected parts, the same is true for its Fourier transform. It will therefore be sufficient to set up the desired Feynman rules for connected diagrams in momentum space. We shall denote the external and internal points by the symbols $\mathbf{x}_k$ $(k = 1, \ldots, n)$ and $\mathbf{z}_i$ $(i = 1, \ldots, p)$. Each momentum can be represented by a line. Depending on the endpoints, we distinguish external lines $(k = 1, \ldots, n)$ and internal lines $\mathbf{p}_i$ $(i = 1, \ldots, I)$. The number of external lines is obviously n. The number of internal lines I is determined by n and the number of vertices p as follows:

$$I = (4p - n)/2. \tag{4.11}$$

This number is a direct consequence of each line having two ends, coinciding either with one of the n external points $\mathbf{x}_k$ or with one of the p internal points $\mathbf{z}_i$ at which four lines meet. Subtracting from the resulting total number $(4p + n)/2$ of lines the number n of external lines, we obtain (4.11).

Let $G_c^{(n)}(\mathbf{x}_1, \ldots, \mathbf{x}_n)$ stand for the Feynman integral symbolized by a connected diagram in $G_p^{(n)}(\mathbf{x}_1, \ldots, \mathbf{x}_n)$. Omitting for a moment the coupling and weight factors $(-g)^p$ and W_G, the integral involves a product of free propagators:

$$G_c^{(n)}(\mathbf{x}_1, \ldots, \mathbf{x}_n) = \int d^D z_1 \cdots d^D z_p \prod_{i=1}^{n} G_0(\mathbf{x}_i - \mathbf{z}_i) \times \prod_{i=1}^{I} G_0(\mathbf{z}_j - \mathbf{z}_j). \tag{4.12}$$

The Fourier transform of $G_c^{(n)}(\mathbf{x}_1, \ldots, \mathbf{x}_n)$ is defined as in (4.10):

$$G_c^{(n)}(\mathbf{k}_1, \ldots, \mathbf{k}_n) = \int d^D x_1 \cdots d^D x_n \, e^{-i(\mathbf{k}_1 \cdot \mathbf{x}_1 + \ldots + \mathbf{k}_n \cdot \mathbf{x}_n)} G_c^{(n)}(\mathbf{x}_1, \ldots, \mathbf{x}_n). \tag{4.13}$$

It contains n factors $e^{-i\mathbf{k}_k \cdot \mathbf{x}_k}$, whose momenta $\mathbf{k}_k$ are represented by an external line. Each external point $\mathbf{x}_k$ appears in a free Green function, which has the Fourier representation

$$G_0(\mathbf{x}_k - \mathbf{z}_i) = \int \frac{d^D k_k}{(2\pi)^D} \, e^{i \, \mathbf{k}_k \cdot (\mathbf{x}_k - \mathbf{z}_i)} \, G_0(\mathbf{k}_k). \tag{4.14}$$

These contribute to the integral (4.12) an exponential factor $e^{i\mathbf{k}_k \cdot (\mathbf{x}_k - \mathbf{z}_i)}$. Each pair of internal points $\mathbf{z}_i$ appears in a free Green function

$$G_0(\mathbf{z}_i - \mathbf{z}_j) = \int \frac{d^D p}{(2\pi)^D} \, e^{i \, \mathbf{p} \cdot (\mathbf{z}_i - \mathbf{z}_j)} \, G_0(\mathbf{p}), \tag{4.15}$$

contributing an exponential factor $e^{i\mathbf{p} \cdot (\mathbf{z}_i - \mathbf{z}_j)}$. As each of the external points $\mathbf{x}_i$ appears twice in those phase factors, the integrals over $\mathbf{x}_i$ produce a factor $(2\pi)^D \delta^{(D)}(\mathbf{k} - \mathbf{p})$ for each external line. The subsequent n integrals over the corresponding momenta $\mathbf{p}$ can all be done, leaving us with $(4p - n)/2$ nontrivial momentum integrals, one for each internal line.

Each vertex appears in four exponential factors $e^{i\mathbf{P} \cdot \mathbf{x}}$. The integrals over the internal vertex positions $\mathbf{z}_i$ yield p δ-distributions which express momentum conservation at each vertex. One of these can be chosen to contain the sum over all external momenta. It guarantees overall momentum conservation. The others can trivially be done, thereby removing $p-1$ integrations. We end up with

$$L = I - p + 1 \tag{4.16}$$

nontrivial integrals over momentum variables $\mathbf{l}_i$, the *loop momenta*, which may be associated with the independent loops in the diagrams. The Fourier transform of $G_c^{(n)}(\mathbf{x}_1, \ldots, \mathbf{x}_n)$ has therefore the form

$$G_c^{(n)}(\mathbf{k}_1, \ldots, \mathbf{k}_n) = G_0(\mathbf{k}_1) \cdots G_0(\mathbf{k}_n) (2\pi)^D \, \delta^{(D)} \left(\sum_{i=1}^{n} \mathbf{k}_i \right)$$

$$\times \int \frac{d^D l_1}{(2\pi)^D} \cdots \frac{d^D l_L}{(2\pi)^D} G_0(\mathbf{p}_1(\mathbf{l}, \mathbf{k})) \cdots G_0(\mathbf{p}_I(\mathbf{l}, \mathbf{k})). \tag{4.17}$$

Each line momentum is expressed by a combination of loop momenta $\mathbf{l}_i$ $(i = 1, \ldots, L)$ and external momenta $\mathbf{k}_i$ $(i = 1, \ldots, n)$, abbreviated by $(\mathbf{l}, \mathbf{k})$.

The direction of a loop momentum is a consequence of momentum conservation. The external momenta $\mathbf{k}_k$ all flow out of each diagram into the external point.

Reintroducing the previously omitted factors $-g/4!$ and W_G, we see that a connected Feynman diagram in the $\mathbf{k}$-space contains:

1. a factor $G_0(\mathbf{k}_i)$ for each external line;

2. a factor $G_0(\mathbf{p}_j)$ with $j = 1, \ldots, I$ for each internal line, where each internal momentum $\mathbf{p}_j$ has an orientation and is expressed by a combination of the $L = I - p + 1$ loop momenta and the n external momenta;

3. an integration over each independent loop momentum
   $(1/2\pi)^D \int d^D l_i$, $(i = 1, \ldots, L)$;

4. a factor $(2\pi)^D \delta^{(D)}(\mathbf{k}_1 + \ldots + \mathbf{k}_n)$ to guarantee overall momentum conservation;

5. a factor $-\lambda/4!$ for each vertex; and

6. a weight factor W_G of the diagram.

From Eq. (4.17), we see that the full connected part of the propagator factorizes in the same way as the free propagator in Eq. (4.4):

$$G_c^{(2)}(\mathbf{k}, \mathbf{k}') \equiv G(\mathbf{k}, \mathbf{k}') = (2\pi)^D \delta^{(D)}(\mathbf{k} + \mathbf{k}') G(\mathbf{k}), \tag{4.18}$$

thus defining a full propagator with a single momentum argument $G(\mathbf{k})$.

4.2 One-Particle Irreducible Diagrams and Proper Vertex Functions

Since the integrations associated with a Feynman diagram run only over the loop momenta, it is convenient to introduce a reduced diagram which represents precisely the loop integrations by removing all external lines. The result is a so-called *amputated diagram*. These diagrams are drawn with short, unlabeled external lines indicating the amputation points. The omission of the external lines removes any difference between diagrams with differently labeled external lines.

The weight factor for these unlabeled diagrams differs from those of the previous diagrams with labeled lines. This generally changes the weight factor by an additional factor N_{perm}, which accounts for the permutations of the external lines and the additional permutations of the internal vertices, as discussed in the context of Eq. (3.16).

Amputated diagrams may contain lines without loop momentum, i.e., without additional factors of $G_0(\mathbf{k}_i)$. Such lines connect loop parts of a diagram, and the diagram falls apart when any of these lines is cut. These lines are called *cutlines*. An amputated diagram which possesses a cutline is said to be *one-particle reducible*. Amputated diagrams without cutlines are called *one-particle irreducible* (1PI) diagrams. They represent the smallest nontrivial Feynman diagrams which form the basic building blocks of all diagrams. The detailed laws of composition of these building blocks will be described in Chapter 5.

For each 1PI diagram with $n > 2$ we introduce the product of loop integrals in Eq. (4.17) as the *proper vertex function* , apart from a minus-sign which is a matter of convention

$$\bar{\Gamma}^{(n)}(\mathbf{k}_1, \ldots, \mathbf{k}_n) \equiv - \int \frac{d^D l_1}{(2\pi)^D} \cdots \frac{d^D l_L}{(2\pi)^D} G_0(\mathbf{q}_1(l, k)) \cdots G_0(\mathbf{q}_I(l, k)), \qquad n > 2. \tag{4.19}$$

The 1PI part of a correlation function $G_c^{(n)}(\mathbf{k}_1, \ldots, \mathbf{k}_n)$ is, of course, recovered from the proper vertex function by multiplication with a free two-point function for each external momentum, and with function $(2\pi)^D \delta^{(D)}(\sum_{i=1}^n \mathbf{k}_i)$ enforcing the total momentum conservation:

$$G_c^{(n)}(\mathbf{k}_1, \ldots, \mathbf{k}_n)\Big|_{1\text{PI}} = -G_0(\mathbf{k}_1) \cdots G_0(\mathbf{k}_n) (2\pi)^D \delta^{(D)} \left(\sum_{i=1}^n \mathbf{k}_i \right) \bar{\Gamma}^{(n)}(\mathbf{k}_1, \ldots, \mathbf{k}_n), \qquad n > 2. \tag{4.20}$$

Below, in Subsection 5.6, we shall see how to construct the missing one-particle reducible parts of the connected correlation functions $G_c^{(n)}(\mathbf{k}_1, \ldots, \mathbf{k}_n)$ from the proper vertex functions $\bar{\Gamma}^{(n)}(\mathbf{k}_1, \ldots, \mathbf{k}_n)$. We shall find that for the ϕ^4-theory in the normal phase under study here, the four-point function has no one-particle-reducible parts, so that Eq. (4.20) happens to give the complete connected correlation function:

$$G_c^{(4)}(\mathbf{k}_1, \mathbf{k}_2, \mathbf{k}_3, \mathbf{k}_4) = -G(\mathbf{k}_1)G(\mathbf{k}_2)G(\mathbf{k}_3)G(\mathbf{k}_4)(2\pi)^D \delta^{(D)} \left(\sum_{i=1}^{4} \mathbf{k}_i \right) \bar{\Gamma}^{(4)}(\mathbf{k}_1, \mathbf{k}_2, \mathbf{k}_3, \mathbf{k}_4). \quad (4.21)$$

Note that the proper vertex functions $\bar{\Gamma}^{(n)}(\mathbf{k}_1, \dots, \mathbf{k}_n)$ are only defined for momenta $\mathbf{k}_i$ which add up to zero. They will be represented diagrammatically by

$$\bar{\Gamma}^{(n)}(\mathbf{k}_i) = - \vcenter{\hbox{}} \quad . \quad (4.22)$$

It is useful to introduce also the *proper correlation functions* $\bar{G}^{(n)}(\mathbf{k}_1, \dots, \mathbf{k}_n)$ in which the δ-function for the overall momentum conservation is removed:

$$G^{(n)}(\mathbf{k}_1, \dots, \mathbf{k}_n) \equiv (2\pi)^D \delta^{(D)} \left(\sum_{i=1}^{4} \mathbf{k}_i \right) \bar{G}^{(n)}(\mathbf{k}_1, \dots, \mathbf{k}_n). \quad (4.23)$$

The functions $\bar{G}^{(n)}(\mathbf{k}_1, \dots, \mathbf{k}_n)$ are defined only for momentum arguments which add up to zero. Then we can invert relation (4.21) to

$$\bar{\Gamma}^{(4)}(\mathbf{k}_1, \mathbf{k}_2, \mathbf{k}_3, \mathbf{k}_4) \equiv -G^{-1}(\mathbf{k}_1) G^{-1}(\mathbf{k}_2) G^{-1}(\mathbf{k}_3) G^{-1}(\mathbf{k}_4) \bar{G}_c^{(4)}(\mathbf{k}_1, \mathbf{k}_2, \mathbf{k}_3, \mathbf{k}_4). \quad (4.24)$$

Up to two loops, the diagrammatic expansion of $\bar{\Gamma}^{(4)}$ is

$$\bar{\Gamma}^{(4)}(\mathbf{k}_i) = - \times - \frac{3}{2} \vcenter{\hbox{}} - 3 \vcenter{\hbox{}} - \frac{3}{4} \vcenter{\hbox{}} - \frac{3}{2} \vcenter{\hbox{}} + \mathcal{O}(g^4). \quad (4.25)$$

Note that with the notation (4.23), the relation (4.18) for the propagator implies the identity

$$G(\mathbf{k}) \equiv \bar{G}_c^{(2)}(\mathbf{k}, -\mathbf{k}). \quad (4.26)$$

The decomposition of the diagrams in the two-point function is somewhat different. The sum of all 1PI diagrams in $G(\mathbf{k})$ without the zeroth-order diagram is defined as the *self-energy* $\Sigma(\mathbf{k})$. Up to third order in the coupling strength, the self-energy has the diagrammatic expansion

$$\Sigma(\mathbf{k}) = \frac{1}{2} \vcenter{\hbox{}} + \frac{1}{4} \vcenter{\hbox{}} + \frac{1}{6} \vcenter{\hbox{}} + \frac{1}{4} \vcenter{\hbox{}} + \frac{1}{12} \vcenter{\hbox{}}$$

$$+ \frac{1}{4} \vcenter{\hbox{}} + \frac{1}{8} \vcenter{\hbox{}} + \frac{1}{8} \vcenter{\hbox{}} + \mathcal{O}(g^4). \quad (4.27)$$

Given the self-energy, the full propagator $G(\mathbf{k})$ is found by forming a *chain* consisting of self-energies connected by single lines. If we denote the self-energy by a diagram

$$\Sigma(\mathbf{k}) = \vcenter{\hbox{}}, \quad (4.28)$$

the sum over these chains looks as follows:

$$G(\mathbf{k}) = \vcenter{\hbox{}} + \vcenter{\hbox{}} + \vcenter{\hbox{}} + \cdots . \quad (4.29)$$

Analytically, this sum reads

$$G(\mathbf{k}) = G_0(\mathbf{k}) + G_0(\mathbf{k})\Sigma(\mathbf{k})G_0(\mathbf{k}) + G_0(\mathbf{k})\Sigma(\mathbf{k})G_0(\mathbf{k})\Sigma(\mathbf{k})G_0(\mathbf{k}) + \dots . \tag{4.30}$$

This is a geometric series, which is readily summed by

$$G(\mathbf{k}) = G_0(\mathbf{k})\sum_{l=0}^{\infty}[\Sigma(\mathbf{k})G_0(\mathbf{k})]^l = \left[G_0^{-1}(\mathbf{k}) - \Sigma(\mathbf{k})\right]^{-1} = \left[\mathbf{k}^2 + m^2 - \Sigma(\mathbf{k})\right]^{-1}. \tag{4.31}$$

Since the momenta in a proper vertex function must add up to zero, the proper vertex function for $n = 2$ has the arguments $\bar{\Gamma}^{(2)}(\mathbf{k}, -\mathbf{k})$. By analogy with the Fourier-transformed two-particle correlation function $G(\mathbf{k})$ which carries only one momentum argument, it will be useful to introduce a quantity

$$\bar{\Gamma}^{(2)}(\mathbf{k}) \equiv \bar{\Gamma}^{(2)}(\mathbf{k}, -\mathbf{k}). \tag{4.32}$$

This is set equal to

$$\bar{\Gamma}^{(2)}(\mathbf{k}) \equiv G_0^{-1}(\mathbf{k}) - \Sigma(\mathbf{k}) = \mathbf{k}^2 + m^2 - \Sigma(\mathbf{k}), \tag{4.33}$$

such that

$$G(\mathbf{k}) = \frac{1}{\bar{\Gamma}^{(2)}(\mathbf{k})}. \tag{4.34}$$

Inserted into (4.18), we obtain for the connected two-point function the relation

$$G_c^{(2)}(\mathbf{k}, \mathbf{k}') = (2\pi)^D \delta^{(D)}(\mathbf{k} + \mathbf{k}') \left[\bar{\Gamma}^{(2)}(\mathbf{k})\right]^{-1}. \tag{4.35}$$

This plays the role of the relation (4.20) for $n = 2$. By analogy with (4.24) we may also write this as

$$\bar{\Gamma}^{(2)}(\mathbf{k}) = G^{-1}(\mathbf{k})\bar{G}^{(2)}(\mathbf{k}, -\mathbf{k})_c G^{-1}(-\mathbf{k}). \tag{4.36}$$

Using these composition formulas, the further development will require only the calculation of 1PI diagrams of two- and four-point functions.

4.3 Composite Fields

A Fourier transformation of the correlation functions $G^{(1,n)}(\mathbf{x}; \mathbf{x}_1, \dots, \mathbf{x}_n)$ introduced in Section 3.5 yields

$$G^{(1,n)}(\mathbf{q}, \mathbf{k}_1, \dots, \mathbf{k}_n) = \prod_{i=1}^{n}\left[\int d^D x_i\, e^{-i\,\mathbf{k}_i\cdot\mathbf{x}_i}\right]\int d^D y\, e^{-i\,\mathbf{q}\cdot\mathbf{y}}\, G^{(1,n)}(\mathbf{y}, \mathbf{x}_1, \dots, \mathbf{x}_n).$$

We define the proper vertex functions $\bar{\Gamma}^{(1,n)}(\mathbf{q} = 0, \mathbf{k}_1, \dots, \mathbf{k}_n)$ by analogy to $\bar{\Gamma}^{(n)}(\mathbf{k}_1, \dots, \mathbf{k}_n)$ by selecting the connected 1PI diagrams in the correlation function $\bar{G}^{(n)}(\mathbf{k}_1, \dots, \mathbf{k}_n)$, and by amputating the legs. In $\bar{G}^{(1,2)}(0, \mathbf{k}, \mathbf{k}')$ and $\bar{\Gamma}^{(1,2)}(0, \mathbf{k}, \mathbf{k}')$ we shall omit the second momentum argument $\mathbf{k}' = -\mathbf{k}$, just as in $\bar{\Gamma}^{(2)}(\mathbf{k})$ in Eq. (4.32). The proper vertex function $\bar{\Gamma}^{(1,2)}(0, \mathbf{k})$ is then obtained from $\bar{G}_c^{(1,2)}(0, \mathbf{k})$, by analogy with Eq. (4.36), with the help of the relation

$$\bar{\Gamma}^{(1,2)}(0, \mathbf{k}) \equiv \bar{\Gamma}^{(1,2)}(0, \mathbf{k}, -\mathbf{k}) = G^{-1}(\mathbf{k})\bar{G}_c^{(1,2)}(0, \mathbf{k}, -\mathbf{k})G^{-1}(-\mathbf{k}). \tag{4.37}$$

We now translate the relation (2.57) to momentum space, where it reads

$$G^{(1,n)}(\mathbf{q} = 0, \mathbf{k}_1, \dots, \mathbf{k}_n) = -\frac{\partial}{\partial m^2}G^{(n)}(\mathbf{k}_1, \dots, \mathbf{k}_n), \qquad n \geq 2, \tag{4.38}$$

yielding the relation

$$\frac{\partial}{\partial m^2} G^{-1}(\mathbf{k}) = G^{-1}(\mathbf{k}) \bar{G}_c^{(1,2)}(\mathbf{0}, \mathbf{k}, -\mathbf{k}) G^{-1}(-\mathbf{k}). \tag{4.39}$$

By expressing the right-hand side in terms of $\bar{\Gamma}^{(1,2)}(\mathbf{0}, \mathbf{k})$ using (4.37), and the left-hand side in terms of $\bar{\Gamma}^{(2)}(\mathbf{k})$ using (4.34), we obtain the relation for the proper vertex functions:

$$\bar{\Gamma}^{(1,2)}(\mathbf{0}, \mathbf{k}) = \frac{\partial}{\partial m^2} \bar{\Gamma}^{(2)}(\mathbf{k}). \tag{4.40}$$

Inserting on the right-hand side the decomposition (4.33), we arrive at the formula

$$\bar{\Gamma}^{(1,2)}(\mathbf{0}, \mathbf{k}) = 1 - \frac{\partial}{\partial m^2} \Sigma(\mathbf{k}). \tag{4.41}$$

The derivative with respect to m^2 can be applied directly to each line in the diagrammatic expansion (4.27) of the self-energy. In Eq. (3.28) we indicated a differentiation with respect to m^2 diagrammatically by a fat dot on a line. Here we have the momentum-space version of this operation. For each line, the differentiation yields

$$\frac{\partial}{\partial m^2} \frac{1}{\mathbf{k}^2 + m^2} = -\frac{1}{(\mathbf{k}^2 + m^2)^2} \; \hat{=} \; \frac{\partial}{\partial m^2} \; \text{———} = \; \text{—•—} \; . \tag{4.42}$$

Using Eq. (4.41), we find from (4.27)

$$\tag{4.43}$$

From a similar diagrammatic analysis it is obvious that there exists an analogous relation to (4.40) for any n-point proper vertex function:

$$\bar{\Gamma}^{(1,n)}(\mathbf{0}, \mathbf{k}_1, \ldots, \mathbf{k}_n) = \frac{\partial}{\partial m^2} \bar{\Gamma}^{(n)}(\mathbf{k}_1, \ldots, \mathbf{k}_n), \tag{4.44}$$

as will be proved in general in Section 5.8.

4.4   Theory in Continuous Dimension D

In all the foregoing development we have left the value D of the space dimension open. The phenomena we want to explain take place in $D = 3$ dimensions. For their theoretical explanation, it will be important to be able to define the theory for continuous values of D. In particular, we shall need to connect the theory for $D = 3$ with the theory for $D = 4$ in an analytic way. This will indeed be possible, and the specific mathematical prescription on how to do this will be given in Chapter 8.

Notes and References

See again the textbook cited at the end of last chapter, and
N. Nakanishi, *Graph Theory and Feynman Integrals*, Gordon and Breach, New York, 1971.

5

Structural Properties of Perturbation Theory

The structural properties of all diagrammatic expansions developed so far can be analyzed systematically with the help of functional equations.

5.1 Generating Functionals

In Chapter 3 we have seen that the correlation functions obtained from the functional derivatives of $Z[j]$ via relation (2.14), and the generating functional itself, contain many disconnected parts. Ultimately, however, we shall be interested only in the connected parts of $Z[j]$. Remember that a meaningful description of a very large thermodynamic system can only be given in terms of the free energy which is directly proportional to the total volume. In the limit of an infinite volume, also called *thermodynamic limit*, one has then a well-defined free energy density. The partition function, on the other hand, has no proper infinite-volume limit. We can observe this property directly in the diagrammatic expansion of $Z[j]$. Each component of a disconnected diagram is integrated over the entire space, thus contributing a volume factor. The expansion of $Z[j]$ therefore diverges at an infinite volume. In thermodynamics, we form the free energy from the logarithm of the partition function, which carries only a single overall volume factor and contains only connected diagrams.

Therefore we expect the logarithm of $Z[j]$ to provide us with the desired generating functional $W[j]$:

$$W[j] = \log Z[j]. \tag{5.1}$$

In this chapter we shall see that the functional derivatives of $W[j]$ produce, indeed, precisely the connected parts of the Feynman diagrams in each correlation function.

Consider the connected correlation functions $G_c^{(n)}(\mathbf{x}_1, \ldots, \mathbf{x}_n)$ defined by the functional derivatives

$$G_c^{(n)}(\mathbf{x}_1, \ldots, \mathbf{x}_n) = \frac{\delta}{\delta j(\mathbf{x}_1)} \cdots \frac{\delta}{\delta j(\mathbf{x}_n)}. W[j] \tag{5.2}$$

At the end, we shall be interested only in those functions at zero external current, where they reduce to the physical quantities (2.46) that vanish for odd n in the normal phase under study here. For the general development in this chapter, however, we shall consider them as functionals of $j(\mathbf{x})$, and go over to $j = 0$ only at the very end. The diagrammatic representation of these correlation functions contains only connected diagrams defined in Section 3.3. Moreover, the connected correlation functions $G_c^{(n)}(\mathbf{x}_1, \ldots, \mathbf{x}_n)$ collect *all* connected diagrams of the full correlation functions $G^{(n)}(\mathbf{x}_1, \ldots, \mathbf{x}_n)$, which then can be recovered via simple composition laws from the connected ones. In order to see this clearly, we shall derive the general relationship between the two types of correlation functions in Section 5.3. First, we shall prove the connectedness property of the derivatives (5.2).

5.2   Connectedness Structure of Correlation Functions

In this section, we shall prove that the generating functional $W[j]$ collects *only* connected diagrams in its Taylor coefficients $\delta^n W/\delta j(x_1)\ldots\delta j(x_n)$. Later, after Eq. (5.26), we shall see that *all* connected diagrams of $G_c^{(n)}(x_1,\ldots,x_n)$ occur in $G^{(n)}(x_1,\ldots,x_n)$.

The basis for the following considerations is the fact that the functional integral (2.13) for the generating functional $Z[j]$ satisfies an elementary identity

$$\int \mathcal{D}\phi\, \frac{\delta}{\delta\phi(\mathbf{x})} e^{-E[\phi,j]} = 0, \tag{5.3}$$

which follows from the vanishing of the Boltzmann factor $e^{-E[\phi,j]}$ at infinite field strength. After performing the functional derivative, we have

$$\int \mathcal{D}\phi\, \frac{\delta E[\phi,j]}{\delta\phi(\mathbf{x})} e^{-E[\phi,j]} = 0. \tag{5.4}$$

Inserting (2.12) for the functional $E[\phi,j]$, this reads

$$\int \mathcal{D}\phi\, \left[G_0^{-1}\phi(\mathbf{x}) + \frac{\lambda}{3!}\phi^3(\mathbf{x}) - j(\mathbf{x}) \right] e^{-E[\phi,j]} = 0. \tag{5.5}$$

Expressing the fields $\phi(\mathbf{x})$ as functional derivatives with respect to the source current $j(\mathbf{x})$, the brackets can be taken out of the integral, and we obtain the functional differential equation for the generating functional $Z[j]$:

$$\left\{ G_0^{-1}\frac{\delta}{\delta j(\mathbf{x})} + \frac{\lambda}{3!}\left[\frac{\delta}{\delta j(\mathbf{x})} \right]^3 - j(\mathbf{x}) \right\} Z[j] = 0. \tag{5.6}$$

With the short-hand notation

$$Z_{j(\mathbf{x}_1)j(\mathbf{x}_2)\ldots j(\mathbf{x}_n)}[j] \equiv \frac{\delta}{\delta j(\mathbf{x}_1)}\frac{\delta}{\delta j(\mathbf{x}_2)} \cdots \frac{\delta}{\delta j(\mathbf{x}_n)} Z[j], \tag{5.7}$$

where the arguments of the currents will eventually be suppressed, this can be written as

$$G_0^{-1}Z_{j(\mathbf{x})} + \frac{\lambda}{3!}Z_{j(\mathbf{x})j(\mathbf{x})j(\mathbf{x})} - j(\mathbf{x})Z[j] = 0. \tag{5.8}$$

Inserting here (5.1), we obtain a functional differential equation for $W[j]$:

$$G_0^{-1}W_j + \frac{\lambda}{3!}\left(W_{jjj} + 3W_{jj}W_j + W_j^3 \right) - j = 0. \tag{5.9}$$

We have employed the same short-hand notation for the functional derivatives of $W[j]$ as in (5.7):

$$W_{j(\mathbf{x}_1)j(\mathbf{x}_2)\ldots j(\mathbf{x}_n)}[j] \equiv \frac{\delta}{\delta j(\mathbf{x}_1)}\frac{\delta}{\delta j(\mathbf{x}_2)} \cdots \frac{\delta}{\delta j(\mathbf{x}_n)} W[j], \tag{5.10}$$

suppressing the arguments $\mathbf{x}_1,\ldots,\mathbf{x}_n$ of the currents, for brevity. Multiplying (5.9) functionally by G_0 gives

$$W_j = -\frac{\lambda}{3!}G_0\left(W_{jjj} + 3W_{jj}W_j + W_j^3 \right) + G_0\,j. \tag{5.11}$$

We have omitted the integral over the intermediate space argument, for brevity. More specifically, we have written $G_0\,j$ for $\int d^D y\, G_0(\mathbf{x},\mathbf{y})j(\mathbf{y})$. Similar expressions abbreviate all functional products. This corresponds to a functional version of *Einstein's summation convention*.

Equation (5.11) may now be expressed in terms of the one-point correlation function

$$G_c^{(1)} = W_j(x), \tag{5.12}$$

defined in (5.2), as

$$G_c^{(1)} = -\frac{\lambda}{3!}G_0\left\{G_{c\,jj}^{(1)} + 3G_{c\,j}^{(1)}G_c^{(1)} + \left[G_c^{(1)}\right]^3\right\} + G_0\,j. \tag{5.13}$$

The solution to this equation is conveniently found by a diagrammatic procedure displayed in Fig. 5.1. To lowest, zeroth, order in λ we have

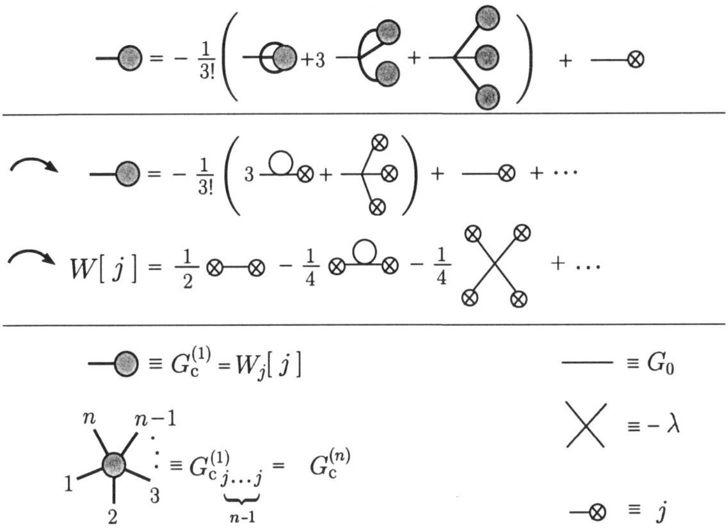

FIGURE 5.1 Diagrammatic solution of recursion relation (5.11) for the generating functional $W[j]$ of all connected correlation functions. First line represents Eq. (5.13), second (5.16), third (5.17). The remaining lines define the diagrammatic symbols.

$$G_c^{(1)} = G_0\,j. \tag{5.14}$$

From this we find by functional integration the zeroth order generating functional $W[j]$

$$W_0[j] = \int \mathcal{D}j\, G_c^{(1)} = \frac{1}{2}jG_0j, \tag{5.15}$$

a result already known from (2.31) and (2.34). As in the perturbation expansions (2.47) of the correlation functions, subscripts of $W[j]$ indicate the order in the interaction strength λ.

Reinserting (5.14) on the right-hand side of (5.13) gives the first-order expression

$$G_c^{(1)} = -G_0\frac{\lambda}{3!}\left[3G_0G_0j + (G_0j)^3\right] + G_0j, \tag{5.16}$$

represented diagrammatically in the second line of Fig. 5.1. The expression (5.16) can be integrated functionally in j to obtain $W[j]$ up to first order in λ. Diagrammatically, this process amounts to multiplying the open line in each diagram by a current j, and dividing each term j^n by n. Thus we arrive at

$$W_0[j] + W_1[j] \;=\; \frac{1}{2} j G_0 j - \frac{\lambda}{4} G_0 \left(G_0 j \right)^2 - \frac{\lambda}{24} \left(G_0 j \right)^4 , \qquad (5.17)$$

as illustrated in the third line of Fig. 5.1. This procedure can be continued to any order in λ.

This diagrammatic procedure allows us to prove that the generating functional $W[j]$ collects *only* connected diagrams in its Taylor coefficients $\delta^n W / \delta j(x_1) \ldots \delta j(x_n)$. For the lowest two orders we can verify the connectedness by inspecting the third line in Fig. 5.1. The diagrammatic form of the recursion relation shows that this topological property remains true for all orders in λ, by induction. Indeed, if we suppose it to be true for some n, then all $G_c^{(1)}$ inserted on the right-hand side are connected, and so are the diagrams constructed from these when forming $G_c^{(1)}$ to the next, $(n+1)$st, order.

Note that this calculation is unable to recover the value of $W[j]$ at $j = 0$ since this is an unknown integration constant of the functional differential equation. For the purpose of generating correlation functions, this constant is irrelevant. We have seen in Section 3.2 that $W[0]$ consists of the sum of all connected vacuum diagrams contained in $Z[0]$.

5.3 Decomposition of Correlation Functions into Connected Correlation Functions

Using the logarithmic relation (5.1) between $W[j]$ and $Z[j]$ we can now derive general relations between the n-point functions and their connected parts. For the one-point function we find

$$G^{(1)}(\mathbf{x}) = Z^{-1}[j] \frac{\delta}{\delta j(\mathbf{x})} Z[j] = \frac{\delta}{\delta j(\mathbf{x})} W[j] = G_c^{(1)}(\mathbf{x}). \qquad (5.18)$$

This equation implies that the one-point function representing the ground state expectation value of the field is always connected:

$$\langle \phi(\mathbf{x}) \rangle \equiv G^{(1)}(\mathbf{x}) = G_c^{(1)}(\mathbf{x}) = \Phi. \qquad (5.19)$$

Consider now the two-point function, which decomposes as follows:

$$
\begin{aligned}
G^{(2)}(\mathbf{x}_1, \mathbf{x}_2) \;&=\; Z^{-1}[j] \frac{\delta}{\delta j(\mathbf{x}_1)} \frac{\delta}{\delta j(\mathbf{x}_2)} Z[j] \\
&=\; Z^{-1}[j] \frac{\delta}{\delta j(\mathbf{x}_1)} \left\{ \left(\frac{\delta}{\delta j(\mathbf{x}_2)} W[j] \right) Z[j] \right\} \\
&=\; Z^{-1}[j] \left\{ W_{j(\mathbf{x}_1) j(\mathbf{x}_2)} + W_{j(\mathbf{x}_1)} W_{j(\mathbf{x}_2)} \right\} Z[j] \\
&=\; G_c^{(2)}(\mathbf{x}_1, \mathbf{x}_2) + G_c^{(1)}(\mathbf{x}_1) G_c^{(1)}(\mathbf{x}_2) .
\end{aligned}
\qquad (5.20)
$$

In addition to the connected diagrams with two ends there are two connected diagrams ending in a single line. These are absent in a ϕ^4-theory with positive m^2 at $j = 0$, in which case the system is in the *normal phase* (recall the discussion in Chapter 1). In that case, the two-point function is automatically connected, as we observed in Eq. (3.23).

For the three-point function we find

$$
\begin{aligned}
G^{(3)}\left(\mathbf{x}_1, \mathbf{x}_2, \mathbf{x}_3\right) & = Z^{-1}[j]\frac{\delta}{\delta j(\mathbf{x}_1)}\frac{\delta}{\delta j(\mathbf{x}_2)}\frac{\delta}{\delta j(\mathbf{x}_3)}Z[j] \\
& = Z^{-1}[j]\frac{\delta}{\delta j(\mathbf{x}_1)}\frac{\delta}{\delta j(\mathbf{x}_2)}\left\{\left[\frac{\delta}{\delta j(\mathbf{x}_3)}W[j]\right]Z[j]\right\} \\
& = Z^{-1}[j]\frac{\delta}{\delta j(\mathbf{x}_1)}\left\{\left[W_{j(\mathbf{x}_3)j(\mathbf{x}_2)} + W_{j(\mathbf{x}_2)}W_{j(\mathbf{x}_3)}\right]Z[j]\right\} \\
& = Z^{-1}[j]\left\{W_{j(\mathbf{x}_1)j(\mathbf{x}_2)j(\mathbf{x}_3)} + \left(W_{j(\mathbf{x}_1)}W_{j(\mathbf{x}_2)j(\mathbf{x}_3)} + W_{j(\mathbf{x}_2)}W_{j(\mathbf{x}_1)j(\mathbf{x}_3)}\right.\right. \\
& \qquad\qquad\qquad \left.\left.+ W_{j(\mathbf{x}_3)}W_{j(\mathbf{x}_1)j(\mathbf{x}_2)}\right) + W_{j(\mathbf{x}_1)}W_{j(\mathbf{x}_2)}W_{j(\mathbf{x}_3)}\right\}Z[j] \\
& = G_c^{(3)}\left(\mathbf{x}_1, \mathbf{x}_2, \mathbf{x}_3\right) + \left[G_c^{(1)}(\mathbf{x}_1)G_c^{(2)}(\mathbf{x}_2, \mathbf{x}_3) + 2\text{ perm}\right] + G_c^{(1)}(\mathbf{x}_1)G_c^{(1)}(\mathbf{x}_2)G_c^{(1)}(\mathbf{x}_3),
\end{aligned} \tag{5.21}
$$

and for the four-point function

$$
\begin{aligned}
G^{(4)}\left(\mathbf{x}_1, \ldots, \mathbf{x}_4\right) & = G_c^{(4)}\left(\mathbf{x}_1, \ldots, \mathbf{x}_4\right) + \left[G_c^{(3)}\left(\mathbf{x}_1, \mathbf{x}_2, \mathbf{x}_3\right)G_c^{(1)}(\mathbf{x}_4) + 3\text{ perm}\right] \\
& \quad + \left[G_c^{(2)}\left(\mathbf{x}_1, \mathbf{x}_2\right)G_c^{(2)}\left(\mathbf{x}_3, \mathbf{x}_4\right) + 2\text{ perm}\right] \\
& \quad + \left[G_c^{(2)}\left(\mathbf{x}_1, \mathbf{x}_2\right)G_c^{(1)}(\mathbf{x}_3)G_c^{(1)}(\mathbf{x}_4) + 5\text{ perm}\right] \\
& \quad + G_c^{(1)}(\mathbf{x}_1)\cdots G_c^{(1)}(\mathbf{x}_4).
\end{aligned} \tag{5.22}
$$

In the pure ϕ^4-theory with positive m^2, i.e., in the *normal phase* of the system, there are no odd correlation functions and we are left with the decomposition (3.24), which was found in Chapter 3 diagrammatically up to second order in the coupling constant λ.

For the general correlation function $G^{(n)}$, the total number of terms is most easily retrieved by dropping all indices and differentiating with respect to j (the arguments $\mathbf{x}_1, \ldots, \mathbf{x}_n$ of the currents are again suppressed):

$$
\begin{aligned}
G^{(1)} & = e^{-W}\left(e^W\right)_j = W_j = G_c^{(1)} \\
G^{(2)} & = e^{-W}\left(e^W\right)_{jj} = W_{jj} + W_j{}^2 = G_c^{(2)} + G_c^{(1)2} \\
G^{(3)} & = e^{-W}\left(e^W\right)_{jjj} = W_{jjj} + 3W_{jj}W_j + W_j{}^3 = G_c^{(3)} + 3G_c^{(2)}G_c^{(1)} + G_c^{(1)3} \\
G^{(4)} & = e^{-W}\left(e^W\right)_{jjjj} = W_{jjjj} + 4W_{jjj}W_j + 3W_{jj}{}^2 + 6W_{jj}W_j{}^2 + W_j{}^4 \\
& = G_c^{(4)} + 4G_c^{(3)}G_c^{(1)} + 3G_c^{(2)2} + 6G_c^{(2)}G_c^{(1)2} + G_c^{(1)4}.
\end{aligned} \tag{5.23}
$$

All relations follow from the recursion relation

$$
G^{(n)} = G_j^{(n-1)} + G^{(n-1)}G_c^{(1)}, \quad n \geq 2, \tag{5.24}
$$

if one uses $G_{cj}^{(n-1)} = G_c^{(n)}$ and the initial relation $G^{(1)} = G_c^{(1)}$. By comparing the first four relations with the explicit forms (5.20)–(5.22) we see that the numerical factors on the right-hand side of (5.23) refer to the permutations of the arguments $\mathbf{x}_1, \mathbf{x}_2, \mathbf{x}_3, \ldots$ of otherwise equal expressions. Since there is no problem in reconstructing the explicit permutations we shall henceforth write all composition laws in the short-hand notation (5.23).

The formula (5.23) and its generalization is often referred to as *cluster decomposition*, or also as the *cumulant expansion*, of the correlation functions.

We can now prove that the connected correlation functions collect precisely all connected diagrams in the n-point functions. For this we observe that the decomposition rules can be inverted by repeatedly differentiating both sides of the equation $W[j] = \log Z[j]$ functionally with respect to the current j:

$$
\begin{aligned}
G_c^{(1)} &= G^{(1)} \\
G_c^{(2)} &= G^{(2)} - G^{(1)}G^{(1)} \\
G_c^{(3)} &= G^{(3)} - 3G^{(2)}G^{(1)} + 2G^{(1)3} \\
G_c^{(4)} &= G^{(4)} - 4G^{(3)}G^{(1)} + 12G^{(2)}G^{(1)2} - 3G^{(2)2} - 6G^{(1)4}.
\end{aligned}
\tag{5.25}
$$

Each equation follows from the previous one by one more derivative with respect to j, and by replacing the derivatives on the right-hand side according to the rule

$$
G_j^{(n)} = G^{(n+1)} - G^{(n)}G^{(1)}.
\tag{5.26}
$$

Again the numerical factors imply different permutations of the arguments and the subscript j denotes functional differentiations with respect to j.

Note that Eqs. (5.25) for the connected correlation functions are valid in the normal phase as well as in the phase with spontaneous symmetry breakdown. In the normal phase, the equations simplify, since all terms involving $G^{(1)} = \Phi = \langle \phi \rangle$ vanish.

It is obvious that any connected diagram contained in $G^{(n)}$ must also be contained in $G_c^{(n)}$, since all the terms added or subtracted in (5.25) are products of $G_j^{(n)}$s, and thus necessarily disconnected. Together with the proof in Section 5.2 that the correlation functions $G_c^{(n)}$ contain *only* the connected parts of $G^{(n)}$, we can now be sure that $G_c^{(n)}$ contains precisely the connected parts of $G^{(n)}$.

5.4   Functional Generation of Vacuum Diagrams

The functional differential equation (5.11) for $W[j]$ contains all information on the connected correlation functions of the system. However, it does not tell us anything about the vacuum diagrams of the theory. These are contained in $W[0]$, which remains an undetermined constant of functional integration of these equations.

In order to gain information on the vacuum diagrams, we consider a modification of the generating functional (2.54), in which we set the external source j equal to zero, but generalize the source $K(\mathbf{x})$ to a bilocal form $K(\mathbf{x}, \mathbf{y})$:

$$
Z[K] = \int \mathcal{D}\phi(\mathbf{x})\, e^{-E[\phi, K]},
\tag{5.27}
$$

where $E[\phi, K]$ is the energy functional:

$$
E[\phi, K] \equiv E_0[\phi] + E_{\text{int}}[\phi] + \frac{1}{2}\int d^D x \int d^D y\, \phi(\mathbf{x})K(\mathbf{x}, \mathbf{y})\phi(\mathbf{y}).
\tag{5.28}
$$

When forming the functional derivative with respect to $K(\mathbf{x}, \mathbf{y})$ we obtain the correlation function in the presence of $K(\mathbf{x}, \mathbf{y})$:

$$
G^{(2)}(\mathbf{x}, \mathbf{y}) = 2Z^{-1}[K]\frac{\delta Z}{\delta K(\mathbf{x}, \mathbf{y})}.
\tag{5.29}
$$

At the end we shall set $K(\mathbf{x}, \mathbf{y}) = 0$, just as previously the source j. When differentiating $Z[K]$ twice, we obtain the four-point function

$$G^{(4)}(\mathbf{x}_1, \mathbf{x}_2, \mathbf{x}_3, \mathbf{x}_4) = 4Z^{-1}[K]\frac{\delta^2 Z}{\delta K(\mathbf{x}_1, \mathbf{x}_2)\delta K(\mathbf{x}_3, \mathbf{x}_4)}. \tag{5.30}$$

As before, we introduce the functional $W[K] \equiv \log Z[K]$. Inserting this into (5.29) and (5.30), we fin

$$G^{(2)}(\mathbf{x}, \mathbf{y}) = 2\frac{\delta W}{\delta K(\mathbf{x}, \mathbf{y})}, \tag{5.31}$$

$$G^{(4)}(\mathbf{x}_1, \mathbf{x}_2, \mathbf{x}_3, \mathbf{x}_4) = 4\left[\frac{\delta^2 W}{\delta K(\mathbf{x}_1, \mathbf{x}_2)\delta K(\mathbf{x}_3, \mathbf{x}_4)} + \frac{\delta W}{\delta K(\mathbf{x}_1, \mathbf{x}_2)}\frac{\delta W}{\delta K(\mathbf{x}_3, \mathbf{x}_4)}\right]. \tag{5.32}$$

With the same short notation as before, we shall use again a subscript K to denote functional differentiation with respect to K, and write

$$G^{(2)} = 2W_K, \qquad G^{(4)} = 4\left[W_{KK} + W_K W_K\right] = 4\left[W_{KK} + G^{(2)}G^{(2)}\right]. \tag{5.33}$$

From Eq. (5.23) we know that in the absence of a source j and in the normal phase, $G^{(4)}$ has the connectedness structure

$$G^{(4)} = G_c^{(4)} + 3G_c^{(2)}G_c^{(2)}. \tag{5.34}$$

This shows that in contrast to W_{jjjj}, the derivative W_{KK} does not directly yield a connected four-point function, but two disconnected parts:

$$W_{KK} = G_c^{(4)} + 2G_c^{(2)}G_c^{(2)}, \tag{5.35}$$

the two-point functions being automatically connected in the normal phase. More explicitly

$$\frac{4\delta^2 W}{\delta K(\mathbf{x}_1, \mathbf{x}_2)\delta K(\mathbf{x}_3, \mathbf{x}_4)} = G_c^{(4)}(\mathbf{x}_1, \mathbf{x}_2, \mathbf{x}_3, \mathbf{x}_4) + G_c^{(2)}(\mathbf{x}_1, \mathbf{x}_3)G_c^{(2)}(\mathbf{x}_2, \mathbf{x}_4) + G_c^{(2)}(\mathbf{x}_1, \mathbf{x}_4)G_c^{(2)}(\mathbf{x}_2, \mathbf{x}_3). \tag{5.36}$$

Let us derive functional differential equations for $Z[K]$ and $W[K]$. By analogy with (5.3) we start out with the trivial functional differential equation

$$\int \mathcal{D}\phi(\mathbf{x}) \, \phi(\mathbf{x})\frac{\delta}{\delta\phi(\mathbf{y})}e^{-E[\phi,K]} = -\delta^{(D)}(\mathbf{x} - \mathbf{y})Z[K], \tag{5.37}$$

which is immediately verified by a functional integration by parts. Performing the functional derivative yields

$$\int \mathcal{D}\phi(\mathbf{x}) \, \phi(\mathbf{x})\frac{\delta E[\phi, K]}{\delta\phi(\mathbf{y})}e^{-E[\phi,K]} = \delta^{(D)}(\mathbf{x} - \mathbf{y})Z[K], \tag{5.38}$$

or

$$\int \mathcal{D}\phi(\mathbf{x}) \int d^D x \int d^D y \left\{\phi(\mathbf{x})G_0^{-1}(\mathbf{x}, \mathbf{y})\phi(\mathbf{y}) + \frac{\lambda}{3!}\phi(\mathbf{x})\phi^3(\mathbf{y})\right\} e^{-E[\phi,K]} = \delta^{(D)}(\mathbf{x} - \mathbf{y})Z[K]. \tag{5.39}$$

For brevity, we have absorbed the source in the free-field correlation function G_0:

$$G_0 \to [G_0^{-1} - K]^{-1}. \tag{5.40}$$

The left-hand side of (5.38) can obviously be expressed in terms of functional derivatives of $Z[K]$, and we obtain the functional differential equation whose short form reads

$$G_0^{-1}Z_K + \frac{\lambda}{3}Z_{KK} = \frac{1}{2}Z. \tag{5.41}$$

Inserting $Z[K] = e^{W[K]}$, this becomes

$$G_0^{-1}W_K + \frac{\lambda}{3}(W_{KK} + W_K W_K) = \frac{1}{2}. \tag{5.42}$$

It is useful to reconsider the functional $W[K]$ as a functional $W[G_0]$. Then $\delta G_0/\delta K = G_0^2$, and the derivatives of $W[K]$ become

$$W_K = G_0^2 W_{G_0}, \quad W_{KK} = 2G_0^3 W_{G_0} + G_0^4 W_{G_0 G_0}, \tag{5.43}$$

and (5.42) takes the form

$$G_0 W_{G_0} + \frac{\lambda}{3}(G_0^4 W_{G_0 G_0} + 2G_0^3 W_{G_0} + G_0^4 W_{G_0} W_{G_0}) = \frac{1}{2}. \tag{5.44}$$

This equation is represented diagrammatically in Fig. 5.2. The zeroth-order solution to this

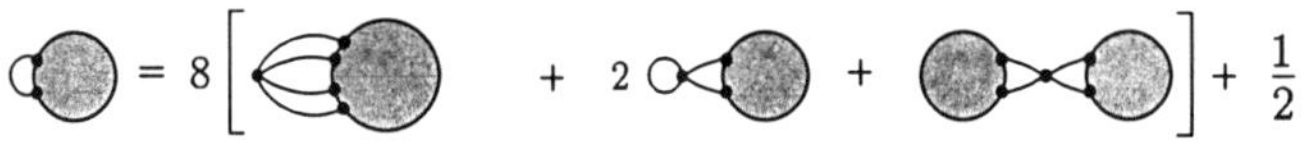

$$G_0 W_{G_0} = 8\frac{-1}{4!}\left[\lambda G_0^4 W_{G_0 G_0} + 2G_0 \lambda G_0^2 W_{G_0} + W_{G_0} G_0^2 \lambda G_0^2 W_{G_0}\right] + \frac{1}{2}$$

FIGURE 5.2 Diagrammatic representation of functional differential equation (5.44). For the purpose of finding the multiplicities of the diagrams, it is convenient to represent here by a vertex the coupling strength $-\lambda/4!$, rather than $-\lambda$ as all other vertices in this book.

equation is obtained by setting $\lambda = 0$:

$$W^{(0)}[G_0] = \frac{1}{2}\text{Tr}\,\log(G_0). \tag{5.45}$$

This is precisely the exponent in the prefactor of the generating functional (2.31) of the free-field theory.

The corrections are found by iteration. For systematic treatment, we write $W[G_0]$ as a sum of a free and an interacting part,

$$W[G_0] = W^{(0)}[G_0] + W^{\text{int}}[G_0], \tag{5.46}$$

insert this into Eq. (5.44), and find the differential equation for the interacting part:

$$G_0 W^{\text{int}}_{G_0} + \frac{\lambda}{3}(G_0^4 W^{\text{int}}_{G_0 G_0} + 3G_0^3 W^{\text{int}}_{G_0} + G_0^4 W^{\text{int}}_{G_0} W^{\text{int}}_{G_0}) = 6\frac{-\lambda}{4!}G_0^2. \tag{5.47}$$

This equation is solved iteratively. Setting $W^{\text{int}}[G_0] = 0$ in all terms proportional to λ, we obtain the first-order contribution to $W^{\text{int}}[G_0]$:

$$W^{\text{int}}[G_0] = 3\frac{-\lambda}{4!}G_0^2. \tag{5.48}$$

This is precisely the contribution of the Feynman diagram. The number 3 is its multiplicity, as defined in Section 3.1.

In order to see how the iteration of Eq. (5.47) may be solved systematically, let us ignore for the moment the functional nature of Eq. (5.47), and treat G_0 as an ordinary real variable rather than a functional matrix. We expand $W[G_0]$ in a Taylor series:

$$W^{\text{int}}[G_0] = \sum_{p=1}^{\infty} \frac{1}{p!} W_p \left(\frac{-\lambda}{4!} \right)^p (G_0)^{2p}, \tag{5.49}$$

and find for the expansion coefficients the recursion relation

$$W_{p+1} = 4 \left\{ [2p(2p-1) + 3(2p)] W_p + \sum_{q=1}^{p-1} \binom{p}{q} 2q \, W_q \times 2(p-q) W_{p-q} \right\}. \tag{5.50}$$

Solving this with the initial number $W_1 = 3$, we obtain the multiplicities of the connected vacuum diagrams of pth order:

$$3, 96, 9504, 1880064, 616108032, 301093355520, 205062331760640, 185587468924354560,$$
$$215430701800551874560, 312052349085504377978880. \tag{5.51}$$

To check these numbers, we go over to $Z[G] = e^{W[G_0]}$, and find the expansion:

$$\begin{aligned} Z[G_0] &= \exp \left[\frac{1}{2} \text{Tr} \log G_0 + \sum_{p=1}^{\infty} \frac{1}{p!} W_p \left(\frac{-\lambda}{4!} \right)^p (G_0)^{2p} \right] \\ &= \text{Det}^{1/2}[G_0] \left[1 + \sum_{p=1}^{\infty} \frac{1}{p!} z_p \left(\frac{-\lambda}{4!} \right)^p (G_0)^{2p} \right] \end{aligned} \tag{5.52}$$

The expansion coefficients z_p count the total number of vacuum diagrams of order p. The exponentiation (5.52) yields $z_p = (4p-1)!!$, which is the correct number of Wick contractions of p interactions ϕ^4.

In fact, by comparing coefficients in the two expansions in (5.52), we may derive another recursion relation for W_p:

$$W_p + 3 \binom{p-1}{1} W_{p-1} + 7 \cdot 5 \cdot 3 \binom{p-1}{2} + \ldots + (4p-5)!! \binom{p-1}{p-1} = (4p-1)!!, \tag{5.53}$$

which is fulfilled by the solutions of (5.50).

In order to find the associated Feynman diagrams, we must perform the differentiations in Eq. (5.47) functionally. The numbers W_p become then a sum of diagrams, for which the recursion relation (5.50) reads

$$W_{p+1} = 4 \left[G_0^4 \frac{d^2}{d\cap^2} W_p + 3 \cdot G_0^3 \frac{d}{d\cap} W_p + \sum_{q=1}^{p-1} \binom{p}{q} \left(\frac{d}{d\cap} W_q \right) G_0^2 \cdot G_0^2 \left(\frac{d}{d\cap} W_{p-q} \right) \right], \tag{5.54}$$

where the differentiation $d/d\cap$ removes one line connecting two vertices in all possible ways. This equation is solved diagrammatically, as shown in Fig. 5.3.

Starting the iteration with $W_1 = 3\,\infty$, we have $dW_p/d\cap = 6\,\bigcirc$ and $d^2 W_p/d\cap^2 = 6\times$. Proceeding to order five loops and going back to the usual vertex notation $-\lambda$, we find the vacuum diagrams with their weight factors as shown in Fig. 5.4. For more than five loops, the

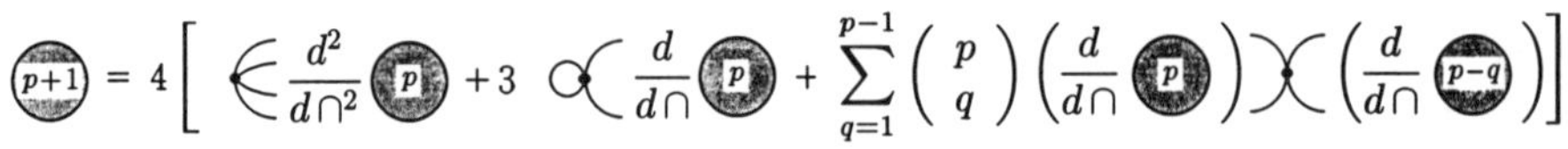

$$\left(\widehat{p{+}1}\right) = 4\left[\;\in\frac{d^2}{d\cap^2}\,\boxed{p} + 3\;\propto\frac{d}{d\cap}\,\boxed{p} + \sum_{q=1}^{p-1}\binom{p}{q}\left(\frac{d}{d\cap}\,\bullet\right)\times\left(\frac{d}{d\cap}\,\widehat{p{-}q}\right)\right]$$

$$W_{p+1} = 4\left[\;G_0^4\frac{d^2}{d\cap^2}\,W_p + 3\cdot G_0^3\,\frac{d}{d\cap}\,W_p + \sum_{q=1}^{p-1}\binom{p}{q}\left(\frac{d}{d\cap}W_q\right)G_0^2\cdot G_0^2\left(\frac{d}{d\cap}W_{p-q}\right)\right]$$

FIGURE 5.3 Diagrammatic representation of functional differential equation (5.54). A vertex represents the coupling strength $-\lambda$.

reader is referred to the paper quoted in Notes and References, and to the internet address from which Mathematica programs can be downloade,d which solve the recursion relations and plot all diagrams of $W[0]$ and the resulting two-and four-point functions.

In Section 14.2 we shall describe a somewhat shorter computer scheme for generating all diagrams used in this text.

order	diagrams and multiplicities	number
g^1	$3\,\text{OO}$	3
g^2	$\dfrac{1}{2!}\left(24\,\ominus \quad 72\,\text{OOO}\right)$	96
g^3	$\dfrac{1}{3!}\left(1728\,\bigtriangledown \quad 3456\,\ominus \quad 2592\,\text{OOOO} \quad 1728\,\text{OOO}\right)$	9504
g^4	$\dfrac{1}{4!}\Big(62208\,\square \quad 66296 \quad 248832 \quad 497664 \quad 165888 \quad 248832$ $165888 \quad 124416\,\text{OOOOO} \quad 248832 \quad 62208\Big)$	1880064

FIGURE 5.4 Vacuum diagrams up to five loops and their multiplicities. In contrast to Fig. 5.3, and the usual diagrammatic notation in (3.5), a vertex stands here for $-\lambda/4!$ for brevity. For more than five loops see the tables on the internet (www.physik.fu-berlin/~kleinert/294/programs).

5.5   Correlation Functions From Vacuum Diagrams

The vacuum diagrams contain information on all correlation functions of the theory. One may rightly say that the vacuum is the world. The two- and four-point functions are given by the functional derivatives (5.33) of the vacuum functional $W[K]$. Diagrammatically, a derivative with respect to K corresponds to cutting one line of a vacuum diagram in all possible ways. Thus, all diagrams of the two-point function $G^{(2)}$ can be derived from such cuts, multiplied by a factor 2. As an example, consider the first-order vacuum diagram of $W[K]$ in Table 5.4

(compare also Fig. 3.5). Cutting one line, which is possible in two ways, and recalling that in Table 5.4 a vertex stands for $-\lambda/4!$ rather than $-\lambda$, as in the other diagrams, we find

$$W_1[0] = \frac{1}{8} \; \bigcirc\!\!\bigcirc \qquad \longrightarrow \qquad G_1^{(2)}(\mathbf{x}_1, \mathbf{x}_2) = 2 \times \frac{1}{8} \; 2 \; \underset{\mathbf{x}_1 \quad \mathbf{x}_2}{\bigcirc} . \tag{5.55}$$

The right-hand side is the correct first-order contribution to the two-point function [recall Eq. (3.23)].

The second equation in (5.33) tells us that all connected contributions to the four-point function $G^{(4)}$ may be obtained by cutting two lines in all combinations, and multiplying the result by a factor 4. As an example, take the second-order vacuum diagrams of $W[0]$ with the proper translation of vertices by a factor 4! (compare again Fig. 3.5), which are

$$W_2[0] = \frac{1}{16} \; \bigcirc\!\!\bigcirc\!\!\bigcirc + \frac{1}{48} \; \ominus . \tag{5.56}$$

Cutting two lines in all possible ways yields the following contributions to the connected diagrams of the two-point function:

$$G^{(4)} = 4 \times \left(2 \cdot 1 \cdot \frac{1}{16} + 4 \cdot 3 \cdot \frac{1}{48} \right) \; \asymp\!\!\asymp . \tag{5.57}$$

This agrees with the first-order contribution calculated in Eq. (3.25).

It is also possible to find all diagrams of the four-point function from the vacuum diagrams by forming a derivative of $W[0]$ with respect to the coupling constant $-\lambda$, and multiplying the result by a factor 4!. This follows directly from the fact that this differentiation applied to $Z[0]$ yields the correlation function $\int d^D x \langle \phi^4 \rangle$. As an example, take the first diagram of order g^3 in Table 5.4 [with the vertex normalization (3.5)]:

$$W_2[0] = \frac{1}{48} \; \bigtriangledown . \tag{5.58}$$

Removing one vertex in the three possible ways and multiplying by a factor 4! yields

$$G^{(4)} = 4! \times \frac{1}{48} \; 3 \; \asymp\!\!\asymp . \tag{5.59}$$

which agrees with the contribution of this diagram in Eq. (3.25).

These relations will be used in Chapter 14 to generate all diagrams by computer methods.

5.6   Generating Functional for Vertex Functions

Apart from the connectedness structure, the most important step in economizing the calculation of Feynman diagrams consists in the decomposition of higher connected correlation functions into 1PI vertex functions and 1PI two-particle correlation functions, as shown in Section 4.2. There is, in fact, a simple algorithm which supplies us in general with such a decomposition. For this purpose let us introduce a new generating functional $\Gamma[\Phi]$, to be called the *effective energy* of the theory. It is defined via a Legendre transformation of $W[j]$:

$$-\Gamma[\Phi] \equiv W[j] - W_j \, j. \tag{5.60}$$

Here and in the following, we use a short-hand notation for the functional multiplication, $W_j\, j = \int d^D x\, W_j(\mathbf{x}) j(\mathbf{x})$, which considers fields as vectors with a continuous index $\mathbf{x}$. The new variable Φ is the functional derivative of $W[j]$ with respect to $j(\mathbf{x})$ [recall (5.10)]:

$$\Phi(\mathbf{x}) \equiv \frac{\delta W[j]}{\delta j(\mathbf{x})} \equiv W_{j(\mathbf{x})} = \langle \phi \rangle_{j(\mathbf{x})}, \tag{5.61}$$

and thus gives the ground state expectation of the field operator in the presence of the current j. When rewriting (5.60) as

$$-\Gamma[\Phi] \equiv W[j] - \Phi\, j, \tag{5.62}$$

and functionally differentiating this with respect to Φ, we obtain the equation

$$\Gamma_\Phi[\Phi] = j. \tag{5.63}$$

This equation shows that the physical field expectation $\Phi(\mathbf{x}) = \langle \phi(\mathbf{x}) \rangle$, where the external current is zero, extremizes the effective energy:

$$\Gamma_\Phi[\Phi] = 0. \tag{5.64}$$

In this text, we shall only study physical systems whose ordered low-temperature phase has a uniform field expectation value $\Phi(\mathbf{x}) \equiv \Phi_0$. Thus we shall not consider systems such as cholesteric or smectic liquid crystals, which possess a space dependent $\Phi_0(\mathbf{x})$, although such systems can also be described by ϕ^4-theories by admitting more general types of gradient terms, for instance $\phi(\partial^2 - k_0^2)^2 \phi$. The ensuing space dependence of $\Phi_0(\mathbf{x})$ may be crystal- or quasicrystal-like [1]. Thus we shall assume a constant

$$\Phi_0 = \langle \phi \rangle|_{j=0}, \tag{5.65}$$

which may be zero or non-zero, depending on the phase of the system.

Let us now demonstrate that the effective energy contains all the information on the proper vertex functions of the theory. These can be found directly from the functional derivatives:

$$\Gamma^{(n)}(\mathbf{x}_1, \ldots, \mathbf{x}_n) \equiv \frac{\delta}{\delta \Phi(\mathbf{x}_1)} \cdots \frac{\delta}{\delta \Phi(\mathbf{x}_n)} \Gamma[\Phi] . \tag{5.66}$$

We shall see that the proper vertex functions of Section 4.2 are obtained from these functions by a Fourier transform and a simple removal of an overall factor $(2\pi)^D \delta^{(D)} \left(\sum_{i=1}^n \mathbf{k}_i \right)$ to ensure momentum conservation. The functions $\Gamma^{(n)}(\mathbf{x}_1, \ldots, \mathbf{x}_n)$ will therefore be called *vertex functions*, without the adjective *proper* which indicates the absence of the δ-function. In particular, the Fourier transforms of the vertex functions $\Gamma^{(2)}(\mathbf{x}_1, \mathbf{x}_2)$ and $\Gamma^{(4)}(\mathbf{x}_1, \mathbf{x}_2, \mathbf{x}_3, \mathbf{x}_4)$ are related to their proper versions by

$$\Gamma^{(2)}(\mathbf{k}_1, \mathbf{k}_2) = (2\pi)^D \delta^{(D)}(\mathbf{k}_1 + \mathbf{k}_2)\, \bar{\Gamma}^{(2)}(\mathbf{k}_1), \tag{5.67}$$

$$\Gamma^{(4)}(\mathbf{k}_1, \mathbf{k}_2, \mathbf{k}_3, \mathbf{k}_4) = (2\pi)^D \delta^{(D)}\left(\sum_{i=1}^4 \mathbf{k}_i \right) \bar{\Gamma}^{(4)}(\mathbf{k}_1, \mathbf{k}_2, \mathbf{k}_3, \mathbf{k}_4). \tag{5.68}$$

For the functional derivatives (5.66) we shall use the same short-hand notation as for the functional derivatives (5.10) of $W[j]$, setting

$$\Gamma_{\Phi(\mathbf{x}_1) \ldots \Phi(\mathbf{x}_n)} \equiv \frac{\delta}{\delta \Phi(\mathbf{x}_1)} \cdots \frac{\delta}{\delta \Phi(\mathbf{x}_n)} \Gamma[\Phi] . \tag{5.69}$$

The arguments $\mathbf{x}_1, \ldots, \mathbf{x}_n$ will usually be suppressed.

In order to derive relations between the derivatives of the effective energy and the connected correlation functions, we first observe that the connected one-point function $G_c^{(1)}$ at a nonzero source j is simply the field expectation Φ [recall (5.19)]:

$$G_c^{(1)} = \Phi. \tag{5.70}$$

Second, we see that the connected two-point function at a nonzero source j is given by

$$G_c^{(2)} = G_j^{(1)} = W_{jj} = \frac{\delta \Phi}{\delta j} = \left(\frac{\delta j}{\delta \Phi} \right)^{-1} = \Gamma_{\Phi\Phi}^{-1}. \tag{5.71}$$

The inverse symbols on the right-hand side are to be understood in the functional sense, i.e., $\Gamma_{\Phi\Phi}^{-1}$ denotes the functional matrix:

$$\Gamma_{\Phi(\mathbf{x})\Phi(\mathbf{y})}^{-1} \equiv \left[\frac{\delta^2 \Gamma}{\delta\Phi(\mathbf{x})\delta\Phi(\mathbf{y})} \right]^{-1}, \tag{5.72}$$

which satisfies

$$\int d^D y \, \Gamma_{\Phi(\mathbf{x})\Phi(\mathbf{y})}^{-1} \Gamma_{\Phi(\mathbf{y})\Phi(\mathbf{z})} = \delta^{(D)}(\mathbf{x} - \mathbf{z}). \tag{5.73}$$

Relation (5.71) states that the second derivative of the effective energy determines directly the connected correlation function $G_c^{(2)}(\mathbf{k})$ of the interacting theory in the presence of the external source j. Since j is an auxiliary quantity, which eventually be set equal to zero thus making Φ equal to Φ_0, the actual physical propagator is given by

$$G_c^{(2)} \Big|_{j=0} = \Gamma_{\Phi\Phi}^{-1} \Big|_{\Phi=\Phi_0}. \tag{5.74}$$

By Fourier-transforming this relation and removing a δ-function for the overall momentum conservation, the propagator $G(\mathbf{k})$ in Eq. (4.18) is related to the vertex function $\Gamma^{(2)}(\mathbf{k})$, defined in (5.67) by

$$G(\mathbf{k}) \equiv \bar{G}^{(2)}(\mathbf{k}) = \frac{1}{\bar{\Gamma}^{(2)}(\mathbf{k})}, \tag{5.75}$$

as observed before on diagrammatic grounds in Eq. (4.34).

The third derivative of the generating functional $W[j]$ is obtained by functionally differentiating W_{jj} in Eq. (5.71) once more with respect to j, and applying the chain rule:

$$W_{jjj} = -\Gamma_{\Phi\Phi}^{-2}\Gamma_{\Phi\Phi\Phi}\frac{\delta\Phi}{\delta j} = -\Gamma_{\Phi\Phi}^{-3}\Gamma_{\Phi\Phi\Phi} = -G^3\Gamma_{\Phi\Phi\Phi}. \tag{5.76}$$

This equation has a simple physical meaning. The third derivative of $W[j]$ on the left-hand side is the full three-point function at a nonzero source j, so that

$$G_c^{(3)} = W_{jjj} = -G_c^{(2)^3}\Gamma_{\Phi\Phi\Phi}. \tag{5.77}$$

This equation states that the full three point function arises from a third derivative of $\Gamma[\Phi]$ by attaching to each derivation a full propagator, apart from a minus sign. This structure was observed empirically in the low-order diagrammatic expansion (4.21) for the four-point function.

We shall express Eq. (5.77) diagrammatically as follows:

where

denotes the connected n-point function, and

the negative n-point vertex function.

   For the general analysis of the diagrammatic content of the effective energy, we observe that according to Eq. (5.76), the functional derivative of the correlation function G with respect to the current j satisfies

$$G^{(2)}_c{}_j = W_{jjj} = G^{(3)}_c = -G^{(2)}_c{}^3 \Gamma_{\Phi\Phi\Phi}. \tag{5.78}$$

This is pictured diagrammatically as follows:

(5.79)

This equation may be differentiated further with respect to j in a diagrammatic way. From the definition (5.2) we deduce the trivial recursion relation

$$G^{(n)}_c\left(\mathbf{x}_1,\ldots,\mathbf{x}_n\right) = \frac{\delta}{\delta j(\mathbf{x}_n)} G^{(n-1)}_c\left(\mathbf{x}_1,\ldots,\mathbf{x}_{n-1}\right), \tag{5.80}$$

which is represented diagrammatically as

$n > 2$.

By applying $\delta/\delta j$ repeatedly to the left-hand side of Eq. (5.78), we generate all higher connected correlation functions. On the right-hand side of (5.78), the chain rule leads to a derivative of all correlation functions $G = G^{(2)}_c$ with respect to j, thereby changing a line into a line with an extra three-point vertex as indicated in the diagrammatic equation (5.79). On the other hand, the vertex function $\Gamma_{\Phi\Phi\Phi}$ must be differentiated with respect to j. Using the chain rule, we obtain for any n-point vertex function:

$$\Gamma_{\Phi\ldots\Phi j} = \Gamma_{\Phi\ldots\Phi\Phi} \frac{\delta\Phi}{\delta j} = \Gamma_{\Phi\ldots\Phi\Phi} G^{(2)}_c, \tag{5.81}$$

which may be represented diagrammatically as

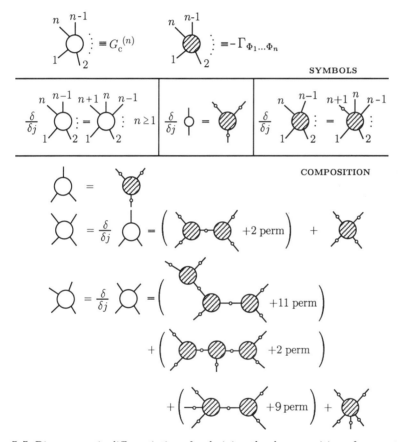

With these diagrammatic rules, we can differentiate (5.76) any number of times, and derive the diagrammatic structure of the connected correlation functions with an arbitrary number of external legs. The result up to $n = 5$ is shown in Fig. 5.5.

FIGURE 5.5 Diagrammatic differentiations for deriving the decomposition of connected correlation functions into trees of 1PI diagrams. The last term in each decomposition contains, after amputation and removal of an overall δ-function of momentum conservation, precisely all 1PI diagrams of Eqs. (4.19) and (4.22).

The diagrams generated in this way have a tree-like structure, and for this reason they are called *tree diagrams*. The tree decomposition reduces all diagrams to their one-particle irreducible contents. This proves our earlier statement that the vertex functions contain precisely the same Feynman diagrams as the proper vertex functions defined diagrammatically in Section 4.2, apart from the δ-function that ensures overall momentum conservation.

The effective energy $\Gamma[\Phi]$ can be used to prove an important composition theorem: The full propagator G can be expressed as a geometric series involving the so-called *self-energy*, a fact that was observed diagrammatically for low orders earlier in the explicit expansion (4.30). Let us decompose the vertex function as

$$\bar{\Gamma}^{(2)} = G_0^{-1} + \bar{\Gamma}_{\Phi\Phi}^{\text{int}}, \tag{5.82}$$

such that the full propagator (5.74) can be rewritten as

$$G = \left(1 + G_0 \bar{\Gamma}_{\Phi\Phi}^{\text{int}}\right)^{-1} G_0. \tag{5.83}$$

Expanding the denominator, this can also be expressed in the form of an integral equation:

$$G = G_0 - G_0 \bar{\Gamma}_{\Phi\Phi}^{\text{int}} G_0 + G_0 \bar{\Gamma}_{\Phi\Phi}^{\text{int}} G_0 \bar{\Gamma}_{\Phi\Phi}^{\text{int}} G_0 - \dots . \tag{5.84}$$

In this equation we identify the self-energy introduced diagrammatically in Eq. (4.30) as

$$\Sigma \equiv -\bar{\Gamma}_{\Phi\Phi}^{\text{int}}, \tag{5.85}$$

i.e., the self-energy is given by the interacting part of the second functional derivative of the effective energy, except for an opposite sign.

Equation (5.84) is the analytic proof of the chain decomposition (4.30) of the full propagator G. All diagrams can be obtained from a repetition of self-energy diagrams connected by a single line. The corresponding Eq. (5.83) confirms the earlier observation, in diagrams of lower orders, that the full propagator can be expressed in terms of Σ as [recall Eq. (4.31)]:

$$G \equiv [G_0^{-1} - \Sigma]^{-1}. \tag{5.86}$$

This equation can, incidentally, be rewritten in the form of an integral equation for the correlation function G:

$$G = G_0 - G_0 \bar{\Gamma}_{\Phi\Phi}^{\text{int}} G. \tag{5.87}$$

5.7 Landau Approximation to Generating Functional

Since the vertex functions are the functional derivatives of the effective energy [see (5.66)], we can expand the effective energy into a functional Taylor series

$$\Gamma[\Phi] = \sum_{n=0}^{\infty} \frac{1}{n!} \int d^D x_1 \dots d^D x_n \Gamma^{(n)}(\mathbf{x}_1, \dots, \mathbf{x}_n) \Phi(\mathbf{x}_1) \dots \Phi(\mathbf{x}_n). \tag{5.88}$$

The expansion in the number of loops of the generating functional $\Gamma[\Phi]$ collects systematically the contributions of fluctuations. To zeroth order, all fluctuations are neglected, and the effective energy reduces to the initial energy, which is the Landau approximation to the Gibbs functional [2] described in Chapter 1. In fact, in the absence of loop diagrams, the vertex functions contain only the lowest-order terms in $\Gamma^{(2)}$ and $\Gamma^{(4)}$:

$$\Gamma_0^{(2)}(\mathbf{x}_1, \mathbf{x}_2) = \left(-\partial_{\mathbf{x}_1}^2 + m^2\right) \delta^{(D)}(\mathbf{x}_1 - \mathbf{x}_2), \tag{5.89}$$

$$\Gamma_0^{(4)}(\mathbf{x}_1, \mathbf{x}_2, \mathbf{x}_3, \mathbf{x}_4) = \lambda \delta^{(D)}(\mathbf{x}_1 - \mathbf{x}_2) \delta^{(D)}(\mathbf{x}_1 - \mathbf{x}_3) \delta^{(D)}(\mathbf{x}_1 - \mathbf{x}_4). \tag{5.90}$$

Inserted into (5.88), this yields the zero-loop approximation to $\Gamma[\Phi]$:

$$\Gamma_0[\Phi] = \frac{1}{2!} \int d^D x \left[(\partial_{\mathbf{x}}\Phi)^2 + m^2\Phi^2\right] + \frac{\lambda}{4!} \int d^D x \, \Phi^4. \tag{5.91}$$

This is precisely the original energy functional (2.1). By allowing $\Phi(\mathbf{x})$ to be a vector $\boldsymbol{\Phi}(\mathbf{x})$, we recover the original phenomenological Ginzburg-Landau energy functional (1.83). Upon replacing the fluctuating field $\phi(\mathbf{x})$ further by its constant expectation value Φ_0, and by identifying this with the magnetic order parameter $\mathbf{M}$, we find the Gibbs free energy (1.38) used by Landau to explain the magnetic phase transition in the mean-field approximation:

$$\Gamma_0[\mathbf{M}] = V\left(\frac{m^2}{2!}\mathbf{M}^2 + \frac{\lambda}{4!}\mathbf{M}^4\right). \tag{5.92}$$

5.8 Composite Fields

In Sections 2.4, 3.5, and 4.3, we encountered a correlation function in which two fields coincide at one point, to be denoted by

$$G^{(1,n)}(\mathbf{x}, \mathbf{x}_1, \ldots, \mathbf{x}_n) = \frac{1}{2}\langle \phi^2(\mathbf{x})\phi(\mathbf{x}_1)\cdots\phi(\mathbf{x}_n)\rangle. \tag{5.93}$$

If multiplied by a factor m^2, the composite operator $m^2\phi^2(\mathbf{x})/2$ is precisely the mass term in the energy functional (2.2). For this reason one speaks of a *mass insertion* into the correlation function $G^{(n)}(\mathbf{x}_1, \ldots, \mathbf{x}_n)$. The negative sign is chosen for convenience.

Actually, we shall never make use of the full correlation function (5.93), but only of the integral over $\mathbf{x}$ in (5.93). This can be obtained directly from the generating functional $Z[j]$ of all correlation functions by differentiation with respect to the square mass in addition to the source terms

$$\int d^D x\, G^{(1,n)}(\mathbf{x}, \mathbf{x}_1, \ldots, \mathbf{x}_n) = -\left. Z^{-1}\frac{\partial}{\partial m^2}\frac{\delta}{\delta j(\mathbf{x}_1)}\cdots\frac{\delta}{\delta j(\mathbf{x}_n)}Z[j]\right|_{j=0}. \tag{5.94}$$

The desire to have a positive sign on the right-hand side was the reason for choosing a minus sign in the definition of the mass insertion (5.93). By going over to the generating functional $W[j]$, we obtain in a similar way the connected parts:

$$\int d^D x\, G_c^{(1,n)}(\mathbf{x}, \mathbf{x}_1, \ldots, \mathbf{x}_n) = -\left. \frac{\partial}{\partial m^2}\frac{\delta}{\delta j(\mathbf{x}_1)}\cdots\frac{\delta}{\delta j(\mathbf{x}_n)}W[j]\right|_{j=0}. \tag{5.95}$$

The right-hand side can be rewritten as

$$\int d^D x\, G_c^{(1,n)}(\mathbf{x}, \mathbf{x}_1, \ldots, \mathbf{x}_n) = -\frac{\partial}{\partial m^2}G_c^{(n)}(\mathbf{x}_1, \ldots, \mathbf{x}_n). \tag{5.96}$$

The connected correlation functions $G_c^{(1,n)}(\mathbf{x}, \mathbf{x}_1, \ldots, \mathbf{x}_n)$ can be decomposed into tree diagrams consisting of lines and one-particle irreducible vertex functions $\Gamma^{(1,n)}(\mathbf{x}, \mathbf{x}_1, \ldots, \mathbf{x}_n)$. The integral over $\mathbf{x}$ of these diagrams is obtained from the Legendre transform (5.60) by a further differentiation with respect to m^2:

$$\int d^D x\, \Gamma^{(1,n)}(\mathbf{x}, \mathbf{x}_1, \ldots, \mathbf{x}_n) = -\left. \frac{\partial}{\partial m^2}\frac{\delta}{\delta\Phi(\mathbf{x}_1)}\cdots\frac{\delta}{\delta\Phi(\mathbf{x}_n)}\Gamma[\Phi]\right|_{\Phi_0}, \tag{5.97}$$

implying the relation

$$\int d^D x\, \Gamma^{(1,n)}(\mathbf{x}, \mathbf{x}_1, \ldots, \mathbf{x}_n) = -\frac{\partial}{\partial m^2}\Gamma^{(n)}(\mathbf{x}_1, \ldots, \mathbf{x}_n), \tag{5.98}$$

which was derived by diagrammatic arguments for the proper vertex functions in Eq. (4.44), and which will be needed later in Section 10.1.

Notes and References

The derivation of the graphical recursion relation in Fig. 5.2 was given in
H. Kleinert, Fortschr. Phys. **30**, 187 (1982) (www.physik.fu-berlin/~kleinert/82); also in
Fortschr. Phys. **30**, 351 (1982) (www.physik.fu-berlin/~kleinert/84).
Its evaluation is discussed in detail in
H. Kleinert, A. Pelster, B. Kastening,. M. Bachmann, Phys. Rev. E **62**, 1537 (2000) (hep-th/9907168).
Diagrams beyond five loops can be found on the internet (www.physik.fu-berlin/~kleinert/294/programs).

The individual citations in the text refer to:

[1] See, for example, H. Kleinert and K. Maki, Fortschr. Phys. **29**, 1 (1981);
H. Kleinert, Phys. Lett. A **90**, 259 (1982);
H. Kleinert and F. Langhammer, Phys. Rev. A **40**, 5988 (1989).
The 1981 paper was the first to investigate icosahedral quasicrystalline structures discovered later in aluminum.

[2] L.D. Landau, J.E.T.P. **7**, 627 (1937).

6

Diagrams for Multicomponent Fields

So far, we have considered only a single real field ϕ with a ϕ^4-interaction. The theory can, however, easily be extended to a set of N identical real fields ϕ_α with $\alpha = 1, \ldots, N$. This extension does not produce any new Feynman diagrams. Additional work arises from the fact that the coupling constant g becomes now a tensor $g_{\alpha\beta\gamma\delta}$, and each momentum integral in a Feynman diagram is accompanied by a corresponding sum over indices.

If the coupling tensor is decomposed into basis tensors which satisfy certain symmetry and completeness properties, the result of each index contraction can again be decomposed into these basis tensors, with invariant factors called *symmetry factors* S_G. In this chapter we show how these are calculated.

6.1 Interactions with O(N) and Cubic Symmetry

In Section 1.1 we have explained that many physical systems are described by an O(N)-symmetric ϕ^4-theory with N identical fields. The simplest example is superfluid ^{4}He, whose local fluctuations are described by a complex field $\phi = (\phi_1 + i\phi_2)/\sqrt{2}$, with an energy functional

$$E[\phi] = \int d^D x \left\{ \frac{1}{2} |\partial_i \phi(\mathbf{x})|^2 + \frac{m^2}{2} |\phi(\mathbf{x})|^2 + \frac{\lambda}{4!} \left(|\phi(\mathbf{x})|^2 \right)^2 \right\}. \tag{6.1}$$

This may be written as an O(2)-symmetric functional of a two-component vector field $\boldsymbol{\phi}(\mathbf{x}) = (\phi_1(\mathbf{x}), \phi_1(\mathbf{x}))$, of the form (1.83):

$$E[\boldsymbol{\phi}] = \int d^D x \left\{ \frac{1}{2} \partial_i \boldsymbol{\phi}(\mathbf{x}) \partial_i \boldsymbol{\phi}(\mathbf{x}) + \frac{m^2}{2} \boldsymbol{\phi}^2(\mathbf{x}) + \frac{\lambda}{4!} [\boldsymbol{\phi}^2(\mathbf{x})]^2 \right\}. \tag{6.2}$$

As explained in Chapter 1, the same energy functional with a three-component vector field (ϕ_1, ϕ_2, ϕ_3) describes the ferromagnetic phase transition, in which case the mass and interaction terms are proportional to

$$\begin{aligned} \boldsymbol{\phi}^2 &\to \phi_1^2 + \phi_2^2 + \phi_3^2, \\ |\boldsymbol{\phi}^2|^2 &\to \left(\phi_1^2 + \phi_2^2 + \phi_3^2 \right)^2. \end{aligned} \tag{6.3}$$

Beside the O(N)-symmetric interaction term, another fourth-order term is frequently encountered in physical systems with cubic crystalline structure, which will also be treated in this text. All our results up to five loops will be derived for an interaction energy with a mixture of an O(N)-symmetric and a cubic-symmetric interaction energy with two coupling constants, where the fourth-order term in (6.2) is replaced by

$$E_{\text{int}}[\boldsymbol{\phi}] = \int d^D x \left[\lambda_1 \left(\sum_{\alpha=1}^{N} \phi_\alpha^2 \right)^2 + \lambda_2 \sum_{\alpha=1}^{N} \phi_\alpha^4 \right]. \tag{6.4}$$

This energy exhibits a broken $O(N)$ symmetry, with physical consequences to be investigated in Chapter 18. In particular, we shall answer the important physical question as to the circumstances under which the broken $O(N)$ symmetry can be restored by the violent fluctuations near the phase transition. In this case a system with a cubic-symmetric interaction energy has the same critical exponents as an isotropic system.

6.2   Free Generating Functional for N Fields

The energy functional for N free real fields is

$$E_0[\boldsymbol{\phi}] = \frac{1}{2} \int d^D x \sum_{\alpha=1}^{N} \left\{ [\partial_{\mathbf{x}} \phi_\alpha(\mathbf{x})]^2 + m^2 \phi_\alpha^2(\mathbf{x}) \right\}. \tag{6.5}$$

The equal mass for all field components makes the energy invariant under $O(N)$-rotations. The path integral for the free partition function factorizes into a product of N identical path integrals [see (2.7)]:

$$Z_0^{\text{phys}} = \prod_{\alpha=1}^{N} \left[\int \mathcal{D}\phi_\alpha \exp \left(\int d^D x \left\{ \frac{1}{2} [\partial \phi_\alpha(\mathbf{x})]^2 + \frac{1}{2} m^2 \phi_\alpha^2(\mathbf{x}) \right\} \right) \right]. \tag{6.6}$$

When adding an $O(N)$-symmetric linear source term

$$E_{\text{source}} = - \int d^D x \sum_{\alpha=1}^{N} \phi_\alpha(\mathbf{x}) j_\alpha(\mathbf{x}) \tag{6.7}$$

to the energy functional, the factorization property of (6.6) remains unchanged, and we obtain, with the normalization (2.9),

$$Z_0[\mathbf{j}] = \prod_{\alpha=1}^{N} \exp \left\{ \frac{1}{2} \int d^D x \, d^D y \, j_\alpha(\mathbf{x}) G_0(\mathbf{x}, \mathbf{y}) j_\alpha(\mathbf{y}) \right\}. \tag{6.8}$$

The basic functional differentiation with respect to the current contains now an additional Kronecker symbol $\delta_{\alpha\beta}^{(N)}$ for the N field indices:

$$\frac{\delta}{\delta j_\alpha(\mathbf{x})} \int d^D z \, j_\beta(\mathbf{z}) = \delta_{\alpha\beta}^{(N)} \int d^D z \, \delta^{(D)}(\mathbf{x} - \mathbf{z}). \tag{6.9}$$

The same Kronecker symbol appears now as a factor in the propagator of the free theory:

$$G_{0;\alpha\beta}^{(2)}(\mathbf{x}_1, \mathbf{x}_2) \equiv \langle \phi_\alpha(\mathbf{x}_1) \phi_\beta(\mathbf{x}_2) \rangle_0, \tag{6.10}$$

which reads

$$\begin{aligned}
G_{0;\alpha\beta}^{(2)}(\mathbf{x}, \mathbf{y}) &= \frac{\delta}{\delta j_\alpha(\mathbf{x})} \frac{\delta}{\delta j_\beta(\mathbf{y})} Z_0[\mathbf{j}] \bigg|_{\mathbf{j}=0} \\
&= \frac{\delta}{\delta j_\alpha(\mathbf{x})} \frac{\delta}{\delta j_\beta(\mathbf{y})} \prod_{\gamma=1}^{N} \exp \left\{ \frac{1}{2} \int d^D x_1 \, d^D x_2 \, j_\gamma(\mathbf{x}_1) G_0(\mathbf{x}_1, \mathbf{x}_2) j_\gamma(\mathbf{x}_2) \right\} \bigg|_{\mathbf{j}=0} \\
&= \delta_{\alpha\beta}^{(N)} G_0(\mathbf{x}, \mathbf{y}),
\end{aligned} \tag{6.11}$$

where $G_0(\mathbf{x}, \mathbf{y})$ is the propagator of the scalar field without labels α. The Fourier transformation yields the momentum space representation of the propagator

$$G^{(2)}_{0;\alpha\beta}(\mathbf{p}_1, \mathbf{p}_2) = (2\pi)^D \delta^{(D)}(\mathbf{p}_1 + \mathbf{p}_2)\, G_{0;\alpha\beta}(\mathbf{p}_1), \tag{6.12}$$

with

$$G_{0;\alpha\beta}(\mathbf{p}) = \delta^{(N)}_{\alpha\beta}\, G_0(\mathbf{p}) = \delta^{(N)}_{\alpha\beta} \frac{1}{\mathbf{p}^2 + m^2}. \tag{6.13}$$

By Wick's theorem, the n-point correlation functions of the free theory

$$G^{(2n)}_{0;\alpha_1\ldots\alpha_{2n}}(\mathbf{x}_1, \ldots, \mathbf{x}_{2n})$$

are sums over products of n two-point functions corresponding to all $(2n-1)!!$ pair contractions. Each contraction involves a free propagator $G_0(\mathbf{p})$ and a Kronecker symbol $\delta^{(N)}_{\alpha\beta}$.

6.3 Perturbation Expansion for N Fields and Symmetry Factors

The interaction energy (6.4) is a special case of the most general local fourth-order expression

$$E_{\text{int}}[\boldsymbol{\phi}] = \frac{1}{4!} \int d^D x \sum_{\alpha,\beta,\gamma,\delta=1}^{N} \lambda_{\alpha\beta\gamma\delta}\, \phi_\alpha(\mathbf{x})\phi_\beta(\mathbf{x})\phi_\gamma(\mathbf{x})\phi_\delta(\mathbf{x}), \tag{6.14}$$

where $\lambda_{\alpha\beta\gamma\delta}$ is some combination of basis tensors:

$$\lambda_{\alpha\beta\gamma\delta} = \sum_i \lambda_i\, T^{(i)}_{\alpha\beta\gamma\delta}. \tag{6.15}$$

The basis tensors may be chosen to be symmetric in all indices. This greatly reduces their number. If they are not initially, we may replace them by

$$T^{(i)}_{\alpha\beta\gamma\delta} \to \frac{1}{24}\left[T^{(i)}_{\alpha\beta\gamma\delta} + \{23 \text{ perm.}\}\right]. \tag{6.16}$$

For systems with a symmetry, there are further limitations, which are analyzed systematically in the literature [1]. Further limitations come from the structure of the theory. First, the perturbative corrections to the free part of the energy functional should maintain the initial $O(N)$ symmetry. This is only possible if the contraction of any two indices of all tensors $T^{(i)}_{\alpha\beta\gamma\delta}$ produces a Kronecker symbol $\delta^{(N)}_{\alpha\beta}$. Second, the perturbative corrections to the interaction energy should not produce new interactions that are not contained in the initial energy functional. The tensors $T^{(i)}_{\alpha\beta\gamma\delta}$ must therefore be complete in the sense that the contraction in two indices, of a product of any two linear combinations of $T^{(i)}_{\alpha\beta\gamma\delta}$ must yield again a linear combination of $T^{(i)}_{\alpha\beta\gamma\delta}$. This property will be important for the renormalizability of the theory. Tensors $T^{(i)}_{\alpha\beta\gamma\delta}$ with these properties will be referred to as *symmetry tensors*.

The expansion of the generating functional in powers of g reads

$$Z[\mathbf{j}] = \sum_{p=0}^{\infty} \frac{1}{p!}\left(\frac{-1}{4!}\right)^p \int \mathcal{D}\boldsymbol{\phi}(\mathbf{x}) \int d^D z_1 \cdots d^D z_p$$

$$\times \prod_{i=1}^{p} \sum_{\alpha_1^{(i)},\ldots,\alpha_4^{(i)}=1}^{N} \lambda_{\alpha_1^{(i)}\alpha_2^{(i)}\alpha_3^{(i)}\alpha_4^{(i)}}\, \phi_{\alpha_1^{(i)}}(\mathbf{z}_i) \cdots \phi_{\alpha_4^{(i)}}(\mathbf{z}_i). \tag{6.17}$$

The n-point correlation functions of the theory with interaction are obtained from the higher functional derivatives

$$
\begin{aligned}
G^{(n)}_{0;\alpha_1\ldots\alpha_n}(\mathbf{x}_1,\ldots,\mathbf{x}_n) &\equiv \langle \phi_{\alpha_1}(\mathbf{x}_1)\cdots\phi_{\alpha_n}(\mathbf{x}_n)\rangle \\
&= Z^{-1}\left[\int \frac{\delta}{\delta j_{\alpha_1}(\mathbf{x}_1)}\cdots\frac{\delta}{\delta j_{\alpha_n}(\mathbf{x}_n)}Z[j]\right]_{j\equiv 0}.
\end{aligned}
\tag{6.18}
$$

The pth order term of the perturbation expansion for the n-point function reads, by analogy with Eq. (3.2):

$$
\begin{aligned}
G^{(n)}_{p\,\alpha_1\ldots\alpha_n}(\mathbf{x}_1,\ldots,\mathbf{x}_n) &= \left(\frac{-1}{4!}\right)^p\frac{1}{p!}\frac{1}{(2p+n/2)!\,2^{2p+n/2}}\frac{\delta}{\delta j_{\alpha_1}(\mathbf{x}_1)}\cdots\frac{\delta}{\delta j_{\alpha_n}(\mathbf{x}_n)} \\
&\times \prod_{i=1}^{p}\left[\int d^D z_i \sum_{\gamma_1^{(i)},\ldots,\gamma_4^{(i)}}\lambda_{\gamma_1^{(i)}\gamma_2^{(i)}\gamma_3^{(i)}\gamma_4^{(i)}}\frac{\delta}{\delta j_{\gamma_1^{(i)}}(\mathbf{z}_i)}\frac{\delta}{\delta j_{\gamma_2^{(i)}}(\mathbf{z}_i)}\frac{\delta}{\delta j_{\gamma_3^{(i)}}(\mathbf{z}_i)}\frac{\delta}{\delta j_{\gamma_4^{(i)}}(\mathbf{z}_i)}\right] \\
&\times \left[\sum_{\beta=1}^{N}\int d^D z\, d^D z'\, j_\beta(\mathbf{z})G_0(\mathbf{z},\mathbf{z}')j_\beta(\mathbf{z}')\right]^{2p+n/2}.
\end{aligned}
\tag{6.19}
$$

As before, this yields a sum over $(4p+n-1)!!$ terms. We write them down in the notation of Eq. (3.3), in which the tensor indices associated with different line ends are first distinguished, and their contractions with the tensor indices of the vertices are enforced by additional Kronecker symbols:

$$
\begin{aligned}
G^{(n)}_{p\,\alpha_1\ldots\alpha_n}(\mathbf{x}_1,\ldots,\mathbf{x}_n) &= \left(\frac{-1}{4!}\right)^p\frac{1}{p!}\int d^D z_1\cdots d^D z_p\int d^D y_1\cdots d^D y_{4p+n} \\[2mm]
&\times \prod_{l=1}^{n}\left[\delta^{(D)}(\mathbf{y}_{4p+l}-\mathbf{x}_l)\right]\prod_{k=1}^{p}\left[\delta^{(D)}(\mathbf{y}_{4k-3}-\mathbf{z}_k)\delta^{(D)}(\mathbf{y}_{4k-2}-\mathbf{z}_k)\delta^{(D)}(\mathbf{y}_{4k-1}-\mathbf{z}_k)\delta^{(D)}(\mathbf{y}_{4k}-\mathbf{z}_k)\right] \\[2mm]
&\times \sum_{\gamma_1^{(1)},\ldots,\gamma_4^{(p)},\beta_1,\ldots,\beta_{4p+n}=1}^{N}\delta^{(N)}_{\beta_1\gamma_1^{(1)}}\cdots\delta^{(N)}_{\beta_{4p}\gamma_4^{(p)}}\cdot\delta^{(N)}_{\beta_{4p+1}\alpha_1}\cdots\delta^{(N)}_{\beta_{4p+n}\alpha_n}\cdot\lambda_{\beta_1\beta_2\beta_3\beta_4}\cdots\lambda_{\beta_{4p-3}\ldots\beta_{4p}} \\[2mm]
&\times \sum_{i=1}^{(4p+n-1)!!}G_0(\mathbf{y}_{\pi_i(1)},\mathbf{y}_{\pi_i(2)})\cdots G_0(\mathbf{y}_{\pi_i(4p+n-1)},\mathbf{y}_{\pi_i(4p+n)})\delta_{\beta_{\pi_i(1)}\beta_{\pi_i(2)}}\cdots\delta_{\beta_{\pi_i(4p+n-1)}\beta_{\pi_i(4p+n)}}.
\end{aligned}
\tag{6.20}
$$

6.4 Symmetry Factors

The last sum over the index i in Eq. (6.20) runs over the relevant index permutations π_i defined for $N=1$ in Eq. (3.3). These expansion terms differ from those for $N=1$ only by a factor arising from a contraction of p symmetric tensors $\lambda_{\beta_1\beta_2\beta_3\beta_4}\cdots\lambda_{\beta_{4p-3}\ldots\beta_{4p}}$. This product of symmetry tensors will be called the *tensor factor* F_G of a diagram.

Since $\lambda_{\alpha\beta\gamma\delta}$ is symmetric in all its indices, the tensor factor F_G is invariant under all index permutations which merely relabel the lines of a vertex. It is also invariant under permutations which relabel the vertices, as this amounts only to a change of the order of the factors in the tensor product. As a consequence, the tensor contractions associated with a specific Feynman integral can also be represented by the same Feynman diagram, in which lines indicate index contractions rather than integrals. Such contractions yield sums of tensors $T^{(i)}_{\alpha\beta\gamma\delta}$ for $n=4$, or simply $\delta_{\alpha\beta}$ for $n = 2$, and a factor consisting of a sum over all possible distributions of coupling constants. For two coupling constants and an L-loop diagram with V vertices, we have as tensor factors:

$$F_G = \sum_{k=0}^{V} \lambda_1^{V-k} \lambda_2^k \left[S_{4_1;(V-k,k)} T^{(1)}_{\alpha\beta\gamma\delta} + S_{4_2;(k,V-k)} T^{(2)}_{\alpha\beta\gamma\delta} \right], \qquad \text{for } n = 4, \qquad (6.21)$$

$$F_G = \sum_{k=0}^{V} \lambda_1^{V-k} \lambda_2^k \left[S_{2;(V-k,k)} \delta_{\alpha\beta} \right], \qquad \text{for } n = 2. \qquad (6.22)$$

In these expressions we can replace V by L, since $V = L$ for a two-point diagram and $V = L + 1$ for a four-point diagram. This will be done in Appendix B where all symmetry factors up to five loops are listed and organized by the number of loops L. The factors $S_{2;(i,j)}$, $S_{4_1;(i,j)}$, and $S_{4_2;(i,j)}$ in (6.21) and (6.22) are called the *symmetry factors* of a diagram. The subscript records the number of external lines n, and the combination of coupling constants associated with the vertices. The label i is the power of λ_1, while j is the power of λ_2. Note that $S_{4_1;(0,V)} = S_{4_2;(V,0)} = 0$ for mixed $O(N)$ and cubic symmetry, since contractions of $T^{(1)}$-type tensors cannot produce a $T^{(2)}$-type tensor, and vice versa. This fact will be important for the renormalizability of the theory with pure $O(N)$ or cubic symmetry.

For one coupling constant and an L-loop diagram, the tensor factors reduce to:

$$F_G = \lambda^V S_4 T^{(1)}_{\alpha\beta\gamma\delta}, \qquad \text{for } n = 4, \qquad (6.23)$$

$$F_G = \lambda^V S_2 \delta_{\alpha\beta}, \qquad \text{for } n = 2. \qquad (6.24)$$

In this case the index of the symmetry factor consists only of the number of lines, since there is only one symmetry factor for each diagram. Since $T^{(i)}_{\alpha\beta\gamma\delta}$ and $\delta_{\alpha\beta}$ are normalized to unity for $N = 1$, the symmetry factors will all reduce to unity for $N = 1$. This is not true for more than one coupling constant. For two coupling constants, $S_{2;(V-k,k)}$ and $S_{4_1;(V-k,k)} + S_{4_2;(V-k,k)}$ reduce to $V!/(V-k)!k!$, the number of ways the coupling constants $\lambda_1^{V-k}\lambda_2^k$ can be arranged on the V vertices. We shall see this explicitly in Section 6.4.2 when calculating the symmetry factors for such a theory.

6.4.1 Symmetry Factors for $O(N)$ Symmetry

The tensor $\lambda_{\alpha\beta\gamma\delta}$ introduced in Eq. (6.14) takes the following form for the $O(N)$-symmetric interaction energy of Eq. (6.4):

$$\lambda^{O(N)}_{\alpha\beta\gamma\delta} = \frac{\lambda}{3} \left(\delta_{\alpha\beta}\delta_{\gamma\delta} + \delta_{\alpha\gamma}\delta_{\beta\delta} + \delta_{\alpha\delta}\delta_{\beta\gamma} \right) = \lambda T^{(1)}_{\alpha\beta\gamma\delta}. \qquad (6.25)$$

The tensor $T^{(1)}$ is one of the basis tensors which also appears in the mixed $O(N)$-cubic-symmetric theory with two coupling constants.

For the Feynman diagrams occurring in the perturbation expansions of the self-energy (4.27) and the vertex function (4.25) up to two loops, we have to form the corresponding tensor contractions. The symmetry factors for the $O(N)$ symmetry will, in general, be denoted by an upper index $S^{O(N)}$. Since we only consider $O(N)$-symmetric tensors in this section, we shall omit this index.

$$\underset{\alpha}{}\overset{\bigcirc}{\rule{0pt}{0pt}}\,\underset{\beta}{} \quad : \quad \lambda^{O(N)}_{\alpha\beta\sigma\sigma} \qquad\qquad\qquad = \lambda S_{\bigcirc} \, \delta_{\alpha\beta},$$

$$\text{(diagram)} \quad : \quad \lambda^{O(N)}_{\alpha\beta\sigma_1\sigma_2}\lambda^{O(N)}_{\sigma_1\sigma_2\sigma_3\sigma_3} \qquad = \lambda^2 S_8\, \delta_{\alpha\beta} = \lambda^2 S_Q^2\, \delta_{\alpha\beta},$$

$$\text{(diagram)} \quad : \quad \lambda^{O(N)}_{\alpha\sigma_1\sigma_2\sigma_3}\lambda^{O(N)}_{\sigma_1\sigma_2\sigma_3\beta} \qquad = \lambda^2 S_\ominus\, \delta_{\alpha\beta},$$

$$\text{(diagram)} \quad : \quad \lambda^{O(N)}_{\alpha\beta\sigma_1\sigma_2}\lambda^{O(N)}_{\sigma_1\sigma_2\gamma\delta}\Big|_{\text{sym}} \qquad = \lambda^2 S_\circlearrowright\, T^{(1)}_{\alpha\beta\gamma\delta}, \tag{6.26}$$

$$\text{(diagram)} \quad : \quad \lambda^{O(N)}_{\alpha\sigma_1\sigma_2\sigma_3}\lambda^{O(N)}_{\beta\sigma_1\sigma_2\sigma_4}\lambda^{O(N)}_{\sigma_3\sigma_4\gamma\delta}\Big|_{\text{sym}} \qquad = \lambda^3 S_\ominus\, T^{(1)}_{\alpha\beta\gamma\delta},$$

$$\text{(diagram)} \quad : \quad \lambda^{O(N)}_{\alpha\beta\sigma_1\sigma_2}\lambda^{O(N)}_{\sigma_1\sigma_2\sigma_3\sigma_4}\lambda^{O(N)}_{\sigma_3\sigma_4\gamma\delta}\Big|_{\text{sym}} \qquad = \lambda^3 S_{\infty}\, T^{(1)}_{\alpha\beta\gamma\delta},$$

$$\text{(diagram)} \quad : \quad \lambda^{O(N)}_{\alpha\beta\sigma_1\sigma_2}\lambda^{O(N)}_{\sigma_2\sigma_3\sigma_3\sigma_4}\lambda^{O(N)}_{\gamma\delta\sigma_1\sigma_4}\Big|_{\text{sym}} \qquad = \lambda^3 S_8\, T^{(1)}_{\alpha\beta\gamma\delta} = \lambda^3 S_Q\, S_Q\, T^{(1)}_{\alpha\beta\gamma\delta}.$$

The external indices will always be labeled $\alpha, \beta, \gamma,$ and δ. The symbol $|_{\text{sym}}$ denotes symmetrization of these indices, which replaces each product $\delta_{\alpha\beta}\delta_{\gamma\delta}$ by $T^{(1)}_{\alpha\beta\gamma\delta}$.

In order to calculate the irreducible matrix elements S_n we observe that

$$\lambda^{O(N)}_{\alpha\beta\sigma\sigma} = \lambda\frac{N+2}{3}\delta_{\alpha\beta}, \tag{6.27}$$

$$\lambda^{O(N)}_{\alpha\sigma_1\sigma_2\sigma_3}\lambda^{O(N)}_{\sigma_1\sigma_2\sigma_3\beta} = \lambda^2\frac{N+2}{3}\delta_{\alpha\beta}, \tag{6.28}$$

so that

$$S_Q = \frac{N+2}{3}, \tag{6.29}$$

$$S_\ominus = \frac{N+2}{3}. \tag{6.30}$$

For the evaluation of $S_\circlearrowright$, we first note that

$$\lambda^{O(N)}_{\alpha\beta\sigma_1\sigma_2}\lambda^{O(N)}_{\sigma_1\sigma_2\gamma\delta} = \frac{\lambda^2}{9}\left[(N+4)\,\delta_{\alpha\beta}\delta_{\gamma\delta} + 2\delta_{\alpha\gamma}\delta_{\beta\delta} + 2\delta_{\alpha\delta}\delta_{\beta\gamma}\right]. \tag{6.31}$$

It is useful to abbreviate the three tensors in the brackets as A_1, A_2, A_3, i.e., we write

$$\lambda^{O(N)}_{\alpha\beta\sigma_1\sigma_2}\lambda^{O(N)}_{\sigma_1\sigma_2\gamma\delta} = \lambda^2\left[\frac{N+4}{9}A_1 + \frac{2}{9}A_2 + \frac{2}{9}A_3\right]_{\alpha\beta\gamma\delta}. \tag{6.32}$$

The symmetrization of the tensors A_1, A_2, A_3 gives $\lambda^{O(N)}_{\alpha\beta\gamma\delta}$ each time, so that

$$\lambda^{O(N)}_{\alpha\beta\sigma_1\sigma_2}\lambda^{O(N)}_{\sigma_1\sigma_2\gamma\delta}\big|_{\text{sym}} = \lambda^2\frac{N+8}{9}T^{(1)}_{\alpha\beta\gamma\delta}, \tag{6.33}$$

and hence

$$S_\circlearrowright = \frac{N+8}{9}. \tag{6.34}$$

For the other products (6.26) we observe the multiplication rules

$$(p_1A_1 + p_2A_2 + p_3A_3)_{\alpha\beta\sigma_1\sigma_2} (q_1A_1 + q_2A_2 + q_3A_3)_{\sigma_1\sigma_2\gamma\delta} \tag{6.35}$$
$$= \{[p_1(q_1N + q_2 + q_3) + p_2q_1 + p_3q_1]A_1 + (p_2q_2 + p_3q_3)A_2 + (p_2q_3 + p_3q_2)A_3\}_{\alpha\beta\gamma\delta}.$$

If we want to form $\lambda^{O(N)}_{\alpha\beta\sigma_1\sigma_2}\lambda^{O(N)}_{\sigma_1\sigma_2\sigma_3\sigma_4}\lambda^{O(N)}_{\sigma_3\sigma_4\gamma\delta}$, we have to multiply $A_1+A_2+A_3$ by $\frac{N+4}{9}A_1+\frac{2}{9}A_2+\frac{2}{9}A_3$ and to symmetrize the result, which gives

$$\lambda^3 \left[\frac{1}{3}\left(\frac{N+4}{9}N + \frac{4}{9}\right) + \frac{2}{3}\left(\frac{N+4}{9} + \frac{2}{9} + \frac{2}{9}\right)\right] T^{(1)}_{\alpha\beta\gamma\delta} \tag{6.36}$$

and hence

$$S_{\infty} = \frac{1}{27}\left(N^2 + 6N + 20\right). \tag{6.37}$$

The last product in (6.26) is found by using the general rule

$$(p_1A_1 + p_2A_2 + p_3A_3)_{\alpha\beta\sigma_1\sigma_2} (q_1A_1 + q_2A_2 + q_3A_3)_{\sigma_1\gamma\sigma_2\delta}$$
$$= \{[p_1(q_1 + Nq_2 + q_3) + (p_2 + p_3)q_2]A_1$$
$$(p_2q_1 + p_3q_3)A_2 + (p_3q_1 + p_2q_3)A_3\}_{\alpha\beta\gamma\delta}. \tag{6.38}$$

Applying this to $\frac{1}{3}(A_1 + A_2 + A_3)$ and $\frac{N+4}{9}A_1 + \frac{2}{9}A_2 + \frac{2}{9}A_3$ we find, after symmetrization,

$$S_{\ominus} = \frac{1}{3}\left(\frac{N+4}{9} + N\frac{2}{9} + \frac{2}{9}\right) + \left(\frac{1}{3} + \frac{1}{3}\right)\frac{2}{9} + 2\frac{1}{3}\left(\frac{N+4}{9} + \frac{2}{9}\right)$$
$$= \frac{1}{27}(5N + 22). \tag{6.39}$$

A summary of all reduced matrix elements up to two loops is given by the equations

$$S_{\mathcal{Q}} = \frac{N+2}{3}, \tag{6.40}$$

$$S_{8} = \left(\frac{N+2}{3}\right)^2, \tag{6.41}$$

$$S_{\ominus} = \frac{N+2}{3}, \tag{6.42}$$

$$S_{\mathcal{Q}} = \frac{N+8}{9}, \tag{6.43}$$

$$S_{\infty} = \frac{N^2 + 6N + 20}{27}, \tag{6.44}$$

$$S_{8} = \frac{N+2}{3}\frac{N+8}{9}, \tag{6.45}$$

$$S_{\ominus} = \frac{5N + 22}{27}. \tag{6.46}$$

The results up to five loops are shown in Appendix B.3.

6.4.2 Symmetry Factors for Mixed $O(N)$ and Cubic Symmetry

The tensor $\lambda_{\alpha\beta\gamma\delta}$ introduced in Eq. (6.14) takes the following form for the mixed $O(N)$-cubic-symmetric interaction energy of Eq. (6.4):

$$
\begin{aligned}
\lambda^{\text{cub}}_{\alpha\beta\gamma\delta} &= \frac{\lambda_1}{3}\left(\delta_{\alpha\beta}\delta_{\gamma\delta} + 2\text{ perm}\right) + \lambda_2\,\delta_{\alpha\beta\gamma\delta} \\
&= \lambda_1 T^{(1)}_{\alpha\beta\gamma\delta} + \lambda_2 T^{(2)}_{\alpha\beta\gamma\delta}.
\end{aligned}
\tag{6.47}
$$

The generalized δ-tensors are defined by:

$$
\delta_{\alpha_1\ldots\alpha_n} \equiv
\begin{cases}
1, & \alpha_1 = \cdots = \alpha_n, \\
0, & \text{otherwise.}
\end{cases}
\tag{6.48}
$$

They satisfy the identities

$$
\sum_\gamma \delta_{\alpha_1\cdots\alpha_n\gamma\gamma} = \delta_{\alpha_1\cdots\alpha_n},
\tag{6.49}
$$

$$
\sum_\gamma \delta_{\alpha_1\cdots\alpha_n\gamma}\delta_{\beta_1\cdots\beta_n\gamma} = \delta_{\alpha_1\cdots\alpha_n\beta_1\cdots\beta_n},
\tag{6.50}
$$

$$
\sum_\gamma \delta_{\gamma\gamma} = N.
\tag{6.51}
$$

The tensor contractions for the first two orders in the perturbation expansion are [recall the diagrams in (3.24) and (3.24)]

$$
\lambda^{\text{cub}}_{\alpha\beta\sigma\sigma} = \left(\lambda_1 S^{\text{cub}}_{2;(1,0)\,\bigcirc} + \lambda_2 S^{\text{cub}}_{2;(0,1)\,\bigcirc}\right)\delta_{\alpha\beta},
\tag{6.52}
$$

$$
\lambda^{\text{cub}}_{\alpha\beta\sigma_1\sigma_2}\lambda^{\text{cub}}_{\sigma_1\sigma_2\sigma_3\sigma_3} = \left(\lambda_1^2 S^{\text{cub}}_{2;(2,0)\,8} + \lambda_1\lambda_2 S^{\text{cub}}_{2;(1,1)\,8} + \lambda_2^2 S^{\text{cub}}_{2;(0,2)\,8}\right)\delta_{\alpha\beta},
\tag{6.53}
$$

$$
\lambda^{\text{cub}}_{\alpha\sigma_1\sigma_2\sigma_3}\lambda^{\text{cub}}_{\sigma_1\sigma_2\sigma_3\beta} = \left(\lambda_1^2 S_{2;(2,0)\,\ominus} + \lambda_1\lambda_2 S_{2;(1,1)\,\ominus} + \lambda_2^2 S^{\text{cub}}_{2;(0,2)\,\ominus}\right)\delta_{\alpha\beta},
\tag{6.54}
$$

$$
\begin{aligned}
\lambda^{\text{cub}}_{\alpha\beta\sigma_1\sigma_2}\lambda^{\text{cub}}_{\sigma_1\sigma_2\gamma\delta}\Big|_{\text{sym}} &= \left(\lambda_1 S^{\text{cub}}_{41;(2,0)\,\bigcirc} + \lambda_2 S^{\text{cub}}_{41;(1,1)\,\bigcirc}\right)\lambda_1 T^{(1)}_{\alpha\beta\gamma\delta} \\
&+ \left(\lambda_1 S^{\text{cub}}_{42;(1,1)\,\bigcirc} + \lambda_2 S^{\text{cub}}_{42;(0,2)\,\bigcirc}\right)\lambda_2 T^{(2)}_{\alpha\beta\gamma\delta},
\end{aligned}
\tag{6.55}
$$

$$
\begin{aligned}
\lambda^{\text{cub}}_{\alpha\sigma_1\sigma_2\sigma_3}\lambda^{\text{cub}}_{\beta\sigma_1\sigma_2\sigma_4}\lambda^{\text{cub}}_{\sigma_3\sigma_4\gamma\delta}\Big|_{\text{sym}} &= \\
\left(\lambda_1^2 S^{\text{cub}}_{41;(3,0)\,\ominus} + \lambda_1\lambda_2 S^{\text{cub}}_{41;(2,1)\,\ominus}\right. & \left. + \lambda_2^2 S^{\text{cub}}_{41;(1,2)\,\ominus}\right)\lambda_1 T^{(1)}_{\alpha\beta\gamma\delta} \\
+ \left(\lambda_1^2 S^{\text{cub}}_{42;(2,1)\,\ominus} + \lambda_1\lambda_2 S^{\text{cub}}_{42;(1,2)\,\ominus}\right. & \left. + \lambda_2^2 S^{\text{cub}}_{42;(0,3)\,\ominus}\right)\lambda_2 T^{(2)}_{\alpha\beta\gamma\delta},
\end{aligned}
\tag{6.56}
$$

$$
\begin{aligned}
\lambda^{\text{cub}}_{\alpha\beta\sigma_1\sigma_2}\lambda^{\text{cub}}_{\sigma_1\sigma_2\sigma_3\sigma_4}\lambda^{\text{cub}}_{\sigma_3\sigma_4\gamma\delta}\Big|_{\text{sym}} &= \\
\left(\lambda_1^2 S^{\text{cub}}_{41;(3,0)\,\infty} + \lambda_1\lambda_2 S^{\text{cub}}_{41;(2,1)\,\infty}\right. & \left. + \lambda_2^2 S^{\text{cub}}_{41;(1,2)\,\infty}\right)\lambda_1 T^{(1)}_{\alpha\beta\gamma\delta} \\
+ \left(\lambda_1^2 S^{\text{cub}}_{42;(2,1)\,\infty} + \lambda_1\lambda_2 S^{\text{cub}}_{42;(1,2)\,\infty}\right. & \left. + \lambda_2^2 S^{\text{cub}}_{42;(0,3)\,\infty}\right)\lambda_2 T^{(2)}_{\alpha\beta\gamma\delta},
\end{aligned}
\tag{6.57}
$$

$$
\lambda^{\text{cub}}_{\alpha\beta\sigma_1\sigma_2}\lambda^{\text{cub}}_{\sigma_2\sigma_3\sigma_3\sigma_4}\lambda^{\text{cub}}_{\gamma\delta\sigma_1\sigma_4}\Big|_{\text{sym}}
$$

$$
= \left(\lambda_1^2 S^{\text{cub}}_{41;(3,0)\,\text{\AA}} + \lambda_1\lambda_2 S^{\text{cub}}_{41;(2,1)\,\text{\AA}} + \lambda_2^2 S^{\text{cub}}_{41;(1,2)\,\text{\AA}} \right) \lambda_1 \, T^{(1)}_{\alpha\beta\gamma\delta}
$$
$$
+ \left(\lambda_1^2 S^{\text{cub}}_{42;(2,1)\,\text{\AA}} + \lambda_1\lambda_2 S^{\text{cub}}_{42;(1,2)\,\text{\AA}} + \lambda_2^2 S^{\text{cub}}_{42;(0,3)\,\text{\AA}} \right) \lambda_2 \, T^{(2)}_{\alpha\beta\gamma\delta}. \tag{6.58}
$$

We shall now examine which symmetry factors arise from tensor contractions, without presenting the calculations in full detail as before. It will turn out that the symmetry factor $S^{\text{cub}}_{(V,0)}$ is obtained as a numerical factor resulting from a contraction of $T^{(1)}$-type tensors only. This makes it equal to the corresponding symmetry factor for a pure $O(N)$ symmetry. The symmetry factor $S^{\text{cub}}_{(0,V)}$ is given by a contraction of $T^{(2)}$-type tensors only, yielding always unity.

The result for the first diagram $\mathcal{Q}$ in (6.52) is immediately obtained. Since

$$
\lambda^{\text{cub}}_{\alpha\beta\sigma\sigma} = \lambda_1 T^{(1)}_{\alpha\beta\sigma\sigma} + \lambda_2 T^{(2)}_{\alpha\beta\sigma\sigma} = \lambda_1 S^{O(N)}_{\mathcal{Q}} \, \delta_{\alpha\beta} + \lambda_2 \delta_{\alpha\beta} = \left(\lambda_1 \frac{N+2}{3} + \lambda_2 \right) \delta_{\alpha\beta},
$$

we see directly that

$$
S^{\text{cub}}_{2;(1,0)\,\mathcal{Q}} = \frac{N+2}{3}, \qquad S^{\text{cub}}_{2;(0,1)\,\mathcal{Q}} = 1. \tag{6.59}
$$

For the first two-loop diagram $\text{\small\bf 8}$ in (6.53), we find:

$$
\lambda^{\text{cub}}_{\alpha\beta\sigma_1\sigma_2} \lambda^{\text{cub}}_{\sigma_1\sigma_2\sigma_3\sigma_3}
$$
$$
= \lambda_1^2 \, T^{(1)}_{\alpha\beta\sigma_1\sigma_2} T^{(1)}_{\sigma_1\sigma_2\sigma_3\sigma_3} + \lambda_1\lambda_2 \left[T^{(1)}_{\alpha\beta\sigma_1\sigma_2} T^{(2)}_{\sigma_1\sigma_2\sigma_3\sigma_3} + T^{(2)}_{\alpha\beta\sigma_1\sigma_2} T^{(1)}_{\sigma_1\sigma_2\sigma_3\sigma_3} \right] + \lambda_2^2 \, T^{(2)}_{\alpha\beta\sigma_1\sigma_2} T^{(2)}_{\sigma_1\sigma_2\sigma_3\sigma_3}
$$
$$
= \left[\lambda_1^2 S^{O(N)}_{2\,\text{\small 8}} + \lambda_1\lambda_2 \, 2 \left(\frac{N+2}{3} \right) + \lambda_2^2 \right] \delta_{\alpha\beta},
$$

and we extract:

$$
S^{\text{cub}}_{2;(2,0)\,\text{\small 8}} = \left(\frac{N+2}{3} \right)^2, \qquad S^{\text{cub}}_{2;(1,1)\,\text{\small 8}} = 2 \left(\frac{N+2}{3} \right), \qquad S^{\text{cub}}_{2;(0,2)\,\text{\small 8}} = 1. \tag{6.60}
$$

For the second two-loop diagram $\ominus$ in (6.54), the calculation proceeds analogously:

$$
\lambda^{\text{cub}}_{\alpha\sigma_1\sigma_2\sigma_3} \lambda^{\text{cub}}_{\sigma_1\sigma_2\sigma_3\beta}
$$
$$
= \lambda_1^2 \, T^{(1)}_{\alpha\sigma_1\sigma_2\sigma_3} T^{(1)}_{\sigma_1\sigma_2\sigma_3\beta} + \lambda_1\lambda_2 \left(T^{(1)}_{\alpha\sigma_1\sigma_2\sigma_3} T^{(2)}_{\sigma_1\sigma_2\sigma_3\beta} + T^{(2)}_{\alpha\sigma_1\sigma_2\sigma_3} T^{(1)}_{\sigma_1\sigma_2\sigma_3\beta} \right) + \lambda_2^2 \, T^{(2)}_{\alpha\sigma_1\sigma_2\sigma_3} T^{(2)}_{\sigma_1\sigma_2\sigma_3\beta}
$$
$$
= \left[\lambda_1^2 S^{O(N)}_{2\,\ominus} + \lambda_1\lambda_2 \, 2 + \lambda_2^2 \right] \delta_{\alpha\beta},
$$

and we get

$$
S^{\text{cub}}_{2;(2,0)\,\ominus} = \frac{N+2}{3}, \qquad S^{\text{cub}}_{2;(1,1)\,\ominus} = 2, \qquad S^{\text{cub}}_{2;(0,2)\,\ominus} = 1. \tag{6.61}
$$

Now we turn to the four-point diagram $\rtimes\!\!\ltimes$ in (6.55)

$$
\lambda^{\text{cub}}_{\alpha\beta\sigma_1\sigma_2} \lambda^{\text{cub}}_{\sigma_1\sigma_2\gamma\delta} \Big|_{\text{sym}}
$$
$$
= \lambda_1^2 \, T^{(1)}_{\alpha\beta\sigma_1\sigma_2} T^{(1)}_{\sigma_1\sigma_2\gamma\delta} \Big|_{\text{sym}} + \lambda_1\lambda_2 \left(T^{(1)}_{\alpha\beta\sigma_1\sigma_2} T^{(2)}_{\sigma_1\sigma_2\gamma\delta} \Big|_{\text{sym}} + T^{(2)}_{\alpha\beta\sigma_1\sigma_2} T^{(1)}_{\sigma_1\sigma_2\gamma\delta} \Big|_{\text{sym}} \right) + \lambda_2^2 \, T^{(2)}_{\alpha\beta\sigma_1\sigma_2} T^{(2)}_{\sigma_1\sigma_2\gamma\delta} \Big|_{\text{sym}}
$$
$$
= \lambda_1^2 \, S^{O(N)}_{4\,\rtimes\!\ltimes} \, T^{(1)}_{\alpha\beta\gamma\delta} + \lambda_1\lambda_2 \left(\frac{2}{3} \, T^{(1)}_{\alpha\beta\gamma\delta} + \frac{4}{3} \, T^{(2)}_{\alpha\beta\gamma\delta} \right) + \lambda_2^2 \, T^{(2)}_{\alpha\beta\gamma\delta},
$$

yielding the symmetry factors:

$$S^{\text{cub}}_{4_1;(2,0)\,\text{O}} = \frac{N+8}{9}, \quad S^{\text{cub}}_{4_1;(1,1)\,\text{O}} = \frac{2}{3}, \quad S^{\text{cub}}_{4_2;(1,1)\,\text{O}} = \frac{4}{3}, \quad S^{\text{cub}}_{4_2;(0,2)\,\text{O}} = 1. \tag{6.62}$$

The calculation of the symmetry factors of the four-point diagram with two loops $\ominus$ in (6.56) produces many terms:

$$\lambda^{\text{cub}}_{\alpha\sigma_1\sigma_2\sigma_3}\lambda^{\text{cub}}_{\beta\sigma_1\sigma_2\sigma_4}\lambda^{\text{cub}}_{\sigma_3\sigma_4\gamma\delta}\Big|_{\text{sym}}$$

$$= \lambda_1^3\, T^{(1)}_{\alpha\sigma_1\sigma_2\sigma_3}T^{(1)}_{\beta\sigma_1\sigma_2\sigma_4}T^{(1)}_{\sigma_3\sigma_4\gamma\delta}\Big|_{\text{sym}}$$

$$+ \lambda_1^2\lambda_2\left(T^{(1)}_{\alpha\sigma_1\sigma_2\sigma_3}T^{(1)}_{\beta\sigma_1\sigma_2\sigma_4}T^{(2)}_{\sigma_3\sigma_4\gamma\delta}\Big|_{\text{sym}} + T^{(1)}_{\alpha\sigma_1\sigma_2\sigma_3}T^{(2)}_{\beta\sigma_1\sigma_2\sigma_4}T^{(1)}_{\sigma_3\sigma_4\gamma\delta}\Big|_{\text{sym}} + T^{(2)}_{\alpha\sigma_1\sigma_2\sigma_3}T^{(1)}_{\beta\sigma_1\sigma_2\sigma_4}T^{(1)}_{\sigma_3\sigma_4\gamma\delta}\Big|_{\text{sym}}\right)$$

$$+ \lambda_1\lambda_2^2\left(T^{(1)}_{\alpha\sigma_1\sigma_2\sigma_3}T^{(2)}_{\beta\sigma_1\sigma_2\sigma_4}T^{(2)}_{\sigma_3\sigma_4\gamma\delta}\Big|_{\text{sym}} + T^{(2)}_{\alpha\sigma_1\sigma_2\sigma_3}T^{(1)}_{\beta\sigma_1\sigma_2\sigma_4}T^{(2)}_{\sigma_3\sigma_4\gamma\delta}\Big|_{\text{sym}} + T^{(2)}_{\alpha\sigma_1\sigma_2\sigma_3}T^{(2)}_{\beta\sigma_1\sigma_2\sigma_4}T^{(1)}_{\sigma_3\sigma_4\gamma\delta}\Big|_{\text{sym}}\right)$$

$$+ \lambda_2^3\, T^{(2)}_{\alpha\sigma_1\sigma_2\sigma_3}T^{(2)}_{\beta\sigma_1\sigma_2\sigma_4}T^{(2)}_{\sigma_3\sigma_4\gamma\delta}\Big|_{\text{sym}}$$

$$= \lambda_1^3\, S^{\text{O}(N)}_{4\,\ominus}\,T^{(1)}_{\alpha\beta\gamma\delta} + \lambda_1^2\lambda_2\left(\frac{12}{9}T^{(1)}_{\alpha\beta\gamma\delta} + \frac{N+14}{9}T^{(2)}_{\alpha\beta\gamma\delta}\right) + \lambda_2\lambda_2^2\left(\frac{1}{3}T^{(1)}_{\alpha\beta\gamma\delta} + \frac{8}{3}T^{(2)}_{\alpha\beta\gamma\delta}\right) + \lambda_2^3\, T^{(2)}_{\alpha\beta\gamma\delta}\,.$$

From these we read off the following symmetry factors:

$$S^{\text{cub}}_{4_1;(3,0)\,\ominus} = \frac{5N+22}{27}, \quad S^{\text{cub}}_{4_1;(2,1)\,\ominus} = \frac{12}{9}, \quad S^{\text{cub}}_{4_1;(1,2)\,\ominus} = \frac{1}{3},$$

$$S^{\text{cub}}_{4_2;(0,3)\,\ominus} = 1, \quad S^{\text{cub}}_{4_2;(2,1)\,\ominus} = \frac{N+14}{9}, \quad S^{\text{cub}}_{4_2;(1,2)\,\ominus} = \frac{8}{3}. \tag{6.63}$$

The symmetry factors for the remaining two four-point diagrams (6.57) and (6.58) are calculated in the same way.

Let us collect the results for the reduced matrix elements up to two loops for the mixed $O(N)$ and cubic symmetry:

$$S^{\text{cub}}_{2;(1,0)\,\text{O}} = \frac{N+2}{3}, \qquad S^{\text{cub}}_{2;(0,1)\,\text{O}} = 1, \tag{6.64}$$

$$S^{\text{cub}}_{2;(2,0)\,\text{8}} = \left(\frac{N+2}{3}\right)^2, \qquad S^{\text{cub}}_{2;(1,1)\,\text{8}} = \frac{2(N+2)}{3}, \qquad S^{\text{cub}}_{2;(0,2)\,\text{8}} = 1,$$

$$S^{\text{cub}}_{2;(2,0)\,\ominus} = \frac{N+2}{3}, \qquad S^{\text{cub}}_{2;(1,1)\,\ominus} = 2, \qquad S^{\text{cub}}_{2;(0,2)\,\ominus} = 1,$$

$$S^{\text{cub}}_{4_1;(2,0)\,\text{O}} = \frac{N+8}{9}, \qquad S^{\text{cub}}_{4_1;(1,1)\,\text{O}} = \frac{2}{3}, \qquad S^{\text{cub}}_{4_2;(1,1)\,\text{O}} = \frac{4}{3}, \qquad S^{\text{cub}}_{4_2;(0,2)\,\text{O}} = 1,$$

$$S^{\text{cub}}_{4_1;(3,0)\,\ominus} = \frac{5N+22}{27}, \qquad S^{\text{cub}}_{4_1;(2,1)\,\ominus} = \frac{4}{3}, \qquad S^{\text{cub}}_{4_1;(1,2)\,\ominus} = \frac{1}{3},$$

$$S^{\text{cub}}_{4_2;(0,3)\,\ominus} = 1, \qquad S^{\text{cub}}_{4_2;(2,1)\,\ominus} = \frac{N+14}{9}, \qquad S^{\text{cub}}_{4_2;(1,2)\,\ominus} = \frac{8}{3},$$

$$S^{\text{cub}}_{4_1;(3,0)\,\text{OO}} = \frac{N^2+6N+20}{27}, \qquad S^{\text{cub}}_{4_1;(2,1)\,\text{OO}} = \frac{4+N}{3}, \qquad S^{\text{cub}}_{4_1;(1,2)\,\text{OO}} = 1,$$

$$S^{\text{cub}}_{4_2;(0,3)\,\text{OO}} = 1, \qquad S^{\text{cub}}_{4_2;(2,1)\,\text{OO}} = \frac{4}{3}, \qquad S^{\text{cub}}_{4_2;(1,2)\,\text{OO}} = 2,$$

$$S^{\text{cub}}_{4_1;(3,0)\,\text{8}} = \frac{N+2}{3}\frac{N+8}{9}, \qquad S^{\text{cub}}_{4_1;(2,1)\,\text{8}} = \frac{N+4}{3}, \qquad S^{\text{cub}}_{4_1;(1,2)\,\text{8}} = \frac{2}{3},$$

$$S^{\text{cub}}_{42;(0,3)\Theta} = 1, \qquad S^{\text{cub}}_{42;(2,1)\Theta} = \frac{4(N+2)}{9}, \qquad S^{\text{cub}}_{42;(1,2)\Theta} = \frac{N+6}{3}. \tag{6.65}$$

Note that for $N = 1$, $S_{2;(V,0)}$ and $S_{4_1;(V,0)}$ reduce to unity, whereas $S_{2;(V-k,k)}$ and $S_{4_1;(V-k,k)} + S_{4_2;(V-k,k)}$ reduce to $V!/(V-k)!k!$, which is the number of ways the coupling constants $\lambda_1^{V-k}\lambda_2^k$ can be arranged on V vertices.

A complete list of all symmetry factors up to five loops is found in Appendix B.4.

6.4.3 Other Symmetries

A further generalization sometimes encountered in the literature that is useful for applications is the case of n fields ϕ^i_α ($\alpha = 1, \ldots, n$), each of which occurring in an $O(q)$-symmetric combination:

$$E = \int d^D x \left\{ \frac{1}{2} \sum_{\alpha,i}^{n,q} \left[(\partial \phi^i_\alpha)^2 + m^2 (\phi^i_\alpha)^2 \right] + \frac{1}{4!} \sum_{\alpha,\alpha'=1}^{n} \lambda_{\alpha\alpha'} \sum_{i=1}^{q} (\phi^i_\alpha)^2 \sum_{j=1}^{q} (\phi^j_{\alpha'})^2 \right\}. \tag{6.66}$$

Among these energies, the case of a cubic symmetry in the indices is of special interest. Its interacting part contains only two independent couplings

$$E_{\text{int}} = \int d^D x \left\{ \lambda_1 \left[\sum_{\alpha=1}^{n} \sum_{i=1}^{q} (\phi^i_\alpha)^2 \right]^2 + \lambda_2 \sum_{\alpha=1}^{n} \left[\sum_{i=1}^{q} (\phi^i_\alpha)^2 \right]^2 \right\}. \tag{6.67}$$

The special case of n complex fields, i.e., of fields ϕ^i_α with $i = 1, 2$, governs the statistical mechanics of ensembles of dislocation lines in a crystal [2]. Furthermore, the case of n fields with $q = 0$ belongs to the same universality class as the so-called random n-vector model [3]. The model with mixed $O(N)$-cubic-symmetric interactions studied before is a special case of (6.66) with $n = N$ and $q = 1$.

The interaction (6.67) corresponds to the tensor (6.25) being replaced by

$$\begin{aligned} \lambda_{\alpha i,\beta j,\gamma k,\delta l} &= \lambda_1 S_{\alpha i,\beta j,\gamma k,\delta l} + \lambda_2 F_{\alpha i,\beta j,\gamma k,\delta l} \\ &= \frac{\lambda_1}{3} \left(\delta_{\alpha\beta}\delta_{\gamma\delta}\delta_{ij}\delta_{kl} + \delta_{\alpha\gamma}\delta_{\beta\delta}\delta_{ik}\delta_{jl} + \delta_{\alpha\delta}\delta_{\beta\gamma}\delta_{il}\delta_{jk} \right) \\ &\quad + \frac{\lambda_2}{3} \delta_{\alpha\beta\gamma\delta} \left(\delta_{ij}\delta_{kl} + \delta_{ik}\delta_{jl} + \delta_{il}\delta_{jk} \right). \end{aligned} \tag{6.68}$$

We have changed the notation for the tensors from $T^{(1)}$ and $T^{(2)}$ to S and F, in order to avoid any confusion.

It is straightforward to generalize the previous results to the present interaction. Take, for example, the symmetry factor $S_{\bigcirc}$ arising from the contraction

$$\left. T_{\alpha\beta\sigma_1\sigma_2} T_{\sigma_1\sigma_2\gamma\delta} \right|_{\text{sym}} = T_{\alpha\beta\gamma\delta} S_{\bigcirc}. \tag{6.69}$$

Multiplying the tensors S, F with each other, we find

$$\begin{aligned} SS|_{\text{sym}} &= \frac{nq+8}{9} S, \\ FF|_{\text{sym}} &= \frac{q+8}{9} F, \\ FS|_{\text{sym}} &= \frac{2}{3} F + \frac{q+2}{9} S, \end{aligned} \tag{6.70}$$

such that the result can again be decomposed into the tensors S and F.

6.4.4   General Symmetry Factors

The formation of the symmetry factor for any diagram can be visualized by the modified Feynman rules:

$$\alpha, \mathbf{x} \longmapsto \beta, \mathbf{y} \quad \hat{=} \quad G^{(2)}_{0;\alpha\beta}(\mathbf{x}, \mathbf{y}) \; = \; \delta_{\alpha\beta} G^{(2)}_0(\mathbf{x}, \mathbf{y}) \; ; \tag{6.71}$$

$$\underset{\alpha}{\overset{\beta}{\times}}\underset{\gamma}{\overset{\delta}{}} \quad \hat{=} \quad - \sum_{\alpha,\beta,\gamma,\delta=1}^{N} T_{\alpha\beta\gamma\delta} \int d^D z \; . \tag{6.72}$$

There are $4p$ summations for a total of $4p + n$ indices, two for each of the $(4p + n)/2$ lines. A subset of $(4p + n)/2$ index summations can be carried out immediately by contracting the indices of the tensors $T_{\beta_i\beta_j\beta_k\beta_l}$ with the Kronecker symbols $\delta_{\alpha\beta}$ for each line, as specified by a particular diagram. The number of the $(4p - n)/2$ remaining summations equals the number of internal lines $I = (4p - n)/2$ [recall (4.11)]. The symmetry tensor F_G is thus calculated as a sum over the indices of the internal lines of the product of p tensors T and depends on the indices of the external lines which are the only noncontracted indices.

In Section 4.2, we introduced the so-called amputated diagrams, in which the external lines are removed. The calculation of the symmetry tensor of these diagram in momentum space contains only $4p - n$ summations for $4p + n$ indices. It is not summed over the indices of the amputated lines, since their factors $\delta_{\alpha\beta}$ are now also absent. Only a subset of $(4p - n)/2$ summations is now carried out immediately, and the number of the nontrivial summations again equals the number of the internal lines. The symmetry factor is finally averaged in the indices of the amputated external lines.

As will be explained in Chapter 14, all two- and four-point diagrams are generated from vacuum diagrams. The symmetry factor S_0 of the vacuum diagram determines also the symmetry factor of the generated diagram. Since the process of generation of two- and four-point diagrams changes the number of loops L, the symmetry factors in this subsection have an additional upper index indicating the number of loops of the corresponding diagram.

The two-point diagrams with L loops are generated by cutting a line in a vacuum diagram with $L + 1$ loops. Conversely, by connecting the two end points of a two-point diagram with L loops we obtain a vacuum diagram with $L+1$ loops. The symmetry factors of the two diagrams are therefore related by

$$\sum_{\alpha\beta} S_2^L \delta_{\alpha\beta} = S_0^{L+1}. \tag{6.73}$$

Since $\sum_{\alpha\beta} \delta_{\alpha\beta}\delta_{\alpha\beta} = N$, the relation becomes

$$S_2^L = \frac{S_0^{L+1}}{N} \; . \tag{6.74}$$

The symmetry factor is therefore the same for all two-point diagrams generated from a single vacuum diagram. This is true for all symmetries that have only one quadratic invariant $\delta_{\alpha\beta}$, which is the case in systems with $O(N)$ symmetry and mixture of $O(N)$ and cubic symmetry. If there are two coupling constants, the symmetry factors appear in a sum over all combinations of coupling constants, and the symmetry factors for the vacuum diagrams are related to the symmetry factors of the two-point diagrams as follows:

$$\sum_{k=0}^{V} \lambda_1^{V-k}\lambda_2^k \, S_{0;(V-k,k)}^{L+1} = \sum_{\alpha\beta}\sum_{k=0}^{V} \lambda_1^{V-k}\lambda_2^k \, S_{2;(V-k,k)}^{L} \, \delta_{\alpha\beta} = N \sum_{k=0}^{V} \lambda_1^{V-k}\lambda_2^k \, S_{2;(V-k,k)}^{L}. \tag{6.75}$$

The four-point diagrams are generated by cutting out a vertex. Closing a four-point diagram to a vacuum diagram with an additional vertex implies that the associated symmetry factor is multiplied by $T_{\alpha\beta\gamma\delta}$. In the case of $O(N)$ symmetry, this gives the following factor:

$$\delta_{\alpha\beta}\delta_{\gamma\delta}\,\lambda_{\alpha\beta\gamma\delta}^{O(N)} = \delta_{\alpha\beta}\delta_{\gamma\delta}\frac{1}{3}(\delta_{\alpha\beta}\delta_{\gamma\delta} + \delta_{\alpha\gamma}\delta_{\beta\delta} + \delta_{\alpha\delta}\delta_{\beta\gamma}) = \frac{N(N+2)}{3}, \tag{6.76}$$

since the unsymmetrized four-point diagram is proportional to $\delta_{\alpha\beta}\delta_{\gamma\delta}$. For a vacuum diagram with $L+3$ loops and the symmetry factor S_0^{L+3}, the symmetry factor S_4^L of the generated four-point diagram with L loops is in $O(N)$ symmetry given by

$$S_4^L = 3\frac{S_0^{L+3}}{N(N+2)}. \tag{6.77}$$

All four-point diagrams in $O(N)$ symmetry derived from one vacuum diagram have the same symmetry factor.

Thus, the number of different symmetry factors equals the number of vacuum diagrams. By combining (6.74) with (6.77) we find a relation between the symmetry factors of the four-point diagrams with L loops S_4^L and the symmetry factors of the two-point diagrams with $L+2$ loops S_2^{L+2}:

$$S_2^{L+2} = \frac{N+2}{3}S_4^L. \tag{6.78}$$

The statements for the four-point diagrams are much more complicated in systems with mixed $O(N)$ and cubic symmetry, since there are two quartic invariants. In this case, Eq. (6.76) becomes:

$$\sum_{k=0}^{V+1} S_{0;(V+1-k,k)}^{L+3}\lambda_1^{V+1-k}\lambda_2^k$$

$$= \sum_{k=0}^{V}\left(S_{4_1;(V-k,k)}^L\,\lambda_1^{V-k}\lambda_2^k\,\delta_{\alpha\beta}\delta_{\gamma\delta} + S_{4_2;(V-k,k)}^L\,\lambda_1^{V-k}\lambda_2^k\,\delta_{\alpha\beta\gamma\delta}\right)\left(\lambda_1 T_{\alpha\beta\gamma\delta}^{(1)} + \lambda_2 T_{\alpha\beta\gamma\delta}^{(2)}\right) \tag{6.79}$$

$$= \sum_{k=0}^{V}\left[S_{4_1;(V-k,k)}^L\,\lambda_1^{V-k}\lambda_2^k\left(\lambda_1\frac{N(N+2)}{3} + \lambda_2 N\right) + S_{4_2;(V-k,i)}^L\,\lambda_1^{V-k}\lambda_2^k\left(\lambda_1 N + \lambda_2 N\right)\right]. \tag{6.80}$$

Remember that $S_{4_1;(0,V)} = S_{4_2;(V,0)} = 0$. Again, the symmetry factors of all four-point diagrams which are derived from the same vacuum diagram are related to the symmetry factors of the vacuum diagram. But this time the relation holds only for the sum of $S_{4_1}^L$ and $S_{4_2}^L$. In fact, there are many more different symmetry factors $S_{4_1}^L$ and $S_{4_2}^L$ than vacuum diagrams. The reason is that it makes a difference whether a λ_1-vertex or a λ_2-vertex is cut out of a vacuum diagram. Not all combinations of λ_1- and λ_2-vertices result in the two quartic invariants, and thus contribute to $S_{4_1}^L$ and $S_{4_2}^L$. This may be verified by inspecting the tables in Appendix B.4.2 where all symmetry factors for the mixed $O(N)$ and cubic symmetry are listed.

Note that for $\lambda_2 = 0$, Eq. (6.80) reduces to

$$S_{0;(V+1,0)}^{L+3} = S_{4_1;(V,0)}^L\frac{N(N+2)}{3}. \tag{6.81}$$

This is the same relation found for $O(N)$ symmetry with only one coupling constant. The symmetry factors $S_{4_1;(V,0)}^L$ in the case of mixed $O(N)$ and cubic symmetry equal the factors S_4^L in the pure $O(N)$ case.

Notes and References

There are various articles containing discussions of the tensor structures of the interactions, for example,

A. Aharony, in *Phase Transitions and Critical Phenomena*, Vol.6, ed. C. Domb and M.S. Green, Academic Press, New York, 1976.
D.J. Wallace, J. Phys. C **6**, 1390 (1973);
I.J. Ketley and D.J. Wallace, J. Phys. A **6**, 1667 (1973);
A. Aharony, Phys. Rev. B **8**, 3349 (1973), and Phys. Rev. Lett. **31**, 1494 (1974);
E. Brézin, J.C. Le Guillou, and J. Zinn-Justin, Phys. Rev. B **10**, 892 (1974);
T. Nattermann and S. Trimper, J. Phys. A **8**, 2000 (1975);
I.F. Lyuksutsov and V. Pokrovskii, JETP Letters **21**, 9 (1975);
J. Rudnick, Phys. Rev. B **18**, 1406 (1978);
H. Jacobson and D.J. Amit, Ann. Phys. (N.Y.) **133**, 57 (1981).

The individual citations in the text refer to:

[1] J.-C. Toledano, L. Michel, P. Toledano, E. Brézin, Phys. Rev. B **31**, 7171 (1985).

[2] H. Kleinert, *Gauge Fields in Condensed Matter*, Vol. II, *Stresses and Defects, Differential Geometry and Crystal Melting*, World Scientific, Singapore 1989, pp. 744-1443 (www.physik.fu-berlin.de/~kleinert/re.html#b2).

[3] G. Grinstein and A. Luther, Phys. Rev. B **13**, 1329 (79).
    A detailed discussion of this and other interesting cases can be found in the article ny A. Aharony cited above.

7

Scale Transformations of Fields and Correlation Functions

We now turn to the properties of ϕ^4-field theories which form the mathematical basis of the phenomena observed in second-order phase transitions. These phenomena are a consequence of a nontrivial behavior of fields and correlation functions under scale transformations, to be discussed in this chapter.

7.1 Free Massless Fields

Consider first a free massless scalar field theory, with an energy functional in D-dimensions:

$$E_0[\phi] = \int d^D x \frac{1}{2} [\partial \phi(\mathbf{x})]^2, \tag{7.1}$$

This is invariant under *scale transformations*, which change the coordinates by a scale factor

$$\mathbf{x} \to \mathbf{x}' = e^\alpha \mathbf{x}, \tag{7.2}$$

and transform the fields simultaneously as follows:

$$\phi(\mathbf{x}) \to \phi'_\alpha(\mathbf{x}) = e^{d^0_\phi \alpha} \phi(e^\alpha \mathbf{x}). \tag{7.3}$$

From the point of view of representation theory of Lie groups, the prefactor d^0_ϕ of the parameter α plays the role of a *generator* of the scale transformations on the field ϕ. Its value is

$$d^0_\phi = \frac{D}{2} - 1. \tag{7.4}$$

Under the scale transformations (7.2) and (7.3), the energy (7.1) is invariant:

$$E_0[\phi'_\alpha] = \int d^D x \frac{1}{2} [\partial \phi'_\alpha(\mathbf{x})]^2 = \int d^D x \, e^{2d^0_\phi \alpha} \frac{1}{2} [\partial \phi(e^\alpha \mathbf{x})]^2 = \int d^D x' \frac{1}{2} [\partial' \phi(\mathbf{x}')]^2 = E_0[\phi]. \tag{7.5}$$

The number d^0_ϕ is called the *field dimension* of the free field $\phi(\mathbf{x})$. Its value (7.4) is a direct consequence of the naive dimensional properties of $\mathbf{x}$ and ϕ. In the units used in this text, in which we have set $k_B T = 1$ [recall the convention stated before Eq. (2.6)], the exponential in the partition function is $e^{-E[\phi]}$, so that $E[\phi]$ is a dimensionless quantity. The coordinates have the dimension of a length. This property is expressed by the equation $[\mathbf{x}] = L$. The field in (7.5) has then a *naive dimension* (also called *engineering* or *technical dimension*) $[\phi] = L^{-d^0_\phi}$. To establish contact with the field theories of elementary particle physics, we shall use further natural units in which $c = \hbar = 1$. Then the length L is equal to an inverse mass μ^{-1} (more precisely, L is the Compton wavelength $L = \hbar/mc$ associated with the mass m). It is conventional to specify the dimension of every quantity in *units of mass* μ, rewriting $[\mathbf{x}] = L$ and $[\phi] = L^{-(D/2-1)}$ as

$[\mathbf{x}] = \mu^{-1}$ and $[\phi] = \mu^{d_\phi^0}$. Hereafter, we shall refer to the power d_ϕ^0 of μ as the *free-field dimension*, rather than the power. This will shorten many statements without danger of confusion.

The trivial scale invariance (7.5) of the free-field theory is the reason for a trivial power behavior of the free-field correlation functions. In momentum space, the two-point function reads [recall (4.9)]:

$$G(\mathbf{k}) = \frac{1}{\mathbf{k}^2}, \tag{7.6}$$

which becomes in $\mathbf{x}$-space

$$G(\mathbf{x}) = \int \frac{d^D k}{(2\pi)^D} \frac{e^{i\mathbf{kx}}}{\mathbf{k}^2} = \frac{\Gamma(D/2 - 1)}{(4\pi)^{D/2}} \frac{2^{D-2}}{|\mathbf{x}|^{D-2}}. \tag{7.7}$$

It is instructive to rederive this power behavior as a consequence of scale invariance (7.5) of the energy functional. This implies that the correlation function of the transformed fields $\phi'_\alpha(\mathbf{x})$ in (7.3) must be the same as those of the initial fields $\phi(\mathbf{x})$:

$$\langle \phi'_\alpha(\mathbf{x}) \phi'_\alpha(\mathbf{y}) \rangle = \langle \phi(\mathbf{x}) \phi(\mathbf{y}) \rangle. \tag{7.8}$$

Inserting on the left-hand side the transformed fields from (7.3), we see that

$$e^{2 d_\phi^0 \alpha} \langle \phi(e^\alpha \mathbf{x}) \phi(e^\alpha \mathbf{y}) \rangle = \langle \phi(\mathbf{x}) \phi(\mathbf{y}) \rangle. \tag{7.9}$$

Translational invariance makes this a function of $\mathbf{x} - \mathbf{y}$ only, which has then the property

$$e^{2 d_\phi^0 \alpha} G(e^\alpha (\mathbf{x} - \mathbf{y})) = G(\mathbf{x} - \mathbf{y}). \tag{7.10}$$

Since $G(\mathbf{x} - \mathbf{y})$ is also rotationally invariant, Eq. (7.10) implies the power behavior

$$G(\mathbf{x} - \mathbf{y}) = \text{const} \times |\mathbf{x} - \mathbf{y}|^{-2 d_\phi^0}, \tag{7.11}$$

which is precisely the form (7.7). Note that this power behavior is of the general type (1.11) observed in critical phenomena, with a trivial critical exponent $\eta = 0$.

In order to study the consequences of a continuous symmetry and its violations, it is convenient to first consider infinitesimal transformations. Then the defining transformations (7.2) and (7.3) read:

$$\delta \mathbf{x} = \alpha \mathbf{x}, \tag{7.12}$$
$$\delta \phi(\mathbf{x}) = \alpha (d_\phi^0 + \mathbf{x} \partial_\mathbf{x}) \phi(\mathbf{x}), \tag{7.13}$$

and the invariance law (7.9) takes the form of a differential equation:

$$(d_\phi^0 + \mathbf{x} \partial_\mathbf{x}) \langle \phi(\mathbf{x}) \phi(\mathbf{y}) \rangle + (d_\phi^0 + \mathbf{y} \partial_\mathbf{y}) \langle \phi(\mathbf{x}) \phi(\mathbf{y}) \rangle = 0. \tag{7.14}$$

For an n-point function, this reads

$$\sum_{i=1}^{n} (d_\phi^0 + \mathbf{x}_i \partial_{\mathbf{x}_i}) \langle \phi(\mathbf{x}_1) \cdots \phi(\mathbf{x}_n) \rangle = 0. \tag{7.15}$$

In the notation (2.10) for the correlation functions, this may be written as

$$\sum_{i=1}^{n} \mathbf{x}_i \partial_{\mathbf{x}_i} G^{(n)}(\mathbf{x}_1, \ldots, \mathbf{x}_n) = -n d_\phi^0 G^{(n)}(\mathbf{x}_1, \ldots, \mathbf{x}_n). \tag{7.16}$$

The differential operator on the left-hand side,

$$H_{\mathbf{x}} \equiv \sum_{i=1}^{n} \mathbf{x}_i \partial_{\mathbf{x}_i}, \tag{7.17}$$

measures the degree of homogeneity in the spatial variables $\mathbf{x}_i$, which is $-nd_\phi^0$. The power behavior (7.11) shows this directly.

If we multiply Eq. (7.16) by $e^{-i\Sigma_i \mathbf{k}_i \mathbf{x}_i}$, and integrate over all spatial coordinates [recall (4.10)], we find the momentum-space equation

$$\left(\sum_{i=1}^{n} \mathbf{k}_i \partial_{\mathbf{k}_i} \right) G^{(n)}(\mathbf{k}_1, \ldots, \mathbf{k}_n) = (nd_\phi^0 - D) G^{(n)}(\mathbf{k}_1, \ldots, \mathbf{k}_n). \tag{7.18}$$

The differential operator on the left-hand side,

$$H_{\mathbf{k}} \equiv \sum_{i=1}^{n} \mathbf{k}_i \partial_{\mathbf{k}_i}, \tag{7.19}$$

measures the degree of homogeneity in the momentum variables $\mathbf{x}_i$, which is $H_{\mathbf{k}} = -H_{\mathbf{x}} - D = nd_\phi^0 - D$.

Turning to the connected one-particle irreducible Green functions $G_c^{(n)}(\mathbf{k}_1, \ldots, \mathbf{k}_n)$, and further to the vertex functions $\bar{\Gamma}^{(n)}(\mathbf{k}_1, \ldots, \mathbf{k}_n)$, by removing an overall δ-function and multiplying each leg by a factor $G^{-1}(\mathbf{k}_i)$, [recall Eq. (4.20)], we find for the vertex functions $\bar{\Gamma}^{(n)}(\mathbf{k}_1, \ldots, \mathbf{k}_n)$ the differential equation

$$\left(\sum_{i=1}^{n} \mathbf{k}_i \partial_{\mathbf{k}_i} \right) \bar{\Gamma}^{(n)}(\mathbf{k}_1, \ldots, \mathbf{k}_n) = (-nd_\phi^0 + D) \bar{\Gamma}^{(n)}(\mathbf{k}_1, \ldots, \mathbf{k}_n). \tag{7.20}$$

For the two-point vertex function, this implies the homogeneity property

$$\bar{\Gamma}^{(2)}(\mathbf{k}) = \text{const} \times \mathbf{k}^2, \tag{7.21}$$

which is, of course, satisfied by the explicit form (7.6) [recall (4.34)].

7.2 Free Massive Fields

We now introduce a mass term to the energy functional, and consider a field ϕ whose fluctuations are governed by

$$E[\phi] = \int d^D x \left[\frac{1}{2}(\partial \phi)^2 + \frac{m^2}{2} \phi^2 \right]. \tag{7.22}$$

In the presence of m, the Fourier transform of the two-point function is [recall (4.9)]

$$G(\mathbf{k}) = \frac{1}{\mathbf{k}^2 + m^2}, \tag{7.23}$$

which becomes, in $\mathbf{x}$-space,

$$G(\mathbf{x}) = \int \frac{d^D k}{(2\pi)^D} \frac{e^{i\mathbf{k}\mathbf{x}}}{\mathbf{k}^2 + m^2} = \frac{1}{(4\pi)^{D/2}} \frac{2^{D-2}}{|\mathbf{x}|^{D-2}} \times \frac{1}{2^{D/2-2}} |m\mathbf{x}|^{D/2-1} K_{D/2-1}(|m\mathbf{x}|). \tag{7.24}$$

This possesses the scaling form of the general phenomenologically observed type (1.8), with a length scale $\xi = 1/m$. Since m^2 is proportional to the reduced temperature variable $t = T/T_c - 1$, the expression (7.24) has also the general scaling form (1.28), with a critical exponent $\nu = 1/2$.

In the presence of a mass term, the energy functional (7.22) is no longer invariant under the scale transformations (7.2) and (7.3). Whereas the free part $E_0[\phi]$ of the energy is invariant as in (7.5), the energy of the mass term of the transformed field $\phi'(\mathbf{x})$ is now related to that of the original field by

$$E_m[\phi'_\alpha] = \int d^D x \frac{m^2}{2} \phi'^2_\alpha(\mathbf{x}) = e^{2d^0_\phi \alpha} \int d^D x \frac{m^2}{2} \phi^2(e^\alpha \mathbf{x}) = e^{(2d^0_\phi - D)\alpha} \int d^D x \frac{m^2}{2} \phi^2(\mathbf{x}) = e^{-2\alpha} E_m[\phi].$$

(7.25)

7.3 Interacting Fields

Let us now add a ϕ^4-interaction to a massless energy functional $E_0[\phi]$, and consider the energy

$$E_{0,\lambda}[\phi] = \int d^D x \left[\frac{1}{2}(\partial\phi)^2 + \frac{\lambda}{4!}\phi^4 \right].$$

(7.26)

This remains invariant under the scale transformations (7.2) and (7.3) only for $D = 4$. In less than four dimensions, where the critical phenomena in the laboratory take place, the invariance is broken by the ϕ^4-term. The interaction energy behaves under scale transformations (7.2) and (7.3) as follows:

$$E_\lambda[\phi'_\alpha] = \int d^D x \frac{\lambda}{4!} \phi'^4_\alpha(\mathbf{x}) = e^{4d^0_\phi \alpha} \int d^D x \frac{\lambda}{4!} \phi^4(e^\alpha \mathbf{x}) = e^{(4d^0_\phi - D)\alpha} \int d^D x \frac{\lambda}{4!} \phi^4(\mathbf{x}) = e^{(D-4)\alpha} E_\lambda[\phi].$$

(7.27)

This symmetry breakdown may be viewed as a consequence of the fact that, in $D \neq 4$ dimensions, the coupling constant is not dimensionless, but has the naive dimension $[\lambda] = \mu^{4-D}$. With a new dimensionless coupling constant

$$\hat{\lambda} \equiv \mu^{D-4} \lambda,$$

(7.28)

this dimension is made explicit by re-expressing the energy functional (7.26) as

$$E_{0,\lambda}[\phi] = \int d^D x \left[\frac{1}{2}(\partial\phi)^2 + \frac{\mu^{4-D}\hat{\lambda}}{4!}\phi^4 \right]$$

(7.29)

with some mass parameter μ.

The theory we want to investigate contains both mass and interaction terms in the energy functional:

$$E[\phi] = \int d^D x \left[\frac{1}{2}(\partial\phi)^2 + \frac{m^2}{2}\phi^2 + \frac{\mu^{4-D}\hat{\lambda}}{4!}\phi^4 \right].$$

(7.30)

The scale invariance of the free-field term is broken in four dimensions by the mass term, and in less than four dimensions by the mass and interaction terms.

Although the complete energy functional (7.30) is no longer invariant under the scale transformations (7.2) and (7.3), one can derive consequences from the nontrivial transformation behavior of its parts (7.25) and (7.27). These consequences may be formulated in the form of Ward identities characterizing the breakdown of scale invariance.

7.3.1   Ward Identities for Broken Scale Invariance

There exists a functional formalism for deriving the invariance property (7.15) of arbitrary n-point functions from the scale invariance of the free-field theory. Consider the generating functional (2.13), normalized to $Z[0] = 1$:

$$Z[j] = (Z_0^{\mathrm{phys}})^{-1} \int \mathcal{D}\phi \, e^{-E_0[\phi] + \int d^D x \, j\phi}, \qquad (7.31)$$

and change the coordinates and the field in the functional integrand by the scale transformations (7.2) and (7.3). Since the free-field energy functional $E_0[\phi]$ is invariant, and since the measures of functional integration in numerator and denominator transform in the same way, we find the property

$$Z[j] = Z[j'_{-\alpha}], \qquad (7.32)$$

where

$$j'_{-\alpha}(\mathbf{x}) = e^{(d_\phi^0 - D)\alpha} j(e^{-\alpha}\mathbf{x}) \qquad (7.33)$$

is the scale-transformed current. Infinitesimally, the equality (7.32) becomes

$$Z[j] = Z[j - \alpha(D - d_0^\phi + \mathbf{x}\partial_\mathbf{x})j], \qquad (7.34)$$

which can be written as

$$\int d^D x [(D - d_\phi^0 + \mathbf{x}\partial_\mathbf{x})j(\mathbf{x})] \frac{\delta Z[j]}{\delta j(\mathbf{x})} = 0. \qquad (7.35)$$

A partial integration brings this to the form

$$-\int d^D x \, j(\mathbf{x})(d_\phi^0 + \mathbf{x}\partial_\mathbf{x}) \frac{\delta Z[j]}{\delta j(\mathbf{x})} = 0. \qquad (7.36)$$

Differentiating this functionally $n-1$ times with respect to the current $j(\mathbf{x})$ yields the invariance property (7.15) for the correlation functions.

Let us now see what happens if a mass term is added to the energy functional. According to (7.25), the massive energy functional transforms infinitesimally like

$$\delta E[\phi] = -2\alpha E_m[\phi]. \qquad (7.37)$$

In the presence of a mass term, we consider now the extended generating functional

$$Z[j, K] = (Z_0^{\mathrm{phys}})^{-1} \int \mathcal{D}\phi \, e^{-E_0[\phi] + \int dx(j\phi + K\phi^2/2)}. \qquad (7.38)$$

The additional source $K(\mathbf{x})$ permits us to generate n-point functions with additional insertions of quadratic terms $\phi^2(\mathbf{x})$. Recall their introduction in Section 2.4. A mass term in the energy corresponds to a constant background source $K(\mathbf{x}) \equiv -m^2$.

Under an infinitesimal scale transformation (7.13), the generating functional (7.38) behaves like

$$Z[j, K] = Z[j - \alpha(D - d_\phi^0 + \mathbf{x}\partial_\mathbf{x})j, \, K - \alpha(D - 2d_\phi^0 + \mathbf{x}\partial_\mathbf{x})K], \qquad (7.39)$$

implying the relation

$$\int d^D x \left\{ (D - d_\phi^0 + \mathbf{x}\partial_\mathbf{x})j(\mathbf{x}) \frac{\delta Z[j, K]}{\delta j(\mathbf{x})} + (D - 2d_\phi^0 + \mathbf{x}\partial_\mathbf{x})K(\mathbf{x}) \frac{\delta Z[j, K]}{\delta K(\mathbf{x})} \right\} = 0. \qquad (7.40)$$

After a partial integration, this becomes

$$-\int d^D x \left\{ j(\mathbf{x})(d_\phi^0 + \mathbf{x}\partial_\mathbf{x})\frac{\delta Z[j,K]}{\delta j(\mathbf{x})} + K(\mathbf{x})(2d_\phi^0 + \mathbf{x}\partial_\mathbf{x})\frac{\delta Z[j,K]}{\delta K(\mathbf{x})} \right\} = 0. \qquad (7.41)$$

Differentiating this n times with respect to $j(\mathbf{x})$ and setting $K = -m^2$, we find the *Ward identity*

$$\sum_{i=1}^{n}(d_\phi^0 + \mathbf{x}_i\partial_{\mathbf{x}_i})\langle\phi(\mathbf{x}_1)\cdots\phi(\mathbf{x}_n)\rangle - \frac{m^2}{2}\int d^D x (2d_\phi^0 + \mathbf{x}\partial_\mathbf{x})\langle\phi^2(\mathbf{x})\phi(\mathbf{x}_1)\cdots\phi(\mathbf{x}_n)\rangle = 0. \qquad (7.42)$$

A partial integration brings this to the form

$$\sum_{i=1}^{n}(d_\phi^0 + \mathbf{x}_i\partial_{\mathbf{x}_i})\langle\phi(\mathbf{x}_1)\cdots\phi(\mathbf{x}_n)\rangle + m^2\int d^D x \langle\phi^2(\mathbf{x})\phi(\mathbf{x}_1)\cdots\phi(\mathbf{x}_n)\rangle = 0. \qquad (7.43)$$

It was shown in (2.58) that the integral over the $\phi^2/2$-insertion in an n-point function can be generated by differentiating an n-point function without insertion with respect to $-m^2$. Thus we arrive at the simple differential equation

$$\sum_{i=1}^{n}(d_\phi^0 + \mathbf{x}_i\partial_{\mathbf{x}_i} - m\,\partial_m)\langle\phi(\mathbf{x}_1)\cdots\phi(\mathbf{x}_n)\rangle = 0. \qquad (7.44)$$

By analogy with (7.16), this may be written as

$$\left(\sum_{i=1}^{n}\mathbf{x}_i\partial_{\mathbf{x}_i} - m\,\partial_m\right)G^{(n)}(\mathbf{x}_1,\ldots,\mathbf{x}_n) = -nd_\phi^0\,G^{(n)}(\mathbf{x}_1,\ldots,\mathbf{x}_n). \qquad (7.45)$$

This equation implies that the correlation functions $G^{(n)}(\mathbf{x}_1,\ldots,\mathbf{x}_n)$ are homogeneous of degree $-nd_\phi^0$ in $\mathbf{x}_i$ and $1/m$. They can therefore be written as

$$G^{(n)}(\mathbf{x}_1,\ldots,\mathbf{x}_n) = m^{nd_\phi^0}f(m\mathbf{x}_i), \qquad (7.46)$$

where $f(\mathbf{x}_i m)$ is some function of its dimensionless arguments. The two-point function (7.24) is an example of this general statement.

The interaction term can be included similarly by extending the generating functional $Z[j,K]$ in (7.38) to

$$Z[j,K,L] = (Z_0^{\text{phys}})^{-1}\int \mathcal{D}\phi\, e^{-E_0[\phi]+\int d^D x(j\phi + K\phi^2/2 + L\phi^4/4!)}. \qquad (7.47)$$

By going through the same derivations as before, we arrive at the differential equation

$$\left[\sum_{i=1}^{n}\mathbf{x}_i\partial_{\mathbf{x}_i} - m\,\partial_m - (4-D)\lambda\,\partial_\lambda\right]G^{(n)}(\mathbf{x}_1,\ldots,\mathbf{x}_n) = -nd_\phi^0\,G^{(n)}(\mathbf{x}_1,\ldots,\mathbf{x}_n), \qquad (7.48)$$

implying the general homogeneity property of the correlation function

$$G^{(n)}(\mathbf{x}_1,\ldots,\mathbf{x}_n) = m^{nd_\phi^0}f(m\mathbf{x}_i, m^{D-4}\lambda). \qquad (7.49)$$

Upon multiplying Eq. (7.48) by $e^{-i\Sigma_i\mathbf{k}_i\mathbf{x}_i}$, and integrating over all spatial coordinates as in the treatment of Eq. (7.16), we find the momentum-space equation analogous to (7.18)

$$\left[\sum_{i=1}^{n}\mathbf{k}_i\partial_{\mathbf{k}_i} + m\,\partial_m + (4-D)\lambda\,\partial_\lambda\right]G^{(n)}(\mathbf{k}_1,\ldots,\mathbf{k}_n) = n(d_\phi^0 - D)\,G^{(n)}(\mathbf{k}_1,\ldots,\mathbf{k}_n). \qquad (7.50)$$

Taking $n = 2$ and removing an overall δ-function guaranteeing momentum conservation [recall (4.4)], we find the homogeneity of the Fourier-transformed correlation function $G(\mathbf{k})$:

$$[\mathbf{k}\partial_{\mathbf{k}} + m\,\partial_m + (4 - D)\lambda\,\partial_\lambda]\,G(\mathbf{k}) = (2d_\phi^0 - D)\,G(\mathbf{k}) = -2\,G(\mathbf{k}). \qquad (7.51)$$

For the vertex functions $\bar{\Gamma}(\mathbf{k}_1, \ldots, \mathbf{k}_n)$ defined in (4.20), we obtain

$$\left[\sum_{i=1}^{n}\mathbf{k}_i\partial_{\mathbf{k}_i} + m\,\partial_m + (4 - D)\lambda\,\partial_\lambda\right]\bar{\Gamma}^{(n)}(\mathbf{k}_1, \ldots, \mathbf{k}_n) = (-nd_\phi^0 + D)\,\bar{\Gamma}^{(n)}(\mathbf{k}_1, \ldots, \mathbf{k}_n), \qquad (7.52)$$

implying the homogeneity property

$$\bar{\Gamma}^{(n)}(\mathbf{k}_1, \ldots, \mathbf{k}_n) = m^{-nd_\phi^0 + D}f(\mathbf{k}_i/m, m^{D-4}g). \qquad (7.53)$$

After expressing the correlation functions in terms of the dimensionless coupling $\hat{\lambda}$ of Eq. (7.28), the differential equation (7.52) looses its derivative term with respect to λ and becomes

$$\left[\sum_{i=1}^{n-1}\mathbf{k}_i\partial_{\mathbf{k}_i} + m\,\partial_m\right]\bar{\Gamma}^{(n)}(\mathbf{k}_1, \ldots, \mathbf{k}_n)\Big|_{\hat{\lambda}} = (-nd_\phi^0 + D)\,\bar{\Gamma}^{(n)}(\mathbf{k}_1, \ldots, \mathbf{k}_n). \qquad (7.54)$$

Note that the homogeneity relations derived in this way are actually a trivial consequence of the invariance of the theory under *trivial scale transformations* (also referred to as *engineering* or *technical scale transformations*) in which one changes $\mathbf{x}$ as in (7.2), $\phi(\mathbf{x})$ as in (7.3), and simultaneously

$$m \to e^{-\alpha}m, \qquad \lambda \to e^{-(4-D)\alpha}\lambda. \qquad (7.55)$$

Every physical quantity changes by a phase factor $e^{-\alpha}$ for each of its mass dimension found by a naive dimensional analysis. These mass dimensions of correlation and vertex functions can be read off directly from Eqs. (7.49) and (7.53). These trivial scale transformations are *not* the origin of the physical scaling properties of a system near a second-order phase transition.

7.4   Anomaly in the Ward Identities

The above results would be correct if all correlation functions were finite. In a fluctuating field system, however, the infinitely many degrees of freedom give rise to infinities in the Feynman diagrams, which require a regularization. This leads to a correction term in the Ward identities (7.48) and (7.52).

Consider the case of four dimensions, where the interaction is scale invariant under (7.2) and (7.3). The Feynman integrals can be made finite only with the help of regularization procedures. These will be introduced in the next chapter. In the present context we use only the simplest of these based on a restriction of all momentum space integrals to a sphere of radius Λ. This is called a *cutoff regularization*. With a cutoff, all correlation functions depend on Λ in addition to $\mathbf{k}_i, m$, and λ. Now, a field theory such as the ϕ^4-theory under study has an important property which will be discussed in detail in Chapter 9: it is *renormalizable*. This will allow us to multiply the parameters of the theory m^2 and λ and all vertex functions $\bar{\Gamma}^{(n)}(\mathbf{k}_1, \ldots, \mathbf{k}_n)$ by Λ-dependent *renormalization constants*, leading to new renormalized quantities which remain finite in the limit of an infinite cutoff Λ. To describe this procedure quantitatively, we shall call all quantities in the original energy density more specifically *bare quantities*, and emphasize this by attaching a subscript B to them. The quantities without this subscript will henceforth denote renormalized quantities. The multiplicative renormalization establishes the following relations between bare and renormalized quantities:

$$\bar{\Gamma}_B^{(2)}(\mathbf{0}) = Z_\phi^{-1}(\lambda, m, \Lambda)\bar{\Gamma}^{(2)}(\mathbf{0}), \tag{7.56}$$

$$m_B^2 = \frac{Z_{m^2}(\lambda, m, \Lambda)}{Z_\phi(\lambda, m, \Lambda)}m^2, \tag{7.57}$$

$$m_B^{D-4}\lambda_B = \frac{Z_\lambda(\lambda, m, \Lambda)}{Z_\phi^2(\lambda, m, \Lambda)}m^{D-4}\lambda. \tag{7.58}$$

The renormalizability of the ϕ^4-theory implies that the renormalized quantities $\bar{\Gamma}^{(2)}(\mathbf{0}), m^2, \lambda$ have a finite value in the limit of an infinite cutoff Λ.

Equation (7.58) contains factors of mass to the power $D-4$ which is zero in four dimensions. However, as we have announced in Section 4.4, it will be important to study the theory in a continuous number of dimensions, in particular in the neighborhood of $D = 4$. The equations (7.56)–(7.58) are then applicable for any D close to four.

Since the cutoff has the same dimension as the mass m, the function f in (7.49), which we shall now call f_B to emphasize its bare nature, and the similar function f in (7.53) depend in general also on Λ/m_B. In the differential equations (7.48) and (7.52), this corresponds to an extra derivative term $\mp\Lambda\partial_\Lambda$, respectively. Then the latter equation reads

$$\left[\sum_{i=1}^n \mathbf{k}_i\partial_{\mathbf{k}_i} + m_B\partial_{m_B} + \Lambda\partial_\Lambda + (4-D)\lambda_B\partial_{\lambda_B}\right]\bar{\Gamma}_B^{(n)}(\mathbf{k}_1,\ldots,\mathbf{k}_n)$$
$$= (-nd_\phi^0 + D)\,\bar{\Gamma}_B^{(n)}(\mathbf{k}_1,\ldots,\mathbf{k}_n). \tag{7.59}$$

The extra term $\Lambda\partial_\Lambda$ ruins the naive Ward identities, and is therefore called an *anomaly*. The general solution of Eq. (7.59) is

$$\bar{\Gamma}_B^{(n)}(\mathbf{k}_1,\ldots,\mathbf{k}_n) = m_B^{-nd_\phi^0+D}f_B(\mathbf{k}_i/m_B, m_B^{D-4}\lambda_B, \Lambda/m_B). \tag{7.60}$$

The understanding of the scaling properties of the theory will make it necessary to find the dependence of bare quantities in (7.56)–(7.58) on the cutoff Λ.

We have stated above that the renormalized vertex functions $\bar{\Gamma}^{(n)}(\mathbf{k}_1,\ldots,\mathbf{k}_n)$ have a finite limit for $\Lambda \to \infty$. Using the same naive dimensional arguments by which we interpreted the functional form (7.53) derived from the Ward identity (7.52), we can now conclude that the original Ward identity (7.54) remains valid for the renormalized vertex functions, if the derivatives refer to renormalized masses and coupling constants.

The above statements are independent of the way the theory is regularized. Since we are going to define the theory for a continuous number of dimensions, another regularization will be possible, called *analytic regularization*. It will turn out that after an analytic continuation, the integrals which require a cutoff in four dimensions no longer need a cutoff in $D \leq 4$ dimensions. Nevertheless, the detailed specification of the infinities requires a mass parameter, which in a massive theory can be the physical mass m, but in a massless theory must be introduced separately, usually under the name μ. In this case, the derivative $\mu\partial_\mu$ replaces the term $\Lambda\partial_\Lambda$ in (7.59).

The anomaly in the Ward identity (7.59) will turn out to be the origin of the nontrivial critical exponents in the scaling laws observed in critical phenomena, as we shall see in Chapter 10. The reason is that in the limit $m \to 0$, the renormalization factors $Z(\lambda, m, \Lambda)$ in Eqs. (7.56)–(7.58) behave like powers $(m/\Lambda)^{\text{power}}$. Moreover, for small bare mass m_B, the renormalized quantity $m^{D-4}\lambda$ tends to a constant, as we shall see in Section 10.5, where this limit is denoted by g^*.

In order to illustrate the nontrivial scaling behavior more specifically, consider the two-point vertex function (7.53). Before renormalization, it has the general form

$$\bar{\Gamma}_B^{(2)}(\mathbf{k}) = m_B^2 f_B(\mathbf{k}/m_B, m_B^{D-4}\lambda_B, \Lambda/m_B). \tag{7.61}$$

At zero momentum this becomes

$$\bar{\Gamma}_B^{(2)}(\mathbf{0}) = m_B^2 f_B(m_B^{D-4}\lambda_B, \Lambda/m_B). \tag{7.62}$$

After renormalization, the Λ-dependence disappears for large Λ, so that the functional dependence becomes [see (7.53)]

$$\bar{\Gamma}^{(2)}(\mathbf{k}) = m^2 f(\mathbf{k}/m, m^{D-4}g). \tag{7.63}$$

Since the renormalized quantity $m^{D-4}\lambda$ tends to a constant g^* for $m_B \to 0$, the renormalized vertex function tends towards $m^2 f(\mathbf{k}/m)$. The function may be expanded in powers of $\mathbf{k}/m$. Rotational invariance leads to $f(\mathbf{k}/m) = c_1 + c_2 \mathbf{k}^2/m^2 + \dots$. The renormalization constants Z_ϕ, Z_{m^2} may be chosen such that the constants c_1 and c_2 are both equal to 1, so that the vertex function starts out for small $\mathbf{k}$ like

$$\bar{\Gamma}^{(2)}(\mathbf{k}) = m^2 + \mathbf{k}^2 + \dots . \tag{7.64}$$

From these first two terms we identify the coherence length as being equal to the inverse mass:

$$\xi = 1/m. \tag{7.65}$$

In the critical regime, the renormalization constants show power behavior, so that the renormalization equations (7.56)—(7.58) read for small m:

$$\bar{\Gamma}_B^{(2)}(\mathbf{k}) \underset{m_B \approx 0}{\approx} \left(\frac{\Lambda}{m}\right)^\eta \bar{\Gamma}^{(2)}(\mathbf{k}), \tag{7.66}$$

$$m_B^2 \underset{m_B \approx 0}{\approx} \left(\frac{\Lambda}{m}\right)^{\eta_m} m^2, \tag{7.67}$$

$$\lambda_B \underset{m_B \approx 0}{\approx} \left(\frac{\Lambda}{m}\right)^{\beta+4-D} \lambda, \tag{7.68}$$

where η, η_m, and β are constants depending on g^*. The constancy of $m^{D-4}\lambda$ in the limit $m_B \to 0$ implies that β vanishes at g^*.

For the functions f_B and f in Eqs. (7.61) and (7.63), equation (7.66) implies that

$$m_B^2 f_B(0, m_B^{D-4}\lambda_B, \Lambda/m_B) \underset{m_B \approx 0}{\approx} \left(\frac{\Lambda}{m}\right)^\eta m^2 f(0, g^*). \tag{7.69}$$

By assumption, the bare mass parameter in the initial energy density behaves near the critical temperature T_c like $m_B^2 \propto t$ where $t \equiv T/T_c - 1$. For the renormalized mass we find from (7.67)

$$m \propto t^{1/(2-\eta_m)}, \tag{7.70}$$

implying that the coherence length diverges like $t^{-1/(2-\eta_m)}$. This fixes the critical exponents ν defined in (1.10) as being

$$\nu = \frac{1}{2 - \eta_m}. \tag{7.71}$$

The bare vertex function $\bar{\Gamma}_B^{(2)}(\mathbf{0})$ behaves therefore like $t^{(2-\eta)/(2-\eta_m)}$. This quantity is observable in magnetic susceptibility experiments:

$$\chi \propto G(\mathbf{0}) = \frac{1}{\bar{\Gamma}_B^{(2)}(\mathbf{0})} \propto m^{-(2-\eta)} \propto t^{-(2-\eta)/(2-\eta_m)}. \tag{7.72}$$

Recalling (1.17), we identify the associated critical exponent as

$$\gamma = \frac{2-\eta}{2-\eta_m} = \nu(2-\eta). \tag{7.73}$$

The existence of such power laws may be interpreted as a consequence of modified exact scale invariance, which are analogous to the free scaling equations (7.20), but hold now for the renormalized correlation functions and the corresponding vertex functions. In fact, in Section 10.1 we shall prove that at the critical point where $m = 0$ and $m^{D-4}g = g^*$, the renormalized vertex functions satisfy a scaling equation

$$\left(\sum_{i=1}^{n-1} \mathbf{k}_i \partial_{\mathbf{k}_i}\right) \bar{\Gamma}^{(n)}(\mathbf{k}_1, \ldots, \mathbf{k}_n) = [-n(d_\phi^0 + \eta/2) + D] \bar{\Gamma}^{(n)}(\mathbf{k}_1, \ldots, \mathbf{k}_n). \tag{7.74}$$

Thus we recover, at the critical point, the free massless scaling equations (7.20) with only a small modification: the *value* of the free-field dimension d_ϕ^0 on the right-hand side is changed to the new value

$$d_\phi = d_\phi^0 + \eta/2. \tag{7.75}$$

The number $\eta/2$ is called the *anomalous dimension* of the interacting field ϕ. Remembering the derivation of the differential equation (7.20) from the invariance (7.9) under the transformation (7.3), we see that the renormalized correlation functions of the interacting theory are obviously invariant under the modified scale transformations of the renormalized fields

$$\phi(\mathbf{x}) \to \phi'_\alpha(\mathbf{x}) = e^{d_\phi \alpha}\phi(e^\alpha \mathbf{x}). \tag{7.76}$$

Thus, at the critical point, the scale-breaking interactions in the energy functional do not lead to a destruction of the scale invariance of the free-field correlation functions, but to a new type of scale invariance with a different field dimension.

The behavior of the coherence length $\xi \propto t^{-\nu}$ may also be interpreted as a consequence of such modified transformation laws. Initially, the coherence length is given by the inverse bare mass $m_B^{-1} \propto t^{-1/2}$. After the interaction is turned on, the role is taken over by the inverse renormalized mass, whose dimension is m_B^ν rather than m_B.

These general considerations will be made specific in the subsequent chapters. Our goal is to calculate the properties of interacting ϕ^4-theories in the scaling regime. Since these theories are naively scale invariant only in $D = 4$ dimensions, it was realized by Wilson that the modified scaling laws in the physical dimension $D = 3$ become accessible by considering the correlation functions as analytic functions in D. By expanding all results around $D = 4$ in powers of $\epsilon = 4 - D$, the scale invariance can be maintained at every level of calculations. This is the approach to be followed in this text.

Notes and References

For supplementary reading see the introduction into four-dimensional dilation invariance of quantum field theory by

S. Coleman, *Dilatations*, Erice Lectures 1971, Ed. A. Zichichi, Editrice Compositori, Bologna, 1973.
Scale invariance in the context of critical phenomena is discussed by
E. Brézin, J.C. Le Guillou, J. Zinn-Justin, in *Phase Transitions and Critical Phenomena*, Vol. 6, edited by C. Domb and M.S. Green (Academic Press, New York, 1976).

8

Regularization of Feynman Integrals

For dimensions close to $D = 4$, the Feynman integrals in momentum space derived in Chapter 4 do not converge since their integrands fall off too slowly at large momenta. Divergences arising from this short-wavelength region of the integrals are called *ultraviolet (UV)-divergences*. For massive fields, these are the only divergences of the integrals. In the zero-mass limit relevant for critical phenomena, there exists further divergences at small momenta and long wavelength. These so-called *infrared (IR)-divergences* will be discussed in Chapter 12.

In this chapter we shall consider only UV-divergences. We shall therefore often omit specifying their UV character. The divergences can be controlled by various mathematical methods whose advantages and disadvantages will be pointed out, and from which we shall select the best method for our purposes.

In principle, all masses and coupling constants occurring here ought to carry a subscript B indicating that the perturbative calculations are done starting from the *bare energy functional* $E_B[\phi_B]$ with *bare mass* m_B and *bare field* ϕ_B, introduced earlier in Section 7.3.1:

$$E_B[\phi_B] = \int d^D x \left[\frac{1}{2} (\partial \phi_B)^2 + \frac{m_B^2}{2} \phi_B^2 \right], \tag{8.1}$$

and perturbing it with the bare interaction

$$E_B^{\text{int}}[\phi_B] \equiv \int d^D x \frac{\lambda_B}{4!} \phi_B^4(\mathbf{x}). \tag{8.2}$$

However, in Chapter 9 we shall see that the renormalized quantities can eventually be calculated from the same Feynman integrals with the experimentally observable mass m and coupling constant λ. For this reason, the subscripts B will be omitted in all integrals.

8.1 Regularization

In four dimensions, the integrals of two- and four-point functions diverge. With the help of so-called *regularization procedures* they can be made finite. A regularization parameter is introduced, so that all divergences of the integrals appear as singularities in this parameter. There are various possible regularization procedures:

(a) Momentum *cutoff* Λ regularization

In field descriptions of condensed matter systems, Feynman diagrams are regularized naturally at length scales a, where the field description breaks down. All momentum integrals are limited naturally to a region $|\mathbf{p}| < \Lambda = \pi/a$, so that no UV-divergences can occur. Examples of Feynman integrals in a four-dimensional ϕ^4-theory with cutoff regularization are

$$\frac{1}{2} \bigcirc_\Lambda = -\frac{\lambda}{2} \int_\Lambda \frac{d^4 p}{(2\pi)^4} \frac{1}{\mathbf{p}^2 + m^2} = -\frac{gm^2}{32\pi^2} \left[\frac{\Lambda^2}{m^2} - \log \frac{\Lambda^2}{m^2} \right] + \mathcal{O}\left[\left(\Lambda^{-1} \right)^0 \right],$$

$$\frac{3}{2} \; \text{\Large $\bigotimes$}_{\Lambda} \; = \; -\frac{3\lambda}{2} \int_{\Lambda} \frac{d^4 p}{(2\pi)^2} \frac{1}{[(\mathbf{p} - \mathbf{k})^2 + m^2](\mathbf{p}^2 + m^2)} = \frac{3\lambda^2}{32\pi^2} \left[\log \frac{\Lambda^2}{m^2} \right] + \mathcal{O}\left[\left(\Lambda^{-1} \right)^0 \right].$$

The subscript Λ of the diagrams emphasizes that the momentum integrals are carried out only up to $\mathbf{p}^2 = \Lambda^2$. Both integrals are divergent for $\Lambda \longrightarrow \infty$. The first behaves for large Λ like Λ^2, and is called *quadratically divergent*. The second behaves like $\log \Lambda$ and is called *logarithmically divergent*.

Phase transitions do not depend on the properties of the system at short distances and should therefore not depend on the cutoff. If a field theory is to give a correct description of the phase transition, it must be possible to go to the limit $\Lambda \to \infty$ at the end without changing the critical behavior.

The cutoff regularization has an undesirable feature of destroying the translational invariance. Methods that do not suffer from this are

(b) Pauli-Villars regularization [1]

In this case, convergence is enforced by changing the propagator in such a way that it decreases for $|\mathbf{p}| \to \infty$ faster than before. This is done by the replacement

$$(\mathbf{p}^2 + m^2)^{-1} \to (\mathbf{p}^2 + m^2)^{-1} - (\mathbf{p}^2 + M^2)^{-1}. \tag{8.3}$$

in which M^2 plays the role of a cutoff.

(c) Analytic regularization [2]

The propagator is substituted by $(\mathbf{p}^2 + m^2)^{-z}$ where z is a complex number with $\text{Re}(z)$ large enough to make the integrals convergent. The result is then continued analytically to a region around the physical value $z = 1$. All divergences manifest themselves as simple poles for $z = 1$. Finite physical quantities for $z = 1$ can be defined by subtracting the pole terms.

(d) Dimensional regularization [3, 4, 5]

Instead of changing the power of the propagator, the measure of momentum integration is changed by allowing the dimension D in the integrals to be an arbitrary complex number. This regularization will be introduced in detail in the next section and used in all our calculations. It was invented by 't Hooft and Veltman to regularize nonabelian gauge theories where all previous cutoff methods failed. There are several attractive features of dimensional regularization. First, it preserves all symmetries of the theory, in particular gauge symmetry. Second, it allows an easy identification of the divergences. Third, it suggests in a natural way a *minimal subtraction scheme* (MS scheme), that greatly simplifies the calculations. Fourth, it regularizes at the same time IR-divergences in massless theories, as will be discussed in Section 12.3.

A difficulty with dimensional regularization is the treatment of certain tensors whose definition does not permit an analytic extrapolation to an arbitrary complex number of spatial dimensions, the most prominent example being the completely antisymmetric tensor $\varepsilon_{\alpha\beta\gamma\delta}$. Fortunately, this tensor does not appear in the theories to be discussed in this text, so that dimensional regularization can be applied without problem.

8.2 Dimensional Regularization

The integrand in the diagram $\bigcirc$ behaves for large loop momenta $\mathbf{p}$ like $|\mathbf{p}|^{-2}$, and the integrand in the diagram $\times\!\!\bigcirc\!\!\times$ like $|\mathbf{p}|^{-4}$. The momentum integrals are therefore defined only for dimensions $D < 2$ and $D < 4$ respectively. The idea is to calculate a Feynman integral for a continuous-valued number of dimensions D for which convergence is assured. For this the Feynman integrals for integer D must be extrapolated analytically to complex D.

The concept of a continuous dimension was introduced by Wilson and Fisher [6], who first calculated physical quantities in $D = 4 - \varepsilon$ dimensions with $\mathrm{Re}\,\varepsilon > 0$, and expanded them in powers of the deviation ε from the dimension $D = 4$. This concept was subsequently incorporated into quantum field theory [7], giving rise to many applications in statistical physics. The dimensional regularization by 't Hooft and Veltman was the appropriate mathematical tool for such expansions.

We shall first derive all formulas needed for the upcoming calculations. The analytic extrapolation to noninteger dimension will be based on the extrapolation of the factorial of integer numbers to real numbers by the Gamma function. In Subsection 8.2.2 we shall describe another approach in which D-dimensional Gaussian integrals are used to arrive at the same formulas.

For completeness, the original procedure of 't Hooft and Veltman will be reviewed in Subsection 8.2.4, to be followed in Subsection 8.2.5 by a slightly different method of Collins [8], who introduced integrals in continuous dimensions D via a certain subtraction method.

8.2.1 Calculation in Dimensional Regularization

To explain the method of dimensional regularization, consider first the simplest Feynman integral

$$I(D) = \int \frac{d^D p}{(2\pi)^D} \frac{1}{\mathbf{p}^2 + m^2}. \tag{8.4}$$

It is UV-divergent for $D \geq 2$, and IR-divergent for $D \leq 0$. After introducing polar coordinates as explained in Appendix 8A, it can be rewritten as

$$I(D) = \int \frac{d^D p}{(2\pi)^D} \frac{1}{\mathbf{p}^2 + m^2} = \frac{2\pi}{(2\pi)^D} \prod_{k=1}^{D-2} \int_0^\pi \sin^k \vartheta_k \, d\vartheta_k \int_0^\infty dp \, p^{D-1} \frac{1}{p^2 + m^2} \tag{8.5}$$

$$= \frac{S_D}{(2\pi)^D} \int_0^\infty dp \, p^{D-1} \frac{1}{p^2 + m^2}, \tag{8.6}$$

where

$$S_D = \frac{2\pi^{D/2}}{\Gamma(D/2)} \tag{8.7}$$

is the surface of a unit sphere in D dimensions [recall (1.98)]. The resulting one-dimensional integral can, after the substitution $p^2 = ym^2$, be cast into the form of an integral for the Beta function

$$B(\alpha, \gamma) \equiv \frac{\Gamma(\alpha)\Gamma(\gamma)}{\Gamma(\alpha + \gamma)} = \int_0^\infty dy \, y^{\alpha-1}(1+y)^{-\alpha-\gamma}, \tag{8.8}$$

where $\Gamma(z)$ is the Gamma function with the integral representation

$$\Gamma(z) = \int_0^\infty dt \, t^{z-1} e^{-t}. \tag{8.9}$$

We then find

$$
\begin{aligned}
I(D) &= \frac{S_D}{(2\pi)^D} \int_0^\infty dp\, p^{D-1} \frac{1}{p^2+m^2} = \frac{S_D}{2(2\pi)^D}(m^2)^{D/2-1} \int_0^\infty dy\, y^{D/2-1}(1-y)^{-1} \\
&= \frac{1}{(4\pi)^{D/2}\Gamma(D/2)}(m^2)^{D/2-1}\frac{\Gamma(D/2)\Gamma(1-D/2)}{\Gamma(1)} = \frac{(m^2)^{D/2-1}}{(4\pi)^{D/2}}\Gamma(1-D/2). \quad (8.10)
\end{aligned}
$$

The Gamma function provides us with an analytical extrapolation of the integrals in integer dimensions D to any complex D.

In general, Feynman integrals contain more complicated denominators than just $\mathbf{p}^2+m^2$. They may, for instance, contain another momentum $\mathbf{q}$:

$$
I(D;\mathbf{q}) = \int \frac{d^D p}{(2\pi)^D} \frac{1}{\mathbf{p}^2+2\mathbf{pq}+m^2}. \quad (8.11)
$$

This integral can be reduced to the previous one in (8.4) by completing the squares in the denominator, yielding

$$
I(D;\mathbf{q}) = \int \frac{d^D p}{(2\pi)^D} \frac{1}{\mathbf{p}^2+2\mathbf{pq}+m^2} = \int \frac{d^D p}{(2\pi)^D} \frac{1}{\mathbf{p}^2+m^2-\mathbf{q}^2} = \frac{1}{(4\pi)^{D/2}}(m^2-\mathbf{q}^2)^{D/2-1}\Gamma(1-D/2). \quad (8.12)
$$

By differentiating this with respect to the mass, we get a formula for arbitrary integer powers of such propagators:

$$
I(D,a;\mathbf{q}) = \int \frac{d^D \mathbf{p}}{(2\pi)^D} \frac{1}{(\mathbf{p}^2+2\mathbf{pq}+m^2)^a} = \frac{1}{(4\pi)^{D/2}}\frac{\Gamma(a-D/2)}{\Gamma(a)}(m^2-\mathbf{q}^2)^{D/2-a}, \quad (8.13)
$$

which can be extended analytically to arbitrary complex powers a.

We may differentiate this equation with respect to the external momentum $\mathbf{q}$, and obtain a further formula

$$
I^\mu(D,a;\mathbf{q}) \equiv \int \frac{d^D p}{(2\pi)^D} \frac{p_\mu}{(\mathbf{p}^2+2\mathbf{pq}+m^2)^a} = \frac{1}{(4\pi)^{D/2}}\frac{\Gamma(a-D/2)}{\Gamma(a)}\frac{q_\mu}{(m^2-\mathbf{q}^2)^{a-D/2}}. \quad (8.14)
$$

More differentiations with respect to $\mathbf{q}$ yield formulas with higher tensors $q^{\mu_1}\cdots q^{\mu_n}$ in the integrand.

For products of different propagators, the integrals are reduced to the above form with the help of *Feynman's parametric integral formula*:

$$
\frac{1}{A^a B^b} = \frac{\Gamma(a+b)}{\Gamma(a)\Gamma(b)} \int_0^1 dx \frac{x^{a-1}(1-x)^{b-1}}{[Ax+B(1-x)]^{a+b}}, \quad (8.15)
$$

which is a straightforward generalization of the obvious identity

$$
\frac{1}{AB} = \frac{1}{B-A}\left(\frac{1}{A}-\frac{1}{B}\right) = \frac{1}{B-A}\int_A^B dz \frac{1}{z^2} = \int_0^1 dx \frac{1}{[Ax+B(1-x)]^2}. \quad (8.16)
$$

Differentiation with respect to A and B yields (8.15) for integer values of the powers a and b, and the resulting equation can be extrapolated analytically to arbitrary complex powers. More generally, Feynman's formula reads:

$$
\frac{1}{A_1\cdots A_n} = \Gamma(n) \int_0^1 dx_1\cdots\int_0^1 dx_n \frac{\delta(1-x_1+\cdots+x_n)}{(x_1A_1+\ldots+x_nA_n)^n}, \quad (8.17)
$$

as can easily be proved by induction [10]. By differentiating both sides a_i times with respect to A_i, one finds

$$\frac{1}{A_1^{a_1} \cdots A_n^{a_n}} = \frac{\Gamma(a_1 + \ldots + a_n)}{\Gamma(a_1) \cdots \Gamma(a_n)} \int_0^1 dx_1 \cdots \int_0^1 dx_n \frac{\delta(1 - x_1 + \cdots + x_n) x_1^{a_1 - 1} \cdots x_n^{a_n - 1}}{(x_1 A_1 + \ldots + x_n A_n)^{a_1 + \ldots + a_n}}, \quad (8.18)$$

valid for integer $a_1, \ldots, a_n$. By analytic extrapolation, this formula remains valid for complex values of a_i. The formula is, of course, only true as long as the integrals over x_i converge.

As an example for the use of the Feynman parametrization, take the D-dimensional momentum integral containing two propagators of arbitrary power. This reduces to the single integral

$$\int \frac{d^D p}{(2\pi)^D} \frac{1}{(\mathbf{p}^2 + m^2)^a [(\mathbf{p} - \mathbf{k})^2 + m^2]^b} = \frac{\Gamma(a+b)}{\Gamma(a)\Gamma(b)} \int_0^1 dx \int \frac{d^D p}{(2\pi)^D} \frac{(1-x)^{a-1} x^{b-1}}{[\mathbf{p}^2 + m^2 - 2\mathbf{p}\mathbf{k}\, x + \mathbf{k}^2\, x]^{a+b}}$$

$$= \frac{1}{(4\pi)^{D/2}} \frac{\Gamma(a+b)}{\Gamma(a)\Gamma(b)} \int_0^1 dx \frac{(1-x)^{a-1} x^{b-1}}{[m^2 + \mathbf{k}^2\, x\, (1-x)]^{a+b-D/2}}. \quad (8.19)$$

8.2.2   Dimensional Regularization via Proper Time Representation

Schwinger observed that all propagators may be rewritten as Gaussian integrals by using a so-called *proper time representation* (see Appendix 8C) of Feynman integrals, also referred to as *parametric representation*. This permits defining momentum integrals in D complex dimensions via a generalization of the Gaussian integral to D complex dimensions [4, 11, 12, 13], since the latter possesses a rather straightforward analytic continuation with the help of Gamma functions. This method is closely related to what is called *analytic regularization*, and many properties are common to the two approaches.

In Schwinger's proper time representation, each scalar propagator is rewritten as an integral

$$\frac{1}{\mathbf{p}^2 + 2\mathbf{p}\mathbf{q} + m^2} = \int_0^\infty d\tau\, e^{-\tau(\mathbf{p}^2 + 2\mathbf{p}\mathbf{q} + m^2)}. \quad (8.20)$$

The variable τ is called *proper time* for reasons which are irrelevant to the present development. With the integral representation of the Gamma function (8.9), this can be generalized to

$$\frac{1}{(\mathbf{p}^2 + 2\mathbf{p}\mathbf{q} + m^2)^a} = \frac{1}{\Gamma(a)} \int_0^\infty d\tau\, \tau^{a-1} e^{-\tau(\mathbf{p}^2 + 2\mathbf{p}\mathbf{q} + m^2)}, \quad (8.21)$$

valid for $a > 0$. We now assume that in a typical D-dimensional Feynman integral

$$\int \frac{d^D p}{(2\pi)^D} \frac{1}{(\mathbf{p}^2 + 2\mathbf{p}\mathbf{q} + m^2)^a} = \frac{1}{\Gamma(a)} \int \frac{d^D p}{(2\pi)^D} \int_0^\infty d\tau\, \tau^{a-1} e^{-\tau(\mathbf{p}^2 + 2\mathbf{p}\mathbf{q} + m^2)}, \quad (8.22)$$

the integral over the proper time can be exchanged with the momentum integral, leading to

$$\frac{1}{\Gamma(a)} \int_0^\infty d\tau\, \tau^{a-1} \int \frac{d^D p}{(2\pi)^D} e^{-\tau(\mathbf{p}^2 + 2\mathbf{p}\mathbf{q} + m^2)} = \frac{1}{\Gamma(a)} \int_0^\infty d\tau\, \tau^{a-1} e^{-\tau(m^2 - \mathbf{q}^2)} \int \frac{d^D p}{(2\pi)^D} e^{-\tau \mathbf{p}^2}. \quad (8.23)$$

The D-dimensional Gaussian integral on the right-hand side can easily be done and yields:

$$\int \frac{d^D p}{(2\pi)^D}\, e^{-\tau \mathbf{p}^2} = \left[\int \frac{dp}{2\pi}\, e^{-\tau p^2} \right]^D = \left(\frac{1}{4\pi\,\tau} \right)^{D/2}. \quad (8.24)$$

The important point is now that this integral can easily be generalized from integer values of D to any complex number D of dimensions. While the left-hand side of (8.24) makes initially sense only if D is an integer number, the right-hand side exists for any complex D, so that Eq. (8.24) can be used as a *definition* of the Gaussian integral on the left-hand side for complex dimensions D. Inserting (8.24) into (8.23), we are left with a proper-time integral which can be evaluated using the integral representation (8.9) for the Gamma function:

$$\int_0^\infty d\tau\, \tau^{a-D/2-1} e^{-\tau(m^2-\mathbf{q}^2)} = \Gamma(a-D/2)\frac{1}{(m^2-\mathbf{q}^2)^{a-D/2}}. \tag{8.25}$$

The left-hand side is defined only for $D < 2a$, but the Gamma function possesses a unique analytic continuation to larger D. In this way we find for the Feynman integral (8.22):

$$\int \frac{d^D p}{(2\pi)^D} \frac{1}{(\mathbf{p}^2 + 2\mathbf{p}\mathbf{q} + m^2)^a} = \frac{1}{(4\pi)^{D/2}} \frac{\Gamma(a-D/2)}{\Gamma(a)} \frac{1}{(m^2-\mathbf{q}^2)^{a-D/2}}. \tag{8.26}$$

By expanding this integral in powers of a up to order $\mathcal{O}(a)$ and comparing the coefficients of a we find a further important integral which will be needed for the calculation of vacuum energies in Eqs. (8.116) and (10.131):

$$\int \frac{d^D p}{(2\pi)^D} \log(\mathbf{k}^2 + 2\mathbf{k}\mathbf{q} + m^2) = \frac{1}{(4\pi)^{D/2}} \frac{2}{D} \Gamma(1-D/2)\, (m^2-\mathbf{q}^2)^{D/2}, \tag{8.27}$$

Strictly speaking, the logarithm on the left-hand side does not make sense since its argument has the dimension of a square mass. It should therefore always be written as $\log[(\mathbf{k}^2+2\mathbf{k}\mathbf{q}+m^2)/\mu^2]$ with some auxiliary mass μ. If $m \neq 0$, the auxiliary mass μ can, of course, be taken to be m itself. In dimensional regularization, however, this proper way of writing the logarithm does not change the integral at all, as we shall soon demonstrate. The reason is that it merely adds an integral over a constant, and this vanishes by the so-called Veltman formula, to be derived in Eq. (8.33).

A proper-time representation exists for any Feynman integral. Each propagator is replaced by an integral over τ_i. As a result, all momentum integrals are of the type (8.26), (8.14), and their straightforward generalizations. The UV-divergences arise then from the integration region $\tau_i \approx 0$, which is regulated by the analytic continuation of the Gamma function.

Massless Tadpole Integrals

Let us now derive a result which will turn out to be important for the evaluation of massless Feynman integrals associated with so-called *tadpole diagrams*. Tadpole diagrams are quadratically divergent diagrams which have only one external vertex. They contain therefore no external momenta. In the ϕ^4-theory, the simplest tadpole diagram is Q . The detailed discussion of these diagrams will take place in Section 11.4, in particular their role in the renormalization process.

Consider the following Feynman integral:

$$\int \frac{d^D p}{(2\pi)^D} \frac{m^2}{\mathbf{p}^2(\mathbf{p}^2+m^2)} = \int \frac{d^D p}{(2\pi)^D} \left[\frac{1}{\mathbf{p}^2} - \frac{1}{(\mathbf{p}^2+m^2)} \right]. \tag{8.28}$$

The left hand side is calculated with the help of Schwinger's proper-time integral as

$$m^2 \int \frac{d^D p}{(2\pi)^D} \int_0^\infty d\tau_1\, d\tau_2\, e^{-(\tau_1+\tau_2)\mathbf{p}^2 - \tau_2 m^2} = \frac{m^2}{(4\pi)^{D/2}} \int_0^\infty d\tau_1\, d\tau_2\, (\tau_1+\tau_2)^{-D/2} e^{-\tau_2 m^2}$$

$$= \frac{m^2}{(4\pi)^{D/2}} \int_0^\infty d\tau_2 \int_{\tau_2}^\infty d\tau_{12} \, \tau_{12}^{-D/2} \, e^{-\tau_2 m^2} \quad = \quad \frac{m^2}{(4\pi)^{D/2}} \int_0^\infty d\tau_2 \, \frac{-\tau_2^{-D/2+1}}{1 - D/2} \, e^{-\tau_2 m^2}$$

$$= -\frac{(m^2)^{D/2-1}}{(4\pi)^{D/2}} \, \Gamma(1 - D/2). \tag{8.29}$$

The same result is obtained by applying Eq. (8.26) to the second term on the right-hand side of (8.28):

$$-\int \frac{d^D p}{(2\pi)^D} \frac{1}{\mathbf{p}^2 + m^2} = -\frac{(m^2)^{D/2-1}}{(4\pi)^{D/2}} \, \Gamma(1 - D/2). \tag{8.30}$$

This implies that for any complex D, the first term on the right-hand side of (8.28) must vanish:

$$\int \frac{d^D p}{(2\pi)^D} \frac{1}{\mathbf{p}^2} = 0. \tag{8.31}$$

After a polar decomposition of the measure of integration as in Eq. (8.6), the integral (8.31) reads

$$\frac{S_D}{(2\pi)^D} \int_0^\infty dk \, k^{D-3} = 0. \tag{8.32}$$

The vanishing of this for all D implies, again via a polar decomposition, the vanishing of the more general integral

$$\int \frac{d^D p}{(2\pi)^D} (\mathbf{p}^2)^\kappa = 0 \quad \text{for } \kappa, \ D \text{ complex.} \tag{8.33}$$

This is known as *Veltman's formula* [9]. Actually, it could have been derived before from Eq. (8.26) by taking the limit $a \to 0$, which is well defined for all $D \neq 2, 4, \dots$:

$$\int d^D k = 0. \tag{8.34}$$

By analytic interpolation, it is expected to hold for all D.

The consistency of (8.33) has been discussed in various ways [8, 9]. It is nontrivial owing to the absence of any regime of convergence as a function of D, so that no proper mathematical analytic continuation can be invoked. Leibbrandt introduced an extended Gaussian integral containing an auxiliary mass term which permits taking simultaneously the limits $m^2 \to 0$ and $D \to 4$. Below we shall give further arguments for the validity of (8.33) on the basis of Collins' subtraction method in Subsections 8.2.4 and 8.2.5.

The vanishing of the integrals over simple powers of the momentum has the pleasant consequence that in the course of renormalization via the so-called minimal subtraction procedure it will be superfluous to calculate diagrams which contain a tadpole, as will be explained in Sections 11.4 and 11.7–11.8 (see also on page 207).

8.2.3   Tensor Structures

The generalization of tensors from integer space dimension D to complex values of D proceeds by replacing four-vectors by D-vectors. The integrations are performed and the results interpolated analytically. Consider for instance the integral

$$\int d^D p \, p_\mu e^{-\tau \mathbf{p}^2} = 0,$$

which vanishes in integer dimension. This integral is *defined* to be zero for all D. As another example, we evaluate

$$
\int \frac{d^D p}{(2\pi)^D} \frac{p_\mu}{(\mathbf{p}^2 + m^2)^a} \frac{1}{[(\mathbf{p} - \mathbf{k})^2 + m^2]^b} \tag{8.35}
$$
$$
= \frac{k_\mu}{(4\pi)^{D/2}} \frac{\Gamma(a + b - D/2)}{\Gamma(a)\Gamma(b)} \int_0^1 d\tau \, \tau^b (1 - \tau)^{a-1} \left[\mathbf{k}^2 \tau (1 - \tau) + m^2 \right]^{D/2 - a - b} .
$$

When dealing with Feynman integrals of second or higher order we frequently encounter the unit tensor. For consistency, its trace which is defined initially only for integer values of D must be assumed to have a continuous value D for arbitrary complex D:

$$
\sum_\mu \delta_{\mu\mu} = D. \tag{8.36}
$$

This relation can be used to simplify integrals over tensors in momentum space:

$$
\int \frac{d^D p}{(2\pi)^D} p_\mu p_\nu \, f(\mathbf{p}^2) = \frac{1}{D} \int \frac{d^D p}{(2\pi)^D} \delta_{\mu\nu} \mathbf{p}^2 \, f(\mathbf{p}^2). \tag{8.37}
$$

The correctness of this can be verified by contraction with $\delta_{\mu\nu}$ and using the previous integration rules in a complex number D of dimensions.

8.2.4   Dimensional Regularization of 't Hooft and Veltman

We are now going to review the construction of an analytical continuation in the number of dimensions originally used by 't Hooft and Veltman [3, 4, 5]. This regularization procedure may be exemplified using the integral (8.4), which is obviously UV-divergent for $D \geq 2$ and IR-divergent for $D \leq 0$. It can be used to *define* the integration in a continuous number of dimensions D in the region $D < 2$. In order to extrapolate the integral to a larger domain, 't Hooft and Veltman introduced a procedure called *"partial p"*. This is based on inserting into the integrand the unit differential operator

$$
\frac{1}{D} \frac{\partial p}{\partial p} = 1, \tag{8.38}
$$

and carrying out a partial integration in the region $0 < D < 2$. The surface term is explicitly zero for $0 < D < 2$, and we remain with:

$$
I(D) = -\frac{1}{D} \int d^D p \, p_i \frac{\partial}{\partial p_i} \frac{1}{\mathbf{p}^2 + m^2} \tag{8.39}
$$
$$
= -\frac{S_D}{D} \int dp \, p^{D-1} \frac{2p^2}{(p^2 + m^2)^2}. \tag{8.40}
$$

Inserting $m^2 - m^2$ and re-expressing the right-hand side by $I(D)$ leads to

$$
I(D) = \frac{2 S_D \, m^2}{D - 2} \int dp \, p^{D-1} \frac{1}{(p^2 + m^2)^2}. \tag{8.41}
$$

The region of convergence of the integral is now extended to $0 < D < 4$. There is a pole at $D = 2$ as a consequence of the UV-divergence. This expression provides us with the desired analytic extrapolation of (8.6). The procedure can be repeated to yield a convergent integral in the region $0 < D < 6$. All momentum integrals can be performed and the results expressed in terms of Beta functions (8.8). Thus we find for the integral (8.41):

$$\int_0^\infty dp\, p^{D-1} \frac{1}{(p^2 + m^2)^2} = \frac{(m^2)^{D/2-2}}{2} \frac{\Gamma(D/2)\Gamma(2 - D/2)}{\Gamma(2)}. \tag{8.42}$$

The UV-divergence of the integral is reflected in the pole of the Gamma function at $D = 4$. After including the prefactor of (8.41) and the definition (8.7) of S_D, we obtain

$$I(D) = \int d^D p\, \frac{1}{\mathbf{p}^2 + m^2} = \pi^{D/2}(m^2)^{D/2-1}\Gamma(1 - D/2). \tag{8.43}$$

The Gamma function is analytic in the entire complex-D plane, except for isolated poles at $D = 2, 4, \ldots$. Recall that the same analytic expression was obtained by a naive evaluation of (8.6) in (8.10), without an explicit consideration of ranges of convergence of the momentum integrals. The considerations of 't Hooft and Veltman justify further the simple direct result.

The above analytic continuation in the dimension of a specific class of integrals can now easily be extended to more general Feynman integrals. There the integrand will not only depend on the magnitude of the momentum $\mathbf{p}$ but also on its direction, which may point in some n-dimensional subspace of the D-dimensional space, where n is an integer number $n \leq 4$. Let this space be spanned by unit vectors $\mathbf{q}_i$, $(i = 1, \ldots, n)$. The integrand is then some function $f(\mathbf{p}^2, \mathbf{p} \cdot \mathbf{q}_1, \mathbf{p} \cdot \mathbf{q}_2, \ldots, \mathbf{p} \cdot \mathbf{q}_n)$. Now 't Hooft and Veltman split the measure of integration into an n-dimensional part $d^n p_\parallel$ and a remainder $d^{D-n} p_\perp$, and write the integral as

$$\int d^D p\, f(\mathbf{p}^2, \mathbf{p} \cdot \mathbf{q}_1, \mathbf{p} \cdot \mathbf{q}_2, \ldots, \mathbf{p} \cdot \mathbf{q}_n) \tag{8.44}$$

$$= \int_{-\infty}^\infty dp_1 \cdots dp_n \int d^{D-n} p_\perp f(\mathbf{p}_\parallel^2, \mathbf{p}_\perp^2, \mathbf{p} \cdot \mathbf{q}_1, \mathbf{p} \cdot \mathbf{q}_2, \ldots, \mathbf{p} \cdot \mathbf{q}_n).$$

The integration over the n-dimensional subspace is an ordinary integration in integer dimensions. The remaining $D - n$ -dimensional integration is integration in continuous dimensions. The function f is independent of the direction of $\mathbf{p}_\perp$ because the scalar products $\mathbf{p} \cdot \mathbf{q}_1, \ldots, \mathbf{p} \cdot \mathbf{q}_n$ depend only on $\mathbf{p}_1, \ldots, \mathbf{p}_n$, such that $\mathbf{p}_\perp$ appears only in the argument $\mathbf{p}^2 = p_1^2 + \ldots + p_n^2 + p_\perp^2$. The $D - n$ -dimensional integration over $\mathbf{p}_\perp$ is therefore rotationally invariant. The angular integration can now be carried out in arbitrary dimensions $D - n$, leaving only a one-dimensional radial integral:

$$\int d^{D-n} p_\perp f(\mathbf{p}_\parallel^2, \mathbf{p}_\perp^2, \mathbf{p} \cdot \mathbf{q}_1, \mathbf{p} \cdot \mathbf{q}_2, \ldots, \mathbf{p} \cdot \mathbf{q}_n)$$

$$= S_{D-n} \int_0^\infty dp_\perp\, p_\perp^{D-n-1} f(\mathbf{p}_\parallel^2, \mathbf{p}_\perp^2, \mathbf{p} \cdot \mathbf{q}_1, \mathbf{p} \cdot \mathbf{q}_2, \ldots, \mathbf{p} \cdot \mathbf{q}_n), \tag{8.45}$$

where S_{D-n} is the surface of a unit sphere in $D - n$ dimensions [see (8.7)].

Together with the remaining integrations in (8.44) we have

$$\int d^D p\, f(\mathbf{p}^2, \mathbf{p} \cdot \mathbf{q}_1, \ldots, \mathbf{p} \cdot \mathbf{q}_n) = \frac{2\pi^{(D-n)/2}}{\Gamma((D - n)/2)} \int_{-\infty}^\infty dp_1 \ldots dp_n$$

$$\times \int dp_\perp p_\perp^{D-n-1} f(\mathbf{p}_\parallel^2, \mathbf{p}_\perp^2, \mathbf{p} \cdot \mathbf{q}_1, \ldots, \mathbf{p} \cdot \mathbf{q}_n). \tag{8.46}$$

The splitting of the measure produces artificial IR-divergences in the remaining radial integral via the factor $p_\perp^{D-n-1}$. In 't Hooft and Veltman's approach, these are eliminated by partial integration throwing away the surface terms. This procedure is shown in Appendix 8B for the simple integral of Eq. (8.4) in which the splitting of the measure also generates artificial IR-divergences. Only after generating a finite domain of convergence can the procedure "partial p" be used to go to higher D. The results show again that the Gamma function provides us with a universal analytic continuation in D.

Note that, for brevity, we have omitted here the typical factor $1/(2\pi)^D$ in the measure of all momentum integrations since they are irrelevant to the above arguments. This factor will again be present in all subsequent calculations.

8.2.5 Subtraction Method

Another procedure for the analytic extrapolation of Feynman integrals in D dimensions was presented by Collins [8]. He differs from 't Hooft and Veltman by giving an explicit procedure to subtract the artificial IR-divergences which disappear in the previous approach by discarding the surface terms.

For simplicity, consider a rotationally invariant integrand $f(\mathbf{p}^2)$, in which case we may take $n = 0$ in Eq. (8.44). The integrand is supposed to fall off sufficiently fast for large $\mathbf{p}$ to give the one-dimensional integral in Eq. (8.45) a finite region of convergence $0 < D < D'$. For the analytic extrapolation to smaller $D < 0$, the subtraction procedure consists in adding and subtracting the leading orders of an expansion of the integrand around $\mathbf{p}^2 = 0$, which contain the IR-divergences. The continuation to $-2 < \operatorname{Re} D < D'$ is given by

$$\int \frac{d^D p}{(2\pi)^D} f(\mathbf{p}^2) = \frac{2(4\pi)^{-D/2}}{\Gamma(D/2)} \left\{ \int_C^\infty dp\, p^{D-1} f(p^2) \right. \tag{8.47}$$

$$\left. + \int_0^C dp\, p^{D-1} \left[f(p^2) - f(0) \right] + f(0) \frac{C^D}{D} \right\}.$$

The integral over $f(p^2) - f(0) \sim p^2 f^{(1)}(0) + \mathcal{O}(p^4)$ obviously converges at the origin for $-2 < \operatorname{Re} D < 0$. This formula holds initially only for $0 < D < D'$. We now derive the left-hand side by extrapolating the right-hand side analytically to $D < 0$. There, the limit $C \to \infty$ can be taken since $f(0) C^D / D \to 0$, leaving

$$\int \frac{d^D p}{(2\pi)^D} f(\mathbf{p}^2) = \frac{2(4\pi)^{-D/2}}{\Gamma(D/2)} \int_0^\infty dp\, p^{D-1} \left[f(p^2) - f(0) \right], \tag{8.48}$$

which is a convergent definition for the left-hand side valid for $-2 < \operatorname{Re} D < 0$.

Repeated application of the subtraction procedure leads to a continuation formula for $-2l - 2 < \operatorname{Re} D < -2l$ $(l = 0, 1, \ldots)$:

$$\int d^D p\, f(\mathbf{p}^2) = \frac{2(4\pi)^{-D/2}}{\Gamma(D/2)} \int_0^\infty dp\, p^{D-1} \left[f(p^2) - f(0) - p^2 f'(0) - \cdots - p^{2l} \frac{f^{(l)}(0)}{l!} \right]. \tag{8.49}$$

After a sufficient number of steps, the one-dimensional integral will be convergent and can be evaluated with the help of the integral representation of the Beta function (8.8) and the Gamma function (8.9). This, in turn, can be extrapolated analytically to higher D, thereby exhibiting poles at isolated values of D. This continuation to higher D is found to be consistent with the subtraction formula as all subtraction levels lead to the same Gamma function. We demonstrate this by treating the one-loop integral (8.4):

$$I(D) = \int \frac{d^D p}{(2\pi)^D} \frac{1}{(\mathbf{p}^2 + m^2)}, \tag{8.50}$$

which is UV-divergent for $D \geq 2$ and becomes IR-divergent for $D \leq 0$ upon splitting the measure of integration in Eq. (8.6). Expressing the integral directly in terms of Gamma functions as in (8.10) gives

$$I(D) = \int \frac{d^D p}{(2\pi)^D} \frac{1}{\mathbf{p}^2 + m^2} = \frac{(m^2)^{D/2-1}}{(4\pi)^{D/2}} \Gamma(1 - D/2). \tag{8.51}$$

To find a definition for the integral for $-2 < \operatorname{Re} D < 0$, we use the subtraction formula (8.48). We derive

$$
\begin{aligned}
I(D) &= \int \frac{d^D p}{(2\pi)^D} \frac{1}{(\mathbf{p}^2 + m^2)} = \frac{S_D}{(2\pi)^D} \int dp\, p^{D-1} \left(\frac{1}{p^2 + m^2} - \frac{1}{m^2} \right) \\
&= \frac{S_D}{(2\pi)^D} \int dp\, p^{D-1} \frac{-p^2}{m^2(p^2 + m^2)} = \frac{-(m^2)^{D/2-1}}{(4\pi)^{D/2}\Gamma(D/2)} \int_0^\infty dy\, y^{D/2} (1+y)^{-1} \\
&= \frac{-(m^2)^{D/2-1}}{(4\pi)^{D/2}\Gamma(D/2)} B(D/2 + 1, -D/2) = \frac{(m^2)^{D/2-1}}{(4\pi)^{D/2}} \frac{\Gamma(1 - D/2)}{\Gamma(1)}.
\end{aligned}
\tag{8.52}
$$

This is the same result as for the unsubtracted formula (8.51). The same is true for higher subtractions like $(\mathbf{p}^2/m^2)^2$, implying that the integration over any pure power of $\mathbf{p}^2$ gives zero in this regularization.

These manipulations show again that the naive integration in Subsection 8.2.1 performed without any subtractions provides us with the desired analytic extrapolation to any complex dimension D.

8.3   Calculation of One-Particle-Irreducible Diagrams up to Two Loops in Dimensional Regularization

In order to illustrate dimensional regularization, we shall now calculate explicitly the diagrams in $\Gamma^{(2)}(\mathbf{k})$ and $\Gamma^{(4)}(\mathbf{k}_i)$, and $\Gamma^{(0)}$ up to two loops:

$$\Gamma^{(2)}(\mathbf{k}) = \mathbf{k}^2 + m^2 - \left(\frac{1}{2} \, \bigcirc\!\!\!\!\!\bullet + \frac{1}{4} \, \bigcirc\!\!\!\!\bigcirc + \frac{1}{6} \, \ominus \right), \tag{8.53}$$

$$\Gamma^{(4)}(\mathbf{k}_i) = - \times - \frac{3}{2} \, \times\!\!\bigcirc\!\!\times - 3 \, \ominus\!\!\!- \frac{3}{4} \, \times\!\!\bigcirc\!\!\bigcirc\!\!\times - \frac{3}{2} \, \times\!\!\bigcirc\!\!\times. \tag{8.54}$$

$$\Gamma^{(0)} = \frac{1}{2} \, \bigcirc + \frac{1}{8} \, \bigcirc\!\!\bigcirc. \tag{8.55}$$

In Subsection 8.3.4, we shall also do the calculation for the vacuum diagrams. The results will be given a power series in ε, with the divergences appearing as pole terms in ε.

8.3.1   One-Loop Diagrams

On the one-loop level, there are only two divergent diagrams. The Feynman integral associated with the diagram $\bigcirc$ is divergent for $D \geq 2$.

$$\bigcirc = -\lambda \int \frac{d^D p}{(2\pi)^D} \frac{1}{\mathbf{p}^2 + m^2}. \tag{8.56}$$

Using Eq. (8.27), we find

$$\bigcirc = -\lambda \frac{(m^2)^{D/2-1}}{(4\pi)^{D/2}} \Gamma(1 - D/2). \tag{8.57}$$

The Feynman integral is UV-divergent in two, four, six, ... dimensions, which is reflected by poles in the Gamma function at $D = 2, 4, 6, \ldots$. The poles can be subtracted in various ways, parametrized by an arbitrary mass parameter μ. Introducing the *dimensionless coupling constant g* [similar to the dimensionless $\hat{\lambda} = \lambda m^{D-4}$ of Eq. (7.28)]

$$g \equiv \lambda \mu^{D-4} = \lambda \mu^{-\varepsilon}, \tag{8.58}$$

the integral reads in terms of g and ε

$$\bigcirc = -m^2 \frac{g}{(4\pi)^2} \left(\frac{4\pi\mu^2}{m^2} \right)^{\varepsilon/2} \Gamma(\varepsilon/2 - 1). \tag{8.59}$$

The arbitrary mass parameter μ appears in a dimensionless ratio with the mass. It is this kind of terms which contains IR-divergences in the limit $m^2 \to 0$. They are expanded in powers of ε:

$$\left(\frac{4\pi\mu^2}{m^2} \right)^{\varepsilon/2} = 1 + \frac{\varepsilon}{2} \log \left(\frac{4\pi\mu^2}{m^2} \right) + \mathcal{O}(\varepsilon^2). \tag{8.60}$$

The ε-expansion of the Gamma function in (8.59) reads (see Appendix 8D):

$$\Gamma(-n + \varepsilon) = \frac{(-1)^n}{n!} \left\{ \frac{1}{\varepsilon} + \psi(n+1) + \frac{\varepsilon}{2} \left[\frac{\pi^2}{3} + \psi^2(n+1) - \psi'(n+1) \right] + \mathcal{O}(\varepsilon^2) \right\}, \tag{8.61}$$

where $\psi(z) \equiv \Gamma'(z)/\Gamma(z)$ is the *Euler Digamma function*. Inserting these into (8.59) we find the Laurent expansion in ε:

$$\bigcirc = m^2 \frac{g}{(4\pi)^2} \left[\frac{2}{\varepsilon} + \psi(2) + \log \left(\frac{4\pi\mu^2}{m^2} \right) + \mathcal{O}(\varepsilon) \right]. \tag{8.62}$$

The residue of the pole is proportional to m^2 and *independent* of μ.

The integration over two propagators in $\times\!\!\bigcirc\!\!\times$ is convergent for $D < 4$:

$$\times\!\!\bigcirc\!\!\times = \lambda^2 \int \frac{d^D p}{(2\pi)^D} \frac{1}{\mathbf{p}^2 + m^2} \frac{1}{(\mathbf{p} + \mathbf{k})^2 + m^2}, \tag{8.63}$$

where the external momentum $\mathbf{k}$ is the sum of the incoming momenta: $\mathbf{k} = \mathbf{k}_1 + \mathbf{k}_2$. Using formula (8.16), we can rewrite the Feynman integral (8.63) with Feynman parameters as

$$\times\!\!\bigcirc\!\!\times = \lambda^2 \int_0^1 dx \int \frac{d^D p}{(2\pi)^D} \frac{1}{\{(\mathbf{p}^2 + m^2)(1-x) + [(\mathbf{p} + \mathbf{k})^2 + m^2] x\}^2} \tag{8.64}$$

$$= \lambda^2 \int_0^1 dx \int \frac{d^D p}{(2\pi)^D} \frac{1}{(\mathbf{p}^2 + 2\mathbf{p}\mathbf{k}x + \mathbf{k}^2 x + m^2)^2} \tag{8.65}$$

$$= \frac{\lambda^2}{(4\pi)^{D/2}} \frac{\Gamma(2 - D/2)}{\Gamma(2)} \int_0^1 dx \frac{1}{[\mathbf{k}^2 x(1-x) + m^2]^{2-D/2}}. \tag{8.66}$$

The divergence for $D = 4$ is contained in the Gamma function which possesses poles at $D = 4, 6, \ldots$. The remaining parameter integral is finite for any D as long as $m^2 \neq 0$. In terms of ε and g, the expression for the simple loops reads

$$\times\!\!\bigcirc\!\!\times = g\mu^\varepsilon \frac{g}{(4\pi)^2} \Gamma(\varepsilon/2) \int_0^1 dx \left[\frac{4\pi\mu^2}{\mathbf{k}^2 x(1-x) + m^2} \right]^{\varepsilon/2}. \tag{8.67}$$

In order to separate the pole terms, we expand each term in powers of ε. The Gamma function is expanded with (8.61) to yield

$$
\bigtimes = g\mu^\varepsilon \frac{g}{(4\pi)^2} \left[\frac{2}{\varepsilon} + \psi(1) + \mathcal{O}(\varepsilon)\right] \left\{1 + \frac{\varepsilon}{2} \int_0^1 dx \log\left[\frac{4\pi\mu^2}{\mathbf{k}^2 x(1-x) + m^2}\right] + \mathcal{O}(\varepsilon^2)\right\}
$$

$$
= g\mu^\varepsilon \frac{g}{(4\pi)^2} \left\{\frac{2}{\varepsilon} + \psi(1) + \int_0^1 dx \log\left[\frac{4\pi\mu^2}{\mathbf{k}^2 x(1-x) + m^2}\right] + \mathcal{O}(\varepsilon)\right\}. \tag{8.68}
$$

The prefactor $g\mu^\varepsilon$ is, in fact, the coupling constant λ which will be renormalized, and is not expanded in powers of ε. Only the expression multiplying it contributes to the renormalization constant Z_g with a pole term that is independent of the arbitrary mass parameter μ. This mass parameter appears only in the finite part, where the freedom of its choice exhibits a degree of freedom in the renormalization procedure.

The result (8.66) can be generalized to Feynman integrals of the type (8.63), in which the denominators appear with an arbitrary power. Using Eqs. (8.15) and (8.26), we find

$$
\int \frac{d^D p}{(2\pi)^D} \frac{1}{(\mathbf{p}^2 + m^2)^a} \frac{1}{[(\mathbf{p}-\mathbf{k})^2 + m^2]^b} \tag{8.69}
$$

$$
= \frac{\Gamma(a+b-D/2)}{(4\pi)^{D/2}\Gamma(a)\Gamma(b)} \int_0^1 dx\, x^{b-1}(1-x)^{a-1} \left[\mathbf{k}^2 x(1-x) + m^2\right]^{D/2-a-b}.
$$

The parameter integral on the right-hand side develops IR-divergences for $m^2 = 0$ and $D = 4$ if a or $b \geq 2$.

8.3.2   Two-Loop Self-Energy Diagrams

On the two-loop level, two diagrams contribute to the self-energy: $\bigotimes$ and $\ominus$. The Feynman integral associated with the first factorizes into two independent momentum integrals:

$$
\bigotimes = \lambda^2 \int \frac{d^D p_1}{(2\pi)^D} \frac{d^D p_2}{(2\pi)^D} \frac{1}{\mathbf{p}_1^2 + m^2} \frac{1}{(\mathbf{p}_2^2 + m^2)^2}. \tag{8.70}
$$

The integral over $\mathbf{p}_1$ coincides with the previous integral (8.56), which is expanded in powers of ε in Eq. (8.62). The integral over $\mathbf{p}_2$ is calculated using formula (8.26):

$$
\lambda \int \frac{d^D p_2}{(2\pi)^D} \frac{1}{(\mathbf{p}_2^2 + m^2)^2} = \frac{g\mu^\varepsilon}{(4\pi)^{D/2}} \frac{\Gamma(2-D/2)}{\Gamma(2)} \frac{1}{(m^2)^{2-D/2}}
$$

$$
= \frac{g}{(4\pi)^2} \left(\frac{4\pi\mu^2}{m^2}\right)^{\varepsilon/2} \Gamma(\varepsilon/2)
$$

$$
= \frac{g}{(4\pi)^2} \left[\frac{2}{\varepsilon} + \psi(1) + \log\frac{4\pi\mu^2}{m^2} + \mathcal{O}(\varepsilon)\right], \tag{8.71}
$$

after expanding the Gamma function according to formula (8.61). This result can also be found directly in (8.68) for $\mathbf{k}^2 = 0$. The result of the integration over $\mathbf{p}_1$ is given in (8.62). The product of the integrations over $\mathbf{p}_1$ and $\mathbf{p}_2$ gives

$$
\bigotimes = -\frac{m^2 g^2}{(4\pi)^4} \left[\frac{4}{\varepsilon^2} + 2\frac{\psi(1) + \psi(2)}{\varepsilon} - \frac{4}{\varepsilon}\log\left(\frac{m^2}{4\pi\mu^2}\right) + \mathcal{O}(\varepsilon^0)\right]. \tag{8.72}
$$

We come now to the calculation of the so-called *sunset diagram*:

$$\ominus = \lambda^2 \int \frac{d^D p_1}{(2\pi)^D} \frac{d^D p_2}{(2\pi)^D} \frac{1}{\mathbf{p}_1^2 + m^2} \frac{1}{\mathbf{p}_2^2 + m^2} \frac{1}{(\mathbf{q} + \mathbf{p}_1 + \mathbf{p}_2)^2 + m^2}, \tag{8.73}$$

in which $\mathbf{q}$ is the incoming momentum. The calculation of $\ominus$ is rather difficult, because a naive introduction of the parameter integrals results in divergences of the parameter integral. The problem is solved by lowering the degree of divergence via partial integration. Using the trivial identity $\partial p^\mu / \partial p^\nu = \delta^{\mu\nu}$ and the trace property (8.37), we see that

$$1 = \frac{1}{2D} \left(\frac{\partial p_1^\mu}{\partial p_1^\mu} + \frac{\partial p_2^\mu}{\partial p_2^\mu} \right). \tag{8.74}$$

Inserting this identity into (8.73), and performing a partial integration in which the surface term is discarded, we obtain a sum of two integrals:

$$\ominus = -\frac{\lambda^2}{D-3} \int \frac{d^D p_1}{(2\pi)^D} \frac{d^D p_2}{(2\pi)^D} \frac{3m^2 + \mathbf{q}(\mathbf{q} + \mathbf{p}_1 + \mathbf{p}_2)}{(\mathbf{p}_1^2 + m^2)(\mathbf{p}_2^2 + m^2)[(\mathbf{q} + \mathbf{p}_1 + \mathbf{p}_2)^2 + m^2]^2}$$

$$= -\frac{\lambda^2}{D-3} \left[3m^2 A(\mathbf{q}) + B(\mathbf{q}) \right]. \tag{8.75}$$

In this way, the original quadratically divergent integral is decomposed into a logarithmically divergent integral $A(\mathbf{q})$ and a linearly divergent integral $B(\mathbf{q})$. Let us first evaluate the integral for $A(\mathbf{q})$:

$$A(\mathbf{q}) = \int \frac{d^D p_1}{(2\pi)^D} \frac{d^D p_2}{(2\pi)^D} \frac{1}{(\mathbf{p}_1^2 + m^2)(\mathbf{p}_2^2 + m^2)[(\mathbf{q} + \mathbf{p}_1 + \mathbf{p}_2)^2 + m^2]^2}. \tag{8.76}$$

Replacing the momentum $\mathbf{p}_2$ by $-\mathbf{p} - \mathbf{q} - \mathbf{p}_1$, the integral over $\mathbf{p}_1$ coincides with that in the diagram $\times\!\!\bigcirc\!\!\times$ in Eq. (8.63). It can be performed in the same way, and we obtain

$$A(\mathbf{q}) = \frac{1}{(4\pi)^{D/2}} \frac{\Gamma(2 - D/2)}{\Gamma(2)} \int_0^1 dx \int \frac{d^D p}{(2\pi)^D} \frac{1}{[(\mathbf{q} + \mathbf{p})^2 x(1 - x) + m^2]^{2-D/2}(\mathbf{p}^2 + m^2)^2}. \tag{8.77}$$

Applying the Feynman formula (8.15), this becomes

$$A(\mathbf{q}) = \frac{1}{(4\pi)^{D/2}} \Gamma(4 - D/2) \int_0^1 dx [x(1 - x)]^{D/2-2} \int_0^1 dy \int \frac{d^D p}{(2\pi)^D} \frac{y(1 - y)^{1-D/2}}{[f(\mathbf{q}, \mathbf{p}, x, y)]^{4-D/2}}, \tag{8.78}$$

with

$$f(\mathbf{q}, \mathbf{p}, x, y) = (\mathbf{p}^2 + m^2)y + \left[(\mathbf{p} + \mathbf{q})^2 + \frac{m^2}{x(1 - x)} \right] (1 - y)$$

$$= \mathbf{p}^2 + 2\mathbf{p}\mathbf{q}(1 - y) + \mathbf{q}^2(1 - y) + m^2 \left[y + \frac{1 - y}{x(1 - x)} \right]. \tag{8.79}$$

The momentum integral in (8.78) is now carried out with the help of formula (8.26), leading to

$$A(\mathbf{q}) = \frac{\Gamma(4 - D)}{(4\pi)^D} \int_0^1 dx [x(1 - x)]^{D/2-2} \int_0^1 dy \frac{y(1 - y)^{1-D/2}}{\left[\mathbf{q}^2 y(1 - y) + m^2 \left(y + \frac{1-y}{x(1-x)} \right) \right]^{4-D}}. \tag{8.80}$$

Inserting $D = 4 - \varepsilon$ and expanding the denominator in ε, we find

$$A(\mathbf{q}) = \frac{\Gamma(\varepsilon)}{(4\pi)^4} \left(\frac{4\pi}{m^2}\right)^\varepsilon \int_0^1 dx [x(1-x)]^{-\varepsilon/2} \int_0^1 dy\, y(1-y)^{\varepsilon/2-1}$$
$$\times \left\{1 - \varepsilon \log\left[\frac{\mathbf{q}^2}{m^2} y(1-y) + \left(y + \frac{1-y}{x(1-x)}\right)\right] + \mathcal{O}(\varepsilon^2)\right\}. \tag{8.81}$$

The parameter integrals without the curly brackets can be evaluated with the help of the integral formula [note the difference with respect to (8.8)]:

$$B(\alpha, \beta) \equiv \frac{\Gamma(\alpha)\Gamma(\beta)}{\Gamma(\alpha + \beta)} = \int_0^1 dy\, y^{\alpha-1}(1-y)^{\beta-1}, \tag{8.82}$$

and the expansion derived in Appendix 8D:

$$\Gamma(n+1+\varepsilon) = n! \left\{1 + \varepsilon\, \psi(n+1) + \frac{\varepsilon^2}{2}\left[\psi'(n+1) + \psi(n+1)^2\right] + \mathcal{O}(\varepsilon^3)\right\}, \tag{8.83}$$

yielding

$$\int_0^1 dx\, x^{-\varepsilon/2}(1-x)^{-\varepsilon/2} = \frac{\Gamma(1-\varepsilon/2)\Gamma(1-\varepsilon/2)}{(1-\varepsilon)\Gamma(1-\varepsilon)} = 1 + \varepsilon + \mathcal{O}(\varepsilon^2), \tag{8.84}$$

$$\int_0^1 dy\, y(1-y)^{\varepsilon/2-1} = \frac{\Gamma(2)\Gamma(\varepsilon/2)}{\Gamma(2+\varepsilon/2)} = \frac{2}{\varepsilon(1+\varepsilon/2)} = \frac{2}{\varepsilon} - 1 + \mathcal{O}(\varepsilon). \tag{8.85}$$

Only the latter integral is singular for $\varepsilon \to 0$. The singularity comes from the endpoint at $y = 1$. Using this we find for $A(\mathbf{q})$ from the first term of the curly bracket of (8.81):

$$A(\mathbf{q}) = \frac{1}{(4\pi)^4} \left(\frac{4\pi}{m^2}\right)^\varepsilon \frac{\Gamma(1+\varepsilon)}{\varepsilon} \left[\frac{2}{\varepsilon} - 1 + \mathcal{O}(\varepsilon)\right] \left[1 + \varepsilon + \mathcal{O}(\varepsilon^2)\right]. \tag{8.86}$$

The second term in the curly brackets of (8.81) is of order ε. This ε is canceled against a $1/\varepsilon$ coming from the factor $\Gamma(\varepsilon)$. Since the logarithm itself is convergent, a pole term can only appear for $y \to 1$ where the y-integral (8.85) diverges. But for $y \to 1$, the logarithm goes to zero as $\log y$. Therefore, the second term in the bracket yields no pole term for $A(\mathbf{q})$, which is therefore entirely contained in (8.86). Expanding $\Gamma(1+\varepsilon) = \Gamma(1) + \psi(1)\varepsilon + \mathcal{O}(\varepsilon^2)$ according to formula (8.83), and $(4\pi/m^2)^\varepsilon = 1 + \varepsilon \log(4\pi/m^2) + \mathcal{O}(\varepsilon^2)$, we obtain

$$A(\mathbf{q}) = \frac{1}{(4\pi)^4} \left\{\frac{2}{\varepsilon^2} + \frac{1}{\varepsilon}\left[1 + 2\,\psi(1) + 2\log\frac{4\pi}{m^2}\right] + \mathcal{O}(\varepsilon^0)\right\}. \tag{8.87}$$

There are many ways of calculating $B(\mathbf{q})$, most of them involving cumbersome expressions. The easiest way uses the fact that the integrand of $B(\mathbf{q})$ can be rewritten as:

$$\frac{q^\mu(q + p_1 + p_2)^\mu}{(\mathbf{p}_1^2 + m^2)(\mathbf{p}_2^2 + m^2)[(\mathbf{q} + \mathbf{p}_1 + \mathbf{p}_2)^2 + m^2]^2} \tag{8.88}$$
$$= -\frac{q^\mu}{2}\frac{\partial}{\partial q^\mu}\frac{1}{(\mathbf{p}_1^2 + m^2)(\mathbf{p}_2^2 + m^2)[(\mathbf{q} + \mathbf{p}_1 + \mathbf{p}_2)^2 + m^2]^2}.$$

Then the integral for $B(\mathbf{q})$ becomes

$$B(\mathbf{q}) = -\frac{q^\mu}{2}\frac{\partial}{\partial q^\mu}\int \frac{d^D p_1}{(2\pi)^D}\frac{d^D p_2}{(2\pi)^D}\frac{1}{(\mathbf{p}_1^2 + m^2)(\mathbf{p}_2^2 + m^2)[(\mathbf{q} + \mathbf{p}_1 + \mathbf{p}_2)^2 + m^2]}. \tag{8.89}$$

Introducing Feynman parameters as before, we find

$$
\begin{aligned}
B(\mathbf{q}) &= -\frac{q^\mu}{2}\frac{\Gamma(3-D)}{(4\pi)^D}\frac{\partial}{\partial q^\mu}\int_0^1 dx[x(1-x)]^{D/2-2} \\
&\quad \times \int_0^1 dy\, y^{1-D/2}\left\{\mathbf{q}^2 y(1-y)+m^2\left[1-y+\frac{y}{x(1-x)}\right]\right\}^{D-3} \\
&= \mathbf{q}^2\frac{(3-D)\Gamma(3-D)}{(4\pi)^D}\int_0^1 dx[x(1-x)]^{D/2-2} \\
&\quad \times \int_0^1 dy\, y^{2-D/2}(1-y)\left\{\mathbf{q}^2 y(1-y)+m^2\left[1-y+\frac{y}{x(1-x)}\right]\right\}^{D-4}. \quad (8.90)
\end{aligned}
$$

In terms of $D = 4 - \varepsilon$, this reads

$$
\begin{aligned}
B(\mathbf{q}) &= \mathbf{q}^2\frac{\Gamma(\varepsilon)}{(4\pi)^4}\left(\frac{4\pi}{m^2}\right)^\varepsilon\int_0^1 dx[x(1-x)]^{-\varepsilon/2} \\
&\quad \times \int_0^1 dy\, y^{\varepsilon/2}(1-y)\left\{1-\varepsilon\log\left[\frac{\mathbf{q}^2}{m^2}y(1-y)+\left(1-y+\frac{y}{x(1-x)}\right)\right]+\mathcal{O}(\varepsilon^2)\right\}. \quad (8.91)
\end{aligned}
$$

The parameter integrals without the brackets give no pole in ε:

$$
\int_0^1 dx\,[x(1-x)]^{-\varepsilon/2} = \frac{\Gamma(1-\varepsilon/2)\Gamma(1-\varepsilon/2)}{(1-\varepsilon)\Gamma(1-\varepsilon)} = 1+\varepsilon+\mathcal{O}(\varepsilon^2), \quad (8.92)
$$

$$
\begin{aligned}
\int_0^1 dy\,(1-y)\,y^{\varepsilon/2} &= \frac{\Gamma(2)\Gamma(1+\varepsilon/2)}{\Gamma(3+\varepsilon/2)} = \frac{\Gamma(1+\varepsilon/2)}{(2+\varepsilon/2)(1+\varepsilon/2)\Gamma(1+\varepsilon/2)} \\
&= \frac{1}{2}\left[1-\frac{3}{4}\varepsilon+\mathcal{O}(\varepsilon^2)\right]. \quad (8.93)
\end{aligned}
$$

The only pole in ε comes from the prefactor $\Gamma(\varepsilon)$ in (8.91). Since the second term in the brackets of (8.91) carries a factor ε, it does not contribute to the pole term of $B(\mathbf{q})$, and we have

$$
\begin{aligned}
B(\mathbf{q}) &= \frac{\mathbf{q}^2}{(4\pi)^4}\left(\frac{4\pi}{m^2}\right)^\varepsilon\frac{\Gamma(1+\varepsilon)}{\varepsilon}\left[1+\varepsilon+\mathcal{O}(\varepsilon^2)\right]\frac{1}{2}\left[1-\frac{3}{4}\varepsilon+\mathcal{O}(\varepsilon^2)\right] \\
&= \frac{\mathbf{q}^2}{(4\pi)^4}\left(\frac{4\pi}{m^2}\right)^\varepsilon\frac{1}{2\varepsilon}+\mathcal{O}(\varepsilon^0). \quad (8.94)
\end{aligned}
$$

Expanding $(4\pi/m^2)^\varepsilon$ as before, the final expression for the pole term of $B(\mathbf{q})$ is

$$
B(\mathbf{q}) = \frac{\mathbf{q}^2}{(4\pi)^4}\frac{1}{2\varepsilon}+\mathcal{O}(\varepsilon^0). \quad (8.95)
$$

Together with the result for $A(\mathbf{q})$ in (8.87), we find for the sunset diagram in (8.75) with $\lambda = g\mu^\varepsilon$:

$$
\bigcirc\!\!\!\!-\!\!\!\!-\ = -g^2\frac{m^2}{(4\pi)^4}\left\{\frac{6}{\varepsilon^2}+\frac{6}{\varepsilon}\left[\frac{3}{2}+\psi(1)+\log\frac{4\pi\mu^2}{m^2}\right]+\frac{\mathbf{q}^2}{2m^2\,\varepsilon}+\mathcal{O}(\varepsilon^0)\right\}. \quad (8.96)
$$

8.3.3   Two-Loop Diagram of Four-Point Function

There are three two-loop diagrams contributing to the four-point function: ⋈, ⦷, ⊖ .
The first is a product of two independent integrals:

$$
\bowtie = -\lambda^3 \int \frac{d^D p}{(2\pi)^D} \frac{1}{[(\mathbf{p}-\mathbf{k})^2+m^2](\mathbf{p}^2+m^2)} \int \frac{d^D q}{(2\pi)^D} \frac{1}{[(\mathbf{q}-\mathbf{k})^2+m^2](\mathbf{q}^2+m^2)}, \tag{8.97}
$$

where $\mathbf{k}$ denotes either of the three different momentum combinations $\mathbf{k}_1+\mathbf{k}_2$, $\mathbf{k}_1+\mathbf{k}_3$, and $\mathbf{k}_1+\mathbf{k}_4$. The pole term in (8.97) is easily calculated using Eq. (8.68). Setting $\lambda = g\mu^\varepsilon$, we find

$$
\bowtie = -g\mu^\varepsilon \frac{g^2}{(4\pi)^4} \left\{ \frac{2}{\varepsilon} + \psi(1) + \int_0^1 dx \log\left[\frac{4\pi\mu^2}{\mathbf{k}^2 x(1-x)+m^2} \right] + \mathcal{O}(\varepsilon) \right\}^2 \tag{8.98}
$$

$$
= -g\mu^\varepsilon \frac{g^2}{(4\pi)^4} \left\{ \frac{4}{\varepsilon^2} + \frac{4}{\varepsilon}\psi(1) + \frac{4}{\varepsilon} \int_0^1 dx \log\left[\frac{4\pi\mu^2}{\mathbf{k}^2 x(1-x)+m^2} \right] + \mathcal{O}(\varepsilon^0) \right\}. \tag{8.99}
$$

The second diagram is associated with the following integral

$$
\circledcirc = -\lambda^3 \int \frac{d^D p}{(2\pi)^D} \frac{1}{[(\mathbf{p}-\mathbf{k})^2+m^2](\mathbf{p}^2+m^2)^2} \int \frac{d^D q}{(2\pi)^D} \frac{1}{(\mathbf{q}^2+m^2)}. \tag{8.100}
$$

It is calculated using Eqs. (8.69) and (8.62), replacing again λ by $g\mu^\varepsilon$:

$$
\circledcirc = -m^2 \frac{g}{(4\pi)^2} \left[\frac{2}{\varepsilon} + \psi(2) + \log\left(\frac{4\pi\mu^2}{m^2} \right) + \mathcal{O}(\varepsilon) \right] \tag{8.101}
$$

$$
\times g\mu^\varepsilon \frac{g}{(4\pi)^2} \Gamma(1+\varepsilon/2) \left[1 + \frac{\varepsilon}{2}\log 4\pi\mu^2 + \mathcal{O}(\varepsilon^2) \right] \left[\int_0^1 dx \frac{1-x}{\mathbf{k}^2 x(1-x)+m^2} - \mathcal{O}(\varepsilon) \right] \tag{8.102}
$$

$$
= -g\mu^\varepsilon \frac{g^2}{(4\pi)^4} \frac{2}{\varepsilon} \left[\int_0^1 dx \frac{m^2(1-x)}{\mathbf{k}^2 x(1-x)+m^2} + \mathcal{O}(\varepsilon) \right]. \tag{8.103}
$$

The third Feynman integral,

$$
\ominus = -\lambda^3 \int \frac{d^D p}{(2\pi)^D} \frac{d^D q}{(2\pi)^D} \frac{1}{(\mathbf{p}^2+m^2)\left[(\mathbf{k}_1+\mathbf{k}_2-\mathbf{p})^2+m^2\right]} \frac{1}{(\mathbf{q}^2+m^2)\left[(\mathbf{p}-\mathbf{q}+\mathbf{k}_3)^2+m^2\right]}, \tag{8.104}
$$

can be written as

$$
\ominus = -\lambda^3 \int \frac{d^D p}{(2\pi)^D} \frac{1}{\mathbf{p}^2+m^2} \frac{1}{(\mathbf{k}_1+\mathbf{k}_2-\mathbf{p})^2+m^2} I(\mathbf{p}+\mathbf{k}_3), \tag{8.105}
$$

where $I(\mathbf{k})$ is the same integral (8.63) which occurred in the one-loop diagram ⋈. Using the result of that integral in (8.67), we obtain

$$
\ominus = (g\mu^\varepsilon)^2 \frac{g}{(4\pi)^2} \Gamma(\varepsilon/2) \int_0^1 dx \tag{8.106}
$$

$$
\times \int \frac{d^D p}{(2\pi)^D} \frac{1}{\mathbf{p}^2+m^2} \frac{1}{(\mathbf{k}_1+\mathbf{k}_2-\mathbf{p})^2+m^2} \left[\frac{4\pi\mu^2}{(\mathbf{k}_3+\mathbf{p})^2 x(1-x)+m^2} \right]^{\varepsilon/2}.
$$

The denominators are combined via (8.17), and we get

$$
g\mu^{\varepsilon} \frac{g^2}{(4\pi)^4} (4\pi\mu^2)^{\varepsilon} \Gamma(2+\varepsilon/2) \frac{\Gamma(\varepsilon)}{\Gamma(2+\varepsilon/2)} \int_0^1 dx\, [x(1-x)]^{-\varepsilon/2} \int_0^1 dy (1-y)^{\varepsilon/2-1} y
$$

$$
\times \int_0^1 dz \Big\{ yz(1-yz)(\mathbf{k}_1+\mathbf{k}_2)^2 + y(1-y)\mathbf{k}_3^2
$$

$$
-2yz(1-y)\mathbf{k}_3(\mathbf{k}_1+\mathbf{k}_2) + m^2 \Big[y + \frac{1-y}{x(1-x)} \Big] \Big\}^{-\varepsilon}. \tag{8.107}
$$

The only pole term comes from the end-point singularity in the integral at $y = 1$. For $y = 1$, the curly bracket can be expanded as

$$
\{\ldots\} = \Big[(\mathbf{k}_1+\mathbf{k}_2)^2 \Big]^{-\varepsilon} \Big\{ 1 - \varepsilon \log \Big[z(1-z) + \frac{m^2}{(\mathbf{k}_1+\mathbf{k}_2)^2} \Big] + \mathcal{O}\left(\varepsilon^2\right) \Big\}. \tag{8.108}
$$

Using formula (8D.24), the prefactor of (8.107) is seen to have an expansion

$$
\frac{1}{\varepsilon} [1 - \varepsilon\psi(1)] + \mathcal{O}\left(\varepsilon\right),
$$

so that we have

$$
\ominus \;=\; g\mu^{\varepsilon} \frac{g^2}{(4\pi)^4} (4\pi\mu^2)^{\varepsilon} \frac{1}{\varepsilon} [1 + \varepsilon\psi(1)] \int_0^1 dx\, [x(1-x)]^{-\varepsilon/2} \int_0^1 dy (1-y)^{\frac{\varepsilon}{2}-1} y
$$

$$
\times \Big[(\mathbf{k}_1+\mathbf{k}_2)^2 \Big]^{-\varepsilon} \int_0^1 dz \Big\{ 1 - \varepsilon \log \Big[z(1-z) + \frac{m^2}{(\mathbf{k}_1+\mathbf{k}_2)^2} \Big] \Big\} + \mathcal{O}(\varepsilon). \tag{8.109}
$$

The first two terms in the integral are independent of x and y, and give

$$
\frac{\Gamma^2(1-\varepsilon/2)}{\Gamma(2-\varepsilon)} \frac{\Gamma(\varepsilon/2)\Gamma(2)}{\Gamma(2+\varepsilon/2)} \Big\{ 1 - \varepsilon \int_0^1 dx \log \Big[x(1-x) + \frac{m^2}{(\mathbf{k}_1+\mathbf{k}_2)^2} \Big] \Big\}. \tag{8.110}
$$

The Gamma functions are combined using the following formula:

$$
\frac{\prod_n \Gamma(1+a_n\varepsilon)}{\prod_m \Gamma(1+a'_m\varepsilon)} = 1 + \mathcal{O}\left(\varepsilon^2\right), \tag{8.111}
$$

which holds if

$$
\sum_n a_n - \sum_m a'_m = 0. \tag{8.112}
$$

We then find for the prefactor in (8.110):

$$
\frac{\Gamma^2(1-\varepsilon/2)}{\Gamma(2-\varepsilon)} \frac{\Gamma(\varepsilon/2)\Gamma(2)}{\Gamma(2+\varepsilon/2)} = \frac{2}{\varepsilon} \frac{1}{(1-\varepsilon)(1+\varepsilon/2)} [1 + \mathcal{O}(\varepsilon^2)]. \tag{8.113}
$$

Up to this point, the singular part of (8.106) is therefore

$$
\ominus \;=\; g\mu^{\varepsilon} \frac{g^2}{(4\pi)^4} \frac{2}{\varepsilon^2} \Big\{ 1 + \frac{\varepsilon}{2} + \varepsilon\psi(1) - \varepsilon \int_0^1 dx \log \Big[\frac{(\mathbf{k}_1+\mathbf{k}_2)^2 x(1-x) + m^2}{4\pi\mu^2} \Big] \Big\} + \mathcal{O}(\varepsilon^0), \tag{8.114}
$$

where we have used the mass parameter μ to make the logarithm dimensionless. We now look at the effect of the last logarithm in (8.108). It is accompanied by a factor ε and vanishes at $y = 1$, such that it fails to lead to an end-point singularity at $y = 1$, $\varepsilon = 0$. Therefore, it contributes to order $\mathcal{O}(\varepsilon^0)$ and can be neglected as far as the singular parts are concerned, leaving Eq. (8.110) unchanged.

8.3.4   Two-Loop Vacuum Diagrams

For the discussion of the effective potential in Section 10.6 we shall also need the vacuum diagrams in dimensional regularization. Up to two loops, they are

$$\Gamma^{(0)} = \frac{1}{2}\,\bigcirc\; +\; \frac{1}{8}\;\infty\;. \tag{8.115}$$

The analytic expression for the one-loop diagram is, according to Eq. (5.45),

$$\frac{1}{2}\,\bigcirc \;=\; \frac{1}{2}\int \frac{d^D p}{(2\pi)^D}\,\log(\mathbf{p}^2 + m^2). \tag{8.116}$$

Recalling formula Eq. (8.27), and expanding the result of the integration in powers of ε, we obtain

$$
\begin{aligned}
\frac{1}{2}\,\bigcirc \;&=\; \frac{1}{2}\frac{2}{D}\frac{(m^2)^{D/2}}{(4\pi)^{D/2}}\,\Gamma(1 - D/2)\\[2mm]
&=\; \frac{1}{2}\frac{m^4}{\mu^\varepsilon}\frac{1}{(4\pi)^2}\left[-\frac{1}{\varepsilon} + \frac{1}{2}\log\frac{m^2}{4\pi\mu^2} - \frac{1}{4} - \frac{1}{2}\psi(2)\right] + \mathcal{O}(\varepsilon),
\end{aligned}\tag{8.117}
$$

where the value of the Digamma function $\psi(2)$ can be taken from Eq. (8D.12). The one-loop vacuum diagram is the only diagram with no vertex. To go to the second line with a dimensionless argument of the logarithm, we have inserted the mass parameter μ by inserting the identity $1 = \mu^\varepsilon/\mu^\varepsilon$ and expanded $(m^2/\mu^2)^{\varepsilon/2}$ in powers of ε. Note that the parameter μ in the prefactor $1/\mu^\varepsilon$ cannot be absorbed into the coupling constant.

The two-loop diagram corresponds to the Feynman integral

$$\frac{1}{8}\;\infty\;=\; -\frac{\lambda}{8}\left(\int \frac{d^D p}{(2\pi)^D}\frac{1}{\mathbf{p}^2 + m^2}\right)^2. \tag{8.118}$$

Comparison of (8.118) with (8.56) and (8.57) yields

$$\frac{1}{8}\;\infty\;=\; -\frac{\lambda}{8}\frac{(m^2)^{D-2}}{(4\pi)^D}\,\Gamma^2(1 - D/2). \tag{8.119}$$

This has the ε-expansion in terms of $g = \lambda\mu^{-\varepsilon}$:

$$
\begin{aligned}
\frac{1}{8}\;\infty\;=\; &-\frac{1}{2}\frac{m^4}{\mu^\varepsilon}\frac{g}{(4\pi)^4}\Bigg\{\frac{1}{\varepsilon^2} - \frac{1}{\varepsilon}\left(\log\frac{m^2}{4\pi\mu^2} - \psi(2)\right)\\[2mm]
&-\psi(2)\log\frac{m^2}{4\pi\mu^2} + \frac{1}{2}\left(\log\frac{m^2}{4\pi\mu^2}\right)^2 + \frac{1}{4}\left[\frac{\pi^2}{3} + 2\psi^2(2) - \psi'(2)\right] + \mathcal{O}(\varepsilon)\Bigg\}.
\end{aligned}
$$

Appendix 8A   Polar Coordinates and Surface of a Sphere in D Dimensions

The polar coordinates in D dimensions are:

$$(p_1.\ldots,p_D) \;\rightarrow\; (p,\varphi,\vartheta_1,\ldots.\vartheta_{D-2})\,, \quad p^2 = p_\mu p_\mu\,, \tag{8A.1}$$

$$d^D p \;=\; p^{D-1}dp\,d\varphi\,\sin\vartheta_1\,d\vartheta_1\,\sin^2\vartheta_2\,d\vartheta_2\ldots\sin^{D-2}\vartheta_{D-2}d\vartheta_{D-2}\,, \tag{8A.2}$$

where $0 < p < \infty$, $0 < \varphi < 2\pi$, $0 < \vartheta_i < \pi$, $i = 1, \ldots, D - 2$. We shall denote the directional integral as

$$\int d\hat{\mathbf{p}} \equiv \int d\varphi \, \sin \vartheta_1 \, d\vartheta_1 \, \sin^2 \vartheta_2 \, d\vartheta_2 \ldots \sin^{D-2} \vartheta_{D-2} d\vartheta_{D-2}. \tag{8A.3}$$

The result of the directional integral is the surface of a sphere of unit radius in D dimensions:

$$S_D = \int d\hat{\mathbf{p}}. \tag{8A.4}$$

The $D - 1$ angular integrations can be carried out using the integral formula

$$\int_0^{\pi/2} (\sin t)^{2x-1} (\cos t)^{2y-1} \, dt = \frac{1}{2} \frac{\Gamma(x)\Gamma(y)}{\Gamma(x+y)}, \qquad \mathrm{Re}\, x, \, \mathrm{Re}\, y > 0. \tag{8A.5}$$

Inserting here $y = \frac{1}{2}$ and $x = \frac{k+1}{2}$, we obtain

$$\int_0^\pi dt \, (\sin t)^k = 2 \int_0^{\pi/2} dt \, (\sin t)^k = \frac{\Gamma(\frac{k+1}{2})\Gamma(\frac{1}{2})}{\Gamma(\frac{k+2}{2})} = \sqrt{\pi} \, \frac{\Gamma(\frac{k+1}{2})}{\Gamma(\frac{k+2}{2})}, \tag{8A.6}$$

and thus:

$$S_D = 2\pi \prod_{k=1}^{D-2} \int_0^\pi \sin^k \vartheta_k \, d\vartheta_k = 2\pi^{\frac{D}{2}} \frac{\Gamma(1) \prod_{k=2}^{D-2} \Gamma(\frac{k+1}{2})}{\Gamma(\frac{D}{2}) \prod_{k=1}^{D-3} \Gamma(\frac{k+2}{2})} = 2\pi^{\frac{D}{2}} \frac{\Gamma(1)}{\Gamma(\frac{D}{2})}. \tag{8A.7}$$

The surface of the unit sphere in D dimensions is therefore

$$S_D = \frac{2\pi^{\frac{D}{2}}}{\Gamma(\frac{D}{2})}. \tag{8A.8}$$

Integration of a rotationally invariant integrand gives

$$\begin{aligned} I(D) &= \int d^D p \, f(\mathbf{p}^2) = 2\pi \prod_{k=1}^{D-2} \int_0^\pi \sin^k \vartheta_k \, d\vartheta_k \int_0^\infty dp \, p^{D-1} f(p^2) \\ &= S_D \int_0^\infty dp \, p^{D-1} f(p^2). \end{aligned} \tag{8A.9}$$

These calculations hold initially for integer values of the dimension D. By an analytic continuation of S_D in D, the final results make sense also for continuous values of D. As an example, we extrapolate analytically a D-dimensional Gaussian integral to any complex D. Using Eq. (8A.9) we find

$$\int \frac{d^D p}{(2\pi)^D} e^{-\mathbf{p}^2} = \frac{S_D}{(2\pi)^D} \int_0^\infty dp \, p^{D-1} e^{-p^2} = \frac{S_D}{2(2\pi)^D} \int_0^\infty dx \, x^{D/2-1} e^{-x} \tag{8A.10}$$

$$= \frac{S_D}{2(2\pi)^D} \Gamma(D/2) = \left(\frac{1}{4\pi}\right)^{D/2}. \tag{8A.11}$$

Appendix 8B More on Dimensional Regularization of 't Hooft and Veltman

Dimensional Regularization of 't Hooft and Veltman by Splitting the Integration Measure The splitting of the integration measure may be exemplified using the integral

$$I(D) = \int d^D p \, \frac{1}{\mathbf{p}^2 + m^2} = S_D \int_0^\infty dp \, p^{D-1} \frac{1}{p^2 + m^2}, \tag{8B.1}$$

where S_D is the surface of a unit sphere in D dimensions calculated in Eq. (8A.8). The integral in (8B.1) is obviously UV-divergent for $D \geq 2$ and IR-divergent for $D \leq 0$.

The idea of 't Hooft and Veltman is to define Feynman integrals in *noninteger dimensions* by splitting the measure of integration into a physical four- and an additional artificial $D - 4$-dimensional one. Let the four-dimensional momentum variable part be $\mathbf{p}_\parallel$, the remaining $D - 4$-dimensional one $\mathbf{p}_\perp$, so that

$$I(D) \;=\; \int d^4 p_\parallel \int d^{D-4} p_\perp \, \frac{1}{p_\parallel^2 + p_\perp^2 + m^2}. \tag{8B.2}$$

The $D - 4$-dimensional integrals can be reduced to a one-dimensional integral by observing that the integrand depends only on the length of the $D - 4$-dimensional integration variable. Proceeding as in Appendix 8A, we perform the angular integration in noninteger dimension which yields the surface of a unit sphere in $D - 4$ dimensions:

$$S_{D-4} = \frac{2\pi^{(D-4)/2}}{\Gamma((D-4)/2)}. \tag{8B.3}$$

The resulting expression for $I(D)$ contains only ordinary integrals:

$$I(D) = S_{D-4} \int d^4 p_\parallel \int_0^\infty dp_\perp \, \frac{p_\perp^{D-4-1}}{p_\parallel^2 + p_\perp^2 + m^2}. \tag{8B.4}$$

This integral does not converge for any D. For $D \geq 2$ it is UV-divergent, whereas for $D \leq 4$ it is IR-divergent. The reason is the artificial IR-divergence generated by splitting the measure of integration. An integral with the original region of convergence is constructed by *partial integration*. For this we rewrite

$$dp_\perp p_\perp^{D-4-1} = \frac{1}{2} dp_\perp^2 p_\perp^{D-6} = dp_\perp^2 \frac{1}{D-4} \frac{d}{dp_\perp^2} (p_\perp^2)^{D/2-2}, \tag{8B.5}$$

and integrate (8B.4) partially over $p_\perp^2$, yielding

$$I(D) = \frac{S_{D-4}}{D-4} \int d^4 p_\parallel \left[\frac{(p_\perp^2)^{D/2-2}}{p_\parallel^2 + p_\perp^2 + m^2} \bigg|_0^\infty + \int_0^\infty dp_\perp^2 \frac{(p_\perp^2)^{D/2-2}}{(p_\parallel^2 + p_\perp^2 + m^2)^2} \right]. \tag{8B.6}$$

The surface term is UV-divergent for $D \geq 2$, and IR-divergent for $D \leq 4$, and is discarded. The remaining integral is UV-divergent for $D \geq 2$ and IR-divergent for $D \leq 2$. Hence there is still no region of convergence. We must repeat the procedure, and arrive with $\Gamma(D/2 - 2)(D/2 - 2)(D/2 - 1) = \Gamma(D/2)$ at the expression

$$I(D) \;=\; \frac{2\pi^{D/2-2}}{\Gamma(D/2)} \int d^4 p_\parallel \int_0^\infty dp_\perp^2 \, \frac{(p_\perp^2)^{D/2-1}}{(p_\parallel^2 + p_\perp^2 + m^2)^3}. \tag{8B.7}$$

This integral is now convergent for $0 < D < 2$. It is the defining integral for the D-dimensional integral in Eq. (8.4) in this interval and can be used as a starting point for an analytic extrapolation to higher D.

For this purpose, 't Hooft and Veltman used their procedure "*partial p*", inserting into the integrand the unit differential operator

$$\frac{1}{D} \left(\frac{\partial p_{\parallel i}}{\partial p_{\parallel i}} + \frac{\partial p_\perp}{\partial p_\perp} \right) = 1, \tag{8B.8}$$

and carrying out a partial integration in the region $0 < D < 2$. This time, the surface term is explicitly zero for $0 < D < 2$. The region of convergence of the remaining integral is extended to $0 < D < 4$, except for a pole at $D = 2$ as a consequence of the UV-divergence. After re-expressing the right-hand side in terms of I, we find

$$I(D) = -3m^2 \frac{4\pi^{D/2-2}}{(D/2-1)\Gamma(D/2)} \int d^4 p_\parallel \int_0^\infty dp_\perp \frac{p_\perp^{D-1}}{(p_\parallel^2 + p_\perp^2 + m^2)^4}. \tag{8B.9}$$

This expression provides us with the desired analytic extrapolation of (8B.7).

The procedure can be repeated to yield a convergent integral in the region $0 < D < 6$:

$$I(D) = 3 \cdot 4\, m^4 \frac{4\pi^{D/2-2}}{(D/2-2)(D/2-1)\Gamma(D/2)} \int d^4 p_\parallel \int_0^\infty dp_\perp \frac{p_\perp^{D-1}}{(p_\parallel^2 + p_\perp^2 + m^2)^5}. \tag{8B.10}$$

The only traces of the original UV-divergence are the poles for $D = 2$ and $D = 4$. The integrations can be performed and the result expressed in terms of Beta functions (8.8). The four-dimensional momentum integration in (8B.10) yields

$$\int d^4 p_\parallel \int_0^\infty dp_\perp \frac{p_\perp^{D-1}}{(\mathbf{p}_\parallel^2 + p_\perp^2 + m^2)^5} = \pi^2 \frac{\Gamma(3)}{\Gamma(5)} \int_0^\infty dp_\perp \frac{p_\perp^{D-1}}{(p_\perp^2 + m^2)^3}, \tag{8B.11}$$

and the remaining one-dimensional integration gives

$$\pi^2 \frac{(m^2)^{D/2-3}}{2} \frac{\Gamma(D/2)\Gamma(3 - D/2)}{\Gamma(5)}. \tag{8B.12}$$

The UV-divergence of the integrations is reflected by the pole of the Gamma function at $D = 6$. After including the prefactor in (8B.10), we obtain

$$I(D) = \int d^D p \frac{1}{\mathbf{p}^2 + m^2} = \pi^{D/2}(m^2)^{D/2-1}\Gamma(1 - D/2). \tag{8B.13}$$

This is the same result as in (8.43) which was reached without splitting the integration measure. The splitting is useful for integrals that depend on external momentum.

Appendix 8C   Parametric Representation of Feynman Integrals

We follow Itzykson and Zuber [15] in the derivation of a proper-time representation of a Feynman integral with $L = I - V + 1$ loops, I internal lines, and V vertices. It is superficially UV-convergent for $\omega(G) = \mathrm{Re}(D)L - 2I < 0$.

An oriented diagram is defined with the help of a so-called *incidence matrix* ϵ_{vl} with $v = 1, \ldots, V$ and $l = 1, \ldots, I$:

$$\epsilon_{vl} = \begin{cases} +1 & \text{if } v \text{ is starting vertex of line } l \\ -1 & \text{if } v \text{ is ending vertex of line } l \\ 0 & \text{if line } l \text{ is not incident on vertex } v \end{cases}. \tag{8C.1}$$

The integral of an n-point diagram G containing no tadpole part is called $I'_G(\mathbf{k}_1, \ldots, \mathbf{k}_n)$. The sum of the external momenta entering the diagram at a vertex v is denoted by $\bar{\mathbf{k}}_v$. Momentum conservation at each vertex is expressed by $\delta^{(D)}(\bar{\mathbf{k}}_v - \sum_l \epsilon_{vl} \mathbf{p}_l)$, where $\mathbf{p}_l$ are the internal momenta. The integral has the following general form:

$$I'_G(\mathbf{k}_1, \ldots, \mathbf{k}_n) = (-\lambda)^V W_G \int \prod_{l=1}^I \frac{d^D p}{(2\pi)^D} \frac{1}{\mathbf{p}_l^2 + m^2} \prod_{v=1}^V (2\pi)^D \delta^{(D)}\left(\bar{\mathbf{k}}_v - \sum_l \epsilon_{vl} \mathbf{p}_l\right), \tag{8C.2}$$

where W_G is the weight factor. Each propagator and the δ-distributions at each vertex are expressed by an integral representation:

$$\frac{1}{\mathbf{p}^2 + m^2} = \int_0^\infty d\tau\, e^{-\tau(\mathbf{p}^2 + m^2)}, \tag{8C.3}$$

$$(2\pi)^D \delta^{(D)}\left(\bar{\mathbf{k}}_v - \sum_l \epsilon_{vl}\mathbf{p}_l\right) = \int d^D y_v\, e^{-i\mathbf{y}_v \cdot (\bar{\mathbf{k}}_v - \sum_l \epsilon_{vl}\mathbf{p}_l)}. \tag{8C.4}$$

Insertion of these two formulas and interchange of the order of integration in Eq. (8C.2) leads to $\mathbf{p}$-integrations of the following form:

$$\int \frac{d^D p_l}{(2\pi)^D} e^{-\tau_l \left(\mathbf{p}_l^2 - \frac{i}{\tau_l}\sum_v \mathbf{y}_v \epsilon_{vl}\mathbf{p}_l\right)} = \int \frac{d^D p_l}{(2\pi)^D} e^{-\tau_l \left(\mathbf{p}_l - \frac{i}{2\tau_l}\sum_v \mathbf{y}_v \epsilon_{vl}\right)^2 - \frac{1}{4\tau_l}\left(\sum_v \mathbf{y}_v \epsilon_{vl}\right)^2}$$

$$= \frac{1}{(4\pi\tau_l)^{D/2}} e^{-\frac{1}{4\tau_l}\left(\sum_v \mathbf{y}_v \epsilon_{vl}\right)^2}. \tag{8C.5}$$

The interchange is justified if $\mathrm{Re}(D)L < 2I$ for the entire diagram as well as for all subdiagrams, or if the integrals are regularized. The complete expression for I'_G becomes

$$I'_G(\mathbf{k}_i) = \frac{(-\lambda)^V W_G}{(4\pi)^{ID/2}} \int \prod_{v=1}^V d^D y_v \int_0^\infty \prod_{l=1}^I \left[d\tau_l \frac{e^{-\tau_l m^2 - \left(\sum_v \mathbf{y}_v \epsilon_{vl}\right)^2/4\tau_l}}{\tau_l^{D/2}} \right] e^{-i\sum_{v=1}^V \mathbf{y}_v \cdot \bar{\mathbf{k}}_v}. \tag{8C.6}$$

After a variable transformation

$$\mathbf{y}_1 = \mathbf{z}_1 + \mathbf{z}_V, \quad \mathbf{y}_2 = \mathbf{z}_2 + \mathbf{z}_V, \quad \ldots, \quad \mathbf{y}_V = \mathbf{z}_V,$$

and using the fact that

$$\mathbf{z}_V \sum_{v=1}^V \epsilon_{vl} = 0,$$

the integration over $\mathbf{z}_V$ yields the factor

$$\int d^D z_V \exp\left(-i\mathbf{z}_V \sum_{v=1}^V \bar{\mathbf{k}}_v\right) = (2\pi)^D \delta^{(D)}\left(\sum_v \bar{\mathbf{k}}_v\right) = (2\pi)^D \delta^{(D)}\left(\sum_i^n \mathbf{k}_i\right).$$

It is useful to change the notation and remove this factor, since the integrals in $\Gamma^{(n)}$ are defined without it. Thus we introduce

$$I'_G = (2\pi)^D \delta^{(D)}\left(\sum_i^n \mathbf{k}_i\right) I_G, \tag{8C.7}$$

and the remaining integral is rewritten as

$$I_G(\mathbf{k}_i) = (-\lambda)^V W_G \int \prod_{v=1}^{V-1} d^D z_v \int_0^\infty \prod_{l=1}^I \left[d\tau_l \frac{e^{-\tau_l m^2 - \left(\sum_{v_1, v_2}^{V-1} \mathbf{z}_{v_1}\epsilon_{v_1 l}\frac{1}{\tau_l}\epsilon_{v_2 l}\mathbf{z}_{v_2}\right)/4}}{(4\pi\tau_l)^{D/2}} \right] e^{-i\sum_{v=1}^{V-1} \mathbf{z}_v \bar{\mathbf{k}}_v}.$$

The matrix

$$[d_G]_{v_1 v_2} = \sum_l \epsilon_{v_1 l}\frac{1}{\tau_l}\epsilon_{v_2 l}$$

is nonsingular and its determinant is found to be [16]

$$\Delta_G \equiv \det[d_G] = \sum_{\text{trees } T} \prod_{l \in T} \frac{1}{\tau_l},$$
(8C.8)

where $\sum_{\text{trees } T}$ denotes a sum over all tree diagrams contained in G. Recall the definition of tree diagrams on page 73 and in Fig. 5.5 on page 73. They consist of a sum over all maximally connected subsets of lines in G which contain no loop but all vertices of G. As each tree has $V-1$ lines, Δ_G is a homogeneous polynomial in τ_l^{-1} of order $V-1$. Integrating over $\mathbf{z}_i$ gives

$$I_G(\mathbf{k}_i) = \frac{(-\lambda)^V W_G}{(4\pi)^{LD/2}} \int_0^\infty \prod_{l=1}^I d\tau_l \frac{e^{-\tau_l m^2 - \sum_{v_1,v_2}^{V-1} \bar{\mathbf{k}}_{v_1}[d^{-1}]_{v_1 v_2} \bar{\mathbf{k}}_{v_2}}}{\prod_{l=1}^I \tau_l^{D/2} \Delta_G(\tau)^{D/2}}.$$

The denominator in the integrand has the final form $[M_G(\tau)]^{D/2}$, where

$$M_G(\tau) = \tau_1, \ldots, \tau_I \times \sum_{\text{trees } T} \prod_{l \in T} \frac{1}{\tau_l} = \sum_{\text{trees } T} \prod_{l \notin T} \tau_l,$$
(8C.9)

which is a homogeneous polynomial of degree $I - (V - 1) = L$.

Let us denote the quadratic form in the exponent of the integrand by $Q_G(\bar{\mathbf{k}}, \tau)$. It is given by a ratio of two homogeneous polynomials in τ of degree $L + 1$ and L respectively:

$$\sum_{v_1,v_2}^{V-1} \bar{\mathbf{k}}_{v_1}[d^{-1}]_{v_1 v_2} \bar{\mathbf{k}}_{v_2} = \frac{1}{M_G} \sum_{\text{cuts } C} \left(\sum_{v \in G_1(C)} \bar{\mathbf{k}}_v \right)^2 \prod_{l \in C} \tau_l \equiv Q_G(\bar{\mathbf{k}}, \tau).$$
(8C.10)

A cut is a subset of lines of G which, if removed, divides G into two parts, say $G_1(C)$ and $G_2(C)$. A tree becomes a cut if one of its lines is taken away. Therefore a cut has $I - (V - 1) + 1 = L + 1$ lines. With these notations the final form reads

$$I_G(\mathbf{k}_i) = \frac{(-\lambda)^V W_G}{(4\pi)^{LD/2}} \int_0^\infty \prod_{l=1}^I d\tau_l \frac{e^{-\tau_l m^2 - Q_G(\bar{\mathbf{k}}, \tau)}}{M_G(\tau)^{D/2}}.$$
(8C.11)

The dependence on D emerges explicitly from the one-dimensional τ-integrations. For this reason, formula (8C.11) has been used to define Feynman integrals in D dimension with complex D [12]. The UV-divergences are found in the $\tau \to 0$ limit.

The superficial divergence of the parametric integral is seen explicitly if the τ-integration is rescaled by a homogeneity parameter σ: $\tau_l \to \sigma \tau_l$, such that

$$Q_G(\bar{\mathbf{k}}, \sigma\tau) = \sigma Q_G(\bar{\mathbf{k}}, \tau),$$
(8C.12)
$$M_G(\sigma\tau) = \sigma^L M_G(\tau).$$
(8C.13)

Inserting the trivial identity

$$1 = \int_0^\infty d\sigma \, \delta\left(\sigma - \sum_l^I \tau_l\right)$$
(8C.14)

into (8C.11) gives then for I_G:

$$I_G(\mathbf{k}_i) = (-\lambda)^V W_G \int_0^\infty \frac{d\sigma}{\sigma} \sigma^{I - DL/2} \int_0^1 \prod_{l=1}^I d\tau_l \frac{e^{-\sigma[Q_G(\bar{\mathbf{k}}, \tau) + m^2 \sum_l \tau_l]}}{[(4\pi)^L M_G(\tau)]^{D/2}} \delta\left(1 - \sum_l \tau_l\right)$$

$$= (-\lambda)^V W_G \Gamma(I - DL/2) \int_0^1 \prod_{l=1}^I d\tau_l \frac{[Q_G(\bar{\mathbf{k}}, \tau) + m^2 \sum_l \tau_l]^{\frac{DL}{2} - I}}{[(4\pi)^L M_G(\tau)]^{D/2}} \delta\left(1 - \sum_l \tau_l\right).$$
(8C.15)

The integration over σ produces the superficial divergence of G. It is convergent at $\sigma = 0$ only if $2I - \text{Re}(D)L = -\omega(G) > 0$. The integral is analytically continued with the Gamma function $\Gamma((2I - DL)/2) = \Gamma(-\omega(G)/2)$, which gives rise to simple poles for $\omega(G) = 0, 2, \ldots$, where the divergences are logarithmic or quadratic. All possible subdivergences come from the remaining τ-integrations, which produce further Gamma functions [13].

We shall exemplify the above formula for a one-loop diagram which has $V = 2$, $I = 2$, $L = 1$ and therefore $I - DL/2 = 2 - D/2$. The momentum conservation is taken care of by a factor $(2\pi)^D \delta^{(D)}(\sum_i^n \mathbf{k}_i)$ to be omitted.

$$
\begin{aligned}
\text{(diagram)} \; &= \; \lambda^2 \frac{3}{2} \int \frac{d^D p}{(2\pi)^D} \frac{1}{(\mathbf{p}^2 + m^2)[(\mathbf{p} - \bar{\mathbf{k}})^2 + m^2]} \\[2mm]
&= \lambda^2 \frac{3}{2} \int \frac{d\sigma}{\sigma} \sigma^{2-D/2} \int_0^1 d\tau_1 d\tau_2 \frac{e^{-\sigma[Q_G(\bar{\mathbf{k}}, \tau_1, \tau_2) + m^2(\tau_1 + \tau_2)]}}{[4\pi M_G(\tau_1, \tau_2)]^{D/2}} \delta(1 - \tau_1 - \tau_2) \\[2mm]
&= \lambda^2 \frac{3}{2} \Gamma(2 - D/2) \int_0^1 d\tau_1 d\tau_2 \frac{[Q_G(\bar{\mathbf{k}}, \tau_1, \tau_2) + m^2(\tau_1 + \tau_2)]^{\frac{D}{2}-2}}{[4\pi M_G(\tau_1, \tau_2)]^{D/2}} \delta(1 - \tau_1 - \tau_2).
\end{aligned}
\tag{8C.16}
$$

This simple diagram contains only two trees, $\mathcal{T}_1$ and $\mathcal{T}_2$. The first consists of line 1, the second of line 2. There is also one cut $\mathcal{C}$, which consists of line 1 and 2, such that $G_1(\mathcal{C})$ contains vertex 1 and $G_2(\mathcal{C})$ contains vertex 2. The relations for the number of lines of a tree and a cut are fulfilled: $I(\mathcal{T}) = V - 1 = 1$ and $I(\mathcal{C}) = L + 1 = 2$. We then find for M_G and Q_G:

$$
M_G \; = \; \prod_{l \notin \mathcal{T}_1} \tau_l + \prod_{l \notin \mathcal{T}_2} \tau_l = \tau_2 + \tau_1,
\tag{8C.17}
$$

$$
Q_G \; = \; \frac{1}{\tau_1 + \tau_2} \bar{\mathbf{k}}^2 \tau_1 \tau_2,
\tag{8C.18}
$$

and the integral becomes

$$
\begin{aligned}
\text{(diagram)} \; &= \; \lambda^2 \frac{\Gamma(2 - D/2)}{(4\pi)^{D/2}} \int_0^1 d\tau_1 d\tau_2 \frac{\left[\bar{\mathbf{k}}^2 \frac{\tau_1 \tau_2}{\tau_1 + \tau_2} + m^2(\tau_1 + \tau_2)\right]^{D/2-2}}{(\tau_1 + \tau_2)^{D/2}} \delta(1 - \tau_1 - \tau_2) \\[2mm]
&= \; \lambda^2 \frac{\Gamma(2 - D/2)}{(4\pi)^{D/2}} \int_0^1 d\tau \left[\bar{\mathbf{k}}^2 \tau(1 - \tau) + m^2\right]^{D/2-2}.
\end{aligned}
\tag{8C.19}
$$

We have reproduced the result in (8.66), which we obtained with Feynman's parameter integral formula.

Appendix 8D Expansion of Gamma Function

The Gamma function is defined by the integral (8.9). This integral has poles for $z = 0$ and negative integers. Partial integration applied to the integral representation of $\Gamma(z+1)$ leads to the identity

$$
\Gamma(z + 1) = z \int_0^\infty dt\, t^{z-1} e^{-t} = z\, \Gamma(z),
\tag{8D.1}
$$

which shows that the Gamma function is the generalization of the factorial to arbitrary complex variables z.

In the context of dimensional regularization, we need a series expansion of the Gamma function near zero or negative integer values of z. Using (8D.1), we see that

$$
\Gamma(2 + \varepsilon) = (1 + \varepsilon)\, \Gamma(1 + \varepsilon) = (\varepsilon + 1)\, \varepsilon\, (\varepsilon - 1)\, \Gamma(-1 + \varepsilon).
\tag{8D.2}
$$

This relates the expansion of $\Gamma(-1+\varepsilon)$ to the expansion of $\Gamma(2+\varepsilon)$ in powers of ε, which reads

$$\Gamma(2+\varepsilon) \;=\; \Gamma(2) + \Gamma'(2)\,\varepsilon + \frac{1}{2}\Gamma''(2)\,\varepsilon^2 + \mathcal{O}(\varepsilon^3) \tag{8D.3}$$

$$=\; 1 + \psi(2)\,\varepsilon + \frac{1}{2}\frac{\Gamma''(2)}{\Gamma(2)}\,\varepsilon^2 + \mathcal{O}(\varepsilon^3). \tag{8D.4}$$

In the second line, we have inserted $\Gamma(2) = 1$ as well as the definition of the Euler Digamma function $\psi(z)$:

$$\psi(z) \equiv \Gamma'(z)/\Gamma(z). \tag{8D.5}$$

From this follows the recurrence relation:

$$\psi(z) = \frac{d}{dz}\log\Gamma(z) = \frac{d\,\log(z-1)}{dz} + \frac{\log\Gamma(z-1)}{dz} = \frac{1}{z-1} + \psi(z-1), \tag{8D.6}$$

which for integer values of z implies

$$\psi(n) = \psi(1) + \sum_{l=1}^{n-1}\frac{1}{l}. \tag{8D.7}$$

Analogously, we find for $\psi'(z)$:

$$\psi'(z) = \frac{-1}{(z-1)^2} + \psi'(z-1), \tag{8D.8}$$

and

$$\psi'(n) = \psi'(1) - \sum_{l=1}^{n-1}\frac{1}{l^2}. \tag{8D.9}$$

The value of $\psi(z)$ at $z = 1$ is equal to the negative of *Euler's constant* γ:

$$\psi(1) = -\gamma = -0.5772156649\ldots. \tag{8D.10}$$

The derivative of $\psi(z)$ has at $z = 1$ the value

$$\psi'(1) = \pi^2/6. \tag{8D.11}$$

Thus we find:

$$\psi(n) \;=\; -\gamma + \sum_{l=1}^{n-1}\frac{1}{l}, \tag{8D.12}$$

$$\psi'(n) \;=\; \frac{\pi^2}{6} - \sum_{l=1}^{n-1}\frac{1}{l^2}. \tag{8D.13}$$

Differentiating Eq. (8D.5), we obtain

$$\psi'(z) = \Gamma''(z)/\Gamma(z) - [\Gamma'(z)]^2/\Gamma(z)^2, \tag{8D.14}$$

and thus an expression for $\Gamma''(z)/\Gamma(z)$ which may be inserted into the expansion Eq. (8D.4) to find

$$\Gamma(2+\varepsilon) = 1 + \psi(2)\varepsilon + \frac{1}{2}\left[\psi'(2) + \psi(2)^2\right]\varepsilon^2 + \mathcal{O}(\varepsilon^3). \tag{8D.15}$$

For $\Gamma(-1+\varepsilon)$, we derive the expansion

$$\Gamma(-1+\varepsilon) = -\frac{\Gamma(2+\varepsilon)}{\varepsilon(1+\varepsilon)(1-\varepsilon)} = -\frac{1}{\varepsilon}\Gamma(2+\varepsilon)\left[1+\varepsilon^2+\mathcal{O}(\varepsilon^3)\right] \tag{8D.16}$$

$$= -\left\{\frac{1}{\varepsilon}+\psi(2)+\varepsilon\left[1+\frac{1}{2}\psi'(2)+\frac{1}{2}\psi(2)^2\right]\right\}+\mathcal{O}(\varepsilon^2). \tag{8D.17}$$

A generalization of this formula to any integer n is obtained using repeatedly the identity (8D.1):

$$\Gamma(-n+\varepsilon) = \frac{\Gamma(n+1+\varepsilon)}{(n+\varepsilon)(n+\varepsilon-1)\dots(\varepsilon+1)\,\varepsilon\,(\varepsilon-1)\dots(\varepsilon-n)} \tag{8D.18}$$

$$= \frac{\Gamma(n+1+\varepsilon)}{(-1)^n\varepsilon\,(1+\varepsilon)(1-\varepsilon)(2+\varepsilon)(2-\varepsilon)\dots(n+\varepsilon)(n-\varepsilon)} \tag{8D.19}$$

$$= \frac{(-1)^n\,\Gamma(n+1+\varepsilon)}{n!^2\,\varepsilon\,(1-\varepsilon^2)\left[1-\left(\frac{\varepsilon}{2}\right)^2\right]\left[1-\left(\frac{\varepsilon}{3}\right)^2\right]\dots\left[1-\left(\frac{\varepsilon}{n}\right)^2\right]} \tag{8D.20}$$

$$= \frac{(-1)^n}{n!^2\,\varepsilon}\,\Gamma(n+1+\varepsilon)\left[1+\varepsilon^2\sum_{j=1}^{n}\frac{1}{j^2}+\mathcal{O}(\varepsilon^4)\right]. \tag{8D.21}$$

Together with Eq. (8D.13), this becomes

$$\Gamma(-n+\varepsilon) = \frac{(-1)^n}{n!^2\,\varepsilon}\,\Gamma(n+1+\varepsilon)\left\{1+\varepsilon^2\left[\frac{\pi^2}{6}-\psi'(n+1)\right]+\mathcal{O}(\varepsilon^4)\right\}. \tag{8D.22}$$

We further need the Taylor expansion of $\Gamma(n+1+\varepsilon)$ for integer n:

$$\Gamma(n+1+\varepsilon) = \Gamma(n+1)\left[1+\varepsilon\,\psi(n+1)+\frac{1}{2}\frac{\Gamma''(n+1)}{\Gamma(n+1)}\varepsilon^2+\mathcal{O}(\varepsilon^3)\right]$$

$$= n!\left\{1+\varepsilon\,\psi(n+1)+\frac{\varepsilon^2}{2}\left[\psi'(n+1)+\psi(n+1)^2\right]+\mathcal{O}(\varepsilon^3)\right\}. \tag{8D.23}$$

Using (8D.21) and (8D.23), the expansion for $\Gamma(-n+\varepsilon)$ takes the form:

$$\Gamma(-n+\varepsilon) = \frac{(-1)^n}{n!}\left\{\frac{1}{\varepsilon}+\psi(n+1)+\frac{\varepsilon}{2}\left[\frac{\pi^2}{3}+\psi(n+1)^2-\psi'(n+1)\right]+\mathcal{O}(\varepsilon^2)\right\}. \tag{8D.24}$$

Notes and References

For the dimensional regularization see also
M.E. Fisher and K.G. Wilson, Phys. Rev. Lett. **28**, 240 (1972);
K.G. Wilson, Phys. Rev. D **7**, 2911 (1973);
K.G. Wilson and J. Kogut, Phys. Rep. C **12**, 77 (1974).
Many textbooks on quantum field theory treat dimensional regularization. An explicit analytic interpolation is constructed by
J.C. Collins, *Renormalization*, Cambridge University Press, Cambridge, 1984.
Explicit calculations for diagrams of the ϕ^4- theory, for instance the sunset diagram (8.73), are found in Ref. [14].
The development of parametric representations of Feynman integrals in the textbook [15] goes

back to
R.J. Eden, P.V. Landshoff, D.I. Olive, and J.C. Polkinghorne, *The Analytic S-Matrix*, Cambridge University Press, 1966.
See also
J.D. Bjorken and S.D. Drell, *Relativistic Quantum Fields*, McGraw-Hill, New York, 1965.

The individual citations in the text refer to:

[1] W. Pauli and F. Villars, Rev. Mod. Phys. **21**, 434 (1949).

[2] E.R. Speer, *Generalized Feynman Amplitudes*, Princeton University, Princeton, 1969.

[3] G.'t Hooft and M. Veltman, Nucl. Phys. B **44**, 189 (1972).

[4] C.G. Bollini, J.J. Giambiagi, Nuovo Cimento B **12**, 20 (1972).

[5] G.'t Hooft, Nucl. Phys. B **61**, 455 (1973).

[6] M.E. Fisher and K.G. Wilson, Phys. Rev. Lett. **28**, 240 (1972);
K.G. Wilson, Phys. Rev. D **7**, 2911 (1973).

[7] K.G. Wilson and J. Kogut, Phys. Rep. C **12**, 77 (1974).

[8] J.C. Collins, *Renormalization*, Cambridge University Press, Cambridge, 1984.

[9] G. Leibbrandt, Rev. Mod. Phys. **74**, 843 (1975).

[10] See Vol. I of E. Goursat and E.R. Hedrick, *A Course in Mathematical Analysis*, Ginn and Co., Boston, 1904.

[11] G. Cicuta and E. Montaldi, Nuovo Cimento Lett. **4**, 329 (1972);
P. Butera, G. Cicuta, and E. Montaldi, Nuovo Cimento A **19**, 513 (1974).

[12] J.F. Ashmore, Nuovo Cimento Lett. **4**, 289 (1972); Commun. Math. Phys. **29**, 177 (1973);

[13] E.R. Speer, J. Math. Phys. **15**, 1 (1974).

[14] P. Ramond, *Field Theory, A Modern Primer*, Benjamin/Cummings Pub. Comp., 1981.

[15] C. Itzykson and J.-B. Zuber, *Quantum Field Theory*, McGraw-Hill, New York, 1980.

[16] N. Nakanishi, *Graph Theory and Feynman Integrals*, Gordon and Breach, New York, 1971.

9

Renormalization

All Feynman integrals discussed in Chapter 8 are infinite in four dimensions. After dimensional regularization in $4 - \varepsilon$ dimensions, they diverge in a specific way for $\varepsilon \to 0$. For an increasing number of loops L, the Feynman integrals possess singularities of the type $1/\varepsilon^i$ $(i = 1, \ldots, L)$. These divergences turn out to contain all the information on the critical exponents of the theory in $4 - \varepsilon$ dimensions.

If we want to find these divergences, we do not have to calculate the full Feynman integral. Expanding a Feynman integral of any 1PI diagram into a power series in the external momenta, we observe the following general property: except for the lowest expansion coefficient in the four-point function, and the lowest two coefficients in the two-point function, all higher expansion coefficients are convergent for $\varepsilon \to 0$. The lowest expansion coefficient is a constant. Due to the rotational invariance of the theory, the first is proportional to $\mathbf{q}^2$. Precisely such terms are found in a perturbation expansion of a theory with a modified energy functional $E[\phi]$ in Eq. (2.1), which contains additional terms of the form $\int d^D x \, \phi^4(x)$, $\int d^D x \, \phi^2$, and $\int d^D x \, (\partial \phi)^2$, respectively. These terms are of the same form as those in the original energy functional, and it is this property which makes the theory *renormalizable*. Indeed, as anticipated in Section 7.4, finite observables can be obtained by multiplying, in each correlation function, the fields, the mass, and the coupling constant by compensating factors, the renormalization constants Z_ϕ, Z_{m^2}, and Z_g. If we use dimensional regularization in evaluating the Feynman integrals, we shall be able to give these factors the generic form

$$Z = 1 + \sum_{k=1}^{L} g^k \sum_{i=1}^{k} \frac{c_i^k}{\varepsilon^i}, \tag{9.1}$$

in which the coefficients a_k and c_i^k are pure c-numbers. The renormalization constants convert the initial objects in the energy functional, the *bare fields*, *bare mass*, and *bare coupling constant* introduced in Eqs. (7.56)–(7.58), into finite *renormalized fields*, *renormalized mass*, and *renormalized coupling constant*. The compensation takes place order by order in g. The expansion coefficients c_i are determined for each order to cancel the above divergences. If done consistently, all observables become finite in the limit $\varepsilon \to 0$ [1]. In Sections 9.2 and 9.3, we shall demonstrate in detail how this works up to second order in g.

The renormalization procedure may be performed basically in two different ways, which differ in their emphasis on the bare versus the renormalized quantities in the energy functional. Either way will be illustrated up to two loops in Sections 9.2 and 9.3. The more efficient method works with renormalized quantities. Starting out with an energy functional containing immediately the renormalized field, mass, and coupling strength, one determines at each order in g certain divergent counterterms to be added to the field energy to remove the divergences. Completely analogous to this *counterterm method* is the *recursive subtraction method* developed by Bogoliubov and Parasiuk [2]. It proceeds diagram by diagram, a fact which is essential for performing calculations up to five loops. The equivalence between the two methods is nontrivial because of the multiplicities of the diagrams and their symmetry factors. We shall study this equivalence up to two loops in Subsection 9.3.3 as an introduction to the recursive subtraction method which will be the main topic of Chapter 11.

The first conjecture on the renormalizability of quantum field theories was put forward by Dyson [3], stimulating Weinberg [4] to prove an important convergence theorem by which the renormalization program was completed. The recursive procedure of Bogoliubov and Parasiuk gave an independent proof and opened the way to the practical feasibility of higher-order calculations. An error in their work was corrected by K. Hepp [5], and for this reason the proof of the renormalizability for a large class of field theories is commonly referred to as the *H-theorem*.

9.1 Superficial Degree of Divergence

In order to localize the UV-divergence of a diagram, naive power counting is used. According to the Feynman rules in Subsection 4.1.2, a Feynman integral I_G of a diagram G with p vertices contains one integration per loop. A diagram with I internal lines contains

$$L = I - p + 1 \tag{9.2}$$

loop integrations and thus $DL = D(I - p + 1)$ powers of momentum in the numerator. Each of the I internal lines is associated with a propagator, thus contributing $2I$ powers of momentum in the denominator. Thus there are altogether

$$\omega(G) \;=\; DL - 2I \;=\; (D-2)I + D - Dp \tag{9.3}$$

powers of momentum in a Feynman integral. The behavior of the integral at large momenta can be characterized by rescaling all internal momenta as $\mathbf{p} \to \lambda\mathbf{p}$, and observing a power behavior

$$I_G \propto \lambda^{\omega(G)} \quad \text{for} \quad \lambda \to \infty. \tag{9.4}$$

The power $\omega(G)$ is called the *superficial degree of divergence* of the diagram G. For $\omega(G) \geq 0$, a diagram G is said to be *superficially divergent*. For $\omega(G) = 0, 2, \ldots$, the superficial divergence of a diagram is *logarithmic, quadratic, ...* , respectively (see page 103). The superficial divergence arises from regions in momentum space of the Feynman integral where all loop momenta become simultaneously large.

A diagram is said to have *subdivergences* if it contains a superficially divergent *subdiagram*, i.e., a subdiagram γ with $\omega(\gamma) \geq 0$. A subdiagram is any subset of lines and vertices of G which form a ϕ^4-diagram of lower order in the perturbation expansion. Subdivergences come from regions in momentum space where the loop momenta of subdiagrams γ become large. If a diagram G has no subdivergences but if $\omega(G) \geq 0$, the superficial divergence is the only divergence of the integral, and the associated Feynman integral is convergent if one of the loop integrals is omitted, which corresponds to cutting one of the lines. A negative superficial degree of divergence $\omega(G)$, on the other hand, implies convergence only if no subdivergences are present. Some examples are shown in Fig. 9.1.

A Feynman diagram G is absolutely convergent if the superficial degree of divergence $\omega(G)$ is negative, and if the superficial degrees of divergence $\omega(\gamma)$ of *all* subdiagrams γ are negative as well. This is part of the famous *power counting theorem* of Dyson, whose proof was completed by Weinberg [4]. This theorem is also called the *Weinberg-Dyson convergence theorem*. A more elementary proof was given later by Hahn and Zimmermann [6]. It will not be repeated here since it can be found in standard textbooks [7, 8]. An essential part of the theorem is

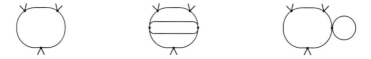

FIGURE 9.1 Three superficially convergent diagrams with $\omega(G) = -2$. The first has no subdivergences. Such a diagram is referred to as a *skeleton diagram*. The second diagram has a logarithmically divergent subdiagram $\gamma = $ ⋈ $[\omega(\gamma) = 0]$. The third diagram has a quadratically divergent subdiagram $\gamma = $ ◯ $[\omega(\gamma) = 2]$.

the elimination of possible extra *overlapping divergences*, which can in principle occur in sets of subdiagrams which have common loop momenta, as shown in Fig. 9.2. In Appendix 9A, the content of the theorem is illustrated by showing explicitly, in a diagram without subdivergences, that no extra divergences are created by overlapping integrations, as stated by the above theorem. All divergences come exclusively from superficial divergences of subdiagrams, and from the superficial divergence of the final integral, but not from overlapping divergences.

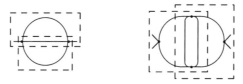

FIGURE 9.2 Two examples for overlapping divergences in ϕ^4-theory. The overlapping subdiagrams are enclosed by dashed boxes.

Historically, overlapping divergences were an obstacle to proving renormalizability of quantum electrodynamics. For the electron self-energy, the problem was solved with the help of the so-called *Ward identity* [9], which expresses the electron self-energy in terms of the vertex function, thereby eliminating all overlapping divergences. The convergence theorem is fundamental to renormalization theory since it enables us to replace all subdivergences by finite subtracted expressions.

A special outcome of the convergence theorem is that, after the subtraction of the divergences, a subdiagram γ behaves as a function of its external momentum like $\gamma(\lambda \mathbf{p}) = \lambda^{\omega(\gamma)} \log^k \lambda$ for $\lambda \to \infty$ with any k. Therefore, after the replacement of the superficially divergent subdiagrams by the corresponding finite subtracted expressions, power counting tells us that any superficially convergent diagram becomes finite. As an example, consider the diagram in Fig. 9.1. If the superficial divergences are subtracted from the subdiagram, it will depend on its external momentum like an ordinary vertex or like an ordinary propagator.

For ϕ^4-theories, the number of the internal lines I in a Feynman diagram may be expressed in terms of the number of vertices p and the number of external lines n as:

$$I = 2p - \frac{n}{2}. \tag{9.5}$$

The superficial degree of divergence of the associated integral becomes therefore

$$\omega(G) = D + n(1 - D/2) + p(D - 4). \tag{9.6}$$

In four dimensions, this simplifies to

$$\omega(G) = 4 - n, \tag{9.7}$$

implying that, in four dimensions, only two- and four-point 1PI diagrams are superficially divergent. Thus the only possible divergent subintegrations are those of two- and four-point subdiagrams. If the integrals of the two- and four-point functions are made finite by some mathematical procedure, any n-point function will be finite, as we know from the convergence theorem. A theory with these properties is said to be *renormalizable*. Hence the ϕ^4-theory is renormalizable in four dimensions.

In three and two dimensions, formula (9.6) yields $\omega(G) = 3 - n/2 - p$ and $\omega(G) = 2 - 2p$, respectively, implying that only a few low-order diagrams possess divergences. A theory with this property is said to be *superrenormalizable*.

In more than four dimensions, the last term $p(D - 4)$ in formula (9.6) is positive, implying that new divergences appear at each higher order in perturbation theory. This property makes the theory *nonrenormalizable*. For a theory with an arbitrary power r of the field in the interaction ϕ^r, there exists the *upper critical dimension*

$$D_c = \frac{r}{r/2 - 1} \tag{9.8}$$

(recall the definition on page 20). For $D > D_c$, the superficial degree of divergence $\omega(G)$ becomes independent of the number p of vertices. In the case of ϕ^4-theory, the upper critical dimension is $D_c = 4$.

The smaller number of the UV-divergences in a superrenormalizable theory for $D < D_c$ makes the Feynman integrals more divergent in the infrared. This is seen as follows. If we let masses and momenta in a Feynman integral go to zero by rescaling them by a factor λ and letting $\lambda \to 0$, we observe a power behavior of the final integrand like $\lambda^{\omega(G)}$. The leading IR-behavior for zero mass is therefore determined by the same power of λ which governs the UV-behavior. The important difference between the two divergences is that the worst UV-divergence is given by the largest $\omega(G) > 0$, whereas the worst IR-divergence is associated with the smallest $\omega(G) < 0$. A nonrenormalizable ϕ^4-theory in $D > 4$ dimensions has, according to Eq. (9.6), the smallest $\omega(G)$ for $p = 0$, which is the case of a free theory, where the critical behavior is mean-field like.

For a superrenormalizable theory, the IR-divergences become worse for increasing order in perturbation theory. In the perturbation expansions, these divergences pile up to give rise to powers of masses in the critical regime. This way of deriving critical exponents will be discussed in detail in Chapters 20 and 21.

Let us end this Section by noting that some authors no longer consider nonrenormalizability as a serious defect of fundamental quantum field theories (even though they did in their own earlier work). It is also possible to give simple calculation rules for extracting experimentally observable properties from such theories [10].

9.2 Normalization Conditions

In a renormalizable theory, Feynman integrals have superficial divergences, in any order of perturbation theory. On account of the power counting theorem, however, these divergences are limited. All superficial divergences are contained in the first coefficients of a Taylor series expansion of self-energy and vertex function around some chosen *normalization point* in the external momentum space. A differentiation with respect to the external momenta lowers the degree of divergence so that the higher coefficients do not contain superficial divergences. After removing all subdivergences of a diagram G, its superficial divergence is contained in the first coefficients of the expansion.

Discussing now the divergences of the theory, we must now clearly distinguish whether we are dealing with *bare or renormalized quantities*. In the first method of renormalizing a field theory, we perform the perturbation starting from the bare energy functional (8.1) with the bare interaction (8.2). On the basis of power counting, we may verify that the superficial divergences are contained in $\Sigma_B(0)$, $\Sigma'_B(0)$, and $\bar{\Gamma}_B^{(4)}(0)$. Here and in what follows we denote the derivative with respect to $\mathbf{k}^2$ by a prime:

$$\Sigma'_B(\mathbf{k}) \equiv \frac{\partial}{\partial \mathbf{k}^2} \Sigma_B(\mathbf{k}). \tag{9.9}$$

The Feynman integrals in $\Sigma_B(0)$ are quadratically divergent; those in $\Sigma'_B(0)$ and $\bar{\Gamma}_B^{(4)}(0)$ are logarithmically divergent. We separate the divergent parts from $\Sigma_B(\mathbf{k})$ and $\bar{\Gamma}_B^{(4)}(\mathbf{k}_i)$ by expanding these quantities in a power series. The superficially convergent parts are contained in the remainder, labeled by a subscript sc:

$$\Sigma_B(\mathbf{k}) = \Sigma_B(0) + \Sigma'_B(0)\mathbf{k}^2 + \Sigma_{Bsc}(\mathbf{k}), \tag{9.10}$$
$$\bar{\Gamma}_B^{(4)}(\mathbf{k}_i) = \bar{\Gamma}_B^{(4)}(0) + \bar{\Gamma}_{Bsc}^{(4)}(\mathbf{k}_i). \tag{9.11}$$

The diagrammatic expansion of these quantities up to two loops was given in Eqs. (8.53) and (8.54) [recalling that $\Sigma_B(\mathbf{k}) = \mathbf{k}^2 + m_B^2 - \bar{\Gamma}(\mathbf{k})$]. The analytic expressions associated with the individual diagrams will be denoted as follows:

$$\Sigma(\mathbf{k}) = \frac{1}{2}\,\text{⊶}\, + \frac{1}{4}\,\text{⊗}\, + \frac{1}{6}\,\text{⬡}\, \equiv -\frac{g_B}{2}Q_1(m_B) + \frac{g_B^2}{4}Q_1(m_B)Q_2(m_B) + \frac{g_B^2}{6}Q_3(\mathbf{k}, m_B),$$

$$\bar{\Gamma}_B^{(4)}(\mathbf{k}_i) = -\,\text{×}\, - \frac{3}{2}\,\text{⋈}\, \equiv g_B - \frac{g_B^2}{2}[L_1(\mathbf{k}_1 + \mathbf{k}_2, m_B) + 2\text{ perm}],$$

$$\tag{9.12}$$

where the notation "2 perm" indicates that the function $L_1(\mathbf{k}_1 + \mathbf{k}_2, m_B)$ contributes also with the permuted arguments $\mathbf{k}_1 + \mathbf{k}_3$ and $\mathbf{k}_1 + \mathbf{k}_4$. The quadratically divergent integrals in $\Sigma_B(\mathbf{k})$ are called Q_i and the logarithmically divergent ones in $\bar{\Gamma}_B^{(4)}(\mathbf{k}_i)$ are called L_i. The following expressions collect all superficial divergences:

$$\Sigma_B(0) = -\frac{g_B}{2}Q_1(m_B) + \frac{g_B^2}{4}Q_1(m_B)Q_2(m_B) + \frac{g_B^2}{6}Q_3(0, m_B) + \mathcal{O}(g_B^3),$$

$$\Sigma'_B(0) = \frac{g_B^2}{6}Q_3'(0, m_B) + \mathcal{O}(g_B^3),$$

$$\bar{\Gamma}_B^{(4)}(0) = g_B - g_B^2\frac{3}{2}L_1(0, m_B) + \mathcal{O}(g_B^3).$$

At this point, the associated Feynman integrals may be made finite either by introducing a momentum cutoff, or by dimensional regularization. The remainders $\Sigma_{Bsc}(\mathbf{k})$ and $\bar{\Gamma}_{Bsc}^{(4)}(\mathbf{k}_i)$ are only superficially convergent and not, in general, finite for $\Lambda \to \infty$ or $\varepsilon \to 0$, due to the possible presence of subdivergences. These appear in second- and higher-order diagrams. We shall see immediately how these are removed, leading ultimately to finite physical results. The number of superficially divergent terms in the expansions (9.10) and (9.11) corresponds to the number of parameters of the theory. Together with the simple momentum dependence of the divergent terms, this fact makes it possible to absorb them into coupling constant, mass, and field normalization.

Let us explain this procedure in detail for the proper two-point vertex function $\bar{\Gamma}_B^{(2)}(\mathbf{k})$, which reads, according to Eq. (9.10),

$$
\begin{aligned}
\bar{\Gamma}_B^{(2)}(\mathbf{k}) &= \mathbf{k}^2 + m_B^2 - \Sigma_B(0) - \Sigma'_B(0)\mathbf{k}^2 - \Sigma_{Bsc}(\mathbf{k}) \\
&= \mathbf{k}^2[1 - \Sigma'_B(0)] + [m_B^2 - \Sigma_B(0)] - \Sigma_{Bsc}(\mathbf{k}) \\
&= [1 - \Sigma'_B(0)]\left\{ \mathbf{k}^2 + m_B^2 \frac{1 - \Sigma_B(0)/m_B^2}{1 - \Sigma'_B(0)} - \frac{\Sigma_{Bsc}(\mathbf{k})}{1 - \Sigma'_B(0)} \right\}.
\end{aligned}
\tag{9.13}
$$

We may now introduce the renormalized field and mass:

$$
\phi \equiv [1 - \Sigma'_B(0)]^{1/2}\, \phi_B,
\tag{9.14}
$$

$$
m^2 \equiv \frac{1 - \Sigma_B(0)/m_B^2}{1 - \Sigma'_B(0)}\, m_B^2,
\tag{9.15}
$$

and a finite renormalized self-energy

$$
\Sigma(\mathbf{k}) \equiv \frac{\Sigma_{Bsc}(\mathbf{k})}{1 - \Sigma'_B(0)},
\tag{9.16}
$$

which starts out like $\mathcal{O}(\mathbf{k}^4)$ for small $\mathbf{k}$. Then the quantity in curly brackets in Eq. (9.13) constitutes a finite renormalized proper two-point vertex function

$$
\bar{\Gamma}^{(2)}(\mathbf{k}) = \mathbf{k}^2 + m^2 - \Sigma(\mathbf{k}).
\tag{9.17}
$$

The relation between bare and renormalized quantities is

$$
\bar{\Gamma}_B^{(2)}(\mathbf{k}) = [1 - \Sigma'_B(0)]\bar{\Gamma}^{(2)}(\mathbf{k}).
\tag{9.18}
$$

As announced at the beginning of this chapter, and earlier in the discussion of scale invariance in Eqs. (7.56)–(7.58), the proper two-point vertex functions can be made finite via a multiplicative renormalization employing three renormalization constants Z_ϕ, Z_{m^2}, and Z_g, to be called *field* or *wave function*, *mass*, and *coupling renormalization constant*, respectively:

$$
\begin{aligned}
\bar{\Gamma}_B^{(2)}(\mathbf{k}) &= Z_\phi^{-1}\bar{\Gamma}^{(2)}(\mathbf{k}), \\
m_B^2 &= Z_{m^2}Z_\phi^{-1}m^2, \\
g_B &= Z_g Z_\phi^{-2} g.
\end{aligned}
\tag{9.19}
\tag{9.20}
\tag{9.21}
$$

Comparing the first two equations with (9.14) and (9.15) yields

$$
Z_\phi = [1 - \Sigma'_B(0)]^{-1}, \qquad Z_{m^2} = [1 - \Sigma_B(0)/m_B^2]^{-1}.
\tag{9.22}
$$

The renormalized quantity $\bar{\Gamma}^{(2)}(\mathbf{k})$ satisfies the equations

$$
\bar{\Gamma}^{(2)}(0) = m^2,
\tag{9.23}
$$

$$
\frac{\partial}{\partial \mathbf{k}^2}\bar{\Gamma}^{(2)}(0) = 1.
\tag{9.24}
$$

The inverse of $\bar{\Gamma}^{(2)}(\mathbf{k})$ is the renormalized propagator of the renormalized field ϕ:

$$
G(\mathbf{k}) = \frac{1}{\bar{\Gamma}^{(2)}(\mathbf{k})} = \frac{1}{\mathbf{k}^2 + m^2 - \Sigma(\mathbf{k})},
\tag{9.25}
$$

which is equal to the Fourier transformation of the correlation function of the renormalized fields defined in (9.14):

$$(2\pi)^D \, \delta^{(D)}(\mathbf{k} + \mathbf{k}')G(\mathbf{k}) = \int d^D x \, d^D x' \, e^{-i\,\mathbf{k}'\cdot\mathbf{x}'-i\,\mathbf{k}\cdot\mathbf{x}} \langle \phi(\mathbf{x})\phi(\mathbf{x}')\rangle. \tag{9.26}$$

The renormalized field is related to the bare field via the *wave function renormalization constant* Z_ϕ in Eq. (9.22):

$$\phi = Z_\phi^{-1/2}\phi_B. \tag{9.27}$$

As we shall see, the same field renormalization makes all n-point functions finite. In the present notation, the Green functions introduced in Eq. (2.10), but expressed in terms of the bare fields corresponding to the energy functionals (8.1) and (8.2), are the *bare Green functions*

$$G_B^{(n)}(\mathbf{x}_1,\ldots,\mathbf{x}_n) \;\equiv\; \langle \phi_B(\mathbf{x}_1)\cdots\phi_B(\mathbf{x}_n)\rangle. \tag{9.28}$$

In quantum field theories formulated in a continuous spacetime [11], physical observations are described with the help of renormalized Green functions, which are the correlation functions of the renormalized fields $\phi(\mathbf{x})$ in Eq. (9.71):

$$G^{(n)}(\mathbf{x}_1,\ldots,\mathbf{x}_n) \equiv \langle \phi(\mathbf{x}_1)\cdots\phi(\mathbf{x}_n)\rangle. \tag{9.29}$$

The two are related by a multiplicative renormalization:

$$G_B^{(n)}(\mathbf{x}_1,\ldots,\mathbf{x}_n) = Z_\phi^{n/2}G^{(n)}(\mathbf{x}_1,\ldots,\mathbf{x}_n). \tag{9.30}$$

The same relation holds between unrenormalized and renormalized Green functions in momentum space.

Remembering the relation (4.21) between correlation functions and proper vertex functions, and its generalization to an arbitrary number n, all proper vertex functions can be made finite by the multiplicative factors inverse to (9.30):

$$\bar{\Gamma}_B^{(n)}(\mathbf{k}_i) = Z_\phi^{-n/2}\bar{\Gamma}^{(n)}(\mathbf{k}_i). \tag{9.31}$$

This renormalization will be performed in detail for the four-point vertex function below, where we shall obtain a finite renormalized quantity $\bar{\Gamma}_B^{(4)}(\mathbf{k}_i)$ from the equation

$$\bar{\Gamma}_B^{(4)}(\mathbf{k}_i) = Z_\phi^{-2}\bar{\Gamma}^{(4)}(\mathbf{k}_i). \tag{9.32}$$

The finite value of $\bar{\Gamma}^{(4)}(\mathbf{k}_i)$ at $\mathbf{k}_i = 0$ will be defined as the renormalized coupling constant g:

$$\bar{\Gamma}^{(4)}(\mathbf{0}) = g. \tag{9.33}$$

The relation between the coupling constants g_B and g can be written precisely in the form (9.21), from which we identify the renormalization constant Z_g. Equations (9.23), (9.24), and (9.33) are called *normalization conditions*. More explicitly, we may write the relation (9.31) as follows:

$$\bar{\Gamma}_B^{(n)}(\mathbf{k}, m_B, g_B, \Lambda) = Z_\phi^{-n/2}\bar{\Gamma}^{(n)}(\mathbf{k}, m, g), \tag{9.34}$$

where we have added mass, coupling constant, and cutoff to the list of arguments of the proper vertex functions. The powers of the renormalization constant Z_ϕ determined by the renormalization of the two-point function absorbs all divergences that remain after having renormalized mass and coupling constant.

The presence of subdivergences complicates the renormalization procedure. Fortunately, the remaining divergences contained in the superficially convergent remainders $\Sigma_{Bsc}(\mathbf{k})$ and $\bar{\Gamma}^{(4)}_{Bsc}(\mathbf{k}_i)$ of Eqs. (9.10) and (9.11) are all removed when re-expressing the bare mass m_B and coupling constant g_B in terms of the renormalized quantities m and g. The renormalization constants of these quantities remove precisely all subdivergences. The normalization conditions (9.23), (9.24), and (9.33) may be used successively in each order of perturbation theory to calculate the renormalized quantities $\bar{\Gamma}^{(2)}(\mathbf{k})$ and $\bar{\Gamma}^{(4)}(\mathbf{k}_i)$, and from these the renormalized mass m and coupling constant g. At each order, the original variables in the energy functional can be re-expressed in terms of the renormalized ones. In this process, we eliminate order by order all subdivergences in $\Sigma_B(\mathbf{k})$, which therefore become finite for $\Lambda \to \infty$ or $\varepsilon \to 0$. We shall now calculate these finite expressions for one- and two-loop diagrams.

9.2.1   One-Loop Mass Renormalization

The one-loop approximation to $\bar{\Gamma}^{(2)}_B(\mathbf{k})$ is

$$\bar{\Gamma}^{(2)}_B(\mathbf{k}) = \mathbf{k}^2 + m_B^2 + \frac{g_B}{2} Q_1(m_B). \tag{9.35}$$

With condition (9.23), we find the (preliminary) renormalized mass

$$m_1^2 = m_B^2 + \frac{g_B}{2} Q_1(m_B). \tag{9.36}$$

Inverting this equation gives $m_B(m_1)$:

$$m_B^2 = m_1^2 - \frac{g_B}{2} Q_1(m_1), \tag{9.37}$$

where the first-order result $m_1^2 = m_B^2$ is inserted into the argument of Q_1, since the error committed in this way is of the order g_B^2. Reexpressing $\bar{\Gamma}^{(2)}_B$ in terms of the renormalized mass yields a finite expression up to the first order in g_B:

$$\bar{\Gamma}^{(2)}_B(\mathbf{k}) = \mathbf{k}^2 + m_1^2 + \mathcal{O}(g_B^2). \tag{9.38}$$

The bare mass m_B is now a function of g_B, m_1, and the cutoff Λ; if ε-regularization is used, it contains a pole term $1/\varepsilon$.

9.2.2   One-Loop Coupling Constant Renormalization

Consider now the one-loop expression for the proper four-point vertex function in Eq. (9.12). In $L_1(\mathbf{k}_1 + \mathbf{k}_2, m_B)$, the mass m_B can again be replaced by m_1 without creating additional terms of order g_B^2. So far there are no subdivergences, and the only divergence occurs in $L_1(\mathbf{0}, m_B)$. A regular expression for $\bar{\Gamma}^{(4)}(\mathbf{k}_i)$ is therefore found by defining

$$g_1 = g_B - \frac{3}{2} g_B^2 L_1(\mathbf{0}, m_1), \tag{9.39}$$

which is inverted to

$$g_B = g_1 + \frac{3}{2} g_1^2 L_1(\mathbf{0}, m_1). \tag{9.40}$$

The error committed in substituting g_B by g_1 in front of L_1 is of the order g_B^3. Expressing the four-point function in terms of g_1, we obtain

$$\bar{\Gamma}^{(4)}_B(\mathbf{k}_i) = g_1 - \frac{g_1^2}{2} [L_1(\mathbf{k}_1 + \mathbf{k}_2, m_1) + 2 \text{ perm} - 3 L_1(\mathbf{0}, m_1)] + \mathcal{O}(g_1^3). \tag{9.41}$$

The subtracted expressions $L_1(\mathbf{k}, m_1) - L_1(\mathbf{0}, m_1)$ are finite for $\Lambda \to \infty$ or $\varepsilon \to 0$. This is seen explicitly by carrying out the subtraction in the integrand for each of the momenta permutations:

$$\frac{1}{(\mathbf{p}^2 + m_1^2)[(\mathbf{p} - \mathbf{k})^2 + m_1^2]} - \frac{1}{(\mathbf{p}^2 + m_1^2)^2} = \frac{-\mathbf{k}^2 + 2\mathbf{pk}}{(\mathbf{p}^2 + m_1^2)^2[(\mathbf{p} - \mathbf{k})^2 + m_1^2]}, \qquad (9.42)$$

which lowers the degree of superficial divergence to a sum of Feynman integrals with $\omega = -1$ and $\omega = -2$.

Assuming g_B to be chosen such that g_1 is finite, the vertex function (9.41) is finite and can be identified directly with the renormalized quantity $\bar{\Gamma}^{(4)}(\mathbf{k}_i)$. The normalization condition (9.33) implies that the quantity g_1 constitutes the renormalized, finite coupling constant g. Thus, at the one-loop level, all divergences have been removed by a redefinition of the coupling constant and the mass.

At the two-loop level, new divergences will appear, and in particular subdivergences. The latter will, however, automatically disappear by the one-loop renormalization of the mass, whereas the superficial two-loop divergences will change further the renormalization constants of mass and coupling constant, and produce a first contribution to the renormalization constant of the field, which is of the order g_B^2.

9.2.3   Two-Loop Mass and Field Renormalization

The diagrammatical two-loop expansion is given by

$$\bar{\Gamma}_B^{(2)}(\mathbf{k}) = (-)^{-1} - \frac{1}{2}\,\mathcal{Q} - \frac{1}{4}\,\mathcal{8} - \frac{1}{6}\,\ominus$$

$$= \mathbf{k}^2 + m_B^2 + \frac{g_B}{2}Q_1(m_B) - \frac{g_B^2}{4}Q_1(m_B)Q_2(m_B) - \frac{g_B^2}{6}Q_3(\mathbf{k}, m_B), \qquad (9.43)$$

and the second-order renormalized mass m_2 is defined by the normalization condition (9.23), implying that the bare vertex function $\bar{\Gamma}_B^{(2)}(\mathbf{0})$ satisfies

$$\bar{\Gamma}_B^{(2)}(\mathbf{0}) = Z_\phi^{-1} m_2^2, \qquad (9.44)$$

where Z_ϕ is yet to be determined.

Let us express m_B as a function of m_2 and g_1. For this we insert into (9.44) the renormalized mass m_1 of Eq. (9.37), and the renormalized coupling constant g_1 of Eq. (9.40). The term of order g_B then gives rise to additional terms of order g_1^2 which contribute at the two-loop level. Specifically, the one-loop integral is re-expanded as

$$\frac{1}{2}g_B\,Q_1(m_B) = \frac{1}{2}g_B \int \frac{d^D p}{(2\pi)^D} \frac{1}{\mathbf{p}^2 + m_B^2}$$

$$= \frac{1}{2}\left[g_1 + \frac{3}{2}g_1^2\,L_1(\mathbf{0}, m_1)\right]\int \frac{d^D p}{(2\pi)^D}\frac{1}{\mathbf{p}^2 + m_1^2 - \frac{1}{2}g_1 Q_1(m_1)}$$

$$= \frac{1}{2}g_1\,Q_1(m_1) + \frac{1}{4}g_1^2\,Q_1(m_1)Q_2(m_1) + \frac{3}{4}g_1^2\,Q_1(m_1)L_1(\mathbf{0}, m_1) + \mathcal{O}(g_1^3). \quad (9.45)$$

Using the fact that $Q_2(m_1) = L_1(\mathbf{0}, m_1)$, we obtain

$$\frac{1}{2}g_B\,Q_1(m_B) = \frac{1}{2}g_1\,Q_1(m_1) + g_1^2\,Q_1(m_1)Q_2(m_1) + \mathcal{O}(g_1^3). \qquad (9.46)$$

The term of order g_1^2 is a manifestation of the subtraction of the subdivergences by the previous first-order renormalizations of mass and coupling constant. Substituting (9.46) into (9.43), and replacing, in the terms quadratic in g_B, the coupling constant g_B and the mass m_B by g_1 and m_1, respectively, we find from (9.44) the equation for m_B^2 as a function of m_1^2:

$$m_B^2 = Z_\phi^{-1} m_2^2 - \frac{g_1}{2} Q_1(m_1) - \frac{3}{4} g_1^2 Q_1(m_1) Q_2(m_1) + \frac{g_1^2}{6} Q_3(0, m_1) + \mathcal{O}(g_1^3). \tag{9.47}$$

Thus we can rewrite (9.43) as

$$\bar{\Gamma}_B^{(2)}(\mathbf{k}) = \mathbf{k}^2 + Z_\phi^{-1} m_2^2 - \frac{g_2^2}{6} \left[Q_3(\mathbf{k}, m_2) - Q_3(0, m_2) \right] + \mathcal{O}(g_2^3), \tag{9.48}$$

where the arguments m_1 and g_1 on the right-hand side have been replaced by m_2 and the second-order renormalized coupling constant g_2 without committing an error of order g_2^2. The first-order terms have disappeared.

The two-point function in Eq. (9.48) is not yet finite in the limit of an infinite cutoff $\Lambda \to \infty$, or $\varepsilon \to 0$, since the subtracted $Q_3(\mathbf{k}, m_2)$ has an integrand

$$\frac{-\mathbf{k}^2 + 2\mathbf{k}(\mathbf{p} + \mathbf{q})}{[(\mathbf{k} - \mathbf{p} - \mathbf{q})^2 + m_2^2](\mathbf{p}^2 + m_2^2)(\mathbf{q}^2 + m_2^2)[(\mathbf{p} + \mathbf{q})^2 + m_2^2]}, \tag{9.49}$$

leading to divergent integrals over $\mathbf{p}$ and $\mathbf{q}$. Thus we expand $Q_3(\mathbf{k}, m_2)$ into the sum of a quadratically and a logarithmically divergent integral, plus a finite subtracted part

$$Q_3(\mathbf{k}, m_2) = Q_3(0, m_2) + \mathbf{k}^2 Q_3'(0, m_2) + Q_{3\,\mathrm{sc}}(\mathbf{k}, m_2)], \tag{9.50}$$

and rewrite (9.48) as

$$\bar{\Gamma}_B^{(2)}(\mathbf{k}) = \left[1 - \frac{g_2^2}{6} Q_3'(0, m_2)) \right] \mathbf{k}^2 + Z_\phi^{-1} m_2^2 - \frac{g_2^2}{6} Q_{\mathrm{sc}}(\mathbf{k}, m_2) + \mathcal{O}(g_2^3). \tag{9.51}$$

Here is the place where the field renormalization comes in. According to the normalization condition (9.24), the renormalized $\bar{\Gamma}^{(2)}(0)$ is supposed to have a unit coefficient of the $\mathbf{k}^2$ term, such that Eq. (9.31) gives for the unrenormalized expression the condition

$$\bar{\Gamma}_B^{(2)\prime}(\mathbf{k})\Big|_0 = Z_\phi^{-1}. \tag{9.52}$$

Thus we identify the wave function renormalization constant as being

$$Z_\phi^{-1} = \bar{\Gamma}_B^{(2)\prime}(0) = 1 - \frac{g_2^2}{6} Q_3'(0, m_2). \tag{9.53}$$

Inserting this into Eq. (9.47), and using Eq. (9.37) to replace m_B by m_1, we find that the renormalized mass m_2 of order g_2^2 differs from m_1 only by terms of order g_2^2:

$$m_2^2 = m_1^2 + \frac{3}{4} g_2^2 Q_1(m_1) Q_2(m_1) - \frac{g_2^2}{6} Q_3(0, m_1) + \frac{g_2^2}{6} m_1^2 Q_3'(0, m_1) + \mathcal{O}(g_2^3). \tag{9.54}$$

We can therefore rewrite (9.47) directly as a relation between m_B^2 and m_2^2 as follows:

$$m_B^2 = m_2^2 - \frac{g_2}{2} Q_1(m_2) - \frac{3}{4} g_2^2 Q_1(m_2) Q_2(m_2) + \frac{g_2^2}{6} Q_3(0, m_2) - \frac{g_2^2}{6} m_2^2 Q_3'(0, m_2) + \mathcal{O}(g_2^3). \tag{9.55}$$

Multiplying $\bar{\Gamma}_B^{(2)}(\mathbf{k})$ in Eq. (9.51) by Z_ϕ, we find the renormalized two-point vertex function

$$\bar{\Gamma}^{(2)}(\mathbf{k}, m, g) = \mathbf{k}^2 + m_2^2 - \frac{g^2}{6} Q_{3\,\mathrm{sc}}(\mathbf{k}, m) + \mathcal{O}(g^3), \qquad (9.56)$$

where we have omitted the subscripts of the renormalized mass and coupling constant to the second-order m and g. The right-hand side has the properly normalized small-$\mathbf{k}$ expansion with unit coefficients of m^2 and $\mathbf{k}^2$:

$$\bar{\Gamma}^{(2)}(\mathbf{k}, m, g) = \mathbf{k}^2 + m^2 + \mathcal{O}(\mathbf{k}^4). \qquad (9.57)$$

This follows from $Q_{3\,\mathrm{sc}}(\mathbf{k}, m) = \mathcal{O}(\mathbf{k}^4)$. Note that Z_ϕ also renormalizes the coupling constant, but not at the two-loop level.

At this place we must add a few remarks concerning the field-theoretic study of critical phenomena directly at the critical temperature T_c, i.e. for $m^2 = 0$. Then the normalization conditions (9.23)–(9.33) possess, in principle, additional IR-divergences. In $D = 4$ dimensions, however, these IR-divergences happen to be absent in logarithmically divergent integrals if the external momenta are *nonexceptional*. Nonexceptional means that none of the partial sums of external momenta vanishes. This will be explained in more detail in Sections 12.1 and 12.3. For quadratically divergent integrals, superficial IR-divergences are absent altogether in $D = 4$ dimensions. Thus at zero mass, the normalization conditions (9.23)–(9.33) have to be modified by employing nonzero external momenta in the two last equations, requiring instead

$$\bar{\Gamma}^{(2)}(\mathbf{0}, 0, g) = 0, \qquad (9.58)$$

$$\left. \frac{\partial}{\partial \mathbf{k}^2} \bar{\Gamma}^{(2)}(\mathbf{k}, 0, g) \right|_{\mathbf{k}^2 = \kappa^2} = 1, \qquad (9.59)$$

$$\left. \bar{\Gamma}^{(4)}(\mathbf{k}_i, 0, g) \right|_{\mathrm{SP}} = g. \qquad (9.60)$$

The subscript SP denotes the symmetric point, where $\mathbf{k}_i \cdot \mathbf{k}_j = (4\delta_{ij} - 1)\kappa^2/4$. The momenta defined by this condition are always *nonexceptional*. In a more mathematical notation, the nonexceptionality means that $\sum_{i \in I} \mathbf{k}_i \neq 0$ for any subset I of the set of indices $\{1, \ldots, n\}$ of the external momenta $\mathbf{k}_i$, $(i = 1, \ldots, n)$.

The above renormalization procedure can be continued to any order in perturbation theory. Obviously, it will be quite difficult to keep track of all involved terms with the repeated re-expansions in terms of the lower-order renormalized quantities. Fortunately, work can be organized more efficiently by using the method of counterterms which will be explained in the next section, where it is carried out explicitly up to second order in the coupling strength. The procedure will be simplified even further by abandoning the normalization conditions and using the so-called minimal subtraction scheme to derive finite expressions for divergent ones.

9.3   Method of Counterterms and Minimal Subtraction

Depending on the regularization scheme, the Feynman integrals are seen to diverge with Λ^2 or $\log \Lambda$ for $\Lambda \to \infty$, or to have poles in ε for $\varepsilon \to 0$. These divergences can also be removed by working with renormalized fields, mass, and coupling constant right from the beginning. The renormalized quantities can be viewed as functions of the bare quantities and of Λ, or of ε. In this test we shall mainly work with dimensional regularization. The details will be explained in Subsection 9.3.2. An analogous treatment exists of course for a cutoff regularization, but this will not be considered here.

The renormalized theory is defined with the help of a renormalized energy functional

$$E[\phi] \;=\; E_0[\phi] + E_{\text{int}}[\phi], \tag{9.61}$$

with a free part

$$E_0[\phi] \;=\; \int d^D x \left[\frac{1}{2}(\partial\phi)^2 + \frac{1}{2}m^2\phi^2 \right] \tag{9.62}$$

and an interaction part which is extended by additional quadratic terms, the so-called *counterterms*, to

$$E_{\text{int}}[\phi] \;=\; \int d^D x \left[\frac{\mu^\varepsilon g}{4!}\phi^4 + c_\phi \frac{1}{2}(\partial\phi)^2 + c_{m^2}\frac{1}{2}m^2\phi^2 + c_g \frac{\mu^\varepsilon g}{4!}\,\phi^4 \right]. \tag{9.63}$$

The additional terms are of the same type as the original ones, such that we can write

$$E[\phi] \;=\; \int d^D x \left[(1+c_\phi)\frac{1}{2}(\partial\phi)^2 + (1+c_{m^2})\frac{1}{2}m^2\phi^2 + (1+c_g)\,\frac{\mu^\varepsilon g}{4!}\,\phi^4 \right]. \tag{9.64}$$

The counterterms c_ϕ, c_{m^2}, and c_g produce additional vertices in the diagrammatic expansion. In momentum space, these have the form

$$\text{—⚹—} \;=\; (-c_{m^2})\, m^2\,, \tag{9.65}$$

$$\text{—◦—} \;=\; (-c_\phi)\, \mathbf{k}^2\,, \tag{9.66}$$

$$\text{✖} \;=\; (-c_g)\, g\, \mu^\varepsilon\,. \tag{9.67}$$

The definition of the original vertex now includes the mass parameter μ, already introduced in Section 8.3 to make g dimensionless:

$$\text{✕} \;=\; (-g)\, \mu^\varepsilon\,. \tag{9.68}$$

The counterterms c_ϕ, c_{m^2}, and c_g are chosen in such a way that all divergent terms are subtracted and the Green functions are finite for $\varepsilon \to 0$, order by order in perturbation theory.

Now, dimensional analysis of Eq. (9.63) shows that the counterterms are dimensionless. In dimensional regularization, they can therefore only depend on the dimensionless coupling constant g or on dimensionless combinations like m^2/μ^2 or $\mathbf{k}^2/\mu^2$. It turns out that the combination $\mathbf{k}^2/\mu^2$ appears only at intermediate steps as $\log \mathbf{k}^2/\mu^2$. It is crucial for the renormalization program that the nonlocal terms all cancel in the final expressions for the counterterms. This implies that the counterterms c_ϕ, c_{m^2}, and c_g depend only on g, ε, and m^2/μ^2. In the minimal subtraction scheme introduced in the next subsection, the dependence on m^2/μ^2 also disappears. Then, the only dimensional dependence of the counterterm diagrams consists in the factors $\mathbf{k}^2$, m^2 and μ^ε in Eqs. (9.65)–(9.67). The cancellation of the logarithms will be observed explicitly in the calculation up to two loops on page 146.

The calculation of the weight factors proceeds as before, but when counting the identical vertex permutations, the different nature of the vertices has to be taken into account. The vertices with two legs require an extension of the previous rules. They carry a factor $1/2!$ by analogy with the factor $1/4!$ for vertices of degree 4.

The quantities ϕ, m, and g in Eq. (9.63) are the renormalized field, renormalized mass, and renormalized coupling constant. The original form of the theory is recovered by a *multiplicative renormalization*. We define the renormalization constants

$$Z_\phi \equiv 1 + c_\phi, \quad Z_{m^2} \equiv 1 + c_{m^2}, \quad Z_g \equiv 1 + c_g, \tag{9.69}$$

and the energy functional (9.64) becomes

$$E[\phi] = \int d^D x \left[\frac{1}{2} Z_\phi (\partial \phi)^2 + \frac{1}{2} m^2 Z_{m^2} \phi^2 + \frac{\mu^\varepsilon g}{4!} Z_g \phi^4 \right], \tag{9.70}$$

with the ε-dependent coefficients. This energy functional still differs from the original one by the factor Z_ϕ in the gradient term. This may be removed by a renormalization of the field, defining the bare field

$$\phi_B \equiv Z_\phi^{1/2} \phi, \tag{9.71}$$

and bare mass and coupling constant

$$m_B^2 \equiv \frac{Z_{m^2}}{Z_\phi} m^2, \qquad g_B \equiv \frac{Z_g}{Z_\phi^2} \mu^\varepsilon g, \tag{9.72}$$

which brings (9.70) to the form:

$$E[\phi] = E[\phi_B] = \int d^D x \left[\frac{1}{2} (\partial \phi_B)^2 + \frac{1}{2} m_B^2 \phi_B^2 + \frac{g_B}{4!} \phi_B^4 \right]. \tag{9.73}$$

This is precisely the initial energy functional in Eqs. (8.1) and (8.2) [or in Eq. (7.30), recalling that the quantities m, g, ϕ in that functional coincide with the presently used bare objects m_B, g_B, and ϕ_B; the subscript B was introduced afterwards in Section 7.4]. The bare quantities are functions of the renormalized quantities m, g, of the mass scale μ, and of $\varepsilon = 4 - D$.

9.3.1   Minimal Subtraction Scheme

The above normalization conditions (9.23), (9.24), and (9.33), [or (9.58)–(9.60)] can be used in connection with any regularization of the divergent integrals. However, if we decide to employ dimensional regularization, the most practical regularization is based on a simultaneous expansion of all Feynman integrals in powers of ε, followed by a subtraction of terms containing ε-poles. This procedure has an important advantage, especially for a study of the critical region in $D < 4$ dimensions. It can be used not only to remove the ultraviolet divergences of the four-dimensional theory, but also at finite small ε-values at the critical point, i.e. in less than four dimensions at zero mass. This is nontrivial, since for $D < 4$, massless diagrams develop IR-divergences, so that the zero-mass condition, Eq. (9.58), cannot be fulfilled, the left-hand side being infinite if the order of the perturbation expansion is sufficiently high. Dimensional analysis tells us that to nth order in the coupling constant g, the two-point vertex function diverges for small $\mathbf{k}$ like $\bar{\Gamma}^{(2)} \approx \mathbf{k}^2 |\mathbf{k}^{-n\varepsilon}| g^n$, so that the prefactor $\mathbf{k}^2$ will eventually turn into a negative power of $\mathbf{k}^2$. In dimensional regularization, this problem disappears since all quantities are expanded in powers of g and ε. Such an ε-expansion looks like [12]

$$\mathbf{k}^2 |\mathbf{k}|^{-n\varepsilon} = \mathbf{k}^2 \mu^{-n\varepsilon} \exp\left[-n\varepsilon \log(\mathbf{k}^2/\mu^2) \right] = \mathbf{k}^2 \mu^{-n\varepsilon} \left[1 - n\varepsilon \log(\mathbf{k}^2/\mu^2) + \ldots \right], \tag{9.74}$$

and the condition (9.58) is fulfilled to all orders in g and ε. Fortunately, the log k-terms are found to cancel in any renormalization scheme, which is necessary for renormalizability, since they could not be canceled by local counterterms in (9.63).

The normalization conditions (9.23)–(9.33) define counterterms which for $m^2 \neq 0$ depend on the mass. For the critical theory with zero mass, we may use the conditions (9.58)–(9.60). Then the counterterms will depend on the mass parameter κ^2 of the symmetry point.

An enormous simplification comes about by the existence of a regularization procedure in which the counterterms become *independent* of the mass m, except for a trivial overall factor m^2

in c_{m^2}. This is known as *minimal subtraction scheme* (MS). It was invented by G.'t Hooft [13] to renormalize nonabelian gauge theories of weak and electromagnetic interactions. In this scheme, the counterterms acquire the generic form (9.1), in which the coefficients of g^n consist of *pure* pole terms $1/\varepsilon^i$, with no finite parts for $\varepsilon \to 0$ [14]. The coefficients are c-numbers and do not contain the mass m or the mass parameter μ introduced in the process of dimensional regularization. In principle, these masses could have appeared in the form of a power series of the dimensionless ratios m^2/μ^2 or its logarithms. The absence of such logarithms is a highly nontrivial virtue of minimal subtraction.

This absence is the origin for another important property of the renormalization constants in the minimal subtraction scheme. They always have the same expansion in powers of the dimensionless coupling constant g, even if this were initially defined to carry an arbitrary analytic function $f(\varepsilon)$ with $f(0) = 1$ as a factor. In other words, if we were to redefine the coupling constant (8.58) to

$$g\mu^\varepsilon \longrightarrow g\mu^\varepsilon f(\varepsilon) = g\mu^\varepsilon \left(1 + f_1\varepsilon + f_2\varepsilon^2 + \dots\right), \tag{9.75}$$

we would find precisely the same expansions (9.1) for the renormalization constants Z_ϕ, Z_{m^2}, and Z_g in powers of the new g. The reason for this is rather obvious: we can always write $f(\varepsilon) = c^\varepsilon$ and absorb the factor c into the mass parameter μ. Since this mass parameter does not appear in the final expansions (9.1), the redefined mass parameter μc cannot appear there as well. The mechanism for this cancellation will be illustrated once more at the end of this section, up to two loops.

Due to the invariance of the final expansions under a rescaling of $g \to gf(\varepsilon)$, there exists an infinite variety of subtraction schemes which may all be called minimal, depending on the choice of the function $f(\varepsilon)$. They all lead to the same counterterms and renormalization constants. In the strict version of MS, one expands *all* functions of ε in each regularized Feynman integral in powers of ε, e.g. the typical common factors $(4\pi)^{\varepsilon/2}$. This will be done in the next section. In a slight modification of this procedure, a common factor $f(\varepsilon) = (4\pi)^{\varepsilon/2}$ may be omitted from each power in g, since it can be thought of as having been absorbed into the irrelevant mass parameter μ. In the five-loop calculation to be presented later, a certain modification of MS will be used, which we shall call $\overline{MS}$-*scheme*, to be explained in more detail in Section 13.1.2 together with some other modified MS-schemes which have been used in the literature.

Formally, the MS-scheme is implemented with the help of an operator $\mathcal{K}$ defined to pick out the pure pole terms of the dimensionally regularized integral:

$$\mathcal{K} \sum_{n=-k}^{\infty} A_i \varepsilon^i = \sum_{n=-k}^{-1} A_i \varepsilon^i = \sum_{i=1}^{k} \frac{A_{-i}}{\varepsilon^i}. \tag{9.76}$$

By definition, $\mathcal{K}$ is a projection operator since

$$\mathcal{K}^2 = \mathcal{K}. \tag{9.77}$$

Application of $\mathcal{K}$ to a diagram means application to the integral associated with the diagram. Take for example the divergent diagrams (8.62) and (8.68), whose pole terms are picked out as follows:

$$\mathcal{K}(\,\underset{}{\mathcal{Q}}\,) = m^2 \left[\frac{g}{(4\pi)^2}\frac{2}{\varepsilon}\right], \qquad \mathcal{K}(\,\times\!\!\times\,) = \mu^\varepsilon g \left[\frac{g}{(4\pi)^2}\frac{2}{\varepsilon}\right]. \tag{9.78}$$

Both one-loop pole terms are local. They are proportional to m^2 for the quadratically divergent diagram, and to $\mu^\varepsilon g$ for the logarithmically divergent diagram. The latter is independent of the mass m. Note that the absence of the external momenta in the pole term on the right-hand side in (9.78) implies the important relation:

$$\mathcal{K}(\,\bigcirc\,) = \mathcal{K}(\,\times\!\!\times\,), \tag{9.79}$$

where the integral on the left-hand side emerges from the one on the right-hand side by setting the external momentum equal to zero. The pole term remains unchanged by this operation, as is obvious from Eq. (8.68). This kind of diagram will be important later on.

In the context of minimal subtraction, the absence of any nonlocal log $\mathbf{k}$-terms in the counterterms has been shown to be an extension of the Dyson-Weinberg convergence theorem in Section 9.1. It can be shown, and we shall observe this explicitly below, that the pole part $\mathcal{K}G$ of any subdivergence-free diagram G is polynomial in its external momenta. This is not only the case in the ϕ^4-theories under study, but also in other renormalizable theories, where the residues of the ε-poles always contain the external momenta and masses as low-order polynomials [13, 14, 15, 16] The proof uses the following properties of the operator $\mathcal{K}$ and the differential operator with respect to the external momentum $\partial \equiv \partial/\partial k$ (omitting the component label):

1. A function vanishes after a finite number of momentum differentiations ∂ if and only if it is a polynomial in the momenta.

2. The $\mathcal{K}$-operation commutes with momentum differentiation, since the two operations act on different spaces.

3. The superficial degree of divergence of a diagram is reduced by one unit for each momentum derivative: $\omega(\partial^s G) = \omega(G) - s$. This is obvious on dimensional grounds, for example:

$$\frac{\partial}{\partial k_\mu} \frac{1}{(\mathbf{k} + \mathbf{p})^2 + m^2} = \frac{-2(k + p)_\mu}{[(\mathbf{k} + \mathbf{p})^2 + m^2]^2}.$$

The argument goes as follows: according to property 3, a subdivergence-free diagram G has $\omega(\partial^{\omega(G)+1} G) < 0$, where $\partial^{\omega(G)+1} G$ is also subdivergence-free [17]. Then the convergence theorem can be invoked stating that $\partial^{\omega(G)+1} G$ is absolutely convergent so that $\mathcal{K}(\partial^{\omega(G)+1} G) = 0$. Using property 2, we deduce that $\partial^{\omega(G)+1} \mathcal{K}G = 0$. Then property 1 implies that $\mathcal{K}G$ is a polynomial of degree lower or equal to $\omega(G)$ in the external momenta.

An analogous statement holds for derivatives with respect to the mass, $\partial_{m^2} G = \partial G/\partial m^2$. As mass and external momenta are the only dimensional parameters of the theory, the pole terms of the integrals are homogeneous polynomials of order $\omega(G)$ in these parameters. Therefore, $\mathcal{K}G$ is proportional to $\mathbf{k}^2$ or m^2 for quadratically divergent diagrams and independent of these dimensional parameters for logarithmically divergent diagrams.

This implies that all $\log(m^2/\mu^2)$-terms arising in the process of subtraction of the pole terms in ε can only survive in the finite parts, but not in the pole terms as required for renormalizability. This will be seen in the next section, where the counterterms for the ϕ^4-theory up to two loops are calculated explicitly. Furthermore , the result ensures that the counterterms can be chosen to be independent of the mass m^2, implying that the limit $m^2 \to 0$ does not produce IR-divergences in the counterterms.

9.3.2 Renormalization in MS-Scheme

We shall now calculate finite two- and four-point vertex functions $\bar{\Gamma}^{(2)}(k^2)$ and $\bar{\Gamma}^{(4)}(k_i)$, starting from the renormalized energy functional (9.61):

$$E[\phi] = \int d^D x \left[\frac{1}{2}(\partial\phi)^2 + c_\phi \frac{1}{2}(\partial\phi)^2 + \frac{m^2}{2}\phi^2 + c_{m^2}\frac{m^2}{2}\phi^2 + \frac{\mu^\varepsilon g}{4!}\phi^4 + c_g \frac{\mu^\varepsilon g}{4!}\phi^4 \right]. \tag{9.80}$$

One-Loop Calculation

To first order in g, the counterterms which are necessary to make the two-point vertex function $\bar{\Gamma}^{(2)}(k)$ finite are also of first order in g. We may write the result in diagrammatical terms as

$$\bar{\Gamma}^{(2)}(k) = k^2 + m^2 - \left(\frac{1}{2}\,\mathcal{Q}\, + \!\!\times\!\! + \!\!-\!\!\circ\!\!- + \mathcal{O}(g^2) \right). \tag{9.81}$$

The cross and the small circle on a line indicate the contribution of the mass and field counterterms c_{m^2} and c_ϕ. The first is chosen to cancel precisely the pole term of $\mathcal{Q}$ proportional to m^2, written down in Eq. (9.78), i.e., we set

$$-\!\!\times\!\!- = -m^2 c_{m^2}^1 = -\frac{1}{2}\mathcal{K}\!\left(\mathcal{Q}\right) = -m^2 \frac{g}{(4\pi)^2}\frac{1}{\varepsilon}, \tag{9.82}$$

where the superscript denotes the order of approximation. Since the counterterm (9.78) contains no contribution proportional to k^2, there is no counterterm c_ϕ to first order in g:

$$-\!\!\circ\!\!- = -k^2 c_\phi^1 = 0. \tag{9.83}$$

Choosing the counterterms in this way, the $1/\varepsilon$-pole in the one-loop diagram is canceled, and the renormalized two-point vertex function

$$\bar{\Gamma}^{(2)}(k) = k^2 + m^2 - \left[\frac{1}{2}\,\mathcal{Q}\, - \frac{1}{2}\mathcal{K}(\mathcal{Q}\,) \right] + \mathcal{O}(g^2) \tag{9.84}$$

remains finite for $\varepsilon \to 0$ up to the first order in g.

The first finite perturbative correction to the vertex function $\bar{\Gamma}^{(4)}(k_i)$ is of the order g^2. The $1/\varepsilon$-pole in the Feynman integral is removed by a counterterm for the coupling constant, to be denoted by a fat dot:

$$\bar{\Gamma}^{(4)} = - \left(\times + \frac{3}{2}\,\rtimes\!\!\ltimes + \,\blacksquare\, \right) + \mathcal{O}(g^3). \tag{9.85}$$

As in Eq. (9.41), the one-loop diagram contributes with three different momentum combinations $k_1 + k_2$, $k_1 + k_3$, and $k_1 + k_4$, indicated by the prefactor 3. Choosing the pole term of $\rtimes\!\!\ltimes$ in Eq. (9.78) as a counterterm, we identify

$$\blacksquare = -\mu^\varepsilon g\, c_g^1 = -\frac{3}{2}\mathcal{K}\!\left(\rtimes\!\!\ltimes\right) = -\mu^\varepsilon g\, \frac{3g}{(4\pi)^2}\frac{1}{\varepsilon}, \tag{9.86}$$

and obtain the finite vertex function

$$\bar{\Gamma}^{(4)} = - \left[\times + \frac{3}{2}\,\rtimes\!\!\ltimes - \frac{3}{2}\mathcal{K}(\rtimes\!\!\ltimes) \right] + \mathcal{O}(g^3). \tag{9.87}$$

Two-Loop Calculation

We now turn to the two-loop counterterms. At this stage the $\log m^2/\mu^2$- and $\log \mathbf{k}^2/\mu^2$-terms will enter at intermediate steps. First we calculate $\bar{\Gamma}^{(2)}(\mathbf{k})$ up to the order g^2. We have to form all previous second-order diagrams plus those which arise from the above-determined first-order counterterms:

$$\bar{\Gamma}^{(2)} = (-)^{-1} - \left[\frac{1}{2}\,\Omega \; + \!\!-\!\!\times\!\!- + \!\!-\!\!\circ\!\!- + \frac{1}{4}\,\mathbb{8} + \frac{1}{2}\,\mathbb{\emptyset} + \frac{1}{6}\,\ominus + \frac{1}{2}\,\mathbb{\bullet} \right] + \mathcal{O}(g^3). \quad (9.88)$$

There are two second-order diagrams, $\mathbb{8}$ and $\ominus$, which contain only $g\phi^4$-interactions. The pole term of $\mathbb{8}$ is given in Eq. (8.72):

$$\frac{1}{4}\mathcal{K}\left(\mathbb{8}\right) = -\frac{m^2 g^2}{(4\pi)^4} \left[\frac{1}{\varepsilon^2} + \frac{\psi(1) + \psi(2)}{2\,\varepsilon} - \frac{1}{\varepsilon}\log\frac{m^2}{4\pi\mu^2} \right]. \quad (9.89)$$

The pole term of $\ominus$ is found in Eq. (8.96):

$$\frac{1}{6}\mathcal{K}\left(\ominus\right) = -\frac{m^2 g^2}{(4\pi)^4} \left[\frac{1}{\varepsilon^2} + \frac{3}{2\,\varepsilon} + \frac{\psi(1)}{\varepsilon} - \frac{1}{\varepsilon}\log\frac{m^2}{4\pi\mu^2} \right] - \frac{g^2}{(4\pi)^4}\frac{\mathbf{k}^2}{12\varepsilon}. \quad (9.90)$$

The divergent term proportional to $\mathbf{k}^2$ will give the first contribution to the wave function counterterm c_ϕ.

Both pole terms (9.89) and (9.90) contain logarithms of the form $\log(m^2/4\pi\mu^2)$. These arise from subdivergences as follows: a regular $\log(m^2/4\pi\mu^2)$ term of one of the loop integrals is multiplied by an ε-pole of the other, and vice versa. The fact that the argument of the logarithm is always $m^2/4\pi\mu^2$ can easily be understood. For dimensional reasons, a two-point diagram is proportional to $m^{2-L\varepsilon}$, where L is the number of loops. Furthermore, each power of g carries a factor μ^ε and each loop integration generates a factor $1/(4\pi)^{2-\varepsilon/2}$. Since the number of loops equals the number of coupling constants, we always run into the combination $(m^2/4\pi\mu^2)^\varepsilon$ whose ε-expansion yields the above logarithms.

The expansion in (9.88) contains in addition two diagrams, $\mathbb{\emptyset}$ and $\mathbb{\bullet}$, arising from the first-order counterterms, to be called *counterterm diagrams*. They are calculated by replacing the coupling constant in the corresponding ϕ^4-diagram by the counterterm. For $\mathbb{\emptyset}$, we replace one of the coupling constants $-\mu^\varepsilon g$ in $\mathbb{X}\big|_{\mathbf{k}^2=0} = \mathbb{Q}$ by $-m^2 c^1_{m^2}$ and find

$$\frac{1}{2}\,\mathbb{\emptyset} = -\frac{m^2 c^1_{m^2} g}{(4\pi)^2} \left[\frac{1}{\varepsilon} + \frac{1}{2}\psi(1) - \frac{1}{2}\int_0^1 d\alpha \, \log\frac{m^2 + \mathbf{k}^2\alpha(1-\alpha)}{4\pi\mu^2} + \mathcal{O}(\varepsilon) \right]\Bigg|_{\mathbf{k}^2=0} \quad (9.91)$$

$$= \frac{m^2 g^2}{(4\pi)^4} \left[\frac{1}{\varepsilon^2} + \frac{1}{2\varepsilon}\psi(1) - \frac{1}{2\varepsilon}\log\frac{m^2}{4\pi\mu^2} + \mathcal{O}(\varepsilon^0) \right]. \quad (9.92)$$

The calculation for $\mathbb{\bullet}$ merely requires replacing the coupling constant $-\mu^\varepsilon g$ in $\mathbb{Q}$ by $-\mu^\varepsilon g\, c^1_g$:

$$\frac{1}{2}\,\mathbb{\bullet} = -\frac{-m^2 g\, c^1_g}{(4\pi)^2} \left[\frac{1}{\varepsilon} + \frac{\psi(2)}{2} - \frac{1}{2}\log\frac{m^2}{4\pi\mu^2} + \mathcal{O}(\varepsilon) \right]$$

$$= \frac{3\, m^2 g^2}{(4\pi)^4} \left[\frac{1}{\varepsilon^2} + \frac{1}{2\varepsilon}\,\psi(2) - \frac{1}{2\varepsilon}\log\frac{m^2}{4\pi\mu^2} + \mathcal{O}(\varepsilon^0) \right]. \quad (9.93)$$

There are again terms of the form $\log(m^2/4\pi\mu^2)$. But this time they come with a prefactor $1/2$ because they are generated only by the single loop integral. The counterterm c_g^1 is free of logarithms and supplies only a factor proportional to $1/\varepsilon$ [see (9.86)]. This is crucial for the ultimate cancellation of all logarithmic terms. The cancellation mechanism will be illustrated in more detail at the end of this section.

Collecting all terms contributing to Eq. (9.88), and using the relation $\psi(n+1)-\psi(n)=1/n$, we find the contributions to the counterterms c_{m^2} and c_ϕ up to the second order in g:

$$
\begin{aligned}
\times\!\!\!\!\!\!-\!\!+-\!\!\!\circ\!\!- \; &= \; -\mathcal{K}\left[\frac{1}{2}\,\bigcirc + \frac{1}{4}\,\bigcirc\!\!\!\bigcirc + \frac{1}{6}\,\bigcirc\!\!\!\!- + \frac{1}{2}\,\bigcirc\!\!\!\!\cdot + \frac{1}{2}\,\bullet\!\!\!\circ\right] \\
&= \; -\left[\frac{g}{(4\pi)^2}\frac{m^2}{\varepsilon} + \frac{g^2}{(4\pi)^4}\left(\frac{2m^2}{\varepsilon^2} - \frac{m^2}{2\varepsilon} - \frac{\mathbf{k}^2}{12\varepsilon}\right)\right].
\end{aligned}
\tag{9.94}
$$

The whole expression is polynomial in m^2 and $\mathbf{k}^2$. The pole terms proportional to m^2 extend the mass counterterms as follows:

$$
m^2\left(c_{m^2}^1 + c_{m^2}^2\right) = m^2\left[\frac{g}{(4\pi)^2}\frac{1}{\varepsilon} + \frac{g^2}{(4\pi)^4}\left(\frac{2}{\varepsilon^2} - \frac{1}{2\varepsilon}\right)\right].
\tag{9.95}
$$

The pole terms proportional to $\mathbf{k}^2$ give the second-order counterterm of the field renormalization

$$
\mathbf{k}^2 c_\phi^2 = \frac{1}{6}\mathcal{K}\left(\bigcirc\!\!\!\!-\right)\Big|_{m=0} = -\mathbf{k}^2\frac{g^2}{(4\pi)^4}\frac{1}{12\,\varepsilon}.
\tag{9.96}
$$

As noted before, c_ϕ possesses no first-order term in g.

We now turn to the two-loop renormalization of the four-point vertex function, whose diagrammatic expansion reads

$$
\bar{\Gamma}^{(4)} \; = \; -\left[\times + \frac{3}{2}\,\times\!\!\!\circ + \bullet\!\!\!\!\ast + 3\,\bigcirc\!\!\!\!- + \frac{3}{4}\,\times\!\!\!\circ\!\!\!\circ\!\!\!\times + \frac{3}{2}\,\bigcirc\!\!\!\!\!\cdot + 3\,\times\!\!\!\circ\!\!\ast + 3\,\times\!\!\!\!\times\right].
\tag{9.97}
$$

As in the one-loop diagram in Eqs. (9.41) and (9.85), we had to sum over all combinations of external momenta, resulting in the factors 3. The pole term in the first two-loop diagram was calculated in Eq. (8.114), yielding

$$
3\,\mathcal{K}\left(\bigcirc\!\!\!\!-\right) = -\mu^\varepsilon g\frac{3g^2}{(4\pi)^4}\left\{\frac{2}{\varepsilon^2} + \frac{1}{\varepsilon} + \frac{2}{\varepsilon}\psi(1) - \frac{2}{\varepsilon}\int_0^1 d\alpha\,\log\left[\frac{m^2 + \mathbf{k}^2\alpha(1-\alpha)}{4\pi\mu^2}\right]\right\},
\tag{9.98}
$$

where $\mathbf{k}$ indicates either of the three different momentum combinations $\mathbf{k}_1 + \mathbf{k}_2$, $\mathbf{k}_1 + \mathbf{k}_3$, and $\mathbf{k}_1 + \mathbf{k}_4$. The pole term in the second two-loop diagram was obtained in Eq. (8.99):

$$
\frac{3}{4}\mathcal{K}\left(\times\!\!\!\circ\!\!\!\circ\!\!\!\times\right) = -\mu^\varepsilon g\frac{3g^2}{(4\pi)^4}\left\{\frac{1}{\varepsilon^2} + \frac{1}{\varepsilon}\psi(1) - \frac{1}{\varepsilon}\int_0^1 d\alpha\,\log\left[\frac{m^2 + \mathbf{k}^2\alpha(1-\alpha)}{4\pi\mu^2}\right]\right\}.
\tag{9.99}
$$

The pole term in the third integral was calculated with the help of the integral formula (8.69) and Eq. (8.62), and reads, according to Eq. (8.103),

$$
\frac{3}{2}\mathcal{K}\left(\bigcirc\!\!\!\!\!\cdot\right) = -\mu^\varepsilon g\frac{3g^2}{(4\pi)^4}\frac{1}{\varepsilon}\int_0^1 d\alpha\,\frac{m^2(1-\alpha)}{\mathbf{k}^2\alpha(1-\alpha)+m^2}.
\tag{9.100}
$$

The pole term of the first counterterm diagram is calculated with the help of Eq. (8.68), where $-\mu^\varepsilon g$ is replaced by $\ast = -\mu^\varepsilon g\,3g/(4\pi)^2\varepsilon$:

$$
3\,\mathcal{K}\left(\times\!\!\!\circ\!\!\ast\right) = \mu^\varepsilon g\frac{3g^2}{(4\pi)^4}\left\{\frac{6}{\varepsilon^2} + \frac{3}{\varepsilon}\psi(1) - \frac{3}{\varepsilon}\int_0^1 d\alpha\,\log\left[\frac{m^2 + \mathbf{k}^2\alpha(1-\alpha)}{4\pi\mu^2}\right]\right\}.
\tag{9.101}
$$

The pole term of the second counterterm diagram is calculated as in (9.98), yielding

$$3\,\mathcal{K}\left(\vcenter{\hbox{$\times\!\!\!\!\times$}}\right) = \mu^\varepsilon g\,\frac{3g^2}{(4\pi)^4}\,\frac{1}{\varepsilon}\int_0^1 d\alpha\,\frac{m^2(1-\alpha)}{\mathbf{k}^2\alpha(1-\alpha)+m^2}\,. \tag{9.102}$$

All pole terms containing either $\psi(1)$ or the parameter integral cancel each other in the counterterm of the coupling constant, which becomes, up to the second order in g,

$$\mu^\varepsilon g\,\left(c_g^1 + c_g^2\right) = \mu^\varepsilon g\,\left[\frac{g}{(4\pi)^2}\frac{3}{\varepsilon} + \frac{g^2}{(4\pi)^4}\left(\frac{9}{\varepsilon^2} - \frac{3}{\varepsilon}\right)\right]. \tag{9.103}$$

The counterterms in Eqs. (9.95), (9.96), and (9.103) have all the local form (9.65)–(9.67). No nonlocal proportional to $\log(\mathbf{k}^2/\mu^2)$ appear, which would have impeded the incorporation of the pole terms into the initial energy functional (9.80).

Up to the order g^2, we thus obtain finite correlation functions by starting out from the initial energy functional (9.80), written in the form:

$$E[\phi] = \int d^D x\,\left[\frac{1}{2}Z_\phi(\partial\phi)^2 + \frac{m^2}{2}Z_{m^2}\phi^2 + \frac{\mu^\varepsilon g}{4!}Z_g\phi^4\right], \tag{9.104}$$

with the renormalization constants

$$Z_\phi(g,\varepsilon^{-1}) = 1 + c_\phi \;\; = \;\; 1 + \frac{1}{\mathbf{k}^2}\frac{1}{6}\mathcal{K}(\ominus)\Big|_{m^2=0} = 1 - \frac{g^2}{(4\pi)^4}\frac{1}{12}\frac{1}{\varepsilon}, \tag{9.105}$$

$$\begin{aligned} Z_{m^2}(g,\varepsilon^{-1}) = 1 + c_{m^2} \;\; &= \;\; 1 + \frac{1}{m^2}\left[\frac{1}{2}\mathcal{K}(\ominus) + \frac{1}{4}\mathcal{K}(\ominus) + \frac{1}{2}\mathcal{K}(\ominus)\right.\\ &\qquad\qquad \left. + \frac{1}{2}\mathcal{K}(\ominus) + \frac{1}{6}\mathcal{K}(\ominus)\Big|_{\mathbf{k}^2=0}\right]\\ &= \;\; 1 + \frac{g}{(4\pi)^2}\frac{1}{\varepsilon} + \frac{g^2}{(4\pi)^4}\left(\frac{2}{\varepsilon^2} - \frac{1}{2\varepsilon}\right), \end{aligned} \tag{9.106}$$

$$\begin{aligned} Z_g(g,\varepsilon^{-1}) = 1 + c_g \;\; &= \;\; 1 + \frac{1}{\mu^\varepsilon g}\left[\frac{3}{2}\mathcal{K}(\bowtie) + 3\,\mathcal{K}\left(\ominus\right) + \frac{3}{4}\mathcal{K}(\bowtie)\right.\\ &\qquad\qquad \left. + \frac{3}{2}\mathcal{K}\left(\ominus\right) + 3\,\mathcal{K}(\bowtie) + 3\,\mathcal{K}(\bowtie)\right]\\ &= \;\; 1 + \frac{g}{(4\pi)^2}\frac{3}{\varepsilon} + \frac{g^2}{(4\pi)^4}\left(\frac{9}{\varepsilon^2} - \frac{3}{\varepsilon}\right). \end{aligned} \tag{9.107}$$

The renormalization constants are expansions in the dimensionless coupling constant g, with expansion coefficients containing only pole terms of the form $1/\varepsilon^i$, where i runs from 1 to n in the term of order g^n. This is precisely the form anticipated in Eq. (9.1).

By writing the energy functional in the form (9.104), it is multiplicatively renormalized. A comparison with the bare energy functional (9.73) allows us to identify the bare field, mass, and coupling constant as in Eqs. (9.71) and (9.72). Note that since the divergences in the ϕ^4-theory come exclusively from the 2- and 4-point 1PI diagrams, all n-point vertex functions are finite for $\varepsilon \to 0$ up to this order in g, if the perturbation expansion proceeds from the energy functional (9.104).

For N field components, the two-loop results are extended by the symmetry factors introduced in Section 6.3 and listed in Eqs. (6.40)–(6.46). They multiply each diagram as follows:

$$Z_\phi(g,\varepsilon^{-1}) \;\; = \;\; 1 + \frac{1}{\mathbf{k}^2}\frac{1}{6}\mathcal{K}(\ominus)\Big|_{m^2=0}S_\ominus\,, \tag{9.108}$$

$$Z_{m^2}(g,\varepsilon^{-1}) = 1 + \frac{1}{m^2}\left[\frac{1}{2}\mathcal{K}(\text{⊘})S_\text{O} + \frac{1}{4}\mathcal{K}(\text{⧖})S_\text{8} + \frac{1}{6}\mathcal{K}(\ominus)\big|_{\mathbf{k}^2=0}S_\ominus\right.$$
$$\left.+\frac{1}{2}\mathcal{K}(\text{⧗})S_\text{⧗} + \frac{1}{2}\mathcal{K}(\text{⦶})S_\text{⦶}\right], \tag{9.109}$$

$$Z_g(g,\varepsilon^{-1}) = 1 + \frac{1}{\mu^\varepsilon g}\left[\frac{3}{2}\mathcal{K}(\text{⋈})S_\text{⋈} + 3\mathcal{K}(\ominus)S_\ominus + \frac{3}{4}\mathcal{K}(\text{∞})S_\text{∞}\right.$$
$$\left.+\frac{3}{2}\mathcal{K}(\text{⦰})S_\text{8} + 3\text{⦿}S_\text{⋈} + 3\text{✕}S_\text{⧖}\right]. \tag{9.110}$$

The symmetry factors associated with the counterterm diagrams must still be calculated, with the following results:

$$S_\text{⧗} = S_\text{O}\,\delta_{\sigma\tau}T_{\alpha\beta\sigma\tau} = S_\text{O}\,S_\text{O} = \left(\frac{N+2}{3}\right)^2, \tag{9.111}$$

$$S_\text{⦶} = S_\text{O}\,T_{\alpha\beta\sigma\tau}\delta_{\sigma\tau} = S_\text{O}\,S_\text{O} = \frac{N+2}{3}\frac{N+8}{9}, \tag{9.112}$$

$$S_\text{⋈} = S_\text{O}\,T_{\alpha\beta\sigma\tau}T_{\sigma\tau\gamma\delta} = S_\text{O}\,S_\text{O} = \left(\frac{N+8}{9}\right)^2, \tag{9.113}$$

$$S_\text{⧖} = S_\text{O}\,\delta_{\sigma\sigma'}T_{\alpha\beta\sigma\tau}T_{\sigma'\tau\gamma\delta} = S_\text{O}\,S_\text{O} = \frac{N+2}{3}\frac{N+8}{9}. \tag{9.114}$$

With these symmetry factors, the renormalization constants up to g^2 are

$$Z_\phi(g,\varepsilon^{-1}) = 1 - \frac{g^2}{(4\pi)^4}\frac{1}{12}\frac{1}{\varepsilon}\frac{N+2}{3}, \tag{9.115}$$

$$Z_{m^2}(g,\varepsilon^{-1}) = 1 + \frac{g}{(4\pi)^2}\frac{1}{\varepsilon}\frac{N+2}{3}$$
$$+\frac{g^2}{(4\pi)^4}\left[\left(-\frac{1}{\varepsilon^2}+\frac{1}{\varepsilon}\log\frac{m^2}{4\pi\mu^2}-\frac{\psi(1)+\psi(2)}{2\varepsilon}\right)\left(\frac{N+2}{3}\right)^2\right.$$
$$+\left(-\frac{1}{\varepsilon^2}-\frac{3}{2\varepsilon}+\frac{1}{\varepsilon}\log\frac{m^2}{4\pi\mu^2}-\frac{\psi(1)}{\varepsilon}\right)\frac{N+2}{3}$$
$$+\left(\frac{1}{\varepsilon^2}-\frac{1}{2\varepsilon}\log\frac{m^2}{4\pi\mu^2}+\frac{\psi(1)}{2\varepsilon}\right)\left(\frac{N+2}{3}\right)^2$$
$$\left.+\left(\frac{3}{\varepsilon^2}-\frac{3}{2\varepsilon}\log\frac{m^2}{4\pi\mu^2}+\frac{3\psi(2)}{2\varepsilon}\right)\frac{N+2}{3}\frac{N+8}{9}\right] \tag{9.116}$$

$$= 1 + \frac{g}{(4\pi)^2}\frac{1}{\varepsilon}\frac{N+2}{3}+\frac{g^2}{(4\pi)^4}\left[-\frac{1}{\varepsilon}\frac{N+2}{6}+\frac{1}{\varepsilon^2}\frac{(N+2)(N+5)}{9}\right], \tag{9.117}$$

$$Z_g(g,\varepsilon^{-1}) = 1 + \frac{g}{(4\pi)^2}\frac{3}{\varepsilon}\frac{N+8}{9}$$
$$+\frac{g^2}{(4\pi)^4}\left\{-\left[\frac{2}{\varepsilon^2}+\frac{1}{\varepsilon}+\frac{2\psi(1)}{\varepsilon}-\frac{2}{\varepsilon}\int_0^1 d\alpha\log\frac{m^2+\mathbf{k}^2\alpha(1-\alpha)}{4\pi\mu^2}\right]\frac{5N+22}{9}\right.$$
$$-\left[\frac{1}{\varepsilon^2}+\frac{\psi(1)}{\varepsilon}-\frac{1}{\varepsilon}\int_0^1 d\alpha\log\frac{m^2+\mathbf{k}^2\alpha(1-\alpha)}{4\pi\mu^2}\right]\frac{N^2+6N+20}{9}$$
$$-\left[\frac{1}{\varepsilon}\int_0^1 d\alpha\frac{m^2(1-\alpha)}{\mathbf{k}^2\alpha(1-\alpha)+m^2}\right]\frac{N^2+10N+16}{27}$$

$$+\left[\frac{6}{\varepsilon^2}+\frac{3\psi(1)}{\varepsilon}-\frac{3}{\varepsilon}\int_0^1 d\alpha\log\frac{m^2+\mathbf{k}^2\alpha(1-\alpha)}{4\pi\mu^2}\right]\frac{(N+8)^2}{27}$$

$$+\left[\frac{1}{\varepsilon}\int_0^1 d\alpha\frac{m^2(1-\alpha)}{\mathbf{k}^2\alpha(1-\alpha)+m^2}\right]\frac{N^2+10N+16}{9}\Bigg\} \tag{9.118}$$

$$=1+\frac{g}{(4\pi)^2}\frac{3}{\varepsilon}\frac{N+8}{9}+\frac{g^2}{(4\pi)^4}\left[\frac{1}{\varepsilon^2}\frac{(N+8)^2}{9}-\frac{1}{\varepsilon}\frac{5N+22}{9}\right]. \tag{9.119}$$

Note that, although the symmetry factors are different for the counterterm diagrams, the combinatorics involved in constructing these are just right to cancel conveniently all logarithms in the final expressions, and lead to local counterterms as necessary for the renormalizability of the theory.

As discussed before, the absence of the mass parameter μ in these expansions offers us the opportunity to redefine the coupling constant g with an arbitrary factor $f(\varepsilon)$ without changing these expansions. It is, however, important to realize that such a redefinition cannot be simply done in the final expressions (9.115), (9.117), (9.119). If we were to replace in these the coupling constant g to $gf(\varepsilon)=g(1+f_1\varepsilon+f_2\varepsilon^2+\dots)$ and delete all positive powers of ε, the coefficients of $1/\varepsilon$ in the g^2-terms would change. The invariance is a consequence of the special preparation of the counterterms, as pointed out earlier after Eq. (9.90). With the prescription for determining the counterterms by minimal subtraction, the redefinition does not modify the coupling constants in the counterterms of the diagrammatic expansion (9.108)–(9.110). The intermediate expressions (9.116) for Z_{m^2} can once more be used to demonstrate this invariance. The powers g^n of the coupling constants, which do not come from a counterterm subdiagram, are multiplied by $(1+f_1\varepsilon+\dots+f_{n-1}\varepsilon^{n-1})^n$, exhibiting only relevant powers up to ε^n. This transforms the graphical expansion (9.109) for Z_{m^2} into

$$Z_{m^2}(g,\varepsilon^{-1})=1+\frac{1}{m^2}\left[\frac{1}{2}\mathcal{K}(\mathcal{Q})S_\mathcal{Q}+\frac{(1+f_1\varepsilon)^2}{4}\mathcal{K}(\mathcal{S})S_\mathcal{S}+\frac{(1+f_1\varepsilon)^2}{6}\mathcal{K}(\ominus)\Big|_{\mathbf{k}^2=0}S_\ominus\right.$$

$$\left.+\frac{(1+f_1\varepsilon)}{2}\mathcal{K}(\mathcal{Q})S_\mathcal{Q}+\frac{(1+f_1\varepsilon)}{2}\mathcal{K}(\mathcal{Q})S_\mathcal{Q}\right]. \tag{9.120}$$

In the analytic expression (9.117), this replacement changes the logarithms as follows:

$$\log\frac{m^2}{4\pi\mu^2}\longrightarrow\log\frac{m^2}{4\pi\mu^2}-2f_1=\log\frac{m^2}{4\pi(\mu e^{f_1})^2},$$

thus multiplying the mass parameter μ by a factor e^{f_1} to this order in g. Since all logarithms disappear in the final renormalization constant (9.117), the redefinition of g leaves no trace in the renormalization constant Z_{m^2}.

9.3.3 Recursive Diagrammatic Subtraction

The subtraction of the ε-poles by the counterterm diagrams can be organized in a different manner. Each counterterm diagram can be associated with one or more ϕ^4-diagrams from which it subtracts the ε-poles coming from subdiagrams of these ϕ^4-diagrams. Only the pole term remaining after all these subtractions contributes to the counterterm in this order. It will be called *superficial pole term*. The main difficulty in these calculations is to find the correct combinatorial factors.

As an example, we calculate recursively the second order contribution to the three counterterms. We begin with c_{m^2} and c_ϕ. To first order in g, there are two counterterm diagrams, which can be written diagrammatically as $\rightarrow\!\!\ast\!\!\leftarrow = -\frac{1}{2}\mathcal{K}(\,\ominus\,)$ and $\boldsymbol{\divideontimes} = -\frac{3}{2}\mathcal{K}(\times\!\!\times)$. With these, the counterterm diagrams of second order in g may be expressed as follows:

$$\frac{1}{2}\,\ominus\!\!\!\!\ominus \;=\; -m^2 c_{m^2} \ast \frac{1}{2}\,\ominus \;=\; -\frac{1}{2}\mathcal{K}(\,\ominus\,)\ast\frac{1}{2}\,\ominus \;, \tag{9.121}$$

$$\frac{1}{2}\,\ominus\!\!\!\bullet \;=\; -\mu^\varepsilon g\, c_g \ast \frac{1}{2}\,\ominus \;=\; -\frac{3}{2}\mathcal{K}(\times\!\!\times)\ast\frac{1}{2}\,\ominus \;, \tag{9.122}$$

where the operation $\ast$ denotes the substitution of the counterterm $-m^2 c_{m^2}$ or $-\mu^\varepsilon g\, c_g$ for the dots in the diagrams $\rightarrow\!\!\leftarrow$ or $\times$, respectively. If the counterterm does not depend on the momentum, this substitution leads simply to the multiplication of the counterterm by the remaining integral, which contains one coupling constant less than the initial diagram. The star operation is less trivial for the counterterms of wave function renormalization which contains a factor $\mathbf{k}^2$. This momentum-dependent factor must be included into the integrand of the remaining loop integrals, thereby complicating its evaluation.

The star operation is now used in the diagrammatic expansion of the sum of the counterterms c_{m^2} and c_ϕ in Eq. (9.94), by splitting the factors and inserting Eqs. (9.121), (9.122), and (9.79):

$$\mathcal{K}\left[\frac{1}{4}\Big(\,\ominus\!\!\ominus + 2\,\ominus\!\!\!\!\ominus + \frac{2}{3}\,\ominus\!\!\!\bullet\,\Big) + \frac{1}{6}\Big(\,\ominus\!\!\!\ominus + 2\,\ominus\!\!\!\bullet\,\Big)\right] \tag{9.123}$$

$$= \mathcal{K}\left[\frac{1}{4}\Big(\,\ominus\!\!\ominus - \mathcal{K}(\,\ominus\,)\ast\ominus - \mathcal{K}(\,\ominus\,)\ast\ominus\,\Big) + \frac{1}{6}\Big(\,\ominus\!\!\!\ominus - 3\mathcal{K}(\times\!\!\times)\ast\ominus\,\Big)\right]$$

$$= \qquad \frac{m^2 g^2}{(4\pi)^4}\frac{1}{\varepsilon^2} \qquad\qquad + \frac{m^2 g^2}{(4\pi)^4}\Big(\frac{1}{\varepsilon^2} - \frac{1}{2\varepsilon}\Big) - \frac{k^2 g^2}{(4\pi)^4}\frac{1}{12\varepsilon}.$$

From the ϕ^4-diagrams $\ominus\!\!\ominus$ and $\ominus\!\!\!\ominus$, all terms are subtracted in which the integration of a superficially divergent subdiagram $\ominus$, $\times\!\!\times$, or $\ominus$ is replaced by its counterterm. This step will be referred to as the *diagrammatic subtraction of subdivergences*. As a result, only the pole terms of the superficial divergences of $\ominus\!\!\ominus$ and $\ominus\!\!\!\ominus$ remain, which are fewer than those in the full counterterms $\mathcal{K}(\,\ominus\!\!\ominus\,)$ and $\mathcal{K}(\,\ominus\!\!\!\ominus\,)$. They provide us with the second-order contributions to the counterterms c_{m^2} and c_ϕ.

The counterterms associated with the vertex function (9.97) can be calculated similarly. After rewriting

$$3\times\!\!\!\bullet = -\frac{3}{2}\mathcal{K}\big(\times\!\!\times\big)\ast 3\times\!\!\times \qquad\text{and}\qquad 3\,\times\!\!\!\!\times = -\frac{1}{2}\mathcal{K}\big(\,\ominus\,\big)\ast 3\times\!\!\times\,, \tag{9.124}$$

the right-hand side of (9.97) yields the following superficial pole terms

$$3\mathcal{K}\left[\,\ominus\!\!\!\ominus - \mathcal{K}\big(\times\!\!\times\big)\ast\times\!\!\times\,\right] + \frac{3}{4}\mathcal{K}\left[\times\!\!\times\!\!\times - 2\mathcal{K}\big(\times\!\!\times\big)\ast\times\!\!\times\,\right] + \frac{3}{2}\mathcal{K}\left[\,\ominus\!\!\!\!\times - \mathcal{K}\big(\,\ominus\,\big)\ast\times\!\!\times\,\right]$$

$$= \frac{g^3}{(4\pi)^4}\left[\Big(\frac{6}{\varepsilon^2} - \frac{3}{\varepsilon}\Big) \qquad + \qquad \frac{3}{\varepsilon^2} \qquad + \qquad 0 \;\right]. \tag{9.125}$$

We have spaced the second line to better indicate the association with the terms above. Using this recursive procedure, Eqs. (9.105)–(9.107) can be calculated diagrammatically as follows:

$$Z_\phi(g,\varepsilon^{-1}) \;=\; 1 + \frac{1}{\mathbf{k}^2}\frac{1}{6}\mathcal{K}(\,\ominus\!\!\!\ominus\,)\Big|_{m^2=0}$$

$$= 1 - \frac{g^2}{(4\pi)^4} \frac{1}{12} \frac{1}{\varepsilon}, \tag{9.126}$$

$$Z_{m^2}(g, \varepsilon^{-1}) = 1 + \frac{1}{m^2} \Big\{ \frac{1}{2} \mathcal{K}(\mathcal{Q}) + \frac{1}{4} \mathcal{K}\Big[\text{⧖} - \mathcal{K}(\mathcal{Q}) * \mathcal{Q} - \mathcal{K}(\mathcal{Q}) * \mathcal{Q} \Big]$$

$$+ \frac{1}{6} \mathcal{K}\Big[\text{⊝} - 3\mathcal{K}(\text{⋈}) * \mathcal{Q} \Big] \Big|_{\mathbf{k}^2 = 0} \Big\}$$

$$= 1 + \frac{g}{(4\pi)^2} \frac{1}{\varepsilon} + \frac{g^2}{(4\pi)^4} \left[\frac{1}{\varepsilon^2} + \left(\frac{1}{\varepsilon^2} - \frac{1}{2\varepsilon} \right) \right], \tag{9.127}$$

$$Z_g(g, \varepsilon^{-1}) = 1 + \frac{1}{\mu^\varepsilon g} \Big\{ \frac{3}{2} \mathcal{K}(\text{⋈}) + 3\mathcal{K}\Big[\text{⊖} - \mathcal{K}(\text{⋈}) * \text{⋈} \Big]$$

$$+ \frac{3}{4} \mathcal{K}\Big[\text{⋈⋈} - 2\mathcal{K}(\text{⋈}) * \text{⋈} \Big]$$

$$+ \frac{3}{2} \mathcal{K}\Big[\text{⧗} - \mathcal{K}(\mathcal{Q}) * \text{⋈} \Big] \Big\}$$

$$= 1 + \frac{g}{(4\pi)^2} \frac{3}{\varepsilon} + \frac{g^2}{(4\pi)^4} \left[\left(\frac{6}{\varepsilon^2} - \frac{3}{\varepsilon} \right) + \frac{3}{\varepsilon^2} + 0 \right]. \tag{9.128}$$

The result is, of course, the same as before.

For N field components, the recursive procedure offers an important advantage over the previous calculation scheme. It saves us from having to calculate the symmetry factors for the counterterm diagrams! This is due to the fact that all subtraction terms of a vertex diagram carry the same symmetry factor. Thus we extend the expansions (9.126)–(9.128) immediately to $O(N)$-symmetric ϕ^4-theory:

$$Z_\phi(g, \varepsilon^{-1}) = 1 - \frac{g^2}{(4\pi)^4} \frac{1}{12} \frac{1}{\varepsilon} S_\ominus, \tag{9.129}$$

$$Z_{m^2}(g, \varepsilon^{-1}) = 1 + \frac{g}{(4\pi)^2} \frac{1}{\varepsilon} S_\bigcirc + \frac{g^2}{(4\pi)^4} \left[\frac{1}{\varepsilon^2} S_8 + \left(\frac{1}{\varepsilon^2} - \frac{1}{2\varepsilon} \right) S_\ominus \right], \tag{9.130}$$

$$Z_g(g, \varepsilon^{-1}) = 1 + 3 \frac{g}{(4\pi)^2} \frac{1}{\varepsilon} S_\bowtie + \frac{g^2}{(4\pi)^4} \left[\frac{3}{\varepsilon^2} S_{\bowtie\bowtie} + \left(\frac{6}{\varepsilon^2} - \frac{3}{\varepsilon} \right) S_\ominus \right]. \tag{9.131}$$

After inserting the symmetry factors of Eqs. (6.40)–(6.46), we obtain more explicitly

$$Z_\phi(g, \varepsilon^{-1}) = 1 - \frac{g^2}{(4\pi)^4} \frac{1}{12} \frac{1}{\varepsilon} \frac{N+2}{3}, \tag{9.132}$$

$$Z_{m^2}(g, \varepsilon^{-1}) = 1 + \frac{g}{(4\pi)^2} \frac{1}{\varepsilon} \frac{N+2}{3}$$

$$+ \frac{g^2}{(4\pi)^4} \left[\frac{1}{\varepsilon^2} \left(\frac{N+2}{3} \right)^2 + \left(\frac{1}{\varepsilon^2} - \frac{1}{2\varepsilon} \right) \frac{N+2}{3} \right], \tag{9.133}$$

$$Z_g(g, \varepsilon^{-1}) = 1 + 3 \frac{g}{(4\pi)^2} \frac{1}{\varepsilon} \frac{N+8}{9}$$

$$+ \frac{g^2}{(4\pi)^4} \left[\frac{3}{\varepsilon^2} \frac{N^2 + 6N + 20}{27} + \left(\frac{6}{\varepsilon^2} - \frac{3}{\varepsilon} \right) \frac{5N+22}{27} \right], \tag{9.134}$$

in agreement with Eqs. (9.115)–(9.119). The simplification brought about by the recursive procedure is nontrivial: it is based on the fact that the symmetry factor of the counterterm

✖ picks up the symmetry factors of the counterterm diagrams ⊖ and ✗✖ in a symmetrized form. For example, $S_{✗✖} \neq S_8$ since ✖ is symmetrized, whereas the subdiagram ✗ in ⊖ is not.

Specifically, the proper symmetry factors in Eq. (9.130) for $Z_{m^2}(g, \varepsilon^{-1})$ are a result of the equality:

$$\frac{1}{4}S_8 + \frac{1}{2}S_⊖ = \frac{3}{4}S_● ,$$

$$\frac{1}{4}\left(\frac{N+2}{3}\right)^2 + \frac{1}{2}\frac{N+2}{3} = \frac{3}{4}\frac{N+2}{3}\frac{N+8}{9}.$$

In Eq. (9.131) for $Z_g(g, \varepsilon^{-1})$, the analogous equalities are

$$3S_⊖ + \frac{3}{2}S_{✗✗} = \frac{9}{2}S_{✗✖} ,$$

$$\frac{5N+22}{9} + \frac{1}{2}\frac{N^2+6N+20}{9} = \frac{1}{2}\frac{(N+8)^2}{9}.$$

The generation of all possible counterterm diagrams via a diagrammatic subtraction of subdivergences in this example can be developed into a systematic technique with the help of the so-called *R*-operation which will be introduced in the Chapter 11.

Appendix 9A Overlapping Divergences

An overlapping divergence could, in principle, arise from the integral over certain directions in the multidimensional space of all loop momenta, even though a diagram has $\omega(G) < 0$ and all $\omega(\gamma) < 0$. According to the convergence theorem, this cannot happen. As an example, consider the following Feynman integral for $D = 4$:

$$\underset{\text{IR}}{\diamond} \; \triangleq \int_{\text{IR}} \frac{d^4k \, d^4p}{\mathbf{p}^4(\mathbf{p}+\mathbf{k})^2\mathbf{k}^4} . \tag{9A.1}$$

Since danger comes only from the large momentum-region, all masses and external momenta have been set equal to zero, for simplicity. The subscript IR on the integral indicates some cutoff at small momenta to prevent IR-divergences. Power counting shows that $\omega(G) = -2$, thus indicating a superficial convergence. Subdivergences are not present, as we can see also by naive power counting. Obviously, naive power counting fails to inform us whether the integral converges in the subspace with fixed $(\mathbf{k}+\mathbf{p})^2$. The above theorem implies that this cannot happen. To verify this, we consider the integral in the eight-dimensional space of the two loop momenta $\mathbf{k}$ and $\mathbf{p}$. A divergence could in principle appear, and this would not be caused by subdivergences, since there are none. Such a divergence could not be predicted by the power counting theorem. The danger of a divergence in the present example is eliminated by the following consideration: the eight-dimensional momentum space is divided into several regions in such a way that, in each region, one of the squared momenta in the denominator is smaller than the others. For the integral in (9A.1), one of the regions to be considered is $U = \{\mathbf{k}|\mathbf{k}^2 \geq 1\}$, $V = \{\mathbf{p}|\mathbf{p}^2 \geq \mathbf{k}^2, (\mathbf{p}+\mathbf{k})^2 \geq \mathbf{k}^2\}$. The momenta are then rescaled by the inverse absolute value of the smallest momentum (here $\mathbf{k}$). In the example, this rescaling is $\mathbf{k} = \hat{\mathbf{k}}k$, where $\hat{\mathbf{k}}$ is a four vector of unit length, and $\mathbf{p} = \mathbf{p}'k$. The resulting integral is

$$\int_U dk \, k^3 \frac{1}{k^6} \int d\hat{\mathbf{k}} \int_{V'} d^4p' \frac{1}{\mathbf{p}'^4(\mathbf{p}'+\hat{\mathbf{k}})^2}, \tag{9A.2}$$

with $V' = \{\mathbf{p}'|\mathbf{p}'^2 \geq 1 = \hat{\mathbf{k}}^2, (\mathbf{p}' + \hat{\mathbf{k}})^2 \geq 1\}$. The integral $\int d\hat{\mathbf{k}}$ covers the surface S_4 of the sphere in four dimensions with unit radius. Now, the integration over $\mathbf{p}'$ is absolutely convergent for each fixed $\hat{\mathbf{k}}$, since (9A.1) is free of subdivergences. The remaining integration over the absolute value of $\mathbf{k}$ is governed by the degree of superficial divergence $\omega(G)$, and is convergent in this region. Similar arguments apply to the other regions. In general, the integrations over the larger momenta are always absolutely convergent if no subdivergences are present. After having carried out the large-momentum integrals, the final integration is always of the form:

$$\int_U dk\, k^3\, \frac{1}{|k|^{2I-4(L-1)}} \sim \int_1^\infty dk\, \frac{1}{|k|^{-\omega(G)+1}}, \tag{9A.3}$$

which is convergent as long as $\omega(G) < 0$.

Notes and References

The possibility to remove all infinities by multiplicative renormalization was first observed by Dyson for quantum electrodynamics in Ref. [3]. For general discussions of renormalization see the textbooks

N.N. Bogoliubov and D.V. Shirkov, *Introduction to the Theory of Quantized Fields*, Wiley Interscience, New York, 1958;
J.C. Collins, *Renormalization*, Cambridge University Press, Cambridge, 1984;
J. Zinn-Justin, *Quantum Field Theory and Critical Phenomena*, Clarendon, Oxford, 1989;
S. Weinberg, *The Quantum Theory of Fields*, Cambridge University Press, New York, 1995;
Vol I: *Foundations*, Vol II: *Modern Applications*;
and Refs. [7], [8], [12].

The individual citations in the text refer to:

[1] J.C. Collins, Nucl. Phys. B **80**, 341 (1974);
    E.R. Speer, J. Math. Phys. **15**, 1 (1974);
    P. Breitenlohner and D. Maison, Commun. Math. Phys. **52**, 11, 39, 55 (1977);
    W.E. Caswell and A.D. Kennedy, Phys. Rev. D **25**, 392 (1982).

[2] N.N. Bogoliubov and O.S. Parasiuk, Acta Math. **97**, 227-266 (1957).

[3] F.J. Dyson, Phys. Rev. **75**, 1736 (1949).

[4] S. Weinberg, Phys. Rev. **118**, 838 (1960).

[5] K. Hepp, Commun. Math. Phys. **2**, 301 (1966).

[6] Y. Hahn and W. Zimmermann, Commun. Math. Phys. **10**, 330 (1968).

[7] J.D. Bjorken and S.D. Drell, *Relativistic Quantum Fields*, McGraw-Hill, New York (1965);

[8] C. Itzykson and J.-B. Zuber, *Quantum Field Theory*, McGraw-Hill, New York (1980).

[9] J.C. Ward, Phys. Rev. **78**, 182 (1959).

[10] The physical content of nonrenormalizable theories is explored in
J. Gomis and S. Weinberg, *Are Nonrenormalizable Gauge Theories Renormalizable?*, Nucl. Phys. B **469**, 473 (1996) (hep-th/9803099);
S. Weinberg, *Non-Renormalization Theorems in Non-Renormalizable Theories*, Phys. Rev. Lett. **80**, 3702 (1998) (hep-th/9803099);
S. Weinberg, *What is Quantum Field Theory, and What Did We Think It Is?*, Talk presented at the conference on *Historical and Philosophical Reflections on the Foundations of Quantum Field Theory*, at Boston University, March 1996.

[11] See page 24 for the relevance of renormalizability in quantum field theories of critical phenomena.

[12] D.J. Amit, *Field Theory, the Renormalization Group and Critical Phenomena*, McGraw-Hill, 1978.

[13] G.'t Hooft, Nuclear Physics B **61**, 455 (1973).

[14] G.'t Hooft and M. Veltman, Nucl. Phys. B **44**, 189 (1972),
and Ref. [1].

[15] J.C. Collins and A.J. MacFarlane, Phys. Rev. D *10*, 1201 (1974).

[16] D.R.T. Jones, Nucl. Phys. B **75**, 531 (1974).

[17] A.A. Vladimirov, Theor. Math. Phys. **36**, 732 (1978);
W.E. Caswell and A.D. Kennedy, Phys. Rev. D **25**, 392 (1982).

10

Renormalization Group

The renormalization procedure in the last chapter has eliminated all UV-divergences from the Feynman integrals arising from large momenta in $D = 4 - \varepsilon$ dimensions. This was necessary to obtain finite correlation functions in the limit $\varepsilon \to 0$. We have seen in Chapter 7 that the dependence on the cutoff or any other mass scale, introduced in the regularization process, changes the Ward identities derived from scale transformations by an additional term—the anomaly of scale invariance. The precise consequences of this term for the renormalized proper vertex functions were first investigated independently by Callan and Symanzik [1].

10.1 Callan-Symanzik Equation

The original derivation of the scaling properties of interacting theories did not quite proceed along the lines discussed in Section 7.4. Callan and Symanzik wanted to find the behavior of the renormalized proper vertex functions $\bar{\Gamma}^{(n)}(\mathbf{k}_1, \ldots, \mathbf{k}_n)$ for $n \geq 1$ under a change of the renormalized mass m. They did this in the context of cutoff regularization and renormalization conditions at a subtraction point as in Eqs. (9.23)–(9.33). To keep track of the parameters of the theory, we shall from now on enter these explicitly into the list of arguments, and write the proper vertex functions as $\bar{\Gamma}^{(n)}(\mathbf{k}_1, \ldots, \mathbf{k}_n; m, g)$. To save space, we abbreviate the list of momenta $\mathbf{k}_1, \ldots, \mathbf{k}_n$ by a momentum symbol $\mathbf{k}_i$, and use the notation $\bar{\Gamma}^{(n)}(\mathbf{k}_i; m, g)$. In addition, the bare proper vertex functions depend on the cutoff, or the deviation $\varepsilon = 4 - D$ from the dimension $D = 4$, and will be denoted by $\bar{\Gamma}_B^{(n)}(\mathbf{k}_i; m_B, \lambda_B, \Lambda)$.

Differentiating the bare proper vertex function $\bar{\Gamma}_B^{(n)}(\mathbf{k}_i; m_B, \lambda_B, \Lambda)$ with respect to the renormalized mass m at a fixed bare coupling constant and cutoff gives

$$m \frac{\partial}{\partial m} \bar{\Gamma}_B^{(n)}(\mathbf{k}_i; m_B, \lambda_B, \Lambda) \bigg|_{\lambda_B, \Lambda} = m \frac{\partial}{\partial m} m_B^2 \bigg|_{\lambda_B, \Lambda} \bar{\Gamma}_B^{(1,n)}(\mathbf{0}, \mathbf{k}_i; m_B, \lambda_B, \Lambda), \qquad (10.1)$$

where

$$\bar{\Gamma}_B^{(1,n)}(\mathbf{0}, \mathbf{k}_i; m_B, \lambda_B, \Lambda) = \frac{\partial}{\partial m_B^2} \bar{\Gamma}_B^{(n)}(\mathbf{k}_i; m_B, \lambda_B, \Lambda) \qquad (10.2)$$

is the proper vertex function associated with the correlation function containing an extra term $-\phi^2(\mathbf{x})/2$ inside the expectation value [recall Eqs. (5.98) and (5.93)]:

$$G^{(1,n)}(\mathbf{x}, \mathbf{x}_1, \ldots, \mathbf{x}_n) = -\frac{1}{2} \langle \phi^2(\mathbf{x}) \phi(\mathbf{x}_1) \cdots \phi(\mathbf{x}_n) \rangle. \qquad (10.3)$$

With the help of the renormalization constants Z_ϕ we go over to renormalized correlation functions as in Eq. (9.31):

$$\bar{\Gamma}_B^{(n)}(\mathbf{k}_i; m_B, \lambda_B, \Lambda) = Z_\phi^{-n/2} \bar{\Gamma}^{(n)}(\mathbf{k}_i; m, g), \qquad n \geq 1. \qquad (10.4)$$

We also introduce a renormalization constant Z_2 which makes the composite vertex function finite in the limit $\lambda_B \to \infty$ via

$$\bar{\Gamma}_B^{(1,n)}(\mathbf{0}, \mathbf{k}_i; m_B, \lambda_B, \Lambda) = Z_\phi^{-n/2} Z_2 \bar{\Gamma}^{(1,n)}(\mathbf{0}, \mathbf{k}_i; m, g), \quad n \geq 1 . \tag{10.5}$$

The constant Z_2 is fixed by the normalization condition

$$\bar{\Gamma}^{(1,n)}(\mathbf{0}, \mathbf{0}; m, g) = 1. \tag{10.6}$$

Then we define auxiliary functions

$$\beta = m\frac{\partial g}{\partial m}\bigg|_{\lambda_B, \Lambda}, \tag{10.7}$$

$$\gamma = \frac{1}{2} Z_\phi^{-1} m \frac{\partial Z_\phi}{\partial m}\bigg|_{\lambda_B, \Lambda}, \tag{10.8}$$

and rewrite the differential equation (10.1) in the form

$$\left(m\frac{\partial}{\partial m} + \beta\frac{\partial}{\partial g} - n\gamma\right)\bar{\Gamma}^{(n)}(\mathbf{k}_i; m, g) = Z_2 m \frac{\partial m_B^2}{\partial m}\bigg|_{\lambda_B, \Lambda} \bar{\Gamma}^{(1,n)}(\mathbf{0}, \mathbf{k}; m, g), \quad n \geq 1 . \tag{10.9}$$

The normalization conditions (9.23) requires that $\bar{\Gamma}^{(2)}(\mathbf{0}; m, g) = m^2$. Inserting this together with (10.6) into Eq. (10.9) for $n = 2$ gives

$$(2 - 2\gamma)m^2 = Z_2 m \frac{\partial m_B^2}{\partial m}\bigg|_{\lambda_B, \Lambda} . \tag{10.10}$$

This permits us to express the right-hand side of (10.9) in terms of renormalized quantities, yielding the *Callan-Symanzik equation*:

$$\left(m\frac{\partial}{\partial m} + \beta\frac{\partial}{\partial g} - n\gamma\right)\bar{\Gamma}^{(n)}(\mathbf{k}_i; m, g) = (2 - 2\gamma)m^2\bar{\Gamma}^{(1,n)}(\mathbf{0}, \mathbf{k}; m, g), \quad n \geq 1 . \tag{10.11}$$

In general, the dimensionless functions β and γ depend on g and m/Λ. But since they govern differential equations for the renormalized proper vertex functions, they do not depend on Λ after all, being functions of g and m. Their properties will be studied in detail in the next section.

The Callan-Symanzik equation makes statements on the scaling properties of correlation functions by going to small masses or, equivalently, to large momenta. In this limit, one may invoke results by Weinberg [2], according to which the right-hand side becomes small compared with the left-hand side. From the resulting approximate homogeneous equation one may deduce the critical behavior of the theory, provided the β-function is zero at some coupling strength $g = g^*$. In the critical regime, the proper vertex functions therefore satisfy the differential equation

$$\left(m\frac{\partial}{\partial m} - n\gamma\right)\bar{\Gamma}^{(n)}(\mathbf{k}_i; m, g) \underset{m \approx 0}{\approx} 0, \quad n \geq 1 . \tag{10.12}$$

By combining this with the original scaling relation (7.54) which is valid for renormalized quantities on the basis of naive scaling arguments, we find

$$\left[\sum_{i=1}^{n} \mathbf{k}_i \partial_{\mathbf{k}_i} + n(d_\phi^0 + \gamma) - D\right]\bar{\Gamma}^{(n)}(\mathbf{k}_i; m, g) \underset{m \approx 0}{\approx} 0, \quad n \geq 1 . \tag{10.13}$$

Recall that our renormalized coupling g is dimensionless by definition, just as the quantity $\hat{\lambda}$ in (7.28).

The scaling equation (10.13) was discussed earlier in Eq. (7.74). It has the same form as the original scaling relation for a massless theory (7.20), except that the free-field dimension d_ϕ^0 is replaced by the modified dimension $d_\phi = d_\phi^0 + \gamma$. By comparison with (7.75), we identify the critical exponent η as

$$\eta = 2\gamma. \tag{10.14}$$

The interacting theory is invariant under scale transformations (7.76) of the renormalized interacting field. The critical scaling equation (10.13) is the origin of the anomalous dimensions observed in Eq. (7.66)–(7.68).

We shall not explore the consequences of the Callan-Symanzik equation further but derive a more powerful equation for studying the critical behavior of the theory.

10.2   Renormalization Group Equation

In Chapter 8 we have decided to regularize the theory by analytic extension of all Feynman integrals from integer values of the dimension D into the complex D-plane, and by subtracting the singularities in $\varepsilon = 4 - D$ in a certain minimal way referred to as minimal subtraction. Adding the dimensional parameter ε to the list of arguments, we shall write the bare correlation functions as

$$G_B^{(n)}(\mathbf{x}_1, \ldots, \mathbf{x}_n; m_B, \lambda_B, \varepsilon) \;\equiv\; \langle \phi_B(\mathbf{x}_1) \cdots \phi_B(\mathbf{x}_n) \rangle. \tag{10.15}$$

They are calculated from a generating functional (2.13), whose Boltzmann factor contains the bare energy functional $E_B[\phi]$ of Eq. (9.73).

In the subtracted terms, one has the freedom of introducing an arbitrary mass parameter μ. The renormalization constants of the theory will therefore depend on μ rather than on a cutoff Λ. The renormalized correlation functions have the form

$$G^{(n)}(\mathbf{x}_1, \ldots, \mathbf{x}_n; m, g, \mu, \varepsilon) \;\equiv\; \langle \phi(\mathbf{x}_1) \cdots \phi(\mathbf{x}_n) \rangle, \tag{10.16}$$

and are related to the bare quantities (10.15) by a multiplicative renormalization:

$$G_B^{(n)}(\mathbf{x}_1, \ldots, \mathbf{x}_n; m_B, \lambda_B, \varepsilon) = Z_\phi^{n/2}(g(\mu), \varepsilon) G^{(n)}(\mathbf{x}_1, \ldots, \mathbf{x}_n; m, g, \mu, \varepsilon), \qquad n \geq 1, \tag{10.17}$$

where $Z_\phi(g(\mu), \varepsilon)$ is the field normalization constant defined by $\phi_B = Z_\phi^{1/2}\phi$. For the propagator with $n = 2$, this implies the relation

$$G_B^{(2)}(\mathbf{x}_1, \mathbf{x}_2; m_B, \lambda_B, \varepsilon) = Z_\phi(g(\mu), \varepsilon)\, G^{(2)}(\mathbf{x}_1, \mathbf{x}_2; m, g, \mu, \varepsilon). \tag{10.18}$$

After a Fourier transform according to Eq. (4.13), the same factors $Z_\phi(g(\mu), \varepsilon)$ renormalize the momentum space n-point functions $G_B^{(n)}(\mathbf{k}_1, \ldots, \mathbf{k}_n; m_B, \lambda_B, \varepsilon)$ to $G^{(n)}(\mathbf{k}_1, \ldots, \mathbf{k}_n; m, g, \mu, \varepsilon)$. For the proper vertex functions $\bar{\Gamma}_B^{(n)}(\mathbf{k}_1, \ldots, \mathbf{k}_n; m_B, \lambda_B, \varepsilon)$, which are obtained from the 1PI parts of the connected n-point functions $G_B^{(n)}(\mathbf{k}_1, \ldots, \mathbf{k}_n; m_B, \lambda_B, \varepsilon)$ by amputating the external lines, i.e., by dividing out n external propagators $G_B^{(2)}(\mathbf{k}_i; m_B, \lambda_B, \varepsilon)$, the renormalized quantities are given by

$$\bar{\Gamma}^{(n)}(\mathbf{k}_1, \ldots, \mathbf{k}_n; m, g, \mu, \varepsilon) = Z_\phi^{n/2}(g(\mu), \varepsilon)\, \bar{\Gamma}_B^{(n)}(\mathbf{k}_1, \ldots, \mathbf{k}_n; m_B, \lambda_B, \varepsilon), \qquad n \geq 1. \tag{10.19}$$

These expressions remain finite in the limit $\varepsilon \to 0$. In the following discussion we shall suppress the obvious ε-dependence in the arguments of all quantities, for brevity, unless it is helpful for a better understanding.

The renormalized parameters g, m, and ϕ defined in Eqs. (9.72) depend on the bare quantities, and on the mass parameter μ:

$$\phi^2 = Z_\phi^{-1}(g(\mu), \varepsilon)\, \phi_B^2, \quad m^2 = m^2(\mu) \equiv \frac{Z_\phi(g(\mu), \varepsilon)}{Z_{m^2}(g(\mu), \varepsilon)}\, m_B^2, \quad g = g(\mu) \equiv \mu^{-\varepsilon} \frac{Z_\phi^2(g(\mu), \varepsilon)}{Z_g(g(\mu), \varepsilon)}\, \lambda_B \; . \tag{10.20}$$

The renormalized proper vertex functions $\bar{\Gamma}^{(n)}(\mathbf{k}_1, \ldots, \mathbf{k}_n; m, g, \mu, \varepsilon)$ depend on μ in two ways: once explicitly, and once via $g(\mu)$ and $m(\mu)$. The explicit dependence comes from factors μ^ε which are generated when replacing λ by $\mu^\varepsilon g$ in (8.58). By contrast, the unrenormalized proper vertex functions $\bar{\Gamma}_B^{(n)}(\mathbf{k}_1, \ldots, \mathbf{k}_n; m_B, \lambda_B, \varepsilon)$ do not depend on μ. On the right-hand side of Eq. (10.19), only Z_ϕ depends on μ via $g(\mu)$.

The bare proper vertex functions are certainly independent of the artificially introduced arbitrary mass parameter μ. When rewriting them via Eq. (10.19) as $Z_\phi^{-n/2}\, \bar{\Gamma}^{(n)}(\mathbf{k}_1, \ldots, \mathbf{k}_n; m, g, \mu, \varepsilon)$, this implies a nontrivial behavior of the renormalized vertex functions under changes of μ. The associated changes of the renormalized proper vertex functions, and the other renormalized parameters, must be related to each other in a specified way. It is this relation which ensures that the physical information in the renormalized functions remains invariant under changes of μ.

Let us calculate these changes. We apply the dimensionless operator $\mu\, \partial/\partial\mu$ to Eq. (10.19) with fixed bare parameters, and obtain for $n \geq 1$:

$$\left[-n\mu \frac{\partial}{\partial\mu} \log Z_\phi^{1/2} \Big|_B + \mu \frac{\partial g}{\partial\mu} \Big|_B \frac{\partial}{\partial g} + \mu \frac{\partial m}{\partial\mu} \Big|_B \frac{\partial}{\partial m} + \mu \frac{\partial}{\partial\mu} \right] \bar{\Gamma}^{(n)}(\mathbf{k}_1, \ldots, \mathbf{k}_n; m, g, \mu) = 0. \tag{10.21}$$

The symbol $|_B$ indicates that the bare parameters m_B, λ_B are kept fixed. This equation expresses the invariance of $\bar{\Gamma}^{(n)}(\mathbf{k}_1, \ldots, \mathbf{k}_n; m, g, \mu)$ under a transformation $(\mu, m(\mu), g(\mu)) \to (\mu', m(\mu'), g(\mu'))$. The observables of the field system are invariant under a change of the mass scale $\mu \to \mu'$ if coupling constant $g(\mu)$ and mass $m(\mu)$ are changed appropriately. The mass scale μ is not an independent parameter.

The appropriate dependence of g, m and Z_ϕ on μ is described by the *renormalization group functions* (RG functions):

$$\gamma(m, g, \mu) = \mu \frac{\partial}{\partial\mu} \log Z_\phi^{1/2} \Big|_B , \tag{10.22}$$

$$\gamma_m(m, g, \mu) = \frac{\mu}{m} \frac{\partial m}{\partial\mu} \Big|_B , \tag{10.23}$$

$$\beta(m, g, \mu) = \mu \frac{\partial g}{\partial\mu} \Big|_B . \tag{10.24}$$

They allow us to rewrite Eq. (10.21) as the *renormalization group equation* (RGE) for the proper vertex functions with $n \geq 1$:

$$\left[\mu \frac{\partial}{\partial\mu} + \beta(m, g, \mu) \frac{\partial}{\partial g} - n\gamma(m, g, \mu) + \gamma_m(m, g, \mu)\, m \frac{\partial}{\partial m} \right] \bar{\Gamma}^{(n)}(\mathbf{k}_1, \ldots, \mathbf{k}_n; m, g, \mu) = 0 . \tag{10.25}$$

The solution of a partial differential equation like (10.25) is generally awkward, since β, γ, γ_m may depend on m, g and μ. It is an important property of 't Hooft's minimal subtraction scheme [3, 4] that the counterterms happen to be independent of the mass m, and that they depend only on the coupling constant g, apart from ε. The renormalization group functions (10.22)–(10.24) are therefore independent of m and μ, and depend only on g:

$$\gamma(g) \overset{\mathrm{MS}}{=} \mu \frac{\partial}{\partial \mu} \log Z_\phi^{1/2} \Big|_B \,, \tag{10.26}$$

$$\gamma_m(g) \overset{\mathrm{MS}}{=} \frac{\mu}{m} \frac{\partial m}{\partial \mu} \Big|_B \,, \tag{10.27}$$

$$\beta(g) \overset{\mathrm{MS}}{=} \mu \frac{\partial g}{\partial \mu} \Big|_B \,. \tag{10.28}$$

With these, the renormalization group equation (10.25) becomes

$$\left[\mu \frac{\partial}{\partial \mu} + \beta(g) \frac{\partial}{\partial g} - n\gamma(g) + \gamma_m(g)\, m \frac{\partial}{\partial m} \right] \bar{\Gamma}^{(n)}(\mathbf{k}_1, \ldots, \mathbf{k}_n; m, g, \mu) = 0 \,, \qquad n \geq 1 \,, \tag{10.29}$$

which is much easier to solve than the general form (10.25).

10.3   Calculation of Coefficient Functions from Counterterms

We now calculate the renormalization group functions taking advantage of the fact that the renormalization constants depend, with minimal subtractions, only on μ via the renormalized coupling constant $g(\mu)$. Consider first the function $\beta(g)$. Inserting the renormalization equation $\lambda_B = \mu^\varepsilon Z_g Z_\phi^{-2} g$ of (10.20) into (10.28), we find

$$\beta(g) = -\mu \frac{(\partial_\mu \lambda_B)_g}{(\partial_g \lambda_B)_\mu} = -\varepsilon \left[\frac{d}{dg} \log(g Z_g Z_\phi^{-2}) \right]^{-1} = -\varepsilon g \left[\frac{d \log g_B(g)}{d \log g} \right]^{-1}. \tag{10.30}$$

By the chain rule of differentiation, we rewrite (10.26) as

$$\gamma(g) = \mu \frac{\partial g}{\partial \mu} \Big|_B \frac{d}{dg} \log Z_\phi^{1/2} = \beta(g) \frac{d}{dg} \log Z_\phi^{1/2}. \tag{10.31}$$

With this, Eq. (10.30) takes the form

$$\beta(g) = \frac{-\varepsilon + 4\gamma(g)}{d \log[g Z_g(g)]/dg} \,. \tag{10.32}$$

Finally, we find from the relation $m_{\mathrm{B}}^2 = m^2\, Z_{m^2}/Z_\phi$ of (10.20) the renormalization group function

$$\gamma_m(g) = -\frac{\beta(g)}{2} \left[\frac{d}{dg} \log Z_{m^2} - \frac{d}{dg} \log Z_\phi \right] = -\frac{\beta(g)}{2} \frac{d}{dg} \log Z_{m^2} + \gamma(g). \tag{10.33}$$

In principle, the right-hand sides still depend on ε, so that we should really write the RG functions as

$$\beta = \beta(g, \varepsilon), \quad \gamma = \gamma(g, \varepsilon), \quad \gamma_m = \gamma_m(g, \varepsilon). \tag{10.34}$$

However, the ε-dependence turns out to be extremely simple. Due to the renormalizability of the theory, the functions $\beta(g, \varepsilon)$, $\gamma(g, \varepsilon)$, $\gamma_m(g, \varepsilon)$ have to remain finite in the limit $\varepsilon \to 0$, and thus free of poles in ε. In fact, an explicit evaluation of the right-hand side of Eqs. (10.30),

(10.31), and (10.33) demonstrates the cancellation of all poles in ε. Thus we can expand these functions in a power series in ε with nonnegative powers ε^n.

Moreover, we can easily convince ourselves that, of the nonnegative powers ε^n, only the first is really present, and this only in the function $\beta(g, \varepsilon)$. In order to show this, we make use of the explicit general form of the $1/\varepsilon$-expansions of the renormalization constants in minimal subtraction, which is

$$Z_\phi(g, \varepsilon) = 1 + \sum_{n=1}^{\infty} Z_{\phi,n}(g)\frac{1}{\varepsilon^n}, \tag{10.35}$$

$$Z_{m^2}(g, \varepsilon) = 1 + \sum_{n=1}^{\infty} Z_{m^2,n}(g)\frac{1}{\varepsilon^n}, \tag{10.36}$$

$$Z_g(g, \varepsilon) = 1 + \sum_{n=1}^{\infty} Z_{g,n}(g)\frac{1}{\varepsilon^n}. \tag{10.37}$$

Then Eqs. (10.31)–(10.33) can be rewritten as

$$\gamma(g, \varepsilon)\left[1 + \sum_{n=1}^{\infty} Z_{\phi,n}(g)\varepsilon^{-n}\right] = \frac{1}{2}\beta(g, \varepsilon)\sum_{n=1}^{\infty} Z'_{\phi,n}(g)\varepsilon^{-n}, \tag{10.38}$$

$$\beta(g, \varepsilon)\left\{1 + \sum_{n=1}^{\infty} [gZ_{g,n}(g)]'\varepsilon^{-n}\right\} = [-\varepsilon + 4\gamma(g, \varepsilon)]\, g\left[1 + \sum_{n=1}^{\infty} Z_{g,n}(g)\varepsilon^{-n}\right], \tag{10.39}$$

$$[-\gamma_m(g, \varepsilon) + \gamma(g, \varepsilon)]\left[1 + \sum_{n=1}^{\infty} Z_{m^2,n}(g)\varepsilon^{-n}\right] = \frac{\beta(g, \varepsilon)}{2}\sum_{n=1}^{\infty} Z'_{m^2,n}(g)\varepsilon^{-n}. \tag{10.40}$$

By inserting (10.38) into (10.39), we see that $\beta(g, \varepsilon)$ can at most contain the following powers of ε:

$$\beta(g, \varepsilon) = \beta_0(g) + \varepsilon\beta_1(g). \tag{10.41}$$

Using this to eliminate $\beta(g, \varepsilon)$ from Eqs. (10.38) and (10.40), we find $\gamma(g, \varepsilon)$ and $\gamma_m(g, \varepsilon)$ as functions of ε. By equating the regular terms in the three equations, we find

$$\begin{aligned}
\beta_0 + \varepsilon\beta_1 + \beta_1(Z_{g,1} + gZ'_{g,1}) &= (-\varepsilon + 4\gamma)g - gZ_{g,1}, \\
\gamma &= \tfrac{1}{2}\beta_1 Z_{\phi,1}, \\
\gamma_m - \gamma &= -\tfrac{1}{2}\beta_1 Z'_{m^2,1}.
\end{aligned} \tag{10.42}$$

The solutions are

$$\begin{aligned}
\beta_1(g) &= -g, \tag{10.43} \\
\beta_0(g) &= gZ'_{g,1}(g) + 4g\gamma(g), \tag{10.44} \\
\gamma(g) &= \tfrac{1}{2}Z'_{\phi,1}(g)\,\beta_1(g), \tag{10.45} \\
\gamma_m(g) &= \tfrac{1}{2}gZ'_{m,1}(g) + \gamma(g). \tag{10.46}
\end{aligned}$$

Thus, amazingly, the three functions $\beta(g), \gamma(g), \gamma_m(g)$ have all been expressed in terms of the derivatives of the three residues $Z_{g,1}(g), Z_{\phi,1}(g), Z_{m^2,1}(g)$ of the simple $1/\varepsilon$ -pole in the counterterms. The dimensional parameter $\varepsilon = 4 - D$ enters the renormalization group function only at a single place: in the $-\varepsilon g$-term of $\beta(g)$:

$$\beta(g) = -\varepsilon g + g^2 \, Z'_{g,1}(g) + 4g\gamma(g) \, . \tag{10.47}$$

The finiteness of the observables β, γ, γ_m at $\varepsilon = 0$ requires that none of the higher residues of Eqs. (10.38)–(10.40) can contribute. Indeed, we can easily verify in the available expansions that there exists an infinite set of relations among the expansion coefficients, useful for checking calculations:

$$\beta_0(gZ_{g,n})' - g(gZ_{g,n+1})' = 4\gamma \, gZ_{g,n} - Z_{g,n+1} \, g, \tag{10.48}$$

$$\gamma Z_{\phi,n} = \tfrac{1}{2}\beta_0 Z'_{\phi,n} - \tfrac{1}{2}gZ'_{\phi,n+1}, \tag{10.49}$$

$$(\gamma_m - \gamma)Z_{m^2,n} = -\tfrac{1}{2}\beta_0 Z'_{m^2,n} + \tfrac{1}{2}gZ'_{m^2,n+1}. \tag{10.50}$$

From the two-loop renormalization constants in Eqs. (9.115)–(9.119), we extract the residues of $1/\varepsilon$:

$$Z_{g,1} = \frac{N+8}{3}\frac{g}{(4\pi)^2} - \frac{5N+22}{9}\frac{g^2}{(4\pi)^4} \, ,$$

$$Z_{\phi,1} = -\frac{N+2}{36}\frac{g^2}{(4\pi)^4} \, , \tag{10.51}$$

$$Z_{m^2,1} = \frac{N+2}{3}\frac{g}{(4\pi)^2} - \frac{N+2}{6}\frac{g^2}{(4\pi)^4} \, ,$$

so that we obtain:

$$\beta_1(g) = -g, \quad \beta_0(g) = g^2 Z'_{g,1}(g) + 4g\gamma(g) = \frac{N+8}{3}\frac{g^2}{(4\pi)^2} - \frac{3N+14}{3}\frac{g^3}{(4\pi)^4} \, , \tag{10.52}$$

$$\gamma(g) = \tfrac{1}{2}Z'_{\phi,1}(g)\,\beta_1(g) = \frac{N+2}{36}\frac{g^2}{(4\pi)^4} \, , \tag{10.53}$$

$$\gamma_m(g) = \tfrac{1}{2}gZ'_{m^2,1}(g) + \gamma(g) = \frac{N+2}{6}\frac{g}{(4\pi)^2} - \frac{5(N+2)}{36}\frac{g^2}{(4\pi)^4} \, . \tag{10.54}$$

The coupling constant always appears with a factor $1/(4\pi)^2$, which is generated by the loop integrations. We therefore introduce a modified coupling constant

$$\bar{g} \equiv \frac{g}{(4\pi)^2}, \tag{10.55}$$

which brings the renormalization group functions to the shorter form:

$$\beta_{\bar{g}}(\bar{g}) = \bar{g}\left(-\varepsilon + \frac{N+8}{3}\,\bar{g} - \frac{3N+14}{3}\,\bar{g}^2\right), \tag{10.56}$$

$$\gamma(\bar{g}) = \frac{N+2}{36}\,\bar{g}^2, \tag{10.57}$$

$$\gamma_m(\bar{g}) = \frac{N+2}{6}\,\bar{g} - \frac{5(N+2)}{36}\,\bar{g}^2. \tag{10.58}$$

It is always possible to introduce a further modified coupling constant defined by an expansion of the generic type $g_H = G(g) = g + a_2 g^2 + \ldots$ with unit coefficient of the first term, which has the property that the function $\beta(g_H)$ consists only of the first three terms

$\beta_H(g_H) = -\varepsilon \bar{g} + b_2 g_H^2 + b_3 g_H^3$ [5]. Since we shall not use this fact, we refer the reader to the original work for a proof.

It is instructive to verify explicitly the cancellation of $1/\varepsilon^n$-singularities in the calculation of the renormalization group functions from Eqs. (10.30)–(10.33). Take, for instance, $\gamma(g)$ of Eq. (10.31). For a more impressive verification, let us anticipate here the five-loop results (15.11) for the renormalization constant $Z_\phi(\bar{g})$ and Eq. (17.5) for the β-function, extending our two-loop expansions (9.115) and (10.56). Selecting the case of $N = 1$ for brevity of the formulas, the five-loop extension of the expansion (10.56) reads

$$\begin{aligned}
\beta_{\bar{g}}(\bar{g}) &= -\varepsilon \bar{g} + 3 \bar{g}^2 - \frac{17}{3} \bar{g}^3 + \left[\frac{145}{8} + 12 \zeta(3) \right] \bar{g}^4 \\
&+ \left[-\frac{3499}{48} + \frac{\pi^4}{5} - 78 \zeta(3) - 120 \zeta(5) \right] \bar{g}^5 \\
&+ \left[\frac{764621}{2304} - \frac{1189 \pi^4}{720} - \frac{5 \pi^6}{14} + \frac{7965 \zeta(3)}{16} + 45 \zeta^2(3) + 987 \zeta(5) + 1323 \zeta(7) \right] \bar{g}^6 .
\end{aligned}$$
(10.59)

From $Z_\phi(\bar{g})$ of Eq. (17.5) for $N = 1$, we find the five-loop expansion of the logarithmic derivative on the right-hand side of Eq. (10.31):

$$\begin{aligned}
[\log Z_\phi(\bar{g})]' &= -\frac{1}{6\varepsilon} \bar{g} + \left(\frac{-1}{2\varepsilon^2} + \frac{1}{8\varepsilon} \right) \bar{g}^2 + \left(\frac{-3}{2\varepsilon^3} + \frac{95}{72\varepsilon^2} - \frac{65}{96\varepsilon} \right) \bar{g}^3 \\
&+ \left[\frac{-9}{2\varepsilon^4} + \frac{163}{24\varepsilon^3} - \frac{553}{96\varepsilon^2} + \frac{3709}{1152\varepsilon} + \frac{\pi^4}{90\varepsilon} - \frac{2\zeta(3)}{\varepsilon^2} - \frac{3\zeta(3)}{8\varepsilon} \right] \bar{g}^4 \\
&+ \left(\frac{-13}{48\varepsilon^4} + \frac{179}{864\varepsilon^3} - \frac{23}{256\varepsilon^2} \right) \bar{g}^5 .
\end{aligned}$$
(10.60)

When forming the product $\beta_{\bar{g}}(\bar{g}) \times [\log Z_\phi(\bar{g})]'$ in (10.31) to obtain $\gamma(\bar{g})$, the contribution to $\gamma(g)$ comes from the product of the ε-term in $\beta_{\bar{g}}(\bar{g})$ with the $1/\varepsilon$ terms in $[\log Z_\phi(\bar{g})]'$. The higher singularities $1/\varepsilon^n$, $1/\varepsilon^{n-1}$, ... in the $\bar{g}^n$-term of $[\log Z_\phi(\bar{g})]'$ are reduced by one power of $1/\varepsilon$ when multiplied with the ε-term of $\beta_{\bar{g}}(\bar{g})$. For $n \geq 2$ the resulting terms are canceled by products of the other terms in $\beta_{\bar{g}}(\bar{g})$ with the singular terms associated with the lower powers $\bar{g}^{n-1}$, $\bar{g}^{n-2}$, ... in $[\log Z_\phi(\bar{g})]'$.

Having determined the renormalization group functions, we shall now solve the renormalization group equations (10.29). In order to avoid rewriting these equations in terms of the new reduced coupling constant $\bar{g}$, we shall rename $\bar{g}$ as g and drop the subscript $\bar{g}$ on the β-function.

10.4 Solution of the Renormalization Group Equation

The renormalization group equation (10.29) is a partial differential equation. Its coefficients depend only on g. Such an equation is solved by the method of characteristics. We introduce a *dimensionless scale parameter* σ and replace μ by $\sigma\mu$, so that variations of the mass scale μ are turned into variations of σ at a fixed mass scale μ. Then we introduce auxiliary functions $g(\sigma)$, $m(\sigma)$, called *running coupling constant* and *running mass*, which satisfy the first-order differential equations:

$$\beta(g(\sigma)) = \sigma \frac{dg(\sigma)}{d\sigma} , \qquad g(1) = g, \tag{10.61}$$

$$\gamma_m(g(\sigma)) = \frac{\sigma}{m(\sigma)} \frac{dm(\sigma)}{d\sigma} , \qquad m(1) = m , \tag{10.62}$$

$$\mu(\sigma) = \sigma \frac{d\mu(\sigma)}{d\sigma} , \qquad \mu(1) = \mu . \tag{10.63}$$

The solutions define trajectories in the (σ, m, g)-space which connect theories renormalized with different mass parameter $\sigma\mu$.

The temperature dependence of the theory is introduced via the mass parameter $m(1)$, i.e. via the renormalized mass $m(\sigma)$ at a fixed mass scale $\sigma\mu$ with $\sigma = 1$. Specifically we assume

$$m^2 = \mu^2 t, \quad t = \frac{T}{T_c} - 1. \tag{10.64}$$

Recall that when setting up the energy density in Section 1.4 to serve as a starting point for the field-theoretic treatment of thermal fluctuations, we followed Landau by taking the squared *bare mass* to be proportional to the temperature deviation from the critical temperature:

$$m_B^2 \propto t, \quad t = \frac{T}{T_c} - 1. \tag{10.65}$$

According to the renormalization equation (10.20), the renormalized mass is proportional to the bare mass, $m^2 = m_B^2 Z_\phi(g, \mu)/Z_{m^2}(g, \mu)$. Thus, at a fixed auxiliary mass scale μ and $\sigma = 1$, the renormalized mass is also proportional to t, as stated in Eq. (10.64). Note that this is a peculiarity of the present regularization procedure. In the earlier procedure used in the general qualitative discussion of scale behavior in Section 7.4, the mass was renormalized in Eq. (7.57) with m-dependent renormalization constants to $m^2 = m_B^2 Z_\phi(\lambda, m, \Lambda)/Z_{m^2}(\lambda, m, \Lambda)$. For small m, these change the linear behavior of the bare square mass $m_B^2 \propto t$ to the power behavior $m \propto t^\nu$ of the renormalized mass [see (7.70), (7.71)]. Such a behavior can be derived within the present scheme if we set the mass scale $\mu(\sigma)$ equal to the running renormalized mass $m(\sigma)$ for some σ_m. This will be done when fixing the parametrization in Eq. (10.78).

The solution of (10.61) is immediately found to be

$$\log \sigma = \int_g^{g(\sigma)} \frac{dg'}{\beta(g')}. \tag{10.66}$$

Inserting this into the second equation (10.62), we obtain

$$m(\sigma) = m \exp\left[\int_1^\sigma \frac{d\sigma'}{\sigma'} \gamma_m(g(\sigma'))\right]. \tag{10.67}$$

The last equation (10.63) is solved by

$$\mu(\sigma) = \mu\sigma. \tag{10.68}$$

With these functions, Eq. (10.29) becomes

$$\left[\sigma\frac{d}{d\sigma} - n\gamma(g(\sigma))\right]\bar{\Gamma}^{(n)}(\mathbf{k}_i; m(\sigma), g(\sigma), \mu(\sigma)) = 0, \quad n \geq 1. \tag{10.69}$$

For brevity, we have written $\mathbf{k}_i$ for $\mathbf{k}_1, \ldots, \mathbf{k}_n$. The solution of Eq. (10.69) is

$$\bar{\Gamma}^{(n)}(\mathbf{k}_i; m, g, \mu) = e^{-n\int_1^\sigma d\sigma' \gamma(g(\sigma'))/\sigma'} \bar{\Gamma}^{(n)}(\mathbf{k}_i; m(\sigma), g(\sigma), \mu\sigma), \quad n \geq 1. \tag{10.70}$$

One set of proper vertex functions specified by the arguments m, g, μ represents an infinite family of vertex functions of the ϕ^4-theory whose parameters are connected by a trajectory $g(\sigma), m(\sigma), \mu\sigma$ in the parameter space. This trajectory is traced out when σ runs from zero to infinity.

The renormalization group trajectory connects can be employed to study the behavior of the theory as the mass parameter approaches zero. For this purpose we take into account the trivial scaling behavior of the proper vertex function in all variables which follow from the dimensional analysis in Subsection 7.3.1. The n-point correlation function is the expectation value of n fields of naive dimension $d_\phi^0 = D/2 - 1$. As such it has a mass dimension [compare (7.49)]

$$[G^{(n)}(\mathbf{x}_1, \ldots, \mathbf{x}_n)] = \mu^{n(D/2-1)}. \tag{10.71}$$

When going to momentum space [see (4.13)], each of the n Fourier integrals adds a number $-D$ to the dimension, so that [compare (7.16)]

$$[G^{(n)}(\mathbf{k}_1, \ldots, \mathbf{k}_n)] = \mu^{n(-D/2-1)}. \tag{10.72}$$

For the two-point function, this implies

$$[G^{(2)}(\mathbf{k}_1, \mathbf{k}_2)] = \mu^{-D-2}. \tag{10.73}$$

The propagator $G(\mathbf{k})$ with a single momentum argument arises from this by removing the D-dimensional δ-function which guarantees overall momentum conservation [recall (4.4)]. Its naive dimension is

$$[G(\mathbf{p})] = \mu^{-2}. \tag{10.74}$$

Equation (10.71) implies for the connected n-point proper vertex functions the dimension

$$[\bar{\Gamma}^{(n)}(\mathbf{k}_i; m, g, \mu)] = \mu^{D-n(D/2-1)}. \tag{10.75}$$

If we now rescale all dimensional parameters by an appropriate power of σ, we obtain the trivial scaling relation [compare (7.53)]

$$\bar{\Gamma}^{(n)}(\mathbf{k}_i; m, g, \mu) = \sigma^{D-n(D/2-1)} \bar{\Gamma}^{(n)}(\mathbf{k}_i/\sigma; m/\sigma, g, \mu/\sigma). \tag{10.76}$$

Inserting (10.70) into the right-hand side, we find for $n \geq 1$:

$$\bar{\Gamma}^{(n)}(\mathbf{k}_i; m, g, \mu) = \sigma^{D-n(D/2-1)} \exp\left[-n \int_1^\sigma d\sigma' \frac{\gamma(g(\sigma'))}{\sigma'}\right] \bar{\Gamma}^{(n)}(\mathbf{k}_i/\sigma; m(\sigma)/\sigma, g(\sigma), \mu). \tag{10.77}$$

We now choose $\sigma = \sigma_m$ in such a way that the running mass $m(\sigma)$ equals the running additional mass scale $\mu(\sigma)$:

$$m^2(\sigma_m) = \mu^2(\sigma_m) = \mu^2 \sigma_m^2. \tag{10.78}$$

For $m^2 > 0$, the rescaled mass $m(\sigma)/\sigma$ is now equal to the mass parameter μ. Then (10.77) becomes

$$\bar{\Gamma}^{(n)}(\mathbf{k}_i; m, g, \mu) = \sigma_m^{D-n(D/2-1)} \exp\left[-n \int_1^{\sigma_m} d\sigma' \frac{\gamma(g(\sigma'))}{\sigma'}\right] \bar{\Gamma}^{(n)}(\mathbf{k}_i/\sigma_m; \mu, g(\sigma_m), \mu). \tag{10.79}$$

This equation relates the renormalized proper vertex functions $\bar{\Gamma}^{(n)}(\mathbf{k}_i, m, g, \mu)$ of an arbitrary mass to those of a fixed mass equal to the mass parameter μ at rescaled momenta $\mathbf{k}_i/\sigma_m$ and a running coupling constant $g(\sigma_m)$. Apart from a trivial overall rescaling factor $\sigma_m^{D-n(D/2-1)}$ due to the naive dimension, there is also a nontrivial exponential function.

Our goal is to study the behavior of the proper vertex functions on the left-hand side of (10.79) in the critical region where $m \to 0$. This is possible with the help of Eq. (10.79), whose right-hand side has a fixed mass equal to the mass parameter μ. All mass dependence of the right-hand side resides in the rescaling parameter σ_m. The index on σ_m indicates that it is related to m. The relation is found from Eq. (10.67), which yields the ratio

$$
\begin{aligned}
\frac{m^2}{\mu^2} &= \frac{m^2(\sigma_m)}{\mu^2} \exp\left\{\int_1^{\sigma_m} \frac{d\sigma'}{\sigma'} \left[-2\gamma_m(g(\sigma'))\right]\right\} \\
&= \frac{m^2(\sigma_m)}{\sigma_m^2\mu^2} \exp\left\{\int_1^{\sigma_m} \frac{d\sigma'}{\sigma'} \left[2 - 2\gamma_m(g(\sigma'))\right]\right\}.
\end{aligned}
\tag{10.80}
$$

Inserting here Eq. (10.78), we obtain

$$
\frac{m^2}{\mu^2} = \exp\left\{\int_1^{\sigma_m} \frac{d\sigma'}{\sigma'} \left[2 - 2\gamma_m(g(\sigma'))\right]\right\}.
\tag{10.81}
$$

Near the critical point, experimental correlation functions show the simple scaling behavior stated in Eq. (1.28). Such a behavior can be reproduced by Eqs. (10.81) and (10.79), if the coupling constant g runs for $m \to 0$ into a *fixed point* g^*, for which the running coupling constant $g(\sigma)$ becomes independent of σ, satisfying

$$
\left[\frac{d\,g(\sigma)}{d\,\sigma}\right]_{g=g^*} = 0.
\tag{10.82}
$$

Assuming that $\gamma_m^* \equiv \gamma_m(g^*) < 1$, which will be found to be true in the present field theory, the integrand in Eq. (10.81) is singular at $\sigma = 0$, and the asymptotic behavior of σ_m for $m \to 0$ can immediately be found:

$$
\frac{m^2}{\mu^2} = t \overset{m\approx0}{\approx} \exp\left\{\int_1^{\sigma_m} \frac{d\sigma'}{\sigma'} \left[2 - 2\gamma_m(g^*)\right]\right\} = \sigma_m^{2-2\gamma_m^*}.
\tag{10.83}
$$

This shows that σ_m goes to zero for $m \to 0$ with the power law

$$
\sigma_m \approx \left(\frac{m^2}{\mu^2}\right)^{1/(2-2\gamma_m^*)} \equiv t^{1/(2-2\gamma_m^*)}.
\tag{10.84}
$$

The power behavior of $\sigma_m \propto t^{1/(2-2\gamma_m^*)}$ enters crucially into the critical behavior of all correlation functions for $T \to T_c$.

In the limit $\sigma_m \to 0$, the exponential prefactor in (10.79) becomes the following power of t:

$$
\exp\left[-n\int_1^{\sigma_m} d\sigma' \frac{\gamma(g(\sigma'))}{\sigma'}\right] \overset{\sigma_m\approx0}{\approx} \sigma_m^{-n\gamma^*} \propto t^{-n\gamma^*/(2-2\gamma_m^*)},
\tag{10.85}
$$

where $\gamma^* \equiv \gamma(g^*)$. The n-point proper vertex function behaves therefore like

$$
\bar{\Gamma}^{(n)}(\mathbf{k}_i; m, g, \mu) \overset{m\approx0}{\approx} \sigma_m^{D-n(D/2-1)-n\gamma^*} \bar{\Gamma}^{(n)}(\mathbf{k}_i/\sigma_m; \mu, g^*, \mu), \qquad n \geq 1.
\tag{10.86}
$$

For the two-point proper vertex function this implies a scaling form

$$
\bar{\Gamma}^{(2)}(\mathbf{k}_i; m, g, \mu) \overset{m\approx0}{\approx} \mu^2 \sigma_m^{2-2\gamma^*} \tilde{g}(\mathbf{k}/\mu\sigma_m),
\tag{10.87}
$$

with some function $\tilde{g}(x)$. Comparison of the argument of $\tilde{g}$ with the general scaling expression in Eq. (1.8), we identify $\mu\sigma_m$ with the inverse correlation length ξ^{-1}. Together with equation

(10.78), this implies that the renormalized running mass is equal to the inverse coherence length: $m(\sigma_m) = \xi^{-1}$.

Comparing further the behavior (10.84) with that in (1.10), we can identify the critical exponent ν as

$$\nu = \frac{1}{2 - 2\gamma_m^*}. \tag{10.88}$$

With this, the relation (10.84) between σ_m and $t = m/\mu^2$ reads simply

$$\sigma_m \approx \left(\frac{m^2}{\mu^2}\right)^\nu \equiv t^\nu. \tag{10.89}$$

The length scale ξ characterizing the spatial behavior of the correlation function (10.87) diverges for $m^2 \to 0$ like

$$\xi(t) = \xi_0 \left|\frac{m^2}{\mu^2}\right|^{-\nu} = \xi_0 |t|^{-\nu}, \qquad t = \left|\frac{T}{T_c} - 1\right|. \tag{10.90}$$

If the expression (10.87) is nonzero for $m \to 0$, the limit must have the momentum dependence

$$\bar{\Gamma}^{(2)}(\mathbf{k}_i; m, g, \mu) \stackrel{m\to 0}{=} \text{const} \times \mu^{2\gamma^*} |\mathbf{k}|^{2-2\gamma^*}, \tag{10.91}$$

with some function $\tilde{f}(\kappa)$. When going over to the two-point function in x-space [recall (4.34)]

$$G^{(2)}(\mathbf{x}; m, g, \mu) = \int \frac{d^D k}{(2\pi)^D} e^{ikx} \frac{1}{\bar{\Gamma}^{(2)}(\mathbf{k}_i; m, g, \mu)} \tag{10.92}$$

this amounts to an x-dependence

$$G^{(2)}(\mathbf{x}; m, g, \mu) \stackrel{m\approx 0}{\propto} \frac{1}{r^{D-2+2\gamma^*}} \tilde{G}(\mathbf{x}\,\mu\sigma_m) = \frac{1}{r^{D-2+2\gamma^*}} \tilde{G}(\mathbf{x}/\xi). \tag{10.93}$$

This expression exhibits precisely the scaling form (1.28) discovered by Kadanoff, such that we identify the critical exponent η as

$$\eta = 2\gamma^*. \tag{10.94}$$

By analogy with this relation, we shall also introduce a critical exponent η_m as

$$\eta_m = 2\gamma_m^*, \tag{10.95}$$

so that (10.88) becomes

$$\nu = \frac{1}{2 - \eta_m}. \tag{10.96}$$

10.5   Fixed Point

Let us now see how such a fixed point with the property Eq. (10.82) is derived from Eq. (10.66). In first-order perturbation theory, the β-function has, from Eqs. (10.41), (10.43), and (10.44), the general form

$$\beta(g) = -\varepsilon g + bg^2, \tag{10.97}$$

where the constant b is, according to Eq. (10.52), equal to $(N+8)/3$ [recall that we have gone over to $\bar{g}$ via (10.55) and dropped the bar over g]. The β-function starts out with negative slope and has a zero at

$$g^* = \varepsilon/b, \tag{10.98}$$

as pictured in Fig. 10.1. For small ε, this statement is reliable even if we know $\beta(g)$ only to order g^2 in perturbation theory. Inserting (10.97) into equation (10.66), we calculate

$$\log \sigma = \int_g^{g(\sigma)} \frac{dg'}{-\varepsilon g' + bg'^2}. \tag{10.99}$$

This equation shows the important consequence of any zero in the β-function: If g is sufficiently close to a zero at $g = g^*$ then the value $g(\sigma)$ always runs into g^* in the limit $\sigma \to 0$, no matter whether $g = g(1)$ lies slightly above or below g^*. The point g^* is the fixed point of the renormalization flow. Since g^* is reached in the limit $\sigma \to 0$, which is the small-mass limit of the theory, one speaks of an *infrared-stable fixed point*. In Fig. 10.1, the flow of $g(\sigma)$ for $\sigma \to 0$ is illustrated by an arrow.

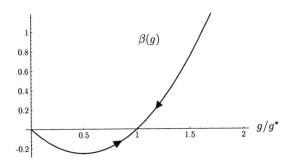

FIGURE 10.1 Flow of the coupling constant $g(\sigma)$ as the scale parameter σ approaches zero, i.e., in the infrared limit. For the opposite scale change, i.e., in the ultraviolet limit, the arrows reverse and the origin is a stable fixed point.

Using the variable $1/g$ instead of g, Eq. (10.99) becomes

$$\log \sigma = -\frac{1}{\varepsilon} \int_{1/g}^{1/g(\sigma)} \frac{dx}{1/g^* - x}. \tag{10.100}$$

This can be integrated directly to

$$\sigma = \frac{|1/g^* - 1/g(\sigma)|^{1/\varepsilon}}{|1/g^* - 1/g|^{1/\varepsilon}}, \tag{10.101}$$

so that

$$g(\sigma) = \frac{g^*}{1 + \sigma^\varepsilon \left(g^*/g - 1\right)}. \tag{10.102}$$

Near the fixed point, the behavior of $g(\sigma_m)$ can be calculated more generally. For this we go back to Eq. (10.66) and expand the denominator around the zero of the β-function:

$$\beta(g) \sim \beta'(g^*)(g - g^*) + \ldots \equiv \omega(g - g^*) + \ldots , \tag{10.103}$$

where we have introduced the slope of the β-function at the fixed point g^*:

$$\omega \equiv \beta'(g^*). \tag{10.104}$$

This is another critical exponent, as we shall see in Section 10.8. The exponent ω governs the leading corrections to the scaling laws. The sign of ω controls the stability of the fixed point. For an infrared stable fixed point, ω must be positive. Then we obtain from (10.62) an equation for $\sigma = \sigma_m$:

$$\log \sigma_m = \int_g^{g(\sigma_m)} \frac{dg'}{\beta(g')} \sim \frac{1}{\omega} \log \left[\frac{g(\sigma_m) - g^*}{g - g^*} \right], \qquad (10.105)$$

implying the following σ_m-dependence of $g(\sigma_m)$, correct to lowest order in $g - g^*$:

$$\frac{g(\sigma_m) - g^*}{g - g^*} = \sigma_m^\omega. \qquad (10.106)$$

This agrees with the specific solution (10.102) derived from the β-function (10.97) which has $\omega = \varepsilon$.

In general, the β-function may behave in many different ways for larger g. In particular, there may be more zeros to the right of g^*. We can see from Eq. (10.99) that, for positive β, the coupling constant $g(\sigma)$ will always run towards zero from the right. For negative β, it will run away from zero to the right.

Note that, in general, the initial coupling $g(\mu) = g(1)$ can flow only into the zero which lies in its *range of attraction*. In the present case this is guaranteed for small ε, if $g(1)$ is sufficiently small.

In the limit $\sigma \to \infty$, we see from Eq. (10.102) that $g(\sigma)$ tends to zero, which is the trivial zero of the β-function. This happens for any zero with a negative slope of $\beta(g)$. The limit $\sigma \to \infty$ corresponds to $m^2 \to \infty$, and for this reason such zeros are called *ultraviolet stable*. In this limit $g(\sigma) \to 0$, $\gamma(g(\sigma)) \to 0$, and $\gamma_m(g(\sigma)) \to 0$. Then scaling relation (10.84) implies that $\sigma_m = m/\mu$, and the correlation functions behave, by (10.86), like those of a free theory:

$$\bar{\Gamma}^{(n)}(\mathbf{k}_i; m, g, \mu) \overset{m \approx 0}{\approx} \left(\frac{m}{\mu} \right)^{D - n(D/2 - 1)} \bar{\Gamma}^{(n)}(\mathbf{k}_i \mu/m; 0, \mu, \mu). \qquad (10.107)$$

This is the behavior of a free-field theory where the fields fluctuate in a trivial purely Gaussian way. The zero in $\beta(g)$ at $g = 0$ is therefore called the *Gaussian* or *trivial fixed point*. In the ϕ^4-theory, the Gaussian fixed point is ultraviolet stable (UV-stable). Since the theory tends for $m \to \infty$ against a free theory, one also says that it is ultraviolet free. Note that this is true only in less than four dimensions.

In $D = 4$ dimensions, where $\varepsilon = 0$, the β-function has only one fixed point, the trivial Gaussian fixed point at the origin.

10.6 Effective Energy and Potential

The above considerations are useful for deriving the critical properties of a system only in the normal phase, where $T \geq T_c$. If we want to study the system in the phase with spontaneous symmetry breakdown, which exists for $T \leq T_c$, we have to perform a renormalization group analysis for the effective energy $\Gamma[\Phi]$ of the system, introduced in Section 5.6, and analyze its behavior as a function of the mass parameter μ. For this purpose we expand the effective energy in a power series in the field expectations in momentum space $\Phi(\mathbf{p})$ as

$$\Gamma[\Phi; m, g, \mu] = \sum_{n=1}^\infty \frac{1}{n!} \int \frac{d^D p_1}{(2\pi)^D} \cdots \frac{d^D p_n}{(2\pi)^D} \Phi(\mathbf{p}_1) \cdots \Phi(\mathbf{p}_n) \bar{\Gamma}^{(n)}(\mathbf{p}_1, \ldots, \mathbf{p}_n; m, g, \mu). \qquad (10.108)$$

The coefficients are the proper vertex functions. We have omitted for a moment the term $n = 0$ in this expansion, since it will require extra treatment. Actually, the omission calls for a new notation for the effective action, but since the reduced sum (10.108) will appear frequently in what follows, while the full effective action appears only in a few equations, we prefer keeping the notation unchanged, and shall instead refer to the full effective action including the $n = 0$-term as $\Gamma_{\rm tot}[\Phi; m, g, \mu]$.

Let us now apply the renormalization group equation (10.29) to each coefficient. Then we observe that the factor n in front of $\gamma(g)$ in Eq. (10.29) can be generated by a functional derivative with respect to the field expectations $\Phi(\mathbf{p})$, replacing

$$n \to \int \frac{d^D p}{(2\pi)^D} \Phi(\mathbf{p}) \frac{\delta}{\delta \Phi(\mathbf{p})}. \tag{10.109}$$

Then we find immediately the renormalization group equation

$$\left[\mu \partial_\mu + \beta(g)\partial_g - \gamma(g) \int \frac{d^D p}{(2\pi)^D} \Phi(\mathbf{p}) \frac{\delta}{\delta \Phi(\mathbf{p})} + \gamma_m(g)m\partial_m\right] \Gamma[\Phi(\mathbf{p}); m, g, \mu] = 0. \tag{10.110}$$

The addition of the missing $n = 0$ term will modify this equation as we shall see in the next section.

A corresponding equation holds for the *effective potential* $v(\Phi)$. This is defined as the negative effective energy density at a constant average field $\Phi(x) \equiv \Phi$:

$$v(\Phi) = -L^{-D}\bar{\Gamma}[\Phi; m, g, \mu]|_{\Phi(x)\equiv\Phi}. \tag{10.111}$$

Here L is the linear size of the D-dimensional box under consideration. The effective potential satisfies the differential equation

$$[\mu \partial_\mu + \beta(g)\partial_g - \gamma(g)\Phi\partial_\Phi + \gamma_m(g)m\partial_m]\, v(\Phi; m, g, \mu) = 0. \tag{10.112}$$

Due to its special relevance to physical applications, we solve here only the latter equation along the lines of Eqs. (10.61)–(10.70). We introduce a running field strength $\Phi(\sigma)$, satisfying the differential equation

$$\frac{1}{\Phi(\sigma)}\sigma\frac{d}{d\sigma}\Phi(\sigma) = -\gamma(g(\sigma)), \tag{10.113}$$

with the initial condition

$$\Phi(1) = \Phi. \tag{10.114}$$

The equation is solved by

$$\frac{\Phi(\sigma)}{\Phi} = \exp\left\{-\int_1^\sigma \frac{d\sigma'}{\sigma'}\gamma(g(\sigma'))\right\} = \exp\left\{-\int_g^{g(\sigma)} dg' \frac{\gamma(g')}{\beta(g')}\right\}. \tag{10.115}$$

Using this function $\Phi(\sigma)$, the effective potential satisfies the renormalization group equation [analogous to (10.70)]:

$$v(\Phi; m, g, \mu) = v(\Phi(\sigma); m(\sigma), g(\sigma), \mu\sigma). \tag{10.116}$$

Note that there is no prefactor as in (10.70).

Since $\Gamma[\Phi]$ is dimensionless and $v(\Phi)$ is related to $\Gamma[\Phi]$ by (10.111), there is a naive scaling relation analogous to (10.76):

$$v(\Phi; m, g, \mu) = \sigma^D v\left(\frac{\Phi}{\sigma^{D/2-1}}; \frac{m}{\sigma}, g, \frac{\mu}{\sigma}\right). \tag{10.117}$$

Together with (10.116), this gives

$$v(\Phi; m, g, \mu) = \sigma^D v\left(\frac{\Phi(\sigma)}{\sigma^{D/2-1}}; \frac{m(\sigma)}{\sigma}, g(\sigma), \mu\right). \tag{10.118}$$

At the mass dependent value $\sigma_m = m(\sigma_m)/\mu$, we obtain the analog of (10.79) for the effective potential

$$v(\Phi; m, g, \mu) = \sigma_m^D v\left(\frac{\Phi(\sigma_m)}{\sigma_m^{D/2-1}}; \mu, g(\sigma_m), \mu\right). \tag{10.119}$$

In the limit $m \to 0$, where $\sigma \to 0$, we see from Eq. (10.115) that the field behaves like

$$\frac{\Phi(\sigma_m)}{\Phi} \approx \sigma_m^{-\gamma^*}. \tag{10.120}$$

The effective potential has therefore the power behavior

$$v(\Phi; m, g, \mu) \overset{m \approx 0}{\approx} \sigma_m^D v(\Phi/\sigma_m^{\gamma^*+D/2-1}; \mu, g^*, \mu), \tag{10.121}$$

where σ_m is related to $t = m^2/\mu^2$ by (10.84).

For applications to many-body systems below T_c, it is most convenient to consider, instead of $\bar{\Gamma}[\Phi; m, g, \mu]$, the proper vertex functions in the presence of an external magnetization. They are obtained by expanding $\bar{\Gamma}[\Phi; m, g, \mu]$ functionally around $\Phi(x) \equiv \Phi_0$:

$$\bar{\Gamma}^{(n)}(\mathbf{x}_1, \ldots, \mathbf{x}_n; \Phi; m, g, \mu) \equiv \left.\frac{\delta^n \bar{\Gamma}[\Phi; m, g, \mu]}{\delta\Phi(\mathbf{x}_1) \ldots \delta\Phi(\mathbf{x}_n)}\right|_{\Phi \equiv \Phi_0}. \tag{10.122}$$

In momentum space, this gives

$$\bar{\Gamma}^{(n)}(\mathbf{k}_1, \ldots, \mathbf{k}_n; \Phi_0; m, g, \mu) = \sum_{n'=0}^{\infty} \frac{\Phi_0^{n'}}{n'!} \bar{\Gamma}^{(n+n')}(\mathbf{k}_1, \ldots \mathbf{k}_n, \mathbf{0}, \ldots, \mathbf{0}; m, g, \mu), \tag{10.123}$$

where the zeros after the arguments $\mathbf{k}_1, \ldots, \mathbf{k}_n$ indicate that there are n' more momentum arguments $\mathbf{k}_{n+1}, \ldots, \mathbf{k}_{n+n'}$ which have been set zero since a constant field $\Phi(\mathbf{x}) \equiv \Phi_0$ has a Fourier transform $\Phi(\mathbf{k}) \equiv \Phi_0 \delta^{(D)}(\mathbf{k})$. Thus the renormalization group equation for the proper vertex function at a nonzero field, $\bar{\Gamma}^{(n)}(\mathbf{k}_1, \ldots, \mathbf{k}_n; \Phi_0; m, g, \mu)$, can be obtained from those at zero field $\bar{\Gamma}^{(n+n')}(\mathbf{k}_1, \ldots, \mathbf{k}_n, \mathbf{k}_{n+1}, \ldots, \mathbf{k}_{n+n'}; m, g, \mu)$ with the last n' momenta set equal to zero, i.e., from

$$[\mu\partial_\mu + \beta(g)\partial_g - (n+n')\gamma(g) + \gamma_m(g)m\partial_m]$$
$$\times\bar{\Gamma}^{(n+n')}(\mathbf{k}_1, \ldots, \mathbf{k}_n, \mathbf{0}, \ldots, \mathbf{0}; m, g, \mu) = 0. \tag{10.124}$$

Inserting this into (10.123), we obtain the renormalization group equation

$$\left[\mu\partial_\mu + \beta(g)\partial_g - \gamma(g)\left(n + \Phi_0\frac{\partial}{\partial\Phi_0}\right) + \gamma_m(g)m\partial_m\right]$$
$$\times\bar{\Gamma}^{(n)}(\mathbf{k}_1, \ldots, \mathbf{k}_n; \Phi_0, m, g, \mu) = 0. \tag{10.125}$$

When treated as above, this leads to the scaling relation

$$\bar{\Gamma}^{(n)}(\mathbf{k}_i; \Phi_0; m, g, \mu) = e^{-n \int_1^\sigma \frac{d\sigma'}{\sigma'} \gamma(g(\sigma'))} \bar{\Gamma}^{(n)}(\mathbf{k}_i; \Phi_0(\sigma); m(\sigma), g(\sigma), \mu\sigma). \tag{10.126}$$

Together with the trivial scaling relation

$$\bar{\Gamma}^{(n)}(\mathbf{k}_i; \Phi_0; m, g, \mu) = \sigma^{D - n(D/2 - 1)} \bar{\Gamma}^{(n)}(\mathbf{k}_i/\sigma; \Phi_0/\sigma^{D/2 - 1}; m/\sigma, g, \mu/\sigma), \tag{10.127}$$

we find

$$\begin{aligned}
\bar{\Gamma}^{(n)}(\mathbf{k}_i; \Phi_0; m, g, \mu) &= \sigma^{D - n(D/2 - 1)} e^{-n \int_1^\sigma \frac{d\sigma'}{\sigma'} \gamma(g(\sigma'))} \\
&\quad \times \bar{\Gamma}^{(n)}(\mathbf{k}_i/\sigma; \Phi_0(\sigma)/\sigma^{D/2 - 1}; m(\sigma)/\sigma, g(\sigma), \mu).
\end{aligned} \tag{10.128}$$

This becomes, at $\sigma = \sigma_m$ of Eq. (10.78),

$$\begin{aligned}
\bar{\Gamma}^{(n)}(\mathbf{k}_i; \Phi_0; m, g, \mu) &= \sigma_m^{D - n(D/2 - 1)} e^{-n \int_1^{\sigma_m} \frac{d\sigma'}{\sigma'} \gamma(g(\sigma'))} \\
&\quad \times \bar{\Gamma}^{(n)}(\mathbf{k}_i/\sigma_m; \Phi_0(\sigma_m)/\sigma_m^{D/2 - 1}; \mu, g(\sigma_m), \mu),
\end{aligned} \tag{10.129}$$

and thus, near the critical point,

$$\bar{\Gamma}^{(n)}(\mathbf{k}_i; \Phi_0; m, g, \mu) = n(\gamma^* + D/2 - 1) \bar{\Gamma}^{(n)}(\mathbf{k}_i/\sigma_m; \Phi_0/\sigma_m^{\gamma^* + D/2 - 1}; \mu, g^*, \mu). \tag{10.130}$$

10.7   Special Properties of Ground State Energy

When deriving the behavior of the proper vertex functions $\bar{\Gamma}^{(n)}(\mathbf{k}_1, \ldots, \mathbf{k}_n; m, g)$ under changes of the scale parameter μ, the number n was restricted to positive integer values $n \geq 1$. The vacuum energy contained in $\bar{\Gamma}^{(0)}(m, g)$ was omitted from the sum in Eq. (10.108). Indeed, the vacuum energy does not follow the regular renormalization pattern. For the above calculation of the critical exponents, this irregularity is irrelevant. But if we want to calculate amplitude ratios (recall the definition in Section 1.2), we have to know the full thermodynamic potential as the temperature approaches the critical point from above and from below. Then the special renormalization properties of the vacuum diagrams can no longer be ignored. The fundamental difference between the ground state energies above and below the transition was seen at the mean-field level in Eq. (1.43). While the vacuum energy is identically zero above the transition, it behaves like $-(T - T_c)^2$ below T_c, which is the condensation energy in mean-field approximation.

More subtleties appear when calculating loop corrections. The lowest-order vacuum diagram shows a peculiar feature: with the help of (8.116) and (8.117), we find the full semiclassical effective potential at zero average field Φ and coupling constant:

$$\begin{aligned}
v_{\text{tot}}(0; m, 0, \mu) &= \frac{N}{2} \int \frac{d^D p}{(2\pi)^D} \log(\mathbf{p}^2 + m^2) \\
&= \frac{N}{2} \frac{2}{D} \frac{(m^2)^{D/2}}{(4\pi)^{D/2}} \Gamma(1 - D/2) = \frac{N}{2} \frac{m^4}{\mu^\varepsilon} \frac{1}{(4\pi)^2} \left[-\frac{1}{\varepsilon} + \frac{1}{2} \left(\log \frac{m^2}{4\pi\mu^2 e^{-\gamma}} - \frac{3}{2} \right) \right] + \mathcal{O}(\varepsilon). \tag{10.131}
\end{aligned}$$

The pole term at $\varepsilon = 0$ can be removed by adding to the potential a counterterm

$$v^{\text{sg}} \equiv \frac{N}{2} \frac{m^4}{\mu^\varepsilon} \frac{1}{(4\pi)^2} \frac{1}{\varepsilon}. \tag{10.132}$$

To find a finite effective potential order by order in perturbation theory, we must perform the perturbation expansion with an additional term in the initial energy functional, which carries an additional set of pole terms, to be written collectively as [6, 7]

$$\Delta E \equiv -L^D \frac{m^4(\mu)}{(4\pi)^2 g(\mu)\mu^\epsilon} Z_v, \tag{10.133}$$

where Z_v is the renormalization constant of the vacuum which has an expansion in powers of $1/\varepsilon$ analogous to the other renormalization constants in (10.35)–(10.37):

$$Z_v(g,\varepsilon) \equiv \sum_{n=1}^\infty Z_{v,n}(g)\frac{1}{\varepsilon^n}. \tag{10.134}$$

The expansion coefficients of Z_v up to five loops will be given in Eq. (15.35).

The analogy of Z_v with the other renormalization constants is not perfect: there is no constant zero-loop term in Z_v. Such a term would be there if we had added to the bare energy functional a term $-L^D m_B^4 h_B/\lambda_B$ with an arbitrary constant h_B. Such a term would have a temperature dependence $\propto m_B^4 \propto t^2$ contributing a constant background term to the specific heat near T_c which is needed to describe experiments. Indeed, the specific heat [see the curves in Fig. 1.1 and their best fit (1.22)] shows a critical power behavior of t superimposed upon a smooth background term. The latter can be fitted by an appropriate constant h_B. The sum over the pole terms in Z_v, on the other hand, diverges for $t \to 0$ and generates the critical power behavior proportional to $|t|^{D\nu}$ which, after two derivatives with respect to t, produces the observed peak in the specific heat $C \propto t^{D\nu-2} \propto |t|^{-\alpha}$. This will be seen explicitly on the next page.

The effective energy of the vacuum is obtained from the sum of all loop diagrams $\Gamma_B^{(0)}$, plus the additional term ΔE. The total sum is the renormalized effective energy of the vacuum:

$$\Gamma^{(0)} = \Gamma_B^{(0)} + \Delta E. \tag{10.135}$$

Now, the bare effective energy at fixed bare quantities is certainly independent of the regularization parameter μ, and therefore satisfies trivially the differential equation

$$\mu \frac{d}{d\mu}\Gamma_B^{(0)}\bigg|_B = 0. \tag{10.136}$$

For the renormalized effective energy (10.135), this implies that

$$\mu \frac{d}{d\mu}\Gamma^{(0)}\bigg|_B = \mu \frac{d}{d\mu}\Delta E\bigg|_B. \tag{10.137}$$

Inserting here the right-hand side of (10.133), this equation can be written as

$$\mu \frac{d}{d\mu}\Gamma^{(0)}\bigg|_B = -L^D \frac{m^4(\mu)}{(4\pi)^2 g(\mu)\mu^\varepsilon}\gamma_v(g), \tag{10.138}$$

with the *renormalization group function of the vacuum*

$$\gamma_v(g) \equiv -\left[\varepsilon + \beta(g,\varepsilon)/g - 4\gamma_m\right] Z_v(g,\varepsilon) + \beta(g,\varepsilon)\partial_g Z_v(g,\varepsilon). \tag{10.139}$$

This function depends only on the renormalized coupling constant g as a consequence of the renormalizability of the theory and the minimal subtraction scheme. The derivative on the left-hand side of (10.137) is converted, via the chain rule, into a sum of differentiations with respect to the renormalized parameters, as in Eq. (10.21). Thus we obtain for $\Gamma^{(0)}(m, g, \mu)$ the renormalization group equation

$$\left[\mu\frac{\partial}{\partial\mu} + \beta(m, g, \mu)\frac{\partial}{\partial g} + \gamma_m(m, g, \mu)\, m\frac{\partial}{\partial m}\right]\Gamma^{(0)}(m, g, \mu) = -\frac{L^D}{(4\pi)^2}\frac{m^4(\mu)}{\mu^\varepsilon g(\mu)}\gamma_v(g(\mu)).\quad(10.140)$$

On the right-hand side we have emphasized the μ-dependence of m and g, to avoid confusion with $m = m(1)$ and $g = g(1)$ defined in (10.61), (10.62). Inserting the expansion (10.134) of Z_v into the right-hand side of (10.140), we find that the differentiations isolate from Z_v precisely the residue of the simple pole term $1/\varepsilon$, yielding

$$\gamma_v = gZ'_{v,1}(g).\quad(10.141)$$

All higher pole terms cancel, since the functions $Z_{v,n}$ satisfy recursion relations similar to (10.48)–(10.50):

$$gZ'_{v,n+1}(g) = [4\gamma_m - \beta_0(g)/g]\, Z_{v,n} + \beta_0(g)Z'_{v,n}(g),\quad(10.142)$$

with $\beta_0(g)$ of Eq. (10.41).

The renormalization group equation (10.140) can now be solved to find the renormalized effective energy of the vacuum [recall (10.119)]:

$$\Gamma^{(0)}(m, g, \mu) = L^D\sigma_m^D\, v\left(\frac{\Phi(\sigma_m)}{\sigma_m^{D/2-1}}; \mu, g(\sigma_m), \mu\right)_{\min} - \frac{L^D}{(4\pi)^2}\frac{m^4 h}{g\mu^\varepsilon}\int_1^{\sigma_m}\frac{d\sigma}{\sigma^{1+\varepsilon}}\gamma_v(g(\sigma))m^4(\sigma).$$

$$(10.143)$$

In the scaling regime, where σ_m is small and the mass goes to zero like

$$m(\sigma_m) \approx m\,\sigma_m^{\gamma_m^*},\quad(10.144)$$

the additional term ΔE in the effective energy of the vacuum in (10.135) is proportional to

$$\Delta E \propto m^4\sigma_m^{4\gamma_m^*-\varepsilon} \propto t^{\nu(4\gamma_m^*-\varepsilon)+2} = t^{D\nu}.\quad(10.145)$$

Thus it has the same scaling behavior as the *incomplete* effective potential $v(\Phi)$ at $\Phi = 0$ [i.e., the effective potential without the $n = 0$ -term in the sum (10.108)], which according to (10.89) and (10.121) behaves like $\sigma_m^D \propto t^{D\nu}$. It is also the same as that for the effective potential at a nontrivial minimum $\Phi = \Phi_0$ in the ordered state, as we shall see from (10.167). This will be important later in Subsection 10.10.3 when calculating the critical exponent of the specific heat defined in (1.16). The calculation of the universal ratios of the amplitudes of the specific heat and other quantities depend crucially on the renormalized vacuum energy $\Gamma^{(0)}(m, g, \mu)$ [8].

10.8   Approach to Scaling

In Eq. (10.93), we derived Kadanoff's scaling law (1.28) from the scaling relation (10.86) for the two-point proper vertex function. From this, we extracted the critical exponents $\nu = 1/(2 - 2\gamma_m^*)$ governing the temperature behavior of the correlation length, and the exponent $\eta = 2\gamma^*$ determining the critical power behavior of the Green function.

In Eq. (10.104), we introduced a further important critical exponent which governs the approach to the scaling law (10.93) for $g \to g^*$. In order to find this, we expand the right-hand side of Eq. (10.79) around g^* and write

$$\bar{\Gamma}^{(n)}(\mathbf{k}_i; m, g, \mu) = \sigma_m^{D-n(D/2-1)}\exp\left[-n\int_1^{\sigma_m}d\sigma'\frac{\gamma(g(\sigma'))}{\sigma'}\right]\bar{\Gamma}^{(n)}(\mathbf{k}_i/\sigma_m, \mu, g^*, \mu)$$

$$\times\, C^{(n)}(\mathbf{k}_i/\sigma_m; \mu, g(\sigma_m), \mu),\quad(10.146)$$

with the correction factor $C'^{(n)}$ given by

$$C'^{(n)}(\mathbf{k}_i/\sigma_m; \mu, g(\sigma_m), \mu) = 1 + [g(\sigma_m) - g^*] \frac{\partial}{\partial g} \log \bar{\Gamma}^{(n)}(\mathbf{k}_i/\sigma_m; \mu, g, \mu)\Big|_{g=g^*} + \dots . \quad (10.147)$$

Using Eq. (10.106), the correction factor is rewritten as

$$C'^{(n)}(\mathbf{k}_i/\sigma_m; \mu, g(\sigma_m), \mu) = 1 + (g - g^*) \sigma_m^\omega \frac{\partial}{\partial g} \log \bar{\Gamma}^{(n)}(\mathbf{k}_i/\sigma_m; \mu, g, \mu)\Big|_{g=g^*} + \dots , \quad (10.148)$$

where ω is the slope of the β-function at $g = g^*$, as defined in Eq. (10.104).

When approaching the critical point $\sigma_m \to 0$, a finite correction to scaling is observed if $\partial \log \bar{\Gamma}^{(n)}/\partial g$ is at $g = g^*$ homogenous of degree ω in the variables $\mathbf{k}_i/\sigma_m$. For the two-point proper vertex function such a behavior implies the following form of the correction factor

$$C^{(2)}(\mathbf{k}/\sigma_m, 1, g^*, 1) \overset{m \approx 0}{\approx} 1 + \text{const} \times (g - g^*)\sigma_m^\omega \times (|\mathbf{k}|/\mu\sigma_m)^\omega + \dots . \quad (10.149)$$

Then

$$\bar{\Gamma}^{(2)}(\mathbf{k}; m, g, \mu) = \sigma_m^2 \exp\left[-2\int_1^{\sigma_m} d\sigma' \frac{\gamma(g(\sigma'))}{\sigma'}\right] \bar{\Gamma}^{(2)}(\mathbf{k}/\sigma_m, \mu, g^*, \mu) C^{(2)}(\mathbf{k}/\sigma_m; \mu, g(\sigma_m), \mu) \quad (10.150)$$

behaves for $t \approx 0$ like

$$\bar{\Gamma}^{(2)}(\mathbf{k}; m, g, \mu) \overset{m \approx 0}{\approx} |\mathbf{k}|^{2-\eta} f(|\mathbf{k}|/\mu t^\nu) \left[1 + (g - g^*) \times \text{const} \times \left(\frac{|\mathbf{k}|}{\mu}\right)^\omega + \dots\right]. \quad (10.151)$$

Thus the correction to scaling is described by the exponent ω which is the slope of the β-function at the fixed point g^*. From the above discussion it is obvious that ω is positive for an infrared stable fixed point.

The most accurately measured approach to scaling comes from space shuttle experiments on the specific heat in superfluid helium, plotted in Fig. 1.2. The correction factor for this approach is obtained from Eq. (10.148) for $n = 0$ to have the general scaling form

$$C^{(0)} = 1 + \text{const} \times \sigma_m^\omega = 1 + \text{const} \times t^{\nu\omega}, \quad (10.152)$$

where we have used (10.84) to express σ_m in terms of $t = T/T_c - 1$. The exponent $\nu\omega$ is usually called Δ [compare Eq. (1.22)].

At this point one may wonder about the universality of this result since, in principle, other corrections to scaling might arise from neglected higher powers of the field of higher gradient terms in the energy functional, for example ϕ^6 or $\phi(\partial\phi)^2$. Fortunately, all such terms can be shown to be irrelevant for the value of ω. This is suggested roughly by dimensional considerations, and proved by studying the flow of these terms towards the critical limit with the help of the renormalization group [9].

10.9 Further Critical Exponents

The critical exponents ν, η, ω determine the critical behavior of all observables and the approach to this behavior. Let us derive the scaling relations for several important thermodynamic quantities and correlation functions.

10.9.1 Specific Heat

Consider the specific heat as a function of temperature. The ground state energy above T_c is given by the effective potential at zero average field $v_h(\Phi = 0)$ which, according to (10.89), (10.121), and (10.145), has the scaling behavior

$$v_h(\Phi = 0) \overset{m \approx 0}{\approx} \sigma_m^D \times \text{const.} + m^4 \sigma_m^{4\gamma_m^* - \varepsilon} \times \text{const.} \approx t^{D\nu}, \tag{10.153}$$

the second term coming from the sum of the vacuum diagrams in Eq. (10.145). Forming the second derivative with respect to t, we find for the specific heat at constant volume

$$C \overset{t \approx 0}{\approx} t^{D/(2-2\gamma_m^*)-2} = t^{D\nu-2}, \quad t > 0. \tag{10.154}$$

This behavior has been observed experimentally, and the critical exponent has been named α [recall (1.16)]:

$$C \overset{t \approx 0}{\approx} t^{-\alpha}; \quad t > 0. \tag{10.155}$$

Thus we can identify

$$\alpha = 2 - D\nu = 2 - D/(2 - 2\gamma_m^*), \tag{10.156}$$

showing that the exponent α is directly related to ν, as stated before in the scaling relation (1.32).

10.9.2 Susceptibility

Suppose the system at $T > T_c$ is coupled to a nonzero external source j, which is the generalization of an external magnetic field in magnetic systems [recall Eq. (1.44) and (1.45)]. The equilibrium value of the magnetization $M \equiv \Phi$ is no longer zero but $M_B \equiv \Phi(j)$. It is determined by the equation of state [the generalization of (1.46)]

$$j = \left. \frac{\partial v(\Phi)}{\partial \Phi} \right|_{\Phi(j)}. \tag{10.157}$$

From (10.121) we see that in the vicinity of the critical point

$$j \overset{t \approx 0}{\approx} \sigma_m^{D-\gamma^*-(D/2-1)} v'(\Phi(j)/\sigma_m^{\gamma^*+D/2-1}; \mu, g^*, \mu), \quad t > 0. \tag{10.158}$$

The *susceptibility* is obtained by an additional differentiation with respect to Φ [see Eq. (1.47)]:

$$\chi^{-1} \equiv \left. \frac{\partial^2 v}{\partial \Phi^2} \right|_{\Phi=\Phi(j)} \overset{t \approx 0}{\approx} \sigma_m^{D-(D-2)} \sigma_m^{-2\gamma^*} v''(\Phi(j)\sigma_m^{-(D-2)/2-\gamma^*}; \mu, g^*, \mu), \quad t > 0. \tag{10.159}$$

For $O(N)$-symmetric systems above T_c, this equation applies to the invariant part of the susceptibility matrix defined by Eq. (1.12). Below T_c, we must distinguish between longitudinal and transverse susceptibilities. This will be done in Subsection 10.10.5.

At $t = 0$ and zero field, one has $\Phi(j) = 0$ and finds

$$\chi^{-1} \propto \sigma_m^{2-2\gamma^*} = t^{(2-2\gamma^*)/(2-2\gamma_m^*)}. \tag{10.160}$$

Experimentally, this critical exponent is called γ [recall (1.17)]:

$$\chi^{-1} \overset{t \approx 0}{\approx} t^\gamma, \quad t > 0. \tag{10.161}$$

The critical exponent γ should not be confused with the renormalization group function $\gamma(g)$ of Eq. (10.26). Comparing (10.161) with (10.160), we identify

$$\gamma = 2\frac{1 - \gamma^*}{2 - 2\gamma_m^*} = \nu(2 - \eta), \tag{10.162}$$

thus reproducing the scaling relation (1.34).

10.9.3 Critical Magnetization

At the critical point, the proportionality of j and Φ (or of H and M) is destroyed by fluctuations. Experimentally, one observes a scaling relation [recall (1.20)]

$$M \approx B^{1/\delta}, \quad t = 0. \tag{10.163}$$

This can be derived from Eq. (10.158) which shows that a finite effective potential at the critical point, where $\sigma_m = 0$, requires the derivative v' to behave like some power for small $\sigma_m = 0$:

$$v' \overset{t \approx 0}{\approx} \text{const.} \times \left(\Phi/\sigma_m^{\gamma^* + D/2 - 1}\right)^\delta, \quad t > 0. \tag{10.164}$$

From the proportionality $j \propto \sigma_m^{D - \gamma^* - (D/2 - 1)} v'$, the power δ which makes j finite in the limit $\sigma_m \to 0$ must satisfy

$$D - \gamma^* - (D/2 - 1) - [\gamma^* + (D/2 - 1)]\delta = 0. \tag{10.165}$$

From this we obtain

$$\delta = \frac{D + 2 - 2\gamma^*}{D - 2 + 2\gamma^*} = \frac{D + 2 - \eta}{D - 2 + \eta}, \tag{10.166}$$

which is the scaling relation (1.35).

10.10 Scaling Relations Below T_c

Let us now turn to scaling results *below* T_c. Since all individual vertex functions in the expansion of the effective energy (10.108) can be calculated for $m^2 < 0$ just as well as for $m^2 > 0$, the main difference lies in $v(\Phi)$ not having a minimum at $\Phi = 0$ but at $\Phi = \Phi_0 \neq 0$ for vanishing external fields.

10.10.1 Spontaneous Magnetization

Consider first the behavior of the spontaneous magnetization $M_0 \equiv \Phi_0$ as the temperature approaches T_c from below. The equilibrium value of Φ is determined by the minimum of the effective potential $v(\Phi)$. According to (10.121), the minimum must have a constant ratio:

$$\Phi_0/\sigma_m^{\gamma^* + D/2 - 1} = \text{const}. \tag{10.167}$$

Hence, Φ_0 depends on m^2 and thus on the reduced temperature t as follows:

$$M_0 \equiv \Phi_0 \propto \sigma_m^{\gamma^* + D/2 - 1} = t^{(\gamma^* + D/2 - 1)/(2 - 2\gamma_m^*)}. \tag{10.168}$$

Thus we derive the experimentally observable relation [compare (1.19)]

$$M_0 \equiv \Phi_0 \propto (-t)^\beta, \tag{10.169}$$

with the critical exponent

$$\beta = \frac{\gamma^* + D/2 - 1}{2 - 2\gamma_m^*} = \frac{\nu}{2}(D - 2 + \eta). \tag{10.170}$$

This relation agrees with (1.33).

10.10.2 Correlation Length

Consider now the temperature dependent correlation length below T_c. From (10.130) we read off that the two-point function satisfies

$$\bar{\Gamma}^{(2)}(\mathbf{k}; \Phi_0; m, g, \mu) \stackrel{t\approx 0}{\approx} \sigma_m^{2-2\gamma^*} \bar{\Gamma}^{(2)}\left(\mathbf{k}/\sigma_m; \Phi_0/\sigma_m^{\gamma^*+D/2-1}; \mu, g^*, \mu\right), \quad t < 0. \qquad (10.171)$$

As in the previous case above T_c [recall (10.86), (10.87)], this is a function of

$$\mathbf{k}/\sigma_m = \mu\,\xi(t)\mathbf{k}, \qquad (10.172)$$

with the same temperature behavior as in (10.90). Thus the same critical exponent governs the divergence of the correlation length below and above T_c.

This *above-below equality* will now also be derived for the critical exponents α and γ of specific heat and susceptibility, respectively. The derivation of their scaling behaviors for $t < 0$ requires keeping track of the change of the average field Φ_0 with temperature.

10.10.3 Specific Heat

The exponent α of the specific heat below T_c follows from the $\Phi_0 \neq 0$ -version of Eq. (10.153) for the effective potential:

$$v_h(\Phi_0) \stackrel{t\approx 0}{\approx} \sigma_m^D v\left(\Phi_0/\sigma_m^{\gamma^*+D/2-1}, \mu, g^*, \mu\right) + \text{const.} \times m^4 \sigma_m^{4\gamma_m^*-\varepsilon}. \qquad (10.173)$$

Since the temperature change of Φ_0 takes place at a constant combination $\Phi_0/\sigma_m^{\gamma^*+D/2-1}$ [see Eq. (10.167)], the presence of $\Phi_0 \neq 0$ can be ignored and we obtain the same result as in (10.153), implying a temperature behavior

$$v(\Phi_0) \propto \sigma_m^D \approx t^{D\nu}. \qquad (10.174)$$

This agrees with the $T > T_c$ -behavior (10.153), leading to the same critical exponent of the specific heat as in (10.156).

10.10.4 Susceptibility

Suppose now that an external magnetic field is switched on in the ordered phase. It will cause a deviation $\delta\Phi \equiv \Phi(j) - \Phi_0$ from Φ_0. From Eqs. (10.157) and (10.158) we obtain for $\delta\Phi$ the scaling relation

$$j \to \sigma_m^{D-(\gamma^*+D/2-1)} v'((\Phi_0 + \delta\Phi)/\sigma_m^{\gamma^*+D/2-1}; \mu, g^*, \mu). \qquad (10.175)$$

Expanding this to first order in $\delta\Phi$ gives

$$\delta j \to \delta\Phi\, \sigma_m^{2-2\gamma^*} v''(\Phi_0/\sigma_m^{\gamma^*+D/2-1}; \mu, g^*, \mu). \qquad (10.176)$$

Since Φ_0 changes with t according to (10.169), the last factor is independent of temperature, and the susceptibility $\chi(t)$ has the same functional form as in (10.159), exhibiting the same critical exponent γ as in (10.161).

In $O(N)$-symmetric systems, this result holds for the longitudinal susceptibility only. The transverse susceptibility requires the following separate discussion.

10.10.5   Transverse Susceptibility and Bending Stiffness

Suppose now that the ground state breaks spontaneously an $O(N)$ symmetry of the system. Then the susceptibility decomposes into a longitudinal part and a transverse part, as shown in Eq. (1.14), and these two parts have completely different scaling properties. Since susceptibilities are proportional to correlation functions according to (1.15), we extract their scaling properties from the lowest gradient term in the effective energy $\bar{\Gamma}[\boldsymbol{\Phi}; m, g, \mu]$ in the deviation of the average field $\boldsymbol{\Phi}(\mathbf{x})$ from the equilibrium value $\boldsymbol{\Phi}_0$. We write bold-face letters for vectors in $O(N)$ field space. The quadratic term in the deviation $\delta\boldsymbol{\Phi}(\mathbf{x}) \equiv \boldsymbol{\Phi}(\mathbf{x}) - \boldsymbol{\Phi}_0$ has the general form

$$\bar{\Gamma}[\boldsymbol{\Phi}; m, g, \mu] \approx \int d^D x \, \delta\boldsymbol{\Phi}(\mathbf{x})\bar{\Gamma}^{(2)}(-i\partial_{\mathbf{x}}; \boldsymbol{\Phi}_0; m, g, \mu))\delta\boldsymbol{\Phi}(\mathbf{x}). \tag{10.177}$$

For smooth field configurations $\delta\boldsymbol{\Phi}(\mathbf{x})$, we expand

$$\Gamma^{(2)}(-i\partial_{\mathbf{x}}; \boldsymbol{\Phi}_0; m, g, \mu) \approx c_1(\boldsymbol{\Phi}_0; m, g, \mu) - c_2(\boldsymbol{\Phi}_0; m, g, \mu)\partial_{\mathbf{x}}^2. \tag{10.178}$$

The temperature dependence of the expansion coefficients $c_{1,2}(\boldsymbol{\Phi}_0; m, g, \mu)$ can be extracted from Eq. (10.171) and (10.167). To ensure the existence of a nontrivial term proportional to $\mathbf{k}^2$ on the right-hand side of (10.171), we see that

$$\bar{\Gamma}^{(2)}(\mathbf{k}_i; \boldsymbol{\Phi}_0; m, g, \mu) \overset{t \approx 0}{\propto} \sigma_m^{2-2\gamma^*} + \text{const} \times \sigma_m^{-2\gamma^*}\mathbf{k}^2. \tag{10.179}$$

Recalling the dependence (10.89) of σ_m on $t = m^2/\mu^2$, and the relation (10.94) for the critical exponent η, we obtain, for smooth field configurations, the temperature dependence of the leading terms in the effective energy

$$\bar{\Gamma}^{(2)}[\boldsymbol{\Phi}; m, g, \mu] \overset{t \approx 0}{\propto} \int d^D x \left\{ t^{(2-\eta)\nu}\boldsymbol{\Phi}^2(\mathbf{x}) + \text{const} \times t^{-\eta\nu}\left[\partial_{\mathbf{x}}\boldsymbol{\Phi}(\mathbf{x})\right]^2 \right\}. \tag{10.180}$$

For an $O(N)$-symmetric system, the order field can be decomposed into size and direction as

$$\boldsymbol{\Phi}(\mathbf{x}) = \Phi(\mathbf{x})\,\mathbf{n}(\mathbf{x}) \tag{10.181}$$

which brings the effective energy for small and smooth deviations $\delta\Phi(\mathbf{x})$ and $\delta\mathbf{n}(\mathbf{x}) \equiv \mathbf{n}(\mathbf{x}) - \mathbf{n}_0$ from the average ordered configurations to the form as

$$\bar{\Gamma}^{(2)}[\boldsymbol{\Phi}; m, g, \mu] \overset{t \approx 0}{\propto} \int d^D x \left(t^{(2-\eta)\nu}\left\{ \Phi_0^2 + [\delta\Phi(\mathbf{x})]^2 \right\} + \text{const} \times t^{-\eta\nu}\left\{ [\partial_{\mathbf{x}}\delta\Phi(\mathbf{x})]^2 + \Phi_0^2[\partial_{\mathbf{x}}\delta\mathbf{n}(\mathbf{x})]^2 \right\} \right). \tag{10.182}$$

In $O(2)$-symmetric systems, the order field has two components, and can be replaced by a complex field $\Phi(\mathbf{x}) = e^{i\theta(\mathbf{x})}\Phi_0(\mathbf{x})$ with a real $\Phi_0(\mathbf{x})$. Then the last term in (10.182) has the form

$$\text{const} \times \int d^D x\, t^{-\eta\nu}\Phi_0^2\left[\partial_{\mathbf{x}}\delta\theta(\mathbf{x})\right]^2. \tag{10.183}$$

In both gradient terms we can, of course, omit the deviation symbols δ. The directional deviation field $\delta\mathbf{n}(\mathbf{x})$ possesses only a gradient term, and describes long-range (massless) excitations whose existence is ensured by the Nambu-Goldstone theorem stated after Eq. (1.50).

From the coefficients of the quadratic terms we extract the scaling behavior of the longitudinal and tranverse correlation functions in momentum space:

$$G_{cL}(\mathbf{k}) \;\propto\; \left[t^{(2-\eta)\nu} + \text{const} \times t^{-\eta\nu}\mathbf{k}^2 \right]^{-1}, \tag{10.184}$$

$$G_{cT}(\mathbf{k}) \;\propto\; \left[\text{const} \times t^{-\eta\nu}\Phi_0^2\mathbf{k}^2 \right]^{-1}. \tag{10.185}$$

The longitudinal and tranverse susceptibilities are proportional to these [recall Eq. (1.15)]. Their critical behavior is given by

$$\chi_{cL}^{-1}(0) \quad \propto \quad t^{(2-\eta)\nu} \tag{10.186}$$

$$\mathbf{k}^{-2}\chi_{cT}^{-1}(\mathbf{k}) \quad \propto \quad \frac{\partial}{\partial \mathbf{k}^2}\chi_{cL}^{-1}(\mathbf{k}) \propto t^{-\eta\nu}\Phi_0^2. \tag{10.187}$$

Recalling the temperature dependence (10.169) of Φ_0, and using the scaling relation for the average field Φ_0 in Eq. (10.169), the second relation becomes

$$\mathbf{k}^{-2}\chi_{cT}^{-1}(\mathbf{k}) \quad \propto \quad \frac{\partial}{\partial \mathbf{k}^2}\chi_{cL}^{-1}(\mathbf{k}) \propto t^{(D-2)\nu}. \tag{10.188}$$

Comparison of the last term in Eq. (10.182) with (1.110) shows that the prefactor supplies us with the temperature behavior of the bending stiffness of the directional field $\mathbf{n}(\mathbf{x})$ near the critical point. In superfluid helium, this is by definition proportional to the experimentally measured superfluid density ρ_s [recall Eq. (1.122)]. The bending stiffness, or the superfluid density, are therefore proportional to those in (10.188), and we obtain the temperature behavior of the superfluid density

$$\rho_s \propto t^{(D-2)\nu}. \tag{10.189}$$

The experimental verification of this scaling behavior was described in Chapter 1, the crucial plots being shown in Fig. 1.3.

10.10.6   Widom's Relation

Finally it is worth noticing that Eq. (10.158) corresponds exactly to Widom's scaling relation (1.26). That relation can be differentiated with respect to M to yield the magnetic equation of state

$$B = t^{3-\alpha}M^{-1-1/\beta}\psi'(t/M^{1/\beta}), \tag{10.190}$$

which may also be written as

$$\frac{B}{M^\delta} = f\left(\frac{t}{M^{1/\beta}}\right) \tag{10.191}$$

with some function $f(x)$. This is easily proven with the help of Griffith's scaling relation $\delta = -1 + (2-\alpha)/\beta$ [recall (1.29)]. In terms of the variables of our field theory, the equation of state (10.191) may be rewritten as

$$j = t^{\delta\beta}\left(\frac{t}{\Phi^{1/\beta}}\right)^{-\delta\beta} f\left(\frac{t}{\Phi^{1/\beta}}\right) = t^{\delta\beta}g\left(\frac{\Phi}{t^\beta}\right), \tag{10.192}$$

where $g(x)$ is some other function. By comparing this with (10.158), we see that

$$\delta\beta = [D - \gamma^* - (D/2 - 1)]\frac{1}{2 - 2\gamma_m^*} = \frac{\nu}{2}(D + 2 - \eta), \tag{10.193}$$

which is in agreement with (10.166), (10.170), or (1.33) and (1.35).

10.11 Comparison of Scaling Relations with Experiment

For a comparison with experiment, we may pick three sets of critical data and extract the values of η, ν, and ω. The remaining critical exponents can then be found from the scaling relations (1.32)–(1.35).

As an example take the magnetic system $CrBr_3$ where one measures

$$\beta \approx 0.368, \quad \delta \approx 4.3, \quad \gamma \approx 1.215. \tag{10.194}$$

Inserting these into Widom's scaling relation [recall (1.30)]

$$\beta = \gamma/(\delta - 1), \tag{10.195}$$

we see that the relation is satisfied excellently. Inserting δ into the relation

$$\eta = \frac{D + 2 - (D - 2)\delta}{\delta + 1}, \tag{10.196}$$

we find for $D = 3$

$$\eta \approx 0.132, \tag{10.197}$$

and from the relation $\nu = \gamma/(2 - \eta)$ [recall (1.34)]:

$$\nu \approx 0.65. \tag{10.198}$$

10.12 Critical Values g^*, η, ν, and ω in Powers of ε

Let us now calculate explicitly the critical properties of the $O(N)$-symmetric ϕ^4-theory in the two-loop approximation. In Eq. (10.56) we gave the β-function

$$\beta(\bar{g}) = -\varepsilon \bar{g} + \frac{N + 8}{3} \bar{g}^2 - \frac{3N + 14}{3} \bar{g}^3. \tag{10.199}$$

In $D = 4$ dimensions, $\beta(\bar{g})$ starts with $\bar{g}^2$, and the only IR-stable fixed point lies at $\bar{g}^* = 0$. Thus the massless ϕ^4-theory behaves asymptotically as a free theory. From Eqs. (10.52) and (10.53) we see that the anomalous dimensions γ and γ_m are zero for $\bar{g}^* = 0$. Hence the critical exponents in Eqs. (10.88), (10.94), and (10.104) possess the mean field values for $D = 4$:

$$\eta = 0, \quad \nu = 1/2, \quad \omega = 0. \tag{10.200}$$

In $D = 4 - \varepsilon$ dimensions, the equation $\beta(\bar{g}^*) = 0$ for the fixed point has the nontrivial solution

$$\bar{g}^* = \frac{3}{N + 8}\varepsilon + \frac{9(3N + 14)}{(N + 8)^3}\varepsilon^2 + \dots . \tag{10.201}$$

If this expansion is inserted into the $\bar{g}$-expansions (10.57)–(10.58), we obtain for the critical exponents ν and η the ε-expansions:

$$\eta = 2\gamma^*(\varepsilon, N) = \frac{N + 2}{2(N + 8)^2}\varepsilon^2 + \dots , \tag{10.202}$$

$$\nu = \frac{1}{2 - 2\gamma_m^*(\varepsilon, N)} = \frac{1}{2} + \frac{N + 2}{4(N + 8)}\varepsilon + \frac{(N + 2)(N^2 + 23N + 60)}{8(N + 8)^3}\varepsilon^2 + \dots . \tag{10.203}$$

The critical exponent ω governing the approach to scaling is found from the derivative of the β-function (10.56) at $\bar{g} = \bar{g}^*$ [recall (10.104)]:

$$\omega = \beta'^*(\varepsilon, N) = \varepsilon - 3\frac{3N + 14}{(N + 8)^2}\varepsilon^2 + \dots . \tag{10.204}$$

All ε-expansions are independent of the choice of the coupling constant. The critical exponents depend via ε and N only on the dimension of space and order parameter space. This is a manifestation of the universality of phase transitions, which states that the critical behavior depends only on the type of interaction, its symmetry, and the space dimensionality.

Let us compare the above ε-expansion with the experimental critical exponents in Section 10.11. The expansion can be used only for infinitesimal ε. For applications to three dimensions we have to evaluate them at $\varepsilon = 1$, which cannot be done by simply inserting this large ε-value, since the series diverge. Let us ignore this problem for the moment, deferring a proper resummation until Chapters 16, 19, and 20. Inserting $\varepsilon = 1$, and estimating the reliability of the result from the size of the last term in each series, we calculate for $N = 0, 1, 2, 3, \infty$:

$$
\begin{aligned}
\nu &= \tfrac{1}{2} + \tfrac{1}{16}\varepsilon + \tfrac{15}{512}\varepsilon^2 + \dots &&= \tfrac{303}{512} + \dots &&\approx 0.5918 \pm 0.0293, &&N = 0, \\
\nu &= \tfrac{1}{2} + \tfrac{1}{12}\varepsilon + \tfrac{7}{162}\varepsilon^2 + \dots &&= \tfrac{203}{324} + \dots &&\approx 0.6265 \pm 0.0432, &&N = 1, \\
\nu &= \tfrac{1}{2} + \tfrac{1}{10}\varepsilon + \tfrac{11}{200}\varepsilon^2 + \dots &&= \tfrac{131}{200} + \dots &&\approx 0.6550 \pm 0.0550, &&N = 2, \\
\nu &= \tfrac{1}{2} + \tfrac{5}{44}\varepsilon + \tfrac{345}{5324}\varepsilon^2 + \dots &&= \tfrac{903}{1331} + \dots &&\approx 0.6874 \pm 0.0648, &&N = 3, \\
\nu &= \tfrac{1}{2} + \tfrac{1}{4}\varepsilon + \tfrac{1}{8}\varepsilon^2 + \dots &&= \tfrac{7}{8} + \dots &&\approx 0.8750 \pm 0.1250, &&N = \infty.
\end{aligned}
\tag{10.205}
$$

The other critical exponents are

$$
\begin{aligned}
\eta &= \tfrac{1}{64} &&\approx 0.016, &&N = 0, \\
\eta &= \tfrac{1}{54} &&\approx 0.019, &&N = 1, \\
\eta &= \tfrac{1}{50} &&\approx 0.02, &&N = 2, \\
\eta &= \tfrac{5}{242} &&\approx 0.021, &&N = 3, \\
\eta &= 0, &&&&N = \infty,
\end{aligned}
\tag{10.206}
$$

and

$$
\begin{aligned}
\omega &= \varepsilon - \tfrac{17}{27}\varepsilon^2 &&= \tfrac{10}{27} &&\approx 0.3704 \pm 0.6296, &&N = 0, \\
\omega &= \varepsilon - \tfrac{17}{27}\varepsilon^2 &&= \tfrac{10}{27} &&\approx 0.3704 \pm 0.6296, &&N = 1, \\
\omega &= \varepsilon - \tfrac{3}{5}\varepsilon^2 &&= \tfrac{2}{5} &&\approx 0.4 \pm 0.6, &&N = 2, \\
\omega &= \varepsilon - \tfrac{69}{121}\varepsilon^2 &&= \tfrac{52}{121} &&\approx 0.4298 \pm 0.5702, &&N = 3, \\
\omega &= \varepsilon &&= 1, &&N = \infty.
\end{aligned}
\tag{10.207}
$$

The ε-expansion for ν has decreasing contributions from higher orders. The value up to order ε^2 is $\nu \approx 0.627$, and agrees reasonably with the experimental value $\nu \approx 0.65$ of Eq. (10.198).

The expansions for η contain only one term, so no convergence can be judged. The agreement with experiment is nevertheless reasonable. The value to order ε^2 at $\varepsilon = 1$ is $\eta \approx 0.019$ which, via the scaling relation $\nu = \gamma/(2 - \eta)$, leads to the exponent $\gamma \approx 1.287$, quite close to the experimental value $\gamma = 1.215$ in Eq. (10.194).

The expansion for ω are obviously useless since the errors are too large.

If we attempt to calculate critical exponents to higher order than ε^2 by inserting $\varepsilon = 1$ into the expansions, we observe that the agreement becomes worse since the series diverge. The rough agreement for ν and ϵ up to order ε^2 is a consequence of the *asymptotic convergence* of the series. In Chapter 16, we shall see how high-precision estimates can still be extracted from asymptotic series. The reader who is curious to see how the direct evaluation of the series becomes worse with higher orders in ε may anticipate the five-loop expansions from Eqs. (17.13)–(17.15) and insert $\varepsilon = 1$ into these.

For a judgment of the reliability of all numbers (10.205)–(10.207), we refer the reader to the most accurate currently available critical exponents in Tables 20.2 and 21.3 .

10.13 Several Coupling Constants

For fields with more than one component, several ϕ^4-couplings are possible which may all become simultaneously relevant in four dimensions. This was discussed in detail in Chapter 6. For each coupling constant, there exists a β-function and there may be two or more fixed points. The stability of the fixed points depends on N and channels the flow in the space of the coupling constants. It can be shown in general [10] that the $O(N)$-symmetric fixed point is the only stable one for $N \leq 4 - \mathcal{O}(\varepsilon)$.

In order to have only a single wave function renormalization constant for the N field components ϕ_α, the following condition has to be fulfilled:

$$\bar{\Gamma}^{(2)}_{\alpha\beta}(k) \sim \bar{\Gamma}^{(2)}(k)\, \delta_{\alpha\beta} . \tag{10.208}$$

This property is guaranteed for all theories which are symmetric under reflection $\phi_\alpha \to -\phi_\alpha$ and under permutations of the N field indices α. The same symmetry ensures that $\bar{\Gamma}^{(4)}$ is, to all orders in perturbation theory, a linear combination of the tensors specifying the ϕ^4-couplings. For two tensors $T^{(1)}_{\alpha\beta\gamma\delta}$ and $T^{(2)}_{\alpha\beta\gamma\delta}$, this condition reads

$$\bar{\Gamma}^{(4)}_{\alpha\beta\gamma\delta} \sim \bar{\Gamma}^{(4)}_1\, T^{(1)}_{\alpha\beta\gamma\delta} + \bar{\Gamma}^{(4)}_2\, T^{(2)}_{\alpha\beta\gamma\delta} . \tag{10.209}$$

If the conditions (10.208) and (10.209) are satisfied, we can find four scalar renormalization constants Z_A $(A = \phi, m^2, g_1, g_2)$ relating the bare mass m_B and the two coupling constants g_{iB} to the corresponding physical parameters by

$$m_B^2 = \frac{Z_{m^2}}{Z_\phi}\, m^2; \qquad g_{iB} = \mu^\varepsilon \frac{Z_{g_i}}{(Z_\phi)^2}\, g_i \qquad \text{for } i = 1, 2 . \tag{10.210}$$

The renormalization group functions are introduced in the usual way:

$$\beta_i(g_1, g_2) = \mu \partial_\mu g_i \big|_{g_{1B}, g_{2B}, m_B, \varepsilon} = \mu \partial_\mu g_i \big|_B , \tag{10.211}$$

$$\gamma(g_1, g_2) = \mu \partial_\mu \log Z_\phi^{1/2} \big|_{g_{1B}, g_{2B}, m_B, \varepsilon} = \mu \partial_\mu \log Z_\phi^{1/2} \big|_B , \tag{10.212}$$

$$\gamma_m(g_1, g_2) = \mu \partial_\mu \log m \big|_{g_{1B}, g_{2B}, m_B, \varepsilon} = \mu \partial_\mu \log m \big|_B . \tag{10.213}$$

We have written Eqs. (10.211)–(10.213) by analogy with Eqs. (10.30)–(10.33). Since the renormalization constants depend on g_1 and g_2, the functions g_1 and g_2 are implicitly given by

$$g_{iB} = \mu^\varepsilon Z_{g_i} Z_\phi^{-2} g_i = g_{iB}(\mu, g_1(\mu), g_2(\mu)). \tag{10.214}$$

The derivatives $\partial_\mu g_1$ and $\partial_\mu g_2$ follow therefore from the two equations

$$\frac{\partial g_{iB}}{\partial \mu} + \frac{\partial g_{iB}}{\partial g_1} \frac{\partial g_1}{\partial \mu} + \frac{\partial g_{iB}}{\partial g_2} \frac{\partial g_2}{\partial \mu} = 0 \quad \text{for } i = 1, 2. \tag{10.215}$$

Using $\partial_\mu g_{iB} = \varepsilon\, g_{iB}/\mu$, we find

$$\frac{\partial \log g_{iB}}{\partial g_1}\, \beta_1 + \frac{\partial \log g_{iB}}{\partial g_2}\, \beta_2 = -\varepsilon \quad \text{for } i = 1, 2. \tag{10.216}$$

The renormalization group function $\gamma(g_1, g_2)$ is given by

$$\gamma(g_1, g_2) = \frac{\beta_1(g_1, g_2)}{2}\frac{\partial \log Z_\phi}{\partial g_1} + \frac{\beta_2(g_1, g_2)}{2}\frac{\partial \log Z_\phi}{\partial g_2}, \tag{10.217}$$

while γ_m is obtained from the equation

$$\gamma_m(g_1, g_2) = -\frac{\beta_1(g_1, g_2)}{2}\frac{\partial \log Z_{m^2}}{\partial g_1} - \frac{\beta_2(g_1, g_2)}{2}\frac{\partial \log Z_{m^2}}{\partial g_2} + \gamma(g_1, g_2). \tag{10.218}$$

Extracting the regular terms of Eqs. (10.216)–(10.218), we find the analog of Eqs. (10.43)–(10.46) for the case with two coupling constants:

$$\begin{aligned}
\beta_1 &= -\varepsilon g_1 + g_1 \left(g_1 \partial_{g_1} Z_{g_1, 1} + g_2 \partial_{g_2} Z_{g_1, 1} + 4\,\gamma \right), \\
\beta_2 &= -\varepsilon g_2 + g_2 \left(g_2 \partial_{g_2} Z_{g_2, 1} + g_1 \partial_{g_1} Z_{g_2, 1} + 4\,\gamma \right), \\
\gamma &= -\tfrac{1}{2} g_1 \partial_{g_1} Z_{\phi, 1} - \tfrac{1}{2} g_2 \partial_{g_2} Z_{\phi, 1}, \\
\gamma_m &= \tfrac{1}{2} g_1 \partial_{g_1} Z_{m^2, 1} + \tfrac{1}{2} g_2 \partial_{g_2} Z_{m^2, 1} + \gamma.
\end{aligned} \tag{10.219}$$

The stability of the fixed points can be examined using the critical exponents ω_1, ω_2, which are the eigenvalues of the matrix $\partial\beta_i / \partial g_j$. They should be positive for an infrared stable fixed point. An example for a system with two coupling constants will be treated in Chapter 18.

10.14   Ultraviolet versus Infrared Properties

Some remarks may be useful concerning the special role of ultraviolet divergences in critical phenomena. In three dimensions, ϕ^4-theories are superrenormalizable and possess finite correlation functions after only a few subtractions. So one may wonder about the relevance of ultraviolet divergences to critical phenomena, in particular, since the system at short distances is not supposed to be represented by the field theory. The explanation of this apparent paradox is the following. Consider some real physical system with a microstructure, such as a lattice, at a temperature very close to the critical temperature at which the correlation length ξ extends over many lattice spacings. There the correlation functions have three regimes. At very long distances $x \gg \xi$, they fall off exponentially like $e^{-x/\xi}$. For distances much larger than the lattice spacing but much smaller than the correlation length, they behave like a power in x. At the critical temperature, this power behavior extends all the way out to infinite distances. In the third regime, where distances are of the order of the lattice spacing, the behavior is nonuniversal and depends crucially on the composition of the material. Nothing can be said about this regime on the basis of field-theoretic studies.

Let us compare these behaviors of correlation functions of real systems with the behaviors found in the present ϕ^4 field theories. Here we can also distinguish three regimes. The third, unphysical regime, lies now at distances which are shorter than the inverse cutoff Λ of the theory. In this regime, the perturbation theory has unphysical singularities, first discussed by Landau, that are completely irrelevant to the critical phenomena to be explained. At length scales much shorter than the correlation length, but much longer than $1/\Lambda$, the correlation functions show power behavior, from which we can extract the critical exponents of the field theory and compare them with experiments made in the above lattice system. In field theory, this is the so-called short-distance behavior. Its properties are governed by the ultraviolet divergences. At the critical point, the short-distance behavior extends all the way to infinity. This is the reason why ultraviolet divergences are relevant for the understanding of long-distance phenomena observed in many-body systems near the critical point, that are independent of the microstructure.

Notes and References

Excellent reviews are found in
E. Brézin, J.C. Le Guillou, J. Zinn-Justin, in *Phase Transitions and Critical Phenomena*, Vol.6, edited by C. Domb and M.S. Green, Academic Press, New York, 1976;
K.G. Wilson, J. Kogut, Phys. Rep. **12**, 75 (1974),
and in the textbook by
D.J. Amit, *Field Theory, the Renormalization Group and Critical Phenomena*, McGraw-Hill, 1978,
and the other textbooks cited in Notes and References of Chapter 1.

The individual citations in the text refer to:

[1] C.G. Callan, Phys. Rev. D **2**, 1541 (1970);
K. Symanzik, Commun. Math. Phys. **18**, 227 (1970).

[2] S. Weinberg, Phys. Rev. **118**, 838 (1960).

[3] G.'t Hooft, Nucl. Phys. B **61**, 455 (1973).

[4] J.C. Collins, A.J. MacFarlane, Phys. Rev. D **10**, 1201 (1974).

[5] G.'t Hooft in *The Whys of Subnuclear Physics*, Ed. A. Zichichi, Proceedings, Erice 1977, Plenum, New York 1979.
See also
N.N. Khuri, Phys. Rev. D **23**, 2285 (1981).

[6] B. Kastening, Phys. Lett. B **283**, 287 (1992).
See also
M. Bando, T. Kugo, N. Maekawa and H. Nakano, Phys. Lett. B **301**, 83 (1993).

[7] B. Kastening, Phys. Rev. D **54**, 3965 (1996); Phys. Rev. D **57**, 3567 (1998).

[8] S.A. Larin, M. Mönnigmann, M. Strösser and V. Dohm (cond-mat/9805028);
S.A. Larin, and V. Dohm, Nucl.Phys. B **540**, 654 (1999) (cond-mat/9806103);
H. Kleinert, B. Van den Bossche, Phys. Rev. E **63**, 056113 (2001) (cond-mat/0011329)

[9] See the review article by E. Brézin et al. and textbook by D. Amit cited above.

[10] E. Brézin, J.C. Le Guillou, J. Zinn-Justin, Phys. Rev. B **10**, 893 (1974).

11

Recursive Subtraction of UV-Divergences by R-Operation

The recursive counterterm procedure developed by Bogoliubov, Parasiuk, and Hepp [1], which was briefly described at the beginning of Chapter 9, yields a finite result for each Feynman diagram, order by order in perturbation theory. It is useful to interpret this procedure as an application of a *R-operation* to the Feynman diagram. All the finite results on higher loops in this book will be derived via the R-operation.

We shall not follow a possible alternative approach based on a general solution of the recursion relation developed by Zimmermann [2], who specified the R-operation by a complete diagrammatic expansion in the form of a so-called *forest formula*. With this formula, the renormalization procedure is known as the *BPHZ formalism*. It was applied by Zimmermann directly to the integrands of the Feynman integrals, making them manifestly convergent without regularization. Subsequently, several authors showed that the R-operation leads to the same finite results, if applied individually to the Feynman integrals regularized by minimal subtraction [3].

The chapter will end with a discussion of the general structure of the counterterms. This will show how the minimal subtraction scheme considerably reduces the number of relevant diagrams to be evaluated explicitly.

In the following manipulations, the negative coupling constant $-\lambda$ associated with each vertex will be omitted, unless otherwise stated, since they will be of no relevance to the analytic expressions represented by the Feynman diagrams.

11.1 Graph-Theoretic Notations

For recursive diagrammatic subtractions of divergences, some graph-theoretic concepts are useful. The first important concept of a *subdiagram* was introduced before, after Eq. (9.3). It will also be useful to admit to the set of all subdiagrams the original diagram itself. If this is excluded, we speak otherwise of a *proper subdiagram*. The notation will be $\gamma \subset G$ for subdiagrams and $\bar{\gamma} \subset G$ for proper subdiagrams of the diagram G. Then we define

- a *shrunk diagram* G/γ, obtained from G by shrinking all lines in γ to zero length, so that their endpoints coincide. Each connected part of γ then collapses to a single vertex.

As an example, consider the five-loop diagram

$$G = \;\;\text{⟨diagram⟩}\;\;,$$

with the two-loop subdiagram

$$\gamma = \;\;\text{⟨diagram⟩}\;.$$

By shrinking all lines of the subdiagram, we find the shrunk diagram

$$G/\gamma = \;\;\text{⟨diagram⟩}\;.$$

Next we define

- UV-divergent 1PI subdiagrams γ_i of G to be *UV-disjoint in G* if they have no internal line (I) and no vertex (V) in common.

Note that a UV-divergent 1PI diagram cannot consist of UV-disjoint components only. Even if the components of a diagram are merely connected by a cutvertex (to be defined in the next item below), the components cannot be UV-disjoint subdiagrams in G because they have a vertex in common. A set of UV-disjoint subdiagrams $\gamma \subset G$ will be denoted by $\Gamma(G)$:

$$\Gamma = \{\gamma | \gamma \in G, \gamma \text{ UV-disjoint}\}. \tag{11.1}$$

For proper UV-disjoint subdiagrams $\bar{\gamma}$, the notation will be $\bar{\Gamma}(G)$. The detailed set-theoretic definition for $\bar{\Gamma}(G)$ is

$$\bar{\Gamma}(G) = \{\bar{\gamma}_i | \bar{\gamma}_i \subset G, \text{ 1PI}, \ \omega(\bar{\gamma}_i) \geq 0, \ I(\bar{\gamma}_i \cap \bar{\gamma}_j) = 0, \ V(\bar{\gamma}_i \cap \bar{\gamma}_j) = 0\}.$$

We furthermore introduce the set of all sets $\Gamma(G)$ or $\bar{\Gamma}(G)$ contained in a diagram G, denoting them by Γ_G or $\bar{\Gamma}_G$:

$$\bar{\Gamma}_G = \{\text{all } \bar{\Gamma}(G)\} \quad \text{and} \quad \Gamma_G = \{\text{all } \Gamma(G)\} \quad \text{with} \quad \Gamma_G = \bar{\Gamma}_G \cup \{G\}. \tag{11.2}$$

Two sets $\Gamma(G)$ of UV-disjoint subdiagrams of G can never contain the same integrations.

Finally, we introduce the concept of a cutvertex:

- A diagram G is said to contain a *cutvertex* if it can be decomposed into two components G_1 and G_2 by cutting it at a vertex. Each of the components is defined to include the vertex where the cut is made. The same definition holds for subdiagrams γ decomposing into components γ_1 and γ_2.

As an example, the complete set $\bar{\Gamma}_G$ of all sets $\bar{\Gamma}(G)$ of proper UV-disjoint subdiagrams $\bar{\gamma} \in G$ for the diagram

$$G = \quad$$

is

$$\bar{\Gamma}_G = \{\bar{\Gamma}(G)\} = \left\{ \emptyset, \{\bigcirc\}, \{\bigcirc\}, \{\bigcirc, \bigcirc\}, \{\bigcirc\}, \{\bigcirc\}, \{\bigcirc\} \right\}.$$

The last set $\bar{\Gamma}(G)$ is forbidden since it contains subdiagrams which have a common vertex. In fact, this last set and the one before contain the same integrations, which would have led to double-counting. Another example will be discussed below in Eqs. (11.16)–(11.19).

11.2 Definition of *R*- and *R̄*-Operation

The R-operation is defined as follows: when applied to a diagram G, it subtracts shrunk diagrams in such a way that RG is free of divergences. In dimensional regularization, all divergences are of the form $1/\varepsilon^i$, such that RG contains, by definition, no such terms, and is finite for $\varepsilon \to 0$. In terms of the $\mathcal{K}$-operation defined in Eq. (9.76), which selects precisely all $1/\varepsilon^i$ terms of a Feynman integral, we may state this property as

$$\mathcal{K}RG = 0 \quad \text{for any} \quad \omega(G). \tag{11.3}$$

Application of the R-operation to G removes all divergences coming from subdiagrams, and the superficial divergence of G itself. In the above set-theoretic notation, the result of the

R-operation may be stated as follows: the first term in RG is G itself. From this, all possible diagrams are subtracted in which the integration over some divergent subdiagrams is substituted by the corresponding counterterm. Among these, there is the term in which the whole integration of G is substituted by the corresponding counterterm. This is the term we are looking for. The remainder consists of a sum over all sets $\bar{\Gamma}(G)$, which contain all possible arrangements of divergent subdiagrams without double-counting. Let us denote by Δ_γ the counterterm associated with a subdiagram γ, i.e., the pole term of the superficial divergence of this subdiagram. This differs, in general, from $\mathcal{K}(\gamma)$ by possible pole terms of subdivergences. With this notation, the result of the R-operation may be written as

$$
\begin{aligned}
RG &= G + \sum_{\Gamma \neq \emptyset} \prod_{\gamma_i \in \Gamma} \Delta_{\gamma_i} * G/\Gamma \\
&= \sum_{\Gamma} \prod_{\gamma_i \in \Gamma} \Delta_{\gamma_i} * G/\Gamma,
\end{aligned}
\tag{11.4}
$$

with $\Delta_\emptyset * G/\emptyset = G$. The sum over Γ runs over all UV-disjoint subdiagrams $\Gamma(G) \in \Gamma_G$. The shrunk integral G/Γ stands for the integral of G without propagators and integrations of the subdiagrams contained in $\Gamma(G)$. Diagrammatically, each subdiagram contained in Γ is shrunk to a vertex in G. Remember that the vertices no longer represent the negative coupling constant. Recall also that, by definition in Eq. (9.122), the operation $*$ implies the substitution of the counterterm into the remaining integrations of the shrunk diagram. For a UV-divergent subdiagram γ, the integrand of G can be written as $I_G(\mathbf{k}) = I_\gamma(\mathbf{p}, \mathbf{q}) I_{G/\gamma}(\mathbf{q}, \mathbf{l}, \mathbf{k})$, where $\mathbf{k}$ and $\mathbf{q}$ are the external momenta of G and γ, respectively, whereas $\mathbf{p}$ stands for the momenta of integration in γ. The momenta $\mathbf{q}$, $\mathbf{l}$ appear in the loop integrals of G/γ. The subtracted terms are then constructed from the original Feynman diagram of G as follows:

$$
G = \int \frac{d^D q}{(2\pi)^D} \gamma(\mathbf{q}) I_{G/\gamma}(\mathbf{q}, \mathbf{l}, \mathbf{k}) \longrightarrow \Delta_\gamma * G/\gamma = \int \frac{d^D q}{(2\pi)^D} \int \frac{d^D l}{(2\pi)^D} \Delta_\gamma(\mathbf{q}) I_{G/\gamma}(\mathbf{q}, \mathbf{l}, \mathbf{k}).
$$

In Section 9.3.2, this construction was carried out explicitly for the renormalization up to two loops.

For a logarithmically divergent subdiagram γ, the counterterm is independent of the external momenta of γ, consisting only of $1/\varepsilon^i$-poles: $\Delta_\gamma = \mathcal{Z}_\gamma(\varepsilon^{-1})$. For this reason, the counterterms of the subdiagram γ do not contribute to the integrals in the full diagram, and the operation $*$ reduces to a simple multiplication:

$$
\Delta_\gamma * G/\gamma = \mathcal{Z}_\gamma(\varepsilon^{-1}) \int \frac{d^D q}{(2\pi)^D} \int \frac{d^D l}{(2\pi)^D} I_{G/\gamma}(\mathbf{q}, \mathbf{l}, \mathbf{k}).
\tag{11.5}
$$

In this case, we usually omit the symbol $*$, unless required for clarity.

The operation $*$ becomes nontrivial for quadratically divergent subdiagrams γ, as we have seen already in Eq. (9.100), and as will be discussed in detail in Section 11.6. Then the counterterms for the two-point vertex function contain contributions proportional to m^2 and to $\mathbf{p}^2$, where $\mathbf{p}$ is the external momentum of the subdiagram. In this case, $\mathbf{p}^2$ will participate in the integration of the total diagram. Writing the associated counterterm in the generic form $\Delta_\gamma = \mathbf{q}^2 \mathcal{Z}'_\gamma(\varepsilon^{-1})$, the counterterm diagram contributes the Feynman integral

$$
\Delta_\gamma * G/\gamma = \mathcal{Z}'_\gamma(\varepsilon^{-1}) \int \frac{d^D q}{(2\pi)^D} \int \frac{d^D l}{(2\pi)^D} \mathbf{q}^2 I_{G/\gamma}(\mathbf{q}, \mathbf{l}, \mathbf{k}).
\tag{11.6}
$$

Actually, the relevant calculations will eventually be carried out at zero mass, such that (11.6) will be the only integral to be evaluated. An example for this procedure will be treated in Eq. (12.18).

Let us abbreviate the product of the counterterms of the subdiagrams contained in Γ by the symbol Δ_Γ:

$$\Delta_\Gamma = \prod_{\gamma_i \in \Gamma} \Delta_{\gamma_i}, \tag{11.7}$$

the R-operation in Eq. (11.4) may be written as

$$RG = \sum_\Gamma \Delta_\Gamma * G/\Gamma. \tag{11.8}$$

Then we separate the term $\Gamma = \{G\}$ from the sum, and apply the operator $\mathcal{K}$ as follows:

$$\mathcal{K}\, \Delta_G G/G + \mathcal{K} \sum_{\bar\Gamma} \Delta_{\bar\Gamma} * G/\bar\Gamma = 0. \tag{11.9}$$

The sum runs now over all $\bar\Gamma(G) \neq \{G\}$. In minimal subtraction, $\mathcal{K}\Delta_G = \Delta_G$, and $G/G = 1$, so that we find the counterterm Δ_G of G as

$$\Delta_G = -\mathcal{K} \sum_{\bar\Gamma} \Delta_{\bar\Gamma} * G/\bar\Gamma. \tag{11.10}$$

It is useful to define the negative of the right-hand side as the result of a so-called *incomplete R-operation*, or *$\bar R$-operation*,

$$\mathcal{K}\bar R G \equiv \mathcal{K} \sum_{\bar\Gamma} \Delta_{\bar\Gamma} * G/\bar\Gamma = \mathcal{K} \sum_{\bar\Gamma} \prod_{\gamma_i \in \bar\Gamma} \Delta_{\gamma_i} * G/\bar\Gamma. \tag{11.11}$$

With the help of Eq. (11.10), we now express the counterterm Δ_γ of any subdiagram γ as the result of an $\bar R$-operation applied to γ:

$$\Delta_\gamma = -\mathcal{K}\bar R \gamma. \tag{11.12}$$

Inserting this into (11.11), we find the counterterm Δ_G as a result of a recursive application of the $\bar R$-operation:

$$\Delta_G = -\mathcal{K}\bar R G = -\mathcal{K} \sum_{\bar\Gamma} \prod_{\gamma \in \bar\Gamma} (-\mathcal{K}\bar R\gamma) * G/\bar\Gamma \tag{11.13}$$

$$= -\mathcal{K}\Big[G + \sum_{\bar\Gamma \neq \emptyset} \prod_{\gamma \in \bar\Gamma} (-\mathcal{K}\bar R\gamma) * G/\bar\Gamma \Big].$$

This formula is used to calculate all counterterms up to five loops, as shown explicitly in Appendix A.

As an illustration of the recursive procedure, let us recalculate the second-order counterterms to $\bar\Gamma^{(2)}(\mathbf{k})$ in Section 9.3.2 in terms of the $\bar R$-operation. Since the vertices do not contain the coupling constants as in the previous chapters, we must write them down explicitly as prefactors of each diagram. Remembering the diagrammatic expansion in the first equation of (9.12), we have the equation for the pole terms

$$\mathcal{K}\bar R\bar\Gamma^{(2)} = \Big[-g\mu^\varepsilon \frac{1}{2}\mathcal{K}\bar R(\,\bigcirc\,) + g^2\mu^{2\varepsilon}\frac{1}{4}\mathcal{K}\bar R(\,\bigotimes\,) + g^2\mu^{2\varepsilon}\frac{1}{6}\mathcal{K}\bar R(\,\ominus\,) \Big]. \tag{11.14}$$

Performing the $\bar{R}$-operation on the two-loop diagrams yields

$$\mathcal{K}\bar{R}(\,\ominus\,) = \mathcal{K}\ominus ,$$

$$\mathcal{K}\bar{R}(\,\ominus\,) = \mathcal{K}\big(\,\ominus\, - \mathcal{K}(\,\ominus\,)*\ominus - \mathcal{K}(\,\ominus\,)*\ominus\,\big),$$

$$\mathcal{K}\bar{R}(\,\ominus\,) = \mathcal{K}\big(\,\ominus\, - 3\mathcal{K}(\,\times\!\times\,)*\ominus\,\big).$$

Note the factor three in the subtraction in the last line. The subdiagram is indeed contained three times in the diagram. This is an example of the treatment of overlapping divergence by the subtraction method. Together with Eqs. (9.121) and (9.122), the right-hand side of Eq. (11.14) can be rewritten in terms of counterterm diagrams:

$$\mathcal{K}\bar{R}\bar{\Gamma}^{(2)} = \mathcal{K}\left[-g\mu^{\varepsilon}\frac{1}{2}\ominus + g^2\mu^{2\varepsilon}\frac{1}{4}\ominus + g^2\mu^{2\varepsilon}\frac{1}{2}\ominus + g^2\mu^{2\varepsilon}\frac{1}{2}\ominus + g^2\mu^{2\varepsilon}\frac{1}{6}\ominus\right]. \quad (11.15)$$

Apart from the explicit prefactors $-g\mu^{\varepsilon}$, this is precisely the previous expression (9.94) for the pole terms.

11.3   Properties of Diagrams with Cutvertices

A cutvertex has the generic diagrammatic form

$$G = \,\bigotimes\!\!\bigotimes\!\!- = G_1 \cdot G_2. \quad (11.16)$$

The integrations in each component G_1 and G_2 are obviously independent, and so are the associated divergences. Hence the counterterm of G must be the product of the counterterms of the two components:

$$\Delta_G = -\mathcal{K}\bar{R}G = (-\mathcal{K}\bar{R}G_1)\cdot(-\mathcal{K}\bar{R}G_2) = \Delta_{G_1}\cdot\Delta_{G_2}. \quad (11.17)$$

Since each nonzero Δ_G contains at least a simple pole in ε, the counterterm of a cutvertex contains a pole ε^{-n} with $n \geq 2$. In Eqs. (10.43)–(10.46), we have learned that the renormalization group functions in the MS-scheme consist only of the simple poles of the counterterms. As a consequence, diagrams with cutvertices do not contribute to the renormalization group functions.

The simplest example for relation (11.17) is the diagram $G = \times\!\times\!\times$. Applying $\mathcal{K}\bar{R}$ to this yields

$$\mathcal{K}\bar{R}(\,\times\!\times\!\times\,) = \mathcal{K}\left[\,\times\!\times\!\times\, - 2\,\mathcal{K}(\,\times\!\times\,)\,\times\!\times\,\right]. \quad (11.18)$$

The first term on the right-hand side can be rewritten as

$$\mathcal{K}(\,\times\!\times\!\times\,) = \mathcal{K}\left[-\mathcal{K}(\,\times\!\times\,)\mathcal{K}(\,\times\!\times\,) + 2\,\mathcal{K}(\,\times\!\times\,)\,\times\!\times\,\right], \quad (11.19)$$

implying that the pole terms factorize:

$$\mathcal{K}\bar{R}(\,\times\!\times\!\times\,) = -\mathcal{K}(\,\times\!\times\,)\mathcal{K}(\,\times\!\times\,).$$

Let us also give an example for the effect of the $\bar{R}$-operation upon a more complicated three-loop diagram, in which these pole terms appear in a subdiagram:

$$\mathcal{K}\bar{R}(\,\times\!\boxminus\,) = \mathcal{K}\left[\,\times\!\boxminus\, - \mathcal{K}(\,\times\!\times\,)\,\times\!\boxminus\, - \mathcal{K}(\,\times\!\times\,)\,\times\!\boxminus\, - \mathcal{K}\bar{R}(\,\times\!\times\!\times\,)\,\times\!\times\,\right]. \quad (11.20)$$

The last subtraction term contains the cutvertex, whose pole terms factorize as

$$\mathcal{K}\bar{R}(\,\text{×○○×}\,) = -\mathcal{K}(\text{×○×})\mathcal{K}(\text{×○×})\,. \tag{11.21}$$

The right-hand side does not appear in (11.20) since the two subdiagrams are not UV-disjoint in G. Remember that the sets $\bar{\Gamma}(G)$ were defined to contain only UV-disjoint subdiagrams, in order to avoid a double subtraction of $(-\mathcal{K}\bar{R}G_1) \cdot (-\mathcal{K}\bar{R}G_2)$ and of $-\mathcal{K}\bar{R}G$.

As a consequence of (11.16), all four-point chain diagrams like

$$\text{×○○×}\,,\quad \text{×○○○×}\,,\quad \text{×○○○○×}\,, \tag{11.22}$$

do not contribute to the RG-functions. Their counterterms are, by formula $\left[-\mathcal{K}(\text{×○×})\right]^n \propto 1/\varepsilon^n$, free of simple poles in ε. But only these simple poles survive in the calculation of β, γ, and γ_m, as we have seen in Section 10.3. In the five-loop calculation in Chapter 15, we shall nevertheless list the full renormalization constants including all higher poles. The cancellation of these poles in the calculation of the renormalization group functions provides us with a useful check of the lower ones.

For the contribution of the chain diagrams to the mass counterterm in the two-point function see the discussion after Eq. (11.42).

11.4 Tadpoles in Diagrams with Superficial Logarithmic Divergence

Diagrams or subdiagrams of the form pictured in Fig. 11.1 are called *tadpole diagrams, tadpole parts* of a diagram, or briefly *tadpoles*. The name tadpole referred originally to diagrams which have the shape of a biological tadpole drawn on the left-hand side in Fig. 11.1. Such diagrams occur in ϕ^3-theories or in perturbation expansions of ϕ^4-theories in the ordered phase, which is not considered here. In the disordered phase under study, analogous diagrams occur with two legs, as shown on the right-hand side of the figure. Similar diagrams occur, incidentally, in quantum electrodynamics, where they are more appropriately called "seagull" diagrams, since they look more like (somewhat fat) seagulls on the ocean sky. More generally, we shall define as tadpole diagrams all diagrams which are quadratically divergent and contain no external momenta. Tadpole diagrams have therefore only one external vertex, which is a cutvertex if the tadpole is a subdiagram. Thus tadpole diagrams renormalize only the mass. In the present theory, the simplest tadpole diagram is $\mathcal{Q}$.

Diagrams with a tadpole part produce counterterms only if they are quadratically divergent by power counting. If they are merely logarithmically divergent, thus possessing an equal number of momenta in numerator and denominator of the integrand, the subdiagram to which the tadpole part is attached has two momentum powers less in the numerator and therefore no superficial divergence by power counting. Thus, if we apply the $\bar{R}$-operation to such a

FIGURE 11.1 Structure of tadpole diagrams. The left figure sketches the shape of a biological tadpole. The right figure shows the shape of the "tadpole diagrams" occurring in the normal phase of ϕ^4-theories, which really look like seagulls. Logarithmically divergent Feynman diagrams containing these as subdiagrams do not contribute in dimensional regularization.

diagram and use the factorization property of diagrams with cutvertices as discussed in the last subsection, we obtain zero. An example is:

$$\mathcal{K}\bar{R}\left(\right) = -\mathcal{K}\left(\right) \cdot \underbrace{\mathcal{K}\left(\right)}_{=0} = 0. \tag{11.23}$$

For quadratically divergent diagrams, the factorized subdiagram is still logarithmically divergent, such that the operation $\mathcal{K}\bar{R}$ does yield a nonzero counterterm.

11.5 Nontrivial Example for $\bar{R}$-Operation

A good nontrivial example for the application of the $\bar{R}$-operation is the five-loop diagram

which appears in the complete list of recursive diagrammatic expansions in Appendix A on page 439 as No. 59. It contains a subdiagram with a cutvertex. The first step in the recursion yields

$$\mathcal{K}\bar{R}\left(\right) = \mathcal{K}\Big[\ -\mathcal{K}( ) \ -\mathcal{K}( ) \\ +\mathcal{K}( )\mathcal{K}( ) \ -\mathcal{K}\bar{R}( ) \\ +\mathcal{K}( )\mathcal{K}\bar{R}( ) \\ -\mathcal{K}\bar{R}( ) \ -\mathcal{K}\bar{R}( ) \Big]. \tag{11.24}$$

A second step is necessary in the fifth and sixth term:

$$\mathcal{K}\bar{R}\left(\right) = \mathcal{K}\Big[\ -\mathcal{K}( ) \Big]. \tag{11.25}$$

The last term in (11.24) factorizes at its cutvertex

$$\mathcal{K}\bar{R}\left(\right) = -\mathcal{K}\bar{R}\Big[\Big]\mathcal{K}\Big[\Big], \tag{11.26}$$

and only the first component, which also appears in the second-last term in (11.24), requires further consideration. It is expanded as

$$\mathcal{K}\bar{R}\left(\right) = \mathcal{K}\Big[\ -\mathcal{K}( ) \ -\mathcal{K}\bar{R}( ) \Big]. \tag{11.27}$$

The right-hand side contains only pole terms of lower-order and shrunk diagrams.

11.6 Counterterms in Minimal Subtraction

Renormalizability requires the counterterms to have the general local form described in Section 9.3. As explained in Subsection 9.3.1, the counterterms are further simplified by using minimal subtraction. For the proofs of the renormalizability in the context of dimensional regularization see Ref. [3].

The first step in such proofs consists in showing that application of the R-operation to a diagram G yields a Feynman integral RG which is indeed free of subdivergences, and that the counterterms subtracted to achieve this are all local. The proof proceeds by induction in the number of loops.

As already mentioned in Subsection 9.3.3 and 11.2, we have to make sure that the combinatorics of the subtractions in each diagram allow for a complete absorption of all $1/\varepsilon^i$ -terms via counterterms by a redefinition of the parameters in the energy functional. This is not obvious since the counterterm diagrams subtract the subdivergences of different diagrams. Some work is necessary to demonstrate that the combinatorics of the subtraction scheme reproduces the correct weight factors of the counterterms. At the end one verifies that the counterterms can indeed be found as the sum of the pole terms $\mathcal{K}\bar{R}G$ of the individual diagrams G. They are polynomials in the external momenta with a maximal power $\mathbf{k}^2$ allowed by locality. In ϕ^4-theory, all divergences appear in the two- and four-point functions. The pole terms $\mathcal{K}\bar{R}\bar{\Gamma}^{(2)}$ contain only terms proportional to m^2 and $\mathbf{k}^2$, whereas $\mathcal{K}\bar{R}\bar{\Gamma}^{(4)}$ contains no momenta at all:

$$-\mathcal{K}\bar{R}\bar{\Gamma}^{(2)}(\mathbf{k}) = \mathcal{K}\bar{R}(\Sigma) \;\longrightarrow\; m^2 c_{m^2}(g,\varepsilon^{-1}) + \mathbf{k}^2 c_\phi(g,\varepsilon^{-1})\,, \tag{11.28}$$

$$-\mathcal{K}\bar{R}\bar{\Gamma}^{(4)}(\mathbf{k}_i) \;\longrightarrow\; \mu^\varepsilon g\, c_g(g,\varepsilon^{-1})\,. \tag{11.29}$$

This is the general form of the counterterms leading to a local counterterm energy functional (9.63) [recall Eqs. (9.65)–(9.67)].

The counterterms are expansions in g and ε^{-1} independent of the mass and the external momentum and have the generic form (9.1). No additional divergences appear for the critical theory with $m^2 = 0$. The trivial mass- and $\mathbf{k}^2$-dependences permit us to rewrite the renormalization constants as follows:

$$Z_{m^2}(g,\varepsilon^{-1}) = 1 - \mathcal{K}\bar{R}\frac{\partial}{\partial m^2}\,\bar{\Gamma}^{(2)}(\mathbf{k})\,, \tag{11.30}$$

$$Z_\phi(g,\varepsilon^{-1}) = 1 - \mathcal{K}\bar{R}\frac{\partial}{\partial \mathbf{k}^2}\,\bar{\Gamma}^{(2)}(\mathbf{k})\,, \tag{11.31}$$

$$Z_g(g,\varepsilon^{-1}) = 1 - \mathcal{K}\bar{R}\frac{1}{\mu^\varepsilon g}\,\bar{\Gamma}^{(4)}(\mathbf{k}_i)\,. \tag{11.32}$$

In writing them down , we have made use of the fact that the operations $\mathcal{K}$ and $\bar{R}$ commute with the differentiations with respect to m^2 and $\mathbf{k}^2$. This follows from the absolute convergence of the initial Feynman integrals in $D = 4 - \varepsilon$ dimensions with $\varepsilon \neq 0$. Performing the differentiations under the integrals, their superficial degree of divergence is lowered by two, and all integrals on the right-hand side of Eqs. (11.30)–(11.32) become logarithmically divergent.

The calculation of all three renormalization constants simplifies in minimal subtraction because diagrams with tadpole subdiagrams do not contribute. For Z_g this is due to the fact that all divergences contributing to the vertex function $\bar{\Gamma}^{(4)}(\mathbf{k}_i)$ are at most logarithmically divergent, and we have shown in Section 11.4 that a tadpole subdiagram causes the remaining diagram to be superficially convergent. Thus the total Feynman integral vanishes under the application of $\mathcal{K}\bar{R}$.

For the renormalization constants Z_{m^2} and Z_ϕ, additional simplifications arise, which we shall now discuss.

11.7   Simplifications for Z_{m^2}

Differentiating a propagator with respect to m^2 raises the power of momenta in the denominator by two. The resulting Feynman integral may be viewed diagrammatically as having arisen from the insertion of an extra $-\phi^2$-vertex with two legs (recall Sections 2.4 and 3.5):

$$\frac{\partial}{\partial m^2} \frac{1}{\mathbf{p}^2 + m^2} = \frac{-1}{(\mathbf{p}^2 + m^2)^2} \; \hat{=} \; \frac{\partial}{\partial m^2} \; -\!\!\!- \; = \; -\!\cdot\!- \; . \tag{11.33}$$

The two additional powers of momenta in the denominator lower the superficial degree of divergence ω by two, changing a quadratic divergence into a logarithmic one. With the product rule of differentiation, application of $\partial/\partial m^2$ to diagrams with several lines leads to a sum of logarithmically divergent diagrams, each containing an extra $-\phi^2$-vertex. If n lines are indistinguishable, the differentiation results in a multiplicity factor n, which is 3 in the following example:

$$\frac{\partial}{\partial m^2} \ominus = 3 \ominus . \tag{11.34}$$

The logarithmically divergent Feynman integral of the diagram on the right-hand side corresponds to a diagram $\ominus$ of the four-point vertex function of $\bar{\Gamma}^{(4)}(\mathbf{k}_i)$, in which the external momenta incoming at the above ϕ^2-vertex have been set equal to zero. For dimensional reasons, the external momenta do not appear in the counterterms $\mathcal{K}\bar{R}G$ of a logarithmically divergent integral G, and the differentiated Feynman integral used for the mass renormalization produces the same counterterm as the corresponding integral in $\bar{\Gamma}^{(4)}(\mathbf{k}_i)$. In the $\mathcal{K}\bar{R}$-notation, we may write:

$$\mathcal{K}\bar{R}\left(\ominus \right) = \mathcal{K}\bar{R}\left(\ominus \right) . \tag{11.35}$$

The superficial divergence of the two Feynman diagrams on the two sides of the equation differ only in the power of the coupling constant associated with the vertex on the top. Having set g equal to unity, we see that the two sides are identical.

Another example illustrates the differentiation of quadratically divergent diagrams with a tadpole part:

$$\frac{\partial}{\partial m^2} \, 8 = 2 \, 8 + 2 \, 8 + \, 8 \; . \tag{11.36}$$

The R-operation applied to the two logarithmically divergent diagrams in Eq. (11.36), which contain a tadpole part, gives zero as explained in Section 11.4:

$$\mathcal{K}\bar{R}\left(8 \right) = -\mathcal{K}\bar{R}(\underbracket{\mathcal{Q}}) \underbrace{\mathcal{K}\bar{R}(\ominus)}_{=0} = 0, \tag{11.37}$$

$$\mathcal{K}\bar{R}\left(8 \right) = -\mathcal{K}\bar{R}(\mathcal{Q}) \underbrace{\mathcal{K}\bar{R}(\ominus)}_{=0} = 0. \tag{11.38}$$

The pole terms in the third diagram of Eq. (11.36), which has a $-\phi^2$-vertex inserted into the tadpole part, is a product of the pole terms of two logarithmically divergent diagrams:

$$\mathcal{K}\bar{R}\left(8 \right) = -\mathcal{K}\bar{R}(\mathcal{Q})\mathcal{K}\bar{R}\left(\ominus \right) . \tag{11.39}$$

The subdiagram $\mathcal{Q}$ is not of the tadpole type since it is not quadratically divergent. The pole terms of this diagram were considered earlier in Eq. (9.79), where we showed that

$$\mathcal{K}(\mathcal{Q}) = \mathcal{K}(\times) . \tag{11.40}$$

Hence the counterterm contribution of the whole diagram on the left-hand side of Eq. (11.36) is the same as that of a four-point diagram renormalizing the coupling constant:

$$\frac{\partial}{\partial m^2}\, \mathcal{K}\bar{R}\, \vcenter{\hbox{⊗}} \; = \mathcal{K}\bar{R}\big(\vcenter{\hbox{⊗}} \big) = \mathcal{K}\bar{R}\big(\vcenter{\hbox{⊗}} \big)\,. \tag{11.41}$$

Another example are all two-point chain diagrams consisting of simple bubbles like the following:

$$\vcenter{\hbox{⧗}} \; , \quad \vcenter{\hbox{⧗}} \; , \quad \vcenter{\hbox{⧗}} \; . \tag{11.42}$$

After the differentiation with respect to the mass, we are left with the contributions of the following four-point diagrams:

$$\vcenter{\hbox{)○○○}} \; , \quad \vcenter{\hbox{)○○○○}} \; , \quad \vcenter{\hbox{)○○○○○}} \; , \tag{11.43}$$

whereas all diagrams with ϕ^2-insertions in any other line give zero upon application of the $\bar{R}$-operation. The remaining diagrams have the same counterterm contribution as the four-point chain diagrams (11.22) and, like those, they contribute to the renormalization constants only with higher poles which cancel when calculating the renormalization group functions.

These examples suggest that the diagrams contributing to the mass renormalization form a subset of the diagrams contributing to the coupling constant renormalization. This can be observed to all orders in perturbation theory. The subset consists of all four-point diagrams without tadpole parts, and with two external lines entering at one common vertex. After dropping a factor $-\mu^\varepsilon g$, and multiplying everything with the appropriate weight factors and symmetry factors, we obtain directly the contribution to the mass renormalization factor $Z_{m^2}(g, \varepsilon^{-1})$.

11.8   Simplifications for Z_ϕ

The calculation of $Z_\phi(g, \varepsilon^{-1})$ is also simplified by the fact that all diagrams with tadpole subdiagrams do not contribute. Here this happens for a different reason than in $Z_{m^2}(g, \varepsilon^{-1})$: The counterterm of any such subdiagram carries now a factor m^2, such that the counterterm of the remaining diagram, which is a pure factor of zero dimension, cannot depend on $\mathbf{k}^2$. Thus such counterterms will not appear in $Z_\phi(g, \varepsilon^{-1})$ obtained from Eq. (11.31) via a momentum differentiation. Diagrams with tadpole parts contribute only to the mass renormalization, but not to $Z_\phi(g, \varepsilon^{-1})$.

The remaining quadratically divergent diagrams without tadpole parts depend on the external momentum. Carrying out the differentiation with respect to this momentum,

$$\frac{\partial}{\partial \mathbf{k}^2}\mathcal{K}\bar{R}\bar{\Gamma}^{(2)}(\mathbf{k}, m^2) \; = \; \mathcal{K}\bar{R}\frac{\partial}{\partial \mathbf{k}^2}\bar{\Gamma}^{(2)}(\mathbf{k}, m^2)\,, \tag{11.44}$$

generates logarithmically divergent integrals. The differentiation can always be interchanged with the R-operation [4]. In a Feynman integral some terms may depend on the individual components k_μ of $\mathbf{k}$ rather than on $\mathbf{k}^2$. When performing the differentiations in such terms, we use the relation

$$\frac{\partial}{\partial \mathbf{k}^2} = \frac{1}{2D}\frac{\partial^2}{\partial k_\mu \partial k_\mu}, \tag{11.45}$$

which follows from the trivial equation

$$\frac{\partial^2}{\partial k_\mu \partial k_\mu}\mathbf{k}^2 = 2D\,. \tag{11.46}$$

In $D \equiv 4 - \varepsilon$ dimensions, the differentiations are done as follows:

$$\frac{\partial}{\partial k_\mu} \frac{1}{(\mathbf{p} - \mathbf{k})^2 + m^2} = -\frac{2(p - k)_\mu}{[(\mathbf{p} - \mathbf{k})^2 + m^2]^2} \tag{11.47}$$

$$\frac{\partial^2}{\partial k_\mu \partial k_\mu} \frac{1}{(\mathbf{p} - \mathbf{k})^2 + m^2} = \frac{-2D}{[(\mathbf{p} - \mathbf{k})^2 + m^2]^2} + \frac{8 (\mathbf{p} - \mathbf{k})^2}{[(\mathbf{p} - \mathbf{k})^2 + m^2]^3}. \tag{11.48}$$

The differentiation raises the power of the denominator by one unit and multiplies the numerator by the loop vector. Diagrammatically, this raising of power may be attributed to the presence of an additional $-\phi^2$-vertex. If we indicate the momentum vector in the numerator by a vertical dash, we can express the operations (11.47) and (11.48) diagrammatically as

$$2 \;\text{-•+} \quad \text{and} \quad 2D \;\text{-•-} + 8 \;\text{-•+}\text{-} , \tag{11.49}$$

respectively. The line with the dash symbolizing the propagator $p_\mu/(\mathbf{p}^2 + m^2)$ and will be referred to as a *line with a vector index*.

Although these operations are quite lengthy, the result has at least the advantage of being less divergent. Moreover, they will lead to real simplifications due to the fact that the calculation of all counterterms can eventually be done with massless Feynman integrals. Then the propagators in the last expression can be combined to $(8 - 2D)/[(\mathbf{p} - \mathbf{k})^2]^2$. All such simplifications will be discussed in detail in Subsection 12.2.2.

Notes and References

Good textbooks explaining the R-operation are
N.N. Bogoliubov and D.V. Shirkov, *Introduction to the Theory of Quantized Fields*, Wiley Interscience, New York, 1958,
and
C. Itzykson and J.-B. Zuber, *Quantum Field Theory*, McGraw-Hill, New York, 1980.

The individual citations in the text refer to:

[1] See, in particular, Refs. [2] and [5] in Chapter 9.

[2] W. Zimmermann, Commun. Math. Phys. *16*, 208 (1969).

[3] J.C. Collins, Nucl. Phys. B **80**, 341 (1974);
    E.R. Speer, J. Math. Phys. **15**, 1 (1974);
    P. Breitenlohner and D. Maison, Commun. Math. Phys. **52**, 11, 39, 55 (1977);
    W.E. Caswell and A.D. Kennedy, Phys. Rev. D **25**, 392 (1982).

[4] For a proof see
    W.E. Caswell, A.D. Kennedy, Phys. Rev. D *25*, 392 (1982).

12

Zero-Mass Approach to Counterterms

Massive Feynman integrals containing more than one loop momentum are hard to evaluate. Fortunately, the understanding of the critical behavior of the field theory requires only knowledge of the divergent counterterms. In the last chapter we have seen that their calculation reduces to the calculation of logarithmically divergent diagrams without tadpole parts for Z_g and Z_{m^2} and of quadratically divergent diagrams without tadpole parts for Z_ϕ. From the superficial divergence of the latter. only the mass-independent part is needed. In addition, the superficial divergences of the logarithmically divergent diagrams are independent of the mass and of the external momenta. These properties have the important consequence that masses and external momenta of Feynman integrals may be modified in a variety of ways without changing the counterterms. In particular, masses and external momenta may be set equal to zero as long as this does not produce unphysical IR-divergences. We shall see that overall IR-divergences do not occur if at least one external momentum is kept nonzero. There are different ways of choosing the nonzero momentum, and the corresponding mathematical modifications of Feynman integrals are called *infrared rearrangement* (IRR) [1]. A suitable rearrangement allows us to simplify considerably the calculation of counterterms in a massive theory.

In many Feynman integrals, the single nonzero external momentum is still an obstacle to an analytical calculation. In this case one employs a more drastic IR-rearrangement by admitting a final nonzero external momentum which *does* generate unphysical *IR-divergences*. These must be properly identified and subtracted to arrive at the desired UV-counterterms.

The minimal subtraction scheme which is so convenient for regularizing the theory has, unfortunately, an unpleasant feature as far as the new unphysical IR-divergences are concerned. The new divergences have the same $1/\varepsilon^i$ -pole form as the UV-divergences. The identification and subtraction of the infrared parts in the total counterterms are therefore nontrivial. These parts will be called *IR-counterterms*, for brevity, and we shall develop a diagrammatic method for calculating them for each Feynman diagram. We shall construct so-called *IR-diagrams* which must, of course, be such that no new UV-divergences arise. This procedure leads, unfortunately, to a proliferation of diagrams, but these have the advantage of containing only massless lines, which greatly simplifies the associated Feynman integrals. In some cases, this is the only way to find an analytic result for the UV-counterterms.

In this chapter we describe a recursive diagrammatic construction and subtraction of IR-divergences which proceeds by close analogy with the subtraction of UV-subdivergences. The combined recursive scheme will eventually be formulated as an extension of the R-operation, called R^*-operation.

As a basis for understanding the IR-divergences of Feynman integrals, in which one or more masses and external momenta are set equal to zero, we introduce the technique of *infrared power counting*. This will help us decide which masses and external momenta can be set equal to zero without changing the counterterms. Subsequently, IR-rearrangement will be explained, and the ensuing additional IR-divergences will be removed.

12.1   Infrared Power Counting

Masses and external momenta ensure the infrared finiteness of all Feynman integrals. If these dimensional parameters are all set equal to zero, IR-divergences will appear. In dimensional regularization, they manifest themselves as pole terms $1/\varepsilon^i$, making them indistinguishable from UV-divergences. To understand the relation between the two types of divergences, we must study the properties of Feynman integrals in the small-momentum regime.

For this purpose, let us rescale all loop momenta in a Feynman integral I_G by a factor σ, and suppose that the integral behaves for $\sigma \to 0$ like

$$I_G \propto \sigma^{-\tilde{\omega}} \qquad \text{for } \sigma \to 0. \tag{12.1}$$

The number $\tilde{\omega}(G)$ is called the *superficial degree of IR-divergence*. An integral is divergent in the infrared if $\tilde{\omega}(G) \geq 0$. Any line containing a mass or an external momentum does not contribute to the superficial IR-divergence. It does, however, reduce the degree of UV-divergence $\omega(G)$. This implies

$$\omega(G) \neq -\tilde{\omega}(G). \tag{12.2}$$

In a massless theory, an external momentum supplies a Feynman integral with an *IR-cutoff* which removes its superficial IR-divergence. This does not yet imply that the entire Feynman integral is IR-convergent. For this to happen, the massless diagrams in four dimensions have to possess *nonexceptional* external momenta. Only then can we rule out IR-subdivergences. Recall that according to the definition on page 140, external momenta are nonexceptional if none of their partial sums vanishes. In a ϕ^4-theory, the external momenta of two- and four-point diagrams are nonexceptional as long as they do not vanish. For this reason, we can always calculate all counterterms from massless diagrams.

If the external momenta of a diagram are exceptional, a subdiagram can be separated from the remaining diagram by cutting lines which do not carry external momenta. Such a subdiagram produces an IR-divergence in the Feynman integral. As an example of an infrared divergence arising in this way, take the Feynman integral

$$\text{\footnotesize XX} \sim \int \frac{d^4p}{(2\pi)^4} \frac{1}{\mathbf{p}^4(\mathbf{p}-\mathbf{k})^2} = \int \frac{d^4p}{(2\pi)^4} \frac{1}{(\mathbf{p}-\mathbf{k})^4 \mathbf{p}^2}. \tag{12.3}$$

The ϕ^2-insertion on the upper line may be considered as a ϕ^4-vertex with two amputated lines of zero external momentum. The momenta are therefore exceptional. The flow of the remaining external momentum can then be chosen as specified in the first integral in (12.3), where it enters and leaves through one of the external lines on the right and left. The integral in (12.3) diverges logarithmically in the infrared, with $\tilde{\omega}(G) = 0$. Note that the original character of the vertex at the top is irrelevant to this divergence, which also would have arisen if the top vertex had been a three-point, five-point, or higher vertex.

In general, the degree of IR-divergence for a massless diagram G is found as follows. One constructs from G a subdiagram $\tilde{\gamma}$ by shrinking all lines which do not contribute to the IR-divergence. These lines are identified by searching for the maximal set of propagators which become singular at different points in momentum space. The shrinking produces a new vertex referred to as *virtual vertex*. In the diagram in (12.3), we may either shrink the upper line or the lower line because the external momenta flowing through them do not contribute to the same IR-divergence. In fact, only the integration over the upper lines with the ϕ^2-vertex is IR-divergent. The subdiagram $\tilde{\gamma}$ is found either by shrinking all lines to a new vertex which carry an external momentum, or by shrinking all lines which do not. In some cases, several subdiagrams $\tilde{\gamma}$ have to be formed and analyzed to find all IR-counterterms. But we shall not need more than one nonzero external momentum. For this reason, the connectedness of the diagram will always lead to a shrunk diagram with only a single virtual vertex.

For the diagram (12.3), the shrunk subdiagram is $\tilde{\gamma} = \mathbb{O}$, omitting the irrelevant external lines.

An arbitrary shrunk diagram satisfies the following topological equations:

$$L(\tilde{\gamma}) = I(\tilde{\gamma}) - [V(\tilde{\gamma}) + 1] + 1, \tag{12.4}$$
$$2I(\tilde{\gamma}) = 4V(\tilde{\gamma}) + n, \tag{12.5}$$

where $L(\tilde{\gamma})$, $I(\tilde{\gamma})$, and $V(\tilde{\gamma})$ are the numbers of loops, internal lines, and vertices of the shrunk diagram. The number $V(\tilde{\gamma})$ does not include the virtual vertex arising from the shrinking. The number n counts the number of lines emerging from this vertex. The degree of IR-divergence of G is now given by the negative degree of UV-divergence of the shrunk diagram $\tilde{\gamma}$, which is

$$\omega(\tilde{\gamma}) = L(\tilde{\gamma})D - 2I(\tilde{\gamma}). \tag{12.6}$$

Thus, a diagram G is IR-divergent if the degree of UV-divergence of $\tilde{\gamma}$ is $\omega(\tilde{\gamma}) \leq 0$. Inserting relations (12.4) and (12.5) for the shrunk diagram, we obtain an equation for the superficial degree of IR-divergence:

$$\tilde{\omega}(G) = -\omega(\tilde{\gamma}) = -L(\tilde{\gamma})D + 2I(\tilde{\gamma}) = L(\tilde{\gamma})(4 - D) - n. \tag{12.7}$$

For 1PI diagrams G, the number n is never smaller than two, such that IR-convergence is guaranteed for $D = 4$. The mass can thus be set equal to zero without generating IR-divergences as long as the external momenta are nonexceptional. This means that we can perform all calculations with massless integrals. This result holds also for 1PI subdiagrams, thus eliminating possible IR-subdivergences. In 1PI subdiagrams, the IR-regulating role of external momenta is played by the loop momenta of the remaining diagram flowing into the subdiagram.

The IR-situation in our calculations is less trivial because these are performed in $D = 4 - \varepsilon$ dimensions. Then, according to Eq. (12.6), the degree of divergence depends on the number of loops $L(\tilde{\gamma})$, and therefore on the order of perturbation theory:

$$\tilde{\omega}(G) = -\omega(\tilde{\gamma}) = L(\tilde{\gamma})\varepsilon - n. \tag{12.8}$$

This gives rise to a small-momentum power behavior $|\mathbf{k}^2|^{n - L(\tilde{\gamma})\varepsilon}$ of the counterterms of the vertex functions. Such a power behavior is of the type discussed earlier in Eq. (9.74), and becomes compatible with the necessary locality of the counterterms only after an expansion in powers of ε.

12.2 Infrared Rearrangement

In the last section we have seen that in 1PI two- and four-point ϕ^4-diagrams with nonzero external momenta, the masses can be set equal to zero without generating IR-divergences. Thus, the four-point diagrams contributing to Z_g and Z_{m^2}, and the two-point diagrams contributing to Z_ϕ can be calculated with zero mass. We also know from Sections 11.7 and 11.8 that the renormalization constants can be calculated from diagrams whose divergence is only logarithmic, if the differentiations are applied directly to the integrands. Since the pole terms of logarithmically divergent diagrams are independent of the external momenta, we may wonder whether the external momentum could not be set to zero as well, thus simplifying the calculations further.

Unfortunately, this is not possible. If we were to do this, all lines of an integral would contribute to the IR-behavior, implying that the superficial degree of UV-divergence $\omega(G)$ also specifies the superficial IR-divergence. A diagram has a superficial UV-divergence for $\omega(G) \geq 0$, and a superficial IR-divergence for $\omega(G) \leq 0$. The last of the $L+1$ integrations in a logarithmically IR-divergent diagram, with zero masses and external momenta, is always of the form:

$$\int \frac{d^D p}{(\mathbf{p}^2)^{2+L\varepsilon/2}} \,. \tag{12.9}$$

This integral is of the massless tadpole type considered in Eq. (8.33), which is equal to zero for any power $2 + L\varepsilon/2$ in dimensional regularization. It is UV-divergent for $D \geq 4 + L\varepsilon$, and IR-divergent for $D \leq 4 + L\varepsilon$. Dimensional regularization provides us with an analytic extrapolation of all Feynman integrals to dimensions $D < 4$, which produces pole terms $1/\varepsilon^i$. The IR-divergences occuring in massless Feynman integrals at $D = 4$ give rise to similar pole terms. In the generic tadpole integral (12.9), the IR-divergence compensates completely the UV-divergence.

In contrast to this, one nonzero external momentum $\mathbf{k}$ entering a massless diagram changes (12.9) to

$$\int \frac{d^D p}{(\mathbf{p} - \mathbf{k})^2 (\mathbf{p}^2)^{1+L\varepsilon/2}} \,. \tag{12.10}$$

Now the integral is IR-convergent, whereas the UV-divergence is unchanged. This is how a nonzero external momentum guarantees superficial IR-convergence of a massless Feynman integral. The lines through which the external momentum flows can be chosen rather arbitrarily. One may even modify the character of the vertices without changing the divergences: they must no longer be all of the ϕ^4-type, but can become ϕ^3-, ϕ^2-, and even ϕ^5-vertices. Examples are shown in Figs. 12.2 and 12.4.

Applying the method of *IR-rearrangement* (IRR), one nonzero external momentum supplies a convenient infrared cutoff, thus making the mass in the Feynman integral superfluous. The resulting zero-mass momentum space integrals consist of a massless $(L-1)$-loop insertion in a one-loop diagram as pictured in Fig. 12.1 [2]. They will generically be referred to as *propagator-type* or shorter *p-integrals*. A detailed discussion will be given in Chapter 13, in particular the powerful algorithms developed [3] for their calculation.

The rearrangement of the momentum is shown explicitly in Fig. 12.2. With IR-rearrangement, five-loop calculations up to $O(\varepsilon^{-1})$ can be reduced to a four-loop calculation up to $O(\varepsilon^0)$. The last integration generates a pole in ε (see 13.1.1).

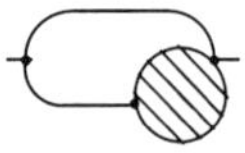

FIGURE 12.1 Generic propagator-type diagram obtained after infrared rearrangement as in Fig. 12.2.

While the introduction of a nonzero external momentum eliminates superficial IR-divergences, it does not in general guarantee the absence of IR-subdivergences. On page 198 we saw that diagrams with nonexceptional external momenta are free of IR-subdivergences. But after the rearrangement of the external momentum, they will in most cases be exceptional. Thus we have to check carefully to see whether an IR-divergence has been created by the rearrangement. This is facilitated by Eq. (12.7), which has to be modified to account for

FIGURE 12.2 Simplification of massless Feynman integral by infrared rearrangement (IRR). Setting the external momenta 2 and 4 equal to zero leads to a diagram of the propagator type, corresponding to a one-loop diagram with an $(L-1)$-loop insertion, pictured in Fig. 12.1. Note the IR-rearrangement produces an artificial ϕ^3-vertex.

the possible existence of artificial ϕ^2-, ϕ^3-, and ϕ^5-vertices generated by the rearrangement. As an example, see the artificial ϕ^3-vertex generated in Fig. 12.2. If the rearranged diagram possesses V_r vertices of the ϕ^r-type for $r = 2, 3, 5$, we find the degree of superficial IR-divergence

$$\tilde{\omega}(\tilde{\gamma}) = -L(\tilde{\gamma})(D-4) - n + 2V_2 + V_3 - V_5, \tag{12.11}$$

where n is the number of external lines of the subdiagram. This expression implies an IR-divergence of the Feynman integral for $D = 4$, if there are n ϕ^3-vertices or $n/2$ ϕ^2-vertices in the shrunk diagram. Remember that in the shrunk diagrams all lines with an IR-cutoff, a mass, or an external momentum are shrunk. A ϕ^2-vertex always implies an IR-subdivergence in $D = 4$ forming a subdiagram with $n = 2$, such that $\omega = 0$. Up to five loops, all but two of the encountered IR-subdivergences are caused by this subdiagram, which is diagrammatically denoted as $\big|$, symbolizing an integral over $1/\mathbf{p}^4$. One subdiagram even occurs which contains two ϕ^2-vertices, which has $\tilde{\omega}(\gamma) = 2$. This subdiagram is $\big|$, corresponding to an integration over $1/\mathbf{p}^6$. Furthermore, IR-divergences are caused by subdiagrams with at least two ϕ^3-vertices for $n = 2$ and four ϕ^3-vertices for $n = 4$. In these cases $\tilde{\omega}(\gamma) = 0$. We shall not encounter subdiagrams with more ϕ^3-vertices than that. Examples containing IR-divergent subdiagrams will be considered in Subsection 12.4.2. The result of these considerations is that propagator-type integrals generated by infrared rearrangement are superficially IR-convergent at $D = 4$ as long as IR-rearrangement generates no IR-subdivergences with virtual vertices of the ϕ^2-type, or subdiagrams with two or four such vertices of the ϕ^3-type.

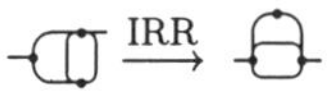

$$-\Box \! \xrightarrow{\text{IRR}} \; \ominus \; .$$

FIGURE 12.3 Simple example for generation of an IR-subdivergence by IR-rearrangement. In the integral on the left-hand side, the external momenta are reduced to a single momentum. The rearrangement produces an IR-divergence.

A simple example for the generation of an IR-subdivergence by IR-rearrangement is shown in Fig. 12.3. The associated Feynman integrals before and after rearrangement can be calculated without infrared rearrangement:

$$\int \frac{d^D p \, d^D q}{(2\pi)^{2D}} \frac{1}{\mathbf{q}^2 (\mathbf{p}-\mathbf{q})^2 (\mathbf{p}-\mathbf{k})^2 \mathbf{p}^2} \xrightarrow{\text{IRR}} \int \frac{d^D p \, d^D q}{(2\pi)^{2D}} \frac{1}{(\mathbf{q}-\mathbf{k})^2 (\mathbf{p}-\mathbf{q})^2 \mathbf{p}^4}. \tag{12.12}$$

The removal of the IR-divergence of the IR-rearranged diagram on the right-hand side will be carried out in detail in Section 12.4.

As mentioned at the beginning of this chapter, sometimes infrared rearrangement generates calculable subdiagrams only if IR-subdivergences are admitted. An example for such a drastic

FIGURE 12.4 Simplification of massless Feynman integral by infrared rearrangement (IRR), generating an IR-divergence in a subdiagram. Setting the external momenta 2, 3, 4 to zero, and importing a nonzero external momentum $1'$, generates a one-loop diagram with a four-loop insertion. Unfortunately, an IR-divergence also arises from the newly created ϕ^2-vertex in the line below $1'$, which must be properly subtracted to obtain the correct UV-divergences.

version of infrared rearrangement [4] is displayed in Fig. 12.4. Here the integration over the massless line with the ϕ^2-vertex introduces the artificial IR-divergence which must be calculated and removed from the total divergence to find the correct UV-divergence. The subtraction will be done recursively in Section 12.4.

12.2.1   The R-Operation for Massless Diagrams

If the masses are set equal to zero, an important simplification occurs in the subtraction of quadratically divergent subdiagrams. The $\bar{R}$-operation involves subtractions of both logarithmically divergent and quadratically divergent subdiagrams, to be denoted by γ_4 and γ_2, respectively:

$$\bar{R}G = \sum_{\bar{\Gamma}} \prod_{\gamma_4 \in \bar{\Gamma}} (-\mathcal{K}\bar{R}\gamma_4) \prod_{\gamma_2 \in \bar{\Gamma}} (-\mathcal{K}\bar{R}\gamma_2) * (G/\bar{\Gamma}) \,. \tag{12.13}$$

For zero mass, the counterterms $\mathcal{K}\bar{R}\gamma_2$ of the quadratically divergent subdiagrams γ_2 are proportional to $\mathbf{p}^2$ only, where $\mathbf{p}$ is the momentum flowing through γ_2. As an example, consider the following Feynman integral where the four external momenta are replaced by only two:

$$G = \text{⬡} \quad = \quad \int \frac{d^D p}{(2\pi)^D} \frac{\gamma_2(\mathbf{p})}{\mathbf{p}^4(\mathbf{p}-\mathbf{k})^2}, \tag{12.14}$$

and which contains the quadratically divergent subdiagram divergence

$$\gamma_2(\mathbf{p}) = \text{⊖} \,, \tag{12.15}$$

whose superficial divergence is given by

$$-\mathcal{K}\bar{R}(\text{⊖}) = \mathcal{Z}_{\gamma_2}(\varepsilon^{-1})\,\mathbf{p}^2. \tag{12.16}$$

The external square momentum $\mathbf{p}^2$ of the subdiagram is part of the integrand of the remaining integral. In the zero-mass case, it cancels a propagator in the denominator of the integrand, i.e., a line in the shrunk integral. The associated counterterm is then calculated as follows:

$$\Delta_{\gamma_2}(\mathbf{p}) * (G/\gamma_2) \quad = \quad -\mathcal{K}\bar{R}(\text{⊖}) * \text{⊖} \tag{12.17}$$

$$\equiv \quad \int \frac{d^D p}{(2\pi)^D} \frac{\mathcal{Z}_{\gamma_2}(\varepsilon^{-1})\,\mathbf{p}^2}{\mathbf{p}^4(\mathbf{p}-\mathbf{k})^2}$$

$$= \quad \mathcal{Z}_{\gamma_2}(\varepsilon^{-1}) \int \frac{d^D p}{(2\pi)^D} \frac{1}{\mathbf{p}^2(\mathbf{p}-\mathbf{k})^2}. \tag{12.18}$$

According to Eq. (12.16), the pole terms $\mathcal{Z}_{\gamma_2}(\varepsilon^{-1})$ may be considered as being obtained from the derivative of $-\mathcal{K}\bar{R}(\,\ominus\,)$ with respect to $\mathbf{p}^2$ as (recall the discussion in Section 11.8)

$$\mathcal{Z}_{\gamma_2}(\varepsilon^{-1}) = \partial_{\mathbf{p}^2}\mathcal{K}\bar{R}\gamma_2\,, \tag{12.19}$$

where the differentiation symbol $\partial_{\mathbf{p}^2} \equiv \partial/\partial\mathbf{p}^2$ acts only on the first expression to its right. Thus we may rewrite (12.18) as

$$\Delta_{\gamma_2}(\mathbf{p}) * (G/\gamma_2) = \mathcal{Z}_{\gamma_2}(\varepsilon^{-1})\ \ominus\ . \tag{12.20}$$

For any massless quadratically divergent diagram, the operation $*$ can be replaced by $\partial_{\mathbf{p}^2}$ acting on the pole term to its left, with a simultaneous removal of an associated propagator in the shrunk diagram to its right. From the point of view of the subdiagram, whose pole term is specified by $\partial_{\mathbf{p}^2}\mathcal{K}\bar{R}(\gamma_2)$, the internal momentum $\mathbf{p}$ of the total diagram is really an external momentum. Since we generally use the notation $\mathbf{k}$ rather than $\mathbf{p}$ to specify external momenta, we shall henceforth denote the pole terms by $\partial_{\mathbf{k}^2}\mathcal{K}\bar{R}(\gamma_2)$ for the sake of a more consistent notation.

Remember that due to the relation $\mathcal{K}\bar{R}\gamma_4 = \mathcal{Z}_{\gamma_4}(\varepsilon^{-1})$, the operation $*$ in logarithmically divergent subdiagrams reduces to a pure multiplication [see Eq. (11.5)]. For this reason, most of our recursive formulas will no longer contain the operation symbols $*$.

As a more complicated example, in which the above diagram appears as a subdiagram, consider the five-loop four-point diagram

This diagram is very hard to calculate even with zero masses, if the external momenta are not rearranged. This will become clear in the next chapter. By setting the external momenta entering through the lower lines equal to zero, the pole terms of this diagram are equal to those of the infrared-rearranged diagram

Application of the $\mathcal{K}\bar{R}$-operation yields the following expansion

$$\mathcal{K}\bar{R}\big(\ \big) = \mathcal{K}\bar{R}\big(\ \big) = \mathcal{K}\Big[\ - \partial_{\mathbf{k}^2}\mathcal{K}(\,\ominus\,)\ - \partial_{\mathbf{k}^2}\mathcal{K}\bar{R}(\,\big)\ \ominus\ \Big]\,.$$

In the complete list of recursive diagrammatic expansions in Appendix A, this appears on page 435 as No. 15.

The propagator-type integrals appearing on the right-hand side are easy to calculate. A nonzero external momentum is assumed to enter through the two remaining external lines, thereby avoiding an IR-divergence in the upper loop. The rearranged diagram is IR-convergent because of the cancellation of one momentum square in the denominator of the quadratically divergent four-loop subdiagram.

12.2.2 Zero-Mass Simplifications for Z_ϕ

The counterterm contribution to Z_ϕ can be calculated from the tadpole-free two-point diagrams with zero mass. Up to four loops, this calculation is a rather straightforward application of the algorithm to be described in Chapter 13. At five loops, however, some non-standard techniques must be used. One of them is based on the possibility of transforming quadratically into logarithmically divergent diagrams by carrying out the differentiation with respect to the external momentum of Section 11.8. Setting the mass equal to zero in the differentiated integrals simplifies the expressions considerably. For $m = 0$, the derivatives (11.47) and (11.48) become

$$
\frac{\partial}{\partial k_\mu} \frac{1}{(\mathbf{p}-\mathbf{k})^2} = -\frac{2(p-k)_\mu}{[(\mathbf{p}-\mathbf{k})^2]^2} = 2 \rightarrowtail , \tag{12.21}
$$

$$
\frac{\partial^2}{\partial k_\mu \partial k_\mu} \frac{1}{(\mathbf{p}-\mathbf{k})^2} = \frac{-2D}{[(\mathbf{p}-\mathbf{k})^2]^2} + \frac{8}{[(\mathbf{p}-\mathbf{k})^2]^2} = \frac{2\varepsilon}{[(\mathbf{p}-\mathbf{k})^2]^2} = -2\varepsilon \rightarrowtail . \tag{12.22}
$$

Application to more than one propagator leads to a sum of terms:

$$
\frac{\partial^2}{\partial k_\mu \partial k_\mu} \frac{1}{(\mathbf{p}-\mathbf{k})^2(\mathbf{q}-\mathbf{k})^2} = 2\frac{\partial}{\partial k_\mu} \left\{ \frac{(p-k)_\mu}{[(\mathbf{p}-\mathbf{k})^2]^2(\mathbf{q}-\mathbf{k})^2} + \frac{(q-k)_\mu}{(\mathbf{p}-\mathbf{k})^2[(\mathbf{q}-\mathbf{k})^2]^2} \right\}
$$

$$
= \frac{-2D+8}{[(\mathbf{p}-\mathbf{k})^2]^2(\mathbf{q}-\mathbf{k})^2} + \frac{-2D+8}{(\mathbf{p}-\mathbf{k})^2[(\mathbf{q}-\mathbf{k})^2]^2} + \frac{8(p-k)_\mu(q-k)_\mu}{[(\mathbf{p}-\mathbf{k})^2]^2[(\mathbf{q}-\mathbf{k})^2]^2}
$$

$$
= 2\varepsilon \left(-\cap - \cap \right) + 8 \left(\curlywedge \right) . \tag{12.23}
$$

The dotted lines indicate propagators $1/(\mathbf{p}-\mathbf{q})^2$ entering into the vertices. These propagators are not written down explicitly to save space. After the differentiation, quadratically divergent diagrams have become logarithmically divergent.

Note that not all of the new logarithmically divergent diagrams are equivalent to four-point diagrams, as shown by an example. Let us calculate the counterterm contribution to Z_ϕ of the following five-loop diagram using $\partial/\partial k^2 = (8-2\varepsilon)^{-1}\partial^2/\partial k_\mu \partial k_\mu$:

$$
\mathcal{K}\bar{R}\left[\frac{\partial}{\partial \mathbf{k}^2}\left(-\!\!\bigtriangleup\!\!\bigtriangledown\!\!- \right)\right] = \mathcal{K}\bar{R}\left[\frac{1}{8-2\varepsilon}\frac{\partial^2}{\partial k_\mu \partial k_\mu}\left(-\!\!\bigtriangleup\!\!\bigtriangledown\!\!- \right)\right]
$$

$$
= \mathcal{K}\bar{R}\left[\frac{1}{8-2\varepsilon}\left(-2 \times 2\varepsilon -\!\!\bigtriangleup\!\!\bigtriangledown\!\!- + 8 -\!\!\bigtriangleup\!\!\bigtriangledown\!\!- \right)\right]. \tag{12.24}
$$

The left of the two final diagrams is a four-point diagram with an artificial IR-divergence for zero mass. This divergence can be avoided by rearranging the flow of external momentum, which brings (12.24) to the form

$$
\mathcal{K}\bar{R}\left[\frac{1}{8-2\varepsilon}\left(-2 \times 2\varepsilon \bigcirc\!\!\!\bigtriangledown + 8 -\!\!\bigtriangleup\!\!\bigtriangledown\!\!- \right)\right].
$$

The resulting propagator-type integral is easy to calculate by the methods to be explained in the Chapter 13.

The second diagram on the right-hand side contains extra vertices and propagators with vector indices. Such diagrams will generically be called *diagrams with vector indices*. Due to the zero mass, the Feynman integrals of such diagrams can often be calculated quite simply with the help of momentum conservation and directional averages. This will now be shown in an example. A further example will be given later in Eqs. (13.36) and (13.37), when illustrating another important calculation technique based on partial integrations of Feynman integrals in momentum space.

The momentum conservation at the vertices allows us to reduce some diagrams with vector indices to ordinary scalar integrals. This requires the introduction of directed lines. Consider the three-loop ϕ^4-diagram

$$
\text{G:} \qquad \text{}
\tag{12.25}
$$

The numbers on the lines label the momenta flowing through them. At the lowest vertex, momentum conservation implies

$$
\mathbf{p}_3 = \mathbf{p}_4 + \mathbf{p}_5 \;\Rightarrow\; 2\,\mathbf{p}_3 \cdot \mathbf{p}_4 = \mathbf{p}_3^2 + \mathbf{p}_4^2 - \mathbf{p}_5^2 .
\tag{12.26}
$$

Inserting this into the integrand gives

$$
\frac{p_{3\mu}p_{4\mu}}{\mathbf{p}_1^2\mathbf{p}_2^2\mathbf{p}_3^2\mathbf{p}_4^2\mathbf{p}_5^2}
= \frac{1}{2}\left(\frac{1}{\mathbf{p}_1^2\mathbf{p}_2^2\mathbf{p}_4^2\mathbf{p}_5^2} + \frac{1}{\mathbf{p}_1^2\mathbf{p}_2^2\mathbf{p}_3^2\mathbf{p}_5^2} - \frac{1}{\mathbf{p}_1^2\mathbf{p}_2^2\mathbf{p}_3^2\mathbf{p}_4^2}\right),
\tag{12.27}
$$

$$
\text{}
= \frac{1}{2}\left(\text{} + \text{} - \text{}\right).
\tag{12.28}
$$

A canceled squared momentum in the denominator corresponds diagrammatically to a shrunk line.

For diagrams containing ϕ^4-vertices, the expansion of the type (12.27) is not always possible. As an example, take the last integral arising in the differentiation

$$
\mathcal{K}\bar{R}\left[\frac{\partial}{\partial \mathbf{k}^2}\left(\text{}\right)\right]
= \mathcal{K}\left[\frac{1}{8-2\varepsilon}\left(-2\varepsilon\bar{R}\,\text{} - 2\varepsilon\bar{R}\,\text{} + 8\bar{R}\,\text{}\right)\right].
\tag{12.29}
$$

The differentiation generates two logarithmically divergent diagrams, which can be viewed as four-point diagrams with a zero external momentum entering at the ϕ^2-vertex. For zero mass, these ϕ^2-vertices create IR-divergences which are avoided by IR-rearrangement. The external momentum is rearranged such that it enters into the ϕ^2-vertex. In this way we replace (12.29) by

$$
\mathcal{K}\left[\frac{1}{4-\varepsilon}\left(-\varepsilon\bar{R}\,\text{} - \varepsilon\bar{R}\,\text{} + 4\bar{R}\,\text{}\right)\right] .
\tag{12.30}
$$

The third diagram on the right-hand side has propagators with vector indices which cannot be expanded as in (12.27), thus requiring a more complicated explicit calculation involving the different vector components.

12.3 Infrared Divergences in Dimensional Regularization

Before entering the general discussion on the subtraction of infrared divergences, let us first discuss their appearance in a few explicit Feynman integrals regularized by analytic continuation. They will illustrate, in particular, that zero masses alone are not sufficient to give rise to $1/\varepsilon$-pole terms, unless the momenta are exceptional. Zero masses lead, in general, to singularities in $D - 2$, D, $D + 2, \dots$. Poles in ε arise only after IR-rearrangement of the massless diagrams.

12.3.1 Nonexceptional External Momenta

As a first example, consider the logarithmically UV-divergent one-loop integral ⟨diagram⟩ , which was calculated in dimensional regularization in Eq. (8.63). Setting $m = 0$ in Eq. (8.67) leads to

$$\left.\text{⟨diagram⟩}\right|_{m=0} = (\mu^\varepsilon g)^2 \frac{\Gamma(\varepsilon/2)}{(4\pi)^{D/2}} (\mathbf{k}^2)^{D/2-2} \int_0^1 dt \, [t \, (1-t)]^{D/2-2} . \tag{12.31}$$

Using the integral representation for the Beta function [compare (8.8)]

$$B(a,b) = \int_0^1 dt \, t^{a-1}(1-t)^{b-1} = \frac{\Gamma(a)\Gamma(b)}{\Gamma(a+b)} \tag{12.32}$$

we obtain

$$\left.\text{⟨diagram⟩}\right|_{m=0} = (\mu^\varepsilon g)^2 \frac{\Gamma(\varepsilon/2)}{(4\pi)^{D/2}} (\mathbf{k}^2)^{D/2-2} \frac{\Gamma(D/2-1)\Gamma(D/2-1)}{\Gamma(D-2)} . \tag{12.33}$$

The first Gamma function $\Gamma(\varepsilon/2)$ contains the $1/\varepsilon$-pole term caused by the UV-divergence. Setting the mass equal to zero leads to additional divergences at the endpoints of the t-integral in (12.31) [which are absent for $m \neq 0$ in Eq. (8.67)]. These divergences appear as poles in the Beta function at $D = 2, 0, -2, \dots$. The integral is IR-convergent for $D = 4$. Note that the single nonzero external momentum in this diagram is certainly nonexceptional (recall the definition on page 140).

12.3.2 Exceptional External Momenta

In the case of exceptional external momenta, we do find IR-divergence for $D = 4$ and $m = 0$. The example in Eq. (12.3) contains an integration over the ϕ^2-vertex:

$$\text{⟨diagram⟩} \sim \int \frac{d^D p}{(2\pi)^D} \frac{1}{(\mathbf{p}^2 + m^2)^2 [(\mathbf{p} - \mathbf{k})^2 + m^2]} . \tag{12.34}$$

Using the integral formula Eq. (8.69) for $a = 2$ and $b = 1$, we find

$$\text{⟨diagram⟩} \propto \frac{\Gamma(3 - D/2)}{(4\pi)^{D/2}\Gamma(2)\Gamma(1)} \int_0^1 dx \, x^0(1-x)^1 \left[\mathbf{k}^2 x(1-x) + m^2\right]^{D/2-3} . \tag{12.35}$$

The integral is UV-divergent for $D \geq 6$. The integration over x is convergent as long as $m \neq 0$. If the mass is zero, it is IR-divergent for $D \leq 4$:

$$\begin{aligned}
\left.\text{⟨diagram⟩}\right|_{m=0} &= -(\mu^\varepsilon g)^2 \frac{\Gamma(3 - D/2)(\mathbf{k}^2)^{D/2-3}}{(4\pi)^{D/2}} \int_0^1 dx \, x^{D/2-3}(1-x)^{D/2-2}. \\
&= -(\mu^\varepsilon g)^2 \frac{\Gamma(3 - D/2)(\mathbf{k}^2)^{D/2-3}}{(4\pi)^{D/2}} \frac{\Gamma(D/2-2)\Gamma(D/2-1)}{\Gamma(D-3)} .
\end{aligned} \tag{12.36}$$

The IR-divergence leads to poles at $D = 4, 2, \dots$; the UV-divergence to poles at $D = 6, 8, \dots$.

12.3.3　Massless Tadpole Diagrams

An important reason for setting all masses equal to zero was the fact that, in the minimal subtraction scheme, all integrals with tadpole parts (8.33) do not contribute to the counterterms (see Sections 11.7-11.8).

As an example, consider the quadratically divergent one-loop diagram $\bigcirc$, which was found in Eq. (8.57) to give

$$\bigcirc = -\mu^\varepsilon g \int \frac{d^D p}{(2\pi)^D} \frac{1}{\mathbf{p}^2 + m^2} = -\mu^\varepsilon g \frac{(m^2)^{D/2-1}}{(4\pi)^{D/2}} \Gamma(1 - D/2). \tag{12.37}$$

For $D > 2$, this vanishes in the limit $m \to 0$, such that we may set the dimensionally regularized integral $\int d^D p \, \mathbf{p}^{-2}$ equal to zero, as stated in Eq. (8.31). The example shows nicely that dimensional regularization controls IR-divergences in the same analytic way as UV-divergences, the two canceling each other for zero mass. This is an important property of the scheme. It allows us to calculate the UV-divergences of massive Feynman integrals from appropriately subtracted massless integrals, which are much easier to perform.

12.4　Subtraction of UV- and IR-divergences: R^*-Operation

From now on we shall deal exclusively with massless, mostly IR-rearranged diagrams. In order to find the correct counterterms produced by such diagrams at each order in perturbation theory, we introduce a generalization of the R-operation, called R^*-operation, which subtracts UV-divergences as well as the artificial IR-divergences by IR-counterterms. The specification of the latter is based on an important difference between the two divergences: a superficial UV-divergence receives contributions from *all* lines of a loop. This is not the case for an IR-divergence where only massless lines contribute, and among these, only those which become singular at the same point in momentum space. We shall refer to such lines as not carrying external momentum. Whereas UV-divergences are subtracted loop by loop, IR-divergences are subtracted line by line, to be cut out of a loop. This fact will somewhat complicate the substitution of the IR-counterterm of the IR-divergent integrations.

In Subsection 12.4.1, an example for a simple R^*-operation is given. It illustrates the definition of the IR-divergent subdiagrams in the Subsection 12.4.2. Note that in the following the coupling constants as well as the factor $1/(2\pi)^2$ accompanying each coupling constant will be omitted.

12.4.1　Example for Subtraction of IR- and UV-Divergences

Consider the following logarithmically UV-divergent diagram $\ominus$, which gives rise to the UV-divergent counterterms

$$\mathcal{K}\bar{R}\left(\ominus\right) = \mathcal{K}\left[\ominus - \mathcal{K}\left(\times\times\right) * \times\times\right]. \tag{12.38}$$

By IR-rearrangement, the diagram can be tranformed into $\ominus$. This contains an additional IR-subdivergence which can be subtracted from $\ominus$ to obtain the correct UV-counterterms. The extra IR-subdivergence in the Feynman integral $\int d^D p \, [\mathbf{p}^4 (\mathbf{p} - \mathbf{q})^2]^{-1}$ is caused by the integral over $1/\mathbf{p}^4$ implied by the upper loop. The propagator $1/(\mathbf{p} - \mathbf{q})^2$ in the lower loop does not produce any IR-divergence. In calculating the IR-pole term, this propagator is nevertheless needed as an UV-cutoff, otherwise the UV-divergence would cancel the IR-divergence and disappear from the integral, since the integral over $1/\mathbf{p}^4$ is a massless tadpole integral giving zero in dimensional integration.

In order to set us an iterative subtraction procedure, the extra IR-divergences will be isolated diagrammatically. The contribution of the upper loop integral in ⊖ has near $D = 4$ dimensions the form:

$$\mathcal{K}(\text{⊖}) = \mathcal{K}\left(\int d^D p \, [\mathbf{p}^4 (\mathbf{p} - \mathbf{q})^2]^{-1}\right) \propto \frac{1}{\varepsilon} \frac{1}{\mathbf{q}^2}, \tag{12.39}$$

with a constant proportionality factor. The resulting $1/\mathbf{q}^2$ on the right-hand side of (12.39) plays the role of a second propagator in the lower loop integral of ⊖ , whose analytic form is $\int d^D q \, [\mathbf{q}^2 (\mathbf{q} - \mathbf{k})^2]^{-1}$. The latter integral is the same as for a Feynman diagram ⊸⊙⊸ evaluated at zero external momentum.

At first one may be tempted to represent the IR-divergence by a diagram such as $\mathcal{K}(\text{⊸⊙⊸})$⊸⊙⊸ . This would be misleading, since the upper line of the right-hand diagram is really the result of the left-hand diagram, so that this line is counted twice. This could be saved by employing the symbolic notation $\mathbf{q}^2 \, \mathcal{K}(\text{⊸⊙⊸})$⊸⊙⊸ for the divergence. This, however, is not recommendable for another reason. The factor in the integrand causing the IR-divergence, here $1/\mathbf{p}^4$, produces $1/\varepsilon^i$-pole terms which are independent of the original Feynman integral. Examples for diagrams in which $1/\mathbf{p}^4$ appears in different loop diagrams will be shown for a different purpose in Fig. 12.5. This independence makes it preferable to set up a diagrammatic notation exhibiting directly the IR-divergent part of the integral as the factor. In the present case, we represent the IR-divergence caused by the integrand $1/\mathbf{p}^4$ as follows: [5]

$$-\mathbf{q}^2 \, \mathcal{K}(\text{⊸⊙⊸}) = \left(\begin{smallmatrix}\circ\\\circ\end{smallmatrix}\right)_{IR} . \tag{12.40}$$

The small open circles at the end of the vertical line indicate vertices at which the line is connected to the initial diagram. With this notation, the diagrammatic formulation of all subtractions is

$$\mathcal{K}\bar{R}^*\left(\text{⊖}\right) = \mathcal{K}\left\{\text{⊖} + \left(\begin{smallmatrix}\circ\\\circ\end{smallmatrix}\right)_{IR} \text{⊸⊙⊸} - \mathcal{K}(\text{⊸⊙⊸})\left[\text{Ω} + \left(\begin{smallmatrix}\circ\\\circ\end{smallmatrix}\right)_{IR}\right]\right\}. \tag{12.41}$$

This produces the same result as in the purely ultraviolet subtraction procedure for massive propagators in Eq. (12.38). There are two IR-subtractions, one for the full two-loop diagram and one for the UV-subtracted subdiagram. The second term on the right-hand side subtracts a pure IR-divergence. The third term vanishes because the remaining diagram is of the massless tadpole type, being zero in dimensional regularization. The last term subtracts an IR- as well as a UV-divergence. Obviously, the calculation of the IR-subtraction terms and their combination with the rest of a diagram are more complicated than in the ultraviolet case.

The calculation of massless diagrams has so far been possible up to five loops, and will be discussed in the next chapter. For the present discussion of IR-subtractions we anticipate the following general formula:

$$\mathbf{k} \, \underset{b}{\overset{a}{\text{⊙}}} \, \sim \frac{\Gamma(D/2 - a)\Gamma(D/2 - b)\Gamma(a + b - D/2)}{(4\pi)^{-\varepsilon/2} \, \Gamma(a)\Gamma(b)\Gamma(D - a - b)} \frac{1}{(\mathbf{k}^2)^{a+b-D/2}}, \tag{12.42}$$

where a and b are the powers of the propagators. For $a = b = 1$, we find the following small-ε expansion:

$$\text{⊗} = \frac{1}{\varepsilon} + 2 - \gamma - \log \frac{\mathbf{k}^2}{4\pi\mu^2} + \mathcal{O}(\varepsilon). \tag{12.43}$$

This expansion agrees with the result for the massive integral in Eq. (8.68), setting $m = 0$ and $\psi(1) = -\gamma$.

The explicit pole term for the IR-divergence is then

$$\mathcal{K}(\,\text{⊸⊙⊸}\,) = -\frac{2}{\varepsilon}\frac{1}{\mathbf{q}^2}. \tag{12.44}$$

As mentioned before, the propagator $1/(\mathbf{p}-\mathbf{q})^2$ does not contribute to the IR-divergence, but acts as UV-cutoff, thereby generating a factor $1/\mathbf{q}^2$:

$$\left(\,\text{⊙}\,\right)_{IR} = -\mathbf{q}^2\mathcal{K}(\,\text{⊸⊙⊸}\,) = \frac{2}{\varepsilon}. \tag{12.45}$$

This pole term multiplies the remaining loop integral ⊸⊙⊸ .

The diagrammatic subtraction of UV- and IR-divergences differs in an important point. A UV-divergence is caused by a loop integral in a subdiagram which is shrunk to a virtual vertex after the integration. For the IR-divergences, on the other hand, not all the lines of a loop integral are relevant. In contrast to the UV-divergence, the lines responsible for the IR-divergence are not shrunk to a point, but removed from the diagram. This process yields the full subtraction term:

$$\mathcal{K}\left[\left(\,\text{⊙}\,\right)_{IR}\,\text{⊸⊙⊸}\,\right] = \mathcal{K}\left[\frac{2}{\varepsilon}\int\frac{d^Dq}{(2\pi)^D}\frac{1}{\mathbf{q}^2(\mathbf{q}-\mathbf{k})^2}\right] = \frac{2}{\varepsilon^2} + \frac{4}{\varepsilon} - \frac{2}{\varepsilon}\gamma - \frac{2}{\varepsilon}\log\frac{\mathbf{k}^2}{4\pi\mu^2}. \tag{12.46}$$

Omitting the vanishing tadpole term in the $\bar{R}^*$-operation in Eq. (12.41), the explicit calculation of the UV-counterterms of the diagram ⊖ involves the following three subtractions, separated by square brackets in the second line:

$$\mathcal{K}\bar{R}^*\left(\,\text{⊖}\,\right) = \mathcal{K}\left[\quad\text{⊖}\quad + \left(\,\text{⊙}\,\right)_{IR}\,\text{⊸⊙⊸}\quad - \left(\,\text{⊙}\,\right)_{IR}\mathcal{K}\left(\,\text{⊸⊙⊸}\,\right)\right]$$

$$= \left[-\frac{2}{\varepsilon^2} - \frac{3}{\varepsilon} + \frac{2}{\varepsilon}\left(\gamma + \log\frac{\mathbf{k}^2}{4\pi\mu^2}\right)\right] + \left[\frac{4}{\varepsilon^2} + \frac{4}{\varepsilon} - \frac{2}{\varepsilon}\left(\gamma + \log\frac{\mathbf{k}^2}{4\pi\mu^2}\right)\right] - \left[\frac{4}{\varepsilon^2}\right]$$

$$= -\frac{2}{\varepsilon^2} + \frac{1}{\varepsilon}. \tag{12.47}$$

For comparison, the calculation of these counterterms from the $\bar{R}$-operation applied to the massive diagram in Eq. (12.38), goes as follows:

$$\mathcal{K}\bar{R}(\,\text{⊖}\,) = \mathcal{K}\left[\quad\text{⊖}\quad - \quad\mathcal{K}\left(\,\text{⋈}\,\right)*\text{⋈}\,\right]$$

$$= \left[\frac{2}{\varepsilon^2} + \frac{5}{\varepsilon} - \frac{2}{\varepsilon}\left(\gamma + \log\frac{\mathbf{k}^2}{4\pi\mu^2}\right)\right] - \left[\frac{4}{\varepsilon^2} + \frac{4}{\varepsilon} - \frac{2}{\varepsilon}\left(\gamma + \log\frac{\mathbf{k}^2}{4\pi\mu^2}\right)\right]$$

$$= -\frac{2}{\varepsilon^2} + \frac{1}{\varepsilon}. \tag{12.48}$$

In order to generalize this diagrammatic subtraction procedure, some new notations must be introduced.

12.4.2 Graph-Theoretic Notations

First, we must properly identify IR-divergent subdiagrams of a diagram G. As observed above, there is an important difference between UV- and IR-divergent subdiagrams. In the first case, subdiagrams are complete 1PI ϕ^4-diagrams. They form closed loops, in which all lines contribute to the UV-divergence. Taking the UV-divergent subintegration out of the diagram is diagrammatically represented by shrinking its loops.

In contrast to this, the IR-divergent subdiagrams are not 1PI subdiagrams of the ϕ^4-theory, Their lines do not form closed loops. All lines which carry momentum of the remaining diagram do not contribute to the IR-divergence. This is why the removal of IR-divergent subintegration from a diagram will be represented diagrammatically by cutting the responsible lines out of the subdiagram. The rules for doing this are:

- A *subdiagram* γ contributing with all of its lines to an IR-divergence is separated from the remaining diagram by cutting one or several loops. The corresponding loop momenta become external momenta in γ. The vertices, at which γ is connected to G, are drawn as hollow circles.

- The *subtracted diagram* or *remaining diagram* $\hat{\gamma} = G\backslash\gamma$ is created from G by taking away the lines and internal vertex points of γ. The integration connecting γ and $G\backslash\gamma$ is omitted, and the corresponding loop momenta play the role of external momenta in γ as well as $\hat{\gamma}$.

- The *contracted subdiagram* $\tilde{\gamma} = G/\hat{\gamma}$ is formed from G by shrinking the lines of $\hat{\gamma}$ to a point. Alternatively, the diagram $\tilde{\gamma}$ is obtained from γ by contracting the vertices which connect γ to $\hat{\gamma}$, and which are represented by open circles, to a single vertex. Note that different subdiagrams γ can produce the same $\tilde{\gamma}$.

Let I_γ be the integrand of γ, and $\mathbf{k} = \mathbf{k}_1, \ldots, \mathbf{k}_n$ the external momenta of G. Further let $\mathbf{p} = \mathbf{p}_1, \ldots, \mathbf{p}_N$ be the common external momenta of γ and $\hat{\gamma}$, which are the momenta of the loops to be cut out of the diagram G, and $\mathbf{l} = \mathbf{l}_1, \ldots, \mathbf{l}_{L(\gamma)}$ and $\mathbf{q} = \mathbf{q}_1, \ldots, \mathbf{q}_m$ be the internal momenta of γ and $\hat{\gamma}$, respectively. Then we find for the above-defined quantities the general analytic expressions:

$$G(\mathbf{k}) \quad \propto \quad \int d^D p\, d^D q\, d^D l\, I_\gamma(\mathbf{l}, \mathbf{p}) I_{\hat{\gamma}}(\mathbf{q}, \mathbf{p}, \mathbf{k}), \tag{12.49}$$

$$\gamma(\mathbf{p}) \quad \propto \quad \int d^D l\, I_\gamma(\mathbf{l}, \mathbf{p}), \tag{12.50}$$

$$\hat{\gamma}(\mathbf{p}, \mathbf{k}) = G\backslash\gamma(\mathbf{p}, \mathbf{k}) \quad \propto \quad \int d^D q\, I_{\hat{\gamma}}(\mathbf{q}, \mathbf{p}, \mathbf{k}), \tag{12.51}$$

$$\tilde{\gamma} \quad \propto \quad \int d^D p\, I_{\gamma(\mathbf{p})}. \tag{12.52}$$

The definitions will be illustrated by the following examples. The hollow circles in γ and in $\tilde{\gamma}$ indicate the vertices connecting γ and $\hat{\gamma}$, which are contracted to a single point in $\tilde{\gamma}$.

Example 1: $G = \bigoplus,$ $\gamma = $ $\Rightarrow \hat{\gamma} = $ $\tilde{\gamma} = $.

Example 2: $G = \bigoplus,$ $\gamma = $ $\Rightarrow \hat{\gamma} = $ $\tilde{\gamma} = $.

Example 3: $G = \bigodot,$ $\gamma = $ $\Rightarrow \hat{\gamma} = $, $\tilde{\gamma} = $.

The three examples illustrate an important role of $\tilde{\gamma}$. The diagrams γ and $\tilde{\gamma}$ contain the same lines. But $\tilde{\gamma}$ contains, in addition, an integral over all external momenta of γ, such that $\tilde{\gamma}$ has the same Feynman integral as a vacuum diagram. Sometimes the diagrams $\tilde{\gamma}$ contain cutvertices, which arise from the contraction of the vertices in $\hat{\gamma}$. The loop diagrams separated by these cutvertices are formed by the lines carrying the momenta of $\hat{\gamma}$, which are the external momenta of the subdiagram γ. The number of external momenta N of γ is thus given by:

$$N = L(\tilde{\gamma}) - L(\gamma). \tag{12.53}$$

The external momenta in $\tilde{\gamma}$ provide an IR-cutoff for the integrations of γ. A cutvertex in $\tilde{\gamma}$ separates loops whose lines contribute to IR-divergences at different points in the momentum space. Something similar happens when external momentum $\mathbf{k}$ of the full diagram G flows into one of the inner vertices of γ, because the lines through which $\mathbf{k}$ flows also contribute to an IR-divergence at a different point in momentum space. The external momentum may well flow through all lines of γ, just as in the shrunk diagram $\tilde{\gamma}$ on page 198. A somewhat artificial example for this case is $G = $ ─◯─ , where both lines, the upper and the lower one, contribute independently to an IR-divergence, but not both together since the external momentum in one of the lines prevents them from getting singular at the same point in momentum space.

We are now ready to define the subdiagrams whose IR-divergences have to be subtracted. A subdiagram $\gamma \subset G$ is called *infrared irreducible* (IRI) if

1. the associated contracted subdiagram $\tilde{\gamma} = G/\hat{\gamma}$ contains no cutvertex,

2. no external momentum of G flows into an internal vertex of the subdiagram γ, i.e., if external momentum flows either through *all* lines of γ or through *none* of them, and

3. the subdiagram γ contains only massless lines.

If all three conditions are satisfied, each line of the subdiagram γ contributes to the IR-divergence of the integrations of $\tilde{\gamma}$.

Among the above examples, only the first one is IRI, and only if the external momentum of G flows entirely through $\hat{\gamma}$. The IRI subdiagrams play the same role for IR-divergences as the 1PI subdiagrams for the UV-divergences. The IR-divergences come exclusively from IR-divergent IRI subdiagrams.

An IRI subdiagram is said to be IR-divergent if its superficial degree of IR-divergence is $\tilde{\omega}(\gamma) \geq 0$.

$$\begin{aligned}
\tilde{\omega}(\gamma) &= \omega(\hat{\gamma}) - \omega(G) = 4L(\hat{\gamma}) - 2I(\hat{\gamma}) - 4L(G) + 2I(G) \\
&= -[4L(\tilde{\gamma}) - 2I(\tilde{\gamma})] = -\omega(\tilde{\gamma}). \tag{12.54}
\end{aligned}$$

This definition coincides with the previous one in Eq. (12.7). In our applications, $\tilde{\omega}$ will be nonzero only once: $\tilde{\omega}\left(\begin{smallmatrix} \vdots \\ \vdots \end{smallmatrix}\right) = 2$. There is one five-loop diagram, in which this subdiagram appears and cannot be avoided by any trick.

A set of IR-divergent IRI subdiagrams $\{\gamma_i\}$ is said to be *IR-disjoint* in G if

(i) the subdiagrams are pairwise non-overlapping, and

(ii) no IRI subdiagram *in* G can be composed from them.

Diagrams which are IR-disjoint contribute independently to IR-divergences, by analogy with UV-disjoint diagrams. Their selection prevents double subtractions in the recursive definition of the IR-counterterms.

By analogy with the definition of the sets in Eq. (11.1),

$$\Gamma = \{\gamma | \gamma \in G, \gamma \text{ UV-disjoint}\},\tag{12.55}$$

we define the sets Γ', which contain the IR-divergent IRI subdiagrams of G:

$$\Gamma'(G) = \{\gamma' | \gamma' \subset G, \gamma' \text{ IR-disjoint}\}.\tag{12.56}$$

Examples of the set of all possible sets Γ' for two different diagrams transformed by IR-rearrangement are shown in Fig. 12.5. The subtraction of all UV- and IR-divergences of a diagram G will involve all sets Γ and Γ' contained in G.

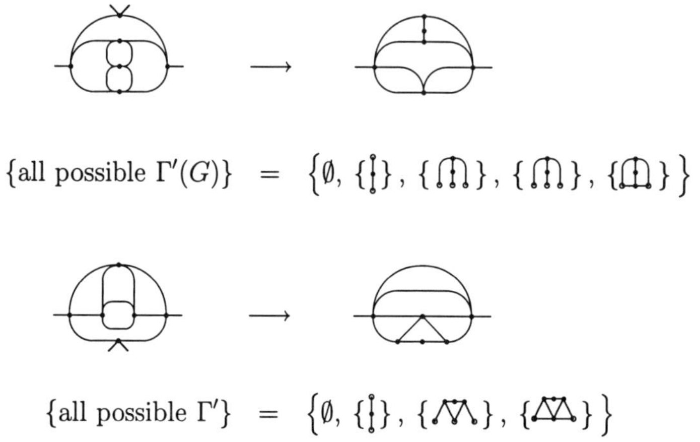

$$\{\text{all possible } \Gamma'(G)\} \quad = \quad \Big\{\emptyset, \{\text{⊡}\}, \{\text{⊞}\}, \{\text{⊞}\}, \{\text{⊞}\}\Big\}$$

$$\{\text{all possible } \Gamma'\} \quad = \quad \Big\{\emptyset, \{\text{⊡}\}, \{\text{∿}\}, \{\text{△}\}\Big\}$$

FIGURE 12.5 All possible sets of infrared-disjoint subdiagrams for two diagrams transformed by infrared rearrangement.

12.4.3   Definition of R^*- and $\bar{R}^*$-Operation

For a systematic recursive subtraction of the artificial IR-divergences in massless Feynman diagrams, the R-operation defined in Eqs. (11.3)–(11.13) is extended to an R^*-operation [4, 6, 7]. The subtraction is possible because the UV- and IR-divergences originate from different regions in momentum space, which yield clearly distinguishable additive contributions. For any diagram G, the R^*-operation yields a finite result:

$$\mathcal{K}R^* G = 0.\tag{12.57}$$

The operation is a sum over all possible terms in which UV- and IR-divergent subintegrations γ and γ' are replaced by the corresponding pole terms with the help of Δ_γ and $\Delta'_{\tilde{\gamma}'}$:

$$\begin{aligned}
\mathcal{K}R^* G &= \sum_{\Gamma \cap \Gamma' = 0} \prod_{\gamma' \in \Gamma'} \Delta'_{\tilde{\gamma}'} * \prod_{\gamma \in \Gamma} \Delta_\gamma * G \backslash \Gamma' / \Gamma \\
&= \sum_{\Gamma \cap \Gamma' = 0} \Delta'_{\Gamma'} * \Delta_\Gamma * G \backslash \Gamma' / \Gamma,
\end{aligned}\tag{12.58}$$

where $\Delta'_{\Gamma'}$ is defined in terms of the IR-disjoint subdiagrams γ' in the same way as Δ_Γ was in terms of the UV-disjoint subdiagrams γ in Eq. (11.8):

$$\Delta'_{\Gamma'} \equiv \prod_{\gamma' \in \Gamma'} \Delta'_{\tilde{\gamma}'}.\tag{12.59}$$

The operations $*$ insert Δ_γ and $\Delta'_{\tilde\gamma'}$ into the remaining integration. We use $\tilde\gamma$ as a subscript of Δ', because it contains all IR-divergent loop integrations, and because different subdiagrams γ may lead to the same $\tilde\gamma$ and Δ'_γ. The sums over all Γ and Γ' with $\Gamma \cap \Gamma' = 0$ guarantee that all subdivergences are subtracted, and that no oversubtractions occur. Since the sum runs over nonoverlapping sets Γ and Γ', the shrinking and the removal of the corresponding subdiagrams in G can be done in an arbitrary order:

$$G\backslash\Gamma'/\Gamma = G/\Gamma\backslash\Gamma'.$$

The Feynman diagrams to be calculated are not superficially IR-divergent because an external momentum is used as an IR-cutoff. Therefore, the term which contains the superficial IR-divergence of G vanishes:

$$\Delta'_G * G\backslash G = \Delta'_G = 0.$$

The superficial UV-divergence, the UV-counterterm, is therefore calculated as follows:

$$\Delta_G = \sum_{\bar\Gamma \cap \bar\Gamma' = 0} \Delta'_{\bar\Gamma'} * \Delta_{\bar\Gamma} * G\backslash\bar\Gamma'/\bar\Gamma \equiv -\mathcal{K}\bar{R}^* G. \tag{12.60}$$

In general, the operators Δ_Γ and $\Delta'_{\Gamma'}$ do not commute because of their momentum dependence. The UV-divergence has to be subtracted first. However, for logarithmically divergent subdiagrams where $\tilde\omega(\gamma) = 0$, the operators commute. This is the most common case. An example with $\tilde\omega(\gamma) > 0$ will be discussed in detail below.

The UV-divergent subdiagrams of an IR-divergent diagram may contain IR-divergences. The counterterm is therefore determined with

$$\Delta_\gamma = -\mathcal{K}\bar{R}^*\gamma. \tag{12.61}$$

The subdiagrams in the five-loop calculation have at most four loops. All their counterterms can be calculated without introducing IR-divergences, such that the extension to the $\bar{R}^*$-operation is superfluous, and we may apply the recursive operation $\bar{R}$ only:

$$\Delta_\gamma = -\mathcal{K}\bar{R}\gamma. \tag{12.62}$$

12.4.4   Construction of Infrared Subtraction Terms of Subdiagrams

Formally, IR-subtraction terms in the sum (12.60) look very similar to UV-subtraction terms:

$$\Delta'_{\tilde\gamma'} * G\backslash\gamma'. \tag{12.63}$$

However, they involve completely different procedures. First, the action of the operator $\Delta'_{\tilde\gamma'}$ is completely different, since it inserts a pole term rather than an IR-divergent subintegration as in the UV-case. Second, the formation of the remaining diagram $G\backslash\gamma'$ is different. The reason for these differences is that the previous UV-divergences are caused by *all* lines of a loop, such that the full loop integral must be subtracted. In contrast, the IR-divergence is caused only by one or a few lines of a loop. Thus the IR-divergent subdiagram γ is only a part of an integrand, which is connected to the remaining diagram by an integration. When replacing the subdiagram by the corresponding pole term, this integration has still to be carried out, and this has an effect on the remaining diagram if $\tilde\omega(\gamma) \neq 0$, unless the result of the integration is a constant.

If the IR-divergent subdiagram consists only of a single line and involves no loop, for example $\gamma = 1/\mathbf{p}^4$ in (12.12) with $\tilde\omega(\gamma) = 0$, the subtraction term is simple. It contains the pole term

$\mathcal{Z}'_{\tilde{\gamma}}(\varepsilon^{-1})$ instead of the integrand γ. The diagrammatic notation for this pole term is $(\gamma)_{IR}$. The integration over the momentum $\mathbf{p}$ is taken care of by inserting a factor $\delta^{(D)}(\mathbf{p})$, thus ensuring that the remaining integral is evaluated for $\mathbf{p} = 0$ where the IR-divergence is generated:

$$\Delta'_{\tilde{\gamma}} * G \backslash \gamma = \mathcal{Z}'_{\tilde{\gamma}}(\varepsilon^{-1}) \int \frac{d^D p}{(2\pi)^D} (2\pi)^D \delta^{(D)}(\mathbf{p}) \, \hat{\gamma}(\mathbf{p}, \mathbf{k}) \,, \qquad \tilde{\omega}(\gamma) = 0, \; L(\gamma) = 0 \,. \qquad (12.64)$$

In the general case, an IRI subdiagram γ of G has internal loops and more than one loop connecting it with the subtracted diagram $\hat{\gamma} = G \backslash \gamma$. The notation for the momenta will be the same as in Section 12.4.2, with loop momenta of γ labeled by $\mathbf{l} = (\mathbf{l}_1, \ldots, \mathbf{l}_{L(\gamma)})$, and the N external momenta of γ and of $\hat{\gamma}$ by $\mathbf{p} = (\mathbf{p}_1, \ldots, \mathbf{p}_N)$. The loop momenta of $\tilde{\gamma}$ are $\mathbf{p}' = (\mathbf{p}'_1, \ldots, \mathbf{p}'_{\tilde{\gamma}})$. By construction, the momenta $\mathbf{p}'$ comprise all $\mathbf{p}$ and $\mathbf{l}$, i.e., $\mathbf{p}' = (\mathbf{p}, \mathbf{l})$. The integrand represented by $\tilde{\gamma}$ or γ is called $I_{\tilde{\gamma}} = I_\gamma$. The integral associated with the full diagram G is, in this notation:

$$G(\mathbf{k}) = \int \frac{d^D p}{(2\pi)^D} \frac{d^D l}{(2\pi)^D} \left[\hat{\gamma}(\mathbf{p}, \mathbf{k}) \right] I_{\tilde{\gamma}}(\mathbf{p}, \mathbf{l}) \,. \qquad (12.65)$$

An example is displayed in Fig. 12.6.

$$G(\mathbf{k}) \qquad\qquad \gamma(\mathbf{p}) \qquad\qquad \hat{\gamma}(\mathbf{k}, \mathbf{p}) \qquad\qquad \tilde{\gamma}$$

FIGURE 12.6 The inner momenta of γ are called $\mathbf{l}_1$, $\mathbf{l}_2$. The contracted subdiagram $\tilde{\gamma}$ contains, in addition, the integration over $\mathbf{p}$. The IR-divergence occurs for vanishing momenta $\mathbf{l}_1$, $\mathbf{l}_2$, $\mathbf{p}$.

In general, the operator $\Delta'_{\tilde{\gamma}}$ acts on G by incorporating the counterterm for γ into the integral:

$$\Delta'_{\tilde{\gamma}} * G \backslash \gamma = \int \frac{d^D p}{(2\pi)^D} \frac{d^D l}{(2\pi)^D} \left[\hat{\gamma}(\mathbf{k}, \mathbf{p}) \right] \Delta'_{\tilde{\gamma}}(\mathbf{p}, \mathbf{l}) \,. \qquad (12.66)$$

The operator $\Delta'_{\tilde{\gamma}}(\mathbf{p}, \mathbf{l})$ is defined as a local expression which isolates the IR-divergences:

$$\Delta'_{\tilde{\gamma}}(\mathbf{p}') = \mathcal{P}'_{\tilde{\gamma}} \left(\frac{\partial}{\partial \mathbf{p}'} \right) \prod_{i=1}^{L(\tilde{\gamma})} (2\pi)^D \delta^{(D)}(p'_i) \,, \qquad (12.67)$$

where $\mathcal{P}'_{\tilde{\gamma}}$ is a homogeneous polynomial of degree $\tilde{\omega}(\gamma)$ in $(\partial/\partial \mathbf{p}'_1, \ldots, \partial/\partial \mathbf{p}'_{L(\tilde{\gamma})})$, with ε-dependent coefficients. They are pure pole terms in the MS-scheme:

$$\mathcal{K} \mathcal{P}_{\tilde{\gamma}} = \mathcal{P}_{\tilde{\gamma}} \,. \qquad (12.68)$$

They are uniquely defined by requiring that $\mathcal{K} R^* G$ in Eq. (12.58) be finite. The differentiations applied to the δ-functions can be moved around in the integral by partial integrations. Since these do not generate surface terms in dimensionally regularized integrals, the differentiations act on those lines of $\hat{\gamma}$ through which flows an external momentum of γ. The differentiations generate the same powers of the associated momenta as would arise from carrying out the integration over the external and internal momenta of γ.

In the calculations of the diagrams in this text, we have encountered only two simple cases: most frequently, the subdiagrams are logarithmically divergent with $\tilde{\omega}(\gamma) = 0$ for an arbitrary number of external momenta N [recall (12.53)]. The operator is then simply given by

$$\Delta'_{\tilde{\gamma}}(\mathbf{p}) = \mathcal{Z}'_{\tilde{\gamma}} \prod_{i=1}^{L(\tilde{\gamma})} (2\pi)^D \delta^{(D)}(\mathbf{p}_i) \,. \qquad (12.69)$$

Only do we encounter a quadratically IR-divergent subdiagram with $\tilde{\omega}(\gamma) = 2$ which has $N = 1$. There the operator is

$$\Delta'_{\tilde{\gamma}}(\mathbf{p}) = \mathcal{Z}'_{\tilde{\gamma}} \left(\frac{\partial}{\partial p_\mu}\right)^2 (2\pi)^D \delta^{(D)}(\mathbf{p}) . \tag{12.70}$$

It has been shown that none of the UV-and IR-pole terms of subdiagrams with L loops produce singularities which are for $\varepsilon \to 0$ more singular than $(\varepsilon^{-1})^L$ [7].

The general structure of the operator (12.67) can be understood better by the following consideration. The IR-divergent integration over γ in $D = 4 - \varepsilon$-dimensions is the integration over a four-dimensional pole in momentum space. The subtraction term is therefore constructed by replacing γ by the corresponding ε-pole term. This replacement is done at the point of the singularity in momentum space. The remaining term in G is the residue of the four-dimensional pole term. For an analytic function $f(z)$ with an isolated singularity at a and a Laurent expansion

$$f(z) = \sum_{n=-m}^{\infty} c_n(z - a)^n , \tag{12.71}$$

the residue c_{-1} is obtained by the operation:

$$(m - 1)! \, c_{-1} = \left(\frac{d}{dz}\right)^{m-1} (z - a)^m f(z) \Bigg|_{z=a} . \tag{12.72}$$

The case of a simple pole with $m = 1$ leads to a logarithmic divergence and corresponds in four dimensions to $1/\mathbf{p}^4$ with $\tilde{\omega} = 0$. The case of $m = 3$ is a quadratic IR-divergence corresponding in four dimensions to $1/\mathbf{p}^6$ with $\tilde{\omega} = 2$. The number of the differentiations is given by the deviation from a logarithmic divergence. For the calculation of the residue in four dimensions, $m - 1$ is thus replaced by $\tilde{\omega}(\gamma)$:

$$\left(\frac{\partial}{\partial p_\mu}\right)^{\tilde{\omega}(\gamma)} \hat{\gamma}(\mathbf{p}, \mathbf{k}) \Bigg|_{\mathbf{p}=0} . \tag{12.73}$$

The IR-divergent poles $1/\mathbf{p}^4$ and $1/\mathbf{p}^6$ with $\tilde{\omega}(\gamma) = 0$ and $\tilde{\omega}(\gamma) = 2$, respectively, are subtracted with the help of the operator

$$\Delta'_{\tilde{\gamma}} * G\backslash\gamma = \mathcal{Z}'_{\tilde{\gamma}}(\varepsilon^{-1}) \int \frac{d^D p}{(2\pi)^D} (2\pi)^D \delta^{(D)}(\mathbf{p}) \left(-\frac{\partial}{\partial p_\mu}\right)^{\tilde{\omega}} \hat{\gamma}(\mathbf{p}, \mathbf{k}) . \tag{12.74}$$

Let us discuss some examples in detail:

Example 1: The diagram

$$G = \quad \reflectbox{\includegraphics{}} \quad \sim \int \frac{d^D p}{(2\pi)^D} \frac{\hat{\gamma}(\mathbf{p}, \mathbf{k})}{\mathbf{p}^4} \tag{12.75}$$

contains the same IR-divergent subdiagram $\gamma = {\textstyle\substack{\bullet \\ |}}$ with $\tilde{\omega}(\gamma) = 0$ as the example discussed in Subsection 12.4.1. The counterterm is constructed with the help of the generic expression (12.64). The resulting IR-pole term was calculated before in (12.45): $\mathcal{Z}'_\gamma(\varepsilon^{-1}) = 2/\varepsilon$. Here the IR-divergent $\mathbf{p}$-integral contains three different propagators:

$$\int \frac{d^D p}{(2\pi)^D} \frac{1}{\mathbf{p}^4(\mathbf{p} - \mathbf{q}_1)^2(\mathbf{p} - \mathbf{q}_2)^2} . \tag{12.76}$$

The δ-function in the **p**-integral of the generic expression (12.64) forces **p** to be zero in $1/(\mathbf{p} - \mathbf{q}_1)^2(\mathbf{p} - \mathbf{q}_2)^2$, such that in the remaining diagram these propagators may simply be replaced by $1/\mathbf{q}_1^2\mathbf{q}_2^2$. The subtraction term and its diagrammatic notation are then

$$\Delta'_\gamma * G\backslash\gamma = \int \frac{d^D p}{(2\pi)^D} \, \mathcal{Z}'(\varepsilon^{-1}) \, (2\pi)^D \delta^{(D)}(\mathbf{p}) \, \hat{\gamma}(\mathbf{p}, \mathbf{k}) = \left(\begin{smallmatrix} \circ \\ \bullet \end{smallmatrix}\right)_{IR} \quad \text{(diagram)} . \tag{12.77}$$

The $\bar{R}^*$-operation of the entire diagram G will be given as an example at the end of the section.

Example 2: The following diagram G contains a subdiagram γ with $\tilde{\omega}(\gamma) = 2$:

$$G = \text{(diagram)} -, \quad \gamma = \text{(diagram)} \quad \Rightarrow \quad \hat{\gamma} = \text{(diagram)} . \tag{12.78}$$

The remaining diagram $\hat{\gamma}$ contains two propagators carrying the momentum of γ. The subtraction term is of the form:

$$\Delta'_\gamma * G\backslash\gamma = \mathcal{Z}'_{\hat{\gamma}}(\varepsilon^{-1}) \left(\frac{\partial^2}{\partial p_\mu \partial p_\mu} G\backslash\gamma \right)\bigg|_{\mathbf{p}=0} . \tag{12.79}$$

The momentum differentiations in Δ'_γ are carried out when incorporating the counterterm in $\hat{\gamma}$. Since

$$\frac{\partial^2}{\partial p_\mu \partial p_\mu} \frac{1}{(\mathbf{p} - \mathbf{q}_1)^2(\mathbf{p} - \mathbf{q}_2)^2} = \tag{12.80}$$

$$\frac{2\varepsilon}{(\mathbf{p} - \mathbf{q}_1)^4(\mathbf{p} - \mathbf{q}_2)^2} + \frac{2\varepsilon}{(\mathbf{p} - \mathbf{q}_1)^2(\mathbf{p} - \mathbf{q}_2)^4} + \frac{8(p - q_1)_\mu(p - q_2)_\mu}{(\mathbf{p} - \mathbf{q}_1)^4(\mathbf{p} - \mathbf{q}_2)^4},$$

the incorporation proceeds diagrammatically as follows:

$$\Delta'_\gamma * G\backslash\gamma = \left(\begin{smallmatrix} \circ \\ \bullet \end{smallmatrix}\right)_{IR} * \text{(diagram)}$$

$$= \mathcal{Z}'_{\hat{\gamma}}(\varepsilon^{-1}) \cdot \left(2\varepsilon \text{ (diagram)} + 2\varepsilon \text{ (diagram)} + 8 \text{ (diagram)} \right) . \tag{12.81}$$

Example 3: Consider a quadratically IR-divergent subdiagram with a UV-divergent remaining diagram. The incorporation of an IR-counterterm with $\tilde{\omega}(\gamma) > 0$ changes drastically the subtracted diagram. Here the order in which the UV- and IR-divergences are subtracted is very important. The calculation of the counterterm of the logarithmically divergent diagram of Eq. (12.18) in the last section via the R-operation has the diagrammatic representation:

$$\mathcal{K}\bar{R}\left(\text{(diagram)}\right) = \mathcal{K}\left[\text{(diagram)} - \partial_{\mathbf{k}^2}\mathcal{K}(\text{(diagram)})\text{(diagram)}\right], \tag{12.82}$$

with the operation $\partial_{\mathbf{k}^2}\mathcal{K}$ defined in Subsection 12.2.1. The UV-divergent subdiagram has $\omega(\gamma)=2$, such that the insertion of $\Delta_\gamma(\mathbf{p})$ into the shrunk diagram cancels a line.

Alternatively, the counterterm can be determined in a somewhat more complicated manner by applying the $\bar{R}^*$-operation to the following diagram G:

$$G = \text{(diagram)}, \quad \gamma = \text{(diagram)} \quad \Rightarrow \quad \hat{\gamma} = \text{(diagram)} . \tag{12.83}$$

It contains the same IRI subdiagram as the last example with $\tilde{\omega}(\gamma) = 2$, and a quadratically UV-divergent remaining diagram $\hat{\gamma}$. If the IR-divergences are subtracted first, the IR-counterterm

$\left(\begin{smallmatrix}\vert\\\vert\end{smallmatrix}\right)_{IR}$ must be incorporated into $\ominus$, leading to the multiplication of the pole term by the diagram $2\,\varepsilon\,\ominus$. The UV-divergence of this diagram

$$-\mathcal{K}\bar{R}^*\left(\,\ominus\,\right) = -\mathcal{K}\bar{R}\left(-\boxminus\right) \tag{12.84}$$

is not a UV-subdivergence of the original diagram. The subtraction of such a term would be wrong:

$$\mathcal{K}\bar{R}^*\left(\,\ominus\,\right) = \mathcal{K}\Big[\ominus + \left(\begin{smallmatrix}\vert\\\vert\end{smallmatrix}\right)_{IR}\big(2\varepsilon\,\ominus\;\underbrace{-\;2\varepsilon\,\mathcal{K}\bar{R}(-\boxminus\,))}_{\text{wrong}}\big)\Big]. \tag{12.85}$$

The subdiagram in G from which the UV-counterterm really originates is $\gamma' = \ominus$. Therefore, if $\tilde{\omega}(\gamma) \neq 0$, the UV-terms have to be subtracted first, thus avoiding a change of the UV-behavior.

The UV-counterterm $\mathcal{K}(\ominus)$ has the explicit diagrammatic form $\mathcal{K}(\ominus) * \mathbb{\bigcirc}$. This vanishes because of the tadpole part. Subtracting both the UV-divergence of $\ominus$ with $\mathcal{K}(\ominus) = \mathcal{Z}_{\gamma'}(\varepsilon^{-1})\mathbf{p}^2$, and the IR-divergence of $\begin{smallmatrix}\vert\\\vert\end{smallmatrix}$ with $\left(\begin{smallmatrix}\vert\\\vert\end{smallmatrix}\right)_{IR} = \mathcal{Z}'_{\tilde{\gamma}}$, gives

$$
\begin{aligned}
\Delta'(\gamma)\,\mathcal{K}(\ominus) &= \mathcal{Z}_{\gamma'}(\varepsilon^{-1})\int \frac{d^D p}{(2\pi)^D}\left[\left(-\frac{\partial}{\partial\mathbf{p}}\right)^2 \mathbf{p}^2\right]\delta^{(D)}(p)\,\mathcal{Z}'_{\tilde{\gamma}}(\varepsilon^{-1})\\
&= \mathcal{Z}_{\gamma'}(\varepsilon^{-1})\,\mathcal{Z}'_{\tilde{\gamma}}(\varepsilon^{-1})\cdot 2D.
\end{aligned} \tag{12.86}
$$

The correct $\bar{R}^*$-operation is

$$\mathcal{K}\bar{R}^*\left(\,\ominus\,\right) = \mathcal{K}\Big[\ominus + \left(\begin{smallmatrix}\vert\\\vert\end{smallmatrix}\right)_{IR}\cdot 2\varepsilon\,\ominus - \left(\begin{smallmatrix}\vert\\\vert\end{smallmatrix}\right)_{IR}\cdot 2D\,\partial_{\mathbf{p}^2}\mathcal{K}(\ominus)\Big]. \tag{12.87}$$

The five-loop calculations contain only one case where this special treatment is necessary: the diagram No. 116 in Appendix A on page 449. It can be calculated analytically only in the IR-rearranged form shown on the right hand side of Fig. 12.7. Here we have dealt only with the IRI subdiagram with $\tilde{\omega} = 2$, i.e., the type examined also in Examples 3 and 4. The full R^*-operation will be presented at the end of the section.

FIGURE 12.7 Infrared rearrangement of the special five-loop diagram No. 116 in Appendix A on page 449, whose IR-counterterm is calculated explicitly in Example 3, and whose complete R^*-operation is constructed in Section 12.5.

12.4.5 IR-Counterterms

The IR-counterterm of a diagram G can be calculated with the help of the general equation (12.60). The contribution from an IR-divergent subdiagram γ can be calculated using an arbitrary auxiliary diagram G_γ which contains γ and is superficially UV-convergent, i.e., which has $\Delta_{G_\gamma} = 0$. The IR-counterterm depends only on the integrand $I_{\tilde{\gamma}}(\mathbf{p}, \mathbf{l})$ in (12.65), such that any G_γ can be used which leaves $I_{\tilde{\gamma}}(\mathbf{p}, \mathbf{l})$ unchanged. The simplest G_γ is constructed by multiplying $I_{\tilde{\gamma}}(\mathbf{p}, \mathbf{l})$ by $\prod_i^N[(\mathbf{p}_i - \mathbf{k})^2]^{-s_i}$ and integrating over $\mathbf{p}_i$ and $\mathbf{l}$. Diagrammatically, each subdiagram $\hat{\gamma}$ is replaced by s_i lines for each loop connecting γ and $\hat{\gamma}$. The number of the connecting loops is N. As an example, we replace for $N = 1$ and $s = 1$:

$$\int \frac{d^D p}{(2\pi)^D} \frac{d^D l}{(2\pi)^D} \, \hat{\gamma}(\mathbf{p}, \mathbf{k}) I_{\hat{\gamma}}(\mathbf{p}, \mathbf{l}) \quad \longrightarrow \quad \int \frac{d^D p}{(2\pi)^D} \frac{d^D l}{(2\pi)^D} \frac{I_{\hat{\gamma}}(\mathbf{p}, \mathbf{l})}{(\mathbf{p} - \mathbf{k})^2} . \qquad (12.88)$$

The numbers s_i are chosen to make G_γ superficially UV-convergent:

$$\omega(G_\gamma) = \omega(\hat{\gamma}) - 2 \, s_i < 0 . \qquad (12.89)$$

For IR-divergent subdiagrams, $s_i = 1$ is in general sufficient. Up to the five-loop level, we shall find only one exception, where $s = 2$ is needed, with the associated IR-divergence complicating the calculation. The term in the R^*-operation of G_γ which subtracts the IR-divergence of γ is then [recall (12.67)]

$$\Delta'_{\hat{\gamma}} * G_\gamma \backslash \gamma \; = \; \mathcal{K} \int \frac{d^D l}{(2\pi)^D} \prod_i^N \left\{ \frac{d^D p_i}{(2\pi)^D} \frac{1}{[(\mathbf{p}_i - \mathbf{k})^2]^{s_i}} \right\} \Delta'_{\hat{\gamma}}(\mathbf{p}_i, \mathbf{l})$$

$$= \; \mathcal{K} \, \mathcal{P}'_\gamma \left(-\frac{\partial}{\partial \mathbf{p}_i} \right) \prod_i^N \frac{1}{[(\mathbf{p}_i - \mathbf{k})^2]^{s_i}} \Bigg|_{\mathbf{p}_i = 0} . \qquad (12.90)$$

For $\omega(\hat{\gamma}) = 0$, this simplifies to

$$\Delta'_{\hat{\gamma}} * G_\gamma \backslash \gamma = \mathcal{Z}'_{\hat{\gamma}}(\varepsilon^{-1}) \frac{1}{(\mathbf{k}^2)^{\Sigma_i^N s_i}}; \qquad (12.91)$$

and for $\omega(\hat{\gamma}) = -2$ and $s = 1$, to

$$\Delta'_{\hat{\gamma}} * G_\gamma \backslash \gamma = \mathcal{Z}'_{\hat{\gamma}}(\varepsilon^{-1}) \frac{2\varepsilon}{(\mathbf{k}^2)^{\Sigma_i^N s_i + 1}} . \qquad (12.92)$$

Now, the R^*-operation of G_γ is used to calculate $\Delta'_{\hat{\gamma}} * G_\gamma \backslash \gamma$. Since G_γ is not superficially UV-divergent, we have $\Delta_{G_\gamma} = 0$, and since G_γ contains an external momentum, we have $\Delta'_{G_\gamma} = 0$. Therefore, we find:

$$\mathcal{K} \bar{R}^* G_\gamma = \mathcal{K} \left[f \Delta'_{\hat{\gamma}} * G_\gamma \backslash \gamma + \sum_{\Gamma, \Gamma' \neq \gamma, G_\gamma} \Delta_\Gamma * \Delta'_{\Gamma'} * G_\gamma \backslash \Gamma' / \Gamma \right] = 0, \qquad (12.93)$$

with a factor f counting how often $\Delta'_{\hat{\gamma}} * G_\gamma \backslash \gamma$ is subtracted in the operation $\mathcal{K} \bar{R}^* G_\gamma$. In most cases, $f = 1$. A case with $f = 2$ occurs in Example 2 below. By construction of G_γ, we have $\mathcal{K}(\Delta'_{\hat{\gamma}} * G_\gamma \backslash \gamma) = \Delta'_{\hat{\gamma}} * G_\gamma \backslash \gamma$. We can thus obtain the IR-counterterms from the following operation:

$$\Delta'_{\hat{\gamma}} * G_\gamma \backslash \gamma = -\frac{1}{f} \mathcal{K} \sum_{\Gamma, \Gamma' \neq \gamma, G_\gamma} \Delta_\Gamma * \Delta'_{\Gamma'} * G_\gamma \backslash \Gamma' / \Gamma \equiv -\frac{1}{f} \mathcal{K} \bar{\bar{R}}^* G_\gamma , \qquad (12.94)$$

where the operator $\bar{\bar{R}}^*$ subtracts the UV- and IR-divergences of all subdiagrams excluding γ.

Combining this with Eq. (12.91), the IR-pole term for $\omega(\hat{\gamma}) = 0$ is then calculated as

$$\mathcal{Z}'_{\hat{\gamma}}(\varepsilon^{-1}) = -\frac{1}{f} (\mathbf{k}^2)^{\Sigma_i^N s_i} \mathcal{K} \bar{\bar{R}}^* G_\gamma . \qquad (12.95)$$

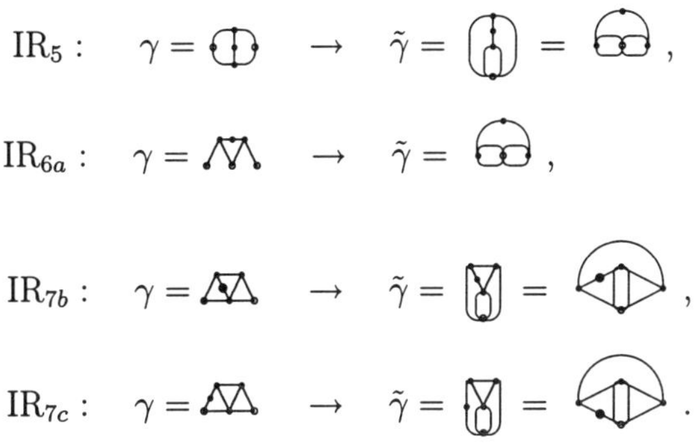

$$\text{IR}_5: \quad \gamma = \;⊕\; \rightarrow \; \tilde{\gamma} = \;⊖\; = \;⊖⊖\;,$$

$$\text{IR}_{6a}: \quad \gamma = \;\triangle\; \rightarrow \; \tilde{\gamma} = \;⊖⊖\;,$$

$$\text{IR}_{7b}: \quad \gamma = \;\triangle\triangle\; \rightarrow \; \tilde{\gamma} = \;⊗\; = \;\triangle\;,$$

$$\text{IR}_{7c}: \quad \gamma = \;\triangle\triangle\; \rightarrow \; \tilde{\gamma} = \;⊗\; = \;\triangle\;.$$

FIGURE 12.8 Typical infrared counterterms depending only on $\tilde{\gamma}$. We see that different IRI subdiagrams γ can have the same infrared counterterms.

In the R^*-operation, only the pole term $\mathcal{Z}'_{\tilde{\gamma}}(\varepsilon^{-1})$ is used. This pole term is called the IR-counterterm, and is denoted by $(\gamma)_{IR}$:

$$(\gamma)_{IR} = \mathcal{Z}'_{\tilde{\gamma}} = -\frac{1}{f}(\mathbf{k}^2)^{\Sigma_i^N s_i}\, \mathcal{K}\bar{\bar{R}}^*\, G_\gamma. \tag{12.96}$$

Although it is $\tilde{\gamma}$ which defines the divergence uniquely (see Fig. 12.8), the counterterm will be written as $(\gamma)_{IR}$ to simplify the notation. We now present a few sample calculations of counterterms.

Example 1: For the calculation of the IR-counterterm of the subdiagram $\gamma = \;|\;$, we use the following G_γ:

$$G_\gamma = \;-\!\bigcirc\!-\; . \tag{12.97}$$

The IR-divergent subdiagram is logarithmically divergent, i.e., $\omega(\tilde{\gamma}) = 0$.

In the ultraviolet, the diagram G_γ is convergent, with $\omega(G_\gamma) = -2$. Hence $\Delta_{G_\gamma} = 0$, and

$$\Delta'_{\tilde{\gamma}} * G_\gamma\backslash\gamma = \mathcal{Z}'_{\tilde{\gamma}}\frac{1}{\mathbf{k}^2} = -\mathcal{K}\bar{\bar{R}}^*\, G_\gamma = -\mathcal{K}\, G_\gamma. \tag{12.98}$$

The pole term is thus given by:

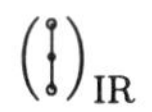

$$\left(|\right)_{IR} = -\mathbf{k}^2\mathcal{K}\left(-\!\bigcirc\!-\right) = -\mathbf{k}^2\mathcal{K}\int\frac{d^D p}{(2\pi)^D}\frac{1}{\mathbf{p}^4(\mathbf{p}-\mathbf{k})^2} = \frac{2}{\varepsilon}. \tag{12.99}$$

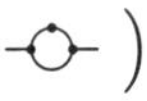

The above IR-counterterm may be used in various $\bar{R}^*$-operations, for instance in

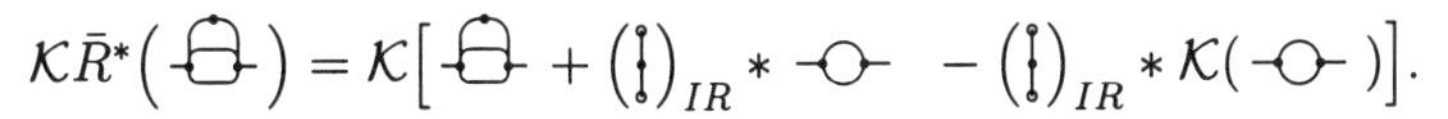

$$\mathcal{K}\bar{R}^*\left(⊖\right) = \mathcal{K}\left[⊖ + \left(|\right)_{IR}*-\!\bigcirc\!- \; - \left(|\right)_{IR}*\mathcal{K}(-\!\bigcirc\!-)\right]. \tag{12.100}$$

The fully subtracted Feynman integral is

$$\mathcal{K}\bar{R}^*\left(⊖\right) = \mathcal{K}\Bigg[\int\frac{d^D p\, d^D q}{(2\pi)^{2D}}\frac{1}{\mathbf{p}^4(\mathbf{p}-\mathbf{q})^2(\mathbf{q}-\mathbf{k})^2} \tag{12.101}$$
$$+\left(-\mathbf{k}^2\mathcal{K}\int\frac{d^D p}{(2\pi)^D}\frac{1}{\mathbf{p}^4(\mathbf{p}-\mathbf{k})^2}\right)\int\frac{d^D q}{(2\pi)^D}\frac{1}{\mathbf{q}^2(\mathbf{q}-\mathbf{k})^2}$$
$$-\left(-\mathbf{k}^2\mathcal{K}\int\frac{d^D p}{(2\pi)^D}\frac{1}{\mathbf{p}^4(\mathbf{p}-\mathbf{k})^2}\right)\mathcal{K}\left(\int\frac{d^D q}{(2\pi)^D}\frac{1}{\mathbf{q}^2(\mathbf{q}-\mathbf{k})^2}\right)\Bigg].$$

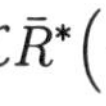

Example 2: Let us also study a diagram with $f \neq 1$ in Eq. (12.94). The IR-counterterms required for the R^*-operation of the diagram in Fig. 12.6 need the determination of the counterterm associated with the following IR-divergent subdiagram:

$$\gamma = \oplus \quad \Rightarrow \quad G_\gamma = \ominus. \tag{12.102}$$

Since $\tilde{\omega} = 0$ and $s = 1$, the counterterm is given by

$$(\oplus)_{IR} = -\mathbf{k}^2 \mathcal{K} \Big[\ominus + \big(\underset{\bullet}{\overset{\bullet}{\mathfrak{l}}} \big)_{IR} \ominus + 2 \big(\mathfrak{N} \big)_{IR} \ominus \Big]. \tag{12.103}$$

Note the factor 2 in the last term, since the corresponding IR-subdiagram appears twice in G_γ. This leads to a factor $f = 2$ in the associated IR-counterterm with the following G_γ:

$$\gamma = \mathfrak{N} \quad \Rightarrow \quad G_\gamma = \ominus. \tag{12.104}$$

We have $N = 2$, and $\tilde{\omega} = 0$. In the $\bar{R}^*$-operation of G_γ, the same factor 2 appears:

$$\mathcal{K}\bar{R}^*(\ominus) = \mathcal{K}\Big[\ominus + 2\big(\mathfrak{N} \big)_{IR} \cdots + \big(\underset{\bullet}{\overset{\bullet}{\mathfrak{l}}} \big)_{IR} \ominus \Big], \tag{12.105}$$

such that $f = 2$. The counterterm for $\gamma = \mathfrak{N}$ is then calculated from

$$\Delta'_{\tilde{\gamma}} * G_\gamma \backslash \gamma = \int \frac{d^D p_1 \, d^D p_2}{(2\pi)^{2D}} \frac{\mathcal{Z}'_{\tilde{\gamma}}(\varepsilon^{-1}) \, (2\pi)^{2D} \, \delta^{(D)}(p_1) \, \delta^{(D)}(p_2)}{(\mathbf{p}_1 - \mathbf{k})^2 (\mathbf{p}_2 - \mathbf{k})^2}$$

$$= \mathcal{Z}'_{\tilde{\gamma}}(\varepsilon^{-1}) \frac{1}{\mathbf{k}^2}. \tag{12.106}$$

This gives

$$(\mathfrak{N})_{IR} = -\frac{\mathbf{k}^4}{2} \mathcal{K}\bar{\bar{R}}^*(\ominus). \tag{12.107}$$

Example 3: For the subdiagram $\gamma = \underset{\bullet}{\overset{\bullet}{\mathfrak{l}}}$, we construct the auxiliary diagram G_γ as follows:

$$\gamma = \big(\underset{\bullet}{\overset{\bullet}{\mathfrak{l}}} \big)_{IR} \quad \Rightarrow \quad G_\gamma = \ominus. \tag{12.108}$$

The IR-divergent subdiagram is quadratically IR-divergent, $\tilde{\omega}(\gamma) = 2$, whereas G_γ is UV-convergent, as it should, with $\omega(G_\gamma) = -4$. Hence $\Delta_{G_\gamma} = 0$, and

$$\Delta'_{\tilde{\gamma}} * G_\gamma = \mathcal{K}\Big[\mathcal{Z}'_{\tilde{\gamma}}(\varepsilon^{-1}) \frac{\partial^2}{\partial p_\mu \partial p_\mu} \frac{1}{[(\mathbf{p} - \mathbf{k})^2]^s} \Big|_{\mathbf{p}=0} \Big] = -\mathcal{K}\bar{\bar{R}}^* G_\gamma. \tag{12.109}$$

For $s = 1$,

$$\frac{\partial^2}{\partial p_\mu \partial p_\mu} \frac{1}{(\mathbf{k} - \mathbf{p})^2} \Big|_{\mathbf{p}=0} = \frac{2\varepsilon}{\mathbf{k}^4}, \tag{12.110}$$

and the pole in ε cancels:

$$-\mathcal{K}(\ominus) = \mathcal{K}\Big[\mathcal{Z}'_{\tilde{\gamma}}(\varepsilon^{-1}) \frac{2\varepsilon}{\mathbf{k}^4} \Big] = 0. \tag{12.111}$$

In the present case, a propagator $1/(\mathbf{p} - \mathbf{k})^2$ cannot be used. We have to choose $s = 2$, giving

$$\frac{\partial^2}{\partial p_\mu \partial p_\mu} \frac{1}{(\mathbf{k} - \mathbf{p})^4}\bigg|_{\mathbf{p}=0} = \frac{8 + 4\varepsilon}{\mathbf{k}^6}, \tag{12.112}$$

leading to

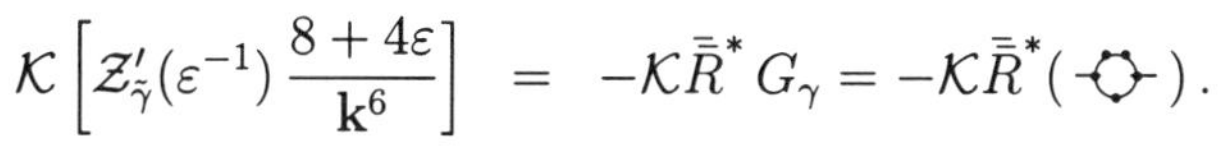

$$\mathcal{K}\left[\mathcal{Z}'_{\bar{\gamma}}(\varepsilon^{-1}) \frac{8 + 4\varepsilon}{\mathbf{k}^6}\right] = -\mathcal{K}\bar{\bar{R}}^* G_\gamma = -\mathcal{K}\bar{\bar{R}}^*(\text{⬡}). \tag{12.113}$$

As $\mathcal{Z}'_{\bar{\gamma}}$ consists of at most a simple pole, it is found by

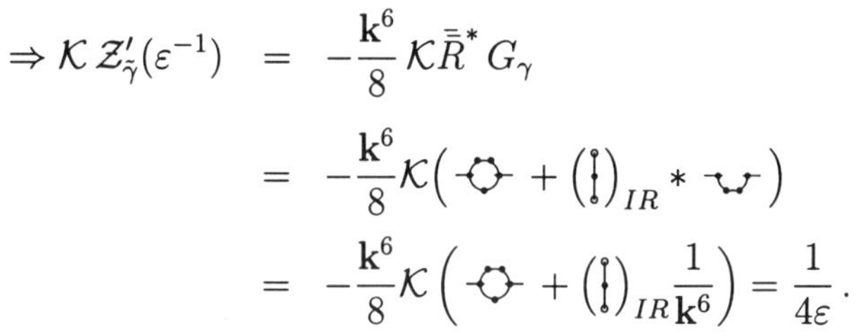

$$\Rightarrow \mathcal{K}\, \mathcal{Z}'_{\bar{\gamma}}(\varepsilon^{-1}) = -\frac{\mathbf{k}^6}{8} \mathcal{K}\bar{\bar{R}}^* G_\gamma \tag{12.114}$$

$$= -\frac{\mathbf{k}^6}{8}\mathcal{K}\left(\text{⬡} + \left(\substack{\circ\\\vert\\\circ}\right)_{IR} * \text{⌣}\right)$$

$$= -\frac{\mathbf{k}^6}{8}\mathcal{K}\left(\text{⬡} + \left(\substack{\circ\\\vert\\\circ}\right)_{IR} \frac{1}{\mathbf{k}^6}\right) = \frac{1}{4\varepsilon}. \tag{12.115}$$

12.5 Examples for the $\bar{R}^*$-Operation

We conclude this chapter by giving two examples for the complete R^*-operation. The first is typical for diagrams containing especially many IRI subdiagrams.

Diagram No. 53

Consider the diagram No. 53 on page 438 of Appendix A. Its IR-structure is rearranged for the calculation as follows:

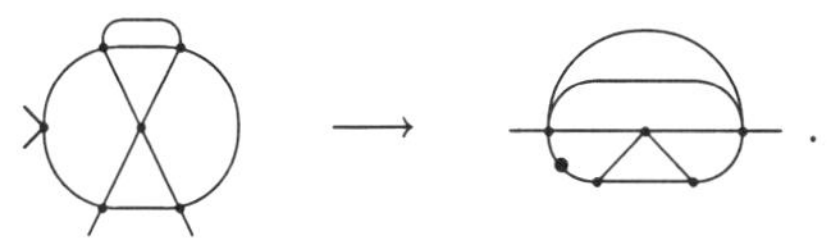

$$\tag{12.116}$$

This gives rise to IR-divergent IRI subdiagrams with up to three loops. The R^*-operation has the diagrammatic representation

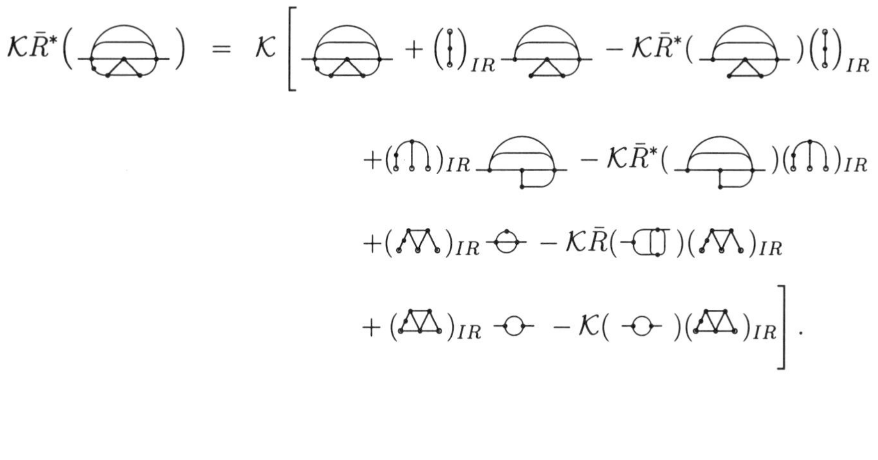

$$\mathcal{K}\bar{R}^*(\text{⬭}) = \mathcal{K}\left[\text{⬭} + \left(\substack{\circ\\\vert}\right)_{IR}\text{⬭} - \mathcal{K}\bar{R}^*(\text{⬭})\left(\substack{\circ\\\vert}\right)_{IR}\right.$$

$$+ \left(\substack{\circ\\\cap}\right)_{IR}\text{⬭} - \mathcal{K}\bar{R}^*(\text{⬭})\left(\substack{\circ\\\cap}\right)_{IR}$$

$$+ (\text{⋀})_{IR}\,\text{⬡} - \mathcal{K}\bar{R}(\text{⊟})(\text{⋀})_{IR}$$

$$\left.+ (\text{⋈})_{IR}\,\text{⊸} - \mathcal{K}(\text{⬡})(\text{⋈})_{IR}\right]. \tag{12.117}$$

There is no term subtracting only a UV-divergence, as the corresponding shrunk diagram has a tadpole form and gives zero. The subtraction of the UV-subdivergences is simplified by IR-rearrangement:

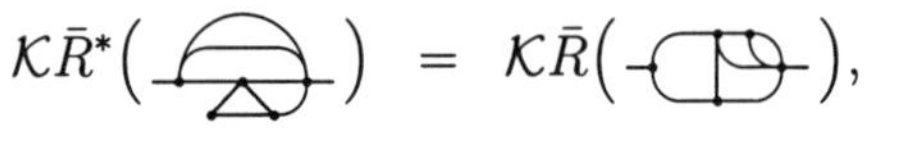

$$\mathcal{K}\bar{R}^*(\,\underset{\triangle}{\ominus}\,) \;=\; \mathcal{K}\bar{R}(\,\ominus\hspace{-3pt}\rhd\,), \tag{12.118}$$

$$\mathcal{K}\bar{R}^*(\,\ominus\,) \;=\; \mathcal{K}\bar{R}(\,\ominus\hspace{-3pt}\rhd\,). \tag{12.119}$$

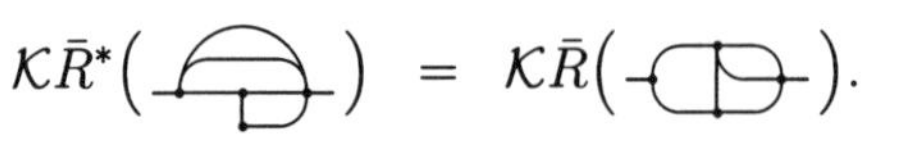

In the following, the subtraction terms of the UV-subdivergences are denoted directly by $\mathcal{K}\bar{R}(\gamma)$. All IR-counterterms have $\tilde{\omega}(\gamma) = 0$. The incorporation is again a simple multiplication, as in the previous Example 2:

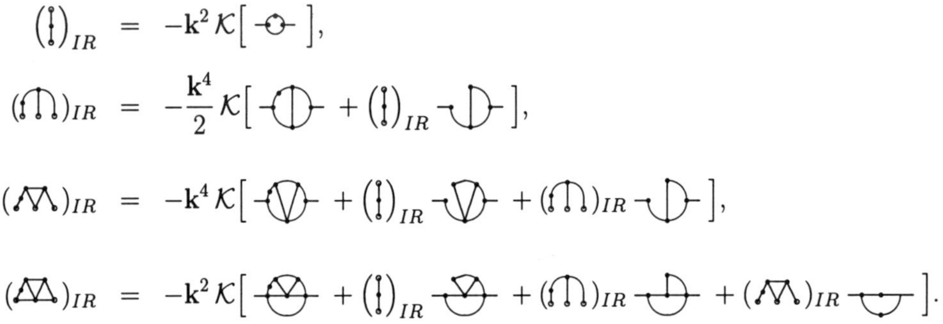

$$\left(\overset{\circ}{\underset{\circ}{\text{\large\textbf{!}}}}\right)_{IR} \;=\; -\mathbf{k}^2\,\mathcal{K}\!\left[\,\multimap\!\!\multimap\,\right], \tag{12.120}$$

$$\left(\text{\small\textrm{(}}\hspace{-2pt}\bigcap\hspace{-2pt}\text{\small\textrm{)}}\right)_{IR} \;=\; -\frac{\mathbf{k}^4}{2}\,\mathcal{K}\!\left[\,\ominus\!\!\ominus + \left(\overset{\circ}{\underset{\circ}{\text{\large\textbf{!}}}}\right)_{IR}\,\ominus\!\!\!\rhd\,\right], \tag{12.121}$$

$$\left(\,\bigwedge\,\right)_{IR} \;=\; -\mathbf{k}^4\,\mathcal{K}\!\left[\,\ominus\!\!\ominus + \left(\overset{\circ}{\underset{\circ}{\text{\large\textbf{!}}}}\right)_{IR}\,\ominus\!\!\!\bigtriangledown + (\text{\small\textrm{(}}\hspace{-2pt}\bigcap\hspace{-2pt}\text{\small\textrm{)}})_{IR}\,\ominus\!\!\!\rhd\,\right], \tag{12.122}$$

$$\left(\,\bigtriangleup\!\!\!\bigtriangleup\,\right)_{IR} \;=\; -\mathbf{k}^2\,\mathcal{K}\!\left[\,\ominus\!\!\!\otimes + \left(\overset{\circ}{\underset{\circ}{\text{\large\textbf{!}}}}\right)_{IR}\,\ominus\!\!\!\otimes + (\text{\small\textrm{(}}\hspace{-2pt}\bigcap\hspace{-2pt}\text{\small\textrm{)}})_{IR}\,\ominus\!\!\!\rhd + (\bigwedge)_{IR}\,\overline{\smile}\,\right]. \tag{12.123}$$

Diagram No. 116

As a second example, we choose the diagram No. 116 of Appendix A, page 449. Its IR-rearrangement was shown in Fig. 12.7. The R^*-operation of this diagram is exceptional, since it is the only case where an IRI subdiagram has $\tilde{\omega} = 2$. The subtraction of such a diagram requires differentiations in the remaining integral, thereby causing the non-commutativity of IR- and UV-subtractions. This problem was already illustrated in the second and third examples on page 216. The full R^*-Operation goes as follows:

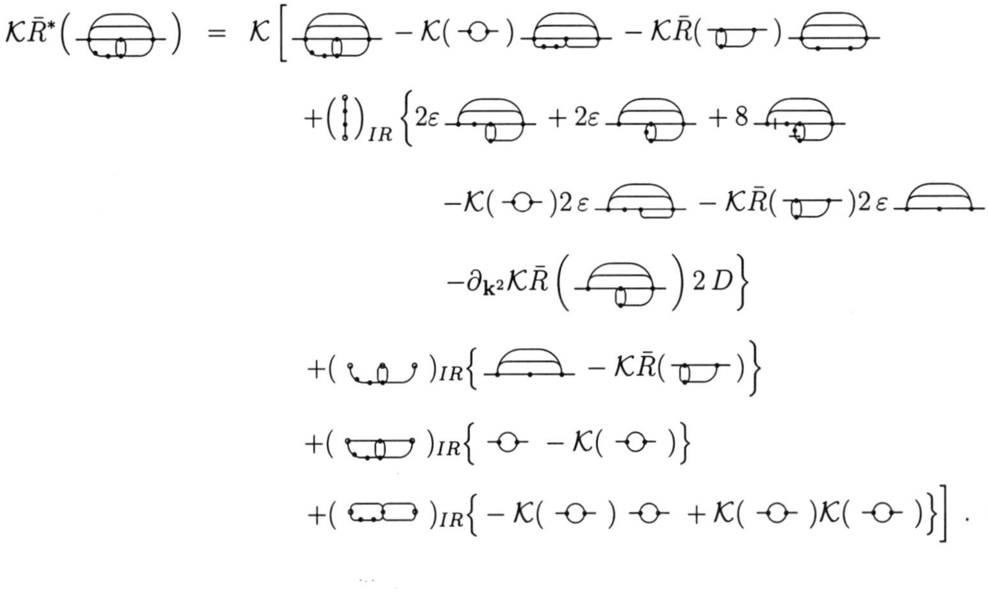

$$\mathcal{K}\bar{R}^*(\,\ominus\!\!\!\ominus\,) \;=\; \mathcal{K}\Bigg[\,\ominus\!\!\!\ominus - \mathcal{K}(\multimap)\,\ominus\!\!\!\ominus - \mathcal{K}\bar{R}(\overline{\smile})\,\ominus\!\!\!\ominus$$

$$+\left(\overset{\circ}{\underset{\circ}{\text{\large\textbf{!}}}}\right)_{IR}\Bigg\{2\varepsilon\,\ominus\!\!\!\ominus + 2\varepsilon\,\ominus\!\!\!\ominus + 8\,\ominus\!\!\!\ominus$$

$$-\mathcal{K}(\multimap)2\varepsilon\,\ominus\!\!\!\ominus - \mathcal{K}\bar{R}(\overline{\smile})2\varepsilon\,\ominus\!\!\!\ominus$$

$$-\partial_{\mathbf{k}^2}\mathcal{K}\bar{R}\left(\,\ominus\!\!\!\ominus\,\right)2\,D\Bigg\}$$

$$+(\,\smile\!\!\smile\,)_{IR}\Big\{\,\ominus\!\!\!\ominus - \mathcal{K}\bar{R}(\overline{\smile})\Big\}$$

$$+(\,\overline{\smile}\,)_{IR}\Big\{\,\multimap - \mathcal{K}(\multimap)\Big\}$$

$$+(\,\underline{\boxminus}\,)_{IR}\Big\{-\mathcal{K}(\multimap)\,\multimap + \mathcal{K}(\multimap)\mathcal{K}(\multimap)\Big\}\Bigg]. \tag{12.124}$$

Note that, by analogy with Subsection 12.4.4, the term

$$2D\left(\substack{\bullet\\\bullet}\right)_{IR}\partial_{\mathbf{k}^2}\mathcal{K}\bar{R}(\;\text{⬡}\;)$$

is subtracted, not

$$2\varepsilon\left(\substack{\bullet\\\bullet}\right)_{IR}\mathcal{K}\bar{R}(\;\text{⬡}\;).$$

Only the first term contains the UV-subdivergence of the diagram. The last term would have to be taken if the IR-subtractions had been carried out first and the UV-subtractions would be applied to the resulting terms (see the examples in Subsection 12.4.4).

The IR-counterterm with $\tilde{\omega} = 2$ was calculated in Eq. (12.115). The other counterterms have $\tilde{\omega} = 0$, and their pole terms are

$$(\text{⌣₀⌣})_{IR} \;=\; -\mathbf{k}^4\,\mathcal{K}\Big[\,\text{⬡}\; -\; \mathcal{K}(\,\text{⊖}\,)\,\text{⬡}\; -\; \mathcal{K}\bar{R}(\,\text{⬡}\,)\,\text{⬡}$$

$$+\left(\substack{\bullet\\\bullet}\right)_{IR}\Big\{2\varepsilon\,\text{⬡}\; +\; 2\varepsilon\,\text{⬡}\; -\; 2\varepsilon\mathcal{K}(\,\text{⊖}\,)\,\text{⬡}$$

$$-\,2\varepsilon\mathcal{K}\bar{R}(\,\text{⬡}\,)\,\text{⋯}\; +\; 8\,\text{⬡}\Big\}\Big] \qquad (12.125)$$

$$(\text{⬡})_{IR} \;=\; -\mathbf{k}^2\,\mathcal{K}\Big[\,\text{⬡}\; -\; \mathcal{K}(\,\text{⊖}\,)\,\text{⬡}\; -\; \mathcal{K}\bar{R}(\,\text{⬡}\,)\,\text{⬡}$$

$$+\left(\substack{\bullet\\\bullet}\right)_{IR}\Big\{2\varepsilon\,\text{⬡}\; +\; 2\varepsilon\,\text{⬡}\; -\; 2\varepsilon\mathcal{K}(\,\text{⊖}\,)\,\text{⬡}$$

$$-\,2\varepsilon\mathcal{K}\bar{R}(\,\text{⬡}\,)\,\text{⌢}\; +\; 8\,\text{⬡}\Big\}$$

$$+\,(\,\text{⬡}\,)_{IR}\Big\{-\mathcal{K}(\,\text{⊖}\,)\,\text{⌢}\Big\}$$

$$+\,(\,\text{⌣₀⌣}\,)_{IR}\,\text{⌢}\Big] \qquad (12.126)$$

$$(\text{⬡})_{IR} \;=\; -\mathbf{k}^2\,\mathcal{K}\Big[\,\text{⬡}\; -\; \mathcal{K}(\,\text{⊖}\,)\,\text{⬡}$$

$$+\left(\substack{\bullet\\\bullet}\right)_{IR}\Big\{2\varepsilon\,\text{⬡}\; -\; 2\varepsilon\mathcal{K}(\,\text{⊖}\,)\,\text{⌢}\Big\}\Big] \qquad (12.127)$$

Appendix 12A   Proof of Interchangeability of Differentiation and $\bar{R}$-Operation

The proof is nontrivial because the operation $\bar{R}$ is defined for diagrams, whereas differentiation operators act on lines [8]. We proceed by induction in the number of loops. By the chain rule, differentiation of a diagram G involves a sum of differentiations for each line l of the diagram:

$$\partial G = \sum_l \partial_l G. \qquad (12A.1)$$

If applied to a subdiagram γ, the line l may lie outside the subdiagram γ under consideration. In this case, we interchange the derivative with γ, i.e., $\partial\gamma = \gamma\partial$. This property will be needed below.

We want to show that $\partial \bar{R} G = \bar{R} \partial G$. The left-hand side is more explicitly

$$
\begin{aligned}
\partial \bar{R} G &= \partial \sum_{\Gamma(G)} \left(\prod_{\gamma \in \Gamma} -\mathcal{K} \bar{R} \gamma \right) G/\Gamma \\
&= \sum_{\Gamma(G)} \left[\partial \left(\prod_{\gamma \in \Gamma} -\mathcal{K} \bar{R} \gamma \right) G/\Gamma + \left(\prod_{\gamma \in \Gamma} -\mathcal{K} \bar{R} \gamma \right) \partial(G/\Gamma) \right].
\end{aligned}
$$

In applying ∂ to the product of factors $-\mathcal{K} \bar{R} \gamma$, we use the product rule, whereby ∂ can be interchanged with the operator $\mathcal{K}$ without problem. As an induction hypothesis, we assume that it also can be interchanged with γ, if γ is a proper subdiagram of G. This implies that $\partial \mathcal{K} \bar{R} \gamma = \mathcal{K} \bar{R} \partial \gamma$, and we find

$$
\partial \bar{R} G = \sum_{\Gamma(G)} \left\{ \sum_{\gamma'} \left[\left(\prod_{\gamma \in \Gamma, \gamma \neq \gamma'} -\mathcal{K} \bar{R} \gamma \right) (-\mathcal{K} \bar{R} \partial \gamma') \right] G/\Gamma + \left(\prod_{\gamma \in \Gamma} -\mathcal{K} \bar{R} \gamma \right) \partial(G/\Gamma) \right\}.
$$

With Eq. (12A.1), this may be rewritten as

$$
\begin{aligned}
\partial \bar{R} G = \sum_{\Gamma(G)} \Bigg\{ &\sum_{\gamma'} \sum_{l \in \gamma'} \left[\left(\prod_{\gamma \in \Gamma, \gamma \neq \gamma'} -\mathcal{K} \bar{R} \gamma \right) (-\mathcal{K} \bar{R} \partial_l \gamma') \right] G/\Gamma \\
&+ \sum_{l \in G/\Gamma} \left(\prod_{\gamma \in \Gamma} -\mathcal{K} \bar{R} \gamma \right) \partial_l G/\Gamma \Bigg\}.
\end{aligned} \tag{12A.2}
$$

The derivative ∂_l can be applied to those terms which do not contain lines l without any effect, leading to

$$
\partial \bar{R} G = \sum_{\Gamma(G)} \left\{ \sum_{l \in \Gamma} \left(\prod_{\gamma \in \Gamma} -(\mathcal{K} \bar{R} \partial_l \gamma) \right) \partial_l(G/\Gamma) + \sum_{l \in G/\Gamma} \left(\prod_{\gamma \in \Gamma} -\mathcal{K} \bar{R} \partial_l \gamma \right) \partial_l(G/\Gamma) \right\}.
$$

Since each line l is either in Γ or in G/Γ, the two terms in the last equation can be combined to a single sum:

$$
\partial \bar{R} G = \sum_{\Gamma(G)} \sum_{l \in G} \left(\prod_{\gamma \in \Gamma} -(\mathcal{K} \bar{R} \partial_l \gamma) \right) \partial_l(G/\Gamma). \tag{12A.3}
$$

The sum over the sets Γ can now be interchanged with the sum over l because the last one is independent of the choice of Γ. Hence

$$
\partial \bar{R} G = \sum_{l \in G} \sum_{\Gamma(G)} \left(\prod_{\gamma \in \Gamma} -(\mathcal{K} \bar{R} \partial_l \gamma) \right) \partial_l(G/\Gamma). \tag{12A.4}
$$

To each set $\Gamma(G)$, there is a corresponding set $\Gamma(\partial G)$, and the sum can be rewritten as running over the sets $\Gamma(\partial G)$:

$$
\partial \bar{R} G = \sum_{l \in G} \sum_{\Gamma(\partial G)} \left(\prod_{\gamma \in \Gamma} -(\mathcal{K} \bar{R} \gamma) \right) (\partial_l G/\Gamma). \tag{12A.5}
$$

The inner sum of this expression corresponds to the $\bar{R}$-operation of $\partial_l G$. Using the linearity of $\bar{R}$, the right-hand side can be brought to the desired form

$$
\partial \bar{R} G = \sum_{l \in G} \bar{R} \partial_l G = \bar{R} \sum_{l \in G} \partial_l G = \bar{R} \partial G, \tag{12A.6}
$$

proving the interchangeability of differentiation and $\bar{R}$-Operation.

Notes and References

The citations in the text refer to:

[1] A.A. Vladimirov, Theor. Math. Phys. **36**, 732 (1978).

[2] D.R.T. Jones, Nucl. Phys. B **75**, 531 (1974).

[3] K.G. Chetyrkin, F.V. Tkachov, Nucl. Phys. B **192**, 159 (1981).

[4] K.G. Chetyrkin, F.V. Tkachov, Phys. Lett. B **114**, 240 (1982).

[5] The equivalent notation in the UV-case would be $(\gamma)_{UV} = \mathcal{K}\bar{R}(\gamma)$. But in the UV-case the implementation of the pole terms into the remaining diagram is much simpler, being always a product of $\mathcal{K}\bar{R}(\gamma)$ with the remaining diagram, and the result can be written down immediately. For this reason there was no advantage in introducing an ultraviolet object $(\gamma)_{UV}$.

[6] K.G. Chetyrkin, F.V. Tkachov, Phys. Lett. B **114**, 240 (1982).

[7] K.G. Chetyrkin, V.A. Smirnov, Phys. Lett. B **144**, 419 (1984).

[8] See also Appendix B of
W.E. Caswell, A.D. Kennedy, Phys. Rev. D **25**, 392 (1982).

13

Calculation of Momentum Space Integrals

In the last chapter we have shown that the problem of determining the counterterms in ϕ^4-theory can be reduced completely to the calculation of massless Feynman integrals in momentum space. Among these, the massless propagator-type integrals generated by the technique of infrared rearrangement in Section 12.2 can be calculated most easily by algebraic methods. The result of a successive application of a generic one-loop integral formula in momentum space yields a Laurent expansion in ε. Some of the integrals can be solved rather directly, others can only be reduced to certain generic two-, three- or four-loop integrals. These can be evaluated by a reduction algorithm in momentum space [1] explained in Section 13.4. In this way, we can find all integrals up to four loops, and most of the five-loop integrals. Some of the five-loop diagrams contain one special type of integrals and a few individual integrals, for which the above reduction algorithm fails. These integrals were initially determined numerically in configuration space, by applying the so-called Gegenbauer-polynomial-**x**-space technique (GPXT) [2]. Later, however, analytic solutions were found by the method of ideal index constellations [3] described in Section 13.5. With these methods, the pole terms of all Feynman integrals up to five loops have been found analytically. In this chapter we shall be concerned only with the momentum integrals associated with the Feynman diagrams. For this reason, we shall ignore the coupling constants attached to the vertices of the diagrams in the Feynman rule (3.5) of the perturbation expansion.

13.1 Simple Loop Integrals

The calculation of all Feynman integrals in momentum space proceeds from the simplest massless loop integral

$$\int \frac{d^D p}{(2\pi)^D} \frac{1}{(\mathbf{p}^2)^a [(\mathbf{p}-\mathbf{k})^2]^b} = \overset{a}{\underset{b}{\mathbf{k} \bigcirc}} \ . \tag{13.1}$$

For $\mathbf{k} = 0$, this integral vanishes by Veltman's formula (8.33). The powers a and b of the massless propagators will be called *line indices*. Initially, the line indices of the simple loop diagram are both equal to unity. However, differentiations with respect to the mass or the external momenta discussed in the last chapter, or successive calculations of nested simple loops, will generate indices greater than one. Working in dimensions $D = 4 - \varepsilon$, the line indices will in general be noninteger, but always close to integer with a typical noninteger form $a = p + q\varepsilon/2$, where p and q are integer.

The D-dimensional Fourier representation of a massless propagator with line index a is, as shown in detail in Appendix 13A,

$$\frac{1}{(\mathbf{p}^2)^a} = \frac{1}{\pi^{D/2} 4^a} \frac{\Gamma(D/2 - a)}{\Gamma(a)} \int d^D x \frac{e^{i\mathbf{p}\mathbf{x}}}{(\mathbf{x}^2)^{D/2-a}} \ . \tag{13.2}$$

We shall also define a line index in configuration space as the power of $1/\mathbf{x}^2$ in a massless integral. In the Fourier integral (13.2), this line index is $a' = D/2 - a$. Inserting (13.2) into (13.1), we find the Fourier representation of the simple loop integral:

$$
\mathbf{k} \overset{a}{\underset{b}{\bigcirc}} = \int \frac{d^D p}{(2\pi)^D} \frac{1}{(\mathbf{p}^2)^a [(\mathbf{p} - \mathbf{k})^2]^b}
$$

$$
= \frac{\Gamma(D/2 - a)\Gamma(D/2 - b)}{\pi^D \Gamma(a)\Gamma(b) 4^{a+b}} \int \frac{d^D p}{(2\pi)^D} \int d^D x \, d^D y \, \frac{e^{i[\mathbf{p}\mathbf{x} + (\mathbf{k}-\mathbf{p})\mathbf{y}]}}{(\mathbf{x}^2)^{D/2-a}(\mathbf{y}^2)^{D/2-b}} . \tag{13.3}
$$

The momentum integral gives rise to a $\delta^{(D)}$-function, forcing $\mathbf{y}$ to be equal to $\mathbf{x}$, so that we obtain

$$
\mathbf{k} \overset{a}{\underset{b}{\bigcirc}} = \frac{\Gamma(D/2 - a)\Gamma(D/2 - b)}{\pi^D \Gamma(a)\Gamma(b) 4^{a+b}} \int d^D x \, \frac{e^{i\mathbf{k}\mathbf{x}}}{(\mathbf{x}^2)^{D-a-b}} . \tag{13.4}
$$

The integral of the right-hand side will be represented graphically by a line marked by its $\mathbf{x}$-space index, leading to the graphical correspondence

$$
\mathbf{k} \overset{a}{\underset{b}{\bigcirc}} \propto \underset{0 \qquad \mathbf{x}}{\overset{D-a-b}{\bullet\!\!-\!\!\bullet}} .
$$

Using once more Formula (13.2), the $\mathbf{x}$-space integral on the right-hand side can be evaluated, and we obtain

$$
\mathbf{k} \overset{a}{\underset{b}{\bigcirc}} = \frac{\Gamma(D/2 - a)\,\Gamma(D/2 - b)}{\pi^D \, \Gamma(a)\,\Gamma(b)\, 4^{(a+b)}} \int d^D x \, \frac{e^{i\mathbf{k}\mathbf{x}}}{(\mathbf{x}^2)^{D/2-(a+b-D/2)}}
$$

$$
= \frac{\Gamma(D/2 - a)\Gamma(D/2 - b)\Gamma(a + b - D/2)}{(4\pi)^{D/2} \, \Gamma(a)\Gamma(b)\Gamma(D - a - b)} \frac{1}{(\mathbf{k}^2)^{a+b-D/2}}
$$

$$
= \frac{1}{(\mathbf{k}^2)^{a+b-D/2}} \frac{1}{(4\pi)^2} L(a, b) . \tag{13.5}
$$

The quantity $L(a, b)$ will be referred to as the *loop function*. Its prefactor $1/(4\pi)^2$ will later be absorbed into the coupling constant (see Subsection 13.1.2). In terms of the Beta function $B(x, y) = \Gamma(x)\Gamma(y)/\Gamma(x + y)$, and the function

$$
\nu(x) \equiv \Gamma(D/2 - x)/\Gamma(x), \tag{13.6}
$$

the loop function can be rewritten conveniently as

$$
L(a, b) = (4\pi)^{\varepsilon/2} \, B(D/2 - a, D/2 - b) \, \frac{\Gamma(a + b - D/2)}{\Gamma(a)\Gamma(b)}
$$

$$
= (4\pi)^{\varepsilon/2} \, \nu(a) \, \nu(b) \, \nu(D - a - b) . \tag{13.7}
$$

Differentiation of a propagator with respect to the momentum generates lines with vector indices of the type introduced in Sections 11.8 and 12.2.2:

$$
\frac{\partial}{\partial p_\mu} \frac{1}{(\mathbf{p}^2)^{a-1}} = -2(a - 1) \frac{p_\mu}{(\mathbf{p}^2)^a} . \tag{13.8}
$$

Using (13.2), we obtain the Fourier representation

$$
\frac{p_\mu}{(\mathbf{p}^2)^a} = \frac{-1/2}{(a-1)} \frac{\partial}{\partial p_\mu} \frac{1}{(\mathbf{p}^2)^{a-1}}
$$

$$
= \frac{-1/2\,\Gamma(D/2-a+1)}{\pi^{D/2} 4^{a-1}\Gamma(a)} \int d^D x \frac{i x_\mu\, e^{i\mathbf{p}\cdot\mathbf{x}}}{(\mathbf{x}^2)^{D/2-a+1}} . \tag{13.9}
$$

A simple loop integral (13.1) containing a line with a vector index has therefore a Fourier representation

$$
\mathbf{k}\,\overset{a}{\underset{b}{\bigcirc}}\, = \int \frac{d^D p}{(2\pi)^D} \frac{p_\mu}{(\mathbf{p}^2)^a[(\mathbf{p}-\mathbf{k})^2]^b}
$$

$$
= -\frac{1}{2} \frac{\Gamma(D/2-a+1)\Gamma(D/2-b)}{\pi^D 4^{(a+b-1)}\Gamma(a)\Gamma(b)} \int d^D x \frac{i\,x_\mu\, e^{i\mathbf{k}\cdot\mathbf{x}}}{(\mathbf{x}^2)^{D-a-b+1}} , \tag{13.10}
$$

where the vertical dash across the upper line of the diagram indicates the vector index [recall Eq. (11.49)]. The right-hand side is evaluated by an inverse Fourier transformation as

$$
\mathbf{k}\,\overset{a}{\underset{b}{\bigcirc}}\, = \frac{\Gamma(D/2-a+1)\Gamma(D/2-b)\Gamma(a+b-D/2)}{(4\pi)^{D/2}\,\Gamma(a)\,\Gamma(b)\,\Gamma(D-a-b+1)} \frac{k_\mu}{(\mathbf{k}^2)^{a+b-D/2}} . \tag{13.11}
$$

We also introduce the generalized loop function

$$
L^{(k)}(a,b) = (4\pi)^{\varepsilon/2}\, B(D/2-a+k, D/2-b) \frac{\Gamma(a+b-D/2)}{\Gamma(a)\Gamma(b)} , \tag{13.12}
$$

of which the original loop function (13.7) is the special case $L^{(0)}(a,b)$. In contrast to $L(a,b) = L^{(0)}(a,b)$, the functions $L^{(k)}(a,b)$ are not symmetric in a and b. In terms of $L^{(k)}(a,b)$, we write the result (13.11) as

$$
\mathbf{k}\,\overset{a}{\underset{b}{\bigcirc}}\, = \frac{1}{(4\pi)^2}\, L^{(1)}(a,b) \frac{k_\mu}{(\mathbf{k}^2)^{a+b-D/2}} . \tag{13.13}
$$

13.1.1   Expansion of Loop Function

A Laurent expansion of the loop function $L(a,b)$ in (13.7) in powers of ε contains at most a simple pole in ε. This is the only singularity caused by the simple loop integral. In order to find its residue, we rewrite (13.7) more explicitly as

$$
L(a,b) = (4\pi)^{\varepsilon/2} \frac{\Gamma(D/2-a)\Gamma(D/2-b)\Gamma(a+b-D/2)}{\Gamma(a)\Gamma(b)\Gamma(D-a-b)} , \tag{13.14}
$$

and expand each Gamma function $\Gamma(n+\alpha\varepsilon)$ as follows:

$$
\Gamma(n+z) = (n-1)!\, \exp\left\{ -z[\gamma-\zeta^{(n)}(1)] + \sum_{j=2}^\infty (-1)^j \frac{z^j}{j} \left[\zeta(j)-\zeta^{(n)}(j)\right] \right\}. \tag{13.15}
$$

This formula is derived in Appendix 13B [see Eq. (13B.12)]. The formula contains Riemann's zeta function

$$
\zeta(z) = \sum_{l=1}^\infty \frac{1}{l^z} , \tag{13.16}
$$

and Euler's constant $\gamma = 0.5772\ldots$ [recall (8D.10)]. The symbol $\zeta^{(n)}(z)$ denotes the *truncated zeta function*

$$\zeta^{(n)}(z) = \sum_{l=1}^{n-1} \frac{1}{l^z}, \tag{13.17}$$

with the special values

$$\zeta^{(1)}(j) = 0, \quad \zeta^{(2)}(j) = 1, \quad \zeta^{(n)}(0) = n - 1. \tag{13.18}$$

Using the expansion (13.15), the Gamma functions in (13.14) may be written as products of exponentials, each expanded in powers of ε. For the simple loop function $L(a, a)$ with $a = b = 1$, the expansion reads

$$L(1,1) = (4\pi)^{\varepsilon/2} \frac{\Gamma(D/2 - 1)\Gamma(D/2 - 1)\Gamma(2 - D/2)}{\Gamma(1)\Gamma(1)\Gamma(D - 2)} = (4\pi)^{\varepsilon/2} \frac{[\Gamma(1 - \varepsilon/2)]^2 \Gamma(1 + \varepsilon/2)}{(\varepsilon/2)\,\Gamma(2 - \varepsilon)}$$

$$= (4\pi)^{\varepsilon/2} \frac{2}{\varepsilon} \exp\left[-\frac{\varepsilon}{2}\gamma + \varepsilon\zeta^{(2)}(1)\right]$$

$$\times \exp\left\{ \sum_{j=2}^{\infty} \frac{[2 + (-1)^j]\,(\varepsilon/2)^j}{j} \left[\zeta(j) - \underbrace{\zeta^{(1)}(j)}_{=0}\right] - \sum_{j=2}^{\infty} \frac{\varepsilon^j}{j}\left[\zeta(j) - \underbrace{\zeta^{(2)}(j)}_{=1}\right]\right\},$$

and can be brought to the form

$$L(1,1) = \frac{2}{\varepsilon} \exp\left\{\frac{\varepsilon}{2}\left[\ln 4\pi - \gamma - \frac{\varepsilon}{4}\zeta(2)\right]\right\} \exp\left[\sum_{j=1}^{\infty} \frac{\varepsilon^j}{j} + \sum_{j=3}^{\infty} \frac{2 + (-1)^j - 2^j}{2^j}\zeta(j)\frac{\varepsilon^j}{j}\right] \equiv \frac{2}{\varepsilon} L_0(\varepsilon). \tag{13.19}$$

On the right-hand side we have factored out a pole term of the form $2/\varepsilon$, defining a finite residue function $L_0(\varepsilon)$ whose Taylor series expansion starts out like $1 + \mathcal{O}(\varepsilon)$:

$$L_0(\varepsilon) \equiv \frac{\varepsilon}{2} L(1,1) = 1 + \frac{\varepsilon}{2}\left(\ln 4\pi - \gamma + 2\right) + \ldots\;. \tag{13.20}$$

Any loop integral $L(a, b)$ can ultimately be reduced to this function $L_0(\varepsilon)$. In a first step, the integer part of the parameters is reduced [4]:

$$L(a, b) = \frac{(a + b - 1 - D/2)(D - 2 - a - b)}{(b - 1)(D/2 - 2 - b)} L(a, b - 1). \tag{13.21}$$

With the help of this formula and the symmetry $L(a, b) = L(b, a)$, the parameters a, b can always be reduced to being near unity up to terms of order ε, say $a = 1 + \alpha\varepsilon$ and $b = 1 + \beta\varepsilon$. For such parameters, the expansion of the loop-function reads

$$L(1 + \alpha\varepsilon, 1 + \beta\varepsilon) = \frac{(4\pi)^{\varepsilon/2}}{\varepsilon(\alpha + b + 1/2)} \frac{\Gamma(1 - (\alpha + \frac{1}{2})\varepsilon)\Gamma(1 - (\beta + 1/2)\varepsilon)\Gamma(1 + (\alpha + \beta + 1/2)\varepsilon)}{\Gamma(1 + \alpha\varepsilon)\Gamma(1 + \beta\varepsilon)\Gamma(2 - (\alpha + \beta + 1)\varepsilon)}$$

$$= \frac{2}{\varepsilon(2\alpha + 2\beta + 1)} \exp\left\{\frac{\varepsilon}{2}\left[\ln 4\pi - \gamma - \frac{\varepsilon}{4}\zeta(2)\right]\right\}$$

$$\times \exp\left[\sum_{j=1}^{\infty} (\alpha + \beta + 1)^j \frac{\varepsilon^j}{j} + \sum_{j=3}^{\infty} F(\alpha, \beta, j)\zeta(j)\frac{\varepsilon^j}{j}\right], \tag{13.22}$$

where

$$F(\alpha, \beta, j) = (\alpha + \tfrac{1}{2})^j + (\beta + \tfrac{1}{2})^j + (-\alpha - \beta - \tfrac{1}{2})^j - (-\alpha)^j - (-\beta)^j - (\alpha + \beta + 1)^j. \tag{13.23}$$

In actual calculations, the direct expansion in Eq. (13.15) turns out to be most convenient. In order to avoid errors, a computer-algebraic program is extremely useful. We have employed the program REDUCE. The calculations up to six loops in this text require an expansion of the loop function up to the seventh order in ε [7].

13.1.2   Modified MS-Scheme and Various Redefinitions of Mass Scale

There are several factors common to all loop integrals. Such factors can be removed by a redefinition of the coupling constant. The reason for this lies in the fixed relationships between the number of loops L and the number of vertices p in the relevant two- and four-point diagrams. From Eqs. (9.2) and (9.5) we see that $L = p$ for the Feynman integrals of the two-point functions, and $L = p - 1$ for those of the four-point functions. Thus we may always go from g to some modified coupling constant $\bar{g} = g \times \text{constant} \times f(\varepsilon)$ with $f(\varepsilon) = 1 + f_1\varepsilon + f_2\varepsilon^2 + \ldots$, and the factor $f(\varepsilon)$ will disappear from the final expansions of the renormalization constants. This observation was made before [recall Eq. (9.75)]. In the above loop calculations in Eq. (13.5), the constant may be chosen to be $1/(4\pi)^2$. This was the reason for expressing all renormalization group functions in powers of

$$\bar{g} = \frac{g}{(4\pi)^2} \tag{13.24}$$

in Eqs. (10.55)–(10.58). This will be done for all higher-loop expansions in Chapter 15.

Although an ε-dependent common factor $f(\varepsilon) = 1 + f_1\varepsilon + f_2\varepsilon^2 + \ldots$ in all loop integrals does not show up in the final expansions, certain ways of choosing them offer different advantages in the calculations. A strict minimal subtraction scheme works with $f(\varepsilon) \equiv 1$, and expands all ε-dependent factors in $L(a, b)$ in powers of ε, as we did with the common factor $f(\varepsilon) = 1/(4\pi)^{-\varepsilon/2}$ in our two-loop calculation in Subsection 9.3.2. Other authors have modified this scheme by keeping some specific nontrivial function $f(\varepsilon)$ unexpanded, and absorbing it into the coupling constant. In fact, if such a function appears as a factor, it may simply be omitted from the calculations by assuming that it has been absorbed into the arbitrary mass scale μ. Such a function $f(\varepsilon)$ constitutes a common portion of all loop functions $L(a, b)$. A so-called $\overline{\text{MS}}$-scheme in the literature omits a factor $(4\pi)^{\varepsilon/2} \exp(-\gamma\varepsilon/2)$. This scheme is favored in quantum chromodynamics (QCD). It permits us to drop all terms 4π and γ in the ε-expansions (13.19) and (13.22). In higher loop-calculations, it is useful to omit also all terms containing $\zeta(2)$ [8]. This amounts to absorbing a function

$$f(\varepsilon) = \exp\left\{\frac{\varepsilon}{2}\left[\ln 4\pi - \gamma - \frac{\varepsilon}{4}\zeta(2)\right]\right\} \tag{13.25}$$

into the mass scale μ. If we remove this factor from the loop integral (13.19), it goes over into the new loop function

$$L(1, 1)_{\text{ours}} \equiv L(1, 1)_{\overline{\text{MS}}} = \frac{2}{\varepsilon}\exp\left\{\sum_{j=1}^{\infty}\frac{\varepsilon^j}{j} + \sum_{j=3}^{\infty}\left[\frac{2 + (-1)^j - 2^j}{j}\left(\frac{\varepsilon}{2}\right)^j \zeta(j)\right]\right\}, \tag{13.26}$$

which we shall use in all our calculations. We shall also refer to our scheme as an $\overline{\text{MS}}$-scheme, even if it is, strictly speaking, an extended version of the standard $\overline{\text{MS}}$-scheme. Thus we shall denote it by $L(1, 1)_{\overline{\text{MS}}}$. Its first terms in an ε-expansion are

$$L(1, 1)_{\overline{\text{MS}}} = \frac{2}{\varepsilon} + 2 + 2\varepsilon - \varepsilon^2\left[\frac{7}{12}\zeta(3) - 2\right] - \varepsilon^3\left[\frac{13}{32}\zeta(4) + \frac{7}{12}\zeta(3) - 2\right] + \mathcal{O}(\varepsilon^4). \tag{13.27}$$

Usually, we shall omit the subscript for notational convenience without much danger of confusion. The loop function $L(1, 1)_{\overline{\text{MS}}}$ can always be distinguished from the original in (13.19) by the absence of all terms containing $\ln 4\pi$, γ, and $\zeta(2)$.

13.1.3   Further Subtraction Schemes

Let us briefly indicate the characteristic features of other subtraction schemes.

a) The *L-scheme* [5] uses the fact that each loop function $L(a, b)$ can be reduced algebraically to $L(1, 1) = 2L_0(\varepsilon)/\varepsilon$ as shown in Subsection 13.1.1. The expansion coefficients contain the ζ-function only with arguments $n > 2$:

$$L(a, b) = L(1, 1) \sum_{n=0}^{\infty} C_n(a, b, \zeta(j > 2)) \, \varepsilon^n = L_0 \frac{2}{\varepsilon} \sum_{n=0}^{\infty} C_n(a, b, \zeta(j > 2)) \, \varepsilon^n . \tag{13.28}$$

This scheme absorbs the complete function $L_0(\varepsilon) = f(\varepsilon)$ into the mass scale μ. An important advantage of this scheme is that the Gamma functions require no longer an expansion in powers of ε. There is, however, a disadvantage to this scheme: all loop functions $L(a, b)$ must be re-expressed explicitly in terms of $L(1, 1)$.

b) The S_D-scheme [6] uses the fact that every loop integration involves the surface of the D-dimensional unit sphere $S_D = 2\pi^{D/2}/\Gamma(D/2)$ and a factor $1/(2\pi)^D$. By analogy with $\hbar \equiv h/2\pi$ we define $\mathcal{S}_D \equiv S_D/(2\pi)^D$, and a new coupling constant

$$g_{\mathcal{S}_D} = \mathcal{S}_D \, g = \frac{2}{(4\pi)^{D/2}\Gamma(D/2)} \, g \tag{13.29}$$

$$= \frac{2}{(4\pi)^{2-\varepsilon/2}} \exp\left\{ \frac{\varepsilon}{2} - \sum_{j=1}^{\infty} \frac{(\varepsilon/2)^j}{j} [\zeta(j) - 1] \right\} g.$$

There appears then a factor 2 which shows up in all expansions. The accompanying function of ε

$$f(\varepsilon) = \frac{1}{(4\pi)^{-\varepsilon/2}} \exp\left\{ \frac{\varepsilon}{2} - \sum_{j=1}^{\infty} \frac{(\varepsilon/2)^j}{j} [\zeta(j) - 1] \right\} \tag{13.30}$$

can again be dropped by absorbing it into the arbitrary mass scale μ. Using $g_{\mathcal{S}_D}$, the effective loop function which remains after factorizing out the ε-expansion for $\mathcal{S}_D$ in (13.26) is

$$L(1, 1)_{\mathcal{S}_D} = \frac{1}{\varepsilon} \exp\left\{ \sum_{j=3}^{\infty} \left[\frac{3 + (-1)^j - 2^j}{j} \left(\frac{\varepsilon}{2} \right)^j \zeta(j) \right] + \sum_{j=1}^{\infty} \left[\frac{\varepsilon^j}{j} - \frac{(\varepsilon/2)^j}{j} \right] \right\}$$

$$= \frac{1}{\varepsilon} + \frac{1}{2} + \frac{1}{2}\varepsilon + \varepsilon^2 \left[\frac{1}{2} - \frac{1}{4}\zeta(3) \right] + \varepsilon^3 \left[\frac{1}{12} - \frac{1}{8}\zeta(3) - \frac{3}{16}\zeta(4) \right] + \mathcal{O}(\varepsilon^4) .$$

As in our $\overline{\text{MS}}$-scheme, the sum over j does not contain any constants $\ln 4\pi$, γ, and $\zeta(2)$, but the remaining expansion is quite different from that in Eq. (13.27).

13.2   Classification of Diagrams

The set of all loop diagrams can be subdivided into three classes:

a) Diagrams with cutvertices, which factorize into independent integrals of lower-order diagrams.

For example:

$$
\text{—} \infty \text{—} \; = \; \int \frac{d^D p}{(2\pi)^D} \frac{1}{\mathbf{p}^2(\mathbf{p}-\mathbf{k})^2} \int \frac{d^D q}{(2\pi)^D} \frac{1}{\mathbf{q}^2(\mathbf{q}-\mathbf{k})^2}
$$

$$
= \; \left[\frac{1}{(4\pi)^2}\right]^2 L(1,1)\, L(1,1) \frac{1}{(\mathbf{k}^2)^{4-D}} \; .
$$

b) *Primitive diagrams*, which consist of nested simple loops and can be calculated recursively.

The integration of a simple loop with line indices a and b raises the line index of its external momentum by $a+b-D/2$, i.e. for $a=1$, $b=1$ by $2-D/2 = \varepsilon/2$. Each integration produces a loop function $L(a,b)$, so that a diagram with L loops generates at most a pole of Lth order in ε. For example:

$$
\text{—} \langle\!\langle \text{—} \; = \; \int \frac{d^D p}{(2\pi)^D} \frac{d^D q}{(2\pi)^D} \frac{d^D r}{(2\pi)^D} \frac{1}{\mathbf{p}^2(\mathbf{p}-\mathbf{r})^2 \mathbf{r}^2(\mathbf{r}-\mathbf{q})^2 \mathbf{q}^2(\mathbf{q}-\mathbf{k})^2}
$$

$$
= \; \left[\frac{1}{(4\pi)^2}\right]^3 L(1,1)\, L(1,1+\varepsilon/2)\, L(1,1+\varepsilon) \frac{1}{(\mathbf{k}^2)^{\frac{3}{2}\varepsilon}} \; .
$$

c) *Generic ϕ^3-diagrams*, which do not belong to the above two classes.

All ϕ^4- and ϕ^3-diagrams may be generated by shrinking lines and choosing appropriate line indices in a generic diagram with the same number of loops. The generic two- and three-loop diagrams are shown in Fig. 13.1. According to their shape, they are called KITE, BENZ, LADDER, and NONPLANAR types.

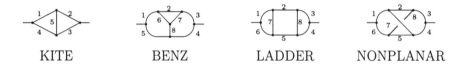

KITE   BENZ   LADDER   NONPLANAR

FIGURE 13.1 Generic two- and three-loop diagrams. Only one generic two-loop diagram occurs, and three with three loops. The numbers label the line indices.

It can be shown [1] that the calculation of all propagator-type integrals up to three loops can ultimately be reduced to the evaluation of one of the generic diagrams in Fig. 13.1, or a primitive diagram. The technique of IR-rearrangement transforms most of the ϕ^4-diagrams up to five loops either into generic two- or three-loop diagrams with insertions of simple loops in their lines, or into simple loops with generic two- or three-loop subdiagrams. The more difficult ones will still contain generic four-loop diagrams, and in one case even a generic five-loop diagram. The five-loop diagrams, to be shown in Fig. 13.10, will pose the main problems in the calculation. Algorithms for the reduction to simple loops exist only for the generic two- and three-loop diagrams, and not even for these with all line index constellations. Those diagrams have to be avoided by IR-rearrangement whenever possible.

13.3   Five-Loop Diagrams

Whenever possible, the five-loop diagrams are IR-rearranged in order to form a simple loop with one or more insertions. The resulting propagator-type integrals may then have ϕ^3- or ϕ^5-vertices. The lower loop insertions must, unfortunately, be calculated up to the finite terms of order $\mathcal{O}(\varepsilon^0)$ because the associated simple loop integral will make it appear in a pole term in ε.

Some examples are drawn in Fig. 13.2 to illustrate the occurrence of the generic ϕ^3-diagrams in the IR-rearranged five-loop ϕ^4-diagrams. The Feynman integrals of the generic types are functions of the line indices, for which we shall use the notation $\mathrm{KI}(a_1, \ldots, a_5)$, $\mathrm{BE}(a_1, \ldots, a_8)$, $\mathrm{LA}(a_1, \ldots, a_8)$, $\mathrm{NO}(a_1, \ldots, a_8)$ for the KITE, BENZ, LADDER, and NONPLANAR types, respectively. The first five examples drawn below are typical of what may be encountered in many five-loop diagrams, generic two- and three-loop diagrams with shrunk lines or combined with simple loops. Full generic three-loop diagrams with no shrunk line appear quite rarely, and each only once. These are of the BENZ type (No. 34), the LADDER type (No. 25), and the NONPLANAR type (No. 28), where the numbers are the running numbers of the diagrams used in Appendix A.

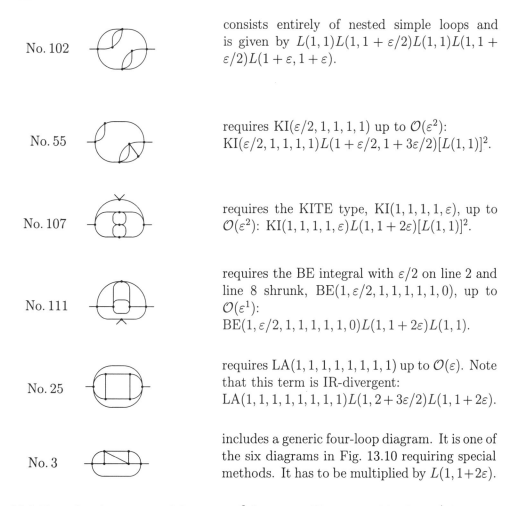

No. 102 — consists entirely of nested simple loops and is given by $L(1,1)L(1,1+\varepsilon/2)L(1,1)L(1,1+\varepsilon/2)L(1+\varepsilon,1+\varepsilon)$.

No. 55 — requires $\mathrm{KI}(\varepsilon/2,1,1,1,1)$ up to $\mathcal{O}(\varepsilon^2)$: $\mathrm{KI}(\varepsilon/2,1,1,1,1)L(1+\varepsilon/2,1+3\varepsilon/2)[L(1,1)]^2$.

No. 107 — requires the KITE type, $\mathrm{KI}(1,1,1,1,\varepsilon)$, up to $\mathcal{O}(\varepsilon^2)$: $\mathrm{KI}(1,1,1,1,\varepsilon)L(1,1+2\varepsilon)[L(1,1)]^2$.

No. 111 — requires the BE integral with $\varepsilon/2$ on line 2 and line 8 shrunk, $\mathrm{BE}(1,\varepsilon/2,1,1,1,1,1,0)$, up to $\mathcal{O}(\varepsilon^1)$: $\mathrm{BE}(1,\varepsilon/2,1,1,1,1,1,0)L(1,1+2\varepsilon)L(1,1)$.

No. 25 — requires $\mathrm{LA}(1,1,1,1,1,1,1,1)$ up to $\mathcal{O}(\varepsilon)$. Note that this term is IR-divergent: $\mathrm{LA}(1,1,1,1,1,1,1,1)L(1,2+3\varepsilon/2)L(1,1+2\varepsilon)$.

No. 3 — includes a generic four-loop diagram. It is one of the six diagrams in Fig. 13.10 requiring special methods. It has to be multiplied by $L(1,1+2\varepsilon)$.

FIGURE 13.2 Examples of occurrence of the generic ϕ^3-diagrams in IR-rearranged five-loop ϕ^4-diagrams.

13.4 Reduction Algorithm based on Partial Integration

A method based on *partial integration* provides us with an algorithm for the reduction of generic two- and three-loop diagrams to primitive diagrams, which can then be calculated to all orders in ε [1]. As announced at the beginning of this chapter, this will enable us to calculate all diagrams of the KITE, BENZ, and LADDER type, if certain line indices are integers. For diagrams with noninteger indices we shall have to resort to the configuration-space methods to be described in Subsection 13.5. Diagrams with simple loop insertions on such a line will have a noninteger line index and therefore cannot be reduced by the algorithm.

The typical loop integral which becomes accessible via the method of partial integration is the triangle diagram in Fig. 13.3. It appears as a subdiagram in all generic diagrams in Fig. 13.1. The associated integrand has the form $1/(\mathbf{p}^2)^a[(\mathbf{p}-\mathbf{q})^2]^b[(\mathbf{p}-\mathbf{k})^2]^c$, and this type of integrals can be reduced to simple loop integrals by partial integration.

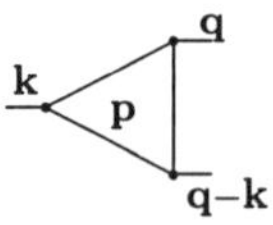

FIGURE 13.3 The triangle-subdiagram which appears in the KITE, LADDER, and BENZ type.

13.4.1   Triangle Diagram

Since surface terms vanish in dimensional regularization, partial integrations are a useful tool in the evaluation of massless Feynman integrals. They lead directly to a reduction algorithm for the integration of the triangle in Fig. 13.3. For this we insert into the integral the identity

$$D \equiv \partial/\partial p_\mu (p-q)_\mu, \tag{13.31}$$

where D is the dimension of space, and perform an integration by parts

$$D \int \frac{d^D p}{(2\pi)^D} \frac{1}{\mathbf{p}^2(\mathbf{p}-\mathbf{q})^2(\mathbf{p}-\mathbf{k})^2} = \int \frac{d^D p}{(2\pi)^D} \left[\frac{\partial}{\partial p_\mu}(p-q)_\mu\right] \frac{1}{\mathbf{p}^2(\mathbf{p}-\mathbf{q})^2(\mathbf{p}-\mathbf{k})^2}$$

$$= -\int \frac{d^D p}{(2\pi)^D} (p-q)_\mu \frac{\partial}{\partial p_\mu} \frac{1}{\mathbf{p}^2(\mathbf{p}-\mathbf{q})^2(\mathbf{p}-\mathbf{k})^2}. \tag{13.32}$$

We may demonstrate explicitly the vanishing of the neglected surface term in dimensional regularization by considering its Fourier representation, obtained with the help of Eqs. (13.2) and (13.9) for $a = 1$:

$$\int \frac{d^D p}{(2\pi)^D} \frac{\partial}{\partial p_\mu} \frac{(p-q)_\mu}{\mathbf{p}^2(\mathbf{p}-\mathbf{q})^2(\mathbf{p}-\mathbf{k})^2} = \int \frac{d^D p}{(2\pi)^D} \frac{\partial}{\partial p_\mu} \int d^D x\, d^D y\, d^D z\, \frac{iy_\mu\, e^{i[\mathbf{p}(\mathbf{x}+\mathbf{y}+\mathbf{z})-\mathbf{q}\mathbf{y}-\mathbf{k}\mathbf{z}]}}{(\mathbf{x}^2)^{D/2-1}(\mathbf{y}^2)^{D/2}(\mathbf{z}^2)^{D/2-1}}. \tag{13.33}$$

Carrying out the differentiation with respect to p_μ on the right-hand side produces a factor $(x+y+z)_\mu$. After the momentum integration, the right-hand side becomes

$$\int d^D x\, d^D y\, d^D z\, \frac{ix_\mu\, (x+y+z)_\mu\, \delta^{(D)}(\mathbf{x}+\mathbf{y}+\mathbf{z})\, e^{-i(\mathbf{q}\mathbf{y}+\mathbf{k}\mathbf{z})}}{(\mathbf{x}^2)^{D/2}(\mathbf{y}^2)^{D/2-1}(\mathbf{z}^2)^{D/2-1}} = 0. \tag{13.34}$$

Thus the surface term is indeed zero, and Eq. (13.32) is correct, which we now proceed to evaluate. Carrying out the momentum differentiation $\partial/\partial p_\mu(1/\mathbf{p}^2) \equiv \partial_{p_\mu}(1/\mathbf{p}^2) = -2p_\mu/\mathbf{p}^4$, we arrive at a sum of three integrals:

$$D \int \frac{d^D p}{(2\pi)^D} \frac{1}{\mathbf{p}^2(\mathbf{p}-\mathbf{q})^2(\mathbf{p}-\mathbf{k})^2} = -\int \frac{d^D p}{(2\pi)^D} \frac{1}{\mathbf{p}^2(\mathbf{p}-\mathbf{q})^2(\mathbf{p}-\mathbf{k})^2} \tag{13.35}$$

$$\times \left[-2\frac{(p-q)_\mu p_\mu}{\mathbf{p}^2} - 2\frac{(p-q)_\mu(p-q)_\mu}{(\mathbf{p}-\mathbf{q})^2} - 2\frac{(p-q)_\mu(p-k)_\mu}{(\mathbf{p}-\mathbf{k})^2}\right].$$

Let us represent the three integrals diagrammatically. A vertical dash on the line denotes a factor p_μ in the numerator, as defined in Eq. (11.49), whereas a dot symbolizes a ϕ^2-vertex insertion with a minus sign, as introduced in Eqs. (11.33) and (11.49), the latter implying an extra factor $\mathbf{p}^2$ in the denominator [recall (11.33)] .

$$
D \;\; \vcenter{\hbox{[triangle diagram with vertices B, A; lines labeled k, p, $p-q$, $p-k$, $q-k$, q]}} \;\; = \; 2\left(\vcenter{\hbox{[diagram]}} + \vcenter{\hbox{[diagram]}} + \vcenter{\hbox{[diagram]}} \right). \tag{13.36}
$$

Momentum conservation at the upper and lower vertices of the diagrams leads to the following equations:

$$
q_\mu = p_\mu - (p-q)_\mu \;\Rightarrow\; -2(p-q)_\mu p_\mu = \mathbf{q}^2 - \mathbf{p}^2 - (\mathbf{p}-\mathbf{q})^2
$$
$$
(p-k)_\mu - (p-q)_\mu = (q-k)_\mu \;\Rightarrow\; 2(p-q)_\mu(p-k)_\mu = (\mathbf{p}-\mathbf{q})^2 + (\mathbf{p}-\mathbf{k})^2 - (\mathbf{q}-\mathbf{k})^2.
$$

With their help we find an example of what will be refereed to as the *triangle rule* or *triangle relation* in momentum space, whose general form will be given in the next subsection:

$$
D\int \frac{d^D p}{(2\pi)^D} \frac{1}{\mathbf{p}^2(\mathbf{p}-\mathbf{q})^2(\mathbf{p}-\mathbf{k})^2} \;=\; -\int \frac{d^D p}{(2\pi)^D} \frac{1}{\mathbf{p}^2(\mathbf{p}-\mathbf{q})^2(\mathbf{p}-\mathbf{k})^2} \tag{13.37}
$$
$$
\times \left[\frac{\mathbf{q}^2 - \mathbf{p}^2 - (\mathbf{p}-\mathbf{q})^2}{\mathbf{p}^2} - 2 - \frac{(\mathbf{p}-\mathbf{q})^2 + (\mathbf{p}-\mathbf{k})^2 - (\mathbf{q}-\mathbf{k})^2}{(\mathbf{p}-\mathbf{k})^2} \right]
$$
$$
= -\int \frac{d^D p}{(2\pi)^D} \frac{1}{\mathbf{p}^2(\mathbf{p}-\mathbf{q})^2(\mathbf{p}-\mathbf{k})^2} \left[-4 + \frac{\mathbf{q}^2}{\mathbf{p}^2} + \frac{(\mathbf{q}-\mathbf{k})^2}{(\mathbf{p}-\mathbf{k})^2} - \frac{(\mathbf{p}-\mathbf{q})^2}{\mathbf{p}^2} - \frac{(\mathbf{p}-\mathbf{q})^2}{(\mathbf{p}-\mathbf{k})^2} \right].
$$

A cancellation of squared momenta corresponds diagrammatically to the shrinking of a line. This equation may be applied to reduce all diagrams with triangles to nested simple loops. Note that the shrinking, and therefore the reduction, cannot be done for noninteger indices of the corresponding line. These arise by the integration of subdiagrams attached to the corresponding line, as mentioned in 13.2.

As an example, we apply the special triangle rule Eq. (13.37) to the KITE-type diagram.

$$
\vcenter{\hbox{[kite diagram with vertices B, A; labels k, p, q, $p-q$, $p-k$, $q-k$]}} \;=\; \int \frac{d^D p}{(2\pi)^D}\frac{d^D q}{(2\pi)^D} \frac{1}{\mathbf{p}^2 \mathbf{q}^2 (\mathbf{q}-\mathbf{k})^2 (\mathbf{p}-\mathbf{k})^2 (\mathbf{p}-\mathbf{q})^2}
$$

$$
= \frac{1}{D-4}\int \frac{d^D p}{(2\pi)^D}\frac{d^D q}{(2\pi)^D} \frac{1}{\mathbf{p}^2 (\mathbf{q})^2 (\mathbf{q}-\mathbf{k})^2 (\mathbf{p}-\mathbf{k})^2 (\mathbf{p}-\mathbf{q})^2}
$$
$$
\times \left[-\frac{\mathbf{q}^2}{\mathbf{p}^2} - \frac{(\mathbf{q}-\mathbf{k})^2}{(\mathbf{p}-\mathbf{k})^2} + \frac{(\mathbf{p}-\mathbf{q})^2}{\mathbf{p}^2} + \frac{(\mathbf{p}-\mathbf{q})^2}{(\mathbf{p}-\mathbf{k})^2} \right]. \tag{13.38}
$$

The symmetry of the diagram allows the replacement of the integration variables $\mathbf{p}$ by $\mathbf{p} - \mathbf{k}$, and $\mathbf{q}$ by $\mathbf{q} - \mathbf{k}$, so that we end up with two integrals

$$
\vcenter{\hbox{[kite diagram with vertices B, A; labels k, p, q, $p-q$, $p-k$, $q-k$]}} \;=\; \frac{2}{D-4}\int \frac{d^D p}{(2\pi)^D}\frac{d^D q}{(2\pi)^D} \left[-\frac{1}{\mathbf{p}^2(\mathbf{p}-\mathbf{k})^4 \mathbf{q}^2 (\mathbf{p}-\mathbf{q})^2} + \frac{1}{\mathbf{p}^2(\mathbf{p}-\mathbf{k})^4 \mathbf{q}^2 (\mathbf{q}-\mathbf{k})^2} \right].
$$
$$
\tag{13.39}
$$

The right-hand side corresponds to two diagrams with shrunk lines:

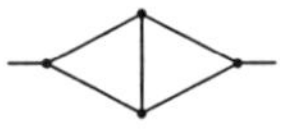

$$\text{} = -\frac{2}{\varepsilon}\left(\quad - \quad \right). \qquad (13.40)$$

Thus the generic two-loop diagram of the KITE type has been reduced to two primitive diagrams with one line index raised by one unit.

13.4.2   General Triangle Rule

The reduction of the triangle integral will be reformulated as a general relation which is applicable to the reduction of all diagrams containing a triangle subdiagram. The general index configuration is specified in Fig. 13.4. Triangle subdiagrams occur in diagrams of the KITE, LADDER, and BENZ type. The method consists in finding all those identities of the type (13.31) which lead to independent reduction formulas of the type (13.37). For this purpose, a generalized differentiation with respect to a loop momentum is introduced. In $\mathbf{x}$-space, the differentiation with respect to the loop momentum generates a factor containing the differences of initial and final coordinates of each line in the loop integral, which add up to zero.

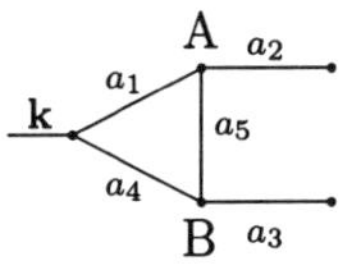

FIGURE 13.4 Subdiagram for the KITE-Type reduction formula.

The integral for the triangle diagram in Fig. 13.4 is denoted by $T(a_1, a_2, a_3, a_4, a_5)$, where a_i are the line indices. Let $\mathbf{p}_i$ be the momenta with line indices a_i. For the moment we assume all momenta to be independent, ignoring momentum conservation at the vertices, i.e., we associate each line with a momentum of its own. In each diagram, there are three integrals to which the three lines of the triangular subdiagram contribute. Now we introduce the operation of differentiation with respect to a closed oriented loop $\mathbf{L}$. For the triangle diagram we define the set of loop momenta and the associated loop derivative:

$$\mathbf{L} = \{\mathbf{p}_1, \mathbf{p}_5, \mathbf{p}_4\}, \qquad \partial_{\mathbf{L}} = \frac{\partial}{\partial \mathbf{p}_1} + \frac{\partial}{\partial \mathbf{p}_5} + \frac{\partial}{\partial \mathbf{p}_4}, \qquad (13.41)$$

where the sign is determined by the orientation of the loop, here clockwise. A vector is chosen

$$\mathbf{P} = \left(\sum_i b_i\, \mathbf{p}_i \right) + b\mathbf{k}, \qquad (13.42)$$

with arbitrary numbers b_i and b, and momenta $\mathbf{p}_i$ of the subdiagram in Fig. 13.4. Since there are three independent loop momenta, or two loop momenta and one external momentum, there is a three-dimensional manifold of choices for $\mathbf{P}$, from which we may pick three linearly independent ones.

A reduction formula for a diagram which contains the triangle is now found from an identity which expresses the vanishing of the surface term:

$$\partial_{\mathbf{L}} \cdot \mathbf{P}\, T\,(a_1, a_2, a_3, a_4, a_5) = 0, \qquad (13.43)$$

this being a generalization of Eq. (13.33). The operation in (13.43) is understood to be performed *inside* the integral over the closed oriented loop L.

This identity holds not only for the triangle diagram itself, but also for any loop integral containing a triangle subdiagram with the loop vector $\mathbf{P}$.

As an example consider the integral of the KITE type. This contains two independent loop momenta and one external momentum, giving three possible independent vectors $\mathbf{P}$. With the two independent loops, we can construct six independent equations. In general, their number is $(L + 1) \times L$. Three other equations are generated by closing the external momenta to form an additional loop $\mathbf{L} = \{\mathbf{k}, \mathbf{p_1}, \mathbf{p_2}\}$. For $\mathbf{P} = \mathbf{k}$, the technical dimension of the integrals $d_t = 2(D - \sum_{i=1}^{I} a_i)$ allows us to conclude that the function $\mathrm{KI}(a_i)$ resulting from the two-loop integral of the KITE type satisfies the identity

$$d_t \mathrm{KI}(a_i) = \left(\mathbf{k} \frac{\partial}{\partial \mathbf{k}}\right) \mathrm{KI}(a_i) = -\mathbf{k} \left(\frac{\partial}{\partial \mathbf{p_1}} + \frac{\partial}{\partial \mathbf{p_2}}\right) \mathrm{KI}(a_i) . \tag{13.44}$$

There exist nine independent equations of this type.

The algorithm for the KITE-type integral treated in detail as an example on page 235, is expressed in this notation by choosing $\mathbf{L} = \{\mathbf{p_1}, \mathbf{p_5}, \mathbf{p_4}\}$, $\mathbf{P} = \mathbf{p_5}$:

$$\mathrm{KI}(a_1, a_2, a_3, a_4, a_5) = (-2a_5 - a_1 - a_4 + D)^{-1} \times \tag{13.45}$$
$$\left[a_1 \Big\{ \mathrm{KI}(a_1 + 1, a_2, a_3, a_4, a_5 - 1) - \mathrm{KI}(a_1 + 1, a_2 - 1, a_3, a_4, a_5) \Big\} \right.$$
$$\left. + a_4 \Big\{ \mathrm{KI}(a_1, a_2, a_3, a_4 + 1, a_5 - 1) - \mathrm{KI}(a_1, a_2, a_3 - 1, a_4 + 1, a_5) \Big\} \right] .$$

This formula leads to a reduction to nested simple loops if the line indices a_2, a_3, and a_5 are integer. It reduces to Eqs. (13.38) and (13.39) for $a_i = 1$. An analogous reduction can be performed with all diagrams of LADDER and BENZ types in Fig. 13.1, since these contain triangular subdiagrams.

For line indices satisfying the relation $2a_5 + a_1 + a_4 = 4 + \mathcal{O}(\varepsilon)$, the above reduction of $\mathrm{KI}(a_1, \ldots)$ produces a pole in ε multiplied by a combination of primitive two-loop diagrams. We may express this schematically as

$$\mathrm{KI}(a_1, \ldots) = \frac{1}{\varepsilon} \sum a_i L(a_k, a_l) \times L(a_p, a_q) , \quad \text{for} \quad 2a_5 + a_1 + a_4 = 4 + \mathcal{O}(\varepsilon) . \tag{13.46}$$

A similar property is shared by all formulas derived by partial integration. The pole requires an expansion of the loop functions $L(a, b)$ to higher orders.

Repeated application of the reduction formula (13.43) allows us to reduce any integer line index a_2, a_3, or a_5 to zero. This is possible for arbitrary line indices a_1, a_4. The restriction to integer line indices is important. A zero index corresponds to shrinking the corresponding line, and results in a primitive diagram. Insertions of primitive loops which give rise to noninteger line indices should therefore lie on the lines 1 and 4. Unfortunately, the so-called special KITE type, $\mathrm{KI}(a_1, \ldots, n + m\varepsilon)$, with $a_1, \ldots, a_4$ integer, cannot be reduced by the algorithm. This type is encountered when reducing generic three-loop diagrams. It has to be calculated by other methods.

For generic four- and five-loop diagrams, no reduction algorithms are at present available. The reason is the appearance of more complicated subdiagrams than triangle diagrams, for example the square subdiagram, which cannot be reduced by partial integration.

The only type of generic diagrams in Fig. 13.1 where the triangle rule cannot be used is the NONPLANAR type, which does not contain any triangle. But it can be applied to the diagrams LADDER and BENZ. The algorithms found for these two correspond to each other if the line indices are mapped as in Fig. 13.5. The so-called *mapping* [1] transforms the external

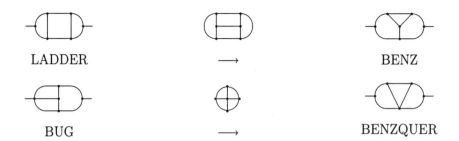

momentum into a loop momentum by identifying its ends to a new loop (see Fig. 13.5). This representation gives full account to the integral's invariance under differentiation with respect to the external momentum as well as with respect to the loop momentum. The momentum conservation at each single vertex is unchanged even if the diagram is cut at a different line. Like the vectors $\mathbf{P}$ and the loops $\mathbf{L}$, the $(L+1)$ equations of the two diagrams can also be transformed into one another by mapping. The same relationship holds for the two diagrams which emerge from one of the generic diagrams listed above by shrinking a line, as can be seen in Fig. 13.5. The reduction formulas for these diagrams, the BEQ and the BUG type, follow from those of the LA or the BE type by setting to zero the corresponding line index.

13.4.3   Reduction Algorithms

The levels of reduction for the generic three-loop diagrams LA and BE will now be displayed explicitly. The algorithms transform the diagrams of level I into those of level II and III (see Fig. 13.6). Those of level II are further reduced to level III and primitive diagrams. The diagrams of level III are reduced to primitive diagrams. The diagram ONE1 cannot be reduced, but can be reduced to the standard expression $KI(1, 1, 1, 1, m\varepsilon/2)L(1, 1)$.

The following reduction algorithms [4] are used in our calculation. The algorithms reduce the integral types KITE, LADDER, BENZ and BENZQUER algebraically to primitive integrals and to the special KITE type. They also reduce the special KITE type $KI\,(n_1, n_2, n_3, n_4, n_5 + m\varepsilon/2)$ with n_i, m integer to the standard form. Other algorithms to bring the NO integral to its standard form with $a_i = 1$ are not listed here. They have not been used in the five-loop calculations because the only time the NO type appears it does so in the standard form.

The notation corresponds to the listing of the diagrams on page 239, the additional

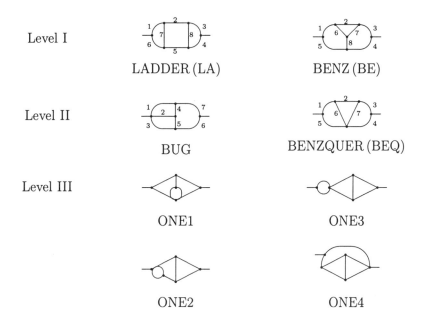

FIGURE 13.6 The diagrams which are generated in the reduction of LADDER- and BUG-type diagrams. The reduction works from Level I down to level III. Application of the reduction algorithms for the KITE-type to the diagrams of level III leads to primitive diagrams.

generation of primitive diagrams of level II and for the KITE type is not explicitly stated. The results are:

<u>LADDER</u>: $\mathrm{LA} = \mathcal{O}(1/\varepsilon)\,\mathrm{BEQ} + \mathcal{O}(1/\varepsilon)\,\mathrm{ONE3}$, $a_2,\ a_5,\ a_8$ integer.

$$
\mathrm{LA}(a_1,\ldots,a_8) \;=\; (D - 2a_8 - a_3 - a_4)^{-1} \times \tag{13.47}
$$

$$
\Big[\; a_4 \mathrm{LA}(a_1, a_2, a_3, a_4 + 1, a_5, a_6, a_7, a_8 - 1)
$$
$$
- a_4 \mathrm{LA}(a_1, a_2, a_3, a_4 + 1, a_5 - 1, a_6, a_7, a_8)
$$
$$
+ a_3 \mathrm{LA}(a_1, a_2, a_3 + 1, a_4, a_5, a_6, a_7, a_8 - 1)
$$
$$
- a_3 \mathrm{LA}(a_1, a_2 - 1, a_3 + 1, a_4, a_5, a_6, a_7, a_8)\Big]
$$

<u>BENZ</u>: $\mathrm{BE} = \mathcal{O}(1/\varepsilon)\,\mathrm{BUG} + \mathcal{O}(1/\varepsilon)\,\mathrm{ONE1}$, $a_1,\ a_2,\ a_3$ integer.

$$
\mathrm{BE}(a_1,\ldots,a_8) \;=\; (D - 2a_2 - a_6 - a_7)^{-1} \times \tag{13.48}
$$

$$
\Big[\; a_6 \mathrm{BE}(a_1, a_2 - 2, a_3, a_4, a_5, a_6 + 1, a_7, a_8)
$$
$$
- a_6 \mathrm{BE}(a_1 - 1, a_2, a_3, a_4, a_5, a_6 + 1, a_7, a_8)
$$
$$
+ a_7 \mathrm{BE}(a_1, a_2 - 1, a_3, a_4, a_5, a_6, a_7 + 1, a_8)
$$
$$
- a_7 \mathrm{BE}(a_1, a_2, a_3 - 1, a_4, a_5, a_6, a_7 + 1, a_8)\Big]
$$

BENZQUER: $\text{BEQ} = \mathcal{O}(1/\varepsilon)\,\text{ONE1} + \mathcal{O}(1/\varepsilon)\,\text{ONE2}$, $a_1,\,a_2,\,a_3$ integer.

$$\text{BENZQUER}(a_1,\dots,a_7) = \text{BENZQUER}(a_1,\dots,a_7,0) \tag{13.49}$$

BUG: $\text{BUG} = \mathcal{O}(1/\varepsilon)\,\text{ONE2} + \mathcal{O}(1/\varepsilon)\,\text{ONE4}$, $a_2,\,a_4,\,a_5,\,a_7$ integer.

$$\text{BUG}(a_1,\dots,a_8) = (D - 2a_5 - a_2 - a_3)^{-1} \times \tag{13.50}$$

$$\left[\; a_2\text{BU}(a_1, a_2+1, a_3, a_4, a_5-1, a_6, a_7, a_8)\right.$$
$$- a_2\text{BU}(a_1, a_2+1, a_3, a_4-1, a_5, a_6, a_7, a_8)$$
$$+ a_3\text{BU}(a_1, a_2, a_3+1, a_4, a_5-1, a_6, a_7, a_8)$$
$$\left. - a_3\text{BU}(a_1, a_2-1, a_3+1, a_4, a_5, a_6, a_7-1, a_8)\right]$$

KITE-TYPE: $\text{KI} = \mathcal{O}(1/\varepsilon) \times L \times L$, $a_2,\,a_3,\,a_5$ integer.

$$\text{KI}(a_1,\dots,a_5) = (D - 2a_5 - a_1 - a_4)^{-1} \times \tag{13.51}$$

$$\left[\; a_1\text{KI}(a_1+1, a_2, a_3, a_4, a_5-1)\right.$$
$$- a_1\text{KI}(a_1+1, a_2-1, a_3, a_4, a_5)$$
$$+ a_4\text{KI}(a_1, a_2, a_3, a_4+1, a_5-1)$$
$$\left. - a_4\text{KI}(a_1, a_2, a_3-1, a_4+1, a_5)\right]$$

SPECIAL KITE-TYPE:

$$\text{KI}\left(a_1,\dots,a_4, n+\tfrac{m}{2}\varepsilon\right) = \mathcal{O}(1/\varepsilon)\,\text{KI}\left(1,1,1,1, n+\tfrac{m}{2}\varepsilon\right),\; a_1,\dots,a_4, m, n \text{ integer.}$$

$$(a_4-1)\text{KI}(a_1, a_2, a_3, a_4, a_5) = \left[(2a_1 + a_4 + a_5 - D - 1)\text{KI}(a_1, a_2, a_3, a_4-1, a_5)\right.$$
$$+ (a_4-1)\text{KI}(a_1-1, a_2, a_3, a_4, a_5)$$
$$+ a_5\Big\{\text{KI}(a_1-1, a_2, a_3, a_4-1, a_5+1)$$
$$\left. - \text{KI}(a_1, a_2-1, a_3, a_4-1, a_5+1)\Big\}\right] \tag{13.52}$$

$$\text{KI}(1,1,1,1, a_5) = (a_5+2-D)(a_5+1-D/2)^{-1}\text{KI}(1,1,1,1, a_5-1)$$
$$+ 2(3D - 10 - 2a_5)L(1, a_5)\,L(1, a_5+2-D/2). \tag{13.53}$$

These algorithms allow us to find analytic expression for the divergent part of all four-loop diagrams, and of most of the five-loop diagrams of the ϕ^4-theory once $\text{KI}(1,1,1,1, m\varepsilon/2)$ is calculated to a sufficient order in ε, which will be done in the next section.

The five-loop diagrams which cannot yet be calculated in this manner contain either generic four-loop diagrams or subdiagrams of the NONPLANAR type. They will be shown in Fig. 13.10. The expansion for $\text{KI}(1,1,1,1, m\varepsilon/2)$ is given in the next section in Eq. (13.75). It cannot be calculated with the momentum space methods presented so far. The same problem exists for the BUG- and BEQ-type integrals with noninteger line indices. Fortunately, up to five loops the calculation of the last two can be circumvented using an appropriate infrared rearrangement, and removing the spurious IR-divergences.

A direct calculation of these diagrams is possible in configuration space. This was first done numerically with the help of the so-called *Gegenbauer polynomial x-space method* [2]. Each propagator is expanded into a set of orthogonal polynomials, the Gegenbauer polynomials. Integration were carried out resulting in multiple series, which were summed numerically on a computer.

Analytical expressions were finally found by the *method of ideal index constellations*, which was developed by Kazakov [3] and which will now be explained.

13.5 Method of Ideal Index Constellations in Configuration Space

The method [3, 9] consists of the combined application of several characteristic identities to diagrams with certain ideal line index constellations. To reach such constellations, the indices are manipulated, after a Fourier transformation, by a configuration space version of the triangle rule of the last section, or by the insertion of a ϕ^2-vertex. This method succeeds in giving analytic expressions for the missing diagrams up to fifth order in ε. An important tool among the following methods is the *duality transformation*.

13.5.1 Dual Diagrams

In Eq. (13.2), we have observed that the Fourier transformation of a massless propagator yields a pure power in $\mathbf{x}$. This has the pleasant consequence that the Fourier transform of an arbitrary massless Feynman integral is again a massless Feynman integral, but has a different mathematical form and is expressed in terms of $\mathbf{x}$ instead of $\mathbf{p}$. If this $\mathbf{x}$-space integral is diagrammatically represented in the same way as a $\mathbf{p}$-space integral, the resulting diagram will be called the *dual Feynman diagram*. It is obvious, that the propagator-type integral of a diagram and the $\mathbf{x}$-space integral of its dual diagram are of the same form. The line indices α of the dual diagram are related to the indices a of the original one by $\alpha = D/2 - a$. Examples for dual diagrams are shown in Fig. 13.7, the integrals of the BUG and the BEQ type are dual to each other, while the integral of the KITE type is dual to itself.

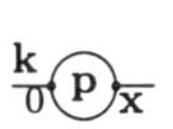

$$\mathbf{k}\underset{}{\bigcirc}\mathbf{x} \quad \sim \int \frac{d^D p}{(2\pi)^D \mathbf{p}^{2a}(\mathbf{p}-\mathbf{k})^{2b}} \qquad\qquad \int \frac{d^D x\, e^{i\mathbf{k}\mathbf{x}}}{(\mathbf{x}^2)^{\alpha+\beta}}$$

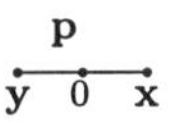

$$\underset{\mathbf{y}\ \ 0\ \ \mathbf{x}}{\xrightarrow{\ \ \mathbf{p}\ \ }} \quad \sim \int \frac{d^D p\, e^{-i\mathbf{p}\mathbf{x}}}{(2\pi)^D (\mathbf{p}^2)^{a+b}} \qquad\qquad \int \frac{d^D x}{\mathbf{x}^{2\alpha}(\mathbf{x}-\mathbf{y})^{2\beta}}$$

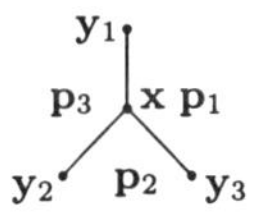

$$\underset{\mathbf{y}_2\ \ \mathbf{p}_2\ \ \mathbf{y}_3}{\overset{\mathbf{y}_1}{\underset{\mathbf{p}_3\ \ \mathbf{x}\ \ \mathbf{p}_1}{}}} \quad \sim \int \frac{(2\pi)^{-3D} d^D p_2\, d^D p_2\, d^D p_3}{(\mathbf{p}_1-\mathbf{p}_3)^{2a}(\mathbf{p}_2-\mathbf{p}_3)^{2b}(\mathbf{p}_1-\mathbf{p}_2)^{2c}} \qquad \int \frac{d^D x}{(\mathbf{x}-\mathbf{y}_1)^{2\alpha}(\mathbf{x}-\mathbf{y}_2)^{2\beta}(\mathbf{x}-\mathbf{y}_3)^{2\gamma}}$$

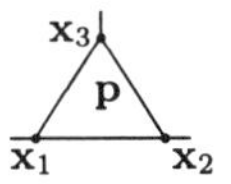

$$\underset{\mathbf{x}_1\ \ \ \ \mathbf{x}_2}{\overset{\mathbf{x}_3}{\triangle}_{\mathbf{p}}} \quad \sim \int \frac{d^D p}{(2\pi)^D \mathbf{p}^{2a}(\mathbf{p}-\mathbf{q})^{2b}(\mathbf{p}-\mathbf{k})^{2c}} \qquad \int \frac{d^D x_1\, d^D x_2\, d^D x_3}{(\mathbf{x}_1-\mathbf{x}_3)^{2\alpha}(\mathbf{x}_2-\mathbf{x}_3)^{2\beta}(\mathbf{x}_1-\mathbf{x}_2)^{2\gamma}}$$

FIGURE 13.7 The $\mathbf{p}$- and $\mathbf{x}$-space representations of the integrals and the corresponding diagrams. The first two diagrams are dual to each other and so are the last two. The $\mathbf{x}$-space integral corresponds mathematically to the $\mathbf{p}$-space integral of the dual diagram and vice versa.

Loops and Chains

Repeating here the Fourier transformation of the simple loop of Eqs. (13.3) and (13.4),

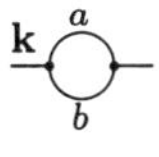

$$\mathbf{k}\underset{b}{\overset{a}{\bigcirc}} \sim \int \frac{d^D p}{(2\pi)^D} \frac{1}{(\mathbf{p}^2)^a[(\mathbf{p}-\mathbf{k})^2]^b} \tag{13.54}$$

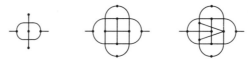

FIGURE 13.8 Construction of dual diagrams

$$= \frac{\Gamma(D/2-a)\Gamma(D/2-b)}{\pi^D \Gamma(a)\Gamma(b) 4^{a+b}} \int d^D x \, \frac{e^{ikx}}{[(\mathbf{x})^2]^{D/2-a}[(\mathbf{x})^2]^{D/2-b}} , \qquad (13.55)$$

we see that the convolution integral with the diagrammatic representation of a loop is transformed in $\mathbf{x}$-space into a product of two identical propagators, causing a collapse of the two lines to a single line whose index is the sum of the original line indices. If the resulting $\mathbf{x}$-space integral is represented diagrammatically as if it were a $\mathbf{p}$-space integral, we have the dual diagram:

$$\underset{\alpha_1 \quad \alpha_2}{\bullet\!\!-\!\!\bullet\!\!-\!\!\bullet} \sim \frac{1}{[(\mathbf{x})^2]^{\alpha_1}[(\mathbf{x})^2]^{\alpha_2}} = \frac{1}{[(\mathbf{x})^2]^{\alpha_1+\alpha_2}} \sim \underset{\alpha_1+\alpha_2}{\bullet\!\!-\!\!\bullet} . \qquad (13.56)$$

The diagram on the left-hand side is called a *chain diagram*. We can also start from the propagator-type integral of the chain diagram, in which case we have to go in the opposite direction from (13.55) to (13.54). If the chain has one link, as in this case, the $\mathbf{x}$-space representation corresponds to a convolution integral of the type (13.54), which can be calculated with the same result as the loop integral (13.4) in momentum space:

$$\int d^D y \frac{1}{[(\mathbf{x}-\mathbf{y})^2]^{\alpha_1}[(\mathbf{y}-\mathbf{z})^2]^{\alpha_2}} = \pi^{D/2} \frac{\nu(\alpha_1)\nu(\alpha_2)\nu(D-\alpha_1-\alpha_2)}{[(\mathbf{x}-\mathbf{z})^2]^{(\alpha_1+\alpha_2-D/2)}} . \qquad (13.57)$$

Another type of diagram arising from the duality transformation of the $\mathbf{p}$-space triangle diagram in Fig. 13.7 is an integral over the variable $\mathbf{x}$ of a ϕ^3-vertex. It is called a *star diagram* and the corresponding integral in $\mathbf{x}$-space reads

$$\hat{=} \int d^D x \, \frac{1}{[(\mathbf{x}-\mathbf{x}_1)^2]^{\alpha_1}[(\mathbf{x}-\mathbf{x}_2)^2]^{\alpha_2}[(\mathbf{x}-\mathbf{x}_3)^2]^{\alpha_3}} . \qquad (13.58)$$

As visible in Fig. 13.7, this $\mathbf{x}$-space integral has the same form as the momentum space triangle integral in Eq. (13.32) except for the indices, and it can also be treated by the reduction formula (13.37) found for diagrams of this form. The star diagram will be considered further in the next subsection.

Generally, the dual diagram is constructed by placing a vertex into each loop and outside of it, and connecting them in such a way that each line of the original diagram is crossed once. In this way one vertex is associated with each loop. This corresponds to the transition from loop integration to $\mathbf{x}$-integration in the Fourier transformation. In momentum space, the lines in chain diagrams collapse when adding the indices. In $\mathbf{x}$-space, the same thing happens for the lines of the simple loop.

We will see that the duality transformation is sometimes helpful in creating calculable diagrams.

13.5.2   Star-Triangle Rule for an Ideal Vertex

The $\mathbf{x}$-space integral for the star diagram

$$\hat{=} \int d^D x \, \frac{1}{[(\mathbf{x}-\mathbf{x}_1)^2]^{\alpha_1}[(\mathbf{x}-\mathbf{x}_2)^2]^{\alpha_2}[(\mathbf{x}-\mathbf{x}_3)^2]^{\alpha_3}} \qquad (13.59)$$

is simplified if one of the line indices is zero, thus producing a chain. Another simplification occurs for index constellations with $\sum \alpha_i = D$, in which case we speak of an *ideal vertex*. The simplification is the content of the *star-triangle rule*, which is derived as follows. Translation of the variable of integration $\mathbf{x} \to \mathbf{x}' = \mathbf{x} + \mathbf{x}_1$ gives

$$\int d^D x \, \frac{1}{(\mathbf{x}^2)^{\alpha_1}[(\mathbf{x} + \mathbf{x}_1 - \mathbf{x}_2)^2]^{\alpha_2}[(\mathbf{x} + \mathbf{x}_1 - \mathbf{x}_3)^2]^{\alpha_3}} \, . \tag{13.60}$$

Inversion of the variables of integration: $\mathbf{x} \to \mathbf{x}' = \mathbf{x}/\mathbf{x}^2$, $d^D x \to d^D x' = d^D x/(\mathbf{x}^2)^D$ such that

$$[(\mathbf{x} - \mathbf{y})^2]^{\alpha} \to \left[\left(\frac{\mathbf{x}}{\mathbf{x}^2} - \mathbf{y}\right)^2\right]^{\alpha} = \frac{(1 - 2\mathbf{x}\mathbf{y} + \mathbf{x}^2\mathbf{y}^2)^{\alpha}}{(\mathbf{x}^2)^{\alpha}} = \frac{[(\mathbf{x} - \frac{\mathbf{y}}{\mathbf{y}^2})^2]^{\alpha}(\mathbf{y}^2)^{\alpha}}{(\mathbf{x}^2)^{\alpha}}, \tag{13.61}$$

results in

$$\int d^D x \, \frac{1}{(\mathbf{x}^2)^D} \left\{ \left(\frac{\mathbf{x}^2}{\mathbf{x}^4}\right)^{\alpha_1} \left[\frac{\left(\mathbf{x} - \frac{\mathbf{x}_1 - \mathbf{x}_2}{(\mathbf{x}_1 - \mathbf{x}_2)^2}\right)^2 (\mathbf{x}_1 - \mathbf{x}_2)^2}{\mathbf{x}^2}\right]^{\alpha_2} \left[\frac{\left(\mathbf{x} - \frac{\mathbf{x}_1 - \mathbf{x}_3}{(\mathbf{x}_1 - \mathbf{x}_3)^2}\right)^2 (\mathbf{x}_1 - \mathbf{x}_3)^2}{\mathbf{x}^2}\right]^{\alpha_3} \right\}^{-1}$$

$$= \int d^D x \left[(\mathbf{x}^2)^{D - \alpha_1 - \alpha_2 - \alpha_3} \left[\left(\mathbf{x} - \frac{\mathbf{x}_1 - \mathbf{x}_2}{(\mathbf{x}_1 - \mathbf{x}_2)^2}\right)^2\right]^{\alpha_2} \left[\left(\mathbf{x} - \frac{\mathbf{x}_1 - \mathbf{x}_3}{(\mathbf{x}_1 - \mathbf{x}_3)^2}\right)^2\right]^{\alpha_3} [(\mathbf{x}_1 - \mathbf{x}_2)^2]^{\alpha_2}[(\mathbf{x}_1 - \mathbf{x}_3)^2]^{\alpha_3}\right]^{-1} \, .$$

For $\sum \alpha_i = D$ the line index of $\mathbf{x}^2$ vanishes, and we are left with

$$\int d^D x \left\{ \left[\left(\mathbf{x} - \frac{\mathbf{x}_1 - \mathbf{x}_2}{(\mathbf{x}_1 - \mathbf{x}_2)^2}\right)^2\right]^{\alpha_2} \left[\left(\mathbf{x} - \frac{\mathbf{x}_1 - \mathbf{x}_3}{(\mathbf{x}_1 - \mathbf{x}_3)^2}\right)^2\right]^{\alpha_3} [(\mathbf{x}_1 - \mathbf{x}_2)^2]^{\alpha_2} [(\mathbf{x}_1 - \mathbf{x}_3)^2]^{\alpha_3} \right\}^{-1} \, .$$

The integration over $\mathbf{x}$ is a chain integration which may be carried out with the help of Eq. (13.57) leading to

$$\frac{\pi^{D/2} \, \nu(\alpha_1) \, \nu(\alpha_2) \, \nu(\alpha_3)}{\left[\left(\frac{\mathbf{x}_1 - \mathbf{x}_2}{(\mathbf{x}_1 - \mathbf{x}_2)^2} - \frac{\mathbf{x}_1 - \mathbf{x}_3}{(\mathbf{x}_1 - \mathbf{x}_3)^2}\right)^2\right]^{\alpha_2 + \alpha_3 - \frac{D}{2}} [(\mathbf{x}_1 - \mathbf{x}_2)^2]^{\alpha_2}[(\mathbf{x}_1 - \mathbf{x}_3)^2]^{\alpha_3}} \, . \tag{13.62}$$

With

$$D = \alpha_1 + \alpha_2 + \alpha_3 \Rightarrow \begin{cases} \alpha_2 &= (\frac{D}{2} - \alpha_3) + (\frac{D}{2} - \alpha_1) \\ \alpha_3 &= (\frac{D}{2} - \alpha_2) + (\frac{D}{2} - \alpha_1) \, , \\ \alpha_2 + \alpha_3 - \frac{D}{2} &= \frac{D}{2} - \alpha_1 \end{cases} \tag{13.63}$$

this can be rewritten to:

$$\frac{\pi^{D/2} \, \nu(\alpha_1) \, \nu(\alpha_2) \, \nu(\alpha_3)}{\left[\left(\frac{\mathbf{x}_1 - \mathbf{x}_2}{(\mathbf{x}_1 - \mathbf{x}_2)^2} - \frac{\mathbf{x}_1 - \mathbf{x}_3}{(\mathbf{x}_1 - \mathbf{x}_3)^2}\right)^2\right]^{\frac{D}{2} - \alpha_1} [(\mathbf{x}_1 - \mathbf{x}_2)^2]^{\frac{D}{2} - \alpha_1}[(\mathbf{x}_1 - \mathbf{x}_3)^2]^{\frac{D}{2} - \alpha_1}[(\mathbf{x}_1 - \mathbf{x}_2)^2]^{\frac{D}{2} - \alpha_3}[(\mathbf{x}_1 - \mathbf{x}_3)^2]^{\frac{D}{2} - \alpha_2}} \, .$$

Finally, we obtain the star-triangle rule:

$$\int \frac{d^D x}{[(\mathbf{x} - \mathbf{x}_1)^2]^{\alpha_1}[(\mathbf{x} - \mathbf{x}_2)^2]^{\alpha_2}[(\mathbf{x} - \mathbf{x}_3)^2]^{\alpha_3}}$$

$$\stackrel{\sum \alpha_i = D}{=} \frac{\pi^{D/2} \, \nu(\alpha_1) \, \nu(\alpha_2) \, \nu(\alpha_3)}{[(\mathbf{x}_2 - \mathbf{x}_3)^2]^{\frac{D}{2} - \alpha_1}[(\mathbf{x}_1 - \mathbf{x}_2)^2]^{\frac{D}{2} - \alpha_3}[(\mathbf{x}_1 - \mathbf{x}_3)^2]^{\frac{D}{2} - \alpha_2}} \, . \tag{13.64}$$

This relation allows us to carry out the integration over ϕ^3-vertices (stars) if their indices satisfy $\sum \alpha_i = D$. The diagrammatic expression is

$$\underset{\mathbf{x}_2 \qquad \mathbf{x}_3}{\overset{\mathbf{x}_1}{\vphantom{|}}} \overset{\sum \alpha_i = D}{=\!=} \pi^{D/2} \nu(\alpha_1)\nu(\alpha_2)\nu(\alpha_3) \underset{\mathbf{x}_2 \quad \frac{D}{2}-\alpha_1 \quad \mathbf{x}_3}{\overset{\frac{D}{2}-\alpha_3 \quad \frac{D}{2}-\alpha_2}{\triangle}} \, .$$

An ideal index constellation admits a simple integration and exists for the star, the triangle, and the chain:

$$A \left\{ \begin{array}{c} \text{line} \\ \text{vertex} \\ \text{triangle} \end{array} \right\} \quad \text{is ideal if} \quad \left\{ \begin{array}{ccc} \alpha & = & 0 \\ \sum \alpha_i & = & D \\ \sum \alpha_i & = & D/2 \end{array} \right. . \tag{13.65}$$

The star-triangle relation transforms an ideal vertex into an ideal triangle. A diagram containing such a vertex can thereby be reduced to primitive diagrams. For example, the application of the star-triangle relation reduces the KITE type to a primitive diagram if a vertex or a triangle is ideal. For the KITE-type diagram in $\mathbf{x}$-space, we find

$$\overset{\alpha_1 \qquad \alpha_2}{\underset{\alpha_4 \qquad \alpha_3}{\boxed{\alpha_5}}} \overset{\alpha_1+\alpha_2+\alpha_5=D}{=\!=} \overset{\frac{D}{2}-\alpha_5}{\underset{\alpha_4 \quad \alpha_3}{\boxed{\frac{D}{2}\alpha_2 \ \frac{D}{2}\alpha_1}}} \quad \nu(\alpha_1)\nu(\alpha_2) \underbrace{\nu(\alpha_5)}_{\nu(D-\alpha_1-\alpha_2)} \tag{13.66}$$

$$= \overset{\frac{D}{2}-\alpha_5}{\underset{\frac{D}{2}-\alpha_2+\alpha_4 \ \frac{D}{2}-\alpha_1+\alpha_3}{\bigcirc}} \quad \underbrace{\nu(\alpha_1)\nu(\alpha_2)\nu(D-\alpha_1-\alpha_2)}_{=L(\alpha_1,\alpha_2)/(4\pi)^{\epsilon/2}} \tag{13.67}$$

$$\overset{\alpha_1 \qquad \alpha_2}{\underset{\alpha_4 \qquad \alpha_3}{\boxed{\alpha_5}}} \overset{\alpha_2+\alpha_3+\alpha_5=\frac{D}{2}}{=\!=} \overset{\alpha_1 \quad \frac{D}{2}-\alpha_3}{\underset{\alpha_4 \quad \frac{D}{2}-\alpha_2}{\bigcirc \frac{D}{2}-\alpha_5}} \quad \nu(\alpha_2)\nu(\alpha_3)\nu(\alpha_5) \, . \tag{13.68}$$

$$\left[L(\tfrac{D}{2}-\alpha_2, \tfrac{D}{2}-\alpha_3) \overset{\alpha_2+\alpha_3+\alpha_5=\frac{D}{2}}{=\!=} \nu(\tfrac{D}{2}-\alpha_2)\nu(\tfrac{D}{2}-\alpha_3)\nu(\tfrac{D}{2}-\alpha_5) = \nu^{-1}(\alpha_2)\nu^{-1}(\alpha_3)\nu^{-1}(\alpha_5) \right] . \tag{13.69}$$

The resulting diagrams are primitive $\mathbf{x}$-space diagrams, and can be calculated using Eqs. (13.56) and (13.57). Transformation to momentum space converts loops into chains and vice versa.

13.5.3 One Step from Ideal Index Constellation

In momentum space the method of partial integration has led to the reduction formula (13.37) for the triangle integral. We are now going to transform this relation into $\mathbf{x}$-space where the triangle integral becomes an integral of the star type:

$$\underset{\mathbf{x}_2 \qquad \mathbf{x}_3}{\overset{\mathbf{x}_1}{\vphantom{|}}} \ \hat{=} \ \int \frac{d^D x}{[(\mathbf{x}-\mathbf{x}_1)^2]^{\alpha_1}[(\mathbf{x}-\mathbf{x}_2)^2]^{\alpha_2}[(\mathbf{x}-\mathbf{x}_3)^2]^{\alpha_3}} \, , \tag{13.70}$$

and the triangle relation (13.37) takes the following form:

$$\overset{\alpha_1}{\underset{\alpha_2 \ \alpha_3}{\triangle}} = \frac{1}{D-2\alpha_1-\alpha_2-\alpha_3} \left(\alpha_2 \overset{\alpha_1-1}{\underset{\alpha_2+1 \ \alpha_3}{\triangle}} + \alpha_3 \overset{\alpha_1-1}{\underset{\alpha_2 \ \alpha_3+1}{\triangle}} - \alpha_2 \overset{\alpha_1}{\underset{\alpha_2+1 \ \alpha_3}{\triangle}}^{-1} - \alpha_3 \overset{\alpha_1}{\underset{\alpha_2 \ \alpha_3+1}{\triangle}}^{-1} \right) . \tag{13.71}$$

This relation can be used for any arrangement of indices, but for particular ones it simplifies considerably. If the vertex on the left-hand side of (13.71) is *one step from ideal*, i.e. if the indices satisfy $\alpha_1 + \alpha_2 + \alpha_3 = D - 1$, the relation turns into

$$
\begin{array}{c}
{}^{\alpha_1}\!\!\!\bigwedge_{\alpha_2\ \ \alpha_3}^{\alpha_1+\alpha_2+\alpha_3=D-1}
- \frac{\alpha_2}{\alpha_1-1}\ {}^{\alpha_1-1}\!\!\!\bigwedge_{\alpha_2+1\ \ \alpha_3}
- \frac{\alpha_3}{\alpha_1-1}\ {}^{\alpha_1-1}\!\!\!\bigwedge_{\alpha_2\ \ \alpha_3+1}
\end{array}
\tag{13.72}
$$

$$
+\ \alpha_2\alpha_3\nu(\alpha_1)\nu(\alpha_2+1)\nu(\alpha_3+1)\ \
\begin{array}{c}
\mathbf{x_1}\\
{}^{\frac{D}{2}-\alpha_3-1}\!\!\!\bigtriangleup^{\frac{D}{2}-\alpha_2-1}\\
\mathbf{x_2}\ \ {}_{\frac{D}{2}-\alpha_1}\ \ \mathbf{x_3}
\end{array}\ ,
$$

where we used $\alpha_2\alpha_3\nu(\alpha_1)\nu(\alpha_2+1)\nu(\alpha_3+1) = \frac{\alpha_2}{\alpha_1-1}\nu(\alpha_1)\nu(\alpha_2+1)\nu(\alpha_3) + \frac{\alpha_3}{\alpha_1-1}\nu(\alpha_1)\nu(\alpha_2)\nu(\alpha_3 + 1)$. If, in addition, both triangles are one step from an ideal index constellation $D/2 + 1$, this relation transforms the KITE-type diagram into integrable diagrams. For indices of the form $\alpha_i = 1 + c_i$ we have $\alpha_3 + \alpha_4 + \alpha_5 = D - 1$, $\alpha_1 + \alpha_4 + \alpha_5 = D/2 + 1$, and $\alpha_2 + \alpha_3 + \alpha_5 = D/2 + 1$, such that:

$$
\begin{array}{c}
{}^{1+c_1}\ \ {}^{1+c_2}\\
\multimap\!\!\!\bigcirc\!\!|_{1+c_5}\!\!\bigcirc\!\!\multimap\\
{}_{1+c_4}\ \ {}_{1+c_3}
\end{array}
=
\frac{1-c_4}{c_5}
\begin{array}{c}
{}^{1+c_1}\ \ {}^{1+c_2}\\
\multimap\!\!\!\bigcirc\!\!|_{c_5}\!\!\bigcirc\!\!\multimap\\
{}_{2+c_4}\ \ {}_{1+c_3}
\end{array}
-
\frac{1+c_3}{c_5}
\begin{array}{c}
{}^{1+c_1}\ \ {}^{1+c_2}\\
\multimap\!\!\!\bigcirc\!\!|_{c_5}\!\!\bigcirc\!\!\multimap\\
{}_{1+c_4}\ \ {}_{2+c_3}
\end{array}
\tag{13.73}
$$

$$
+(1 + c_3)(1 + c_4)\nu(1 + c_5)\nu(2 + c_3)\nu(2 + c_4)\ \
\begin{array}{c}
{}^{1+c_1}\ \ {}^{1+c_2}\\
\bigcirc\!\!\!\bigcirc\\
{}_{\frac{D}{2}-2-c_3}\ \ {}_{\frac{D}{2}-2-c_4}\\
{}_{\frac{D}{2}-1-c_5}
\end{array}\ .
$$

The first diagram on the right-hand side can now be integrated because the right triangle has got an ideal index constellation. In the second diagram, it is the ideal index constellation of the left triangle, which allows application of formula (13.64). The last diagram can be solved easily via Eqs. (13.56) and (13.57).

13.5.4 Transformation of Indices

In general, Eq. (13.71) solves the integral if three elements (lines, vertices or triangles) are one step from an ideal index constellation [9, 3]. For example, if three lines have the index 1, or two triangles have $D/2 + 1$ and one vertex has $D - 1$. Otherwise relation (13.71) is used to manipulate indices in order to reach an ideal index constellation. Simple lines with an index $a = 1$ in momentum space and $\alpha = D/2 - 1 = 1 - \varepsilon/2$ in $\mathbf{x}$-space do not produce ideal index constellations. These properties are summarized in the following table:

	ideal	one step from ideal	simple lines
line	0	1	$1 - \varepsilon/2$
vertex	$4 - \varepsilon$	$3 - \varepsilon$	$3 - 3\varepsilon/2$
triangle	$2 - \varepsilon/2$	$3 - \varepsilon/2$	$3 - 3\varepsilon/2$

An index transformation of order ε is required to reach an ideal index constellation. Apart from Eq. (13.71) the index constellation can also be changed by an insertion of a ϕ^2-vertex using Eq. (13.57), which changes a line into two lines with modified indices, generating an ideal index constellation in the adjacent vertex. Another option is the inversion of the integration variables, $x_\mu \longrightarrow x_\mu/x^2$. Finally, we can use the dual diagram with its transformed indices instead of the original integral.

13.5.5 Construction of Tables

The above rules (13.56), (13.57), (13.64) and (13.71) can be used to calculate the KITE-type and the BEQ- or BUG-type diagram for certain index constellations as in Eq. (13.73). For example, the application of the star-triangle relation reduces the KITE-type to a primitive diagram if a vertex or a triangle has an ideal index constellation. With the results for several index constellations, functional equations for the ε-expansion of the KITE-type or the BEQ-type integral depending on the line indices can be derived [3, 9]. The solutions of these equations are given in tables for indices of the form $1 + a_i \varepsilon$. The KITE-type table was calculated by Kazakov up to ε^4. For the diagram

$$
\begin{array}{c}
1+a_1\varepsilon \quad 1+a_2\varepsilon \\
\text{—}\bigcirc\!\!\!\boxed{1+a_5\varepsilon}\!\!\!\bigcirc\text{—} \\
1+a_4\varepsilon \quad 1+a_3\varepsilon
\end{array}
\;\equiv\; \frac{\text{KI}\,(1+a_1\varepsilon,\ldots,1+a_5\varepsilon)}{(\mathbf{x}^2)^{5+\sum_i a_i\varepsilon - D}}, \tag{13.74}
$$

the expansion has the following form:

$$
\text{KI}\,(1+a_1\varepsilon,\ldots,1+a_5\varepsilon) = \frac{1}{1-\varepsilon}\Bigg[A_0\zeta(3) + \frac{A_1\zeta(4)}{2}\varepsilon + \frac{A_2\zeta(5)}{4}\varepsilon^2 + \frac{A_3\zeta(6)}{8}\varepsilon^3
$$
$$
+ \frac{A_4\zeta^2(3)}{8}\varepsilon^3 - \frac{A_5\zeta(7)}{16}\varepsilon^4 + \frac{A_6\zeta(3)\zeta(4)}{16}\varepsilon^4 + \mathcal{O}(\varepsilon^5)\Bigg], \tag{13.75}
$$

with coefficients A_i depending only on the indices $a_1,\ldots,a_5$ and the combination $a^n = a_1^n + a_2^n + a_3^n + a_4^n$:

$$A_0 = 6, \quad A_1 = 9,$$

$$A_2 = 42 + 30a + 45a_5 + 10a^2 + 15a_5^2 + 15a_5 a + 10(a_1 a_2 + a_3 a_4 + a_1 a_4 + a_2 a_3) + 5(a_1 a_3 + a_2 a_4),$$

$$A_3 = \frac{5}{2}(A_2 - 6),$$

$$
\begin{aligned}
A_4 = {}& 46 + 42a + 45a_5 + 14a^2 + 15a_5^2 + 33a_5 a + 50(a_1 a_2 + a_3 a_4) + 31(a_1 a_3 + a_2 a_4) \\
&+ 14(a_1 a_4 + a_2 a_3) + 6a_5 a^2 + 6a_5^2 a + 24a_5(a_1 a_2 + a_3 a_4) + 12a_5(a_1 a_3 + a_2 a_4) \\
&+ 12(a_1 a_2 a_3 + a_1 a_2 a_4 + a_1 a_3 a_4 + a_2 a_3 a_4) + 12(a_1^2 a_2 + a_2^2 a_1 + a_3^2 a_4 + a_4^2 a_3) \\
&+ 6(a_1^2 a_3 + a_3^2 a_1 + a_2^2 a_4 + a_4^2 a_2),
\end{aligned}
$$

$$
\begin{aligned}
A_5 = {}& 294 + 402a + \frac{2223}{4}a_5 + 260a^2 + \frac{3183}{8}a_5^2 + 516a_5 a + 386(a_1 a_2 + a_3 a_4 + a_1 a_4 + a_2 a_3) \\
&+ \frac{575}{2}(a_1 a_3 + a_2 a_4) + 84a^3 + \frac{567}{4}a_5^3 \\
&+ 168(a_1^2 a_2 + a_2^2 a_1 + a_3^2 a_4 + a_4^2 a_3 + a_4^2 a_1 + a_1^2 a_4 + a_2^2 a_3 + a_3^2 a_2) \\
&+ \frac{441}{4}(a_1^2 a_3 + a_3^2 a_2 + a_2^2 a_4 + a_4^2 a_2) + \frac{945}{4}a_5 a^2 + 252a_5^2 a + \frac{693}{2}a_5(a_1 a_2 + a_3 a_4 + a_1 a_4 + a_2 a_3) \\
&+ \frac{945}{4}(a_1 a_3 + a_2 a_4)a_5 + 210(a_1 a_2 a_3 + a_1 a_2 a_4 + a_1 a_3 a_4 + a_2 a_3 a_4) + 14a^4 + \frac{189}{8}a_5^4 \\
&+ 42a_5 a^3 + \frac{189}{4}a_5^3 a + \frac{525}{8}a_5^2 a^2 + \frac{357}{4}a_5^2(a_1 a_2 + a_3 a_4 + a_1 a_4 + a_2 a_3) + \frac{105}{2}a_5^2(a_1 a_3 + a_2 a_4) \\
&+ 84a_5(a_1^2 a_2 + a_2^2 a_1 + a_3^2 a_4 + a_4^2 a_3 + a_1^2 a_4 + a_4^2 a_1 + a_2^2 a_3 + a_3^2 a_2) \\
&+ \frac{189}{4}a_5(a_1^2 a_3 + a_3^2 a_1 + a_2^2 a_4 + a_4^2 a_2) + \frac{357}{4}a_5(a_1 a_2 a_3 + a_1 a_3 a_4 + a_1 a_2 a_4 + a_2 a_3 a_4) \\
&+ 28(a_1^3 a_2 + a_2^3 a_1 + a_3^3 a_4 + a_4^3 a_3 + a_1^3 a_4 + a_4^3 a_1 + a_2^3 a_3 + a_3^3 a_2) \\
&+ 14(a_1^3 a_3 + a_3^3 a_1 + a_2^3 a_4 + a_4^3 a_2) + 42(a_1^2 a_2^2 + a_3^2 a_4^2 + a_1^2 a_4^2 + a_2^2 a_3^2) + \frac{189}{8}(a_1^2 a_3^2 + a_2^2 a_4^2)
\end{aligned}
$$

$$+42(a_1^2 a_2 a_3 + a_1^2 a_2 a_4 + a_1^2 a_3 a_4 + a_2^2 a_1 a_4 + a_2^2 a_1 a_3 + a_2^2 a_3 a_4$$

$$+a_3^2 a_1 a_4 + a_3^2 a_2 a_4 + a_3^2 a_1 a_2 + a_4^2 a_2 a_3 + a_4^2 a_1 a_3 + a_4^2 a_1 a_2) + \frac{315}{4} a_1 a_2 a_3 a_4,$$

$$A_6 = 3(A_4 - 1), \tag{13.76}$$

This two-loop result can be used for **x**-space and for momentum space integrals since the KITE-type integral is dual to itself. Kazakov also calculated the the tables for the BEQ- and the BUG-type integrals with line indices of the form $\alpha_i = 1 + a_i \varepsilon$ up to order ε^2. Since these two are dual to each other, the **x**-space table for the BEQ-type integral has to be used for the momentum space integral of the BUG type and vice versa. For an integral of the BEQ type with the following form of the indices:

$$\equiv \frac{\text{BEQ}\,(1 + a_1\varepsilon, \ldots, 1 + a_7\varepsilon)}{(\mathbf{x}^2)^{7 + \sum_i a_i\varepsilon - 3D/2}}, \tag{13.77}$$

the **x**-space BEQ-table up to ε^2 is given by [3, 9]

$$\text{BEQ}(1 + a_1\varepsilon, \ldots, 1 + a_7\varepsilon) = \frac{1}{1-\varepsilon}\left[A_0\zeta(5) + \frac{A_1\zeta(6)}{2}\varepsilon + \frac{A_2\zeta^2(3)}{2}\varepsilon \right.$$

$$\left. + \frac{A_3\zeta(7)}{4}\varepsilon^2 + \frac{A_4\zeta(3)\zeta(4)}{4}\varepsilon^2 + \mathcal{O}(\varepsilon^3)\right], \tag{13.78}$$

with the coefficients:

$$A_0 = 20, \quad A_1 = 50, \quad A_2 = 20 + 6(a_4 + a_5 + a_6 + a_7),$$

$$A_3 = 7\left[\frac{380}{7} + 20(a_1 + a_3) + 32a_2 + 17(a_4 + a_5) + 33(a_6 + a_7)\right.$$

$$+6(a_1^2 + a_3^2) + 8a_2^2 + 4(a_4^2 + a_5^2) + 8(a_6^2 + a_7^2) + 8(a_1 + a_3)a_2 + 2(a_1a_4 + a_3a_5)$$

$$+6(a_1a_5 + a_3a_4) + 10(a_1a_6 + a_3a_7) + 6(a_1a_7 + a_3a_6) + 4a_1a_2 + 4(a_4 + a_5)a_2$$

$$+12(a_6 + a_7)a_2 + 2a_4a_5 + 4(a_4a_6 + a_5a_7) + 6(a_4a_7 + a_5a_6) + 10a_6a_7$$

$$\left. +\frac{1}{4}(a_4 + a_5 + a_6 + a_7) + \frac{1}{8}(a_4 + a_5 + a_6 + a_7)^2\right],$$

$$A_4 = 3\left[20 + 6(a_4 + a_5 + a_6 + a_7)\right]. \tag{13.79}$$

The reduction algorithms of partial integration in momentum space solves the BEQ- and the BUG-diagram only if their line indices are integer. If they are not integer, these algorithms can be used to bring the two types to the standard form given by the tables. The diagrams determined by the tables are displayed in Fig. 13.9.

The tables enable us to calculate the N-type diagram with No. 4 in Fig. 13.2, which is the only one that requires the expansion of the BEQ-type diagram up to $\mathcal{O}(\varepsilon^2)$. The calculation is sketched in the next section.

KITE type   BUG type   BEQ type

FIGURE 13.9 Types of diagrams given by the tables for indices of the form $a_i = 1 + a_i\varepsilon$.

13.6 Special Treatment of Generic Four- and Five-Loop Diagrams

There are only six diagrams where IR-rearrangement does not help avoiding the appearance of generic four- and five-loop diagrams and the NO type of Fig. 13.1. They are shown in Fig. 13.10. Even the combined use of the reduction algorithms and the tables is not sufficient to calculate

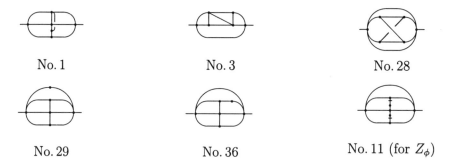

FIGURE 13.10 Diagrams whose calculation requires special methods.

them analytically. They need a special treatment. Three of the diagrams, Nos. 29, 36, and 11, are done in momentum space. The diagrams Nos. 29 and 36 are calculated by a clever application of the R-operation and momentum differentiation [1]. The diagram with vector indices emerging from differentiation of the quadratically divergent diagram No. 11 is solved by applying $\partial^2/\partial k_\mu \partial k_\mu$ to the integrals. The result is obtained by repeated partial integration.

Nos. 1, 3, and 28 have been solved in configuration space, they were first calculated numerically by GPXT. Analytical results up to $\mathcal{O}(\varepsilon^0)$ were found by Kazakov [9], applying the method of ideal index calculation of Section 13.5. The calculation of the N-shaped diagram is shown as an example.

13.6.1 N-Shaped Diagram

The N-shaped four-loop subdiagram has to be calculated up to $\mathcal{O}(\varepsilon^0)$. Being overall convergent without any subdivergences the constant term is the only one needed. The ε-expansion coefficients of the line indices of the subdiagram can thus be chosen arbitrarily. Relation (13.71) applied to the lower left vertex gives for the following index constellation:

$$\text{(diagram)} = -\frac{1}{\varepsilon}\left\{ \text{(diagram)} + \text{(diagram)} - 2\,\text{(diagram)} - \text{(diagram)} \right\}.$$

The chains can be integrated. The right vertex of the last diagram has an ideal index constellation, such that we obtain using (13.64):

$$-\frac{1}{\varepsilon}\left\{\quad 2\,\nu(1)\,\nu(2)\,\nu(1-\varepsilon)\quad \text{(diagram)} \right.$$
$$-\nu(2)\,\nu\!\left(1+\tfrac{1}{2}\varepsilon\right)\nu\!\left(1-\tfrac{3}{2}\varepsilon\right)\quad \text{(diagram)}$$
$$\left. -\nu(1)\,\nu(2)\,\nu(1-\varepsilon)\quad \text{(diagram)}\quad \right\}.$$

The BEQ-type diagram is required up to $\mathcal{O}(\varepsilon^2)$ and can be deduced from Kazakov's expansions (13.75)–(13.79), yielding in **x**-space:

$$\left(-\!\bigotimes\!-\right)_{\varepsilon=0} = \frac{1}{(\mathbf{x}^2)^{1-5\varepsilon/2}} \frac{441}{8} \zeta(7) \tag{13.80}$$

$$\xrightarrow{\text{FT}} \underbrace{\frac{\Gamma(1+2\varepsilon)}{\Gamma(1-5/2\varepsilon)}}_{=1+\mathcal{O}(\varepsilon)} \frac{1}{(\mathbf{k}^2)^{1+2\varepsilon}} \frac{441}{8} \zeta(7) , \tag{13.81}$$

where FT indicates a Fourier transformation. The required pole term of diagram No. 3 is then:

$$\mathcal{K}\left[-\!\bigotimes\!-\right] = \mathcal{K}\left[\frac{441}{8}\zeta(7) \underbrace{L(1+2\varepsilon,1)}_{=2/5\varepsilon+\mathcal{O}(\varepsilon^0)}\right] = \frac{441}{20}\zeta(7)\,\varepsilon^{-1} . \tag{13.82}$$

13.7 Computer-Algebraic Program

The momentum space algorithms of Section 13.4.3 together with the tables for the KITE-type integrals up to ε^4, for the BEQ- and the BUG-type up to ε^2, and the expansion of the loop function $L(a,b)$ up to ε^7 are implemented as a REDUCE program. It is based on the program LOOPS [7] that allows the calculation of primitive two-loop integrals with arbitrary line indices and free vector indices as well as the KITE-type integral with integer line indices without free vector indices. For the five-loop calculation, the expansion of the loop function $L(a,b)$ had to be extended up to $\mathcal{O}(\varepsilon^7)$ to account for extra poles arising in intermediate expressions. The program also contains the formulas in Section 13.4.3 and the tables in Section 13.5.5. The following types of diagrams can then be calculated:

1. Planar generic three-loop diagrams with integer line indices up to $\mathcal{O}(\varepsilon^3)$;

2. BEQ- and BUG-type integrals with $a_i = 1 + m_i\varepsilon/2$ up to $\mathcal{O}(\varepsilon^2)$;

3. KITE-type integrals

   - with arbitrary noninteger indices on the lines 1 and 4 *or* 2 and three up to $\mathcal{O}(\varepsilon^{7-L})$;
   - with arbitrary noninteger index on line 5 up to $\mathcal{O}(\varepsilon^4)$;
   - with indices of the form $a_i = 1 + m_i\varepsilon/2$ on all lines up to $\mathcal{O}(\varepsilon^4)$;

4. Primitive diagrams with L loops up to the order $\mathcal{O}(\varepsilon^{7-L})$.

All but six propagator-type integrals of the ϕ^4-theory up to five loops can be calculated with this program, as they can be reduced by IR-rearrangement and R^*-operation to one of the types above. The results for the 6 diagrams of Fig. 13.10 are inserted by hand.

We now proceed to sort all five-loop diagrams of the ϕ^4-theory according to the above types. The subdivisions depend on the IR-rearrangement and are somewhat arbitrary. There are 124 logarithmically divergent propagator-type integrals for $Z_g(g,\varepsilon^{-1})$ or $Z_{m^2}(g,\varepsilon^{-1})$ and 11 quadratically divergent ones for $Z_{m^2}(g,\varepsilon^{-1})$.

The 124 Logarithmically Divergent Integrals

They consist of (numbering corresponds to that in Appendix A):

a) 27 diagrams with cutvertices not contributing to the renormalization constants, with
$$\mathcal{K}\bar{R}G = \mathcal{K}\bar{R}G_1 \cdot \mathcal{K}\bar{R}G_2 = \sum_{i=1}^{L_1} C_i^1 \varepsilon^{-i} \sum_{i=1}^{5-L_1} C_i^2 \varepsilon^{-i} = \sum_{i=2}^{5} C_i' \varepsilon^{-i};$$

b) Insertions into simple loops. the inserted diagrams contain

   • Generic four-loop diagrams (Nos. 3, 29, 36);

   • Generic three-loop diagrams with integer indices (Nos. 24, 25, 26, 28, 34, 37, 53, 75, 111) or with noninteger indices (No. 27);

   • Generic two-loop diagrams (KITE type) with the index a_5 noninteger (Nos. 32, 40, 42, 47, 58, 65, 73, 93, 98, 109, 112, 116, 123) or with index $a_1, \ldots, a_4$ noninteger (Nos. 38, 39, 44, 48, 51, 55, 59, 64, 67, 69, 74, 101);

   • KITE-type integral with integer indices and primitive diagrams.

c) Insertions in KITE-type diagrams with index a_5 noninteger (Nos. 19, 79) or with index $a_1, \ldots, a_4$ noninteger (No. 57);

d) Generic five-loop diagrams (No. 1) and inserted results of diagrams (Nos. 3, 28, 29, 36).

For 26 diagrams the R^*-operation has to be carried out (Nos. 21, 32, 36, 37, 47, 53, 58, 61, 62, 65, 67, 73, 75, 79, 88, 91, 92, 93, 95, 101, 103, 105, 107, 109, 111, 116).

The 11 Quadratically Divergent p-Integrals

They consist of

a) 5 primitive diagrams (Nos. 1, 2, 3, 4, 5);

b) 6 diagrams require differentiation with respect to external momentum and infrared rearrangement

   • 2 give known logarithmically divergent propagator-type integrals (Nos. 7, 8);

   • 4 give integrals with free vector indices, 3 consisting of primitive or KITE-type integrals (Nos. 6, 9, 10) and 1 of a generic four-loop diagram (No. 11).

Appendix 13A   Fourier Transformation of Simple Powers in D Dimensions

The Fourier transform of a simple power of $\mathbf{x}^2$ is again a simple power:

$$\frac{1}{(\mathbf{p}^2)^a} = \frac{\Gamma(D/2-a)}{\pi^{D/2}\Gamma(a)4^a} \int d^D x \, \frac{e^{i\mathbf{p}\cdot\mathbf{x}}}{(\mathbf{x}^2)^{D/2-a}} \, . \qquad (13A.1)$$

We first split the x-space into a component parallel to $\mathbf{p}$ and a $D-1$-dimensional space perpendicular to it:

$$\int d^D x \, \frac{e^{i\mathbf{p}\cdot\mathbf{x}}}{(\mathbf{x}^2)^{D/2-a}} = \int_{-\infty}^{\infty} dx_0 \cos(|\mathbf{p}|\,x_0) \int d^{D-1}x \frac{1}{(x_0^2 + \mathbf{x}^2)^{D/2-a}}. \qquad (13A.2)$$

The integration over the $D - 1$-dimensional space can be carried out, and the right-hand side becomes

$$2\pi^{(D-1)/2}\frac{\Gamma(1/2-a)}{\Gamma(D/2-a)}\int_0^\infty dx_0\frac{\cos(|\mathbf{p}|\,x_0)}{x_0^{1-2a}}. \tag{13A.3}$$

With the help of the integral formula

$$\int_0^\infty dy\,\frac{\cos(|\mathbf{p}|\,y)}{y^{1-2a}} = \frac{\pi}{2}\frac{|\mathbf{p}|^{-2a}}{\Gamma(1-2a)\,\cos(\frac{1-2a}{2}\pi)}, \tag{13A.4}$$

we find

$$\int d^D x\,\frac{e^{i\mathbf{p}\cdot\mathbf{x}}}{(\mathbf{x}^2)^{D/2-a}} = \frac{\pi^{(D+1)/2}\,\Gamma(1/2-a)}{\Gamma(D/2-a)\,\Gamma(1-2a)\,\cos(\frac{1-2a}{2}\pi)}\,\frac{1}{(p^2)^a}, \tag{13A.5}$$

which can be trivially rewritten as

$$\frac{\pi^{D/2}\,\Gamma(a)4^a}{\Gamma(D/2-a)}\frac{\sqrt{\pi}\,\Gamma(1/2-a)}{4^a\Gamma(a)\,\Gamma(1-2a)\,\cos(\frac{1-2a}{2}\pi)}\,\frac{1}{(p^2)^a}. \tag{13A.6}$$

Using further

$$\cos(\pi/2-a\pi) = \sin(a\pi), \qquad \Gamma(1/2) = \sqrt{\pi}, \tag{13A.7}$$

and

$$\Gamma(1-2a) = \frac{\Gamma(1/2-a)\Gamma(1-a)}{\Gamma(1/2)}2^{-2a}, \tag{13A.8}$$

the right-hand side of (13A.6) can be brought to the form

$$\frac{\pi^{D/2}\Gamma(a)4^a}{\Gamma(D/2-a)}\underbrace{\frac{\pi}{\Gamma(a)\,\Gamma(1-a)\,\sin(a\pi)}}_{=1}\,\frac{1}{p^{2a}}, \tag{13A.9}$$

so that we finally obtain

$$\int d^D x\,\frac{e^{i\mathbf{p}\cdot\mathbf{x}}}{(\mathbf{x}^2)^{D/2-a}} = 4^a\pi^{D/2}\frac{\Gamma(a)}{\Gamma(D/2-a)}\,\frac{1}{p^{2a}}. \tag{13A.10}$$

Appendix 13B   Further Expansions of Gamma Function

In Appendix 8D, the Gamma function is expressed in terms of ψ functions. Here we supplement the formula by another expansion involving Riemann's ζ-functions. For this we recall the well-known expansion of the logarithm of the Gamma function:

$$\ln\Gamma(1+z) = -\gamma z + \sum_{m=2}^\infty(-1)^m\,\zeta(m)\,\frac{z^m}{m}, \tag{13B.1}$$

where γ is the *Euler constant*

$$\gamma = -\psi(1) = \sum_{m=2}^\infty(-1)^m\,\frac{\zeta(m)}{m} = 0.5772..., \tag{13B.2}$$

and $\zeta(x)$ is *Riemann's ζ function*:

$$\zeta(x) = \sum_{k=1}^\infty\frac{1}{k^x}. \tag{13B.3}$$

This representation can be generalized using the identity of the Gamma function $\Gamma(1 + z) = z\Gamma(z)$, and the expansion $\ln(1 + z) = \sum_{n=1}^{\infty} (-1)^{n+1} z^n / n$:

$$
\begin{aligned}
\ln \Gamma(2 + z) &= \ln(1 + z) + \ln \Gamma(1 + z) \\
&= -\sum_{m=1}^{\infty} (-1)^m \frac{z^m}{m} - \gamma z + \sum_{m=2}^{\infty} (-1)^m \frac{z^m}{m} \zeta(m) \\
&= z(1 - \gamma) + \sum_{m=2}^{\infty} (-1)^m \frac{z^m}{m} [\zeta(m) - 1],
\end{aligned}
\tag{13B.4}
$$

$$
\begin{aligned}
\ln \Gamma(3 + z) &= \ln 2 + \ln(1 + z/2) + \ln \Gamma(2 + z) \\
&= \ln 2 - \sum_{m=1}^{\infty} (-1)^m \frac{(z/2)^m}{m} + z(1 - \gamma) + \sum_{m=2}^{\infty} (-1)^m \frac{z^m}{m} [\zeta(m) - 1] \\
&= \ln 2 + z\left(1 + \frac{1}{2} - \gamma\right) + \sum_{m=2}^{\infty} (-1)^m \frac{z^m}{m} \left[\zeta(m) - 1 - \frac{1}{2^m}\right].
\end{aligned}
\tag{13B.5}
$$

In general one has

$$
\ln \Gamma(n + z) = \ln[(n - 1)!] + z \left[\sum_{k=1}^{n-1} \frac{1}{k} - \gamma\right] + \sum_{m=2}^{\infty} (-1)^m \frac{z^m}{m} \left[\zeta(m) - \sum_{k=1}^{n-1} \frac{1}{k^m}\right].
\tag{13B.6}
$$

Inserting here the truncated zeta functions (13.17) with the property $\zeta^{(1)}(x) \equiv 0$ [see (13.18)], this can be written as:

$$
\Gamma(n + z) = (n - 1)! \exp\left\{-z\left[\gamma - \zeta^{(n)}(1)\right] + \sum_{m=2}^{\infty} (-1)^m \frac{z^m}{m} \left[\zeta(m) - \zeta^{(n)}(m)\right]\right\}.
\tag{13B.7}
$$

Note that this is the manifestly finite version of the more symmetric but not well-defined expression:

$$
\Gamma(n + z) \overset{n \geq 0}{=} n! \exp\left\{\sum_{m=1}^{\infty} (-1)^m \frac{z^m}{m} \left[\zeta(m) - \zeta^{(n)}(m)\right]\right\},
\tag{13B.8}
$$

where γ has been set equal to $\zeta(1)$. In fact, the quantity $\zeta(1)$ is divergent, since the truncated sum $\zeta^{(n)}(1)$ grows with n like $\log n + \gamma$.

In Appendix 8D we gave an expansion of the Gamma function near integer arguments in terms of the Digamma functions [see Eq. (8D.23)]:

$$
\Gamma(n + 1 + \varepsilon) = n! \left\{1 + \varepsilon \psi(n + 1) + \frac{\varepsilon^2}{2} \left[\psi'(n + 1) + \psi(n + 1)^2\right] + O(\varepsilon^3)\right\}.
\tag{13B.9}
$$

To see the equivalence of the two expansions, we use the truncated zeta function (13.17) to rewrite (8D.12) and (8D.13) as

$$
\psi(n + 1) = \zeta^{(n+1)}(1) - \gamma,
\tag{13B.10}
$$

$$
\psi'(n + 1) = -\zeta^{(n+1)}(2) + \frac{\pi^2}{6} = -\zeta^{(n+1)}(2) + \zeta(2).
\tag{13B.11}
$$

Inserting this into (13B.9), we obtain

$$
\Gamma(n + \varepsilon + 1) = n!\left(1 + \varepsilon\left\{\zeta^{(n+1)}(1) - \gamma\right\} \right.
$$
$$
\left. + \frac{\varepsilon^2}{2}\left\{-\zeta^{(n+1)}(2) + \zeta(2) + [\zeta^{(n+1)}(1) - \gamma]^2\right\} + \mathcal{O}(\varepsilon^3)\right),
\tag{13B.12}
$$

which is the ε-expansion of Eq. (13B.8) up to second order in ε.

Notes and References

The citations in the text refer to:

[1] K.G. Chetyrkin, F.V. Tkachov, Nucl. Phys. B **192**, 159 (1981).

[2] K.G. Chetyrkin, A.L. Kataev, and F.V. Tkachov, Nucl. Phys. B **174**, 345 (1980).

[3] D.I. Kazakov, Phys. Lett. B **133**, 406 (1983); Theor. Math. Phys. **61**, 84 (1985). The author used the term 'method of uniqueness', but 'method of ideal index constellations' is more descriptive.

[4] S.G. Gorishny, S.A. Larin, L.R. Surguladze, and F.V. Tkachov, Comp. Phys. Comm. **55**, 381, (1989).

[5] See Ref. [2] and [4]. The original name used by those authors was *G-scheme* since the authors denote the loop integrals L by G. The name L-scheme is an adaptation of their name to our notation.

[6] Some authors refer to this as the $\overline{\text{MS}}$-scheme, which is different from ours defined in Eq. (13.24).

[7] For this purpose we extended an available program LOOPS written by L.R. Surguladze and F.V. Tkachov, Comp. Phys. Comm. **55**, 205 (1989), which gave the expansion terms only up to $\mathcal{O}(\varepsilon^4)$.

[8] This is the redefinition of g or μ used in the program LOOPS of Ref. [7].

[9] D.I. Kazakov, Theor. Math. Phys. **58**, 223 (1984).

14

Generation of Diagrams

In Section 5.4, we have developed a functional formalism to find all connected vacuum diagrams and their multiplicities. These serve as a basis for deriving all diagrams contributing to two- and four-point functions. This is done by removing from the connected vacuum diagrams one line and two lines, respectively. Instead of the two lines we may also remove a vertex, as pointed out in Section 5.5, where we also indicated a way to implement the recursive formalism for the generation of all graphs on a computer [1], and the reader is referred to the detailed original paper for a detailed explanation of the techniques.

In this chapter, we shall describe a somewhat shortened way of generating all diagrams which was used in the initial work on the five-loop determinations of the critical exponents [2]. Because of the special features of the regularization by minimal subtraction discussed in Section 11.6, we can restrict the generation to

- all 1PI four-point diagrams without tadpole parts and their weight factors for Z_g,

- all 1PI two-point diagrams without tadpole parts and their weight factors for Z_ϕ,

- all 1PI four-point diagrams without tadpole parts which result from differentiation with respect to m^2, and the weight factors of the corresponding two-point diagrams for Z_{m^2}.

The generation starts with all connected vacuum diagrams. For the generation of tadpole-free two- and four-point diagrams, only vacuum diagrams without cutvertices are needed. From these, the connected two- and four-point diagrams are constructed by cutting lines and vertices, respectively. These diagrams may now contain cutvertices. In the last step, we select the 1PI diagrams.

Since we shall eventually be interested only in the renormalization group functions of the theory, we can reduce the number of diagrams further by selecting only those two- and four-point 1PI diagrams without cutvertices. Recall that the pole part $\mathcal{K}\bar{R}G$ of a diagram G with a cutvertex contains only ε^{-n}-poles with $n \geq 2$, as explained on page 190. It contributes to the renormalization constants, but not to the renormalization group functions due to the cancellation of all higher ε-poles (see Section 10.3). Nevertheless, we find it useful to include diagrams with cutvertices at each stage of the calculations, since they permit us to check the correctness of the ε-poles via the necessary cancellations. They appear among the other diagrams in Appendix A, for example, the first two-, three-, and four-loop diagrams on page 431, or those with numbers 2 to 7 on page 432.

14.1 Algebraic Representation of Diagrams

When generating Feynman diagrams by the method described in Sections 5.4 and 5.5, an important problem is the identification of topologically inequivalent diagrams. For this purpose we set up an algebraic specification of diagrams. With each diagram G, we associate a *diagram matrix* **G**. The row and column indices i and j of its matrix elements G^{ij} run from zero to the number of vertices p. Each element G^{ij} gives the number of lines joining the vertices i and j. In the ϕ^4-theory, the matrix elements all lie in the interval $0 \leq G^{ij} \leq 4$. The diagonal elements G^{ii} count the number of self connections of the ith vertex. External lines of a diagram are labeled as if they were connected to an additional dummy vertex with the number 0. The irrelevant matrix element G^{00} is set equal to zero. The matrix is symmetric and contains $(p+1)(p+2)/2 - 1$ independent elements. Unfortunately, the matrix is not unique. There are obviously several matrices characterizing each diagram, depending on the numbering of the vertices. All of them contain the same information on the numbers S, D, T or F of double, triple, or fourfold connection in a diagram, respectively.

The number S is given by the sum of the numbers on the diagonal, D by the number of times the value 2 appears above the diagonal, and T and F by the number of times the values 3 and 4 appear above the diagonal. For an amputated diagram in which the external lines are not connected to labeled vertices, these numbers include multiple connections of the external lines to an imagined additional dummy vertex (labeled 0 in the matrices). Note that this definition of S, D, T or F differs from the one on page 45 for diagrams whose external lines are connected to labeled external positions.

What does the matrix tell us about the number of identical vertex permutations N_{IVP} which appears in the multiplicity and weight formulas (3.14)–(3.17)? If the diagram is amputated, as it is the case for the diagrams under consideration, N_{IVP} includes all vertex permutations, also those that involve vertices with external lines. The number N_{IVP} is the number of all permutations of lines or rows which leave the matrix unchanged. The dummy vertex 0 is omitted from the permutations. This implies that the number of identical vertex permutations is not equal to that of vacuum diagrams obtained by connecting the external lines to the dummy vertex 0. Given the numbers N_{IVP}, S, D, T, the total weight of an amputated Feynman diagram is given by

$$W_G = \frac{n!}{2^{S+D}3!^T 4!^F\, N_{\text{IVP}}} \ . \tag{14.1}$$

This formula yields the same number as formula (3.17), where we differentiated distinguished between internal and external lines.

A simple example is shown in Fig. 14.1. Table 14.1 lists explicitly the matrix elements and the numbers N_{IVP} for the vacuum diagrams up to eight loops without cutvertices.

The matrix G^{ij} contains some more information about the properties of a diagram. First, the *connectedness of a diagram* may be checked by inspection. If a diagram is not connected, the matrix G^{ij} is a block matrix for a certain vertex numbering. Second, the matrix $G^{(ij)}$ contains information on the presence of cutvertices. Recall their definition on page 187 and the description of its properties in Section 11.3 that if a connected diagram falls into two pieces after cutting a vertex, this vertex is called a cutvertex. If a diagram contains a tadpole part or a self-connection it necessarily contains a cutvertex. For vacuum diagrams, the opposite is also true. In the matrix G^{ij}, a cutvertex with the number i manifests itself by making the G^{ij} approximately a block matrix. The blocks overlap on the diagonal at (i, i), so that the matrix takes block form if the ith row and column are eliminated. Third, the matrix $G^{(ij)}$ helps us to recognize a cutline, a line whose cutting makes a connected diagram fall into two pieces. Recall the definition on page 55. Such a line is identified by cutting it and testing the matrix G^{ij} associated with the resulting modified diagram for connectedness. A connected diagram without cutlines is 1PI. In the normal phase of the ϕ^4-theory, vacuum diagrams do not contain any cutlines.

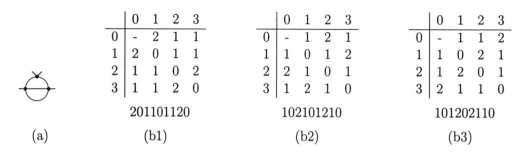

	0	1	2	3
0	-	2	1	1
1	2	0	1	1
2	1	1	0	2
3	1	1	2	0

201101120

(b1)

	0	1	2	3
0	-	1	2	1
1	1	0	1	2
2	2	1	0	1
3	1	2	1	0

102101210

(b2)

	0	1	2	3
0	-	1	1	2
1	1	0	2	1
2	1	2	0	1
3	2	1	1	0

101202110

(b3)

(a)

FIGURE 14.1 The diagram in (a) can be represented by the three matrices (b1), (b2), and (b3). The labeling of the upper vertex is 1 in (b1), 2 in (b2), and 3 in (b3). The number $\hat{G} = G^{10}G^{11}G^{20}G^{21}G^{22}G^{30}G^{31}G^{32}G^{33}$ underneath the matrices in b1)–b3) is constructed as explained in the text. The smallest number is read off from representation (b3) which is chosen as the unique representation of the diagram. The diagram has four external lines ($n = 4$), and two double connections ($D = 2$) including the one of the external lines to the additional dummy vertex 0. The number of identical vertex permutations N_{IVP} is 2, since two of the vertices are symmetric to each other. Hence $W_G = 4!/(2^2\,N_{IVP}) = 24/(4 \cdot 2) = 3$. The weight is therefore $W_G = 3$, as in the second line of the diagrammatic equation (3.25), and in the third entry in the table in Appendix B.1.

Since the vertex numbering is arbitrary, the matrix representation of a diagram is, so far, not unique. Examples for such equivalent matrices are shown in Fig. 14.1. A unique representation is found by associating with each matrix a number whose digits are composed of the matrix elements G^{ij} ($0 \le j \le i \le p$, $j \neq 0$). More explicitly, we form the following number with $[(p + 1)(p + 2)/2 - 1]$-digits:

$$\hat{G} = G^{10}G^{11}G^{20}G^{21}G^{22}G^{30}G^{31}G^{32}G^{33}\ldots G^{pp}. \qquad (14.2)$$

The matrix associated with the smallest number is chosen as the unique representation of the diagrams. In Fig. 14.1 it is the matrix (b3).

It will further be useful to introduce a more condensed notation for diagrams with no self-connections, where all numbers G^{ii} are omitted since they are zero. Then we are left with a reduced number

$$\hat{G}_r = G^{10}G^{20}G^{21}G^{30}G^{31}G^{32}\ldots G^{p+1,p}. \qquad (14.3)$$

If we consider vacuum diagrams, there are no external lines such that $j \neq 0$. Therefore, vacuum diagrams without self-connections may be characterized uniquely by the matrix with the lowest number

$$\hat{G}_V = G^{21}G^{31}G^{32}\ldots G^{p+1,p}. \qquad (14.4)$$

All vacuum diagrams without self-connections are displayed in Fig. 14.3. The numbers (14.4) are listed in Table 14.1 , which shows also the number N_{IVP} of identical vertex permutations and the resulting weight factor. For example, diagram No. 13 with seven vertices has two double connections and two triple connections and there are two identical vertex permutations such that the weight factor is $[(2!)^2 \times (3!)^2 \times 2]^{-1} = 1/288$.

14.2   Generation Procedure

We now describe the shortened recursive graphical scheme starting with the generation of vacuum diagrams. The shortening is possible by observing that in the full graphical recursion relation in Fig. 5.3, the diagrams produced by the last, nonlinear term are no new diagrams, but change only the weights of those generated by the first two linear terms. Since we possess a simple counting formula (3.17) for the weights, we do not have to solve the full recursion relation, but generate the diagrams by a graphical recursion based only on the first two operations in Fig. 5.3, and ignoring the weights. The two- and four-point diagrams are then obtained by cutting one line and a vertex, respectively, as described in Section 5.5.

14.2.1 Vacuum Diagrams

The generation of all diagrams of the theory begins with ∞, the only vacuum diagram with one vertex. The addition of a further vertex to a diagram with p vertices is performed in two ways, both illustrated in Fig. 14.2. First, we cut a line and insert a new vertex with a

(a)   (b)

FIGURE 14.2 Generation of diagrams by cutting one (a) or two (b) lines, and joining the ends in a new vertex.

self-connection. This is done for all lines in each vacuum diagram with p vertices. Cutting topologically equivalent lines generates equivalent diagrams, and only one of such equivalent lines needs to be cut. Next, we cut two lines at a time and join the four ends in a new vertex. The matrices G^{ij} of each newly generated diagram and their associated numbers of Eq. (14.2) have to be identified, and compared to those of previously obtained diagrams to avoid repeated generation of topologically equivalent diagrams. The occurrence of cutvertices during the generation of the vacuum diagrams is registered for later use.

The vacuum diagrams obtained in this way are automatically connected. For the calculation of the renormalization constants, we need only diagrams without tadpole parts which are generated by vacuum diagrams without cutvertices. All vacuum diagrams without cutvertices up to seven vertices are pictured in Fig. 14.3.

14.2.2 Two-Point Diagrams

A connected two-point diagram can be generated from a vacuum diagram with the same number of vertices by cutting a line. All connected two-point diagrams may be found in this way. Repeated generation is again avoided by taking into account symmetries among the lines, i.e. by selecting only one out of a set of topologically equivalent lines as illustrated in Fig. 14.4. All generated diagrams are connected because the ϕ^4-theory does not admit vacuum diagrams with cutlines. They are, however, not necessarily 1PI since they may have cutlines.

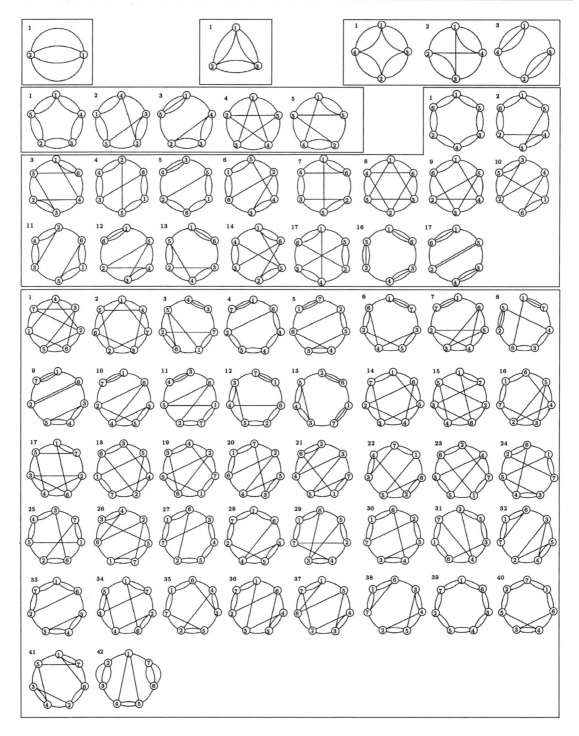

FIGURE 14.3 Vacuum diagrams without cutvertices up to seven vertices. The displayed diagrams allow the generation of all two- and four-point diagrams without tadpole parts (but with cutvertices) up to five loops. The numbers in the upper left corners are running labels, generally denoted by the symbol n_L^{Vac}. Each vacuum diagram possesses a unique classification number $\hat{G}_V = G^{21}G^{31}G^{32}G^{41}G^{42}G^{43}\ldots$ listed in Table 14.1 .

TABLE 14.1   Matrix representation for all vacuum diagrams of Fig. 14.3 specified by the numbers $\hat{G}_V = G^{21}G^{31}G^{32}G^{41}G^{42}G^{43}\ldots$ of Eq. (14.4). To guide the eye, we have inserted bars separating the rows of the triangular part of the matrix below the diagonal. The vertices are labeled as in Fig. 14.3, with the order chosen to achieve the smallest number $\hat{G}$. The first column gives the running number of the diagrams in Fig. 14.3. For each pair $i > j$ in the two top rows, the number G^{ij} underneath gives the number of lines connecting the vertices i and j. The second last column lists the number of single, double, triple, and quadruple connections and the number of identical vertex permutations (S,D,T,Q,N_{IVP}). The last column contains the weight factor W_G calculated from Eq. (3.6). For the corresponding list for the two- and four-point diagrams, see Table 14.3 .

2 vertices, 3 loops: 1 diagram				
j	2			
i	1			
No.	G^{ij}	$(S,D,T,Q;N_{\text{IVP}})$	M_G	W_G
2-1	4	(0,0,0,1;2)	24	1/48

3 vertices, 4 loops: 1 diagram					
j	2	33			
i	1	12			
No.	G^{ij}		$(S,D,T,Q;N_{\text{IVP}})$	M_G	W_G
3-1	2	22	(0,3,0,0;6)	1728	1/48

4 vertices, 5 loops: 3 diagrams					
j	2	33	444		
i	1	12	123		
No.	G^{ij}			$(S,D,T,Q;N_{\text{IVP}})$	M_G	W_G
4-1	0	22	220	(0,4,0,0;8)	62208	1/128
4-2	1	12	211	(4,2,0,0;8)	248832	1/32
4-3	0	13	310	(2,0,2,0;4)	55296	1/144

5 vertices, 6 loops: 5 diagrams						
j	2	33	444	5555		
i	1	12	123	1234		
No.	G^{ij}				$(S,D,T,Q;N_{\text{IVP}})$	M_G	W_G
5-1	0	02	202	2200	(0,5,0,0;10)	2985984	1/320
5-2	0	02	211	2110	(4,3,0,0;4)	29859840	1/32
5-3	0	02	112	3100	(3,2,1,0;2)	19906560	1/48
5-4	1	11	111	1111	(10,0,0,0;120)	7962624	1/120
5-5	0	11	121	2110	(6,2,0,0;4)	59719680	1/16

6 vertices, 7 loops: 17 diagrams							
j	2	33	444	5555	66666		
i	1	12	123	1234	12345		
No.	G^{ij}					$(S,D,T,Q;N_{\text{IVP}})$	M_G	W_G
6-1	0	00	022	2020	22000	(0,6,0,0;12)	$1.7916\,10^8$	1/768
6-2	0	01	012	2011	22000	(4,4,0,0;4)	$2.1499\,10^9$	1/64
6-3	0	02	111	1110	20011	(8,2,0,0;16)	$2.1499\,10^9$	1/64
6-4	0	00	022	2110	21100	(4,4,0,0;8)	$1.0750\,10^9$	1/128
6-5	0	00	013	2110	22000	(3,3,1,0;2)	$1.4333\,10^9$	1/96
6-6	0	01	012	2101	21100	(6,3,0,0;2)	$8.5996\,10^9$	1/16
6-7	0	01	102	1210	21010	(6,3,0,0;4)	$4.2998\,10^9$	1/32
6-8	0	11	110	1111	11110	(12,0,0,0;48)	$2.8665\,10^9$	1/48
6-9	0	01	111	1111	21100	(10,1,0,0;8)	$8.5996\,10^9$	1/16
6-10	0	01	111	1120	21010	(8,2,0,0;2)	$1.7199\,10^{10}$	1/8
6-11	0	01	022	2010	21001	(4,4,0,0;4)	$2.1499\,10^9$	1/64
6-12	0	01	012	1111	31000	(7,1,1,0;4)	$2.8665\,10^9$	1/48
6-13	0	01	012	1201	30100	(5,2,1,0,;2)	$2.8665\,10^9$	1/48
6-14	0	01	102	1201	21100	(6,3,0,0;12)	$1.4333\,10^9$	1/96
6-15	0	00	112	1210	21100	(6,3,0,0;12)	$1.4333\,10^9$	1/96
6-16	0	00	013	1300	30100	(3,0,3,0;6)	$1.0617\,10^8$	1/1296
6-17	0	00	013	1210	31000	(4,1,2,0;4)	$4.7776\,10^8$	1/288

7 vertices, 8 loops: 41 diagrams								
j	2	33	444	5555	66666	777777		
i	1	12	123	1234	12345	123456		
No.	G^{ij}						$(S,D,T,Q;N_{\mathrm{IVP}})$	M_G	W_G
7-1	0	00	111	1110	11101	111100	(14,0,0,0;48)	$4.8158\,10^{11}$	1/48
7-2	0	01	101	1101	11101	111100	(14,0,0,0;14)	$1.6511\,10^{12}$	1/14
7-3	0	00	003	1101	12001	211000	(7,2,1,0;1)	$9.6316\,10^{11}$	1/24
7-4	0	00	002	0202	11200	310000	(3,4,1,0;2)	$1.2039\,10^{11}$	1/192
7-5	0	00	002	0211	11110	310000	(7,2,1,0;2)	$4.8158\,10^{11}$	1/48
7-6	0	00	011	0121	12010	301000	(7,2,1,0;2)	$4.8158\,10^{11}$	1/48
7-7	0	01	011	0111	10111	310000	(11,0,1,0;12)	$3.2105\,10^{11}$	1/72
7-8	0	00	002	0301	11200	300100	(4,2,2,0;4)	$4.0132\,10^{10}$	1/576
7-9	0	00	002	0112	12100	310000	(5,3,1,0;2)	$2.4079\,10^{11}$	1/96
7-10	0	00	011	0121	11110	310000	(9,1,1,0;2)	$9.6316\,10^{11}$	1/24
7-11	0	00	003	1101	11101	220000	(7,2,1,0;4)	$2.4079\,10^{11}$	1/96
7-12	0	00	002	0211	12010	301000	(5,3,1,0;1)	$4.8158\,10^{11}$	1/48
7-13	0	00	002	0112	13000	301000	(4,2,2,0;2)	$8.0263\,10^{10}$	1/288
7-14	0	01	011	1011	11011	211000	(12,1,0,0;4)	$2.8895\,10^{12}$	1/8
7-15	0	00	011	1111	11110	211000	(12,1,0,0;8)	$1.4447\,10^{12}$	1/16
7-16	0	01	011	0111	20011	211000	(10,2,0,0;8)	$7.2237\,10^{11}$	1/32
7-17	0	01	011	1011	11110	210010	(12,1,0,0;4)	$2.8895\,10^{12}$	1/8
7-18	0	00	011	1111	11200	210100	(10,2,0,0;2)	$2.8895\,10^{12}$	1/8
7-19	0	00	012	1110	11101	210100	(10,2,0,0;4)	$1.4447\,10^{12}$	1/16
7-20	0	00	011	0121	21010	211000	(8,3,0,0;1)	$2.8895\,10^{12}$	1/8
7-21	0	01	011	1011	12010	201010	(10,2,0,0;2)	$2.8895\,10^{12}$	1/8
7-22	0	00	012	1110	12001	201100	(8,3,0,0;1)	$2.8895\,10^{12}$	1/8
7-23	0	00	011	1021	12100	210100	(8,3,0,0;2)	$1.4447\,10^{12}$	1/16
7-24	0	00	002	1111	12010	211000	(8,3,0,0;2)	$1.4447\,10^{12}$	1/16
7-25	0	00	002	1102	12100	211000	(6,4,0,0;4)	$3.6118\,10^{11}$	1/64
7-26	0	00	012	1110	11110	210010	(10,2,0,0;2)	$2.8895\,10^{12}$	1/8
7-27	0	00	002	0202	21100	211000	(4,5,0,0;4)	$1.8059\,10^{11}$	1/128
7-28	0	00	011	0121	20110	220000	(6,4,0,0;2)	$7.2237\,10^{11}$	1/32
7-29	0	00	012	0210	20101	210100	(6,4,0,0;4)	$3.6118\,10^{11}$	1/64
7-30	0	00	002	0211	21010	211000	(6,4,0,0;2)	$7.2237\,10^{11}$	1/32
7-31	0	00	011	0220	20110	210100	(6,4,0,0;4)	$3.6118\,10^{11}$	1/64
7-32	0	01	011	0112	20100	210001	(8,3,0,0;4)	$7.2237\,10^{11}$	1/32
7-33	0	00	002	0112	21100	220000	(4,5,0,0;2)	$3.6118\,10^{11}$	1/64
7-34	0	00	011	1021	12010	211000	(8,3,0,0;4)	$7.2237\,10^{11}$	1/32
7-35	0	00	012	0210	20110	210010	(6,4,0,0;4)	$3.6118\,10^{11}$	1/64
7-36	0	00	002	1111	11110	220000	(8,3,0,0;8)	$3.6118\,10^{11}$	1/64
7-37	0	01	012	1011	11001	210001	(10,2,0,0;4)	$1.4447\,10^{12}$	1/16
7-38	0	00	012	0211	20100	210001	(6,4,0,0;2)	$7.2237\,10^{11}$	1/32
7-39	0	00	002	0202	20200	220000	(0,7,0,0;14)	$1.2899\,10^{10}$	1/1792
7-40	0	00	002	0211	20110	220000	(4,5,0,0;4)	$1.8059\,10^{11}$	1/128
7-41	0	01	012	1011	12000	200011	(8,3,0,0;8)	$3.6118\,10^{11}$	1/64

To exclude diagrams with cutlines each line is taken away, and the modified diagram is checked for connectedness. In this way, all 1PI two-point diagrams can be obtained.

Two-point diagrams with tadpole parts are generated from vacuum diagrams with cutvertices, whereas vacuum diagrams without cutvertices lead to tadpole-free two-point diagrams.

(a)   (b1)   (b2)   (b3)

FIGURE 14.4 Identical vertex permutations transform all lines of the vacuum diagram (a) into one of the lines 1, 2, or 3. The three diagrams (b1,b2,b3) of the connected two-point function $G_{\mathrm{c}}^{(2)}$ are generated by cutting the line 1, 2, or 3, respectively.

All 1PI two-point diagrams, with and without tadpole parts, contribute to the mass renormalization. They are shown in Fig. 14.7. The weight factors and the unique matrix representations for all 1PI two-point diagrams are listed on page 265 in Table 14.3 . These diagrams can also be generated from the four-point diagrams as explained in Subsection 14.2.4. For the field renormalization, on the other hand, we need only the two-point diagrams without tadpoles. These diagrams can of course also be generated from the four-point diagrams, but they can more easily be generated directly from the vacuum diagrams without cutvertices listed in Fig 14.3. The resulting two-point diagrams are then automatically tadpole-free. The weight factors and the unique matrix representations for all tadpole-free 1PI two-point diagrams are listed in Table 14.3 .

A two-point diagram with L loops is generated out of a vacuum diagram with $L + 1$ loops, such that for the five-loop calculation the vacuum diagrams up to six loops are needed. The only two-point diagram with one loop is the tadpole diagram itself. The lowest-order vacuum diagram leading to a non-trivial two-point function has therefore two loops. In Table 14.2 we list the numbers of diagrams up to six and eight loops.

L	Field	Coupling	Mass
1	-	1	1
2	1	2	2
3	1	8	6
4	4	26	21
5	11	124	90
6	50	627	444

TABLE 14.2   Number of two- and four-point diagrams with L loops contributing to counterterms. The field renormalization requires all 1PI two-point diagrams without tadpole parts. The coupling constant requires renormalization of all 1PI four-point diagrams without tadpoles. Out of the latter, those with only three external vertices contribute to the mass renormalization.

14.2.3   Four-Point Diagrams

A four-point diagram with p vertices is generated by removing a vertex from a vacuum diagram with $p + 1$ vertices. The resulting four-point diagram is not connected if the removed vertex is either a cutvertex or a vertex with self-connections. The resulting four-point diagram will have no tadpole part if the vacuum diagram from which it is obtained has no cutvertex. All vacuum diagrams of this type up to seven vertices are listed in Fig. 14.3. Cutting vertices in these diagrams generates all connected four-point diagrams without tadpole parts. Removing vertices which participate in identical vertex permutations leads to equivalent diagrams. An example is drawn in Fig. 14.5. Choosing only one of these vertices avoids repeated generation of the same diagram [3].

Finally, the 1PI diagrams have to be selected. Each line of the four-point diagrams is taken away, and the modified diagram is checked for connectedness. For example, if in Fig. 14.3 the four-vertex vacuum diagram No. 3 is turned into a four-point diagram by deleting any of the vertices we obtain a connected diagram which is not 1PI:

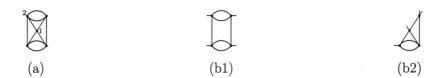

(a)   (b1)   (b2)

FIGURE 14.5 The vertices at the four outer corners of the vacuum diagram (a) are transformed into one another by permutations which keep the figure unchanged. Only by removing the vertex 1 or 2 can we obtain different four-point diagrams shown in (b1) and (b2).

and therefore not connected after deleting the line connecting the vertex to the rest. A detailed list of the matrix representations and the weight factors of the 1PI four-point functions without tadpoles is given in Table 14.3 .

14.2.4   Four-Point Diagrams for Mass Renormalization

Recall that the diagrams relevant for the mass renormalization are a subset of the connected four-point diagrams with p vertices. They emerge by differentiating the connected two-point diagrams with $p - 1$ vertices with respect to m^2. The differentiation generates a ϕ^2-vertex insertion, as explained in Section 2.4. The resulting four-point diagrams possess a vertex with two external lines carrying no momenta. Such four-point diagrams are found by inspecting the matrix representation. There must exist a double connection between an internal vertex and zero, which codes the external vertex. The matrix representation for the two-point diagram follows simply by deleting the double connection to the external vertex. The weight factor for the two-point diagram can then be read off from this matrix. Note that there is no one-to-one correspondence between two-point and four-point diagrams. The differentiation generates several four-point diagrams out of one two-point diagram. Differentiation with respect to the mass of topologically equivalent lines leads to equivalent four-point diagrams. The differentiation therefore generates a factor giving the number of lines in the two-point diagram which are topologically equivalent to the differentiated line. This factor is then combined with the weight factor of the two-point diagram $W^{m^2}_{G^{(2)}}$ to give the mass weight factor of the four-point diagram $W^{m^2}_{\partial_{m^2} G^{(2)}}$. Differentiating an n-fold connection gives the factor n. If a two-point diagram has only one external vertex, the two internal lines connected to it are topologically equivalent lines, and the differentiation of each of them leads to the same four-point diagram. Such a diagram always receives an extra factor 2. Two examples are shown in Fig. 14.6.

14.2.5   Check for Number of Connected Diagrams

The sum of the weight factors of all diagrams with a certain number of vertices and external lines is given by Wick's rule as discussed in Chapter 3. For checking the reduced number of selected 1PI diagrams without tadpole parts, however, the rule is of no use.

The sums of the weight factors of the diagrams appearing in the expansion coefficients of the partition function Z_p, the two-point function $G^{(2)}_p$, and the four-point function $G^{(4)}_p$ are called w_{Z_p}, $w_{G^{(2)}_p}$, and $w_{G^{(4)}_p}$. According to the Wick rule in Chapter 3, the sum of the weight factors for all diagrams with p vertices and n external lines is

$$w_{G^{(n)}_p} = \frac{(4p + n - 1)!!}{(4!)^p p!}.$$

(a)   (b)

FIGURE 14.6 Weight factors of four-point diagrams relevant for coupling constant and for mass renormalization. (a) The weight factor for this four-point diagram is $W^g_{G^{(4)}} = 3/4$, the factor 3 coming from the symmetrization of the external lines. In the mass renormalization, the same four-point diagram is used with two external lines set to zero. The weight factor $W^{m^2}_{\partial_{m^2}G^{(4)}}$ of this diagram is related to the two-point diagram from which it emerges by differentiation as follows: the weight factor of the two-point diagram is 1/8. Differentiation with respect to the mass in the lines labeled by 1 generates a combinatorial factor 2 since the lines are topologically equivalent. The weight factor to be used for the mass renormalization together with the four-point diagram is $W^{m^2}_{\partial_{m^2}G^{(4)}} = 1/4$. (b) The weight factor of the two-point diagram is 1/4. Differentiation generates again a factor 2 even if the lines 1 and 2 are not topologically equivalent. Hence $W^{m^2}_{G^{(4)}} = 1/2$, but $W^g_{G^{(4)}} = 3/2$.

If only the connected diagrams in the cumulants Z_{pc}, $G^{(2)}_{pc}$, and $G^{(4)}_{pc}$ are considered, the sums of the weight factors are called $w_{Z_{pc}}$, $w_{G^{(2)}_{pc}}$, and $w_{G^{(4)}_{pc}}$. These were calculated from recursion relations on p. 67, but they may also be found by the technique explained in the previous sections. A relation between the weight factors for all diagrams and those for the connected diagrams is needed. It is obtained from the defining equation $Z[j] = e^{W[j]}$ of the generating functional $W[j]$ of all connected correlation functions [recall Eq. (3.19)]. For zero external source this equation relates all disconnected and connected vacuum diagrams with each other. Inserting $\sum_{p=0} w_{Z_p} x^p$ for the sum of the weights of all vacuum diagrams, and $\sum_{p=0} w_{Z_{pc}} x^p$ for the sum of the weights of all connected vacuum diagrams, we find

$$\sum_{p=0}^{\infty} w_{Z_p} x^p = \sum_{m=0}^{\infty} \frac{1}{m!} \left[\sum_{p=0}^{\infty} w_{Z_{pc}} x^p \right]^m . \tag{14.5}$$

Composing the coefficients of each order in x yields equations between $w_{Z_{pc}}$ and w_{Z_p}. To calculate the weight factors of the two- and four-point diagrams, the results of Section 3.3.2 or 5.3 are used, especially Eqs. (3.21), (3.23), and (3.24). Replacing in Eq. (3.21) the sum over all n-point functions $\sum_{p=0} G^{(n)}_p$ by the sum over their weight factors $\sum_{p=0} w_{G^{(n)}_p} x^p$ and $\sum_{p=0} G^{(n)}_{p\,\text{ext}}$ for $n = 2$ according to Eq. (3.23) by $\sum_{p=0} w_{G^{(2)}_{pc}} x^p$, and for $n = 4$ according to Eq. (3.24) by $\sum_{p=0} w_{G^{(4)}_{pc}} x^p + 3(\sum_{p=0} w_{G^{(2)}_{pc}} x^p)^2$, where $w_{G^{(2)}_0} = 1$, $w_{G^{(2)}_0} = 3$, $w_{G^{(4)}_{0c}} = 1$, $w_{G^{(4)}_{0c}} = 0$, we find:

$$\sum_{p=0}^{\infty} w_{G^{(2)}_p} x^p = \left(\sum_{p=0}^{\infty} w_{G^{(0)}_p} x^p \right) \times \left(\sum_{p=0}^{\infty} w_{G^{(2)}_p} x^p \right) , \tag{14.6}$$

$$\sum_{p=0}^{\infty} w_{G^{(4)}_p} x^p = \left(\sum_{p=0}^{\infty} w_{G^{(0)}_p} x^p \right) \times \left[3 \left(\sum_{p=0}^{\infty} w_{G^{(2)}_p} x^p \right)^2 + \sum_{p=0}^{\infty} w_{G^{(4)}_p} x^p \right] . \tag{14.7}$$

TABLE 14.3 Matrix representation for all diagrams contributing to Z_ϕ, Z_m and Z_g specified by the associated numbers $\hat{G} = G^{10}G^{11}G^{20}G^{21}G^{22}G^{30}G^{31}G^{32}G^{33}\ldots$ of Eq. (14.2). They are ordered lexicographically in $\hat{G}$. The vertical lines separating groups of digits in $\hat{G}$ mark the successive rows of the below-diagonal part of the matrix G^{ij}. The entries for i and j show which matrix element G^{ij} is displayed below, The matrix elements specify the number of lines connecting the two vertices i and j. The tables display further the numbers of self-, double, and triple connections, and of irreducible vertex permutations ($S, D, T, N_{\mathrm{IVP}}$), the multiplicities M_G, and the weights W_G. The last column in the table for Z_g gives the running number of the diagrams used in the lists in Appendices A.1 and A.2. In these lists the two-point diagrams for the calculation of Z_m do not appear because Z_m is obtained from four-point diagrams. The two-point diagrams are therefore shown separately in Fig. 14.7 of this chapter.

Z_ϕ, 2 vertices, 2 loops: 1 diagram		
i	11	222
j	01	012
#	G^{ij}		$(S,D,T;N_{\mathrm{IVP}})$	M_G	W_G	No.
2-1	10	130	(0,0,1;2)	192	1/6	1

Z_ϕ, 3 vertices, 3 loops: 1 diagram			
i	11	222	3333
j	01	012	0123
#	G^{ij}			$(S,D,T;N_{\mathrm{IVP}})$	M_G	W_G	No.
3-1	00	120	1210	(0,2,0;2)	20736	1/4	1

Z_ϕ, 4 vertices, 4 loops: 4 diagrams				
i	11	222	3333	44444
j	01	012	0123	01234
#	G^{ij}				$(S,D,T;N_{\mathrm{IVP}})$	M_G	W_G	No.
4-1	00	010	1120	12100	(0,2,0;2)	1990656	1/4	4
4-2	00	020	1020	12010	(0,3,0;2)	995328	1/8	1
4-3	00	020	1110	11110	(0,1,0;4)	1990656	1/4	3
4-4	00	030	1010	11020	(0,1,1;2)	663552	1/12	2

Z_ϕ, 5 vertices, 5 loops: 11 diagrams					
i	11	222	3333	44444	555555
j	01	012	0123	01234	012345
#	G^{ij}					$(S,D,T;N_{\mathrm{IVP}})$	M_G	W_G	No.
5-1	00	000	0030	12010	121000	(0,2,1;2)	39813120	1/24	3
5-2	00	000	0120	11110	121000	(0,2,0;1)	477757440	1/2	5
5-3	00	000	0120	11200	120100	(0,3,0;1)	238878720	1/4	6
5-4	00	000	0130	11100	120010	(0,1,1;1)	159252480	1/6	2
5-5	00	000	0220	10200	120010	(0,4,0;2)	59719680	1/16	1
5-6	00	000	0220	11100	111010	(0,2,0;4)	119439360	1/8	7
5-7	00	010	0110	10120	121000	(0,2,0;2)	238878720	1/4	9
5-8	00	010	0110	11110	111100	(0,0,0;12)	159252480	1/6	11
5-9	00	010	0120	10110	120010	(0,2,0;2)	238878720	1/4	10
5-10	00	010	0120	11010	111010	(0,1,0;2)	477757440	1/2	8
5-11	00	010	0220	10100	110020	(0,3,0;2)	119439360	1/8	4

Z_m, 2 vertices, 2 loops: 2 diagrams					
i	11	222			
j	01	012			
#	G^{ij}		$(S,D,T;N_{\text{IVP}})$	M_G	$W_{\bar{G}}$
2-1	10	130	(0,0,1;2)	192	1/6
2-2	01	220	(1,2,0;1)	288	1/4

Z_m, 3 vertices, 3 loops: 5 diagrams						
i	11	222	3333			
j	01	012	0123			
#	G^{ij}			$(S,D,T;N_{\text{IVP}})$	M_G	$W_{\bar{G}}$
3-1	00	030	2110	(0,1,1;2)	6912	1/12
3-2	00	120	1210	(0,2,0;2)	20736	1/4
3-3	01	110	1120	(1,1,0;2)	20736	1/4
3-4	00	021	2200	(1,3,0;1)	10368	1/8
3-5	01	011	2110	(2,1,0;2)	10368	1/8

Z_m, 4 vertices, 4 loops: 17 diagrams							
i	11	222	3333	44444			
j	01	012	0123	01234			
#	G^{ij}				$(S,D,T;N_{\text{IVP}})$	M_G	$W_{\bar{G}}$
4-1	00	001	0220	22000	(1,4,0;1)	497664	1/16
4-2	00	001	0310	21100	(1,1,1;1)	663552	1/12
4-3	00	001	1210	12100	(1,2,0;2)	995328	1/8
4-4	00	010	0130	22000	(0,2,1;2)	331776	1/24
4-5	00	010	0220	21100	(0,3,0;2)	995328	1/8
4-6	00	010	1120	12100	(0,2,0;2)	1990656	1/4
4-7	00	010	1120	12100	(0,2,0;2)	1990656	1/4
4-8	00	011	0111	22000	(2,2,0;2)	497664	1/16
4-9	00	011	0201	21100	(2,2,0;1)	995328	1/8
4-10	00	011	0210	21010	(1,2,0;2)	995328	1/8
4-11	00	011	1110	12010	(1,1,0;1)	3981312	1/2
4-12	00	020	1020	12010	(0,3,0;2)	995328	1/8
4-13	00	021	1100	11020	(1,2,0;2)	995328	1/8
4-14	00	030	1010	11020	(0,1,1;2)	663552	1/12
4-15	01	001	0111	21100	(3,1,0;2)	497664	1/16
4-16	01	001	1110	11110	(2,0,0;4)	995328	1/8
4-17	01	011	1010	11020	(2,1,0;2)	995328	1/8

Z_m, 5 vertices, 5 loops: 68 diagrams								
i	11	222	3333	44444	555555			
j	01	012	0123	01234	012345			
#	G^{ij}					$(S,D,T;N_{\text{IVP}})$	M_G	$W_{\bar{G}}$
5-1	00	000	0021	02200	220000	(1,5,0;1)	29859840	1/32
5-2	00	000	0021	03100	211000	(1,2,1;1)	39813120	1/24
5-3	00	000	0021	12100	121000	(1,3,0;2)	59719680	1/16
5-4	00	000	0030	02110	220000	(0,3,1;2)	19906560	1/48
5-5	00	000	0030	03010	211000	(0,1,2;2)	13271040	1/72
5-6	00	000	0030	12010	121000	(0,2,1;2)	39813120	1/24
5-7	00	000	0111	01300	220000	(1,2,1;1)	39813120	1/24
5-8	00	000	0111	11200	121000	(1,2,0;2)	119439360	1/8
5-9	00	000	0120	01210	220000	(0,4,0;2)	59719680	1/16
5-10	00	000	0120	02110	211000	(0,3,0;2)	119439360	1/8

Z_m, 5 vertices, 5 loops: 68 diagrams								
i	11	222	3333	44444	555555			
j	01	012	0123	01234	012345			
#			G^{ij}			$(S,D,T;N_{\mathrm{IVP}})$	M_G	$W_{\bar{G}}$
5-11	00	000	0120	02200	210100	(0,4,0;2)	59719680	1/16
5-12	00	000	0120	03100	201100	(0,2,1;2)	39813120	1/24
5-13	00	000	0120	11110	121000	(0,2,0;1)	477757440	1/2
5-14	00	000	0012	01120	0120100	(0,3,0;1)	238878720	1/4
5-15	00	000	0130	11100	120010	(0,1,1;1)	159252480	1/6
5-16	00	000	0220	01101	211000	(1,3,0;2)	59719680	1/16
5-17	00	000	0220	10200	120010	(0,4,0;2)	59719680	1/16
5-18	00	000	0220	11100	111010	(0,2,0;4)	119439360	1/8
5-19	00	001	0011	02110	220000	(2,3,0;2)	29859840	1/32
5-20	00	001	0011	03010	211000	(2,1,1;1)	39813120	1/24
5-21	00	001	0011	12010	121000	(2,2,0;2)	59719680	1/16
5-22	00	001	0101	01210	220000	(2,3,0;1)	59719680	1/16
5-23	00	001	0101	02110	211000	(2,1,0;1)	238878720	1/4
5-24	00	001	0101	02200	210100	(2,3,0;1)	59719680	1/16
5-25	00	001	0101	03100	201100	(2,1,1;2)	19906560	1/48
5-26	00	001	0110	01120	220000	(1,3,0;2)	59719680	1/16
5-27	00	001	0110	02020	211000	(1,3,0;1)	119439360	1/8
5-28	00	001	0110	02110	210100	(1,2,0;1)	238878720	1/4
5-29	00	001	0110	03010	201100	(1,1,1;2)	39813120	1/24
5-30	00	001	0110	11020	121000	(1,2,0;1)	238878720	1/4
5-31	00	001	0110	11110	120100	(1,1,0;1)	477757440	1/2
5-32	00	001	0111	01101	220000	(3,2,0;2)	29859840	1/32
5-33	00	001	0111	02001	211000	(3,2,0;1)	59719680	1/16
5-34	00	001	0111	02100	210010	(2,2,0;2)	59719680	1/16
5-35	00	001	0111	11100	120010	(2,1,0;1)	238878720	1/4
5-36	00	001	0120	02001	210100	(2,3,0;2)	29859840	1/32
5-37	00	001	0120	02010	210010	(1,3,0;2)	59719680	1/16
5-38	00	001	0120	11010	120010	(1,2,0;1)	238878720	1/4
5-39	00	001	0200	10120	121000	(1,3,0;2)	59719680	1/16
5-40	00	001	0200	11110	111100	(1,1,0;4)	119439360	1/8
5-41	00	001	0201	11100	111010	(2,1,0;2)	119439360	1/8
5-42	00	001	0210	10110	120010	(1,2,0;1)	238878720	1/4
5-43	00	001	0210	11010	111010	(1,1,0;1)	477757440	1/2
5-44	00	001	0220	11000	110020	(1,3,0;2)	59719680	1/16
5-45	00	001	0300	10110	111010	(1,0,1;2)	79626240	1/12
5-46	00	001	0310	10100	110020	(1,1,1;1)	79626240	1/12
5-47	00	010	0110	01120	211000	(0,2,0;4)	119439360	1/8
5-48	00	010	0110	11020	112000	(0,2,0;2)	238878720	1/4
5-49	00	010	0110	11110	111100	(0,0,0;12)	159252480	1/6
5-50	00	010	0111	01101	211000	(2,1,0;4)	59719680	1/16
5-51	00	010	0111	10200	120010	(1,2,0;2)	119439360	1/8
5-52	00	010	0111	11100	111010	(1,0,0;4)	238878720	1/4
5-53	00	010	0120	10110	120010	(0,2,0;2)	238878720	1/4
5-54	00	010	0120	11010	111010	(0,1,0;2)	477757440	1/2
5-55	00	010	0130	11000	110020	(0,1,1;4)	39813020	1/24
5-56	00	010	0220	10100	110020	(0,3,0;2)	119439360	1/8
5-57	00	011	0101	01011	211000	(3,1,0;2)	59719680	1/16
5-58	00	011	0101	02001	201100	(3,2,0;2)	29859840	1/32
5-59	00	011	0101	10110	120010	(2,1,0;2)	119439360	1/8
5-60	00	011	0101	11010	111010	(2,0,0;2)	238878720	1/4
5-61	00	011	0111	11000	110020	(2,1,0;4)	59719680	1/16
5-62	00	011	0201	10100	110020	(2,2,0;1)	119439360	1/8
5-63	00	020	0011	01101	210100	(2,2,0;1)	119439360	1/8
5-64	00	020	0111	10100	110020	(1,2,0;2)	119439360	1/8
5-65	01	001	0001	11110	111100	(3,0,0;12)	19906560	1/48
5-66	01	001	0011	01011	211000	(4,1,0;2)	29859840	1/32
5-67	01	001	0011	11010	110100	(3,0,0;2)	119439360	1/8
5-68	01	001	0111	10100	110020	(3,1,0;2)	59719680	1/16

Z_g, 2 vertices, 1 loop: 1 diagram					
i	11	222			
j	01	012			
#	G^{ij}		$(S,D,T;N_{IVP})$	M_G	$W_{\bar{G}}$	No.
2-1	20	220	(0,3,0;2)	1728	3/2	1

Z_g, 3 vertices, 2 loops: 2 diagrams						
i	11	222	3333			
j	01	012	0123			
#	G^{ij}			$e(S,D,T;N_{IVP})$	M_G	$W_{\bar{G}}$	No.
3-1	00	220	2200	(0,4,0;2)	62208	3/4	1
3-2	10	120	2110	(0,2,0;2)	248832	3	2

Z_g, 4 vertices, 3 loops: 8 diagrams							
i	11	222	3333	44444			
j	01	012	0123	01234			
#	G^{ij}				$(S,D,T;N_{IVP})$	M_G	$W_{\bar{G}}$	No.
4-1	00	020	2020	22000	(0,5,0;2)	2985984	3/8	1
4-2	00	020	2110	21100	(0,3,0;4)	5971968	3/4	8
4-3	00	030	2010	21010	(0,2,1;2)	3981312	1/2	3
4-4	00	110	1120	22000	(0,3,0;2)	11943936	3/2	2
4-5	00	110	1210	21100	(0,2,0;1)	47775744	6	7
4-6	00	120	1200	20110	(0,3,0;2)	11943936	3/2	5
4-7	10	100	1120	12100	(0,2,0;4)	11943936	3/2	6
4-8	10	110	1110	11110	(0,0,0;24)	7962624	1	4

Z_g, 5 vertices, 4 loops: 26 diagrams								
i	11	222	3333	44444	555555			
j	01	012	0123	01234	012345			
#	G^{ij}					$(S,D,T;N_{IVP})$	M_G	$W_{\bar{G}}$	No.
5-1	00	000	0130	21100	220000	(0,3,1;1)	477757440	1/2	5
5-2	00	000	0220	20200	220000	(0,6,0;2)	179159040	3/16	1
5-3	00	000	0220	21100	211000	(0,4,0;4)	358318080	3/8	14
5-4	00	000	1120	11200	220000	(0,4,0;2)	716636160	3/4	4
5-5	00	000	1120	12100	211000	(0,3,0;2)	1433272320	3/2	22
5-6	00	010	0120	20110	220000	(0,4,0;2)	716636160	3/4	7
5-7	00	010	0120	21010	211000	(0,3,0;2)	1433272320	3/2	8
5-8	00	010	0220	20100	210010	(0,4,0;2)	716636160	3/4	15
5-9	00	010	1020	11110	220000	(0,3,0;1)	2866544640	3	6
5-10	00	010	1020	12010	211000	(0,3,0;2)	1433272320	3/2	20
5-11	00	010	1020	12100	210100	(0,3,0;1)	2866544640	3	23
5-12	00	010	1110	11110	211000	(0,1,0;4)	2866544640	3	21
5-13	00	010	1110	11200	2101 00	(0,2,0;1)	5733089280	6	26
5-14	00	020	1010	10120	220000	(0,4,0;2)	716636160	3/4	2
5-15	00	020	1010	11020	211000	(0,3,0;2)	1433272320	3/2	9
5-16	00	020	1010	11110	210100	(0,2,0;1)	5733089280	6	25
5-17	00	020	1010	12010	201100	(0,3,0;1)	2866544640	3	24
5-18	00	020	1020	12000	200110	(0,4,0;2)	716636160	3/4	18
5-19	00	020	1110	11100	200110	(0,2,0;4)	1433272320	3/2	17
5-20	00	030	1000	10120	210100	(0,2,1;1)	955514880	1	19
5-21	00	030	1010	11010	200110	(0,1,1;2)	955514880	1	16
5-22	00	100	1020	12010	121000	(0,3,0;2)	1433272320	3/2	13
5-23	00	100	1110	11110	121000	(0,1,0;2)	5733089280	6	11
5-24	00	100	1110	11200	120100	(0,2,0;1)	5733089280	6	12
5-25	00	110	1100	11020	112000	(0,2,0;8)	716636160	3/4	3
5-26	00	110	1100	11110	111100	(0,0,0;8)	2866544640	3	10

#	11	222	3333	44444	555555	6666666				
	01	012	0123	01234	012345	0123456				
				G^{ij}			$(S,D,T;N_{IVP})$	M_G	$W_{\bar{G}}$	No.
6-1	00	000	0020	01120	211000	2200000	(0,5,0;1)	103195607040	3/4	99
6-2	00	000	0020	02020	202000	2200000	(0,7,0;2)	12899450880	3/32	117
6-3	00	000	0020	02020	211000	2110000	(0,5,0;4)	25798901760	3/16	79
6-4	00	000	0020	02110	201100	2200000	(0,5,0;2)	51597803520	3/8	120
6-5	00	000	0020	02110	210100	2110000	(0,4,0;2)	103195607040	3/4	93
6-6	00	000	0020	03010	201100	2110000	(0,3,1;1)	68797071360	1/2	21
6-7	00	000	0020	03010	202000	2101000	(0,4,1;1)	34398535680	1/4	7
6-8	00	000	0020	11020	112000	2200000	(0,5,0;2)	51597803520	3/8	78
6-9	00	000	0020	11020	121000	2110000	(0,4,0;1)	206391214080	3/2	71
6-10	00	000	0020	11110	111100	2200000	(0,3,0;4)	103195607040	3/4	108
6-11	00	000	0020	11110	120100	2110000	(0,3,0;1)	412782428160	3	69
6-12	00	000	0020	12010	120100	2020000	(0,5,0;2)	51 597803520	3/8	80
6-13	00	000	0020	12010	121000	2011000	(0,4,0;2)	103195607040	3/4	92
6-14	00	000	0030	03000	200110	2110000	(0,2,2;4)	5733089280	1/24	14
6-15	00	000	0030	03010	201000	2100010	(0,2,2;2)	11466178560	1/12	23
6-16	00	000	0030	11010	111010	2200000	(0,2,1;2)	68797071360	1/2	18
6-17	00	000	0030	11010	120100	2110000	(0,2,1;1)	137594142720	1	6
6-18	00	000	0030	11010	121000	2100100	(0,2,1;1)	137594142720	1	4
6-19	00	000	0030	12000	120100	2010100	(0,3,1;1)	68797071360	1/2	20
6-20	00	000	0110	01120	202000	2200000	(0,5,0;4)	25798901760	3/16	121
6-21	00	000	0110	01120	211000	2110000	(0,3,0;8)	51597803520	3/8	109
6-22	00	000	0110	01210	201100	2200000	(0,4,0;1)	206391214080	3/2	84
6-23	00	000	0110	01210	210100	2110000	(0,3,0;1)	412782428160	3	47
6-24	00	000	0110	01300	200200	2200000	(0,4,1;2)	17199267840	1/8	8
6-25	00	000	0110	01300	210100	2101000	(0,2,1;4)	34398535680	1/4	19
6-26	00	000	0110	02200	200200	2110000	(0,5,0;2)	51597803520	3/8	81
6-27	00	000	0110	02200	201100	2101000	(0,4,0;2)	103195607040	3/4	95
6-28	00	000	0110	10120	112000	2200000	(0,4,0;1)	206391214080	3/2	90
6-29	00	000	0110	10120	121000	2110000	(0,3,0;1)	412782428160	3,	70
6-30	00	000	0110	10210	111100	2200000	(0,3,0;1)	412782428160	3	46
6-31	00	000	0110	10210	120100	2110000	(0,3,0;2)	206391214080	3/2	102
6-32	00	000	0110	10210	121000	2101000	(0,3,0;1)	412782428160	3	63
6-33	00	000	0110	11110	111100	2110000	(0,1,0;4)	412782428160	3	29
6-34	00	000	0110	11110	112000	2101000	(0,2,0;1)	825564856320	6	38
6-35	00	000	0110	11200	121000	2002000	(0,4,0;2)	103195607040	3/4	72
6-36	00	000	0120	02100	200110	2110000	(0,4,0;4)	51597803520	3/8	73
6-37	00	000	0120	02100	201010	2101000	(0,4,0;4)	51597803520	3/8	88
6-38	00	000	0120	02100	201100	2100100	(0,4,0;4)	51597803520	3/8	106
6-39	00	000	0120	02110	201000	2100010	(0,4,0;2)	103195607040	3/4	116
6-40	00	000	0120	02200	200100	2100010	(0,5,0;2)	51597803520	3/8	100
6-41	00	000	0120	03100	200100	2010010	(0,3,1;2)	34398535680	1/4	15
6-42	00	000	0120	10110	111010	2200000	(0,3,0;1)	412782428160	3	48
6-43	00	000	0120	10110	120010	2110000	(0,3,0;1)	412782428160	3	65
6-44	00	000	0120	10110	121000	2100100	(0,3,0;1)	412782428160	3	101
6-45	00	000	0120	10200	110110	2200000	(0,4,0;1)	206391214080	3/2	94
6-46	00	000	0120	10200	120010	2101000	(0,4,0;1)	206391214080	3/2	86
6-47	00	000	0120	10200	120100	2100100	(0,4,0;1)	206391214080	3/2	104
6-48	00	000	0120	11010	111010	2110000	(0,2,0;1)	825564856320	6	37
6-49	00	000	0120	11010	112000	2100100	(0,2,0;1)	412782428160	3	50
6-50	00	000	0120	11010	120010	2020000	(0,4,0;1)	206391214080	3/2	83
6-51	00	000	0120	11010	121000	2010100	(0,3,0;1)	412782428160	3	64
6-52	00	000	0120	11100	111010	2101000	(0,2,0;2)	412782428160	3	43
6-53	00	000	0120	11100	111100	2100100	(0,2,0;1)	825564856320	6	74
6-54	00	000	0120	11100	120010	2011000	(0,3,0;1)	412782428160	3	62
6-55	00	000	0120	11100	120100	2010100	(0,3,0;1)	412782428160	3	57
6-56	00	000	0120	11100	121000	2001100	(0,3,0;1)	412782428160	3	67
6-57	00	000	0120	11110	120000	2010100	(0,3,0;1)	412782428160	3	59
6-58	00	000	0120	11200	120000	2001010	(0,4,0;1)	206391214080	3/2	89
6-59	00	000	0130	10100	110020	2200000	(0,3,1;1)	68797071360	1/2	22
6-60	00	000	0130	10100	120010	2100100	(0,2,1;1)	137594142720	1	5
6-61	00	000	0130	11000	110020	2110000	(0,2,1;2)	68797071360	1/2	9
6-62	00	000	0130	11000	111010	2100100	(0,1,1;1)	275188285440	2	16

Z_g, 6 vertices, 5 loops: 124 diagrams

			Z_g, 6 vertices, 5 loops: 124 diagrams							
i	11	222	3333	44444	555555	6666666				
j	01	012	0123	01234	012345	0123456				
#			G^{ij}				$(S,D,T;N_{\text{IVP}})$	M_G	$W_{\bar{G}}$	No.
6-63	00	000	0130	11000	120010	2010100	(0,2,1;1)	137594142720	1	12
6-64	00	000	0130	11100	120000	2000110	(0,2,1;1)	137594142720	1	10
6-65	00	000	0220	10100	101020	2200000	(0,5,0;2)	51597803520	3/8	119
6-66	0 0	000	0220	10100	110020	2110000	(0,4,0;2)	103195607040	3/4	91
6-67	00	000	0220	10100	111010	2100100	(0,3,0;1)	412782428160	3	51
6-68	00	000	0220	10100	120010	2010100	(0,4,0;1)	206391214080	3/2	82
6-69	00	000	0220	10200	120000	2000110	(0,5,0;2)	51597803520	3/8	118
6-70	00	000	0220	11100	111000	2000110	(0,3,0;4)	103195607040	3/4	107
6-71	00	000	1010	10120	121000	1210000	(0,3,0;4)	103195607040	3/4	110
6-72	00	000	1010	10210	120100	1210000	(0,3,0;1)	412782428160	3	45
6-73	00	000	1010	11020	112000	1210000	(0,3,0;2)	206391214080	3/2	66
6-74	00	000	1010	11110	111100	1210000	(0,1,0;2)	825564856320	6	27
6-75	00	000	1010	11110	112000	1201000	(0,2,0;1)	825564856320	6	39
6-76	00	000	1020	10200	120010	1201000	(0,4,0;4)	51597803520	3/8	96
6-77	00	000	1020	11100	111010	1201000	(0,2,0;4)	206391214080	3/2	41
6-78	00	000	1020	11100	111100	1200100	(0,2,0;2)	412782428160	3	52
6-79	00	000	1110	11100	111010	1111000	(0,0,0;16)	206391214080	3/2	2
6-80	00	010	0110	01110	200110	2110000	(0,2,0;8)	103195607040	3/4	32
6-81	00	010	0110	01120	201000	2100010	(0,3,0;4)	103195607040	3/4	98
6-82	00	010	0110	10020	102010	2200000	(0,4,0;2)	103195607040	3/4	87
6-83	00	010	0110	10020	111010	2110000	(0,2,0;2)	412782428160	3	42
6-84	00	010	0110	10020	112000	2100100	(0,3,0;1)	412782428160	3	56
6-85	00	010	0110	10110	101110	2200000	(0,2,0;4)	206391214080	3/2	31
6-86	00	010	0110	10110	110110	2110000	(0,1,0;2)	825564856320	6	26
6-87	00	010	0110	10110	111100	2100100	(0,1,0;2)	825564856320	6	36
6-88	00	010	0110	10110	120100	2010100	(0,2,0;1)	825564856320	6	54
6-89	00	010	0120	10010	101020	2200000	(0,4,0;2)	103195607040	3/4	105
6-90	00	010	0120	10010	110020	2110000	(0,3,0;1)	412782428160	3	61
6-91	00	010	0120	10010	111010	2100100	(0,2,0;1)	825564856320	6	75
6-92	00	010	0120	10110	120010	2010100	(0,3,0;1)	412782428160	1/2	58
6-93	00	010	0120	10110	110110	2100100	(0,2,0;2)	412782428160	3	113
6-94	00	010	0120	10110	120000	2000110	(0,3,0;2)	206391214080	3/2	123
6-95	00	010	0120	11 000	110020	2011000	(0,3,0;4)	103195607040	3/4	124
6-96	00	010	0120	11000	110110	2010100	(0,2,0;1)	825564856320	6	53
6-97	00	010	0120	11010	111000	2000110	(0,2,0;2)	412782428160	3	111
6-98	00	010	0220	10000	101020	2100100	(0,4,0;1)	206391214080	3/2	115
6-99	00	010	0220	10100	110010	2000110	(0,3,0;2)	206391214080	3/2	97
6-100	00	010	1000	10120	111100	1210000	(0,2,0;1)	825564856320	6	76
6-101	00	010	1000	10120	112000	1201000	(0,3,0;1)	412782428160	3	60
6-102	00	010	1000	10210	111100	1201000	(0,2,0;2)	412782428160	3	44
6-103	00	010	1000	11110	111100	1111000	(0,0,0;12)	275188285440	2	1
6-104	00	010	1010	10110	110110	1210000	(0,1,0;2)	825564856320	6	33
6-105	00	010	1010	10110	111010	1201000	(0,1,0;1)	1651129712640	12	24
6-106	00	010	1010	10120	111000	1200010	(0,2,0;2)	412782428160	3	114
6-107	00	010	1010	10200	110110	1201000	(0,2,0;2)	412782428160	3	40
6-108	00	010	1010	10200	110200	1200100	(0,3,0;2)	206391214080	3/2	68
6-109	00	010	1010	10210	110100	1200010	(0,2,0;2)	412782428160	3	77
6-110	00	010	1010	11010	111010	1111000	(0,0,0;2)	1651129712640	12	3
6-111	00	010	1010	11020	111000	1110010	(0,1,0;4)	412782428160	3	35
6-112	00	020	1000	10020	102010	1201000	(0,4,0;2)	103195607040	3/4	85
6-113	00	020	1000	10020	111010	1111000	(0,2,0;4)	206391214080	3/2	112
6-114	00	020	1000	10110	101110	1201000	(0,2,0;2)	412782428160	3	30
6-115	00	020	1000	10110	101200	1200100	(0,3,0;1)	412782428160	3	49
6-116	00	020	1000	10110	110110	1111000	(0,1,0;2)	825564856320	6	34
6-117	00	020	1000	10110	110200	1110100	(0,2,0;1)	825564856320	6	55
6-118	00	020	1010	10100	110020	1102000	(0,3,0;4)	103195607040	3/4	103
6-119	00	020	1010	10100	110110	1101100	(0,1,0;8)	206391214080	3/2	28
6-120	00	020	1010	10110	110010	1101010	(0,1,0;4)	412782428160	3	25
6-121	00	020	1010	10120	110000	1100020	(0,3,0;8)	51597803520	3/8	122
6-122	00	030	1000	10010	101020	1102000	(0,2,1;2)	68797071360	1/2	11
6-123	00	030	1000	10010	101110	1101100	(0,0,1;4)	137594142720	1	13
6-124	00	030	1 000	10020	101010	1101010	(0,1,1;2)	137594142720	1	17

In Table 14.4 we list the number of the diagrams appearing in Z_{pc}, $G^{(2)}_{pc}$, and $G^{(4)}_{pc}$, and the sum of the corresponding weight factors.

TABLE 14.4 Number of connected diagrams with 0, 2 and 4 external lines and the sums of their weight factors.

	Z_{pc}		$G^{(2)}_{pc}$		$G^{(4)}_{pc}$	
p	number	$w_{Z_{pc}}$	number	$w_{G^{(2)}_{pc}}$	number	$w_{G^{(4)}_{pc}}$
1	1	$\frac{1}{8}$	1	$\frac{1}{2}$	1	1
2	2	$\frac{1}{12}$	3	$\frac{2}{3}$	2	$\frac{7}{2}$
3	4	$\frac{11}{96}$	8	$\frac{11}{8}$	8	$\frac{149}{12}$
4	10	$\frac{17}{72}$	30	$\frac{34}{9}$	37	$\frac{197}{4}$
5	28	$\frac{619}{960}$	118	$\frac{619}{48}$	181	$\frac{15905}{72}$
6	97	$\frac{709}{324}$	548	$\frac{1418}{27}$	1010	$\frac{107113}{96}$

FIGURE 14.7 One-particle irreducible two-point diagrams up to five loops contributing to the mass renormalization constant Z_m. The order is the same as in the list on page 265 in Table 14.3 which lists the weight factors and matrix representations of the diagrams. We show these diagrams here, because they will not appear in the calculations and accompanying tables. For the actual calculation of Z_m we will use the integrals of the four-point diagrams as explained in Section 11.7. These 1PI four-point diagrams, which are also needed when calculating Z_g, are shown in Appendix A. There, we also show some 1PI two-point diagrams again, but only those with no tadpoles, since they contribute to Z_ϕ, and this contribution has to be calculated from the two-point diagrams themselves.

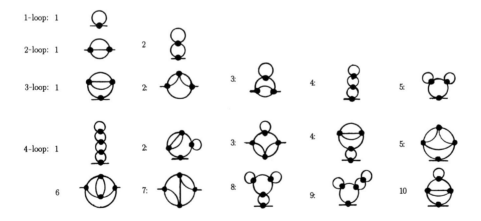

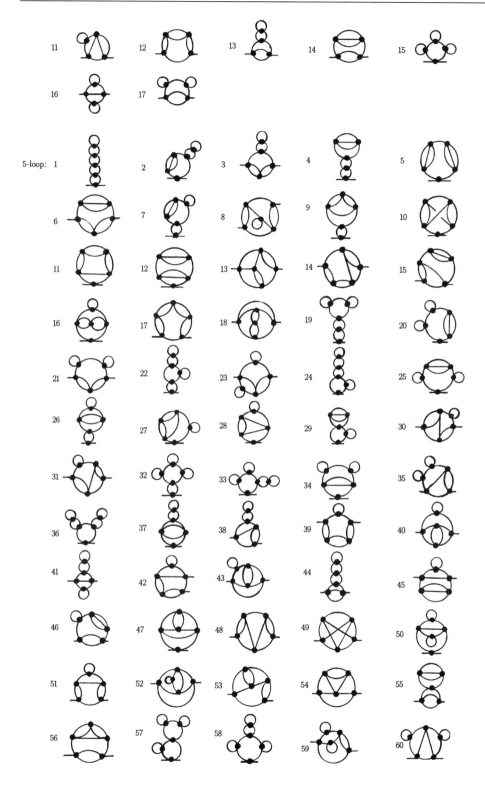

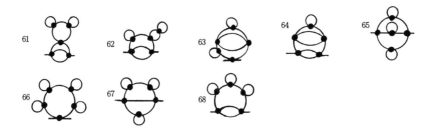

Notes and References

An overview on graph-theoretical notations can be found in
N. Nakanishi, *Graph Theory and Feynman Integrals*, Gordon and Breach, New York, 1971.

The individual citations in the text refer to:

[1] The construction of all diagrams and their multiplicities based on the recursion relation
in Fig. 5.2 is described in
H. Kleinert, A. Pelster, B. Kastening, M. Bachmann, Phys. Rev. E **62**, 1537 (2000) (hep-th/9907168).
The method was developed in
H. Kleinert, Fortschr. Phys. **30**, 187 (1982); ibid. **30**, 351 (1982).
The Mathematica program and its output are available on the internet (www.physik.fu-berlin/~kleinert/294/programs/index.html#5).

The diagrams in the original five-loop paper [2] were generated by a computer-algebraic
program written in our Berlin group in 1990 by J. Neu in his M.S. thesis.
There exist various other computer programs to generate diagrams and count multi-plicites, for instance *FeynArts* by
J. Külbeck, M. Böhm, and A. Denner, Comp. Phys. Comm. **60**, 165 (1991);
T. Hahn, Nucl. Phys. Proc. Suppl. **89**, 231 (2000) (hep-ph/0005029).
The programs can be downloaded from http://www-itp.physik.uni-karlsruhe.de/feynarts.
Another procedure is *QGRAF* by
P. Nogueira, J. Comput. Phys. **105**, 279 (1993),
available from ftp://gtae2.ist.utl.pt/pub/qgraf. These programs are based on a combi-natorial enumeration of all possible ways of connecting vertices by lines according to
Feynman's rules.

Important early enumeration work was done by
B.R. Heap, J. Math. Phys. **7**, 1582 (1966); J.F. Nagle, J. Math. Phys. **7**, 1588 (1966).

[2] H. Kleinert, J. Neu, V. Schulte-Frohlinde, K.G. Chetyrkin, S.A. Larin, Phys. Lett. B **272**,
39 (1991) (hep-th/9503230); Erratum ibid. **319**, 545 (1993).

[3] Note that this would be hard to achieve if we were to generate all four-point diagrams by
another procedure: by cutting two lines in a connected vacuum diagram with p vertices
rather than by removing a vertex.

15

Results of the Five-Loop Calculation

In this chapter, we summarize the results of the five-loop calculations for the power series expansions of the renormalization constants governing the relations between bare and renormalized parameters. These expansions were first calculated more than 15 years ago [1]. Several errors were corrected in a more recent recalculation [2]. An essential error had occurred in diagram No.116 (see page 449) fro which the R^* operation was explained in detail on page 222.

The renormalization constants receive contributions from the quadratically and logarithmically divergent 1PI-diagrams without tadpole parts, to be collectively called G_2 and G_4 for two- and four-point functions. Each diagram contributes a polynomial in $1/\varepsilon$, multiplied by a weight factor W and a symmetry factor S. The renormalized coupling constant in these expansions is $\bar{g}$ whose normalization was specified in Eq. (13.24). The loop expansions are:

$$Z_\phi = 1 - \frac{\partial}{\partial \mathbf{k}^2} \mathcal{K}\bar{R}\bar{\Gamma}^{(2)} = 1 + \sum_{L=1}^{5} \sum_{n_L=1}^{N_L} (-\bar{g})^L I_2(n_L) W_2(n_L) S_2(i_2(n_L)), \tag{15.1}$$

$$Z_{m^2} = 1 - \mathcal{K}\bar{R}\frac{\partial}{\partial m^2}\bar{\Gamma}^{(2)} = 1 - \sum_{L=1}^{5} \sum_{n_L=1}^{N_L} (-\bar{g})^L I_4(n_L) W_{m^2}(n_L) S_2(i_2(n_L)), \tag{15.2}$$

$$Z_g = 1 - \frac{1}{g}\mathcal{K}\bar{R}\bar{\Gamma}^{(4)} = 1 - \sum_{L=1}^{5} \sum_{n_L=1}^{N_L} (-\bar{g})^L I_4(n_L) W_4(n_L) S_4(i_4(n_L)). \tag{15.3}$$

The double sum runs over the number of loops L, and for each L over the different diagrams numbered by $n_L = 1, \ldots, N_L$. The symbols $I_2(n_L)$ and $I_4(n_L)$ denote the polynomials in $1/\varepsilon$ calculated from the diagrams $G_2(n_L)$ and $G_4(n_L)$, respectively. The calculation is done with the help of R-operations:

$$I_2(n_L) = 1/(-\bar{g})^L \, \mathcal{K}\bar{R}\, G_2(n_L), \tag{15.4}$$

$$I_4(n_L) = 1/(-\bar{g})^{L+1} \, \mathcal{K}\bar{R}\, G_4(n_L). \tag{15.5}$$

The steps in the calculation are listed diagrammatically in Appendix A. The zero-mass diagrams require the subtraction of several IR-counterterms. Their diagrammatic construction via the R^*-operation of Section 12.4 is displayed in Appendix A.3, where we have also tabulated all $1/\varepsilon^n$-expansions.

The final pole terms I_4 and I_2 are tabulated in Appendix B, together with the weight factors W_{m^2}, W_2, and W_4. The weight factors W_{m^2} are obtained as sums $\sum_i (1 + F_i)(W_2)_i$ over the weigth factors $(W_2)_i$ of the two-point diagrams from which the corresponding four-point diagrams I_4 have been obtained by graphical differentiation with respect to a line. The factors F_i counts the number of equivalent differentiated lines producing the same four-point diagrams (see Section 11.7). The symmetry factors $S^{O(N)}$ for the $O(N)$-symmetric interaction, and S^{cub} for a mixture of $O(N)$ and cubic symmetry are listed in additional tables in Appendices B.3 and

B.4. The number of different symmetry factors is much smaller than the number of diagrams, as shown at the end of Section 6.4.3.

More details on the tables and the numbering can be found at the beginning of Appendix B.

15.1 Renormalization Constants for O(N)-Symmetric Theory

In order to exemplify the use of the tables in Appendix B, We now show how to compose explicitly the renormalization constants of the O(N)-symmetric theory using the tables in Appendices B.1–B.3.

15.1.1 Renormalization Constants up to Two Loops

The information relevant for two-loops taken from Appendices B.1–B.3. is summarized in Table 15.1 . At the one-loop level, there is only one diagram contributing to Z_g and Z_{m^2} with

Z_g				
$n_1 = 1$	$I_4 = 2\varepsilon^{-1}$	$W_4 = 3/2$	$S_4(i_4 = 1) = (N + 8)/3^2$	$-\bar{g}$
$n_2 = 1$	$I_4 = -4\varepsilon^{-2}$	$W_4 = 3/4$	$S_4(i_4 = 1) = (20 + 6N + N^2)/3^3$	$\bar{g}^2$
$n_2 = 2$	$I_4 = \varepsilon^{-1} - 2\varepsilon^{-2}$	$W_4 = 3$	$S_4(i_4 = 2) = (22 + 5N)/3^3$	$\bar{g}^2$
Z_{m^2}				
$n_1 = 1$	$I_4 = 2\varepsilon^{-1}$	$W_{m^2} = 1/2$	$S_2(i_2 = 1) = (2 + N)/3$	$-\bar{g}$
$n_2 = 1$	$I_4 = -4\varepsilon^{-2}$	$W_{m^2} = 1/4$	$S_2(i_2 = 1) = (2 + N)^2/3^2$	$\bar{g}^2$
$n_2 = 2$	$I_4 = \varepsilon^{-1} - 2\varepsilon^{-2}$	$W_{m^2} = 1/2$	$S_2(i_2 = 2) = 3(N + 2)/3^2$	$\bar{g}^2$
Z_ϕ				
$n_1 = 0$	—	—	—	—
$n_2 = 1$	$I_2 = -\varepsilon^{-1}/2$	$W_2 = 1/6$	$S_2(i_2 = 2) = 3(N + 2)/3^2$	$\bar{g}^2$

TABLE 15.1   Contributions to renormalization constants up to two loops, extracted from the tables in Appendices B.1–B.3.containing all $1/\epsilon$-poles up to five loops.

the pole term $2/\varepsilon$, and the weight factors $3/2$ and $1/2$, respectively. The running numbers of the symmetry factor are $i_4 = i_2 = 1$. The symmetry factors are taken from the tables in Appendix B.3. They are $S_4 = (8 + N)/9$ and $S_2 = (2 + N)/3$. These entries in Table 15.1 must be multiplied with each other and with $(-\bar{g})^L$, and give contributions to Z_g and Z_ϕ in the sums (15.2) and (15.3) with a positive sign. Explicitly, the one-loop results are

$$Z_g = 1 + \bar{g} \frac{2}{\varepsilon} \frac{3}{2} \frac{8 + N}{9} = 1 + \frac{\bar{g}}{\varepsilon} \frac{(8 + N)}{3}, \tag{15.6}$$

$$Z_{m^2} = 1 + g \frac{2}{\varepsilon} \frac{1}{2} \frac{2 + N}{3} = 1 + \frac{\bar{g}}{\varepsilon} \frac{(2 + N)}{3}. \tag{15.7}$$

The two-loop contribution are extracted quite similarly:

$$\Delta Z_g^{L=2} = -\bar{g}^2 \frac{-4}{\varepsilon^2} \frac{3}{4} \frac{20 + 6N + N^2}{27} - \bar{g}^2 \left(\frac{1}{\varepsilon} + \frac{-2}{\varepsilon^2} \right) 3 \frac{22 + 5N}{27}$$

$$= \bar{g}^2 \left[\frac{1}{\varepsilon^2} \frac{(20 + 6N + N^2)}{9} + \left(\frac{2}{\varepsilon^2} - \frac{1}{\varepsilon} \right) \frac{22 + 5N}{9} \right], \tag{15.8}$$

$$\Delta Z_{m^2}^{L=2} = -\bar{g}^2 \frac{-4}{\varepsilon^2} \frac{1}{4} \frac{(2 + N)^2}{9} - \bar{g}^2 \left(\frac{1}{\varepsilon} + \frac{-2}{\varepsilon^2} \right) \frac{1}{2} \frac{2(2 + N)}{9}$$

$$= \bar{g}^2 \left[\frac{1}{\varepsilon^2} \frac{(2+N)^2}{9} + \left(\frac{1}{\varepsilon^2} - \frac{1}{2\varepsilon} \right) \frac{2+N}{3} \right]. \tag{15.9}$$

The contributions to Z_ϕ in Table 15.1 come from Appendix B.2. The one-loop diagram gives no pole term. The lowest pole term arises from the two-loop diagram. The symmetry factors S_2 are listed in the first table of Appendix B.3. For $n_2 = 1$ we have $S_2 = (2+N)/3$. The corresponding entries in Table 15.1 must be multiplied with each other and with $(-\bar{g})^L$, leading to

$$Z_\phi = 1 + \bar{g}^2 \frac{-1}{2\varepsilon} \frac{1}{6} \frac{3(2+N)}{9} = 1 - \bar{g}^2 \frac{1}{12\varepsilon} \frac{2+N}{3}. \tag{15.10}$$

These expressions agree, of course, with the results of the explicit calculations of the renormalization constants in Eqs. (9.115)–(9.119).

15.1.2 Renormalization Constants up to Five Loops

For the O(N)-symmetric theory, the renormalization constants up to five loops are:

$$\begin{aligned}
Z_\phi = {} & 1 - \bar{g}^2 \, (2+N) \, \tfrac{1}{36} \, \varepsilon^{-1} + \bar{g}^3 \, (2+N) \, (8+N) \left\{ -\tfrac{1}{162} \varepsilon^{-2} + \tfrac{1}{648} \, \varepsilon^{-1} \right\} \\
& + \bar{g}^4 \, (2+N) \Big\{ -(8+N)^2 \, \tfrac{1}{648} \, \varepsilon^{-3} \\
& \qquad + [234 + 53\,N + N^2] \, \tfrac{1}{2592} \varepsilon^{-2} + 5 \, [-100 - 18\,N + N^2] \, \tfrac{1}{10368} \, \varepsilon^{-1} \Big\} \\
& + \bar{g}^5 \, (2+N) \Big\{ -(8+N)^3 \, \tfrac{1}{2430} \, \varepsilon^{-4} + (8+N) \, [1210 + 269\,N + 3\,N^2] \tfrac{1}{29160} \, \varepsilon^{-3} \\
& \qquad + [-33872 - 10610\,N - 461\,N^2 + 15\,N^3 - (12672 + 2880\,N) \, \zeta(3)] \tfrac{1}{116640} \, \varepsilon^{-2} \\
& \qquad + [77056 + 22752\,N + 296\,N^2 + 39\,N^3 - (8832 + 3072\,N - 288\,N^2 + 48\,N^3) \, \zeta(3) \\
& \qquad + (25344 + 5760\,N) \, \zeta(4)] \, \tfrac{1}{466560} \, \varepsilon^{-1} \Big\}. \tag{15.11}
\end{aligned}$$

$$\begin{aligned}
Z_{m^2} = {} & 1 + \bar{g} \, (2+N) \tfrac{1}{3} \varepsilon^{-1} + \bar{g}^2 \, (2+N) \left\{ (5+N) \tfrac{1}{9} \varepsilon^{-2} - \tfrac{1}{6} \varepsilon^{-1} \right\} \\
& + \bar{g}^3 \, (2+N) \Big\{ (5+N) \, (6+N) \tfrac{1}{27} \, \varepsilon^{-3} - (50 + 11\,N) \tfrac{1}{54} \, \varepsilon^{-2} \\
& \qquad + (230 + 31\,N) \tfrac{1}{648} \, \varepsilon^{-1} \Big\} \\
& + \bar{g}^4 \, (2+N) \Big\{ (5+N) \, (6+N) \, (13 + 2\,N) \tfrac{1}{162} \, \varepsilon^{-4} \\
& \qquad - [1100 + 438\,N + 43\,N^2] \tfrac{1}{324} \, \varepsilon^{-3} \\
& \qquad + [4870 + 1531\,N + 79\,N^2 + (1584 + 360\,N) \, \zeta(3)] \tfrac{1}{1944} \, \varepsilon^{-2} \\
& \qquad + [-4070 - 981\,N + 2\,N^2 - (408 + 60\,N + 18\,N^2) \, \zeta(3) \\
& \qquad\quad - (792 + 180\,N) \, \zeta(4)] \tfrac{1}{3888} \, \varepsilon^{-1} \Big\} \\
& + \bar{g}^5 \, (2+N) \Big\{ (5+N) \, (6+N) \, (13 + 2\,N) \, (34 + 5\,N) \tfrac{1}{2430} \, \varepsilon^{-5} \\
& \qquad - [10648 + 5960\,N + 1100\,N^2 + 67\,N^3] \tfrac{1}{972} \, \varepsilon^{-4} \\
& \qquad + [170240 + 81504\,N + 11082\,N^2 + 343\,N^3 \\
& \qquad\quad + (58608 + 23616\,N + 2340\,N^2) \, \zeta(3)] \tfrac{1}{14580} \, \varepsilon^{-3} \\
& \qquad + [-485728 - 199496\,N - 17798\,N^2 + 23\,N^3 \\
& \qquad\quad - (279024 + 88176\,N + 7344\,N^2 + 162\,N^3) \, \zeta(3) \\
& \qquad\quad + (17424 + 3168\,N - 180\,N^2) \, \zeta(4) \\
& \qquad\quad - (357120 + 105600\,N + 3840\,N^2) \, \zeta(5)] \tfrac{1}{58320} \, \varepsilon^{-2}
\end{aligned}$$

$$+ [1106880 + 374208\,N + 15282\,N^2 + 33\,N^3$$
$$+ (503680 + 129280\,N + 15232\,N^2 + 240\,N^3)\,\zeta(3)$$
$$- (148992 + 37120\,N + 512\,N^2)\,\zeta(3)^2 + (273024 + 82176\,N$$
$$+ 2784\,N^2 - 288\,N^3)\,\zeta(4) + (18432 + 3584\,N - 1280\,N^2)\,\zeta(5)$$
$$+ (595200 + 176000\,N + 6400\,N^2)\,\zeta(6)]\,\tfrac{1}{311040}\,\varepsilon^{-1}\big\}. \tag{15.12}$$

$$Z_g = 1 + \bar{g}\,(8+N)\,\tfrac{1}{3}\varepsilon^{-1} + \bar{g}^2\,\Big\{(8+N)^2\,\tfrac{1}{9}\varepsilon^{-2} - (22+5\,N)\,\tfrac{1}{9}\varepsilon^{-1}\Big\}$$
$$+ \bar{g}^3\,\Big\{(8+N)^3\,\tfrac{1}{27}\,\varepsilon^{-3} - 2\,(8+N)\,(76+17\,N)\,\tfrac{1}{81}\,\varepsilon^{-2}$$
$$+ [2992 + 942\,N + 35\,N^2 + (2112 + 480\,N)\,\zeta(3)]\,\tfrac{1}{648}\,\varepsilon^{-1}\Big\}$$
$$+ \bar{g}^4\,\Big\{(8+N)^4\,\tfrac{1}{81}\,\varepsilon^{-4} - (8+N)^2\,(248+55\,N)\,\tfrac{1}{243}\,\varepsilon^{-3}$$
$$+ [38272 + 16856\,N + 1934\,N^2 + 43\,N^3$$
$$+ (21120 + 7440\,N + 600\,N^2)\,\zeta(3)]\,\tfrac{1}{972}\,\varepsilon^{-2}$$
$$+ [-49912 - 20624\,N - 1640\,N^2 + 5\,N^3$$
$$- (55968 + 18336\,N + 1512\,N^2)\,\zeta(3) + (12672 + 4464\,N$$
$$+ 360\,N^2)\,\zeta(4) - (89280 + 26400\,N + 960\,N^2)\,\zeta(5)]\,\tfrac{1}{3888}\,\varepsilon^{-1}\Big\}$$
$$+ \bar{g}^5\,\Big\{(8+N)^5\,\tfrac{1}{243}\,\varepsilon^{-5} - 2\,(8+N)^3\,(875+193\,N)\,\tfrac{1}{3645}\,\varepsilon^{-4}$$
$$+ (8+N)\,[803328 + 353392\,N + 40186\,N^2 + 733\,N^3$$
$$+ (363264 + 127968\,N + 10320\,N^2)\,\zeta(3)]\,\tfrac{1}{29160}\varepsilon^{-3}$$
$$+ [-3680256 - 1975616\,N - 306616\,N^2 - 13150\,N^3 + 25\,N^4$$
$$- (3687552 + 1667088\,N + 237168\,N^2 + 9828\,N^3)\,\zeta(3)$$
$$+ (658944 + 314496\,N + 47736\,N^2 + 2340\,N^3)\,\zeta(4)$$
$$- (4642560 + 1953120\,N + 221520\,N^2 + 6240\,N^3)\,\zeta(5)]\,\tfrac{1}{29160}\varepsilon^{-2}$$
$$+ [40148480 + 20429248\,N + 2518864\,N^2 + 39230\,N^3 + 195\,N^4$$
$$+ (63017472 + 26449536\,N + 3236736\,N^2 + 60672\,N^3 - 624\,N^4)\,\zeta(3)$$
$$+ (7520256 + 1027584\,N - 135936\,N^2 - 13824\,N^3)\,\zeta(3)^2$$
$$- (18044928 + 8088192\,N + 1176192\,N^2 + 54432\,N^3)\,\zeta(4)$$
$$+ (126784512 + 51445248\,N + 5733888\,N^2 + 234240\,N^3)\,\zeta(5)$$
$$- (42854400 + 18028800\,N + 2044800\,N^2 + 57600\,N^3)\,\zeta(6)$$
$$+ (178149888 + 64012032\,N + 4741632\,N^2)\,\zeta(7)]\,\tfrac{1}{933120}\,\varepsilon^{-1}\Big\}. \tag{15.13}$$

For the convenience of the reader who wants to derive consequences of these series, we have placed the corresponding files on the World Wide Web [3].

15.1.3 Ratios between Bare and Renormalized Quantities up to Five Loops

The above renormalization constants determine the ratios between bare and renormalized quantities according to Eqs. (9.71) and (9.72):

$$\frac{\phi_B^2}{\phi^2} = Z_\phi, \qquad \frac{m_B^2}{m^2} = \frac{Z_{m^2}}{Z_\phi}, \qquad \frac{\bar{g}_B}{\bar{g}} = \frac{Z_g}{Z_\phi^2}, \tag{15.14}$$

where $\bar{g}_B \equiv \bar{\lambda}_B/\mu^\varepsilon$. For completeness, we list the explicit expansions for these three ratios. The expansion for ϕ_B^2/ϕ^2 is directly given by Eq. (15.11). For the mass ratio, the expansion in powers of the renormalized coupling constant $\bar{g}$ reads, from (15.12) and (15.11),

$$
\begin{aligned}
m_B^2/m^2 = {} & 1 + \bar{g}\,(2+N)\,\tfrac{1}{3}\,\varepsilon^{-1} \\
& + \bar{g}^2\,(2+N)\left\{-5\,\tfrac{1}{36}\,\varepsilon^{-1} + (5+N)\,\tfrac{1}{9}\,\varepsilon^{-2}\right\} \\
& + \bar{g}^3\,(2+N)\left\{(5+N)\,(6+N)\,\tfrac{1}{27}\,\varepsilon^{-3} + (37+5\,N)\,\tfrac{1}{108}\,\varepsilon^{-1}\right. \\
& \qquad\left. - (278+61\,N)\,\tfrac{1}{324}\,\varepsilon^{-2}\right\} \\
& + \bar{g}^4\,(2+N)\left\{(5+N)\,(6+N)\,(13+2\,N)\,\tfrac{1}{162}\,\varepsilon^{-4}\right. \\
& \qquad - \left[6284 + 2498\,N + 245\,N^2\right]\tfrac{1}{1944}\,\varepsilon^{-3} \\
& \qquad + \left[6218 + 1965\,N + 103\,N^2 + (2112 + 480\,N)\,\zeta(3)\right]\tfrac{1}{2592}\,\varepsilon^{-2} \\
& \qquad + \left[-31060 - 7578\,N + N^2 - (3264 + 480\,N + 144\,N^2)\,\zeta(3)\right. \\
& \qquad\quad\left.\left. - (6336 + 1440\,N)\,\zeta(4)\right]\tfrac{1}{31104}\,\varepsilon^{-1}\right\} \\
& + \bar{g}^5\,(2+N)\left\{(5+N)\,(6+N)\,(13+2\,N)\,(34+5\,N)\,\tfrac{1}{2430}\,\varepsilon^{-5}\right. \\
& \qquad - (307976 + 172176\,N + 31752\,N^2 + 1933\,N^3)\,\tfrac{1}{29160}\,\varepsilon^{-4} \\
& \qquad + \left[1307420 + 627164\,N + 85649\,N^2 + 2697\,N^3\right. \\
& \qquad\quad\left. + (468864 + 188928\,N + 18720\,N^2)\,\zeta(3)\right]\tfrac{1}{116640}\,\varepsilon^{-3} \\
& \qquad + \left[-3724856 - 1536688\,N - 138640\,N^2 + 49\,N^3\right. \\
& \qquad\quad - (2181504 + 693888\,N + 58752\,N^2 + 1296\,N^3)\,\zeta(3) \\
& \qquad\quad + (139392 + 25344\,N - 1440\,N^2)\,\zeta(4) \\
& \qquad\quad\left. - (2856960 + 844800\,N + 30720\,N^2)\,\zeta(5)\right]\tfrac{1}{466560}\,\varepsilon^{-2} \\
& \qquad + \left[3166528 + 1077120\,N + 45254\,N^2 + 21\,N^3\right. \\
& \qquad\quad + (1528704 + 393984\,N + 45120\,N^2 + 816\,N^3)\,\zeta(3) \\
& \qquad\quad - (446976 + 111360\,N + 1536\,N^2)\,\zeta(3)^2 \\
& \qquad\quad + (768384 + 235008\,N + 8352\,N^2 - 864\,N^3)\,\zeta(4) \\
& \qquad\quad + (55296 + 10752\,N - 3840\,N^2)\,\zeta(5) \\
& \qquad\quad\left.\left. + (1785600 + 528000\,N + 19200\,N^2)\,\zeta(6)\right]\tfrac{1}{933120}\,\varepsilon^{-1}\right\}. \qquad (15.15)
\end{aligned}
$$

For the ratio between bare and renormalized coupling constants, we obtain from Eqs. (15.11) and (15.13):

$$
\begin{aligned}
\bar{g}_B/\bar{g} = {} & 1 + \bar{g}\,(8+N)\,\tfrac{1}{3}\,\varepsilon^{-1} + \bar{g}^2\left\{(8+N)^2\,\tfrac{1}{9}\,\varepsilon^{-2} - (14+3\,N)\,\tfrac{1}{6}\,\varepsilon^{-1}\right\} \\
& + \bar{g}^3\left\{(8+N)^3\,\tfrac{1}{27}\,\varepsilon^{-3} - 7\,(8+N)\,(14+3\,N)\,\tfrac{1}{54}\,\varepsilon^{-2}\right. \\
& \qquad\left. + \left[2960 + 922\,N + 33\,N^2 + (2112 + 480\,N)\,\zeta(3)\right]\tfrac{1}{648}\,\varepsilon^{-1}\right\} \\
& + \bar{g}^4\left\{(8+N)^4\,\tfrac{1}{81}\,\varepsilon^{-4} - 23\,(8+N)^2\,(14+3\,N)\,\tfrac{1}{324}\,\varepsilon^{-3}\right. \\
& \qquad + \left[150152 + 65288\,N + 7388\,N^2 + 165\,N^3\right. \\
& \qquad\quad\left. + (84480 + 29760\,N + 2400\,N^2)\,\zeta(3)\right]\tfrac{1}{3888}\,\varepsilon^{-2} \\
& \qquad + (-196648 - 80456\,N - 6320\,N^2 + 5\,N^3 \\
& \qquad\quad - (223872 + 73344\,N + 6048\,N^2)\,\zeta(3) \\
& \qquad\quad + (50688 + 17856\,N + 1440\,N^2)\,\zeta(4)
\end{aligned}
$$

$$- (357120 + 105600\,N + 3840\,N^2)\,\zeta(5)]\,\tfrac{1}{15552}\,\varepsilon^{-1}\Big\}$$
$$+ \bar{g}^5\,\Big\{(8+N)^5\,\tfrac{1}{243}\,\varepsilon^{-5} - 163\,(8+N)^3\,(14+3\,N)\,\tfrac{1}{4860}\,\varepsilon^{-4}$$
$$+ (8+N)\,[1572136 + 681832\,N + 76432\,N^2 + 1419\,N^3$$
$$+ (726528 + 255936\,N + 20640\,N^2)\,\zeta(3)]\,\tfrac{1}{58320}\,\varepsilon^{-3}$$
$$+ [-28905152 - 15368600\,N - 2361720\,N^2 - 101836\,N^3 + 65\,N^4$$
$$- (29314560 + 13201536\,N + 1876224\,N^2 + 78624\,N^3)\,\zeta(3)$$
$$+ (5271552 + 2515968\,N + 381888\,N^2 + 18720\,N^3)\,\zeta(4)$$
$$- (37140480 + 15624960\,N + 1772160\,N^2 + 49920\,N^3)\,\zeta(5)]\,\tfrac{1}{233280}\,\varepsilon^{-2}$$
$$+ [13177344 + 6646336\,N + 808496\,N^2 + 12578\,N^3 + 13\,N^4$$
$$+ (21029376 + 8836480\,N + 1082240\,N^2 + 19968\,N^3 - 144\,N^4)\,\zeta(3)$$
$$+ (2506752 + 342528\,N - 45312\,N^2 - 4608\,N^3)\,\zeta(3)^2$$
$$- (6082560 + 2745216\,N + 399744\,N^2 + 18144\,N^3)\,\zeta(4)$$
$$+ (42261504 + 17148416\,N + 1911296\,N^2 + 78080\,N^3)\,\zeta(5)$$
$$- (14284800 + 6009600\,N + 681600\,N^2 + 19200\,N^3)\,\zeta(6)$$
$$+ (59383296 + 21337344\,N + 1580544\,N^2)\,\zeta(7)]\,\tfrac{1}{311040}\,\varepsilon^{-1}\Big\}. \tag{15.16}$$

For $\varepsilon = 1$ corresponding to $D = 3$ dimensions, this becomes numerically:

$$
\begin{aligned}
\bar{g}_B/\bar{g} = \; & 1 + (2.666666 + 0.333333\,N)\,\bar{g} \\
& + (4.77778 + 1.27778\,N + 0.111111\,N^2)\,\bar{g}^2 \\
& + (12.9302 + 4.49844\,N + 0.550926\,N^2 + 0.037037\,N^3)\,\bar{g}^3 \\
& + (1.46948 + 5.10568\,N + 1.95206\,N^2 + 0.224859\,N^3 + 0.0123457\,N^4)\,\bar{g}^4 \\
& + (215.097 + 82.8298\,N + 9.8859\,N^2 + 0.933862\,N^3 + 0.088087\,N^4 + 0.004115\,N^5)\,\bar{g}^5 .
\end{aligned}
\tag{15.17}
$$

Note that the right-hand side has no alternating sign. From the large-order behavior one can deduce that the signs will eventually alternate. For the three-dimensional perturbation expansions this behavior was shown in Ref. [4] to set in surprisingly late, at order 11 (see Table 20.6).

The resulting renormalization group functions will be derived from these expansions in Chapters 17 and 18. For later use in Chapter 19, we also write down the inverse series expansion of the renormalized coupling constant in powers of the dimensionless bare coupling constant $\bar{g}_B = \bar{\lambda}_B/\mu^\varepsilon$:

$$\bar{g}/\bar{g}_B = 1 - \bar{g}_B\,(8+N)\,\tfrac{1}{3}\,\varepsilon^{-1} + \bar{g}_B^2\,\Big\{(8+N)^2\,\tfrac{1}{9}\,\varepsilon^{-2} + (14+3\,N)\,\tfrac{1}{6}\,\varepsilon^{-1}\Big\}$$
$$+ \bar{g}_B^3\,\Big\{-(8+N)^3\,\tfrac{1}{27}\,\varepsilon^{-3} - 4\,(8+N)\,(14+3\,N)\,\tfrac{1}{27}\,\varepsilon^{-2}$$
$$- [2960 + 922\,N + 33\,N^2 + (2112 + 480\,N)\,\zeta(3)]\,\tfrac{1}{648}\,\varepsilon^{-1}\Big\}$$
$$+ \bar{g}_B^4\,\Big\{(8+N)^4\,\tfrac{1}{81}\,\varepsilon^{-4} + 29\,(8+N)^2\,(14+3\,N)\,\tfrac{1}{324}\,\varepsilon^{-3}$$
$$+ [197512 + 85960\,N + 9760\,N^2 + 231\,N^3$$
$$+ (118272 + 41664\,N + 3360\,N^2)\,\zeta(3)]\,\tfrac{1}{3888}\,\varepsilon^{-2}$$
$$+ [196648 + 80456\,N + 6320\,N^2 - 5\,N^3$$
$$+ (223872 + 73344\,N + 6048\,N^2)\,\zeta(3)$$

$$- (50688 + 17856\,N + 1440\,N^2)\,\zeta(4)$$
$$+ (357120 + 105600\,N + 3840\,N^2)\,\zeta(5)]\,\tfrac{1}{15552}\,\varepsilon^{-1}\}$$
$$+ \bar{g}_B^5\,\{- (8+N)^5\,\tfrac{1}{243}\,\varepsilon^{-5} - 37\,(8+N)^3\,(14+3\,N)\,\tfrac{1}{810}\,\varepsilon^{-4}$$
$$- (8+N)\,[420736 + 182752\,N + 20602\,N^2 + 429\,N^3$$
$$+ (219648 + 77376\,N + 6240\,N^2)\,\zeta(3)]\,\tfrac{1}{9720}\,\varepsilon^{-3}$$
$$+ [-21780544 - 11596360\,N - 1789800\,N^2 - 79772\,N^3 + 55\,N^4$$
$$- (22894080 + 10326912\,N + 1494528\,N^2 + 66528\,N^3)\,\zeta(3)$$
$$+ (4460544 + 2128896\,N + 323136\,N^2 + 15840\,N^3)\,\zeta(4)$$
$$- (31426560 + 13221120\,N + 1499520\,N^2 + 42240\,N^3)\,\zeta(5)]\,\tfrac{1}{116640}\,\varepsilon^{-2}$$
$$+ [-13177344 - 6646336\,N - 808496\,N^2 - 12578\,N^3 - 13\,N^4$$
$$- (21029376 + 8836480\,N + 1082240\,N^2 + 19968\,N^3 - 144\,N^4)\,\zeta(3)$$
$$+ (-2506752 - 342528\,N + 45312\,N^2 + 4608\,N^3)\,\zeta(3)^2$$
$$+ (6082560 + 2745216\,N + 399744\,N^2 + 18144\,N^3)\,\zeta(4)$$
$$- (42261504 + 17148416\,N + 1911296\,N^2 + 78080\,N^3)\,\zeta(5)$$
$$+ (14284800 + 6009600\,N + 681600\,N^2 + 19200\,N^3)\,\zeta(6)$$
$$- (59383296 + 21337344\,N + 1580544\,N^2)\,\zeta(7)]\,\tfrac{1}{311040}\,\varepsilon^{-1}\}. \tag{15.18}$$

In three dimensions, this becomes numerically:

$$\begin{aligned}
\bar{g}/\bar{g}_B \;=\; & 1 - (2.666666 + 0.333333\,N)\,\bar{g}_B \\
& + (9.44444 + 2.27778\,N + 0.111111\,N^2)\,\bar{g}_B^2 \\
& - (44.0413 + 15.054\,N + 1.38426\,N^2 + 0.037037\,N^3)\,\bar{g}_B^3 \\
& + (268.364 + 114.149\,N + 14.8689\,N^2 + 0.722672\,N^3 + 0.012346\,N^4)\,\bar{g}_B^4 \\
& - (2087.33 + 1032.51\,N + 169.755\,N^2 + 11.6351\,N^3 + \\
& \quad 0.344796\,N^4 + 0.004115\,N^5)\,\bar{g}_B^5. \tag{15.19}
\end{aligned}$$

In contrast to the inverse expansion (15.17), this does have alternating signs.

15.2 Renormalization Constants for Theory with Mixed O(N) and Cubic Symmetry

In the case of two coupling constants, the expression $(-\bar{g})^L S(i)$ in Eqs. (15.1)–(15.3) is replaced by a sum over symmetry factors, as explained in Section 6.4.2. The symmetry factors in that sum are listed in Appendix B.4. Here we compose the contributions to the renormalization constants up to two loops.

15.2.1 Renormalization Constants up to Two Loops

Table 15.2 shows the relevant extraction from the tables of Section B.4. For two-point diagrams, the combined symmetry factor is given by the sum $\sum_{k=0}^{L}(\bar{g}_1)^{L-k}\,(\bar{g}_2)^k S_{2;(L-k,k)}^{\text{cub}}(i_2)$. Up to two loops, it contains the individual symmetry factors listed in Table 15.2 . The last entry is the symmetry factor associated with each diagram. For the calculation of Z_{m^2}, we need the contributions of all two-point diagrams and combine them with the weight factors and the pole terms in the same way as for O(N) symmetry.

$S_{2;(L-k,k)}$					
$n_1 = 1$	$i_2 = 1$	$S_{2;(1,0)} = \frac{2+N}{3}$		$S_{2;(0,1)} = 1$	$\bar{g}_1 \frac{N+2}{3}3 + \bar{g}_2$
$n_2 = 1$	$i_2 = 1$	$S_{2;(2,0)} = \left(\frac{2+N}{3}\right)^2$	$S_{2;(1,1)} = \frac{2(2+N)}{3}$	$S_{2;(0,2)} = 1$	$\bar{g}_1^2 \frac{(2+N)^2}{9} + \bar{g}_1 \bar{g}_2 \frac{2(2+N)}{3} + \bar{g}_2^2$
$n_2 = 2$	$i_2 = 2$	$S_{2;(2,0)} = \frac{2+N}{3}$	$S_{2;(1,1)} = 2$	$S_{2;(0,2)} = 1$	$\bar{g}_1^2 \frac{2+N}{3} + \bar{g}_1 \bar{g}_2 2 + \bar{g}_2^2$

TABLE 15.2   Symmetry factors $S_{2;(L-k,k)}^{\mathrm{cub}}$ for the two-point diagrams up to two loops. The sum $\sum_{k=0}^L (\bar{g}_1)^{L-k} (\bar{g}_2)^k S_{2;(L-k,k)}^{\mathrm{cub}}(i_2)$ is listed in the last column. This table is extracted from those in Appendix B.4, which contain the symmetry factors for a mixture of $O(N)$-symmetric and cubic interactions up to five loops.

Z_{g_1}			
n_L	I_4	W_4	$\sum_{k=0}^L (\bar{g}_1)^{L-k} (\bar{g}_2)^k S_{41;(L+1-k,k)}^{\mathrm{cub}}(i_4)$
$n_1 = 1$	$2\varepsilon^{-1}$	$3/2$	$\bar{g}_1 \frac{N+8}{9} + \bar{g}_2 \frac{2}{3}$
$n_2 = 1$	$-4\varepsilon^{-2}$	$3/4$	$\bar{g}_1^2 \frac{20+6N+N^2}{27} + \bar{g}_1 \bar{g}_2 \frac{4+N}{3} + g_2^2$
$n_2 = 2$	$\varepsilon^{-1} - 2\varepsilon^{-2}$	3	$\bar{g}_1^2 \frac{22+5N}{27} + \bar{g}_1 \bar{g}_2 \frac{4}{3} + \bar{g}_2^2 \frac{1}{3}$

Z_{g_2}			
n_L	I_4	W_4	$\sum_{k=0}^L (\bar{g}_1)^{L-k} (\bar{g}_2)^k S_{42;(L-k,k+1)}^{\mathrm{cub}}(i_4)$
$n_1 = 1$	$2\varepsilon^{-1}$	$3/2$	$\bar{g}_1 \frac{4}{3} + \bar{g}_2$
$n_2 = 1$	$-4\varepsilon^{-2}$	$3/4$	$\bar{g}_1^2 \frac{4}{3} + \bar{g}_1 \bar{g}_2 2 + g_2^2$
$n_2 = 2$	$\varepsilon^{-1} - 2\varepsilon^{-2}$	3	$\bar{g}_1^2 \frac{14+N}{9} + \bar{g}_1 \bar{g}_2 \frac{8}{3} + \bar{g}_2^2 \frac{1}{3}$

Z_{m^2}			
n_L	I_4	W_{m^2}	$\sum_{k=0}^L (\bar{g}_1)^{L-k} (\bar{g}_2)^k S_{2;(L-k,k)}^{\mathrm{cub}}(i_2)$
$n_1 = 1$	$2\varepsilon^{-1}$	$1/2$	$\bar{g}_1 \frac{N+2}{3}3 + g_2$
$n_2 = 1$	$-4\varepsilon^{-2}$	$1/4$	$\bar{g}_1^2 \frac{(2+N)^2}{9} + \bar{g}_1 \bar{g}_2 \frac{2(2+N)}{3} + g_2^2$
$n_2 = 2$	$\varepsilon^{-1} - 2\varepsilon^{-2}$	$1/2$	$\bar{g}_1^2 \frac{2+N}{3} + \bar{g}_1 \bar{g}_2 2 + \bar{g}_2^2$

Z_{ϕ}			
n_L	I_2	W_2	$\sum_{k=0}^L (\bar{g}_1)^{L-k} (\bar{g}_2)^k S_{2;(L-k,k)}^{\mathrm{cub}}(i_2)$
$n_1 = 0$	$-$	$-$	$-$
$n_2 = 1$	$-\varepsilon^{-1}/2$	$1/6$	$\bar{g}_1^2 \frac{2+N}{3} + \bar{g}_1 \bar{g}_2 2 + \bar{g}_2^2$

TABLE 15.3   Contributions to renormalization constants up to two loops in a theory with mixed $O(N)$ and cubic symmetry. This table is extracted from those in Sections B.1, B.2, and B.4 of Appendix B, which contain the contributions up to five loops. The individual symmetry factors in the last entry are calculated separately in Table 15.4 .

The contributions to the renormalization constant Z_{m^2} are collected in Table 15.3 up to two loops and add up to

$$Z_{m^2} = 1 + \frac{1}{\varepsilon}\left[\bar{g}_1 \frac{(2+N)}{3} + \bar{g}_2\right] + \frac{1}{\varepsilon^2}\left[\bar{g}_1^2 \frac{(2+N)^2}{9} + \bar{g}_1 \bar{g}_2 \frac{2(2+N)}{3} + \bar{g}_2^2\right]$$
$$+ \left(\frac{1}{\varepsilon^2} + \frac{1}{2\varepsilon}\right)\left[\bar{g}_1^2 \frac{(2+N)}{3} + 2\,\bar{g}_1\,\bar{g}_2 + \bar{g}_2^2\right]. \tag{15.20}$$

The renormalization constant Z_ϕ up to two loops is obtained from only one diagram. It carries the number $n_2 = 2$, and Table 15.3 shows that

$$Z_\phi = 1 - \frac{1}{12\varepsilon}\left(\bar{g}_1^2 \frac{2+N}{3} + 2\,\bar{g}_1\,\bar{g}_2 + \bar{g}_2^2\right). \tag{15.21}$$

Let us finally calculate the sum $\sum_{k=0}^{L+1}(\bar{g}_1)^{L+1-k}(\bar{g}_2)^k S^{\text{cub}}_{4_i;(L+1-k,k)}(i_4)$ for the four-point diagrams. The individual symmetry factors are listed in Table 15.4 for $i = 1, 2$ up to two loops,

$S_{4_1;(L-k,k)}$					
$n_1 = 1$	$i_4 = 1$	$S_{4_1;(2,0)} = \frac{8+N}{9}$		$S_{4_1;(1,1)} = \frac{2}{3}$	$\bar{g}_1^2 \frac{N+8}{9} + \bar{g}_1\bar{g}_2 \frac{2}{3}$
$n_2 = 1$	$i_4 = 1$	$S_{4_1;(3,0)} = \frac{20+6N+N^2}{27}$	$S_{4_1;(2,1)} = \frac{4+N}{3}$	$S_{4_1;(1,2)} = 1$	$\bar{g}_1^3 \frac{20+6N+N^2}{27}$
					$+\bar{g}_1^2\bar{g}_2 \frac{4+N}{3} + \bar{g}_1\bar{g}_2^2$
$n_2 = 2$	$i_4 = 2$	$S_{4_1;(3,0)} = \frac{22+5N}{27}$	$S_{4_1;(2,1)} = \frac{4}{3}$	$S_{4_1;(1,2)} = \frac{1}{3}$	$\bar{g}_1^3 \frac{22+5N}{27} + \bar{g}_1^2\bar{g}_2 \frac{4}{3} + \bar{g}_1\bar{g}_2^2 \frac{1}{3}$
$S_{4_2;(L-k,k)}$					
$n_1 = 1$	$i_4 = 1$	$S_{4_2;(1,1)} = \frac{4}{3}$		$S_{4_2;(0,2)} = 1$	$\bar{g}_1\bar{g}_2 \frac{4}{3} + \bar{g}_2^2$
$n_2 = 1$	$i_4 = 1$	$S_{4_2;(2,1)} = \frac{4}{3}$	$S_{4_2;(1,2)} = 2$	$S_{4_2;(0,3)} = 1$	$\bar{g}_1^2\bar{g}_2 \frac{4}{3} + \bar{g}_1\,\bar{g}_2^2 2 + \bar{g}_2^3$
$n_2 = 2$	$i_4 = 2$	$S_{4_2;(2,1)} = \frac{14+N}{9}$	$S_{4_2;(1,2)} = \frac{8}{3}$	$S_{4_2;(0,3)} = 1$	$\bar{g}_1^2\bar{g}_2 \frac{14+N}{9} + \bar{g}_1\,\bar{g}_2^2 \frac{8}{3} + \bar{g}_2^3$

TABLE 15.4 Symmetry factors of four-point diagrams $S^{\text{cub}}_{4_1;(L-k,k+1)}$ and $S^{\text{cub}}_{4_2;(L-k,k+1)}$ up to two loops. The sum $\sum_{k=0}^{L+1}(\bar{g}_1)^{L+1-k}(\bar{g}_2)^k S^{\text{cub}}_{4_i;(L+1-k,k)}(i_4)$ is listed in the last column. It is needed in Table 15.3 . The sum over S_{4_1} is to be divided by $\bar{g}_1$ and the sum over S_{4_2} by $\bar{g}_2$ for use in Z_{g_1} or Z_{g_2}, respectively. This table is extracted from those in Appendix B.4, which contain all contributions up to five loops.

the result being again displayed in the last column. This sum must be divided by $\bar{g}_1$ for the contribution of the diagram to Z_{g_1}, and by $\bar{g}_2$ for the contribution to Z_{g_2}. The results are

$$Z_{g_1} = 1 + \frac{1}{\varepsilon}\left(\bar{g}_1 \frac{8+N}{9} + \bar{g}_2 \frac{2}{3}\right)$$
$$+ \frac{1}{\varepsilon^2}\left(\bar{g}_1^2 \frac{20+6N+N^2}{27} + \bar{g}_1\bar{g}_2 \frac{4+N}{3} + \bar{g}_2^2\right)$$
$$+ \left(\frac{2}{\varepsilon^2} - \frac{1}{\varepsilon}\right)\left(\bar{g}_1^2 \frac{22+5N}{27} + \bar{g}_1\bar{g}_2 \frac{4}{3} + \bar{g}_2^2 \frac{1}{3}\right), \tag{15.22}$$

and

$$Z_{g_2} = 1 + \frac{1}{\varepsilon}\left(\bar{g}_1 \frac{4}{3} + \bar{g}_2\right) + \frac{1}{\varepsilon^2}\left(\bar{g}_1^2 \frac{4}{3} + 2\,\bar{g}_1\,g_2 + \bar{g}_2^2\right)$$
$$+ \left(\frac{2}{\varepsilon^2} - \frac{1}{\varepsilon}\right)\left(\bar{g}_1^2 \frac{14+N}{9} + \bar{g}_1\,\bar{g}_2 \frac{8}{3} + \bar{g}_2^2\right). \tag{15.23}$$

15.2.2   Renormalization Constants up to Three Loops

The expansions for the renormalization constants for a mixture of $O(N)$-symmetric and cubic interactions will be written down here only up to three loops, to save space. In the presence of two coupling constants the number of terms becomes rapidly large. The full five-loop series can be downloaded from the World Wide Web [3]. Up to three loops, they read:

$$
\begin{aligned}
Z_\phi \; = \; & 1 - \bar g_1^2\,(2+N)\tfrac{1}{36}\,\varepsilon^{-1} - \bar g_1\bar g_2\tfrac{1}{6}\,\varepsilon^{-1} - \bar g_2^2\tfrac{1}{12}\,\varepsilon^{-1} \\
& - \bar g_1^3\,(2+N)\left[(8+N)\tfrac{1}{162}\,\varepsilon^{-2} - (8+N)\tfrac{1}{648}\,\varepsilon^{-1}\right] \\
& - \bar g_1^2\bar g_2\left[(8+N)\tfrac{1}{18}\,\varepsilon^{-2} - (8+N)\tfrac{1}{72}\,\varepsilon^{-1}\right] - \bar g_1\bar g_2^2\left[\tfrac{1}{2}\,\varepsilon^{-2} - \tfrac{1}{8}\,\varepsilon^{-1}\right] \\
& - \bar g_2^3\left[\tfrac{1}{6}\,\varepsilon^{-2} - \tfrac{1}{24}\,\varepsilon^{-1}\right],
\end{aligned}
\tag{15.24}
$$

$$
\begin{aligned}
Z_{m^2} \; = \; & 1 + \bar g_2\,\varepsilon^{-1} + \bar g_1(2+N)\tfrac{1}{3}\,\varepsilon^{-1} + \bar g_1^2\,(2+N)\left[(5+N)\tfrac{1}{9}\,\varepsilon^{-2} - \tfrac{1}{6}\,\varepsilon^{-1}\right] \\
& + \bar g_1\bar g_2\left[(5+N)\tfrac{2}{3}\,\varepsilon^{-2} - \varepsilon^{-1}\right] + \bar g_2^2\left[2\,\varepsilon^{-2} - \tfrac{1}{2}\,\varepsilon^{-1}\right] \\
& + \bar g_1^3\,(2+N)\left[(5+N)(6+N)\tfrac{1}{27}\,\varepsilon^{-3} - (50+11N)\tfrac{1}{54}\,\varepsilon^{-2}\right. \\
& \qquad\qquad\qquad \left. + (230+31N)\tfrac{1}{648}\,\varepsilon^{-1}\right] \\
& + \bar g_1\bar g_2^2\left[(8+N)\tfrac{14}{9}\,\varepsilon^{-3} - (530+19N)\tfrac{1}{54}\,\varepsilon^{-2} + (260+N)\tfrac{1}{72}\,\varepsilon^{-1}\right] \\
& + \bar g_1^2\bar g_2\left[(5+N)(6+N)\tfrac{1}{3}\,\varepsilon^{-3} - (50+11N)\tfrac{1}{6}\,\varepsilon^{-2} + (230+31N)\tfrac{1}{72}\,\varepsilon^{-1}\right] \\
& + \bar g_2^3\left[\tfrac{14}{3}\,\varepsilon^{-3} - \tfrac{61}{18}\,\varepsilon^{-2} + \tfrac{29}{24}\,\varepsilon^{-1}\right],
\end{aligned}
\tag{15.25}
$$

$$
\begin{aligned}
Z_{g_1} \; = \; & 1 + \bar g_1\,(8+N)\tfrac{1}{3}\,\varepsilon^{-1} + \bar g_2\,2\,\varepsilon^{-1} + \bar g_1^2\left[(8+N)^2\tfrac{1}{9}\,\varepsilon^{-2} - (22+5N)\tfrac{1}{9}\,\varepsilon^{-1}\right] \\
& + \bar g_1\bar g_2\left[(12+N)\varepsilon^{-2} - 4\,\varepsilon^{-1}\right] + \bar g_2^2\left[5\,\varepsilon^{-2} - \varepsilon^{-1}\right] \\
& + \bar g_1^3\left[(8+N)^3\tfrac{1}{27}\,\varepsilon^{-3} - (8+N)(76+17N)\tfrac{2}{81}\,\varepsilon^{-2}\right. \\
& \qquad\quad \left. + (2992+942N+35\,N^2+2112\,\zeta(3)+480\,N\,\zeta(3))\tfrac{1}{648}\,\varepsilon^{-1}\right] \\
& + \bar g_1^2\bar g_2\left[(116+21\,N+N^2)\tfrac{4}{9}\,\varepsilon^{-3} - (1088+130\,N)\tfrac{1}{27}\,\varepsilon^{-2}\right. \\
& \qquad\quad \left. + (671+41\,N+384\,\zeta(3))\tfrac{1}{54}\,\varepsilon^{-1}\right] \\
& + \bar g_1\bar g_2^2\left[(206+13\,N)\tfrac{2}{9}\,\varepsilon^{-3} - (428+7\,N)\tfrac{2}{27}\,\varepsilon^{-2} + (660+N+192\zeta(3))\tfrac{1}{72}\,\varepsilon^{-1}\right] \\
& + \bar g_2^3\left[\tfrac{40}{3}\,\varepsilon^{-3} - \tfrac{70}{9}\,\varepsilon^{-2} + \tfrac{29}{12}\,\varepsilon^{-1}\right],
\end{aligned}
\tag{15.26}
$$

$$
\begin{aligned}
Z_{g_2} \; = \; & 1 + \bar g_1\,4\,\varepsilon^{-1} + \bar g_2\,3\,\varepsilon^{-1} + \bar g_1^2\left[(40+2N)\tfrac{1}{3}\,\varepsilon^{-2} - (14+N)\tfrac{1}{3}\,\varepsilon^{-1}\right] \\
& + \bar g_1\bar g_2\left[22\,\varepsilon^{-2} - 8\,\varepsilon^{-1}\right] + \bar g_2^2\left[9\,\varepsilon^{-2} - 3\,\varepsilon^{-1}\right] \\
& + \bar g_1^3\left[+(1120+136N+4N^2)\tfrac{1}{27}\,\varepsilon^{-3} - (448+58N+N^2)\tfrac{2}{27}\,\varepsilon^{-2}\right. \\
& \qquad\quad \left. + (550+63N-2\,N^2+448\zeta(3)+32N\zeta(3))\tfrac{1}{54}\,\varepsilon^{-1}\right] \\
& + \bar g_1^2\bar g_2\left[(328+12N)\tfrac{1}{3}\,\varepsilon^{-3} - (400+13N)\tfrac{2}{9}\,\varepsilon^{-2} + (1966+19\,N+1536\zeta(3))\tfrac{1}{72}\,\varepsilon^{-1}\right] \\
& + \bar g_1\bar g_2^2\left[284\tfrac{1}{3}\,\varepsilon^{-3} - 674\tfrac{1}{9}\,\varepsilon^{-2} + (265+192\zeta(3))\tfrac{1}{12}\,\varepsilon^{-1}\right] \\
& + \bar g_2^3\left[27\,\varepsilon^{-3} - 62\tfrac{1}{3}\,\varepsilon^{-2} + (49+32\zeta(3))\tfrac{1}{8}\,\varepsilon^{-1}\right].
\end{aligned}
\tag{15.27}
$$

15.2.3 Ratios between Bare and Renormalized Quantities up to Three Loops

The relations between the renormalized and the bare quantities for more than one coupling constant are defined in Eqs. (10.210) and are repeated here:

$$\frac{\phi_B^2}{\phi^2} = Z_\phi , \qquad \frac{m_B^2}{m^2} = \frac{Z_{m^2}}{Z_\phi} , \qquad \frac{\bar{g}_{1B}}{\bar{g}_2} = \frac{Z_{g_1}}{Z_\phi^2} , \qquad \frac{\bar{g}_{2B}}{\bar{g}_2} = \frac{Z_{g_2}}{Z_\phi^2} . \tag{15.28}$$

The first ratio follows again directly from (15.24). For mass and coupling constant ratios, we find up to three loops:

$$
\begin{aligned}
m_B^2/m^2 =\ & 1 + \bar{g}_1\,(2+N)\tfrac{1}{3}\,\varepsilon^{-1} + \bar{g}_2\,\varepsilon^{-1} + \bar{g}_1^2\left[(2+N)(5+N)\tfrac{1}{9}\,\varepsilon^{-2} - 5(2+N)\tfrac{1}{36}\,\varepsilon^{-1}\right] \\
& + \bar{g}_1\,\bar{g}_2\left[2(5+N)\tfrac{1}{3}\,\varepsilon^{-2} - \tfrac{5}{6}\,\varepsilon^{-1}\right] + \bar{g}_2^2\left[2\,\varepsilon^{-2} - \tfrac{5}{12}\,\varepsilon^{-1}\right] \\
& + \bar{g}_1^3\left[(60 + 52\,N + 13\,N^2 + N^3)\tfrac{1}{27}\,\varepsilon^{-3} - (556 + 400\,N + 61\,N^2)\tfrac{1}{324}\,\varepsilon^{-2}\right. \\
& \qquad \left. + (74 + 47\,N + 5\,N^2)\tfrac{1}{108}\,\varepsilon^{-1}\right] \\
& + \bar{g}_1^2\,\bar{g}_2\left[(5+N)(6+N)\tfrac{1}{3}\,\varepsilon^{-3} - (278 + 61\,N)\tfrac{1}{36}\,\varepsilon^{-2} + (37 + 5\,N)\tfrac{1}{12}\,\varepsilon^{-1}\right] \\
& + \bar{g}_1\,\bar{g}_2^2\left[(112 + 14\,N)\tfrac{1}{9}\,\varepsilon^{-3} - (982 + 35\,N)\tfrac{1}{108}\,\varepsilon^{-2} + (251 + N)\tfrac{1}{72}\,\varepsilon^{-1}\right] \\
& + \bar{g}_2^3\left[\tfrac{14}{3}\,\varepsilon^{-3} - \tfrac{113}{36}\,\varepsilon^{-2} + \tfrac{7}{6}\,\varepsilon^{-1}\right] ,
\end{aligned}
\tag{15.29}
$$

$$
\begin{aligned}
\bar{g}_{1B}/\bar{g}_1 =\ & 1 + \bar{g}_1\,(8+N)\tfrac{1}{3}\,\varepsilon^{-1} + \bar{g}_2\,2\,\varepsilon^{-1} + \bar{g}_1^2\left[(8+N)^2\tfrac{1}{9}\,\varepsilon^{-2} - (14 + 3\,N)\tfrac{1}{6}\,\varepsilon^{-1}\right] \\
& + \bar{g}_1\,\bar{g}_2\left[(12+N)\varepsilon^{-2} - \tfrac{11}{3}\,\varepsilon^{-1}\right] + \bar{g}_2^2\left[5\,\varepsilon^{-2} - \tfrac{5}{6}\,\varepsilon^{-1}\right] \\
& + \bar{g}_1^3\left[(8+N)^3\tfrac{1}{27}\,\varepsilon^{-3} - 7(8+N)(14+3N)\tfrac{1}{54}\,\varepsilon^{-2}\right. \\
& \qquad \left. + (1480 + 461\,N + 22\,N^2 + 1056\,\zeta(3) + 240\,N\,\zeta(3))\tfrac{1}{324}\,\varepsilon^{-1}\right] \\
& + \bar{g}_1^2\,\bar{g}_2\left[4(116 + 16\,N + N^2)\tfrac{1}{9}\,\varepsilon^{-3} - 11(94 + 11\,N)\tfrac{1}{27}\,\varepsilon^{-2}\right. \\
& \qquad \left. + (1318 + 79\,N + 768\,\zeta(3))\tfrac{1}{108}\,\varepsilon^{-1}\right] \\
& + \bar{g}_1\,\bar{g}_2^2\left[2(206 + 13\,N)\tfrac{1}{9}\,\varepsilon^{-3} - (1598 + 25\,N)\tfrac{1}{54}\,\varepsilon^{-2} + (642 + N + 192\,\zeta(3))\tfrac{1}{72}\,\varepsilon^{-1}\right] \\
& + \bar{g}_2^3\left[\tfrac{40}{3}\,\varepsilon^{-3} - \tfrac{64}{9}\,\varepsilon^{-2} + \tfrac{7}{3}\,\varepsilon^{-1}\right] ,
\end{aligned}
\tag{15.30}
$$

$$
\begin{aligned}
\bar{g}_{2B}/\bar{g}_2 =\ & 1 + \bar{g}_1\,4\,\varepsilon^{-1} + \bar{g}_2\,3\,\varepsilon^{-1} + \bar{g}_1^2\left[2(20+N)\tfrac{1}{3}\,\varepsilon^{-2} - (82 + 5\,N)\tfrac{1}{18}\,\varepsilon^{-1}\right] \\
& + \bar{g}_1\,\bar{g}_2\left[22\,\varepsilon^{-2} - \tfrac{23}{3}\,\varepsilon^{-1}\right] + \bar{g}_2^2\left[9\,\varepsilon^{-2} - \tfrac{17}{6}\,\varepsilon^{-1}\right] \\
& + \bar{g}_1^3\left[(1120 + 136\,N + 4\,N^2)\tfrac{1}{27}\,\varepsilon^{-3} - (2636 + 320\,N + 5\,N^2)\tfrac{1}{81}\,\varepsilon^{-2}\right. \\
& \qquad \left. + (6568 + 736\,N - 13\,N^2 + 2688\,\zeta(3) + 192\,N\,\zeta(3))\tfrac{1}{324}\,\varepsilon^{-1}\right] \\
& + \bar{g}_1^2\,\bar{g}_2\left[(328 + 12\,N)\tfrac{1}{3}\,\varepsilon^{-3} - (1554 + 47\,N)\tfrac{1}{18}\,\varepsilon^{-2} + (1950 + 17\,N + 1536\,\zeta(3))\tfrac{1}{72}\,\varepsilon^{-1}\right] \\
& + \bar{g}_1\,\bar{g}_2^2\left[\tfrac{284}{3}\,\varepsilon^{-3} - \tfrac{650}{9}\,\varepsilon^{-2} + (131 + 96\,\zeta(3))\tfrac{1}{6}\,\varepsilon^{-1}\right] \\
& + \bar{g}_2^3\left[27\,\varepsilon^{-3} - \tfrac{119}{6}\,\varepsilon^{-2} + (145 + 96\,\zeta(3))\tfrac{1}{24}\,\varepsilon^{-1}\right] .
\end{aligned}
\tag{15.31}
$$

A resummation procedure to be presented in Chapter 19 requires the renormalized coupling constant as an expansion in powers of the bare coupling constant. Although we shall not apply this scheme to the theory with mixed $O(N)$ and cubic symmetry, the following three-loop expansions will be useful for further investigation. The full five-loop series can be downloaded from the World Wide Web [3].

$$
\begin{aligned}
\bar{g}_1/\bar{g}_{1B} = {} & 1 - \bar{g}_{1B}\,(8+N)\tfrac{1}{3}\,\varepsilon^{-1} - \bar{g}_{2B}\,2\,\varepsilon^{-1} \\
& + \bar{g}_{1B}^2\,\left[(8+N)^2\tfrac{1}{9}\,\varepsilon^{-2} + (14+3N)\tfrac{1}{6}\,\varepsilon^{-1}\right] \\
& + \bar{g}_{1B}\,\bar{g}_{2B}\,\left[(12+N)\,\varepsilon^{-2} + \tfrac{11}{3}\,\varepsilon^{-1}\right] + \bar{g}_{2B}^2\,\left[5\,\varepsilon^{-2} + \tfrac{5}{6}\,\varepsilon^{-1}\right] \\
& + \bar{g}_{1B}^3\,\left[-(8+N)^3\tfrac{1}{27}\,\varepsilon^{-3} - 4(8+N)(14+3N)\tfrac{1}{27}\,\varepsilon^{-2}\right. \\
& \qquad \left. -(2960 + 922\,N + 33\,N^2 + 2112\,\zeta(3) + 480\,N\,\zeta(3))\tfrac{1}{648}\,\varepsilon^{-1}\right] \\
& + \bar{g}_{1B}^2\,\bar{g}_{2B}\,\left[-(464 + 84\,N + 4\,N^2)\tfrac{1}{9}\,\varepsilon^{-3} - (1168 + 134\,N)\tfrac{1}{27}\,\varepsilon^{-2}\right. \\
& \qquad \left. -(1318 + 79\,N + 768\,\zeta(3))\tfrac{1}{108}\,\varepsilon^{-1}\right] \\
& + \bar{g}_{1B}\,\bar{g}_{2B}^2\,\left[-(412 + 26\,N)\tfrac{1}{9}\,\varepsilon^{-3} - (866 + 10\,N)\tfrac{1}{27}\,\varepsilon^{-2} - (642 + N + 192\,\zeta(3))\tfrac{1}{72}\,\varepsilon^{-1}\right] \\
& + \bar{g}_{2B}^3\,\left[-\tfrac{40}{3}\,\varepsilon^{-3} - \tfrac{62}{9}\,\varepsilon^{-2} - \tfrac{7}{3}\,\varepsilon^{-1}\right],
\end{aligned}
\tag{15.32}
$$

$$
\begin{aligned}
\bar{g}_2/\bar{g}_{2B} = {} & 1 - \bar{g}_{1B}\,4\,\varepsilon^{-1} - \bar{g}_{2B}\,3\,\varepsilon^{-1} + \bar{g}_{1B}^2\,\left[2(20+N)\tfrac{1}{3}\,\varepsilon^{-2} + (82 + 5\,N)\tfrac{1}{18}\,\varepsilon^{-1}\right] \\
& + \bar{g}_{1B}\,\bar{g}_{2B}\,\left[22\,\varepsilon^{-2} + \tfrac{23}{3}\,\varepsilon^{-1}\right] + \bar{g}_{2B}^2\,\left[9\,\varepsilon^{-2} + \tfrac{17}{6}\,\varepsilon^{-1}\right] \\
& + \bar{g}_{1B}^3\,\left[-4(14+N)(20+N)\tfrac{1}{27}\,\varepsilon^{-3} - (3040 + 388\,N + 10\,N^2)\tfrac{1}{81}\,\varepsilon^{-2}\right. \\
& \qquad \left. -(3284 + 368\,N - 13\,N^2 + 2688\,\zeta(3) + 192\,N\,\zeta(3))\tfrac{1}{324}\,\varepsilon^{-1}\right] \\
& + \bar{g}_{1B}^2\,\bar{g}_{2B}\,\left[-(328 + 12\,N)\tfrac{1}{3}\,\varepsilon^{-3} - (900 + 32\,N)\tfrac{1}{9}\,\varepsilon^{-2}\right. \\
& \qquad \left. -(1950 + 17\,N + 1536\,\zeta(3))\tfrac{1}{72}\,\varepsilon^{-1}\right] \\
& + \bar{g}_{1B}\,\bar{g}_{2B}^2\,\left[-\tfrac{284}{3}\,\varepsilon^{-3} - \tfrac{754}{9}\,\varepsilon^{-2} - (131 + 96\,\zeta(3))\tfrac{1}{6}\,\varepsilon^{-1}\right] \\
& + \bar{g}_{2B}^3\,\left[-27\,\varepsilon^{-3} - \tfrac{68}{3}\,\varepsilon^{-2} - (145 + 96\,\zeta(3))\tfrac{1}{24}\,\varepsilon^{-1}\right].
\end{aligned}
\tag{15.33}
$$

15.3 Renormalization Constant for Vacuum Energy

For completeness, let us also state without calculation the renormalization constant Z_v introduced in Section 10.7. Recall that the vacuum energy becomes finite by adding to the initial energy functional a term

$$
\Delta E = \frac{m^4}{(4\pi)^2 g \mu^\varepsilon}\,Z_v,
\tag{15.34}
$$

where Z_v collects all pole terms in ε. Up to five loops, it has the expansion [5].

$$
\begin{aligned}
Z_v = {} & \frac{N}{2\varepsilon}\,g + \frac{N(N+2)}{6\varepsilon^2}\,g^2 + \left[\frac{N(N+2)}{144\varepsilon} - \frac{5N(N+2)}{54\varepsilon^2} + \frac{N(N+2)(N+4)}{18\varepsilon^3}\right]g^3 \\
& + \left\{\frac{N(N+2)(N+8)[13\zeta(3) - 25]}{2592\varepsilon} + \frac{N(N+2)(4N+29)}{108\varepsilon^2}\right. \\
& \qquad \left. -\frac{N(N+2)(31N+128)}{324\varepsilon^3} + \frac{N(N+2)(N+4)(N+5)}{54\varepsilon^4}\right\}g^4
\end{aligned}
$$

$$
+\bigg\{ N(N+2)[-319N^2 + 13968N + 64864 + 16(3N^2 - 382N - 1700)\zeta(3)
$$

$$
+96(4N^2 + 39N + 146)\zeta(4) - 1024(5N + 22)\zeta(5)]\frac{1}{207360\varepsilon}
$$

$$
-N(N+2)[31N^2 + 2354N + 9306 + 3(7N^2 - 28N + 48)\zeta(3) + 72(5N + 22)\zeta(4)]\frac{1}{9720\varepsilon^2}
$$

$$
+N(N+2)[519N^2 + 8462N + 25048 + 288(5N + 22)\zeta(3)]\frac{1}{19440\varepsilon^3}
$$

$$
-N(N+2)(293N^2 + 2624N + 5840)\frac{1}{4860\varepsilon^4}
$$

$$
+N(N+2)(N+4)(N+5)(5N+28)810\varepsilon^5 \bigg\} g^5. \tag{15.35}
$$

Notes and References

As indicated in the text, the lengthy expressions in this Chapter are all available on the World Wide Web under the address www.physik.fu-berlin.de/~kleinert/b8/programs/programs.html.

The individual citations in the text refer to:

[1] K.G. Chetyrkin, A.L. Kataev, F.V. Tkachov, Phys. Lett. B **99**, 147-150 (1981); ibid. **101**, 457 (1981)(Erratum);
K.G. Chetyrkin, S.G. Gorishny, S.A. Larin, F.V. Tkachov, Phys. Lett. B *132* 351 (1983);
Preprint INR P-0453, Moscow (1986).

[2] H. Kleinert, J. Neu, V. Schulte-Frohlinde, K.G. Chetyrkin, S.A. Larin, Phys. Lett. B **272**, 39 (1991) (hep-th/9503230); Erratum ibid. **319**, 545 (1993).

[3] Download the full five-loop result from the World Wide Web address given in the above note.

[4] H. Kleinert, Phys. Rev. D **60**, 085001 (1999) (hep-th/9812197).

[5] B. Kastening, Phys. Rev. D **54**, 3965 (1996); Phys. Rev. D **57**, 3567 (1998).

16

Basic Resummation Theory

The power series in the coupling constant g or in ε derived so far are unsuitable for numeric calculations since they are divergent for any coupling strength. They are called *asymptotic series*. Resummation procedures are necessary to extract physical results. The crudest method to approximate a divergent function $f(g)$ whose first L expansion coefficients are known employs *Padé approximants* [1]. These are rational functions with the same power series expansions as $f(g)$. A better approximation can be found by taking into account the knowledge of the *large-order behavior* of the expansion coefficients. This is provided by a semiclassical analysis of the functional integral (2.6) of the ϕ^4-theory, from which it follows that the coefficients grow factorially with the order. Such a factorial growth can be taken into account by means of *Borel transformations*. The *Padé-Borel method* applies the Padé approximation to the Borel transform. We shall see below that the Borel transform has a left-hand cut in the complex g-plane. The Padé approximation replaces the cut by a string of poles. This approximation can be improved further by a *conformal mapping technique* in which the complex g-plane is mapped into a unit circle which contains the original left-hand cut on its circumference. More efficient resummation techniques are based on re-expansions of the asymptotic truncated series for $f(g)$ in terms of special basis functions $I_n(g)$. These can be chosen in different ways to possess precisely the analytic behavior responsible for the divergence of the original series. Thus the basis functions contain from the outset the necessary information on the large-order behavior of the expansion coefficients, thus containing information beyond the expansion coefficients of the perturbation series. All these methods will be explained in this chapter. The most efficient technique will be applied in Chapters 17 and 18 to calculate critical exponents. A completely different approach to the critical exponents and the resummation problem will be developed in Chapters 19 and 20. There we shall consider the problem of calculating the critical behavior of ϕ^4-theory as a strong-coupling problem in the *bare* coupling constant g_B.

16.1 Asymptotic Series

At large orders, the perturbation coefficients of the renormalization group functions grow like a factorial of the order, with alternating signs. This behavior has a simple origin. If we plot the potential part of the energy density in (2.1) for small negative values of the coupling constant g as shown in Fig. 16.1, we see that the fluctuations around the origin are metastable, since they can carry the field with a nonzero probability all across the barrier into the energy abyss to the right or the left. For very high barriers, the probability of such an event is given by the Boltzmann factor associated with a spherically symmetric classical solution to the field equation:

$$\frac{\delta}{\delta \phi} E[\phi] = (-\partial_{\mathbf{x}}^2 + m^2)\phi^2(\mathbf{x}) + \frac{g}{3!}\phi^3(\mathbf{x}) = 0. \tag{16.1}$$

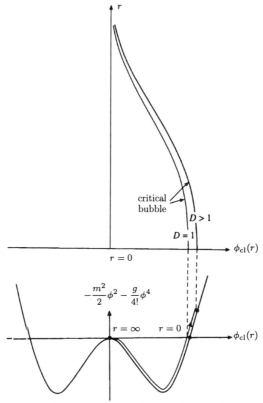

FIGURE 16.1 Fluctuations around the origin are metastable for $g < 0$ since they can reach beyond the barrier (drawn upside down) with ϕ depending on r as indicated by the curve with arrow. The solution of the field equation traversing the barrier looks like the trajectory of a mechanical point particle moving with negative friction in the potential well. It has an activation energy $1/\alpha g$ producing a left-hand cut in the complex g-plane in all correlation functions, with a discontinuity proportional to the Boltzmann factor $e^{-1/\alpha g}$ [in the natural units used in the functional integral (2.6)].

This has the form of a bubble described by a function $\phi_{cl}(r)$, solving the radial equation in D dimensions

$$\left(-\frac{d^2}{dr^2} - \frac{D-1}{r}\frac{d}{dr}\right)\phi_{cl}(r) + m^2\phi_{cl} + \frac{g}{3!}\phi_{cl}^3(r) = 0. \tag{16.2}$$

If we interpret $-r$ as a time t and $\phi_{cl}(r)$ as a position $x(-t)$ of a point particle as a function of time, we obtain an equation of motion for this point particle in the inverse potential $V(x) = -(m^2/2)x - (g/4!)x^4$:

$$\ddot{x}(t) - \frac{D-1}{t}\dot{x}(t) - m^2x(t) - \frac{g}{3!}x^3(t) = 0. \tag{16.3}$$

In point mechanics, the second term proportional to $\dot{x}(t)$ corresponds to a negative friction whose strength decreases with time like $1/t$. The trajectory is indicated in Fig. 16.1 by the curve with an arrow, from which one deduces the properties of the bubble solution. For $D = 1$, the friction is absent and a first integral of the differential equation (16.3) is given by energy conservation, fixing the maximal value of $x = \phi_{cl}$ by the right-hand zero of the potential. In $D > 1$ dimensions, the negative friction makes the particle overshoot on the right-hand side. For more details see the discussion in the textbooks [2]. The Boltzmann factor is proportional to $e^{-1/\alpha g}$, where $1/\alpha g$ is the action of such a solution, playing the role of an activation energy. The instability of the fluctuations gives rise to an imaginary part in the functional integral (2.6) for the partition function and (2.10) for the correlation functions. This property will go over to all renormalization group functions such as $\beta(g)$, $\nu(g)$, $\eta(g)$. Since all these expressions are real for positive g, they have a cut from $g = 0$ to $-\infty$ with a discontinuity across it which behaves like $e^{-1/\alpha g}$ near the tip of the cut.

Functions with such a cut have a vanishing radius of convergence. Their expansion coefficients grow factorially, as we can easily see as follows: let $f(g)$ be such a function with a discontinuity along the left-hand cut:

$$\text{disc } f(g) = f(g + i\varepsilon) - f(g - i\varepsilon) = 2 i \operatorname{Im} f(g + i\varepsilon) \quad \text{for } g < 0. \tag{16.4}$$

The discontinuity determines all expansion coefficients. To see this, we use a dispersion relation relating $f(g)$ on the positive real axis to the discontinuity at the cut:

$$f(g) = \frac{1}{2\pi i} \int_{-\infty}^{0} dg' \, \frac{\text{disc } f(g')}{g' - g} . \tag{16.5}$$

Expansion in powers of g as $f(g) = \sum_k f_k g^k$ yields coefficients f_k expressed as integrals over the discontinuity of the cut:

$$f_k = \frac{1}{2\pi i} \int_{-\infty}^{0} dg' \, \frac{\text{disc } f(g')}{g'^{k+1}} . \tag{16.6}$$

The main contribution to the integral in Eq. (16.6) for $k \to \infty$ comes from the region of small negative g', i.e. from the tip of the cut. In this limit, quasiclassical methods can be used to calculate the imaginary part of the function $f(g)$ [3]. Near the tip of the cut, the functional integral describes a system which decays through a very high barrier. The functional integral can therefore be approximated by the saddle point approximation. The leading term is produced by the classical solution. The fluctuation corrections correspond to small distortions and translations of this solution. As a result we find for $g \to 0^-$ a discontinuity of the form

$$\text{disc } f(-|g|) = 2\pi i \, \frac{\gamma}{(\alpha|g|)^{\beta+1}} \, e^{-1/\alpha|g|} \, [1 + \mathcal{O}(\alpha|g|)] . \tag{16.7}$$

The exponential is a quantum version of the Boltzmann factor for the activation energy of the classical solution. The prefactor accounts for the fluctuation entropy. Inserting (16.7) into Eq. (16.6) yields

$$f_k = \frac{(-1)^k}{2\pi i} \int_0^\infty dg' \, \frac{\text{disc } f(-g')}{g'^{k+1}} \approx \frac{\gamma(-1)^k}{\alpha^{\beta+1}} \int_0^\infty dg' (g')^{-(\beta+k+2)} e^{-1/\alpha g'} , \tag{16.8}$$

which leads via the integral formula

$$\int_0^\infty dg \, g^{-(\beta+k+2)} e^{-1/\alpha g} = \alpha^{\beta+k+1} \Gamma(\beta + k + 1) \tag{16.9}$$

to the large-order behavior of the coefficients:

$$f_k = \gamma(-\alpha)^k \Gamma(k + \beta + 1) . \tag{16.10}$$

For large k, the Gamma function $\Gamma(k + \beta + 1)$ can be approximated with the help *Stirling's formula*

$$\Gamma(pk + q) \sim \sqrt{2\pi}^{1-p} p^{q-1/2} k^{-1/2+q-p/2} p^{pk} (k!)^p , \tag{16.11}$$

from which we find that $\Gamma(k + \beta + 1) = (k + \beta)!$ grows like $k^\beta k!$, and we may write just as well

$$f_k = \gamma(-\alpha)^k k^\beta k![1 + \mathcal{O}(1/k)] \quad \text{for } k \to \infty. \tag{16.12}$$

In a similar way, the large-order behavior can be determined for all correlation functions of the $O(N)$-symmetric ϕ^4-theory [4]. This will be considered in Chapter 17.

The factorial growth of the perturbation expansion implies that we are confronted with asymptotic series which require special resummation procedures if we want to extract reliable results. Let us first recall some of their basic properties. According to Poincaré [5], a divergent series is an asymptotic expansion of a function $f(g)$ if

$$\lim_{|g| \to 0} \left[\frac{1}{g^L} \left| f(g) - \sum_{k=0}^{L} f_k g^k \right| \right] = 0, \quad \text{for } L \geq 0. \tag{16.13}$$

This definition implies that an asymptotic series does not define a function uniquely. The expansion coefficients of a function of the type $e^{-1/\alpha g} \sum_{k=0}^{L} f_k g^k$ being identically zero, such a function can be added to $f(g)$ without changing (16.13):

$$\lim_{|g| \to 0} \left[\frac{1}{g^L} \left(e^{-1/\alpha g} \sum_{k=0}^{L} f_k g^k \right) \right] = 0 \quad \text{for all } L. \tag{16.14}$$

The existence of such an ambiguity can be excluded under certain conditions by the theorem of Carleman [6], which is based on theorems of Phragmen-Lindelöf. Suppose a function $f(g)$ is analytic for $|g| < g_0$ with $|\arg(g)| \leq \pi\delta/2$ and has an asymptotic expansion in this region, so that

$$\Delta(L, g) = |f(g) - \sum_{k=0}^{L-1} f_k g^k| \leq M\alpha^L (L!)^\rho |g|^L, \quad |\arg g| \leq \frac{\pi\delta}{2}, \tag{16.15}$$

where for all k:

$$|f_k| \leq \gamma \, k! \, A^k, \tag{16.16}$$

with some real number A. Such a function is defined uniquely by its asymptotic expansion if $\delta > \rho \geq 1$. According to Nevanlinna [7], this is true even for $\delta \geq \rho \geq 1$. In this case, it makes sense to reconstruct a function from its asymptotic series expansion.

The error of the $(L - 1)$th partial sum of an asymptotic series is, according to Eq. (16.15) for $\rho = 1$, bounded by $\alpha^L L! |g|^L$. For small g, this error decreases for some initial orders and reaches a minimum at some $L_{\min}$. Indeed, from Stirling's formula (16.11) we know that

$$L! \approx L^{L+1/2} \sqrt{2\pi} e^{-L}, \tag{16.17}$$

so that we may estimate

$$\alpha^L L! g^L \approx \exp[-L + L \ln \alpha g + (L + 1/2) \ln L + 1/2 \ln 2\pi]. \tag{16.18}$$

This has a minimum at $L_{\min} \approx 1/\alpha g$, where the error becomes roughly

$$\Delta_{\min}(g) \equiv \Delta(L_{\min}, g) \approx e^{-1/\alpha g}. \tag{16.19}$$

Calculating higher and higher partial sums does not improve the approximation. The best result is reached by the partial sum of order $L_{\min}$, where the error is of the order $\mathcal{O}(e^{-1/\alpha g})$ such that even this result is useful provided g is very small. For larger g, more complicated methods have to be used to extract reliable information from an asymptotic series.

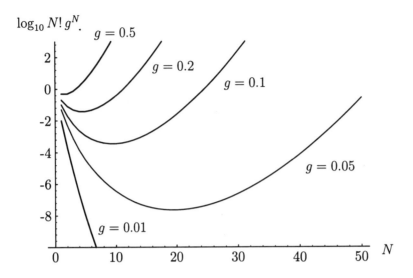

FIGURE 16.2 Minimal error $\Delta_{\min}(g)$ of an expansion with a behavior of the form $L!\,g^L$ is shifted towards larger L for smaller and smaller g.

16.2   Padé Approximants

The *Padé approximant* for a series expansion $f(z) = \sum_{i=0}^{k} f_i z^i + \ldots$ up to the order k is given by

$$[M/N] = \frac{a_0 + a_1 z + \ldots + a_M z^M}{1 + b_1 z + \ldots + b_N z^N}, \quad M + N = k, \tag{16.20}$$

where a_i and b_i are chosen such that the series expansion of $[M/N]$ up to the order k equals the original series

$$\sum_{i=0}^{k} f_i z^i = [M/N] + O[z^{M+N+1}]. \tag{16.21}$$

The $M + N + 1$ unknown coefficients are given uniquely by the $k + 1$ coefficients f_i. The calculation of the b_i proceeds in principle by multiplication of Eq. (16.21) with the denominator of $[M/N]$ and comparison of the coefficients of the equations for z^n with $M < n < M + N + 1$:

$$\begin{aligned}
b_N f_{M-N+1} + b_{N-1} f_{M-N+2} + \ldots + b0 f_{M+1} &= 0 \\
b_N f_{M-N+2} + b_{N-1} f_{M-N+3} + \ldots + b0 f_{M+2} &= 0 \\
&\vdots \\
b_N f_M + b_{N-1} f_{M+1} + \ldots + b0 f_{M+N} &= 0,
\end{aligned} \tag{16.22}$$

where $f_j = 0$ for $j < 0$. The coefficients of the numerator follow after inserting the results for the b_i into Eq. (16.21). For actual calculations, there are many algorithms which can be found in Ref. [1]. The symmetric $[N/N]$ Padé approximant usually gives the fastest approximation for increasing N although this depends on the function.

Being quotients of polynomials, the Padé approximants can describe functions with poles even in lowest order, in contrast to power series which are good approximations only inside some circle of convergence $|z| < R$. Therefore, the method can also be used to approximate functions outside of the region of convergence of their power series. A sequence of Padé approximants which converge outside the region of $|z| \leq R$ can be used to define an analytical continuation of the function for $|z| > R$. An example is shown in Fig. 16.3. The function has a power series with a convergence radius $R = 1/2$:

$$f(z) = \sqrt{\frac{1 + z/2}{1 + 2z}} = 1 - \frac{3}{4}z + \frac{39}{32}z^2 - \ldots . \tag{16.23}$$

Its [1/1] Padé approximant is

$$[1/1] = \frac{1 + 7z/8}{1 + 13z/8}, \tag{16.24}$$

which deviates by less than 8% from the exact function for $z \to \infty$, where $f(\infty) = 0.5$, whereas $[1/1](\infty) \approx 0.54$.

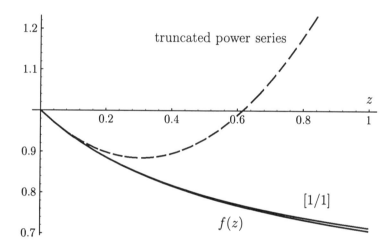

FIGURE 16.3 Plot of the function $f(z) = \sqrt{(1 + z/2)/(1 + 2z)}$, its truncated power series expansion $1 - 3z/4 + 39z^2/32 - \ldots$, and its Padé approximant $[1/1] = (1 + 7z/8)/(1 + 13z/8)$.

For $R = 0$, an asymptotic power series has only a formal character. With the help of a converging sequence of Padé approximants, however, it is possible to resum such a series.

The method has been extended [8] to two-variable approximants $[M, M]/[N, N]$, which could not be used in our applications since they lead to poles in the positive coupling constant plane. For the purpose of extracting reliable results from five-loop calculations of quantum field theory, the Padé approximants converge too slowly. It is necessary to include the knowledge of the large-order behavior which, as argued before, is possible using a Borel transformation of the asymptotic series.

16.3 Borel Transformation

Let the asymptotic series which describes a function $f(g)$ have a zero radius of convergence. Let this divergence be caused by a factorial growth of the coefficients f_k. If we divide each term in the expansion by a factor $k!$, we can obtain a series with a nonzero radius of convergence. This series is called a *Borel sum* [9]. The function $f(g)$ is called *Borel summable*, if the Borel sum can be summed and analytically continued over the whole positive axis. This analytical continuation can be used to recover the original function via a Borel transformation. The Borel transformation restores the factorial in each term of the expansion. For details and citations see Ref. [10].

For a function $f(g) = \sum_k f_k g^k$ with $f_k \propto (-\alpha)^k k!$ for large k, the Borel sum is defined by

$$B(g) \equiv \sum_k^\infty B_k g^k, \quad \text{with} \quad B_k \equiv \frac{f_k}{k!}. \tag{16.25}$$

The denominator in B_k removes the factorial growth of the expansion coefficients f_k and leads to a function $B(g)$ which is analytic in some neighborhood of the origin in the complex g-plane. In fact, for large k, the expansion coefficients B_k behave like

$$B_k \propto (-\alpha)^k, \quad \text{for large } k, \tag{16.26}$$

so that the Borel sum converges for $|g| < 1/\alpha$, and has an expansion

$$\sum_k^\infty B_k g^k = \frac{\text{const}}{1 + \alpha g} [1 + \mathcal{O}(g)], \quad \text{for } |g| < \frac{1}{\alpha}. \tag{16.27}$$

From this convergent expansion, an analytic continuation $B(g)$ can be defined for $g > 1/\alpha$.

Another way of defining the Borel transform (16.25) is via a contour integral:

$$B(t) \equiv \frac{1}{2\pi i} \oint_C \frac{dz}{z} e^z f(t/z), \tag{16.28}$$

where the contour C encircles the origin. Inserting the expansion $f(g) = \sum_n f_n g^n$, the integral produces in each term a factor $1/n!$, thus converting $f(g)$ into $B(t) = \sum_n f_n t^n/n!$.

The factorials can be restored using an integral representation of the Gamma function:

$$k! = \Gamma(k+1) = \int_0^\infty dt\, e^{-t} t^k. \tag{16.29}$$

This shows that the original power series expansion for $f(g)$ is recovered from the Borel sum by the Borel transformation:

$$f(g) = \sum_k^\infty f_k g^k = \sum_k^\infty B_k g^k k! = \sum_k^\infty B_k \int_0^\infty dt\, e^{-t} (gt)^k. \tag{16.30}$$

Interchanging the order of integration and summation, the right-hand side becomes

$$\int_0^\infty dt\, e^{-t} \sum_k^\infty B_k (gt)^k, \tag{16.31}$$

and leads to an integral transformation

$$f(g) = \int_0^\infty dt\, e^{-t} B(gt). \tag{16.32}$$

This is the Borel transformation of the Borel sum $B(g)$. If only a finite number of terms in the Borel sum is known, the integral in Eq. (16.31) just reinserts the factors $k!$ and leads back to the initial diverging series. This can be avoided by trying to find a nonpolynomial function whose initial Taylor coefficients coincide with the known terms in the Borel sum. Such a function could be a Padé approximant. But we can also try to construct an analytical continuation using the knowledge of the large-order behavior.

For finding an analytical continuation it is crucial to know the radius of convergence from the position of the singularity closest to the origin. But this singularity can be deduced from the large-order behavior as is stated by the *theorem of Darboux* [5]. Usually, the singularity will be a cut starting at $g = -1/\alpha$. Indeed, suppose the presence of such a cut in $B(g)$ in the form

$$B(g) = \frac{\gamma}{(1 + \alpha g)^{\beta+1}}. \tag{16.33}$$

Expanding this in powers of g yields

$$\frac{\gamma}{(1 + \alpha g)^{\beta+1}} = \gamma \sum_{k=0}^{\infty} B_k g^k = \gamma \sum_{k=0}^{\infty} \frac{\Gamma(\beta + k + 1)}{\Gamma(\beta + 1)\Gamma(k + 1)} (-\alpha)^k g^k, \tag{16.34}$$

with coefficients behaving for large k like

$$B_k \propto (-\alpha)^k k^\beta. \tag{16.35}$$

The coefficients of the original series $f(g) = \sum_k f_k g^k$ grow therefore like

$$f_k \propto (-\alpha)^k k^\beta k!. \tag{16.36}$$

If $B(g)$ contains a superposition of cuts like (16.34), the cut lying closest to the origin determines the convergence radius of the Borel sum, and dominates the large-k behavior of f_k.

In order to carry out the integration over the Borel sum $B(g)$, it not only has to be analytic in the circle of radius $1/\alpha$, it also has to be regular on the positive real axis. This is assured by a theorem of Watson [11] under conditions which are somewhat stricter than those required for the uniqueness of the asymptotic expansion. The theorem states that a function analytic in $|\arg g| \leq \pi/2 + \delta$ for $\delta > 0$ and $|g| < g_0$ with an asymptotic expansion in this region fulfilling

$$\Delta(L, g) = |f(g) - \sum_{k=0}^{L-1} f_k g^k| = \mathcal{O}(\alpha^L L! |g|^L), \quad |\arg g| \leq \pi/2 + \delta, \tag{16.37}$$

with f_k growing like $k! C^k$ for large k, has a Borel sum $B(t) = \sum_k f_k t^k / k!$ convergent in the circle $|t| < 1/\alpha$ and regular in the angle $|\arg t| < \delta$. The Borel sum can be integrated to give back $f(g)$:

$$f(g) = \int_0^\infty e^{-t} B(gt), \quad \text{for } |g| < g_0, \ |\arg g| < \delta. \tag{16.38}$$

This justifies the interchange of integration and summation under these conditions.

It is not easy to find out whether these conditions are satisfied for various field theories. It has been possible to prove Borel summability for ϕ^4-theory in $D = 0, 1, 2, 3$ dimensions [12], but not yet in $D = 4$ dimensions. We expect it to hold for all $D = 4 - \varepsilon < 4$.

Instead of a Borel transform with a leading cut at $-1/\alpha$, we may also define a generalized Borel transform in which the leading cut becomes a leading pole, called *Borel-Leroy transform* $B^\beta(g)$. It arises from the expansion of $f(g)$ by dividing each coefficient f_k by $\Gamma(k + \beta + 1)$ rather than by $k!$ [13]. The basic inverse Borel-Leroy transform is

$$\Gamma(k + \beta + 1) = \int_0^\infty dt\, e^{-t} t^{k+\beta}, \tag{16.39}$$

which serves to reinsert the growth factor $\Gamma(k + \beta + 1)$ to recover $f(g)$:

$$f(g) = \int_0^\infty dt\, e^{-t} t^\beta B^\beta(gt). \tag{16.40}$$

A pole term $1/(1 + \alpha t g)$ in $B^\beta(gt)$ corresponds obviously to expansion coefficients of $f(g)$ growing like

$$f_k \propto (-\alpha)^k \Gamma(k + \beta + 1) \approx (-\alpha)^k k! k^\beta. \tag{16.41}$$

The simplest way to apply the technique of Borel transformation to the resummation of perturbation expansions for critical exponents proceeds by dividing the expansion coefficients f_k by the leading growth $\Gamma(k + \beta + 1)$, using the Padé approximation to sum the resulting Borel series [14], and recovering $f(g)$ by doing the integral (16.40) numerically. If no high accuracy is required, this method can be applied without knowledge of the large-order behavior. We may try out a few different values of β and optimize the speed of convergence for increasing orders of the expansion.

The Padé approximants are useful to approximate the singularities on the negative axis by a string of poles. This makes them particularly useful for approximating Borel and Borel-Leroy transforms. Unfortunately, they often introduce unphysical poles on the positive axis, in which case the integral (16.40) no longer exists, making it impossible to recover the original function $f(g)$.

For the critical exponents to be evaluated in this book, the radius of convergence of the Borel sum is known from the large-order behavior of the expansions. This knowledge can be exploited to set up conformal mappings to a new complex variable, by which the Borel sum can be continued analytically to a regime outside the circle of convergence, as will now be explained.

16.4   Conformal Mappings

The method of conformal mappings [15, 16] exploits the knowledge of the location of the closest singularity in $B^\beta(g)$ from the large-order behavior of the expansion coefficients of $f(g)$. This knowledge is used to map the entire cut g-plane into the unit circle such that the singularities on the negative axis are moved onto the circumference of the circle. The result is an analytic continuation of the Borel sum to the entire g-plane. For a function $f(g) = \sum_k f_k g^k$ with the large-order behavior

$$f_k = \gamma k^\beta (-\alpha)^k k! [1 + \mathcal{O}(1/k)], \quad k \to \infty, \quad \alpha > 0, \tag{16.42}$$

the Borel-Leroy sum is

$$B^{\beta_0}(g) \;=\; \sum_{k=0}^\infty \frac{f_k}{\Gamma(k + \beta_0 + 1)}\, g^k = \sum_{k=0}^\infty B_k^{\beta_0} g^k. \tag{16.43}$$

Its expansion coefficients grow for large k like

$$B_k^{\beta_0} = \gamma (-\alpha)^k k^{\beta - \beta_0}, \quad \text{for } k \to \infty, \tag{16.44}$$

as a consequence of the singularity in $B^{\beta_0}(g)$ at $g = -1/\alpha$ which is closest to the origin. It is assumed that all other singularities of $B^{\beta_0}(g)$ lie on the negative axis further away from the origin. In ϕ^4-theory, this was proven for $D = 0, 1$ but not for $D = 2, 3$. Such singularities come from subleading instanton contributions.

The following mapping preserves the origin and transforms $g = \infty$ to $w = 1$ such that the whole cut plane will be in a circle of unit radius, and the singularities of the negative axis from $-1/\alpha$ to $-\infty$ will be on the border of the unit circle:

$$w(g) = \frac{\sqrt{1+\alpha g} - 1}{\sqrt{1+\alpha g} + 1}, \qquad g = \frac{4}{\alpha} \frac{w}{(1-w)^2}. \tag{16.45}$$

This implies that the function $W(w) = B^{\beta_0}(g(w))$ has a convergent Taylor expansion for $|w| < 1$ and so will the Borel sum which is re-expanded in the new variable $w(g)$ in the whole cut plane of g:

$$B^{\beta_0}(g) = \sum_{k=0}^{\infty} W_k [w(g)]^k. \tag{16.46}$$

A Borel sum that is known up to the L-th order can be re-expanded in the new variable:

$$B^{\beta_0[L]}(g) = \sum_{k=0}^{L} B_k^{\beta_0} g^k = \sum_{k=0}^{L} W_k [w(g)]^k + \mathcal{O}(g^{L+1}), \tag{16.47}$$

where the coefficients W_k are given by

$$W_k = \sum_{n=0}^{k} B_n^{\beta_0} \left(\frac{4}{\alpha}\right)^n \frac{(n+k-1)!}{(k-n)!(2n-1)!}. \tag{16.48}$$

The re-expanded series on the right-hand side of Eq. (16.47) will give a much better approximation to the Borel sum. The reason is that the analytic behavior which causes the divergent large-order behavior is already incorporated in each of the expansion terms $[w(g)]^k$. This can be seen by expanding $[w(g)]^k$ near $g = -1/\alpha$, which is the starting point of a square root branch cut, as

$$[w(g)]^k = (-1)^k \left[1 - 2k(1+\alpha g)^{1/2} + \mathcal{O}(1+\alpha g)\right], \quad \text{for} \quad g \to -1/\alpha. \tag{16.49}$$

According to the remarks made in the context of Eq. (16.33), each term leads to a growth of f_k of the form (16.42). Inserting (16.46) into the integral transform (16.40) gives an expansion for $f^{[L]}(g)$:

$$f^{[L]}(g) = \sum_{k=0}^{L} W_k \int_0^\infty e^{-t} t^{\beta_0} [w(gt)]^k. \tag{16.50}$$

In the case of ϕ^4-theory in $3, 4$ dimensions it is not known whether the new expansion of $f^{[L]}$ will have a nonzero radius of convergence or is just a new kind of an asymptotic expansion.

The convergence can be improved by an appropriate choice of the parameter β_0. A function with expansion coefficients of the form

$$B_k^{\beta_0} = \frac{\gamma(-\alpha)^k k^\beta k!}{\Gamma(k+\beta_0+1)}[1 + \mathcal{O}(1/k)] \;\to\; \gamma(-\alpha)^k k^{\beta-\beta_0}[1 + \mathcal{O}(1/k)], \; k \to \infty, \tag{16.51}$$

behaves close to its leading singularity at $-1/\alpha$ as:

$$B^{\beta_0}(g) = \gamma\Gamma(1+\beta-\beta_0)(1+\alpha g)^{\beta_0-\beta-1}[1 + \mathcal{O}(1+\alpha g)], \; g \to -1/\alpha. \tag{16.52}$$

The parameter β_0 can be chosen such that the singularity in Eq. (16.52) is of the same form as in Eq. (16.49):

$$\beta_0 = \beta + 3/2. \tag{16.53}$$

Furthermore, we can adapt a possibly known *strong-coupling power behavior* of $f(g)$:

$$f(g) \overset{g \to \infty}{\sim} g^s. \tag{16.54}$$

This is done by re-expanding the Borel sum in a modified manner:

$$B^{\beta_0[L]}(g) \equiv \sum_{k=0}^{L} B_k^{\beta_0} g^k = \frac{1}{[1-w(g)]^{2s}} \sum_{k=0}^{L} W_k(s)[w(g)]^k, \tag{16.55}$$

where the W_k are now depending on s:

$$W_k(s) = \sum_{n=0}^{k} \frac{B_n^{\beta_0}}{\Gamma(n+\beta_0+1)} \left(\frac{4}{\alpha}\right)^n \frac{(k+n-2s-1)!}{(k-n)!(2n-2s-1)!}. \tag{16.56}$$

Noting that for $g \to \infty$, $w(g) \to 1$ for $g \to \infty$ and

$$\frac{1}{[1-w(g)]^{2s}} = \left(\frac{\sqrt{1+\alpha g}+1}{2}\right)^{2s} \overset{g \to \infty}{\to} g^s \left(\frac{\alpha}{4}\right)^s, \tag{16.57}$$

the strong-coupling behavior is obviously reproduced order by order and is unchanged by the integral transform of Eq. (16.50).

16.5 Janke-Kleinert Resummation Algorithm

The conformal mapping results in a re-expansion of the Borel sum into powers of the functions $w(gt)$. The large-order behavior of the Borel sum is included in the functions $[w(gt)]^k$ which have the proper branch cut on the negative axis to generate the correct divergences after the integral transformation. It was realized by Janke and Kleinert [10] that these properties can be implemented in a simple systematic way by re-expanding the asymptotic expansion in a complete set of basis functions $I_n(g)$:

$$f_L(g) = \sum_{k=0}^{L} f_k g^k = \sum_{n=0}^{L} h_n I_n(g) \overset{L \to \infty}{\Rightarrow} f(g). \tag{16.58}$$

The basis functions $I_n(g)$ possess the following properties:

1. They have a simple *Borel-Leroy representation*

$$I_n(g) = \int_0^\infty dt e^{-t} t^{\beta_0} B_{I_n}^{\beta_0}(gt). \tag{16.59}$$

2. They possess the *large-order behavior* of the expansion coefficients f_k:

$$f_k \to \gamma(-\alpha)^k k^\beta k!, \quad \text{for } k \to \infty. \tag{16.60}$$

3. They exhibit a power-like *strong-coupling behavior*

$$f(g) \to g^s, \quad \text{for } g \to \infty. \tag{16.61}$$

4. The expansion coefficients h_n in (16.58) are determined by a triangular system of equations from f_k, so that the first known coefficients $f_1, f_2, \ldots, f_L$ are correctly reproduced.

16.5.1 Reexpansion Functions

The second property implies the existence of the same tip of the branch cut on the negative real axis in the functions $I_n(g)$ as in $f(g)$. Equivalently, the associated basis functions $B_{I_n}^{\beta_0}(gt)$ of the Borel-Leroy transforms have their smallest singularity in absolute size at $t = -1/\alpha|g|$. Janke and Kleinert chose for these a set of *hypergeometric functions* of the argument αgt, which are indeed analytic functions in gt with a cut on the negative axis starting at $-1/\alpha$:

$$
{}_2F_1(a, b; c; -\alpha gt) = \sum_{k=0}^{\infty} \frac{(a)_k (b)_k}{(c)_k} \frac{(-\alpha gt)^k}{k!}, \tag{16.62}
$$

where $(a)_n$ are the *Pochhammer symbols*:

$$
(a)_k = \frac{\Gamma(a+k)}{\Gamma(a)} = a(a+1)(a+2)\ldots(a+k-1), \qquad (a)_0 = 1. \tag{16.63}
$$

The hypergeometric functions are standard special functions whose properties are well-known. The Borel integral can immediately be calculated:

$$
\int_0^{\infty} dt\, e^{-t}\, t^{\beta_0}\, {}_2F_1(a, b, c; -\alpha gt) = \frac{\Gamma(c)}{\Gamma(a)\Gamma(b)} E(a, b, \beta_0 + 1; c; 1/\alpha g). \tag{16.64}
$$

The resulting functions are the *MacRobert functions*. Their asymptotic expansions are [17]

$$
\frac{\Gamma(c)}{\Gamma(a)\Gamma(b)} E(a, b, \beta_0 + 1; c; 1/\alpha g) \equiv \sum_{k=0}^{\infty} e_k g^k, \tag{16.65}
$$

with coefficients of the form:

$$
e_k \longrightarrow \frac{\Gamma(c)}{\Gamma(a)\Gamma(b)}(-1)^k k! k^{a+b-c+\beta_0-1} \alpha^k, \qquad \text{for } k \to \infty, \tag{16.66}
$$

thus reproducing the large-order behavior of (16.60). This behavior is unchanged if the hypergeometric function is multiplied by $(\alpha gt)^n$. A set of functions is therefore

$$
B_{I_n}^{\beta_0}(gt) = (\alpha gt)^n\, {}_2F_1(a, b; c; -\alpha gt) = (\alpha gt)^n \sum_{k=0}^{\infty} \frac{(a)_k (b)_k}{(c)_k} \frac{(-\alpha gt)^k}{k!}. \tag{16.67}
$$

The parameters a, b, and c are not completely fixed by the large-order behavior. The parameter β in (16.60) imposes the following relation upon a, b, c, and β_0 of (16.66):

$$
a + b + \beta_0 - c - 1 = \beta. \tag{16.68}
$$

This leaves three parameters undetermined. They may be adjusted to reduce the hypergeometric function to simple algebraic expressions. Two such choices are

$$
b = a + \frac{1}{2}, \quad c = 2a + 1: \quad F^{(1)}(a, a + \tfrac{1}{2}; 2a + 1; -\alpha gt) = 4^a \left(1 + \sqrt{1 + \alpha gt}\right)^{-2a}, \tag{16.69}
$$

$$
b = a + \frac{1}{2}, \quad c = 2a: \quad F^{(2)}(a, a + \tfrac{1}{2}; 2a; -\alpha gt) = 4^a \left(1 + \sqrt{1 + \alpha gt}\right)^{-2a} \frac{1 + \sqrt{1 + \alpha gt}}{2\sqrt{1 + \alpha gt}}, \tag{16.70}
$$

leaving only one free parameter. From Eq. (16.68) we see that the parameter β_0 is in these two cases equal to $\beta_0 = \beta + 3/2$ and $\beta_0 = \beta + 1/2$, respectively. The two hypergeometric functions possess precisely the cut of square root type in the complex g-plane that was previously generated by the analytic mapping procedure. The hypergeometric functions $_2F_1(a, b; c; -\alpha gt)$ are real for real arguments smaller than one, i.e., for $-\alpha gt < 1$. The imaginary part of the function for negative gt is found by rewriting the hypergeometric function as

$$_2F_1(a, b; c; -\alpha gt) = C_1\,_2F_1(a, b; a + b - c + 1; 1 + \alpha gt) \tag{16.71}$$

$$+(1 + \alpha gt)^{c-a-b}C_2\,_2F_1(c - a, c - b; c - a - b + 1; 1 + \alpha gt),$$

$$\text{with} \quad C_1 = \frac{\Gamma(c - a - b)}{\Gamma(c - a)\Gamma(c - b)} \quad \text{and} \quad C_2 = \frac{\Gamma(a + b - c)}{\Gamma(a)\Gamma(b)}. \tag{16.72}$$

The hypergeometric functions on the right-hand side are real for real negative gt as $1 + \alpha gt < 1$. Since the parameters a, b, c satisfy the relation $c - a - b = \pm 1/2$ for the two sets of functions, the term $(1 + \alpha gt)^{\pm 1/2}$ possesses a branch cut of the square root type. Its explicit form reads, for $g = -|g| + i\varepsilon$,

$$(1 + \alpha gt)^{\pm\frac{1}{2}} = \begin{cases} (1 - \alpha|g|t)^{\pm\frac{1}{2}} & \text{for } t < 1/\alpha|g|, \\ \pm i\,(\alpha|g|t - 1)^{\pm\frac{1}{2}} & \text{for } t > 1/\alpha|g|. \end{cases} \tag{16.73}$$

Inserting the special values of a, b, c, β_0 and the two functions (16.69) and (16.70) into (16.59), and carrying out the t-integration we obtain for $I_n(g)$ an imaginary part of the correct form $e^{-1/\alpha|g|}(1/\alpha|g|)^{\beta_0}$ responsible for the large-order behavior of $f(g)$. This ensures that the re-expansion (16.58) has an improved convergence.

The functions Eq. (16.69) and (16.70) contain a free parameter a which determines the strong-coupling exponent or strong-coupling parameter s in (16.61). For large g, the functions $F^{(1)}$ and $F^{(2)}$, multiplied by the factor $(\alpha g)^n$ to get $B_{I_n}^{\beta_0}(gt)$, grow like g^{n-a}, so that a is related to s by

$$a = n - s. \tag{16.74}$$

Including for later convenience a factor $1/4^n\Gamma(\beta_0 + 1)$ and $2/4^n\Gamma(\beta_0 + 1)$, we arrive at the basis functions for the Borel-Leroy transforms:

$$B_{I_n}^{\beta_0(1)}(gt) = \frac{(\alpha gt)^n}{\Gamma(\beta_0 + 1)4^n}\,_2F_1(n - s, n - s + \tfrac{1}{2}, 2(n - s) + 1, -\alpha gt) \tag{16.75}$$

$$= \frac{1}{\Gamma(\beta_0 + 1)}\left(\tfrac{1}{2} + \tfrac{1}{2}\sqrt{1 + \alpha gt}\right)^{2s}\frac{(\alpha gt)^n}{(1 + \sqrt{1 + \alpha gt})^{2n}}, \qquad \text{with } \beta_0 = \beta + 3/2,$$

$$B_{I_n}^{\beta_0(2)}(gt) = \frac{2(\alpha gt)^n}{\Gamma(\beta_0 + 1)4^n}\,_2F_1(n - s, n - s + \tfrac{1}{2}, 2(n - s), -\alpha gt) \tag{16.76}$$

$$= \frac{1}{\Gamma(\beta_0 + 1)}\left(\tfrac{1}{2} + \tfrac{1}{2}\sqrt{1 + \alpha gt}\right)^{2s}\frac{1 + \sqrt{1 + \alpha gt}}{\sqrt{1 + \alpha gt}}\frac{(\alpha gt)^n}{(1 + \sqrt{1 + \alpha gt})^{2n}}, \quad \text{with } \beta_0 = \beta + 1/2;$$

with the associated basis functions

$$I_n^{(1)}(g) = \int_0^\infty dt\, e^{-t}t^{\beta_0}B_{I_n}^{\beta_0(1)}(gt), \tag{16.77}$$

$$I_n^{(2)}(g) = \int_0^\infty dt\, e^{-t}t^{\beta_0}B_{I_n}^{\beta_0(2)}(gt). \tag{16.78}$$

We now turn to the determination of the re-expansion coefficients h_n in (16.58). They are found by comparing the coefficients of g^k in $f(g)$ and $\sum_{n=0}^{L} h_n I_n(g)$. For this we need the coefficients of the expansion of $I_n(g)$. Inserting the expansion of the hypergeometric function in (16.59) and performing the Laplace transforms in (16.77) and (16.78), we obtain

$$
\begin{aligned}
I_n^{(1)} &= \int_0^\infty dt \, \frac{e^{-t} t^{\beta_0} (\alpha g t)^n}{\Gamma(\beta_0 + 1) 4^n} \, {}_2F_1(n - s, n - s + \tfrac{1}{2}; 2(n - s) + 1; -\alpha g t) \\
&= \sum_{k=0}^\infty \frac{\Gamma(\beta_0 + n + 1)}{\Gamma(\beta_0 + 1) 4^n} \frac{(n - s)_k (n - s + \tfrac{1}{2})_k (\beta_0 + n + 1)_k}{(2n - 2s + 1)_k} \frac{(-1)^n (-\alpha g)^{k+n}}{k!} \\
&\equiv \sum_{k'=n}^\infty I_{k',n}^{(1)} \, g^{k'} .
\end{aligned}
\tag{16.79}
$$

Expressing the Pochhammer symbols in terms of Gamma functions, and using the identity where

$$
\Gamma(2a) = \frac{\Gamma(a)\Gamma(a + 1/2)}{\Gamma(1/2)} 2^{2a-1} ,
\tag{16.80}
$$

the expansion coefficients $I_{k,n}^{(1)}$ become

$$
I_{k,n}^{(1)} = \frac{(-1)^n (-\alpha)^k (n - s)}{4^s \sqrt{\pi}} \frac{\Gamma(k - s)\Gamma(k - s + \tfrac{1}{2})\Gamma(\beta_0 + k + 1)}{\Gamma(\beta_0 + 1)\Gamma(n + k - 2s + 1)\Gamma(k - n + 1)}.
\tag{16.81}
$$

They have the important property that $I_{k,n}^{(1)} = 0$ for $n > k$, i.e., the matrix $I_{k,n}^{(1)}$ has a convenient triangular form. This is due to the fact that the basis functions $B_{I_n}^{\beta_0 (1)}(gt)$ in Eq. (16.75) start out with powers $(gt)^n$. The triangular form will make the inversion of these matrices trivial.

For the second set of functions we find similarly

$$
I_{k,n}^{(2)} = \frac{(-1)^n (-\alpha)^k}{4^s \sqrt{\pi}} \frac{\Gamma(k - s)\Gamma(k - s + 1/2)\Gamma(\beta_0 + k + 1)}{\Gamma(\beta_0 + 1)\Gamma(n + k - 2s)\Gamma(k - n + 1)},
\tag{16.82}
$$

where again $I_{k,n}^{(2)} = 0$ for $n > k$, i.e., the matrix $I_{k,n}^{(2)}$ has once again the convenient triangular form.

The behavior of $I_{k,n}$ for large k follows from $\Gamma(k + 1 + a) \to k! k^a (1 + \mathcal{O}(1/k))$ as:

$$
I_{k,n}^{(1)} \to c_1 (-1)^n (-\alpha)^k k! k^{\beta_0 - 3/2} = c_1 (-1)^n (-\alpha)^k k! k^\beta
\tag{16.83}
$$

$$
I_{k,n}^{(2)} \to c_2 (-1)^n (-\alpha)^k k! k^{\beta_0 - 1/2} = c_2 (-1)^n (-\alpha)^k k! k^\beta,
\tag{16.84}
$$

in accordance with (16.60). We now compare the expansion coefficients in the re-expansion

$$
\sum_{k=0}^L f_k g^k = \sum_{n=0}^L h_n I_n = \sum_{k=0}^L g^k \sum_{n=0}^k h_n I_{k,n}.
\tag{16.85}
$$

Inverting the relation

$$
f_k = \sum_{n=0}^k h_n I_{k,n} ,
\tag{16.86}
$$

the triangular form of the matrix $I_{k,n}$ leads to a recursion relation for the expansion coefficients h_n:

$$
h_k = \frac{1}{I_{k,k}} \left(f_k - \sum_{n=0}^{k-1} h_n I_{k,n} \right).
\tag{16.87}
$$

16.5.2   Convergent Strong-Coupling Expansion

Having re-expanded the function $f(g)$ as $h_n I_n(g)$, we must find an efficient way of evaluating the functions $I_n(g)$. The power series for $I_n(g)$ is useless since it has the same vanishing radius of convergence as the original expansion of $f(g)$. However, there exists a convergent strong-coupling expansion in powers of $1/g$. Since we shall ultimately need the values of $f(g)$ with g of the order of unity, this expansion is expected to converge fast enough for practical calculations. In order to find the strong-coupling expansion, we rewrite once more the Laplace transform of the hypergeometric function $_2F_1$ in terms of MacRobert's function:

$$\int_0^\infty dt\, e^{-t}\, t^{\beta_0}\, _2F_1(a, b, c; -\alpha g t) = \frac{\Gamma(c)}{\Gamma(a)\Gamma(b)}\, E(a, b, \beta_0 + 1; c; 1/\alpha g)\,. \tag{16.88}$$

For MacRobert's function, a convergent $1/g$-expansion is well-known. Thus we write

$$I_n^{(1)} = \int_0^\infty dt \frac{e^{-t} t^{\beta_0} (\alpha g t)^n}{\Gamma(\beta_0 + 1)4^n}\, _2F_1(n-s, n-s+\tfrac{1}{2}; 2(n-s) + 1; -\alpha g t)$$

$$= \frac{\Gamma(2(n-s) + 1)(\alpha g)^n}{\Gamma(n-s)\Gamma(n-s+\tfrac{1}{2})\Gamma(\beta_0+1)4^n} E(n-s, n-s+\tfrac{1}{2}, \beta_0+n+1; 2(n-s)+1; 1/\alpha g)\,. \tag{16.89}$$

Expressing MacRobert's function in terms of generalized hypergeometric functions $_2F_2(a, b; c, d; 1/\alpha g)$, we obtain

$$I_n^{(1)} = \frac{(n-s)(\alpha g)^n}{\sqrt{\pi}\Gamma(\beta_0 + 1)4^s} \times \left[\frac{\Gamma(1/2)\Gamma(\beta_0 + s + 1)\Gamma(n-s)}{\Gamma(n-s+1)(\alpha g)^{n-s}}\, _2F_2(n-s, s-n, \tfrac{1}{2}, -s-\beta_0; 1/\alpha g) \right.$$

$$+ \frac{\Gamma(-1/2)\Gamma(\beta_0 + s + \tfrac{1}{2})\Gamma(n-s+\tfrac{1}{2})}{\Gamma(n-s+1/2)(\alpha g)^{n-s+1/2}}\, _2F_2(n-s+\tfrac{1}{2}, s-n+\tfrac{1}{2}; \tfrac{3}{2}, \tfrac{1}{2} - s - \beta_0; 1/\alpha g)$$

$$\left. + \frac{\Gamma(-s-\beta_0-1)\Gamma(-\beta_0-s-\tfrac{1}{2})\Gamma(\beta_0+n+1)}{\Gamma(n-2s-\beta_0)(\alpha g)^{\beta_0+n+1}}\, _2F_2(\beta_0+n+1, \beta_0+2s-n-1; s+\beta_0+2, s+\beta_0+\tfrac{3}{4}; 1/\alpha g) \right].$$

The expansion of $_2F_2(a, b; c, d; 1/\alpha g)$ in powers of $1/\alpha g$,

$$_2F_2(a, b; c, d; 1/\alpha g) = \sum_{k=0}^\infty \frac{(a)_k (b)_k}{(c)_k (d)_k} \frac{1}{k!} \left(\frac{1}{\alpha g} \right)^k, \tag{16.90}$$

is convergent for sufficiently large g.

16.5.3   Relation with Conformal Mapping Technique

The present algorithm is mathematically equivalent to the conformal mapping technique. Its main advantage lies in its systematic formal nature, which allows us to write a simple resummation program. The above re-expansion corresponds to choosing the analytic mapping in Eq. (16.45) as

$$w(\alpha g t) = \frac{\alpha g t}{(1 + \sqrt{1 + \alpha g t})^2} = \frac{\sqrt{1 + \alpha g t} - 1}{\sqrt{1 + \alpha g t} + 1}\,, \tag{16.91}$$

which implies that

$$\frac{1}{(1 - w)^{2s}} = \left(\frac{\sqrt{1 + \alpha g t} + 1}{2} \right)^{2s}, \tag{16.92}$$

and

$$\alpha g t = \frac{4w}{(1-w)^2}.$$ (16.93)

We now take the re-expansion equation $\sum_n h_n I_n(g) = \sum_k f_k g^k$ and rewrite both sides as Borel integrals using (16.59) and (16.40). Extracting $B_{I_n}^{\beta_0}$ from (16.75), the integrands satisfy the equation

$$\sum_{n=0}^{\infty} \frac{h_n}{\Gamma(\beta_0+1)} \frac{(\frac{1}{2}+\frac{1}{2}\sqrt{1+\alpha g t})^{2s} (\alpha g t)^n}{(1+\sqrt{1+\alpha g t})^{2n}} = \sum_{k=0}^{\infty} \frac{f_k(gt)^k}{\Gamma(k+\beta_0+1)}.$$ (16.94)

Expressing g in terms of w via (16.93), this becomes

$$\sum_{n=0}^{\infty} h_n \frac{w^n}{(1-w)^{2s}} = \sum_{k=0}^{\infty} \frac{f_k}{(\beta_0+1)_k} \left(\frac{4}{\alpha}\right)^k \left[\frac{w}{(1-w)^2}\right]^k,$$ (16.95)

or

$$\sum_{n=0}^{\infty} h_n w^n = \sum_{k=0}^{\infty} \frac{f_k}{(\beta_0+1)_k} \left(\frac{4}{\alpha}\right)^k \frac{w^k}{(1-w)^{2k-2s}}.$$ (16.96)

If we expand $1/(1-w)^{2k-2s}$ in powers of w as

$$\frac{w^k}{(1-w)^{2k-2s}} = \sum_{l=0}^{\infty} \frac{\Gamma(l+2k-2s)}{\Gamma(l+1)\Gamma(2k-2s)} w^{l+k},$$ (16.97)

and replace $l+k \to n$, Eq. (16.96) becomes

$$\sum_{n=0}^{\infty} h_n w^n = \sum_{n=0}^{\infty} \left[\sum_{k=0}^{n} \frac{f_k}{(\beta_0+1)_k} \left(\frac{4}{\alpha}\right)^k \frac{\Gamma(n+k-2s)}{\Gamma(n-k+1)\Gamma(2k-2s)}\right] w^n.$$ (16.98)

This is precisely the expansion (16.56) obtained via the conformal mapping technique up to an additional factor $\Gamma(\beta_0+1)$. Comparison of equal powers of w gives

$$h_n = \sum_{k=0}^{n} \frac{f_k}{(\beta_0+1)_k} \left(\frac{4}{\alpha}\right)^k \frac{(n+k-2s-1)!}{(n-k)!(2k-2s-1)!},$$ (16.99)

which coincides with the present re-expansion coefficients in (16.87).

16.6 Modified Reexpansions

There exists a simple modification of the algorithm based on the observation in Eqs. (16.87) or (16.99), that changing only one coefficient f_k in the expansion of $f(g)$ changes the coefficients h_n for all $n \geq k$. Suppose, for example, that we subtract the constant term from the expansion of $f(g)$. Then, we obtain a new expansion $\hat{f}(g) \equiv f(g) - f_0 = \sum_k \hat{f}_k g^k$ whose coefficients are related to f_k as follows:

$$\hat{f}_0 = 0, \quad \hat{f}_k = f_k, \quad \text{for } k \neq 0.$$ (16.100)

Alternatively, we may write $f(g) = f_0 + g\tilde{f}(g)$ with expansion coefficients

$$\tilde{f}_k = f_{k+1}.$$ (16.101)

The function $f(g)$ is recovered from

$$f(g) = \sum_{k=0}^{\infty} f_k g^k = f_0 + \sum_{k=0}^{\infty} \hat{f}_k g^k = f_0 + g \sum_{k=0}^{\infty} \tilde{f}_k g^k. \tag{16.102}$$

The parameters $\hat{s}$ and $\hat{\beta}$ in the re-expansion functions of $\hat{f}(g)$ have obviously the same values as those of $f(g)$. In contrast, the parameters $\tilde{s}$ and $\tilde{\beta}$ in the re-expansion functions of $\tilde{f}(g)$ are related to those of $f(g)$ by $\tilde{s} = s - 1$ and $\tilde{\beta} = \beta + 1$. Let us denote the associated expansion functions by $\tilde{I}_n(g)$. They possess the modified large-order behavior $\tilde{f}_k \sim -\alpha(-\alpha)^k \Gamma(k + 1 + \beta + 1)$. If $f(g)$ behaves for large g as g^s, then $\tilde{f}$ behaves as g^{s-1}, and so do the re-expansion functions $\tilde{I}_n(g)$. With $\tilde{s} - 1$ and $\tilde{b}_0 = \tilde{\beta} + 3/2 = \beta + 5/2$, Eq. (16.99) for $\tilde{f}_n$ takes the same form as for $\hat{f}_{n+1}$, the relation between the two expansion coefficients being

$$\tilde{f}_n = (\beta_0 + 1)\frac{\alpha}{4} \sum_{l=1}^{n+1} \frac{f_l}{(\beta_0 + 1)_l} \left(\frac{4}{\alpha}\right)^l \binom{n+1+l-1-2s}{n+1-l} = (\beta_0 + 1)\frac{\alpha}{4}\hat{f}_{n+1}. \tag{16.103}$$

An analogous relation holds for the expansion functions:

$$\tilde{I}_n(g) = \left(\frac{4}{\alpha g}\right) \frac{1}{\beta_0 + 1} I_{n+1}(g), \tag{16.104}$$

so that the two re-expansions satisfy

$$\sum_{n=0}^{L} \hat{f}_n I_n(g) = g \sum_{n=0}^{L-1} \tilde{f}_n \tilde{I}_n(g). \tag{16.105}$$

16.6.1 Choosing the Strong-Coupling Growth Parameter s

In the ϕ^4-theories at hand, the leading strong-coupling power s is not known. It will be chosen to ensure the best convergence of the resummation procedure. We may start with $s = 0$, say, and plot the resummed partial sums against the order in g. This usually yields a smooth curve approaching the final value. When raising s, the convergence is improved. After a certain value of s, the points begin to jump around the smooth curve. The optimal value of s is selected by the condition that the difference between the last two points is minimal. The smallest difference serves as an estimate for the systematic error, and the last value is taken as the result.

Another possibility of selecting an optimal value of s is to raise s until the last and the third-last points have the same value. In this case the average of the last two points is taken as the final result, the error being estimated by half the difference between the last two points. Both procedures give about the same result, as we shall see in Section 17.4.

Notes and References

For a survey of resummation techniques see
D.I. Kazakov and D.V. Shirkov, Fortschr. Phys. **28**, 465 (1980);
J. Zinn-Justin, Phys. Rep. **70**, 3 (1981); Phys. Rep. **70**, 109 (1981),
W. Janke, H. Kleinert, *Resummation of Divergent Perturbation Series*, World Scientific, Singapore (to be published), the last reference being the source of the resummation method in Section 16.5. For applications of this method see
H. Kleinert and S. Thoms, Phys. Rev. D **15**, 5926 (1995).

The individual citations in the text refer to:

[1] G.A. Baker, *Essentials of Padé Approximants*, Academic, New York, 1975.

[2] See for instance Sections 17.10 and 17.12 in
H. Kleinert, *Path Integrals in Quantum Mechanics, Statistics and Polymer Physics,* World Scientific, Singapore, 1995, second extended edition (www.physik.fu-berlin.de/~kleinert/b5).

[3] A.I. Vainshtein, *Decaying Systems and the Divergence of Perturbation Series*, Novosibirsk Report (1964), in Russian (unpublished).
L.N. Lipatov, JETP Lett. **25**, 104 (1977); JETP **45**, 216 (1977);
E.B. Bogomolny,V.A. Fateev, L.N. Lipatov, Sov. Sci. Rev. A **2**, 247 (1980).

[4] E. Brézin, J.C. Le Guillou, J. Zinn-Justin, Phys. Rev. D **15**, 1544, 1558 (1977).

[5] R.B. Dingle, *Asymptotic Expansions: Their Derivation and Interpretation*, Academic Press, New York, 1973.

[6] T. Carleman, *Les Fonctions Quasianalytiques*, Gauthiers Villars, Paris, 1926.

[7] F. Nevanlinna, *Zur Theorie der asymptotischen Potenzreihen*, Akademische Abhandlungen, Helsingfors, 1918.

[8] P.R. Graves-Morris, R. Hughes-Jones, G.J. Makinson, J. Inst. Maths. Applics. **13**, 311 (1974).

[9] E. Borel, *Leccedon sur les series divergentes*, Gauthier Villars, Paris, 1928.

[10] W. Janke, H. Kleinert, *Resummation of Divergent Perturbation Series*, World Scientific, Singapore (to be published).

[11] G.N. Watson, Phil. Trans. A **211**, 279 (1911).

[12] J.P. Eckmann, J. Magnen, R. Seneor, Commun. Math. Phys. **39**, 251 (1975);
J.S. Feldman, K. Osterwalder, Ann. Phys. (N.Y.) **97**, 80 (1976);
J. Magnen, R. Seneor, Commun. Math. Phys. **56**, 237 (1977).

[13] E. Leroy, Ann. Fac. Sci. Toulouse **2**, 317 (1900).

[14] G.A. Baker, B.G. Nickel, M.S. Green, D.I. Meiron, Phys. Rev. Lett. **36**, 1351 (1976).

[15] D.I. Kazakov, O.V. Tarasov, D.V. Shirkov, Teo. Mat. Fiz. **38**, 15 (1979).

[16] J.C. Le Guillou, J. Zinn-Justin, Phys. Rev. B **21**, 3976 (1980); J. Phys. Lett. (Paris) **46**, L137 (1985); J. Phys. (Paris) **48**, 19 (1987); ibid. **50**, 1365 (1987); J. Phys. Lett. **46**, 137 (1985).
See also
J.J. Loeffel, in *Large-Order Behaviour of Perturbation Theory*, ed. by J.C. Le Guillou and J. Zinn-Justin, North-Holland, Amsterdam, 1990.

[17] A. Erdely, F. Oberhettinger, and F.G. Tricomi, *Higher Transcendental Functions*, Vol. 1, McGraw-Hill, New-York, 1953.

17

Critical Exponents of O(N)-Symmetric Theory

We shall now use the results of the previous chapters to derive explicitly the critical exponents of the O(N)-symmetric ϕ^4-theory. The unique tensor multiplying the product of fields $\phi_\alpha \phi_\beta \phi_\gamma \phi_\delta$ in the interaction is [recall (6.25)]

$$\lambda^{O(N)}_{\alpha\beta\gamma\delta} \equiv \frac{\lambda}{3} \left[\delta_{\alpha\beta}\delta_{\gamma\delta} + \delta_{\alpha\gamma}\delta_{\beta\delta} + \delta_{\alpha\delta}\delta_{\beta\gamma} \right]. \tag{17.1}$$

As discussed in Chapter 1, different choices of N allow us to calculate the critical behavior of various physical systems. For $N = 1$, the field theory lies in the same universality class as a magnetic system with two preferred orientations, which can also be described by the Ising-model. For $N = 2$, the field theory matches the critical behavior of a magnet with preferred orientations in a plane, which can also be described by an XY-model. The same critical behavior is found in the superfluid transition of liquid helium near the λ-point. For $N = 3$, we obtain the critical behavior of a rotationally invariant ferromagnet, also described by the Heisenberg-model. Finally, the case $N = 0$ explains the critical behavior of polymers [1].

This tensor (17.1) appears in all vertex functions obtained from Feynman diagrams. This property is essential for the renormalizability of the theory (recall the discussion in Sections 6.4.4. and 10.13).

17.1 Series Expansions for Renormalization Group Functions

The renormalization constants are calculated via Eqs. (15.1)–(15.3). All required terms, the pole terms of the integrals, the weight factors, and the symmetry factors for O(N) symmetry are displayed in the tables of Appendix B. Only the simple poles of the renormalization constants contribute to the β-function and the anomalous dimensions γ_m and γ as we saw in Eqs. (10.46) of Chapter 10:

$$\gamma(g) = -\frac{g}{2} \frac{\partial Z_{\phi,1}}{\partial g}, \tag{17.2}$$

$$\gamma_m(g) = \frac{g}{2} \frac{\partial Z_{m^2,1}}{\partial g} + \gamma, \tag{17.3}$$

$$\beta(g) = -\varepsilon g + g^2 \frac{\partial Z_{g,1}}{\partial g} + 4g\gamma \equiv -\varepsilon g - 2g^2\gamma_g + 4g\gamma. \tag{17.4}$$

The expansions in g up to 5th order are [2]

$$\beta(g) = -\varepsilon g + \frac{g^2}{3}(N+8) - \frac{g^3}{3}(3N+14)$$
$$+ \frac{g^4}{216}\left[33N^2 + 922N + 2960 + \zeta(3) \cdot 96(5N + 22)\right]$$
$$- \frac{g^5}{3888}\left[-5N^3 + 6320N^2 + 80456N + 196648\right.$$

304

$$+ \zeta(3) \cdot 96(63N^2 + 764N + 2332)$$
$$- \zeta(4) \cdot 288(5N + 22)(N + 8)$$
$$+ \zeta(5) \cdot 1920(2N^2 + 55N + 186)]$$
$$+ \tfrac{g^6}{62208} \; [13N^4 + 12578N^3 + 808496N^2 + 6646336N + 13177344$$
$$+ \zeta(3) \cdot 16(-9N^4 + 1248N^3 + 67640N^2 + 552280N + 1314336)$$
$$+ \zeta^2(3) \cdot 768(-6N^3 - 59N^2 + 446N + 3264)$$
$$- \zeta(4) \cdot 288(63N^3 + 1388N^2 + 9532N + 21120)$$
$$+ \zeta(5) \cdot 256(305N^3 + 7466N^2 + 66986N + 165084)$$
$$- \zeta(6)(N + 8) \cdot 9600(2N^2 + 55N + 186)$$
$$+ \zeta(7) \cdot 112896(14N^2 + 189N + 526)] \; , \tag{17.5}$$

$$\gamma(g) = \tfrac{g^2}{36}(N + 2) - \tfrac{g^3}{432}(N + 2)[N + 8]$$
$$+ \tfrac{g^4}{5184}(N + 2) \;\; [5(-N^2 + 18N + 100)]$$
$$- \tfrac{g^5}{186624}(N + 2) \; [39N^3 + 296N^2 + 22752N + 77056$$
$$- \zeta(3) \cdot 48(N^3 - 6N^2 + 64N + 184)$$
$$+ \zeta(4) \cdot 1152(5N + 22)] \; , \tag{17.6}$$

$$\gamma_m(g) = \tfrac{g}{6}(N + 2) - \tfrac{g^2}{36}(N + 2)[5] + \tfrac{g^3}{72}(N + 2)[5N + 37]$$
$$- \tfrac{g^4}{15552}(N + 2) \; [-N^2 + 7578N + 31060$$
$$+ \zeta(3) \cdot 48(3N^2 + 10N + 68)$$
$$+ \zeta(4) \cdot 288(5N + 22)]$$
$$+ \tfrac{g^5}{373248}(N + 2) [21N^3 + 45254N^2 + 1077120N + 3166528$$
$$+ \zeta(3) \cdot 48(17N^3 + 940N^2 + 8208N + 31848)$$
$$- \zeta^2(3) \cdot 768(2N^2 + 145N + 582)$$
$$+ \zeta(4) \cdot 288(-3N^3 + 29N^2 + 816N + 2668)$$
$$+ \zeta(5) \cdot 768(-5N^2 + 14N + 72)$$
$$+ \zeta(6) \cdot 9600(2N^2 + 55N + 186)] \; . \tag{17.7}$$

To have an idea of the growth behavior of the expansion coefficients, we write down the series numerically for $N = 1$:

$$\beta(g) = -\varepsilon g + 3.0 \, g^2 - 5.67 \, g^3 + 32.55 \, g^4 - 271.6 \, g^5 + 2848.57 \, g^6 \; , \tag{17.8}$$
$$\gamma(g) = 0.0833 \, g^2 - 0.0625 \, g^3 + 0.3385 \, g^4 - 1.9256 \, g^5 \; , \tag{17.9}$$
$$\gamma_m(g) = 0.5 \, g - 0.42 \, g^2 + 1.75 \, g^3 - 9.98 \, g^4 + 75.38 \, g^5 \; . \tag{17.10}$$

17.2 Fixed Point and Critical Exponents

Beside the Gaussian fixed point at $g^* = 0$ (recall Section 10.5), there is a nontrivial fixed point called Heisenberg or isotropic fixed point. Its ε-expansion reads

$$g^*(\varepsilon) = \frac{3\varepsilon}{8+N} + \frac{9\varepsilon^2}{(8+N)^3}[14+3N]$$

$$+\frac{\varepsilon^3}{(8+N)^5}[\tfrac{3}{8}(4544+1760N+110N^2-33N^3)-\zeta(3)\cdot36(8+N)(22+5N)]$$

$$+\frac{\varepsilon^4}{(8+N)^7}[\tfrac{1}{16}(529792+309312N+52784N^2-5584N^3-2670N^4-5N^5)$$

$$+\zeta(3)(8+N)\cdot6(-9064-3796N-82N^2+63N^3)$$

$$-\zeta(4)(8+N)^3\cdot18(22+5N)$$

$$+\zeta(5)(8+N)^2\cdot120(186+55N+2N^2)]$$

$$+\frac{\varepsilon^5}{(8+N)^9}[\tfrac{3}{256}(-21159936-8425472N+3595520N^2+758144N^3$$
$$-625104N^4-179408N^5-1262N^6-13N^7)$$

$$+\zeta(3)(8+N)\cdot\tfrac{3}{16}(-15131136-8873728N-890208N^2$$
$$+310248N^3+45592N^4-1104N^5+9N^6)$$

$$+\zeta(3)^2(8+N)^2\cdot9(43584+24848N+3626N^2+107N^3+6N^4)$$

$$+\zeta(4)(8+N)^3\cdot\tfrac{27}{8}(-8448-3524N-52N^2+63N^3)$$

$$+\zeta(5)(8+N)^2\cdot3(554208+255188N+12246N^2-5586N^3-305N^4)$$

$$+\zeta(6)(8+N)^4\cdot\tfrac{225}{2}(186+55N+2N^2)$$

$$-\zeta(7)(8+N)^3\cdot1323(526+189N+14N^2)]. \tag{17.11}$$

The critical exponents η, ν, and ω are obtained by evaluating the renormalization group functions at the fixed point:

$$\eta = 2\gamma^*, \quad \nu^{-1} = 2(1-\gamma_m^*), \quad \omega = \beta'(g^*). \tag{17.12}$$

Inserting the ε-expansion of the fixed point (17.11) into (17.6), (17.7), and the derivative $d\beta/dg(g)$ of Eq. (17.5), we obtain the ε-expansions for the critical exponents up to ε^5 [2]

$$\eta(\varepsilon) = \frac{(N+2)\varepsilon^2}{2(N+8)^2}\left\{1+\frac{\varepsilon}{4(N+8)^2}[-N^2+56N+272]\right.$$

$$-\frac{\varepsilon^2}{16(N+8)^4}[5N^4+230N^3-1124N^2-17920N-46144\,\zeta(3)(N+8)384(5N+22)]$$

$$-\frac{\varepsilon^3}{64(N+8)^6}[13N^6+946N^5+27620N^4+121472N^3-262528N^2-2912768N-5655552$$

$$-\zeta(3)(N+8)16(N^5+10N^4+1220N^3-1136N^2-68672N-171264)$$

$$\left.+\zeta(4)(N+8)^31152(5N+22)-\zeta(5)(N+8)^2\cdot5120(2N^2+55N+186)]\right\}, \tag{17.13}$$

$$\nu^{-1}(\varepsilon) = 2+\frac{(N+2)\varepsilon}{N+8}\left\{-1-\frac{\varepsilon}{2(N+8)^2}[13N+44]\right.$$

$$+\frac{\varepsilon^2}{8(N+8)^4}[3N^3-452N^2-2672N-5312$$

$$+\zeta(3)(N+8)\cdot96(5N+22)]$$

$$+\frac{\varepsilon^3}{32(N+8)^6}[3N^5+398N^4-12900N^3-81552N^2-219968N-357120$$

$$+\zeta(3)(N+8)\cdot16(3N^4-194N^3+148N^2+9472N+19488)$$

$$+\zeta(4)(N+8)^3\cdot288(5N+22)\zeta(5)(N+8)^2\cdot1280(2N^2+55N+186)]$$

$$+\frac{\varepsilon^4}{128(N+8)^8}[3N^7-1198N^6-27484N^5-1055344N^4$$

$$-5242112N^3-5256704N^2+6999040N-626688$$

$$-\zeta(3)(N+8)\cdot16(13N^6-310N^5+19004N^4+102400N^3$$
$$-381536N^2-2792576N-4240640)$$

$$-\zeta^2(3)(N+8)^2\cdot1024(2N^4+18N^3+981N^2+6994N+11688)$$

$$+\zeta(4)(N+8)^3 \cdot 48(3N^4 - 194N^3 + 148N^2 + 9472N + 19488)$$
$$+\zeta(5)(N+8)^2 \cdot 256(155N^4 + 3026N^3 + 989N^2 - 66018N - 130608)$$
$$-\zeta(6)(N+8)^4 \cdot 6400(2N^2 + 55N + 186)$$
$$+\zeta(7)(N+8)^3 \cdot 56448(14N^2 + 189N + 526)]\Big\} \,, \qquad (17.14)$$

$$
\begin{aligned}
\omega(\varepsilon) = \ & \varepsilon - \tfrac{3\varepsilon^2}{(N+8)^2} \ [3N + 14] \\
& + \tfrac{\varepsilon^3}{4(N+8)^4} \ [33N^3 + 538N^2 + 4288N + 9568 + \zeta(3)(N+8) \cdot 96(5N + 22)] \\
& + \tfrac{\varepsilon^4}{16(N+8)^6} \ [5N^5 - 1488N^4 - 46616N^3 - 419528N^2 - 1750080N - 2599552 \\
& \qquad - \zeta(3)(N+8) \cdot 96(63N^3 + 548N^2 + 1916N + 3872) \\
& \qquad + \zeta(4)(N+8)^3 \cdot 288(5N + 22) \\
& \qquad - \zeta(5)(N+8)^2 \cdot 1920(2N^2 + 55N + 186)] \\
& + \tfrac{\varepsilon^5}{64(N+8)^8} \ [13N^7 + 7196N^6 + 240328N^5 + 3760776N^4 \\
& \qquad + 38877056N^3 + 223778048N^2 + 660389888N + 752420864 \\
& \qquad - \zeta(3)(N+8) \cdot 16(9N^6 - 1104N^5 - 11648N^4 - 243864N^3 \\
& \qquad\qquad\qquad - 2413248N^2 - 9603328N - 14734080) \\
& \qquad - \zeta^2(3)(N+8)^2 \cdot 768(6N^4 + 107N^3 + 1826N^2 + 9008N + 8736) \\
& \qquad - \zeta(4)(N+8)^3 \cdot 288(63N^3 + 548N^2 + 1916N + 3872) \\
& \qquad + \zeta(5)(N+8)^2 \cdot 256(305N^4 + 7386N^3 + 45654N^2 + 143212N + 226992) \\
& \qquad - \zeta(6)(N+8)^4 \cdot 9600(2N^2 + 55N + 186) \\
& \qquad + \zeta(7)(N+8)^3 \cdot 112896(14N^2 + 189N + 526)] \,. \qquad (17.15)
\end{aligned}
$$

For $N = 1$, these expansions reduce to

$$\eta = 0.0185\,\varepsilon^2 + 0.0187\,\varepsilon^3 - 0.0083\,\varepsilon^4 + 0.0257\,\varepsilon^5 \,, \qquad (17.16)$$
$$\nu^{-1} = 2 - 0.333\,\varepsilon - 0.117\,\varepsilon^2 + 0.124\,\varepsilon^3 - 0.307\,\varepsilon^4 + 0.951\,\varepsilon^5 \,, \qquad (17.17)$$
$$\omega = \varepsilon - 0.63\,\varepsilon^2 + 1.62\,\varepsilon^3 - 5.24\,\varepsilon^4 + 20.75\,\varepsilon^5 \,. \qquad (17.18)$$

These ε-expansions are asymptotic series and, as explained in the last chapter, resummation techniques [3] have to be applied to obtain reliable estimates of the critical exponents. For this we need the large-order behavior of the ϵ-expansions, to be summarized in the next section.

17.3 Large-Order Behavior

At the tip of the left-hand cut in the complex coupling constant plane, i.e., for $g \to 0^-$, the $O(N)$-symmetric β-function has the behavior (16.7):

$$\beta(g) = \beta^p(g) - iC2^7 3\pi^4 \left(-\frac{1}{\alpha g}\right)^{\beta+1} \exp\left(\frac{1}{\alpha g}\right) \,, \qquad (17.19)$$

where $\beta^p(g)$ denotes the perturbative part, the growth parameter β [not to be confused with the β-function of the renormalization group defined in Eq. (10.24)] is equal to $(N + 2 + D)/2$, and $1/\alpha g$ is the energy of the nontrivial solution of the field equation, the instanton [4]. In our normalization, $\alpha = 1$. Application of the dispersion integral (16.5) to the imaginary part leads to the large-order behavior of the expansion coefficients β_k defined by $\beta(g) = \sum_{k=1}^{\infty} \beta_k g^k$ [recall (16.12)]:

$$\beta_k = \gamma \left(-\alpha\right)^k k^\beta k! [1 + \mathcal{O}(1/k)]. \tag{17.20}$$

The other renormalization group functions $\gamma(g)$ and $\gamma_m(g)$ have the same general form for $g \to 0^-$, but with different prefactors and exponents β [5]:

$$\beta = \begin{cases} 3 + N/2 & \overset{N=3}{=} & 9/2 & \text{for} & \beta(g), \\[2ex] 2 + N/2 & \overset{N=3}{=} & 7/2 & \text{for} & \eta(g) = 2\gamma(g), \\[2ex] 3 + N/2 & \overset{N=3}{=} & 9/2 & \text{for} & \nu^{-1}(g) = 2 - 2\gamma_m(g), \\[2ex] 4 + N/2 & \overset{N=3}{=} & 11/2 & \text{for} & \omega(g) = \beta'(g). \end{cases} \tag{17.21}$$

From this we can easily derive the large-order behavior of the corresponding ε-series. The ε-expansion of the fixed point $g^*(\varepsilon)$ is defined by $\beta(g^*(\varepsilon)) = 0$. This implies that for negative ε, $g^*(\varepsilon)$ has an imaginary part, so that the cut of the β-function along the negative axis gives rise to a cut of $g^*(\varepsilon)$ along the negative ε-axis. At the tip of the cut, the imaginary part can be found by inserting a complex g into the full expression for β in (17.19) and separating the imaginary and real parts of the equation $\beta(g^*) = 0$. The real part of $g^*(\varepsilon)$ is given by the real part of the equation $\beta(g^*) = 0$, which is the perturbative part. The imaginary part of $[g^*(\varepsilon)]$ comes from the imaginary part of $\beta(g^*) = 0$, and reads

$$\text{Im } g^*(\varepsilon) = -C' 2^7 \pi^4 \left(-\frac{1}{\alpha\varepsilon}\right)^{(N+6+D)/2} \exp\left(\frac{1}{\alpha\varepsilon}\right). \tag{17.22}$$

Inserting this complex $g^*(\varepsilon)$ into $\gamma(g)$ and $\gamma_m(g)$ yields the imaginary parts of $\gamma(\varepsilon)$ and $\gamma_m(\varepsilon)$. Application of the dispersion integral leads again to a large order behavior of the type (17.20)

$$f(\varepsilon) = \sum_{k=0}^{\infty} f_k \varepsilon^k, \quad \text{with} \quad f_k = \gamma(-\alpha)^k k^\beta k! [1 + O(1/k)], \tag{17.23}$$

where the growth parameter α is now equal to $3/(N+8)$, while the powers β are

$$\beta = \begin{cases} 4 + N/2 & \overset{N=3}{=} & 11/2 & \text{for} & g^*(\varepsilon), \\[2ex] 3 + N/2 & \overset{N=3}{=} & 9/2 & \text{for} & \eta(\varepsilon), \\[2ex] 4 + N/2 & \overset{N=3}{=} & 11/2 & \text{for} & \nu^{-1}(\varepsilon), \\[2ex] 5 + N/2 & \overset{N=3}{=} & 13/2 & \text{for} & \omega(\varepsilon). \end{cases} \tag{17.24}$$

The prefactor γ differs for each expansion, but will not be used in the sequel.

17.4   Resummation

The resummation techniques developed in Chapter 16 can be applied either to the expansions in g or to the expansions in ε. The most accurate results are obtained from the latter, which we shall use here, concentrating only upon the Janke-Kleinert algorithm described in Section 16.5. The growth parameters of the basis functions (16.59) are adjusted to those of the perturbation expansions given in the last section. By comparing (17.20) with (16.60), we identify σ of $I_n(g)$ with α, which in the present normalization of the coupling constant is equal to unity. The growth parameter β of $I_n(g)$ in (16.60) is given by Eqs. (17.24) for the different series and universality classes with $O(N)$ symmetry.   The resulting critical exponents are listed in Table 17.1 . The error in parentheses comes about as follows: for various values of the strong-coupling growth parameter s we plot the successive resummation results from the first, second, ... , fifth order of the ε expansions. This produces the curves shown in Fig. 17.1–17.4. The value of s where the curves just begin to oscillate makes the difference between the last two approximations as small as possible. The critical exponents associated with this s are listed in Table 17.1 . To see the typical systematic error, we vary s symmetrically around this optimal value over a range indicated in the figure headings. The resulting range of critical exponents is given on the top of each figure. For comparison, we list in Tables 17.3 and 17.2 the most recent results of Guida and Zinn-Justin [6].

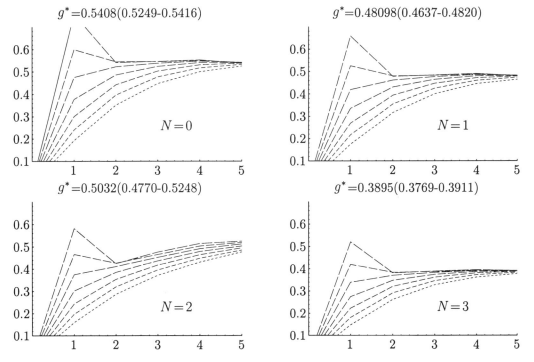

FIGURE 17.1 Resummed ε-expansion of g^* versus the number of loops for strong-coupling parameter s, which increases from 0.1 to 1.9 with increasing dash length. Further explanations are given in the text.

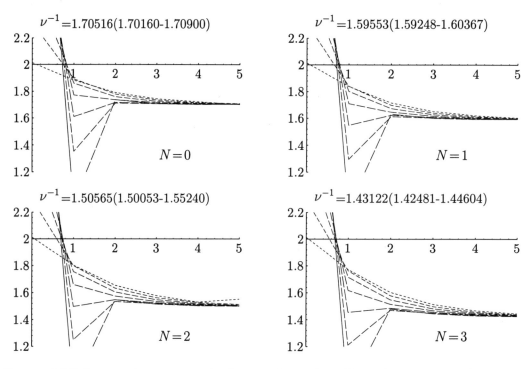

FIGURE 17.2 Resummed ε-expansion of ν^{-1} versus the number of loops. The strong-coupling parameters s changes from 0.0 to 1.2 with increasing dash length. Further explanations are given in the text.

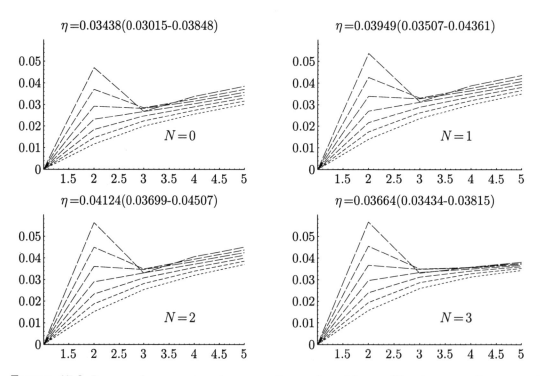

FIGURE 17.3 Resummed ε-expansion of η versus the number of loops. The strong-coupling parameter s changes from 1.6 to 3.4 with increasing dash length. Further explanations are given in the text.

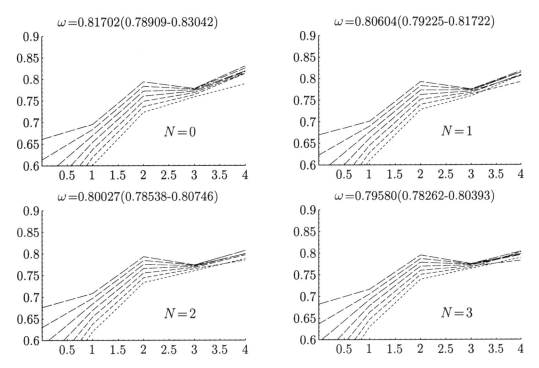

FIGURE 17.4 Resummed ε-expansion of ω versus the number of loops. The strong-coupling parameter s changes from -1.15 to -0.55 with increasing dash length. Further explanations are given in the text.

N	0	1	2	3
$g^*(\varepsilon)$	0.5408(83)	0.4810(91)	0.5032(239)	0.3895(71)
$\nu(\varepsilon)$	0.5865(13)	0.6268(22)	0.6642(111)	0.6987(51)
$\eta(\varepsilon)$	0.0344(42)	0.0395(43)	0.0412(41)	0.0366(20)
$\omega(\varepsilon)$	0.817(21)	0.806(13)	0.800(13)	0.796(11)

TABLE 17.1 Estimates of the critical exponents for $N = 0, 1, 2, 3$ and $\varepsilon = 1$. Resummation with the algorithm of Section 16.5 for the ε-expansions as shown in Figs. 17.1-17.4. The errors are estimated from the variation in dependence on s, as explained in the text.

TABLE 17.2   Critical exponents of the O(N)-models from ε-expansion [6].

N	0	1	2	3
γ (free) γ (bc)	1.1575 ± 0.0050 1.1571 ± 0.0030	1.2360 ± 0.0040 1.2380 ± 0.0045	1.3120 ± 0.0085 1.317	1.3830 ± 0.0135 1.392
ν (free) ν (bc)	0.5875 ± 0.0018 0.5878 ± 0.0011	0.6293 ± 0.0026 0.6304 ± 0.0021	0.6685 ± 0.0040 0.671	0.7050 ± 0.0055 0.708
η (free) η (bc)	0.030 ± 0.006 0.0315 ± 0.0025	0.0360 ± 0.006 0.0365 ± 0.0045	0.0385 ± 0.0065 0.0370	0.0380 ± 0.0060 0.0355
β (free) β (bc)	0.3024 ± 0.0024 0.3032 ± 0.0011	0.3260 ± 0.0020 0.3265 ± 0.0012	0.3472 ± 0.0015	0.3660 ± 0.0015
ω	0.828 ± 0.023	0.814 ± 0.018	0.802 ± 0.018	0.794 ± 0.018
θ	0.486 ± 0.015	0.512 ± 0.013	0.536 ± 0.015	0.560 ± 0.017

TABLE 17.3   Critical exponents of the O(N)-models from $d = 3$ expansion [6].

N	0	1	2	3
g^*_{Ni}	1.413 ± 0.006	1.411 ± 0.004	1.403 ± 0.003	1.391 ± 0.004
g^*	26.63 ± 0.11	23.64 ± 0.07	21.16 ± 0.05	19.07 ± 0.05
γ	1.1598 ± 0.0020	1.2397 ± 0.0013	1.3169 ± 0.0020	1.3895 ± 0.0050
ν	0.5882 ± 0.0011	0.6304 ± 0.0013	0.6703 ± 0.0013	0.7073 ± 0.0030
η	0.0284 ± 0.0025	0.0335 ± 0.0025	0.0354 ± 0.0025	0.0355 ± 0.0025
β	0.3025 ± 0.0008	0.3258 ± 0.0014	0.3470 ± 0.0014	0.3662 ± 0.0025
α	0.235 ± 0.003	0.109 ± 0.004	-0.011 ± 0.004	-0.122 ± 0.009
ω	0.812 ± 0.016	0.799 ± 0.011	0.789 ± 0.011	0.782 ± 0.0013
$\theta = \omega\nu$	0.478 ± 0.010	0.504 ± 0.008	0.529 ± 0.008	0.553 ± 0.012

Notes and References

The ε-expansions for the critical exponents used here are from
H. Kleinert, J. Neu, V. Schulte-Frohlinde, K.G. Chetyrkin, S.A. Larin, Phys. Lett. B **272**, 39-44 (1991); ibid. **319**, 545 (1993) (Erratum).},
V. Schulte-Frohlinde, FU-Berlin Ph.D. Thesis, 1969.

The resummation described here is from
H. Kleinert, V. Schulte-Frohlinde, J. Phys. A **34** 1037 (2001) (cond-mat/9907214).
Among the most accurate resummation data in 4-ε dimensions available in the literature are those in Ref. [6].

The individual citations in the text refer to:

[1] P.G. de Gennes, Phys. Lett. A **38**, 339 (1972).

[2] The expansions can be downloaded from the internet address www.physik.fu-berlin.de/~kleinert/b8/programs/index.html#1 .

[3] D.I. Kazakov and D.V. Shirkov, Fortschr. Phys. **28**, 465 (1980); J. Zinn-Justin, Phys. Rep. **70**, 3 (1981).

[4] See for instance Sections 17.10 and 17.12 in
H. Kleinert, *Path Integrals in Quantum Mechanics, Statistics and Polymer Physics,* World Scientific, Singapore 1995, Second extended edition (www.physik.fu-berlin.de/~kleinert/b5).

[5] D.I. Kazakov, O.V. Tarasov, D.V. Shirkov, Teo. Mat. Fiz. **38**, 15 (1979);
A.J. McKane and D.J. Wallace, J. Phys. A **11**, 2285 (1978);
I.T. Drummond and G.M. Shore, Ann. Phys. **121**, 204 (1979);
E. Brézin, J.C. Le Guillou, J. Zinn-Justin, Phys. Rev. D **15**, 1544, 1558 (1977).

[6] R. Guida, J. Zinn-Justin, *Critical Exponents of the N-Vector Model,* cond-mat 9803240, (1998).
They are improvements of earlier results in
J.C. Le Guillou, J. Zinn-Justin, Phys. Rev. B **21**, 3976 (1980); J. Physique Lett. **46**, 137 (1985).

18

Cubic Anisotropy

In most magnetic systems existing in nature, the $O(N)$ symmetry is broken by the crystal structure of the materials. The magnetization points in certain preferred directions, for example along the edges or the diagonals of a cubic lattice. In order to describe such situations in general, we generalize the $O(N)$-symmetric model, and extend it by an interaction which prefers these directions on a hypercube in the N-dimensional field space. The general form of the ϕ^4-interaction was given in Eq. (6.4). An interaction term $\lambda_2 \sum_{\alpha=1}^{N} \phi_\alpha^4$ accounting for the *cubic anisotropy* is added to the $O(N)$-symmetric interaction $\lambda_1 (\sum_{\alpha=1}^{N} \phi_\alpha^2)^2$. The extended theory interpolates between an $O(N)$-symmetric and a system with increasing *cubic anisotropy*. The tensor associated with the two ϕ^4-interactions was introduced in Subsection 6.4.2, where the relevant changes in the perturbation expansions were discussed. We are now prepared to calculate the critical properties of the field fluctuations.

18.1 Basic Properties

The tensor in the ϕ^4-interaction of mixed $O(N)$-symmetric and cubic symmetry was written down in Eq. (6.47). It reads

$$T^{\text{cub}}_{\alpha\beta\gamma\delta} = g_1\, T^{(1)}_{\alpha\beta\gamma\delta} + g_2\, T^{(2)}_{\alpha\beta\gamma\delta}, \tag{18.1}$$

where $T^{(1)}_{\alpha\beta\gamma\delta}$ and $T^{(2)}_{\alpha\beta\gamma\delta}$ have the symmetrized form:

$$T^{(1)}_{\alpha\beta\gamma\delta} = \frac{1}{3}(\delta_{\alpha\beta}\delta_{\gamma\delta} + \delta_{\alpha\gamma}\delta_{\beta\delta} + \delta_{\alpha\delta}\delta_{\beta\gamma}) \sim T^{O(N)}_{\alpha\beta\gamma\delta}, \tag{18.2}$$

$$T^{(2)}_{\alpha\beta\gamma\delta} = \delta_{\alpha\beta\gamma\delta}. \tag{18.3}$$

For the definition of $\delta_{\alpha\beta\gamma\delta}$ see Eq. (6.48). The tensor T^{cub} fulfills the conditions (10.208) and (10.209) stated in Section 10.13, which ensure that the theory has only one length scale. For $N = 2, 3$, the combinations of $T^{(1)}_{\alpha\beta\gamma\delta}$ and $T^{(2)}_{\alpha\beta\gamma\delta}$ exhaust all tensors of rank 4 for which the theory has that property [1], i.e., for which Eq. (10.208) holds. For $N \geq 4$, more tensors are admissible without introducing new length scales [2].

The energy calculated with (18.1) is no longer invariant under rotations of the field at constant magnitude. It has minima for an axial order of the ground state of the field with an expectation value $\boldsymbol{\Phi} = \langle \boldsymbol{\phi}(\mathbf{x}) \rangle = (1, 0, \dots, 0)$, or with a diagonal expectation value $\boldsymbol{\Phi} = \langle \boldsymbol{\phi}(\mathbf{x}) \rangle = (1/\sqrt{N}, 1/\sqrt{N}, \dots, 1/\sqrt{N})$ in the N-dimensional hypercube (compare Section 1.1). For the axial order, the magnitude of the interaction is $(g_1 + g_2)|\boldsymbol{\Phi}|^4$; for the diagonal ordering it assumes $(g_1 + g_2/N)|\boldsymbol{\Phi}|^4$, implying that for positive g_2 and $N > 1$ diagonals are favored, whereas for negative g_2 and $N > 1$ the edges are favored. The energy is thus positive definite within the *mean-field stability wedge* bounded by the straight lines $g_1 + g_2 > 0$ and $N g_1 + g_2 > 0$, as shown in Fig. 18.1. This will be changed drastically by the fluctuations, as indicated in Fig. 18.2.

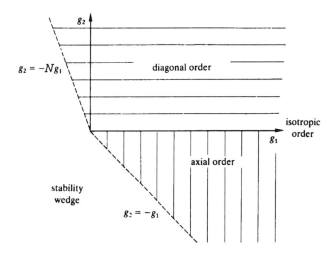

FIGURE 18.1 Stability wedge in mean-field approximation. There exist three stable ordered states below the critical temperature, depending on the two coupling constants g_1 and g_2. The left-hand side of the dashed lines is unstable.

Inside the mean-field domain of stability we have two special cases. For $g_1 = 0$, $g_2 \neq 0$, the field components decouple and we have a simple Ising-type system with an N-dependent overall factor. For $g_1 \neq 0$, $g_2 = 0$ we recover the O(N)-symmetric system.

The system turns out to have four fixed points. The Gaussian fixed point at the origin; the *Ising* fixed point with $g_1^I = 0$, $g_2^I \neq 0$; the *O(N)-symmetric* or *Heisenberg* fixed point with $g_1^H \neq 0$, $g_2^H = 0$; and the *cubic* fixed point with $g_1^C \neq 0$, $g_2^C \neq 0$. The cubic fixed point is the only new one; the others are known from the results in Chapter 17. The first-order results are

$$g_1^H(\varepsilon) \;=\; \frac{3\,\varepsilon}{8+N}, \qquad g_2^H(\varepsilon) \;=\; 0; \tag{18.4}$$

$$g_1^I(\varepsilon) \;=\; 0, \qquad g_2^I(\varepsilon) \;=\; \frac{\varepsilon}{3}; \tag{18.5}$$

$$g_1^C(\varepsilon) \;=\; \frac{\varepsilon}{N}, \qquad g_2^C(\varepsilon) \;=\; \frac{(N-4)\,\varepsilon}{3\,N}. \tag{18.6}$$

The location of the fixed points depends on N. For $N = 1$, there is only one coupling constant $g_1 + g_2$, and the O(N)-symmetric and the Ising fixed points are equivalent, as are the Gaussian and the cubic fixed points. For $N = 2$, a special symmetry comes into play. A rotation in the space of the fields through $\pi/4$:

$$\phi_1' = \frac{\phi_1 + \phi_2}{\sqrt{2}}, \qquad \phi_2' = \frac{\phi_1 - \phi_2}{\sqrt{2}}, \tag{18.7}$$

leaves the fourth-order interaction term invariant if the coupling constants are transformed with

$$g_1' = g_1 + \frac{3}{2}\,g_2, \qquad g_2' = -g_2. \tag{18.8}$$

These equations transform the Ising fixed point into the cubic fixed point for $N = 2$, as can be seen from Eqs. (18.5) and (18.6).

For increasing N, the cubic fixed point moves from the lower half-plane $g_2 < 0$ into the upper one with $g_2 > 0$. For a certain $N = N_c$, the O(N)-symmetric and the cubic fixed point coalesce and interchange stability. For $N < N_c$, the O(N)-symmetric fixed point is stable; for $N > N_c$, the cubic fixed point is stable (see Fig. 18.2). For $N < N_c$, there is symmetry restoration.

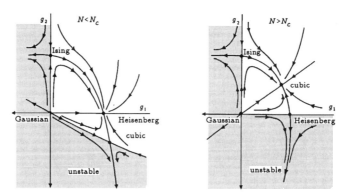

FIGURE 18.2 Stability of the fixed points in ϕ^4-theory with mixed O(N)-symmetric and cubic coupling for $N < N_c$ and $N > N_c$. The results of our analysis are compatible with $N_c = 3$.

Although the O(N) symmetry is broken at the mean field level by the second interaction, the symmetry is restored by the large fluctuations near the critical point. The physically most interesting value of N is 3, where the O(N)-symmetric fixed point characterizes the critical behavior of the classical Heisenberg model of magnetism.

Early work based on three-loop calculations in $D = 4 - \varepsilon$ estimated N_c to lie somewhere between 3 and 4 [3, 4, 5]. This implied that all magnetic systems with cubic symmetry occurring in nature show an O(3)-symmetric critical behavior. In contrast, the five-loop results suggest that the critical value N_c lies *below* $N = 3$, so that the cubic fixed point should govern the critical behavior of magnetic transitions in cubic crystals. A similar result was also found in a four-loop calculation in $D = 3$ dimensions [6]. The five-loop calculations to be presented below will favor this result, although not conclusively. The difficulty of deciding which is the correct fixed point has its parallel in the experimental difficulty of distinguishing the critical exponents of the two fixed points. On account of the vicinity of the Heisenberg fixed point, the difference between the critical exponents of the two fixed points is too small to be measured with present techniques. The cubic universality class for $N = 3$, if it exists, is practically indistinguishable from the O(3)-symmetric Heisenberg class.

18.2   Series Expansions for RG Functions

The general equations for the renormalization group functions of ϕ^4-interactions with two coupling constants were given in Eqs. (10.211)–(10.213). In dimensional regularization, they are given by the simple poles of the renormalization constants [recall Eq. (10.219)]:

$$\beta_i(g_1, g_2) = -\varepsilon g_i - 2 g_i \left[-\frac{g_1}{2} \frac{\partial Z_{g_i,1}}{\partial g_1} - \frac{g_2}{2} \frac{\partial Z_{g_i,1}}{\partial g_2} - 2\gamma(g_1, g_2) \right]$$
$$\equiv -\varepsilon g_i - 2g_i \gamma_{g_i} + 4g_i \gamma(g_1, g_2) , \qquad i = 1, 2, \tag{18.9}$$

$$\gamma(g_1, g_2) = -\frac{g_1}{2} \frac{\partial Z_{\phi^2,1}}{\partial g_1} - \frac{g_2}{2} \frac{\partial Z_{\phi^2,1}}{\partial g_2} , \tag{18.10}$$

$$\gamma_m(g_1, g_2) = \frac{g_1}{2}\frac{\partial Z_{m^2,1}}{\partial g_1} + \frac{g_2}{2}\frac{\partial Z_{m^2,1}}{\partial g_2} + \gamma\,.$$

(18.11)

There are now two renormalization constants for the coupling constants: Z_{g_1} and Z_{g_2}. The four renormalization constants are calculated using the general diagrammatic sums (15.1)–(15.3). All expressions occuring in these sums, the pole terms of the integrals, the weight factors, and the symmetry factors for the cubic symmetry have been listed in the tables of Appendix B.

The resulting expansions for the renormalization group functions up to five loops are [7]:

$$\beta_1(g_1, g_2) = -\varepsilon\,g_1 + g_1{}^2\,\frac{N+8}{3} + g_1\,g_2\,2 + g_1{}^3\left[-N-\frac{14}{3}\right] - g_1{}^2\,g_2\,\frac{22}{3} - g_1\,g_2{}^2\,\frac{5}{3}$$
$$+g_1{}^4\left[N^2\,\frac{11}{72} + N\left(\frac{461}{108}+\frac{20\,\zeta(3)}{9}\right)+\frac{370}{27}+\frac{88\,\zeta(3)}{9}\right]+g_1{}^3\,g_2\left[\frac{79\,N}{36}+\frac{659}{18}+\frac{64\,\zeta(3)}{3}\right]$$
$$+g_1{}^2\,g_2{}^2\left[\frac{N}{24}+\frac{107}{4}+8\,\zeta(3)\right]+g_1\,g_2{}^3\,7$$
$$+g_1{}^5\left[N^3\,\frac{5}{3888}+N^2\left(-\frac{395}{243}-\frac{14\,\zeta(3)}{9}+\frac{10\,\zeta(4)}{27}-\frac{80\,\zeta(5)}{81}\right)+N\left(-\frac{10057}{486}-\frac{1528\,\zeta(3)}{81}+\frac{124\,\zeta(4)}{27}-\frac{2200\,\zeta(5)}{81}\right)\right.$$
$$\left.-\frac{24581}{486}-\frac{4664\,\zeta(3)}{81}+\frac{352\,\zeta(4)}{27}-\frac{2480\,\zeta(5)}{27}\right]$$
$$+g_1{}^4\,g_2\left[N^2\left(\frac{7}{81}-\frac{\zeta(3)}{9}\right)+N\left(-\frac{1319}{54}-\frac{184\,\zeta(3)}{9}+\frac{38\,\zeta(4)}{9}-\frac{400\,\zeta(5)}{27}\right)-\frac{15967}{81}-\frac{4856\,\zeta(3)}{27}+\frac{340\,\zeta(4)}{9}-\frac{2560\,\zeta(5)}{9}\right]$$
$$+g_1{}^3\,g_2{}^2\left[N\left(-\frac{301}{72}-\frac{35\,\zeta(3)}{9}\right)-\frac{13433}{54}-\frac{1456\,\zeta(3)}{9}+\frac{64\,\zeta(4)}{3}-\frac{2000\,\zeta(5)}{9}\right]$$
$$+g_1{}^2\,g_2{}^3\left[N\left(-\frac{25}{36}+\frac{\zeta(3)}{3}\right)-\frac{4867}{36}-50\,\zeta(3)-8\,\zeta(4)-\frac{160\,\zeta(5)}{3}\right]+g_1\,g_2{}^4\left[-\frac{477}{16}-3\,\zeta(3)-6\,\zeta(4)\right]$$
$$+g_1{}^6\left[N^4\left(\frac{13}{62208}-\frac{\zeta(3)}{432}\right)+N^3\left(\frac{6289}{31104}+\frac{26\,\zeta(3)}{81}-\frac{2\,\zeta^2(3)}{27}-\frac{7\,\zeta(4)}{24}+\frac{305\,\zeta(5)}{243}-\frac{25\,\zeta(6)}{81}\right)\right.$$
$$+N^2\left(\frac{50531}{3888}+\frac{8455\,\zeta(3)}{486}-\frac{59\,\zeta^2(3)}{81}-\frac{347\,\zeta(4)}{54}+\frac{7466\,\zeta(5)}{243}-\frac{1775\,\zeta(6)}{162}+\frac{686\,\zeta(7)}{27}\right)$$
$$+N\left(\frac{103849}{972}+\frac{69035\,\zeta(3)}{486}+\frac{446\,\zeta^2(3)}{81}-\frac{2383\,\zeta(4)}{54}+\frac{66986\,\zeta(5)}{243}-\frac{7825\,\zeta(6)}{81}+343\,\zeta(7)\right)$$
$$\left.+\frac{17158}{81}+\frac{27382\,\zeta(3)}{81}+\frac{1088\,\zeta^2(3)}{27}-\frac{880\,\zeta(4)}{9}+\frac{55028\,\zeta(5)}{81}-\frac{6200\,\zeta(6)}{27}+\frac{25774\,\zeta(7)}{27}\right]$$
$$+g_1{}^5\,g_2\left[N^3\left(\frac{161}{10368}-\frac{17\,\zeta(3)}{648}-\frac{\zeta(4)}{36}\right)+N^2\left(\frac{59675}{15552}+\frac{170\,\zeta(3)}{27}-\frac{4\,\zeta^2(3)}{3}-\frac{19\,\zeta(4)}{4}+\frac{602\,\zeta(5)}{27}-\frac{50\,\zeta(6)}{9}\right)\right.$$
$$+N\left(\frac{5723}{27}+\frac{21560\,\zeta(3)}{81}-\frac{190\,\zeta^2(3)}{27}-\frac{2339\,\zeta(4)}{27}+\frac{4046\,\zeta(5)}{9}-\frac{4075\,\zeta(6)}{27}+\frac{1274\,\zeta(7)}{3}\right)$$
$$\left.+\frac{537437}{486}+\frac{116759\,\zeta(3)}{81}+\frac{3148\,\zeta^2(3)}{27}-\frac{10177\,\zeta(4)}{27}+\frac{75236\,\zeta(5)}{27}-\frac{24050\,\zeta(6)}{27}+\frac{11564\,\zeta(7)}{3}\right]$$
$$+g_1{}^4\,g_2{}^2\left[N^2\left(-\frac{1921}{10368}+\frac{763\,\zeta(3)}{648}-\frac{17\,\zeta(4)}{36}+\frac{5\,\zeta(5)}{9}\right)\right.$$
$$+N\left(\frac{270749}{2592}+\frac{9230\,\zeta(3)}{81}-\frac{232\,\zeta^2(3)}{27}-\frac{4841\,\zeta(4)}{108}+\frac{2045\,\zeta(5)}{9}-\frac{1450\,\zeta(6)}{27}+\frac{245\,\zeta(7)}{3}\right)$$
$$\left.+\frac{1314497}{648}+\frac{171533\,\zeta(3)}{81}+\frac{1384\,\zeta^2(3)}{27}-\frac{23105\,\zeta(4)}{54}+\frac{96794\,\zeta(5)}{27}-\frac{25400\,\zeta(6)}{27}+\frac{14210\,\zeta(7)}{3}\right]$$
$$+g_1{}^3\,g_2{}^3\left[N\left(\frac{30277}{1296}+\frac{344\,\zeta(3)}{27}-\frac{25\,\zeta(4)}{6}+\frac{208\,\zeta(5)}{9}\right)\right.$$
$$\left.+\frac{2281727}{1296}+\frac{37789\,\zeta(3)}{27}-\frac{544\,\zeta^2(3)}{9}-\frac{337\,\zeta(4)}{3}+\frac{17444\,\zeta(5)}{9}-\frac{1600\,\zeta(6)}{9}+2352\,\zeta(7)\right]$$
$$+g_1{}^2\,g_2{}^4\left[N\left(\frac{26171}{6912}-\frac{77\,\zeta(3)}{48}+\frac{7\,\zeta(4)}{8}-\frac{4\,\zeta(5)}{3}\right)+\frac{1336801}{1728}+\frac{5495\,\zeta(3)}{12}-\frac{190\,\zeta^2(3)}{3}\right.$$
$$\left.+\frac{141\,\zeta(4)}{2}+\frac{1145\,\zeta(5)}{3}+\frac{575\,\zeta(6)}{3}+441\,\zeta(7)\right]$$
$$+g_1\,g_2{}^5\left[\frac{158849}{1152}+\frac{1519\,\zeta(3)}{24}-18\,\zeta^2(3)+\frac{65\,\zeta(4)}{2}+2\,\zeta(5)+75\,\zeta(6)\right]\,,$$

(18.12)

$$\beta_2(g_1, g_2) = -\varepsilon\,g_2 + 3\,g_2{}^2 + 4\,g_1\,g_2 - g_2{}^3\,\frac{17}{3} - g_1\,g_2{}^2\,\frac{46}{3} - g_1{}^2\,g_2\,\frac{5\,N+82}{9} + g_2{}^4\left[\frac{145}{8}+12\,\zeta(3)\right]$$
$$+g_1\,g_2{}^3\left[\frac{131}{2}+48\,\zeta(3)\right]+g_1{}^2\,g_2{}^2\left[\frac{17\,N}{24}+\frac{325}{4}+64\,\zeta(3)\right]+g_1{}^3\,g_2\left[-N^2\,\frac{13}{108}+N\left(\frac{92}{27}+\frac{16\,\zeta(3)}{9}\right)+\frac{821}{27}+\frac{224\,\zeta(3)}{9}\right]$$
$$+g_2{}^5\left[-\frac{3499}{48}-78\,\zeta(3)+18\,\zeta(4)-120\,\zeta(5)\right]+g_1\,g_2{}^4\left[-\frac{1004}{3}-387\,\zeta(3)+96\,\zeta(4)-600\,\zeta(5)\right]$$
$$+g_1{}^2\,g_2{}^3\left[N\left(-\frac{19}{24}-\frac{19\,\zeta(3)}{3}+4\,\zeta(4)\right)-\frac{10661}{18}-724\,\zeta(3)+184\,\zeta(4)-\frac{3440\,\zeta(5)}{3}\right]$$
$$+g_1{}^3\,g_2{}^2\left[N^2\left(\frac{1}{6}+\frac{\zeta(3)}{9}\right)+N\left(-\frac{508}{27}-\frac{218\,\zeta(3)}{9}+12\,\zeta(4)-\frac{160\,\zeta(5)}{9}\right)-\frac{12349}{27}-\frac{5312\,\zeta(3)}{9}+\frac{440\,\zeta(4)}{3}-960\,\zeta(5)\right]$$
$$+g_1{}^4\,g_2\left[N^3\left(-\frac{29}{1296}+\frac{\zeta(3)}{27}\right)+N^2\left(-\frac{7}{162}-\frac{8\,\zeta(3)}{9}+\frac{4\,\zeta(4)}{9}\right)\right.$$
$$\left.+N\left(-\frac{3479}{162}-\frac{560\,\zeta(3)}{27}+\frac{68\,\zeta(4)}{9}-\frac{280\,\zeta(5)}{9}\right)-\frac{19679}{162}-168\,\zeta(3)+40\,\zeta(4)-\frac{7280\,\zeta(5)}{27}\right]$$

$$+g_2{}^6\left[\frac{764621}{2304}+\frac{7965\,\zeta(3)}{16}+45\,\zeta^2(3)-\frac{1189\,\zeta(4)}{8}+987\,\zeta(5)-\frac{675\,\zeta(6)}{2}+1323\,\zeta(7)\right]$$

$$+g_1\,g_2{}^5\left[\frac{1067507}{576}+\frac{35083\,\zeta(3)}{12}+288\,\zeta^2(3)-\frac{3697\,\zeta(4)}{4}+5920\,\zeta(5)-2100\,\zeta(6)+7938\,\zeta(7)\right]$$

$$+g_1{}^2\,g_2{}^4\left[N\left(-\frac{16223}{3456}+\frac{2947\,\zeta(3)}{72}-17\,\zeta^2(3)-\frac{151\,\zeta(4)}{4}+\frac{290\,\zeta(5)}{3}-\frac{125\,\zeta(6)}{2}\right)\right.$$
$$\left.+\frac{3633377}{864}+\frac{125459\,\zeta(3)}{18}+\frac{2266\,\zeta^2(3)}{3}-2263\,\zeta(4)+14328\,\zeta(5)-\frac{15575\,\zeta(6)}{3}+19404\,\zeta(7)\right]$$

$$+g_1{}^3\,g_2{}^3\left[N^2\left(\frac{8213}{15552}-\frac{35\,\zeta(3)}{108}-\frac{2\,\zeta(4)}{3}+\frac{14\,\zeta(5)}{9}\right)\right.$$
$$+N\left(\frac{496159}{7776}+\frac{1309\,\zeta(3)}{6}-\frac{452\,\zeta^2(3)}{9}-\frac{4076\,\zeta(4)}{27}+478\,\zeta(5)-\frac{2450\,\zeta(6)}{9}+196\,\zeta(7)\right)$$
$$\left.+\frac{9309907}{1944}+\frac{224804\,\zeta(3)}{27}+\frac{3032\,\zeta^2(3)}{3}-\frac{73018\,\zeta(4)}{27}+\frac{155692\,\zeta(5)}{9}-6300\,\zeta(6)+23912\,\zeta(7)\right]$$

$$+g_1{}^4\,g_2{}^2\left[N^3\left(\frac{127}{20736}-\frac{91\,\zeta(3)}{1296}+\frac{\zeta(4)}{18}\right)+N^2\left(-\frac{43295}{31104}+\zeta(3)-\frac{4\,\zeta^2(3)}{3}-\frac{121\,\zeta(4)}{24}+\frac{364\,\zeta(5)}{27}-\frac{50\,\zeta(6)}{9}\right)\right.$$
$$+N\left(\frac{11495}{54}+\frac{31598\,\zeta(3)}{81}-\frac{1045\,\zeta^2(3)}{27}-\frac{10729\,\zeta(4)}{54}+\frac{20917\,\zeta(5)}{27}-\frac{21425\,\zeta(6)}{54}+\frac{1960\,\zeta(7)}{3}\right)$$
$$\left.+\frac{1279979}{486}+\frac{784621\,\zeta(3)}{162}+\frac{18154\,\zeta^2(3)}{27}-\frac{83837\,\zeta(4)}{54}+\frac{275510\,\zeta(5)}{27}-\frac{98975\,\zeta(6)}{27}+\frac{43120\,\zeta(7)}{3}\right]$$

$$+g_1{}^5\,g_2\left[N^4\left(-\frac{61}{15552}-\frac{5\,\zeta(3)}{972}+\frac{\zeta(4)}{108}\right)+N^3\left(-\frac{3557}{46656}-\frac{151\,\zeta(3)}{972}-\frac{4\,\zeta(4)}{27}+\frac{8\,\zeta(5)}{81}\right)\right.$$
$$+N^2\left(\frac{111217}{23328}+\frac{2785\,\zeta(3)}{243}-\frac{92\,\zeta^2(3)}{81}-\frac{1055\,\zeta(4)}{162}+\frac{530\,\zeta(5)}{27}-\frac{950\,\zeta(6)}{81}+\frac{98\,\zeta(7)}{9}\right)$$
$$+N\left(\frac{95588}{729}+\frac{15742\,\zeta(3)}{81}-\frac{92\,\zeta^2(3)}{81}-\frac{6592\,\zeta(4)}{81}+\frac{34460\,\zeta(5)}{81}-\frac{14750\,\zeta(6)}{81}+490\,\zeta(7)\right)$$
$$\left.+\frac{389095}{729}+\frac{259358\,\zeta(3)}{243}+\frac{13288\,\zeta^2(3)}{81}-\frac{27166\,\zeta(4)}{81}+\frac{179696\,\zeta(5)}{81}-\frac{63500\,\zeta(6)}{81}+\frac{28420\,\zeta(7)}{9}\right]\,,\qquad(18.13)$$

$$\gamma_2(g_1,g_2)=\;\;g_1{}^2\,\frac{N+2}{36}+g_1\,g_2\,\frac{1}{6}+g_2{}^2\,\frac{1}{12}-g_1{}^3\left[\frac{N^2}{432}+\frac{5\,N}{216}+\frac{1}{27}\right]-g_1{}^2\,g_2\left[\frac{N}{48}+\frac{1}{6}\right]-g_1\,g_2{}^2\,\frac{3}{16}-g_2{}^3\,\frac{1}{16}$$
$$+g_1{}^4\left[-\frac{5\,N^3}{5184}+\frac{5\,N^2}{324}+\frac{85\,N}{648}+\frac{125}{648}\right]+g_1{}^3\,g_2\left[-\frac{5\,N^2}{432}+\frac{5\,N}{24}+\frac{125}{108}\right]+g_1{}^2\,g_2{}^2\left[\frac{5\,N}{288}+\frac{145}{72}\right]+g_1\,g_2{}^3\,\frac{65}{48}+g_2{}^4\,\frac{65}{192}$$
$$+g_1{}^5\left[N^4\left(-\frac{13}{62208}+\frac{\zeta(3)}{3888}\right)+N^3\left(-\frac{187}{93312}-\frac{\zeta(3)}{972}\right)+N^2\left(-\frac{1459}{11664}+\frac{13\,\zeta(3)}{972}-\frac{5\,\zeta(4)}{162}\right)\right.$$
$$\left.+N\left(-\frac{1915}{2916}+\frac{13\,\zeta(3)}{162}-\frac{16\,\zeta(4)}{81}\right)-\frac{602}{729}+\frac{23\,\zeta(3)}{243}-\frac{22\,\zeta(4)}{81}\right]$$
$$+g_1{}^4\,g_2\left[N^3\left(-\frac{65}{20736}+\frac{5\,\zeta(3)}{1296}\right)+N^2\left(-\frac{185}{7776}-\frac{5\,\zeta(3)}{216}\right)\right.$$
$$\left.+N\left(-\frac{395}{216}+\frac{20\,\zeta(3)}{81}-\frac{25\,\zeta(4)}{54}\right)-\frac{1505}{243}+\frac{115\,\zeta(3)}{162}-\frac{55\,\zeta(4)}{27}\right]$$
$$+g_1{}^3\,g_2{}^2\left[\frac{325\,N^2}{31104}+N\left(-\frac{4453}{3888}+\frac{23\,\zeta(3)}{216}-\frac{5\,\zeta(4)}{27}\right)-\frac{58177}{3888}+\frac{191\,\zeta(3)}{108}-\frac{130\,\zeta(4)}{27}\right]$$
$$+g_1{}^2\,g_2{}^3\left[N\left(-\frac{671}{3456}+\frac{\zeta(3)}{72}\right)-\frac{13741}{864}+\frac{67\,\zeta(3)}{36}-5\,\zeta(4)\right]+g_1\,g_2{}^4\left[-\frac{18545}{2304}+\frac{15\,\zeta(3)}{16}-\frac{5\,\zeta(4)}{2}\right]$$
$$+g_2{}^5\left[-\frac{3709}{2304}+\frac{3\,\zeta(3)}{16}-\frac{\zeta(4)}{2}\right]\,,\qquad(18.14)$$

$$\gamma_m(g_1,g_2)=\;\;g_1\,\frac{2+N}{6}+g_2\,\frac{1}{2}-g_1{}^2\,\frac{5\,N+10}{36}-g_1\,g_2\,\frac{5}{6}-g_2{}^2\,\frac{5}{12}$$
$$+g_1{}^3\left[\frac{5\,N^2}{72}+\frac{47\,N}{72}+\frac{37}{36}\right]+g_1{}^2\,g_2\,\frac{37+5\,N}{8}+g_1\,g_2{}^2\,\frac{251+N}{48}+g_2{}^3\,\frac{7}{4}$$
$$+g_1{}^4\left[N^3\left(\frac{1}{15552}-\frac{\zeta(3)}{108}\right)+N^2\left(-\frac{947}{1944}-\frac{4\,\zeta(3)}{81}-\frac{5\,\zeta(4)}{54}\right)\right.$$
$$\left.+N\left(-\frac{5777}{1944}-\frac{22\,\zeta(3)}{81}-\frac{16\,\zeta(4)}{27}\right)-\frac{7765}{1944}-\frac{34\,\zeta(3)}{81}-\frac{22\,\zeta(4)}{27}\right]$$
$$+g_1{}^3\,g_2\left[N^2\left(\frac{1}{1296}-\frac{\zeta(3)}{9}\right)+N\left(-\frac{421}{72}-\frac{10\,\zeta(3)}{27}-\frac{10\,\zeta(4)}{9}\right)-\frac{7765}{324}-\frac{68\,\zeta(3)}{27}-\frac{44\,\zeta(4)}{9}\right]$$
$$+g_1{}^2\,g_2{}^2\left[N\left(-\frac{1841}{864}-\frac{\zeta(3)}{6}-\frac{\zeta(4)}{3}\right)-\frac{9199}{216}-\frac{13\,\zeta(3)}{3}-\frac{26\,\zeta(4)}{3}\right]$$
$$+g_1\,g_2{}^3\left[N\left(-\frac{25}{72}+\frac{\zeta(3)}{6}\right)-\frac{4243}{144}-\frac{19\,\zeta(3)}{6}-6\,\zeta(4)\right]+g_2{}^4\left[-\frac{477}{64}-\frac{3\,\zeta(3)}{4}-\frac{3\,\zeta(4)}{2}\right]$$
$$+g_1{}^5\left[N^4\left(\frac{7}{124416}+\frac{17\,\zeta(3)}{7776}-\frac{\zeta(4)}{432}\right)+N^3\left(\frac{2831}{23328}+\frac{487\,\zeta(3)}{3888}-\frac{\zeta^2(3)}{243}+\frac{23\,\zeta(4)}{1296}-\frac{5\,\zeta(5)}{486}+\frac{25\,\zeta(6)}{486}\right)\right.$$
$$+N^2\left(\frac{291907}{93312}+\frac{1261\,\zeta(3)}{972}-\frac{149\,\zeta^2(3)}{486}+\frac{437\,\zeta(4)}{648}+\frac{2\,\zeta(5)}{243}+\frac{1475\,\zeta(6)}{972}\right)$$
$$+N\left(\frac{83137}{5832}+\frac{2011\,\zeta(3)}{324}-\frac{436\,\zeta^2(3)}{243}+\frac{1075\,\zeta(4)}{324}+\frac{50\,\zeta(5)}{243}+\frac{1850\,\zeta(6)}{243}\right)$$
$$\left.+\frac{49477}{2916}+\frac{1327\,\zeta(3)}{162}-\frac{194\,\zeta^2(3)}{81}+\frac{667\,\zeta(4)}{162}+\frac{8\,\zeta(5)}{27}+\frac{775\,\zeta(6)}{81}\right]$$
$$+g_1{}^4\,g_2\left[N^3\left(\frac{35}{41472}+\frac{85\,\zeta(3)}{2592}-\frac{5\,\zeta(4)}{144}\right)+N^2\left(\frac{113135}{62208}+\frac{1175\,\zeta(3)}{648}-\frac{5\,\zeta^2(3)}{81}+\frac{145\,\zeta(4)}{432}-\frac{25\,\zeta(5)}{162}+\frac{125\,\zeta(6)}{162}\right)\right.$$

$$+N\left(\frac{4675}{108}+\frac{95\,\zeta(3)}{6}-\frac{725\,\zeta^2(3)}{162}+\frac{85\,\zeta(4)}{9}+\frac{35\,\zeta(5)}{81}+\frac{6875\,\zeta(6)}{324}\right)$$

$$+\frac{247385}{1944}+\frac{6635\,\zeta(3)}{108}-\frac{485\,\zeta^2(3)}{27}+\frac{3335\,\zeta(4)}{108}+\frac{20\,\zeta(5)}{9}+\frac{3875\,\zeta(6)}{54}\Big]$$

$$+g_1{}^3\,g_2{}^2\left[N^2\left(-\frac{2045}{20736}+\frac{785\,\zeta(3)}{1296}-\frac{7\,\zeta(4)}{72}-\frac{7\,\zeta(5)}{54}\right)\right.$$

$$+N\left(\frac{362281}{10368}+\frac{3743\,\zeta(3)}{324}-\frac{61\,\zeta^2(3)}{27}+\frac{1493\,\zeta(4)}{216}-\frac{11\,\zeta(5)}{18}+\frac{775\,\zeta(6)}{54}\right)$$

$$+\frac{267737}{864}+\frac{47327\,\zeta(3)}{324}-\frac{1154\,\zeta^2(3)}{27}+\frac{8039\,\zeta(4)}{108}+\frac{155\,\zeta(5)}{27}+\frac{4675\,\zeta(6)}{27}\Big]$$

$$+g_1{}^2\,g_2{}^3\left[N\left(\frac{26173}{2304}+\frac{71\,\zeta(3)}{48}-\frac{2\,\zeta^2(3)}{9}+\frac{25\,\zeta(4)}{12}-\frac{14\,\zeta(5)}{9}+\frac{25\,\zeta(6)}{9}\right)\right.$$

$$+\frac{32003}{96}+\frac{627\,\zeta(3)}{4}-\frac{403\,\zeta^2(3)}{9}+\frac{475\,\zeta(4)}{6}+\frac{59\,\zeta(5)}{9}+\frac{3325\,\zeta(6)}{18}\Big]$$

$$+g_1\,g_2{}^4\left[N\left(\frac{26171}{13824}-\frac{77\,\zeta(3)}{96}+\frac{7\,\zeta(4)}{16}-\frac{2\,\zeta(5)}{3}\right)+\frac{589141}{3456}+\frac{959\,\zeta(3)}{12}-\frac{45\,\zeta^2(3)}{2}+\frac{643\,\zeta(4)}{16}+\frac{19\,\zeta(5)}{6}+\frac{375\,\zeta(6)}{4}\right]$$

$$+g_2{}^5\left[\frac{158849}{4608}+\frac{1519\,\zeta(3)}{96}-\frac{9\,\zeta^2(3)}{2}+\frac{65\,\zeta(4)}{8}+\frac{\zeta(5)}{2}+\frac{75\,\zeta(6)}{4}\right].\tag{18.15}$$

18.3 Fixed Points and Critical Exponents

The only new fixed point is the cubic one (g_1^C, g_2^C), which represents a system with mixed $O(N)$ and cubic symmetry. We have calculated the ε-expansion of the fixed-point couplings and critical exponents up to the order ε^5. For $N=2$, we verify the relation (18.8) between the cubic and the Ising fixed points:

$$g_1^C = g_1^I + 3\,g_2^I/2,\qquad g_2^C = -g_2^I,$$

as a consequence of the symmetry in Eq. (18.7). As N increases, the cubic fixed point approaches the Heisenberg fixed point from below, crossing it at $N=N_c$, where g_2^C changes its sign. For $N\to\infty$, the cubic fixed point moves towards the Ising fixed point.

The ε-expansions for the cubic fixed point are [8]

$$g_1^C(\varepsilon) = \varepsilon\,\frac{1}{N}+\varepsilon^2\,\frac{(-106+125\,N-19\,N^2)}{27\,N^3}$$

$$+\varepsilon^3\left[\frac{22472}{729\,N^5}-\frac{45080}{729\,N^4}+\frac{38329}{972\,N^3}-\frac{41971}{5832\,N^2}-\frac{1955}{5832\,N}+\left(\frac{56}{9\,N^4}-\frac{28}{9\,N^3}-\frac{8}{3\,N^2}+\frac{8}{9\,N}\right)\zeta(3)\right]$$

$$+\varepsilon^4\left[-\frac{5955080}{19683\,N^7}+\frac{5623300}{6561\,N^6}-\frac{5934115}{6561\,N^5}+\frac{8315992}{19683\,N^4}-\frac{3955061}{52488\,N^3}+\frac{113779}{104976\,N^2}-\frac{2987}{314928\,N}\right.$$

$$+\left(-\frac{29680}{243\,N^6}+\frac{46696}{243\,N^5}-\frac{12764}{243\,N^4}-\frac{7934}{243\,N^3}+\frac{2744}{243\,N^2}+\frac{110}{243\,N}\right)\zeta(3)$$

$$+\left(\frac{32}{9\,N^4}-\frac{16}{9\,N^3}-\frac{16}{9\,N^2}+\frac{2}{3\,N}\right)\zeta(4)+\left(-\frac{80}{3\,N^5}+\frac{200}{27\,N^4}+\frac{80}{27\,N^3}+\frac{80}{9\,N^2}-\frac{80}{27\,N}\right)\zeta(5)\Big]$$

$$+\varepsilon^5\left[\frac{1767467744}{531441\,N^9}-\frac{6468340480}{531441\,N^8}+\frac{9496212881}{531441\,N^7}-\frac{14088835643}{1062882\,N^6}+\frac{43137004355}{8503056\,N^5}\right.$$

$$-\frac{7400332843}{8503056\,N^4}+\frac{2080479877}{68024448\,N^3}+\frac{337198481}{136048896\,N^2}-\frac{5795035}{136048896\,N}$$

$$+\left(\frac{4404512}{2187\,N^8}-\frac{3678896}{729\,N^7}+\frac{9044242}{2187\,N^6}-\frac{1239931}{1458\,N^5}+\frac{1694161}{4374\,N^4}+\frac{653341}{4374\,N^3}-\frac{182483}{34992\,N^2}+\frac{13883}{34992\,N}\right)\zeta(3)$$

$$+\left(-\frac{18020}{243\,N^6}+\frac{56137}{486\,N^5}-\frac{2213}{81\,N^4}-\frac{47369}{1944\,N^3}+\frac{3929}{486\,N^2}+\frac{217}{648\,N}\right)\zeta(4)$$

$$+\left(\frac{16960}{27\,N^7}-\frac{579260}{729\,N^6}+\frac{107902}{729\,N^5}-\frac{508}{27\,N^4}+\frac{84818}{729\,N^3}-\frac{26296}{729\,N^2}-\frac{340}{243\,N}\right)\zeta(5)$$

$$+\left(-\frac{2225}{81\,N^5}+\frac{1525}{162\,N^4}+\frac{125}{81\,N^3}+\frac{850}{81\,N^2}-\frac{100}{27\,N}\right)\zeta(6)+\left(\frac{-1078}{9\,N^6}+\frac{1225}{3\,N^5}-\frac{2450}{9\,N^4}+\frac{539}{9\,N^3}-\frac{343}{9\,N^2}+\frac{98}{9\,N}\right)\zeta(7)$$

$$+\left(\frac{3136}{27\,N^7}-\frac{112}{N^6}-\frac{2834}{81\,N^5}+\frac{3089}{81\,N^4}+\frac{626}{81\,N^3}-\frac{296}{81\,N^2}-\frac{16}{27\,N}\right)\zeta^2(3)\Big]+\mathcal{O}(\varepsilon^6)\,,\tag{18.16}$$

$$g_2^C(\varepsilon) = \varepsilon\,\frac{(N-4)}{3\,N}+\varepsilon^2\,\frac{(424-534\,N+93\,N^2+17\,N^3)}{81\,N^3}$$

$$+\varepsilon^3\left[-\frac{89888}{2187\,N^5}+\frac{187528}{2187\,N^4}-\frac{123707}{2187\,N^3}+\frac{90281}{8748\,N^2}+\frac{11713}{17496\,N}+\frac{709}{17496}+\left(-\frac{224}{27\,N^4}+\frac{16}{3\,N^3}+\frac{80}{27\,N^2}-\frac{32}{27\,N}-\frac{4}{27}\right)\zeta(3)\right]$$

$$+\varepsilon^4\left[\frac{23820320}{59049\,N^7}-\frac{69389720}{59049\,N^6}+\frac{25018256}{19683\,N^5}-\frac{35478331}{59049\,N^4}+\frac{11944655}{118098\,N^3}+\frac{406721}{157464\,N^2}-\frac{511435}{944784\,N}+\frac{10909}{944784}\right.$$

$$+\left(\frac{118720}{729\,N^6}-\frac{200768}{729\,N^5}+\frac{23752}{243\,N^4}+\frac{2704}{81\,N^3}-\frac{10450}{729\,N^2}-\frac{56}{81\,N}-\frac{106}{729}\right)\zeta(3)$$

$$+\left(-\frac{136}{27\,N^4}+\frac{28}{9\,N^3}+\frac{64}{27\,N^2}-\frac{28}{27\,N}-\frac{2}{27}\right)\zeta(4)+\left(\frac{320}{9\,N^5}-\frac{400}{27\,N^4}-\frac{440}{81\,N^3}-\frac{80}{9\,N^2}+\frac{280}{81\,N}+\frac{40}{81}\right)\zeta(5)\Bigg]$$

$$+\varepsilon^5\Bigg[-\frac{7069870976}{1594323\,N^9}+\frac{2937809504}{177147\,N^8}-\frac{4398801284}{177147\,N^7}+\frac{9923276525}{531441\,N^6}-\frac{5033294725}{708588\,N^5}+\frac{3132906331}{2834352\,N^4}+\frac{256333871}{17006112\,N^3}$$

$$-\frac{264392957}{22674816\,N^2}+\frac{4069429}{45349632\,N}-\frac{321451}{408146688}$$

$$+\left(-\frac{17618048}{6561\,N^8}+\frac{46032704}{6561\,N^7}-\frac{40407016}{6561\,N^6}+\frac{10668718}{6561\,N^5}+\frac{4840987}{13122\,N^4}-\frac{1176529}{6561\,N^3}+\frac{64261}{26244\,N^2}+\frac{10361}{104976\,N}-\frac{11221}{104976}\right)\zeta(3)$$

$$+\left(\frac{75472}{729\,N^6}-\frac{125462}{729\,N^5}+\frac{8347}{162\,N^4}+\frac{7747}{243\,N^3}-\frac{72941}{5832\,N^2}-\frac{323}{972\,N}-\frac{443}{5832}\right)\zeta(4)$$

$$+\left(-\frac{67840}{81\,N^7}-\frac{845200}{729\,N^6}-\frac{202864}{729\,N^5}-\frac{10202}{729\,N^4}-\frac{83693}{729\,N^3}+\frac{29770}{729\,N^2}+\frac{1628}{729\,N}+\frac{373}{729}\right)\zeta(5)$$

$$+\left(\frac{3100}{81\,N^5}-\frac{1525}{81\,N^4}-\frac{425}{162\,N^3}-\frac{1025}{81\,N^2}+\frac{275}{54\,N}+\frac{25}{54}\right)\zeta(6)+\left(\frac{4312}{27\,N^6}-\frac{14896}{27\,N^5}+\frac{3626}{9\,N^4}-\frac{2254}{27\,N^3}+\frac{980}{27\,N^2}-\frac{98}{9\,N}-\frac{49}{27}\right)\zeta(7)$$

$$+\left(-\frac{12544}{81\,N^7}+\frac{13888}{81\,N^6}+\frac{712}{27\,N^5}-\frac{4354}{81\,N^4}-\frac{365}{81\,N^3}+\frac{10}{3\,N^2}+\frac{85}{81\,N}+\frac{11}{81}\right)\zeta^2(3)\Bigg]+\mathcal{O}(\varepsilon^6)\,. \tag{18.17}$$

18.4 Stability

The stability of the fixed points is determined from the eigenvalues ω_1 and ω_2 of the matrix

$$M_{ij}=\left.\frac{\partial\beta_i(g_1,g_2)}{\partial g_j}\right|_{g_1^*,g_2^*}. \tag{18.18}$$

For positive real parts of both eigenvalues, the corresponding fixed point is infrared stable. The Gaussian fixed point is doubly unstable. At the Ising fixed point, one eigenvalue ω_1 is negative. The Heisenberg and the cubic fixed points interchange stability for $N=N_c$, the former being stable for $N<N_c$ where $g_2^C<0$. The stability wedges of the critical theory are visible in Fig. 18.1. They differ from the bare stability wedge. Outside these wedges, the transition is of first order [9].

To find the crucial number N_c determining which fixed point governs the critical behavior in $D=3$ dimensions, we study the eigenvalue ω_2^C of the stability matrix as a function of N (the other eigenvalue ω_1^C remains positive and can be ignored). Its ε-expansion reads

$$\omega_2^C=\varepsilon\,\frac{N-4}{3\,N}+(N-1)\left[\varepsilon^2\,\frac{(-848+660\,N+72\,N^2-19\,N^3)}{81\,N^3\,(2+N)}\right.$$

$$\left.+\varepsilon^3\,\frac{\Sigma_{i=0}^7\,C_i^3\,N^i}{8748\,N^5\,(2+N)^3}+\varepsilon^4\,\frac{\Sigma_{i=0}^{11}\,C_i^4\,N^i}{944784\,N^7\,(2+N)^5}+\varepsilon^5\,\frac{\Sigma_{i=0}^{15}\,C_i^5\,N^i}{102036672\,N^9\,(2+N)^7}\right]+\mathcal{O}(\varepsilon^6)\,. \tag{18.19}$$

The coefficients C_i^j are listed in Table 18.1 . The condition of vanishing ω_2^C gives the ε-expansion of N_c:

$$N_c=4-2\,\varepsilon+\varepsilon^2\left[-\frac{5}{12}+\frac{5\zeta(3)}{2}\right]+\varepsilon^3\left[-\frac{1}{72}+\frac{5\zeta(3)}{8}+\frac{15\zeta(4)}{8}-\frac{25\zeta(5)}{3}\right]$$

$$+\varepsilon^4\left[-\frac{1}{384}+\frac{93\,\zeta(3)}{128}-\frac{229\,\zeta^2(3)}{144}+\frac{15\,\zeta(4)}{32}-\frac{3155\,\zeta(5)}{1728}-\frac{125\,\zeta(6)}{12}+\frac{11515\,\zeta(7)}{384}\right]+\mathcal{O}(\varepsilon^5)\,. \tag{18.20}$$

The same expansion is found from the condition $g_2^C=0$. This expansion is badly divergent, making it difficult to calculate the value of N_c at $\varepsilon=1$. With the help of Padé approximants we obtain the following values:

Padé [1/1] :	N_c	=	3.128	Padé [2/2] :	N_c	=	2.958
Padé [2/1] :	N_c	=	2.792	Padé [1/2] :	N_c	=	2.893
Padé [3/1] :	N_c	=	3.068	Padé [1/3] :	N_c	=	2.972

The approximant Padé [1/3] is unreliable since the Padé denominator has a pole at positive ε,

C^3_0	2876416	C^4_2	$22268920832-5633556480\zeta(3)-1074954240\zeta(5)$
C^3_1	$-1740544+580608\zeta(3)$	C^4_3	$-5244000000+5418233856\zeta(3)+188116992\zeta(4)-2120048640\zeta(5)$
C^3_2	$-3188544+829440\zeta(3)$	C^4_4	$-21313343616+8161855488\zeta(3)+456855552\zeta(4)-2508226560\zeta(5)$
C^3_3	$2340592-62208\zeta(3)$	C^4_5	$11104506624-370593792\zeta(3)+295612416\zeta(4)-1425807360\zeta(5)$
C^3_4	$54656-404352\zeta(3)$	C^4_6	$1087015200-2657484288\zeta(3)-83980800\zeta(4)+432967680\zeta(5)$
C^3_5	$-162696-114048\zeta(3)$	C^4_7	$-1171729344-359023104\zeta(3)-173000448\zeta(4)+901393920\zeta(5)$
C^3_6	$15700+7776\zeta(3)$	C^4_8	$54071304+211455360\zeta(3)-67184640\zeta(4)+343388160\zeta(5)$
C^3_7	$-937+2592\zeta(3)$	C^4_9	$26339632+27133056\zeta(3)-5878656\zeta(4)+20062080\zeta(5)$
C^4_0	-12196003840	C^4_{10}	$-3320774-2164320\zeta(3)+1469664\zeta(4)-9797760\zeta(5)$
C^4_1	$5267640320-4923555840\zeta(3)$	C^4_{11}	$-24857+154224\zeta(3)+209952\zeta(4)-933120\zeta(5)$

C^5_0	57916383035392
C^5_1	$-14984240431104+35071472959488\,\zeta(3)$
C^5_2	$-143277741441024+39325680009216\,\zeta(3)+2022633897984\,\zeta^2(3)+10938734346240\,\zeta(5)$
C^5_3	$1211449212928-58923130945536\,\zeta(3)+5995664769024\,\zeta^2(3)$ $-1595232092160\,\zeta(4)+22263974461440\,\zeta(5)-2085841207296\,\zeta(7)$
C^5_4	$162125071257600-92450858139648\,\zeta(3)+5046265184256\,\zeta^2(3)-3420504391680\,\zeta(4)$ $+9499251179520\,\zeta(5)-580475289600\,\zeta(6)+3887249522688\,\zeta(7)$
C^5_5	$38426107425792+21006186430464\,\zeta(3)-2702434959360\,\zeta^2(3)-468572553216\,\zeta(4)$ $-1137660831744\,\zeta(5)-1725301555200\,\zeta(6)+14600888451072\,\zeta(7)$
C^5_6	$-162464030196224+69899709652992\,\zeta(3)-6924425232384\,\zeta^2(3)+3943630872576\,\zeta(4)$ $-19961756909568\,\zeta(5)-2644387430400\,\zeta(6)+13700184293376\,\zeta(7)$
C^5_7	$53925927185664+4311047245824\,\zeta(3)-3110702579712\,\zeta^2(3)+2963245731840\,\zeta(4)$ $-11389462659072\,\zeta(5)-2410584883200\,\zeta(6)+5736063320064\,\zeta(7)$
C^5_8	$14422298978304-19789519380480\,\zeta(3)+993580204032\,\zeta^2(3)-319987003392\,\zeta(4)$ $+1995616714752\,\zeta(5)-874744012800\,\zeta(6)+592568524800\,\zeta(7)$
C^5_9	$-7438725755776-2538882164736\,\zeta(3)+1199003959296\,\zeta^2(3)-1007366492160\,\zeta(4)$ $+4815696457728\,\zeta(5)+528071270400\,\zeta(6)-1724374407168\,\zeta(7)$
C^5_{10}	$-147321611712+2076145468416\,\zeta(3)+251458670592\,\zeta^2(3)-263068176384\,\zeta(4)$ $+1308067024896\,\zeta(5)+730632960000\,\zeta(6)-2005844456448\,\zeta(7)$
C^5_{11}	$258021138336+211760262144\,\zeta(3)-41520107520\,\zeta^2(3)+48221775360\,\zeta(4)$ $-269590685184\,\zeta(5)+317951308800\,\zeta(6)-960701720832\,\zeta(7)$
C^5_{12}	$-8428414672-68664506112\,\zeta(3)-17716589568\,\zeta^2(3)+25217754624\,\zeta(4)$ $-141247120896\,\zeta(5)+51900134400\,\zeta(6)-170733806208\,\zeta(7)$
C^5_{13}	$-2039889852-1386414144\,\zeta(3)-851565312\,\zeta^2(3)+1546506432\,\zeta(4)$ $-7514135424\,\zeta(5)-3086294400\,\zeta(6)+12221725824\,\zeta(7)$
C^5_{14}	$-138029874+936160416\,\zeta(3)-25194240\,\zeta^2(3)-125341344\,\zeta(4)+1017287424\,\zeta(5)$ $-1826582400\,\zeta(6)+6666395904\,\zeta(7)$
C^5_{15}	$-64327+12966480\,\zeta(3)-20155392\,\zeta^2(3)+12492144\,\zeta(4)-49268736\,\zeta(5)-125971200\,\zeta(6)+370355328\,\zeta(7)$

TABLE 18.1 Constants appearing in expansion (18.19) for ω^C_2.

where the exact result should be regular. The highest symmetric approximant is usually the most accurate one, from which we deduce the estimate:

$$N_c \approx 2.958. \tag{18.21}$$

Before the work of the present authors [7], the ε-expansion for N_c was known only up to the order ε^2 [10], so that only the Padé [1/1]-approximant was available which yielded $N_c \approx 3.128$. Estimates of N_c directly in $D = 3$ dimensions yielded $N_c = 3.4$ [11] and $N_c = 2.9$ [6]. Thus most of the previous results gave N_c-values larger than 3, implying that the critical behavior of magnetic systems with cubic symmetry is governed by the Heisenberg fixed point. The symmetric Padé approximant to our expansion suggests that N_c lies below three, so that the cubic fixed point is the relevant one.

By inserting the expansions (18.16) and (18.17) into (18.14) and (18.15), we find the critical exponents η_C and ν_C of the cubic fixed point:

$$1/\nu^C = 2+(N-1)\left\{-\varepsilon\,\tfrac{2}{3N}+\varepsilon^2\left[\tfrac{1}{162N^3}\left(424-326N+19N^2\right)\right]\right.$$
$$\left.+\varepsilon^3\left[\tfrac{1}{17496N^5}\left(-359552+573728N-264936N^2+28358N^3+937N^4+\tfrac{4(N+2)}{27N^4}\left(-14+11N-N^2\right)\zeta(3)\right)\right]\right.$$

$$+\varepsilon^4\left[\frac{1}{1889568\,N^7}\left(381125120-923268480\,N+798088608\,N^2-284926360\,N^3\right.\right.$$
$$\left.+32693424\,N^4+768780\,N^5+24857\,N^6\right)$$
$$+\frac{1}{1458\,N^6}\left(118720-152032\,N+29816\,N^2+17936\,N^3-4124\,N^4-119\,N^5\right)\zeta(3)$$
$$+\frac{(2+N)}{9\,N^4}\left(-14+11\,N-N^2\right)\zeta(4)+\frac{40}{81\,N^5}\left(36-2\,N-4\,N^2-8\,N^3+N^4\right)\zeta(5)\Big]$$
$$+\varepsilon^5\left[\frac{1}{204073344\,N^9}\left(-452471742464+1470211004416\,N-1869697955840\,N^2+1160186503168\,N^3\right.\right.$$
$$\left.-350446218272\,N^4+41122747144\,N^5+144762448\,N^6-68383472\,N^7+64327\,N^8\right)$$
$$+\frac{1}{52488\,N^8}\left(-70472192+153589248\,N-106996288\,N^2\right.$$
$$\left.+18129888\,N^3+6458072\,N^4-1726592\,N^5-8716\,N^6-3335\,N^7\right)\zeta(3)$$
$$+\frac{1}{1944\,N^6}\left(118720-152032\,N+29816\,N^2+17936\,N^3-4124\,N^4-119\,N^5\right)\zeta(4)$$
$$+\frac{1}{2187\,N^7}\left(-915840+897560\,N-53320\,N^2-2676\,N^3-86879\,N^4+20128\,N^5+528\,N^6\right)\zeta(5)$$
$$+\frac{50}{81\,N^5}\left(36-2\,N-4\,N^2-8\,N^3+N^4\right)\zeta(6)+\frac{49}{27\,N^6}\left(44-132\,N+65\,N^2-9\,N^3+8\,N^4-N^5\right)\zeta(7)$$
$$+\frac{8}{81\,N^7}\left(-784+588\,N+232\,N^2-163\,N^3-23\,N^4+5\,N^5+N^6\right)\zeta^2(3)\Big]\Big\}+\mathcal{O}\!\left(\varepsilon^6\right). \tag{18.22}$$

$$\eta^C=(N-1)\left\{\varepsilon^2\left[\frac{(2+N)}{54\,N^2}\right]+\varepsilon^3\left[\frac{-1696+1728\,N-222\,N^2+109\,N^3}{5832\,N^4}\right]\right.$$
$$+\varepsilon^4\left[\frac{1}{629856\,N^6}\left(1797760-3566912\,N+2292328\,N^2-507952\,N^3+28832\,N^4+7217\,N^5\right)\right.$$
$$\left.+\frac{4}{243\,N^5}\left(28-6\,N-16\,N^2+4\,N^3-N^4\right)\zeta(3)\right]$$
$$+\varepsilon^5\left[\frac{1}{68024448\,N^8}\left(-2134300672+6125897728\,N-6643967232\,N^2+3326175872\,N^3\right.\right.$$
$$\left.-731940728\,N^4+46139232\,N^5+1948700\,N^6+321511\,N^7\right)$$
$$+\frac{1}{17496\,N^7}\left(-189952+266624\,N-31584\,N^2-80376\,N^3+29704\,N^4-2196\,N^5-329\,N^6\right)\zeta(3)$$
$$+\frac{1}{81\,N^5}\left(28-6\,N-16\,N^2+4\,N^3-N^4\right)\zeta(4)+\frac{40}{729\,N^6}\left(-36+N^2+18\,N^3-5\,N^4+N^5\right)\zeta(5)\Big]\Big\}+\mathcal{O}(\varepsilon^6). \tag{18.23}$$

As anticipated, for $N=1$ the exponents for the cubic fixed point $(g_1^C=-g_2^C)$ are degenerate with those of the Gaussian fixed point, taking free-field values. At $N=2$, the exponents take Ising values by virtue of the symmetry of the interaction under (18.7).

18.5　Resummation

Different resummation techniques are applied to get a better estimate for the cubic fixed point and the critical exponents.

18.5.1　Padé Approximations for Critical Exponents

The results of Padé approximations to the ε-expansions of the critical exponents are shown in Table 18.2 for $N=3$. The exponents for the symmetric and the cubic fixed point lie very close to each other. Most of the approximants contain poles at real positive ε, and are therefore useless. This is also true for the Padé approximants of the Borel-transformed series, where only the [4/1] approximant has no such unphysical property. For the Heisenberg fixed point, the approximants can be compared with results of resummations which include information from the large-order behavior.

The stability of the cubic fixed point for $N=3$ is given by the sign of $\omega^{(2)}$. The Padé approximants suggest that the Heisenberg fixed point is the only stable one, as predicted [5]. In contrast, the Borel-Padé method indicates that the sign might change upon resummation.

$[M/N]$	$[2/2]$	$-$	$[4/1]$	Borel$[4/1]$	Lit. [12]
η_H	0.0164		0.03706	0.03685	0.040(3)
η_C	0.0157		0.03713	0.03689	$-$
η_I	0.0127		0.03517	0.03493	$-$

$[M/N]$	$[3/2]$	$[2/3]$	$[4/1]$	Borel$[4/1]$	Lit. [12]
$1/\nu_H$	1.4249	1.4273	1.4251	1.4165	1.408(14)
$1/\nu_C$	1.4235	1.42745	1.4241	1.4139	$-$
$1/\nu_I$	1.5951	1.5994	1.5991	1.5903	$-$

$[M/N]$	$[3/2]$	$-$	$[4/1]$	Borel$[4/1]$	Lit. [12]
$\omega_H^{(1)}$	0.8003		0.8698	0.7641	0.79(4)
$\omega_C^{(1)}$	0.8185		0.8869	0.775	$-$
$\omega_I^{(1)}$	0.8058		0.9339	0.7925	$-$

$[M/N]$	$[3/2]$	$-$	$[4/1]$	Borel$[4/1]$	
$\omega_H^{(2)}$	0.0017		0.0111	-0.011	$-$
$\omega_C^{(2)}$	-0.0012		-0.0048	-0.0034	$-$
$\omega_I^{(2)}$	-0.1902		-0.1981	-0.18	$-$

TABLE 18.2 Padé approximations to ε-expansions of critical exponents for cubic symmetry with $N = 3$ and $\varepsilon = 1$. Only the approximants with no poles for $\varepsilon > 0$ are listed. The second-last column contains the result of a Borel-Padé transformation, in which the Borel sums are evaluated with the $[4/1]$-Padé approximant.

For $N = 4$ the situation is clear. The number $\omega_C^{(2)}$ is definitely positive, whereas $\omega_H^{(2)}$ is negative. Here the cubic fixed point is stable.

Instead of applying the Padé approximant to the series in ε, a two-variable Padé $[M, M]/[N, N]$ can be used to resum the series in g_1 and g_2 [13]. This leads to $N_c = 2.9$ [6], where a $[2, 2]/[2, 2]$-Padé is used to evaluate the Borel sum emerging from a four-loop calculation in fixed dimension. The $[3, 3]/[2, 2]$- and the $[4, 4]/[1, 1]$-Padé approximants of the five-loop expansions (18.12) and (18.13), as well as of the associated Borel sums contain poles at positive coupling constants, which renders them useless.

18.5.2 Resummations for Cubic Fixed Point

The cubic fixed point lies in the upper half of the coupling constant plane ($g_2 > 0$) only if $N > N_C$. Calculating the cubic fixed point for $N = 3$ as a zero of the resummed β-function, verifies whether $g_2^C > 0$ or, equivalently, $N_C < 3$.

Large-Order Behavior

From Padé approximants, we know that the cubic fixed point lies close to the isotropic fixed point with $|g_2| < 0.1$. A large-order calculation [14], designed for the region of weak anisotropy where $g_2 \to 0$, will therefore be used for the resummation. This calculation works with the two coupling constants $g = g_1 + g_2$ and $l = g_2$, $|l| < 0.1$. A continuous transition between $l = 0$ and $l \neq 0$ is assumed. The anisotropy is treated as a perturbation of the isotropic case whose large-order behavior is found by an isotropic solution of the field equation (instanton). The β-functions for g and l are given as expansions in l around the isotropic case. The $l = 0$ -term of β_g consists of the isotropic β-function. The same expansion is performed for the imaginary part of the β-functions for $g \to 0^-$. The general form of the expansion is

$$\beta_{g/l}(g, l) = \beta^p_{g/l} + i \sum_{n=0}^{\infty} l^n a_n^{g/l} \left(\frac{-1}{\sigma g}\right)^{\frac{D+4+N+4n}{2}} \exp\left(\frac{1}{\sigma g}\right) \tag{18.24}$$

$$= \sum_{n=0/1}^{\infty} l^n \left[\sum_{m=0}^{\infty} A_{n,m}^{g/l} g^m + i a_n^{g/l} \left(\frac{-1}{\sigma g}\right)^{\frac{D+4+N+4n}{2}} \exp\left(\frac{1}{\sigma g}\right)\right] \tag{18.25}$$

$$= \sum_{n=0/1}^{\infty} l^n \, B_n^{g/l}(g), \tag{18.26}$$

where $\beta^p_{g/l}$ is the perturbative part and $A_{n,m}^{g/l} g^m$ are the coefficients of l^n in the perturbative part. The parameter σg is again the numerical value of the instanton energy, where $\sigma = 1$ in this parametrization. The coefficients $a_n^{g/l}$ are rather complicated expressions.

There is a cut only along the negative real axis of the complex g-plane. For small l, the imaginary part of $\beta(g)$ for small $-g > 0$ governs the asymptotic series. The parameter l, being very small, is treated as an expansion parameter. The sum over all l^n-terms has a non-zero radius of convergence and the expansion is asymptotic in l. But the error of an asymptotic series decreases up to the order $N_{\min} \approx 1/\sigma l$. With $\sigma = 1$ and $l = 0.1$, we find $N_{\min} \approx 10$ and the error $\Delta \approx 4.54 \cdot 10^{-5}$, so that partial sums in l can yield acceptable approximations. Even for $N = 2$ and $l = 0.1$, the error is still small: $\Delta \approx 0.019$. For $N = 1$ and $l = 0.01$, the error is $\Delta \approx 0.009$.

A resummation is needed only for the expansions B_n in g. Application of the dispersion integral to the imaginary part of each $B_n(g)$ in (18.25) reveals [5] the asymptotic large-order behavior of the expansions $B_n(g) = \sum_k B_{n,k} g^k$:

$$B_{n,k}^g \stackrel{k \to \infty}{\to} \gamma_n^g (-\sigma)^k k! k^{\beta_n} [1 + \mathcal{O}(1/k)], \quad \text{with } \beta_n = \frac{D+5+4n}{2}, \tag{18.27}$$

$$B_{n,k}^l \stackrel{k \to \infty}{\to} \gamma_n^l (-\sigma)^k k! k^{\beta_n} [1 + \mathcal{O}(1/k)], \quad \text{with } \beta_n = \frac{D+5+4n}{2}, \tag{18.28}$$

and allows a resummation by the methods described in Section 16.5.

The structure of Eq. (18.25) does not permit us to find the imaginary part and therefore the large-order behavior of the ε-series for both coupling constants, $g^*(\varepsilon)$ and $l^*(\varepsilon)$. The different treatment of the coupling constants is incompatible with the idea of the ε-expansion, which deals with the two coupling constants at the same level by using one expansion parameter for both of them.

Cubic Fixed Point from Resummed β-Function

For $\varepsilon = 1$ and $N = 3$, each series $B_n(g)$ is resummed separately. The resummation of $B_n(g)$ is getting less accurate for increasing n since the corresponding series in g have fewer and fewer terms. But taking into account the smallness of the parameter, the expansion in l may be truncated. This is seen when using Padé approximants to resum $B_n(g)$ in Eqs. (18.29) and (18.30) and calculating the simultaneous zeros l^*, g^* of the β-functions as shown in Table 18.3 . The β-functions at the cubic fixed point will be approximated by

$$0 = \beta_g(g^*, l^*) \approx \sum_{n=0}^{5} l^{*n} \, \text{res}[B_n^g(g^*)], \tag{18.29}$$

$$
\begin{aligned}
n = 0: \quad & [3/2]B_0^g(g^*) \;=\; 0 & & g^* \;=\; 0.38081 \\
& [3/2]B_0^l(g^*) \;=\; 0 & \Rightarrow \quad & l^* \;=\; 0.0
\end{aligned}
$$

$$
\begin{aligned}
n = 1: \quad & [3/2]B_0^g(g^*) + l^*\,[3/2]B_1^g(g^*) \;=\; 0 & & g^* \;=\; 0.373716 \\
& [3/2]B_0^l(g^*) + l^*\,[2/2]B_1^l(g^*) \;=\; 0 & \Rightarrow \quad & l^* \;=\; -0.01845
\end{aligned}
$$

$$
\begin{aligned}
n = 2: \quad & [3/2]B_0^g(g^*) + \ldots + l^{*2}\,[2/2]B_2^g(g^*) \;=\; 0 & & g^* \;=\; 0.373835 \\
& [3/2]B_0^l(g^*) + \ldots + l^{*2}\,[2/1]B_2^l(g^*) \;=\; 0 & \Rightarrow \quad & l^* \;=\; -0.0177435
\end{aligned}
$$

$$
\begin{aligned}
n = 3: \quad & [3/2]B_0^g(g^*) + \ldots + l^{*3}\,[2/1]B_3^g(g^*) \;=\; 0 & & g^* \;=\; 0.37384 \\
& [3/2]B_0^l(g^*) + \ldots + l^{*3}\,[1/1]B_3^l(g^*) \;=\; 0 & \Rightarrow \quad & l^* \;=\; -0.0177294
\end{aligned}
$$

$$
\begin{aligned}
n = 4: \quad & [3/2]B_0^g(g^*) + \ldots + l^{*4}\,[1/1]B_4^g(g^*) \;=\; 0 & & g^* \;=\; 0.37377 \\
& [3/2]B_0^l(g^*) + \ldots + l^{*4}B_4^l(g^*) \;=\; 0 & \Rightarrow \quad & l^* \;=\; -0.017903
\end{aligned}
$$

TABLE 18.3 Padé approximation in presence of anisotropic contributions to β-function. From zeroth to first order in l, the value for g^* decreases by about 1.9%. Including the second order in l changes g^* further by about 0.032%, whereas l^* changes by about 3.98%. The third order in l, changes this value for l^* only by about 0.079%. The main contribution to the difference between the Heisenberg and the cubic fixed points is contained in the first-order contribution. This gives an error of approximately 0.03% in g^*, and 4% in l^*.

$$
0 \;=\; \beta_l(g^*, l^*) \;\approx\; l^* \sum_{n=0}^{5} l^{*n} \, \mathrm{res}[B_n^l(g^*)], \tag{18.30}
$$

where $\mathrm{res}[\ldots]$ indicates the resummed series.   Neglecting all but the first two terms of each β-function leads to the following simple equation:

$$
l_g^* \equiv \frac{\mathrm{res}[B_0^g(g^*)]}{\mathrm{res}[B_1^g(g^*)]} = \frac{\mathrm{res}[B_0^l(g^*)]}{\mathrm{res}[B_1^l(g^*)]} \equiv l_l^* . \tag{18.31}
$$

The sign of l^* depends trivially on the sign of $\mathrm{res}[B_n]$. Since the strong-coupling behavior of the β-function is unknown for the ϕ^4-theory in 4 dimensions, we have to use some trial value of α. This is chosen to acquire an optimal convergence of the series, as mentioned in Section 16.5. With $\varepsilon = 1$, the following four expansions have to be resummed:

$$
\begin{aligned}
B_0^g &= -g + 11/3g^2 - 23/3g^3 + 47.6514g^4 - 437.6456g^5 + 4998.6184g^6, & \beta &= \tfrac{9}{2}, \\
B_1^g &= -4/3g + 44/9g^2 - 45.8922g^3 + 564.7871g^4 - 8113.7392g^5, & \beta &= \tfrac{13}{2}, \\
B_1^l &= -1 + 4g - 97/9g^2 + 75.8751g^3 - 776.2604g^4 + 9707.3624g^5, & \beta &= \tfrac{13}{2}, \\
B_2^l &= -1 + 56/9g - 67.3187g^2 + 944.0496g^3 - 15030.8879g^4, & \beta &= \tfrac{17}{2}.
\end{aligned}
$$

Here, β is again the growth parameter of the large-order behavior. The parameter α turns out to be different for each term. It is chosen as follows:

$$
B_0^g : \alpha = 1.45, \quad B_1^g : \alpha = 0.35, \quad B_1^l : \alpha = 0.325, \quad B_2^l : \alpha = -1.4. \tag{18.32}
$$

The result is displayed in Fig. 18.3 where l_g and l_l, defined in Eq. (18.31), are plotted against g. The resulting cubic fixed point is

$$
g_C^* \approx 0.399 \qquad l_C^* \approx 0.01 . \tag{18.33}
$$

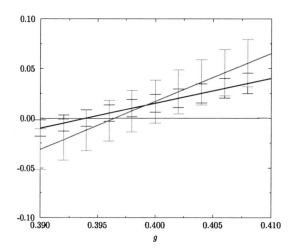

FIGURE 18.3 Determination of $l^* = g_2^*$ by resummation in $g = g_1 + g_2$. The quantities l_g and l_l defined in Eq. (18.31) are plotted against g. Their crossing point determines the values for l^* and g^*.

The errors resulting from variations in α are indicated in Fig. 18.3. They are too large to make a conclusive statement. A further improvement of the resummation of β-functions in two coupling constants is needed to find the correct answer. For more details see the original discussion in Ref. [14].

Notes and References

Textbooks discussing cubic anisotropy are
A. Aharony, in *Phase Transitions and Critical Phenomena*, Vol.6, ed. C. Domb and M.S. Green, Academic Press, New York, 1976.
P. Pfeuty and G. Toulouse, *Introduction to the Renormalization Group and to Critical Phenomena*, John Wiley & Sons, New York 1978.

A six-loop study in $D = 3$ dimensions was performed by
J.M. Carmona, A. Pelissetto, E. Vicari, Phys. Rev. B **61**, 15136 (2000) (cond-mat/9912115).
The results are in agreement with those onbtained in $4 - \varepsilon$ dimensions.
The most recent papers on this subject are
R. Folk, Y. Holovatch, and T. Yavorskii, Phys. Rev. B **61**, 15114 (2000) (cond-mat/9909121); ibid. **62**, 12195 (2000) (cond-mat/0003216).
K.B. Varnashev, Phys. Rev. B **61**, 14660 (2000) (cond-mat/990908); J. Phys. A **33**, 3121 (2000) (cond-mat/0005087).
See also
G. Parisi, F. Ricci-Tersenghi, and J.J. Ruiz-Lorenzo, Phys. Rev. E **60**, 5198 (1999) (cond-mat/9903095).

The individual citations in the text refer to:

[1] R.K. Zia, D.J. Wallace, J. Phys. A **7**, 1089 (1975).

[2] J.-C. Toledano, L. Michel, P. Toledano, E. Brézin, Phys. Rev. B **31**, 7171 (1985).

[3] A. Aharony, Phys. Rev. B **8**, 4270 (1973).

[4] I.J. Ketley, D.J. Wallace, J. Phys. A **6**, 1667 (1973).

[5] E. Brézin, J.C. Le Guillou, J. Zinn-Justin, Phys. Rev. B **10**, 893 (1974).

[6] I.O. Mayer, A.I. Sokolov, B.N. Shalayev, Ferroelectrics, **95**, 93 (1989).

[7] H. Kleinert and V. Schulte-Frohlinde, Phys. Lett. B **342**, 284 (1995).
The expansions can be downloaded from the internet address www.physik.fu-berlin.de/~kleinert/b8/programs/index.html#1 .
V. Schulte-Frohlinde, FU-Berlin Ph.D. Thesis, 1969.

[8] For files of the ε-expansion, see the internet address in Footnote 7.

[9] E. Domany, D. Mukamel, M.E. Fisher, Phys. Rev. B **15**, 5433 (1977);
J. Rudnick, Phys. Rev. B **18**, 1406 (1978).

[10] I.J. Ketley, D.J. Wallace, J. Phys. A **6**, 1667 (1973).

[11] K.E. Newman, E.K. Riedel, Phys. Rev. B **25**, 25 (1982).

[12] J.C. Le Guillou, J. Zinn-Justin, Phys. Rev. B **21**, 3976 (1980); J. Phys. Lett. **46**, 137 (1985).

[13] P.R. Graves-Morris, R. Hughes-Jones, G.J. Makinson, J. Inst. Maths. Applics., **13**, 311 (1974).

[14] H. Kleinert, S. Thoms, Phys. Rev. D **52**, 5926 (1995).
H. Kleinert, S. Thoms, and V. Schulte-Frohlinde, Phys. Rev. B **56**, 14428 (1997).

19

Variational Perturbation Theory

A powerful and yet mathematically simple method of resumming divergent perturbation expansions for critical exponents is provided by *variational perturbation theory*. This is a systematic extension of a variational approximation to path integrals, proposed some time ago by Feynman and Kleinert [1], to any desired order and accuracy [2]. The understanding of its convergence properties was greatly helped by pioneering mathematical work of Seznec and Zinn-Justin [3]. For an anharmonic oscillator, the expansions of energy eigenvalues converge uniformly and exponentially fast, like $e^{-\mathrm{const}\times L^{1/3}}$ in the order L of the approximation. The uniformity of the convergence makes it possible to derive convergent strong-coupling expansions from divergent weak-coupling expansions. It turns out that the speed of convergence of the approach is governed by the finite convergence radius of the strong-coupling expansion: the constant in the above exponential is directly related to the convergence radius [4].

Since convergent strong-coupling expansions can be obtained so easily from divergent weak-coupling expansions, it is straightforward to develop a simple algorithm for finding uniformly convergent optimal interpolations to functions for which both several weak-coupling and strong-coupling expansion coefficients are known [5].

19.1 From Weak- to Strong-Coupling Expansions

We shall apply this algorithm to the calculation of the renormalization constants of ϕ^4-theories for *all* coupling strengths, from which we can extract the limit of infinite bare coupling constant λ_B. As we shall see, this limit corresponds precisely to the fixed point g^* of the renormalized coupling g. The reason for this is that a perturbation expansion of the ϕ^4-theory in $4 - \varepsilon$ dimensions produces series of the form

$$f_L(g_B) = \sum_{n=0}^{L} f_n g_B^n, \tag{19.1}$$

where $g_B = \lambda_B/\mu^\varepsilon$ is the dimensionless bare coupling constant. In a massive theory, the mass parameter μ can be chosen to be equal to the renormalized mass m. The approach to the critical point $m \to 0$ corresponds then to the limit $g_B \to \infty$, which is the same as the strong-coupling limit in the bare coupling constant λ_B. By performing this limit, we shall be able to reproduce the critical exponents obtained in the last chapter from the renormalization group equation.

19.2 Strong-Coupling Theory

Let us first explain in general how a divergent weak-coupling expansion of the type (19.1) may be turned into a strong-coupling expansion

$$f^M(g_B) = g_B^{p/q} \sum_{m=0}^{M} b_m (g_B^{-2/q})^m, \tag{19.2}$$

which has a finite convergence radius g_s. The leading power p/q of g_B is the parameter s in the previous resummation procedure in Eqs. (16.54), (16.61), and in Subsection 16.6.1.

Examples treated in the literature [5] are the anharmonic oscillator with parameters $p = 1/3$, $q = 3$ for the energy eigenvalues; and the Fröhlich polaron with $p = 1$, $q = 1$ for the ground-state energy, and $p = 4$, $q = 1$ for the mass. For the mass of the polaron, the summation gives quite different results from Feynman's, calling for further studies of this system.

The first step is to rewrite the weak-coupling expansion with the help of an auxiliary scale parameter κ [6]:

$$f_L(g_B) = \kappa^p \sum_{n=0}^{L} f_n \left(\frac{g_B}{\kappa^q} \right)^n, \tag{19.3}$$

where κ is eventually set equal to 1. We shall see below that the quotient p/q parametrizes the *leading power behavior* in g_B of the strong-coupling expansion, whereas $2/q$ characterizes the *approach* to the leading power behavior.

In the second step, we replace κ by the identical expression

$$\kappa \to \sqrt{K^2 + \kappa^2 - K^2} \tag{19.4}$$

containing a dummy scaling parameter K. The series (19.3) is then re-expanded in powers of g_B up to the order L, thereby treating $\kappa^2 - K^2$ as a quantity of order g_B. The result is most conveniently expressed in terms of dimensionless parameters $\hat{g}_B \equiv g_B/K^q$ and $\sigma(K) \equiv (1 - \hat{\kappa}^2)/\hat{g}_B$, where $\hat{\kappa} \equiv \kappa/K$ [suppressing g_B and κ in the arguments of $\sigma(K)$]. Then the replacement (19.4) becomes

$$\kappa \longrightarrow K(1 - \sigma\hat{g}_B)^{1/2}, \tag{19.5}$$

so that the re-expanded series reads explicitly

$$W_L(g_B, K) = K^p \sum_{n=0}^{L} \varepsilon_n(\sigma(K))\hat{g}_B^n, \tag{19.6}$$

with the coefficients:

$$\varepsilon_n(\sigma) = \sum_{j=0}^{n} f_j \binom{(p - qj)/2}{n - j} (-\sigma)^{n-j}. \tag{19.7}$$

For any fixed g_B and κ, we form the first and second derivatives of $W_L(g_B, K)$ with respect to K, calculate the K-values of the extrema and the turning points, and select the smallest of these as the optimal scaling parameter K_L. The function $W_L(g_B) \equiv W_L(g_B, K_L)$ constitutes the Lth variational approximation $f_L(g_B)$ to the function $f(g_B)$.

It is easy to take this approximation to the strong-coupling limit $g_B \to \infty$. For this we observe that (19.6) depends on its variables as follows:

$$W_L(g_B, K) = K^p w_L(\hat{g}_B, \hat{\kappa}^2). \tag{19.8}$$

For dimensional reasons, the optimal K_L increases for large g_B like $K_L \approx g_B^{1/q} c_L$, so that $\hat{g}_B$ becomes asymptotically constant, say $\hat{g}_B \to c_L^{-q}$ and $\sigma \to 1/\hat{g}_B \to c_L^q$, implying that they remain finite in the strong-coupling limit. The dimensionless $\hat{\kappa}^2$ tends to zero like $1/[c_L(g_B/\kappa^q)^{1/q}]^2$. Thus $W_L(g_B, K_L)$ behaves for large g_B like

$$W_L(g_B, K_L) \approx g_B^{p/q} c_L^p w_L(c_L^{-q}, 0). \tag{19.9}$$

In this limiting form, c_L plays the role of the variational parameter to be determined by the optimal extremum or turning point of $c_L^p w_L(c_L^{-q}, 0)$.

The full strong-coupling expansion is obtained by expanding $w_L(\hat{g}_B, \hat{\kappa}^2)$ in powers of $\hat{\kappa}^2 = (g_B/\kappa^q \hat{g}_B)^{-2/q}$ at $\hat{g}_B = c_L^{-q}$. The result is

$$W_L(g_B) = g_B^{p/q} \left[\bar{b}_0(\hat{g}_B) + \bar{b}_1(\hat{g}_B) \left(\frac{g_B}{\kappa^q} \right)^{-2/q} + \bar{b}_2(\hat{g}_B) \left(\frac{g_B}{\kappa^q} \right)^{-4/q} + \ldots \right] \qquad (19.10)$$

with

$$\bar{b}_n(\hat{g}_B) = \frac{1}{n!} w_L^{(n)}(\hat{g}_B, 0) \hat{g}_B^{(2n-p)/q} , \qquad (19.11)$$

where $w_L^{(n)}(\hat{g}_B, \hat{\kappa}^2)$ is the nth derivative of $w_L(\hat{g}_B, \hat{\kappa}^2)$ with respect to $\hat{\kappa}^2$. Explicitly:

$$\frac{1}{n!} w_L^{(n)}(\hat{g}_B, 0) = \sum_{l=0}^{L} (-1)^{l+n} \sum_{j=0}^{l-n} f_j \binom{(p-qj)/2}{l-j} \binom{l-j}{n} (-\hat{g}_B)^j . \qquad (19.12)$$

The optimal expansion of the energy (19.10) is found by expanding

$$\hat{g}_B = c_L^{-q} \left[1 + \gamma_1 \left(\frac{g_B}{\kappa^q} \right)^{-2/q} + \gamma_2 \left(\frac{g_B}{\kappa^q} \right)^{-4/q} + \ldots \right] , \qquad (19.13)$$

and finding the optimal extremum (or turning point) in the resulting polynomials of $\gamma_1, \gamma_2, \ldots$. Setting the auxiliary scale parameter equal to unity, we obtain a strong-coupling expansion in powers of $g_B^{-2/q}$:

$$W_L(g_B) = g_B^{p/q} \left[b_0 + b_1 (g_B)^{-2/q} + b_2 (g_B)^{-4/q} + \ldots \right] , \qquad (19.14)$$

which is of the desired type (19.2). In practice, the coefficients b_n are determined successively as follows. First we optimize $\bar{b}_0(\hat{g}_B)$ at $\hat{g}_B = c_L^{-q}$ and obtain $b_0 = \bar{b}_0(c_L^{-q})$. At the same value of $\hat{g}_B$ we calculate $b_1 = \bar{b}_1(c_L^{-q})$ and the coefficients $\bar{b}_i(c_L^{-q})$ $(i = 2, 3, \ldots)$, and their derivatives $\bar{b}_i'(c_L^{-q})$, $\bar{b}_i''(c_L^{-q}) \ldots$. From these we determine the remaining optimized coefficients b_i in the strong-coupling expansion (19.14) by combining $\bar{b}_i(c_L^{-q})$, as specified in Table 19.1 .

This procedure will be applied to renormalization group functions in Sections 19.5 and 20.1. There we shall be dealing only with functions $f(g_B)$ which go to a constant f_L^* in the strong-coupling limit. Thus we have $p = 0$, and the critical exponents follow from the first term in the expansions (19.14). It is given by (19.11) for $n = 0$, which we shall write as

$$f_L^* = \underset{\hat{g}_B}{\mathrm{opt}} \left[\sum_{l=0}^{L} f_l \hat{g}_B^l \sum_{k=0}^{L-l} \binom{-ql/2}{k} (-1)^k \right] , \qquad (19.15)$$

where the expression in square brackets has to be optimized in the variational parameter $\hat{g}_B$. Mnemonically, a better way to express this formula is

$$f_L^* = \underset{\hat{g}_B}{\mathrm{opt}} \left[\sum_{l=0}^{L} f_l \hat{g}_B^l \, [1-1]_{L-l}^{-ql/2} \right] , \qquad (19.16)$$

TABLE 19.1   Combinations of functions $\bar{b}_n(c_L^{-q})$ determining coefficients b_n of strong-coupling expansion (19.14). The arguments c_L^{-q} of $\bar{b}_n, \bar{b}_n', \bar{b}_n''$ are suppressed.

m	b_m	$-\gamma_{m-1}$
2	$\bar{b}_2 + \gamma_1 \bar{b}_1' + \frac{1}{2} \gamma_1^2 \bar{b}_0''$	$\bar{b}_1'/\bar{b}_0''$
3	$\bar{b}_3 + \gamma_2 \bar{b}_1' + \gamma_1 \bar{b}_2' + \gamma_1 \gamma_2 \bar{b}_0'' + \frac{1}{2} \gamma_1^2 \bar{b}_1'' + \frac{1}{6} \gamma_1^3 \bar{b}_0^{(3)}$	$(\bar{b}_2' + \gamma_1 \bar{b}_1'' + \frac{1}{2} \gamma_1^2 \bar{b}_0^{(3)})/\bar{b}_0''$
4	$\bar{b}_4 + \gamma_3 \bar{b}_1' + \gamma_2 \bar{b}_2' + \gamma_1 \bar{b}_3' + (\frac{1}{2} \gamma_2^2 + \gamma_1 \gamma_3) \bar{b}_0''$ $+ \gamma_1 \gamma_2 \bar{b}_1'' + \frac{1}{2} \gamma_1^2 \bar{b}_2'' + \frac{1}{2} \gamma_1^2 \gamma_2 \bar{b}_0^{(3)} + \frac{1}{6} \gamma_1^3 \bar{b}_1^{(3)} + \frac{1}{24} \gamma_1^4 \bar{b}_0^{(4)}$	$(\bar{b}_3' + \gamma_2 \bar{b}_1'' + \gamma_1 \bar{b}_2'' + \gamma_1 \gamma_2 \bar{b}_0^{(3)}$ $+ \frac{1}{2} \gamma_1^2 \bar{b}_1^{(3)} + \frac{1}{6} \gamma_1^3 \bar{b}_0^{(4)})/\bar{b}_0''$

where the symbol $[1 - x]_{L-l}^{-ql/2}$ denotes the binomial expansion of $(1 - x)^{-ql/2}$ up to the power $L - l$ in the second argument x:

$$[1 - x]_{L-l}^{-ql/2} \equiv \sum_{k=0}^{L-l} \binom{-ql/2}{k} (-x)^k. \tag{19.17}$$

19.3 Convergence

The number of weak-coupling coefficients calculated so far for ϕ^4-theories in $4 - \varepsilon$ dimensions is limited to $L = 5$ (see Chapter 17). Thus it will be important to know the specific way in which the approximations $W_L(g_B)$ approach the final result $W_\infty(g_B)$. We shall derive this behavior in general [6, 7] for any strong-coupling parameters p and q. Let us remove the factor κ^p from the function $f(g_B)$, defining the reduced quantity $\tilde{f}(\tilde{g}_B) = f(g_B)/\kappa^p$ as a function of the reduced coupling constant $\tilde{g}_B \equiv g_B/\kappa^q$. We further assume the strong-coupling growth $g_B^{p/q}$ of the function $f(g_B)$ to be less than linear, so that $\tilde{f}(\tilde{g}_B)$ satisfies a once-subtracted dispersion relation (compare the discussion in Section 16.1):

$$\tilde{f}(\tilde{g}_B) = f_0 + \frac{\tilde{g}_B}{2\pi i} \int_0^{-\infty} \frac{d\tilde{g}_B'}{\tilde{g}_B'} \frac{\operatorname{disc} \tilde{f}(\tilde{g}_B')}{\tilde{g}_B' - \tilde{g}_B}, \tag{19.18}$$

where $\operatorname{disc} \tilde{f}(\tilde{g}_B)$ is the discontinuity across the left-hand cut in the complex g_B-plane:

$$\operatorname{disc} \tilde{f}(\tilde{g}_B) = \tilde{f}(\tilde{g}_B + i\eta) - \tilde{f}(\tilde{g}_B - i\eta) = 2i \operatorname{Im} \tilde{f}(\tilde{g}_B + i\eta) \quad \text{for } \tilde{g}_B < 0, \tag{19.19}$$

with η being an infinitesimal positive number. An expansion of the integrand in powers of $\tilde{g}_B$ up to $\tilde{g}_B^L$ reproduces of course the initial perturbation series (19.3), where the expansion coefficients are moment integrals over the discontinuity:

$$a_k = \frac{1}{2\pi i} \int_{-\infty}^0 \frac{d\tilde{g}_B}{\tilde{g}_B^{k+1}} \operatorname{disc} \tilde{f}(\tilde{g}_B). \tag{19.20}$$

The dispersion relation (19.18) can also be used to derive moment integrals for the re-expansion coefficient $\varepsilon_l(\sigma)$ in (19.6) and (19.7). For the dimensionless coupling constant $\tilde{g}_B$, the replacement (19.5) becomes

$$\tilde{g}_B \longrightarrow \tilde{G}_B(\hat{g}_B) \equiv \frac{\hat{g}_B}{(1 - \sigma\hat{g}_B)^{q/2}}. \tag{19.21}$$

Because of the prefactor κ^p in (19.3), the replacement (19.4) also produces a prefactor $K^p/\kappa^p = (1 - \sigma\hat{g}_B)^{p/2}$ to the function $f(g_B)$. For the reduced function $\hat{f}(\hat{g}_B) \equiv f(g_B)/K^p$, which depends only on the reduced coupling constant $\hat{g}_B$, we thus obtain a dispersion relation

$$\hat{f}(\hat{g}_B) = (1 - \sigma\hat{g}_B)^{p/2} \left[f_0 + \frac{\tilde{G}_B(\hat{g}_B)}{2\pi i} \int_{-\infty}^0 \frac{d\tilde{g}_B'}{\tilde{g}_B'} \frac{\operatorname{disc} \tilde{f}(\tilde{g}_B')}{\tilde{g}_B' - \tilde{G}_B(\hat{g}_B)} \right]. \tag{19.22}$$

This function satisfies a dispersion relation in the complex $\hat{g}_B$-plane. If C denotes the image of the left-hand cut in the original dispersion relation (19.18) as it arises from the mapping (19.21), and if $\operatorname{disc}_C f(\hat{g}_B)$ denotes the discontinuity across this cut, the dispersion relation reads

$$\hat{f}(\hat{g}_B) = f_0 + \frac{\hat{g}_B}{2\pi i} \int_C \frac{d\hat{g}_B'}{\hat{g}_B'} \frac{\operatorname{disc}_C \hat{f}(\hat{g}_B')}{\hat{g}_B' - \hat{g}_B}. \tag{19.23}$$

An expansion of the integrand in powers of $\hat{g}_B$ yields moment integrals for the desired re-expansion coefficients $\varepsilon_k(\sigma)$:

$$\varepsilon_k(\sigma) = \frac{1}{2\pi i} \int_C \frac{d\hat{g}_B}{\hat{g}_B^{k+1}} \mathrm{disc}_C \hat{f}(\hat{g}_B). \tag{19.24}$$

The discontinuity in these integrals can be derived from the dispersion relation (19.22). The function $\tilde{G}_B(\hat{g}_B)$ in (19.21) carries the left-hand cut in the complex $\tilde{g}_B$-plane over into several cuts in the $\hat{g}_B$-plane. In Fig. 19.1 we show the image cuts for $q = 3$, where the critical exponent

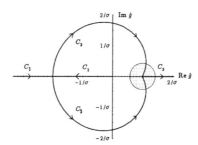

FIGURE 19.1 Image of left-hand cut in complex $\hat{g}_B$-plane for $q = 3$ -mapping (19.21).

of the approach to scaling is $\omega = 2/3$, as in the quantum-mechanical anharmonic oscillator. The cuts run along the contours C_1, $C_{\bar{1}}$, C_2, $C_{\bar{2}}$, C_3, the last being caused by the powers $q/2$ and $p/2$ of $1 - \sigma\hat{g}_B$ in the mapping (19.21) and the prefactor in (19.22), respectively. Let $\tilde{D}(\tilde{g}_B)$ abbreviate the reduced discontinuity (19.19):

$$\tilde{D}(\tilde{g}_B) \equiv \mathrm{disc}\, \tilde{f}(\tilde{g}_B) = 2i\mathrm{Im}\, \tilde{f}(\tilde{g}_B + i\eta), \quad \tilde{g}_B \leq 0. \tag{19.25}$$

Then the discontinuities across the various cuts are

$$\mathrm{disc}_{C_{1,\bar{1},2,\bar{2}}} \hat{f}(\hat{g}_B) = (1 - \sigma\hat{g}_B)^{p/2} \tilde{D}(\hat{g}_B(1 - \sigma\hat{g}_B)^{-q/2}), \tag{19.26}$$

$$\mathrm{disc}_{C_3} \hat{f}(\hat{g}_B) = -2i(\sigma\hat{g}_B - 1)^{p/2} \times \left[f_0 + \int_0^\infty \frac{d\tilde{g}_B'}{2\pi} \frac{\hat{g}_B(\sigma\hat{g}_B - 1)^{-q/2}}{\tilde{g}_B'^2 + \hat{g}_B^2(\sigma\hat{g}_B - 1)^{-q}} \tilde{D}(-\tilde{g}_B') \right]. \tag{19.27}$$

For small negative $\tilde{g}_B$, the discontinuity is given by a standard semiclassical approximation with the typical form

$$\tilde{D}(\tilde{g}_B) \approx i\, \mathrm{const} \times \tilde{g}_B^b\, e^{a/\tilde{g}_B}. \tag{19.28}$$

The exponential plays the role of a Boltzmann factor for the activation of a classical solution to the field equations, whereas the prefactor accounts for the entropy of the field fluctuations around this solution.

Let us denote by $\varepsilon_k(C_i)$ the contributions of the different cuts to the integral (19.24). After inserting (19.28) into (19.26), we obtain from the cut along C_1 the semiclassical approximation

$$\varepsilon_k(C_1) \approx \mathrm{const} \times \int_{C_1} \frac{d\hat{g}_B}{2\pi} \frac{1}{\hat{g}_B^{k+1}} (1 - \sigma\hat{g}_B)^{p/2 - bq/2} \hat{g}_B^b\, e^{a(1 - \sigma\hat{g}_B)^{q/2}/\hat{g}_B}. \tag{19.29}$$

For the kth term of the series $S_k \equiv \varepsilon_k \hat{g}_B^k$, this yields the large-$k$ estimate

$$S_k \propto \left[\int_{C_\gamma} \frac{d\gamma}{2\pi} e^{h_k(\gamma)} \right] \gamma^k, \tag{19.30}$$

where $\gamma \equiv \sigma \hat{g}_B$, and $h_k(\gamma)$ is the function

$$h_k(\gamma) \approx -k \log(-\gamma) + \frac{a\sigma}{\gamma}(1-\gamma)^{q/2} + (b-1)\log(-\gamma) + \ldots \quad (19.31)$$

For large k, the saddle point approximation yields via the extremum at $\gamma \xrightarrow[k\to\infty]{} \gamma_k = -a\sigma/k$:

$$h_k \xrightarrow[k\to\infty]{} k\log(k/ea\sigma) - aq\sigma/2 + (b-1)\log\frac{a\sigma}{k} + \ldots \quad (19.32)$$

The constant $-aq\sigma/2$ in this limiting expression arises when expanding the second term of Eq. (19.31) into a Taylor series, $(a\sigma/\gamma)(1-\gamma)^{q/2} = a\sigma/\gamma_k - aq\sigma/2 + \ldots$. Only the first two terms in (19.32) contribute to the large-k limit. Thus, to leading order in k, the kth term of the re-expanded series becomes

$$S_k \propto e^{-aq\sigma/2}\left(\frac{-k}{e}\right)^k \hat{g}^k. \quad (19.33)$$

The corresponding re-expansion coefficients

$$\varepsilon_k \propto e^{-aq\sigma/2} f_k \quad (19.34)$$

have the remarkable property of growing in precisely the same manner with k as the initial expansion coefficients h_k, except for an overall suppression factor $e^{-aq\sigma/2}$.

To estimate the convergence of the variational perturbation expansion (19.6), we note that $\sigma \hat{g} = 1 - \hat{\kappa}^2$ is smaller than unity for large K, so that the powers $(\sigma \hat{g})^k$ by themselves would yield a convergent series. An optimal re-expansion of the reduced function $\tilde{f}(\hat{g}_B)$ can be achieved by choosing, for a given large maximal order L of the expansion, a parameter σ proportional to L:

$$\sigma \approx \sigma_L \equiv cL. \quad (19.35)$$

Inserting this into (19.31), we obtain for large $k = L$

$$h_L(\gamma) \approx L\left[-\log(-\gamma) + \frac{ac}{\gamma}(1-\gamma)^{q/2}\right]. \quad (19.36)$$

The extremum of this function lies at a γ-value satisfying the equation

$$1 + \frac{ac}{\gamma}(1-\gamma)^{q/2-1}\left[1 + \left(\frac{q}{2}-1\right)\gamma\right] = 0. \quad (19.37)$$

The constant c may be chosen in such a way that the large exponent proportional to L in the exponential function $e^{h_L(\gamma)}$ arising from the first term in (19.36) is canceled by an equally large contribution from the second term, i.e., we require at the extremum

$$h_L(\gamma) = 0. \quad (19.38)$$

The two equations (19.37) and (19.38) are solved by certain constant values of $\gamma < 0$ and c. In contrast to the extremal γ of Eq. (19.31) which dominates the large-k limit, the extremal γ in the present limit, in which k is also large but of the same size as L, remains finite (the previous estimate held for $k \gg L$). Accordingly, the second term $(ac/\gamma)(1-\gamma)^{q/2}$ in $h_L(\gamma)$ contributes in full, not merely via the first two Taylor expansion terms of $(1-\gamma)^{q/2}$ as it did in (19.32).

Since $h_L(\gamma)$ vanishes at the extremum, the Lth term in the re-expansion has the order of magnitude

$$S_L(C_1) \propto (\sigma_L \hat{g}_B)^L = \left(1 - \frac{1}{K_L^2}\right)^L. \tag{19.39}$$

According to (19.35), the scale parameter K_L grows for large L like

$$K_L \sim \sigma_L^{1/q} g_B^{1/q} \sim (cLg_B)^{1/q}. \tag{19.40}$$

As a consequence, the last term of the series decreases for large L like

$$S_L(C_1) \propto \left[1 - \frac{1}{(\sigma_L g_B)^{2/q}}\right]^L \approx e^{-L/(\sigma_N g_B)^{2/q}} \approx e^{-L^{1-2/q}/(cg_B)^{2/q}}. \tag{19.41}$$

This estimate does not yet explain the exponentially fast convergence of the variational perturbation expansion in the strong-coupling limit [8]. For the contribution of the cut C_1 to S_L, the derivation of such a behavior requires including the approach of σ_L to the large-L behavior (19.35), which has the same general form as the strong-coupling expansion (19.10),

$$\sigma_L \sim cL\left(1 + \frac{c'}{L^{2/q}} + \ldots\right), \tag{19.42}$$

as it turns out with positive c'. By inserting this σ_L into $h_L(\gamma)$ of (19.36), we find an extra exponential factor which dominates the large-L behavior at infinite coupling $\hat{g}_B$:

$$e^{\Delta h_L} \approx \exp\left[-L\log(-\gamma)\frac{c'}{L^{2/q}}\right] \approx e^{-c''L^{1-2/q}}. \tag{19.43}$$

What about the contributions of the other cuts? For $C_{\bar{1}}$, the integral in (19.24) runs from $\hat{g} = -2/\sigma$ to $-\infty$ and decrease like $(-2/\sigma)^{-k}$. The associated last term $S_L(C_{\bar{1}})$ is of the negligible order $e^{-L\log L}$. For the cuts $C_{2,\bar{2},3}$, the integral (19.24) starts at $\hat{g} = 1/\sigma$ and has therefore the leading behavior

$$\varepsilon_k(C_{2,\bar{2},3}) \sim \sigma^k, \tag{19.44}$$

yielding at first a contribution to the Lth term in the re-expansion of the order of

$$S_L(C_{2,\bar{2},3}) \sim (\sigma\hat{g})^L, \tag{19.45}$$

which decreases merely like (19.41) and does not explain the empirically observed convergence in the strong-coupling limit. The important additional information [9] is that the cuts in Fig. 19.1 do not really reach the point $\sigma\hat{g} = 1$. There exists a small circle of radius $\Delta\hat{g} > 0$ in which $\hat{f}(\hat{g})$ has no singularities at all. This is a consequence of the fact, unused up to this point, that the strong-coupling expansion (19.10) converges for $g_B > g_s$. For the reduced function $\hat{f}(\hat{g}_B)$, this expansion reads

$$\hat{f}(\hat{g}) = (\hat{g}_B)^{p/q}\left\{b_0 + b_1\left[\frac{\hat{g}_B}{(1 - \sigma\hat{g}_B)^{q/2}}\right]^{-2/q} + b_2\left[\frac{\hat{g}_B}{(1 - \sigma\hat{g}_B)^{q/2}}\right]^{-4/q} + \ldots\right\}. \tag{19.46}$$

The convergence of (19.10) for $g_B > g_s$ implies that (19.46) converges for all $\sigma\hat{g}_B$ in a neighborhood of the point $\sigma\hat{g}_B = 1$ with a radius

$$\Delta(\sigma\hat{g}_B) = \left|\frac{\hat{g}_B}{\tilde{g}_s}\right|^{2/q}, \tag{19.47}$$

where $\tilde{g}_s \equiv g_s/\kappa^q$. For large L, the denominator K^q in $\hat{g}_B$ on the right-hand side makes $\Delta(\sigma\hat{g}_B)$ go to zero like

$$\Delta(\sigma\hat{g}_B) \approx \frac{1}{(L|\tilde{g}_s|c)^{2/q}}. \tag{19.48}$$

Thus the integration contours of the moment integrals (19.24) for the contributions $\varepsilon_k(C_i)$ of the other cuts do not begin at the point $\sigma\hat{g} = 1$, but a little distance $\Delta(\sigma\hat{g})$ away from it. If $q < 4$, i.e., if $\omega > 1/2$, the intersection points of the small circle with the cuts C_2 and $C_{\bar{2}}$ have a real part larger than unity. This produces a suppression factor to the previous result (19.44) of the integral (19.24):

$$(\sigma\hat{g}_B)^{-L} \sim [1 + \Delta(\sigma\hat{g}_B)]^{-L}, \tag{19.49}$$

bringing the last term of the series S_L to

$$S_L(C_{2,\bar{2},3}) \sim (\sigma\hat{g}_B)^L \frac{1}{[1 + \Delta(\sigma\hat{g}_B)]^L} \tag{19.50}$$

instead of (19.45). Inserting (19.48), we find that this goes to zero with the same characteristic behavior (19.43) as the contribution from the cut C_1:

$$S_L(C_{2,\bar{2},3}) \approx e^{-c'''L^{1-\omega}}, \qquad c''' > 0. \tag{19.51}$$

Such a behavior therefore characterizes the convergence for $L \to \infty$, which will be needed in Sections 19.5 and 20.1 to extrapolate finite-L results to $L \to \infty$.

19.4 Strong-Coupling Limit and Critical Exponents

In the examples treated in the original Ref. [6], the strong-coupling parameters p and q were known. In the present field system, only p is known to be zero if we assume the system to have experimentally observed scaling properties. Then there exists an infrared-stable fixed point g^*, and the strong-coupling expansion for the expansion of g in powers of g_B will have a large-g_B behavior of the form [10]

$$g(g_B) = g^* - \frac{\text{const}}{g_B^{\omega/\varepsilon}} + \cdots. \tag{19.52}$$

The assumption of a strong-coupling scaling of this form is equivalent to the assumption that the β-function has a zero g^*. Rewriting the β-function as $-\varepsilon dg/d\log g_B$ and integration with Eq. (10.103) results in (19.52).

In the general framework of Section 19.2, we have to resum the series

$$g_L = \sum_{n=1}^{L} f_n g_B^n \tag{19.53}$$

with the strong-coupling parameters $p = 0$ and $q = 2\varepsilon/\omega$ [compare expansion (19.10)].

Both p and q can be determined to any desired accuracy from the power series which we want to resum. Indeed, the ratios p/q and $2/q$ are obtainable from the strong-coupling limits of the following infinite set of logarithmic derivatives of $W_L(g_B)$:

$$\frac{p}{q} = F_1(\infty), \quad F_1(g_B) \equiv \frac{d\log W_L(g_B)}{d\log g_B} = g_B \frac{W_L'(g_B)}{W_L(g_B)}, \tag{19.54}$$

$$-\frac{2}{q} - 1 = F_2(\infty), \quad F_2(g_B) \equiv \frac{d\log F_1'(g_B)}{d\log g_B} = g_B \frac{F_1''(g_B)}{F_1'(g_B)}. \tag{19.55}$$

Under the assumption that the strong-coupling expansion contains only powers of $g_B^{-\omega}$, there is an infinite set of further equations

$$-\frac{2}{q} - 1 = F_3(\infty), \qquad F_3(g_B) \equiv \frac{d \log F_2'(g_B)}{d \log g_B} = g_B \frac{F_2''(g_B)}{F_2'(g_B)}, \tag{19.56}$$

$$-\frac{2}{q} - 1 = F_4(\infty), \qquad F_4(g_B) \equiv \frac{d \log F_3'(g_B)}{d \log g_B} = g_B \frac{F_3''(g_B)}{F_3'(g_B)}, \tag{19.57}$$

$$\vdots$$

For the anharmonic oscillator, these equations are all satisfied. In the field theoretic perturbation expansions the possible presence of *confluent singularities* [11] containing powers of $g_B^{-\omega'}$, $g_B^{-\omega''}$ with exponents $\omega'' > \omega' > \omega$ makes the higher equations unreliable.

If the parameter p happens to be zero, which is the case in the application to be considered, there is a further formula for the parameter q:

$$-\frac{2}{q} - 1 = G_1(\infty), \qquad G_1(g_B) \equiv \frac{d \log W_L'(g_B)}{d \log g_B} = g_B \frac{W_L''(g_B)}{W_L'(g_B)}. \tag{19.58}$$

and in the absence of confluent singularities a further sequence of equations

$$-\frac{2}{q} - 1 = G_2(\infty), \qquad G_2(g_B) \equiv \frac{d \log G_1'(g_B)}{d \log g_B} = g_B \frac{G_1''(g_B)}{G_1'(g_B)}, \tag{19.59}$$

$$-\frac{2}{q} - 1 = G_3(\infty), \qquad G_3(g_B) \equiv \frac{d \log G_2'(g_B)}{d \log g_B} = g_B \frac{G_2''(g_B)}{G_2'(g_B)}, \tag{19.60}$$

$$\vdots$$

Formulas (19.54) and (19.58) will be crucial to the development in Sections 19.5 and 20.1. They enable us to calculate the critical exponent ω of the approach to scalingfrom the power series expansion (19.53).

When applied to the expansion of $g(g_B)$, the logarithmic derivative (19.54) can be written as

$$\frac{p}{q} = s = g_B \frac{g'(g_B)}{g(g_B)}. \tag{19.61}$$

The function on the right-hand side is, up to a factor $-\varepsilon g(g_B)$, equal to the β-function (10.30) of the renormalization group analysis:

$$\beta(g_B) = -\varepsilon g_B g'(g_B). \tag{19.62}$$

This follows directly from the defining equation (10.30). Thus Eq. (19.61) can be rewritten as

$$\frac{p}{q} = s = -\frac{\beta(g_B)}{\varepsilon g(g_B)}, \tag{19.63}$$

and the property $s = 0$ is equivalent to the vanishing of the β-function in the ordinary renormalization group analysis.

The critical exponent ω will be calculated by requiring the resummed power series expansion for (19.63) to vanish in the strong-coupling limit, thus ensuring a constant limit $g \to g^*$:

$$0 = g_B \frac{g'(g_B)}{g(g_B)} \bigg|_{g_B=\infty}. \tag{19.64}$$

Alternatively we may also use Eq. (19.58) to determine

$$\frac{\omega}{\varepsilon} = -1 - g_B \frac{g''(g_B)}{g'(g_B)}\bigg|_{g_B=\infty}. \tag{19.65}$$

It is easy to verify that this equation is completely equivalent to Eq. (10.104), where ω is defined as the slope of the β-function at $g = g^*$. Indeed, differentiating the Eq. (19.62) for the β-function with respect to g, the chain rule leads directly to (19.65). Equation (19.64) will be used to determine $q = 2\varepsilon/\omega$ directly to make the strong-coupling limit vanish. For comparison, the other equation (19.65) will sometimes be solved as well. This is done recursively, determining the left-hand side using some initial trial ω in the strong-coupling limit of the right-hand side, and repeating the procedure with the new ω until it converges. This procedure will be referred to as the self-consistent determination of ω.

The same exponent ω can be found from the series expansion of any function $f(g)$ of the renormalized coupling constant g in powers of the bare coupling constant g_B, which all behave for large g_B, due to (19.52), like

$$f(g) = f(g^*) + f'(g^*) \times \frac{\text{const}}{g_B^{\omega/\varepsilon}} + \dots . \tag{19.66}$$

From the definitions (10.20), (10.22), (10.23), (10.94), and (10.95) of the renormalization group functions, we deduce the alternative derivative equations for the critical exponents which are useful for the upcoming strong-coupling calculations:

$$\eta_m(g_B) = -\varepsilon \frac{d}{d\log g_B} \log \frac{m^2}{m_B^2}, \qquad \eta(g_B) = \varepsilon \frac{d}{d\log g_B} \log \frac{\phi^2}{\phi_B^2}. \tag{19.67}$$

If we set $\mu = m$, we find that $g_B = \lambda_B \mu^{-\varepsilon}$ goes to infinity like $\text{const} \times m^{-\varepsilon}$ for $m \to 0$. Thus the strong-coupling limit is approached for $m \to 0$, implying the power behavior

$$\frac{m^2}{m_B^2} \propto g_B^{-\eta_m/\varepsilon} \propto m^{\eta_m}, \qquad \frac{\phi^2}{\phi_B^2} \propto g_B^{\eta/\varepsilon} \propto m^{-\eta}, \tag{19.68}$$

where η_m and η are the $g\beta \to \infty$-limits of $\eta_m(g_B)$ and $\eta(g_B)$, the first determining the critical exponent ν via $\nu = 1/(2 - \eta_m)$.

When approaching a second-order phase transition, the bare square mass m_B^2 vanishes like $\tau \equiv T/T_c - 1$. The renormalized mass m^2 will vanish with a different power of τ. This power is obtained from the first equation in (19.68), which shows that $m \propto \tau^\nu$. Experiments observe that the coherence length of fluctuations $\xi = 1/m$ increases near T_c like $\tau^{-\nu}$. Similarly we see from the second equation in (19.68) that the scaling dimension $D/2-1$ of the bare field ϕ_B for $T \to T_c$ is changed, in the strong-coupling limit $g_B \to \infty$, to $D/2 - 1 + \eta/2$, the number η being the so-called anomalous dimension of the field. This implies a change in the large-distance behavior of the correlation functions $\langle\phi(\mathbf{x})\phi(\mathbf{0})\rangle$ at T_c from the free-field behavior r^{-D+2} to $r^{-D+2-\eta}$. The magnetic susceptibility is determined by the integrated correlation function $\langle\phi_B(\mathbf{x})\phi_B(\mathbf{0})\rangle$. At zero coupling constant g_B, this is proportional to $1/m_B^2 \propto \tau^{-1}$, which is changed by fluctuations to $m^{-2}\phi_B^2/\phi^2$. This has a temperature behavior $m^{-(2-\eta)} = \tau^{-\nu(2-\eta)} \equiv \tau^{-\gamma}$, which defines the critical exponent $\gamma = \nu(2 - \eta)$ observable in magnetic experiments.

19.5   Explicit Low-Order Calculations

If the expansions are known to orders $L = 2, 3,$ or 4, we can give analytic expressions for the strong-coupling limits (19.16).

19.5.1   General Formulas

Setting $\rho \equiv 1 + q/2 = 1 + \varepsilon/\omega$, we find for $L = 2$

$$f_2^* = \underset{\hat{g}_B}{\mathrm{opt}} \left[f_0 + f_1 \rho \hat{g}_B + f_2 \hat{g}_B^2 \right] = f_0 - \frac{1}{4} \frac{f_1^2}{f_2} \rho^2. \qquad (19.69)$$

For $L = 3$, we obtain from the extrema

$$\begin{aligned}
f_3^* &= \underset{\hat{g}_B}{\mathrm{opt}} \left[f_0 + \tfrac{1}{2} f_1 \rho(\rho + 1) \hat{g}_B + f_2(2\rho - 1)\hat{g}_B^2 + f_3 \hat{g}_B^3 \right] \\
&= f_0 - \frac{1}{3} \frac{\bar{f}_1 \bar{f}_2}{f_3} \left(1 - \frac{2}{3} r \right) + \frac{2}{27} \frac{\bar{f}_2^3}{f_3^2} (1 - r),
\end{aligned} \qquad (19.70)$$

where $r \equiv \sqrt{1 - 3\bar{f}_1 f_3 / \bar{f}_2^2}$ and $\bar{f}_1 \equiv \tfrac{1}{2} f_1 \rho(\rho + 1)$ and $\bar{f}_2 \equiv f_2(2\rho - 1)$. The positive square root must be taken to connect g_3^* smoothly to g_2^* in the limit of a vanishing coefficient of g_B^3. If the square root is imaginary, the optimum is given by the unique turning point, leading once more to (19.70) but with $r = 0$:

$$f_3^* = f_0 - \frac{1}{3} \frac{\bar{f}_1 \bar{f}_2}{f_3} + \frac{2}{27} \frac{\bar{f}_2^3}{f_3^2}. \qquad (19.71)$$

The parameter $\rho = 1 + \varepsilon/\omega$ can be determined from the expansion coefficients of a function $F(g_B)$ as follows. Assuming $F(g_B)$ to be constant F^* in the strong-coupling limit, the logarithmic derivative $f(g_B) \equiv g_B F'(g_B)/F(g_B)$ must vanish at $g_B = \infty$ [see (19.64)].

If $F(g_B)$ starts out as $F_0 + F_1 g_B + \ldots$ or $F_1 g_B + F_2 g_B^2 + \ldots$, the logarithmic derivative is

$$f(g_B) = F_1' g_B + (2F_2' - F_1'^2)g_B^2 + (F_1'^3 - 3F_1' F_2' + 3F_3')g_B^3 + \ldots, \qquad (19.72)$$

where $F_i' = F_i/F_0$, or

$$f(g_B) = 1 + \hat{F}_2 g_B + (2\hat{F}_3 - \hat{F}_2^2)g_B^2 + (\hat{F}_2^3 - 3\hat{F}_2 \hat{F}_3 + 3\hat{F}_4)g_B^3 + \ldots, \qquad (19.73)$$

where $\hat{F}_i = F_i/F_1$. The expansion coefficients on the right-hand sides are then inserted into (19.69) or (19.70), and the left-hand sides have to vanish to ensure that $F(g_B) \to F^*$.

If the approach $F(g_B) \to F^*$ is of the type (19.66), the function

$$h(g_B) \equiv g_B \frac{F''(g_B)}{F'(g_B)} = 2\hat{F}_2 g_B + (-4\hat{F}_2^2 + 6\hat{F}_3)g_B^2 + (8\hat{F}_2^3 - 18\hat{F}_2 \hat{F}_3 + 12\hat{F}_4)g_B^3 + \ldots (19.74)$$

must have the strong-coupling limit [recall (19.65)]

$$h(g_B) \to h^* = -\frac{\omega}{\varepsilon} - 1. \qquad (19.75)$$

This is true for both expansions $F_0 + F_1 g_B + \ldots$ and $F_1 g_B + F_2 g_B^2 + \ldots$.

19.5.2   Perturbation Series

The above formulas are now applied to the power series expansions of $O(N)$-symmetric ϕ^4-theories in $D = 4 - \varepsilon$ dimensions. Their renormalized coupling constant is related to the unrenormalized ones by Eq. (15.18), which we shall use at first only up to order $L = 3$:

$$\bar{g} = \bar{g}_B - \frac{(N + 8)}{3\varepsilon} \bar{g}_B^2 + \left[\frac{(N + 8)^2}{9\varepsilon^2} + \frac{(3N + 14)}{6\varepsilon} \right] \bar{g}_B^3 + \ldots. \qquad (19.76)$$

Inserting this into the expansions (15.15) and (15.11), we obtain for the mass and field ratios

$$\frac{m^2}{m_B^2} = 1 - \frac{N+2}{3}\frac{\bar{g}_B}{\varepsilon} + \frac{(N+2)}{9}\left[\frac{N+5}{\varepsilon^2} + \frac{5}{4\varepsilon}\right]\bar{g}_B^2 + \dots \, , \tag{19.77}$$

$$\frac{\phi^2}{\phi_B^2} = 1 + \frac{N+2}{36}\frac{\bar{g}_B^2}{\varepsilon} + \dots \, . \tag{19.78}$$

Remember that the dimensionless bare coupling constant $\bar{g}_B$ contains the arbitrary mass scale μ given by

$$\bar{g}_B = \frac{g_B}{(4\pi)^2} = \frac{\lambda_B}{\mu^\varepsilon(4\pi)^2}.$$

The arbitrary mass scale μ will now be set equal to the physical mass m, as discussed at the beginning of this chapter and before Eq. (19.68), considering all quantities as functions of $\bar{g}_B$. In order to describe second-order phase transitions, we let m_B^2 go to zero proportionally to $\tau = T/T_c - 1$ as the temperature T approaches the critical temperature T_c. Then m^2 will also go to zero, and thus $\bar{g}_B$ to infinity.

We use the vanishing of (19.73) at infinite bare coupling $\bar{g}_B$, or Eq. (19.75) with (19.74), to determine the critical exponent of approach to scaling ω. For the other critical exponents we find from formulas (19.67), (19.77), and (19.78) the expansions

$$\eta_m(\bar{g}_B) = \frac{N+2}{3}\bar{g}_B - \frac{N+2}{18}\left(5 + 2\frac{N+8}{\varepsilon}\right)\bar{g}_B^2, \tag{19.79}$$

$$\eta(\bar{g}_B) = \frac{N+2}{18}\bar{g}_B^2. \tag{19.80}$$

The scaling relation $\gamma = \nu(2-\eta)$ with $\nu = 1/(2-\eta_m)$ and the expansions (19.79), (19.80) yield for the critical exponent $\gamma(g_B)$ of the susceptibility defined in Eq. (10.160) the perturbation expansion up to second order in g_B:

$$\gamma(\bar{g}_B) = 1 + \frac{N+2}{6}\bar{g}_B + \frac{N+2}{36}\left(N - 4 - 2\frac{N+8}{\varepsilon}\right)\bar{g}_B^2. \tag{19.81}$$

We shall evaluate this rather than the series (19.80) for $\eta(\bar{g}_B)$, since $\gamma(\bar{g}_B)$ has more than one expansion coefficient at the two-loop level, which is necessary for the existence of an optimum in variational perturbation theory. At the five-loop level, the combination $\eta + \eta_m$ will be used since it seems to converge fastest.

19.5.3   Critical Exponent ω

We begin by calculating the critical exponent ω from the requirement that $g(g_B)$ has a constant strong-coupling limit, implying the vanishing of (19.73) for $g_B \rightarrow \infty$. From the expansion (19.76) we obtain a logarithmic derivative (19.73) up to the term g_B^2, so that Eq. (19.69) can be used to find the scaling condition

$$0 = 1 - \frac{1}{4}\frac{\hat{F}_2^2}{2\hat{F}_3 - \hat{F}_2^2}\rho^2. \tag{19.82}$$

This gives

$$\rho = \sqrt{8\hat{F}_3/\hat{F}_2^2 - 4}. \tag{19.83}$$

Since ω must be greater than zero, only the positive square root is physical. With the explicit coefficients F_1, F_2, F_3 of expansion (19.76), this becomes

$$\rho = 2\sqrt{1 + 3\frac{3N + 14}{(N + 8)^2}\varepsilon}. \tag{19.84}$$

The associated critical exponent $\omega = \varepsilon/(\rho - 1)$ is plotted in Fig. 19.2. It has the ε-expansion

$$\omega = \varepsilon - 3\frac{3N + 14}{(N + 8)^2}\varepsilon^2 + \cdots , \tag{19.85}$$

which is also shown in Fig. 19.2, and agrees with the first two terms obtained from renormalization group calculations in Eq. 17.15.

From Eqs. (19.75), (19.74), and (19.69) we obtain for the critical exponent ω a further equation

$$-\frac{\omega}{\varepsilon} - 1 = -\frac{\rho}{\rho - 1} = -\frac{1}{2}\frac{\hat{F}_2^2\,\rho^2}{3\hat{F}_3 - 2\hat{F}_2^2}, \tag{19.86}$$

which is solved by

$$\rho = \frac{1}{2} + \sqrt{\frac{6\hat{F}_3}{\hat{F}_2^2} - \frac{15}{4}}, \tag{19.87}$$

with the positive sign of the square root ensuring a positive ω. Inserting the coefficients of (19.76), this becomes

$$\rho = \frac{1}{2} + \frac{3}{2}\sqrt{1 + 4\frac{3N + 14}{(N + 8)^2}\varepsilon}. \tag{19.88}$$

The associated critical exponent $\omega = \varepsilon/(\rho - 1)$ has the same ε-expansion (19.85) as the previous approximation (19.84). The full approximation based on (19.88) is indistinguishable from the earlier one in the plot of Fig. 19.2.

Having determined ω, we can now calculate $\bar{g}^*$. Identifying the first three coefficients in the expansion (19.76) with f_0, f_1, f_2 in (19.69) we obtain

$$\bar{g}_2^* = f_0 - \frac{1}{4}\frac{f_1^2}{f_2}\rho^2. \tag{19.89}$$

With (19.84), this reads explicitly

$$\bar{g}_2^* = \frac{3}{N + 8}\varepsilon + 9\frac{3N + 14}{(N + 8)^3}\varepsilon^2, \tag{19.90}$$

which is precisely the well-known ε-expansion of $\bar{g}^*$ in renormalization group calculations up to the second order. Including the next coefficient, we can use formula (19.70) to calculate the next approximation $\bar{g}_3^*$. At $\varepsilon = 1$, the square root turns out to be imaginary, so that it has to be omitted (corresponding to the turning point as optimum). The resulting curve lies slightly ($\approx 8\%$) above the curve (19.90), i.e., represents a worse approximation than (19.90). Indeed, the ε^3-term in $\bar{g}_3^*$ is $81(3N + 14)^2/8(N + 8)^5$ and disagrees in sign with the exact term $\varepsilon^3[3(-33n^3 + 110n^2 + 1760n + 4544)/8 - 36\zeta(3)(N + 8)(5N + 22)]/(N + 8)^5$, which we would find by calculating ρ from an expansion (19.76) with one more power in g_B.

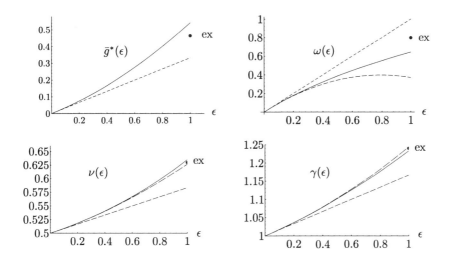

FIGURE 19.2 For Ising universality class ($N = 1$), the first figure shows the renormalized coupling at infinite bare coupling as a function of $\varepsilon = 4 - D$. It is calculated by variational perturbation theory from the first two perturbative expansion terms. The curve coincides with the ε-expansion up to order ε^2. Dashed curves indicate linear and quadratic ε-expansions. The other figures show similarly the critical exponents ω, ν, and γ. The dots mark currently accepted values of $\bar{g}^* \approx 0.466 \pm 0.003$, $\omega \approx 0.802 \pm 0.003$, $\nu = 0.630 \pm 0.002$, and $\gamma = 1.241 \pm 0.004$ obtained from the six-loop calculations in Section 20.2.

19.5.4 Critical Exponent ν

We now turn to the critical exponent ν. Taking the expansion (19.79) to infinite couplings $\bar{g}_B$, we obtain from formula (19.69) the limiting value

$$\eta_m = \frac{\varepsilon}{4} \frac{N+2}{N+8+5\varepsilon/2} \rho^2. \tag{19.91}$$

The corresponding exponent ν is plotted in Fig. 19.2. With the approximation (19.84) for ρ we find for ν the ε-expansion

$$\nu = \frac{1}{2} + \frac{1}{4} \frac{N+2}{N+8} \varepsilon + \frac{(N+2)(N+3)(N+20)}{8(N+8)^3} \varepsilon^2 + \dots , \tag{19.92}$$

which is also shown in Fig. 19.2 and agrees with renormalization group results to this order.

19.5.5 Critical Exponent γ

As a third independent critical exponent we calculate the critical exponent of the susceptibility $\gamma = \nu(2 - \eta) = (2 - \eta)/(2 - \eta_m)$ by inserting the coefficients of the expansion (19.81) into formula (19.69), which yields

$$\gamma = 1 + \frac{\varepsilon}{8} \frac{N+2}{N+8-(N-4)\varepsilon/2} \rho^2, \tag{19.93}$$

plotted in Fig. 19.2. This has an ε-expansion

$$\gamma = 1 + \frac{1}{2} \frac{N+2}{N+8} \varepsilon + \frac{1}{4} \frac{(N+2)(N^2+22N+52)}{(N+8)^3} \varepsilon^2 + \dots , \tag{19.94}$$

shown again in Fig. 19.2 and agreeing with renormalization group results to this order. The full approximation is plotted in Fig. 19.2. The critical exponent $\eta = 2 - \gamma/\nu$ has the ε-expansion $\eta = (N+2)\varepsilon^2/2(N+8)^2 + \ldots$.

Thus we see that variational strong-coupling theory can easily be applied to ϕ^4-theories in $D = 4 - \varepsilon$ dimensions and yields resummed expressions for the ε-dependence of all critical exponents. Their ε-expansions agree with those obtained from renormalization group calculations in Chapter 17.

19.6   Three-Loop Resummation

The three-loop calculations are algebraically more involved [12]. Fortunately, the optima of the variational expressions turn out to be determined by turning points, i.e., by the vanishing of the second derivative. At the three-loop order, this implies that the parameter r in (19.70) is zero, thus leading to the three-loop strong-coupling limit (19.71). It is this feature which renders the calculation analytically manageable, involving only a cubic equation for the determination of ρ (for $r \neq 0$, we would have to solve an eight-order equation). In order to obtain ω to three loops, we apply (19.71) to the logarithmic derivative of the three-loop part of (15.18), and find for ρ the following trigonometric representation of the relevant cubic root:

$$\rho = -\frac{1}{6} + \frac{256}{3}a_0 \left[106 + N(N+25)\right]^2 - a_0 b_0 \cos\left(\frac{-2\pi + \theta}{3}\right). \tag{19.95}$$

The angle θ and the constants $a_0\, b_0$ are given by

$$
\begin{aligned}
\theta &= \arccos\Bigg(\frac{\left[13776 + 4738N + N^2(8N+405) + 96(5N+22)\zeta(3)\right]^2}{2\left[106 + N(N+28)\right]\left\{(N+8)\left[13776 + 4738N + N^2(8N+405) + 96(5N+22)\zeta(3)\right]\right\}^{3/2}} \\
&\quad \times \frac{1}{\left[2209664 + 1040160N + 162982N^2 + 9683N^3 + 184N^4 + 672(N+8)(5N+22)\zeta(3)\right]^{3/2}} \\
&\quad \times \Big\{67181166592 + 64001040384N + 25893312000N^2 + 5641828480N^3 + 713027988N^4 + 54733044N^5 \\
&\quad + 2760157N^6 + 88332N^7 + 1440N^8 \\
&\quad - 192(N+8)(5N+22)\left[4084864 + 1952480N + 323706N^2 + 20021N^3 + 514N^4\right]\zeta(3) \\
&\quad + 746496\left[(N+8)(5N+22)\zeta(3)\right]^2\Big\}\Bigg),
\end{aligned}
\tag{19.96}
$$

$$a_0 = \left[446336 + 213280N + 35334N^2 + 2179N^3 + 56N^4 - 864(N+8)(5N+22)\zeta(3)\right]^{-1} \tag{19.97}$$

$$
\begin{aligned}
b_0 &= 3\sqrt{(N+8)\left[13776 + 4738N + N^2(8N+405) + 96(5N+22)\zeta(3)\right]} \\
&\quad \times \sqrt{2209664 + 1040160N + 162982N^2 + 9683N^3 + 184N^4 + 672(N+8)(5N+22)\zeta(3)}.
\end{aligned}
\tag{19.98}
$$

To save space, we have written down the solution directly for $\varepsilon = 1$. For the physically interesting cases $N = 0, \ldots, 4$, we obtain the values for $D = 3$ dimensions shown in Table 19.2 . Figure 19.3 illustrates the two- and three-loop critical exponent of the approach to scaling $\omega = \varepsilon/(\rho - 1)$ as a function of N.

Given ρ, we can determine the other exponents. The results are

$$
\begin{aligned}
\gamma &= 1 - \frac{\varepsilon(N+2)\left[\varepsilon(N-4) - 2(N+8)\right]\rho(\rho+1)(2\rho-1)}{3\left[8(N+8)^2 - 4\varepsilon(2N^2 - N - 106) + \varepsilon^2(2N^2 + 17N + 194)\right]} \\
&\quad + \frac{8\varepsilon(N+2)\left[\varepsilon(N-4) - 2(N+8)\right]^3(2\rho-1)^3}{27\left[8(N+8)^2 - 4\varepsilon(2N^2 - N - 106) + \varepsilon^2(2N^2 + 17N + 194)\right]^2},
\end{aligned}
\tag{19.99}
$$

TABLE 19.2   Critical exponent ω from three-loop strong-coupling theory in $4 - \varepsilon$ dimensions.

N	0	1	2	3	4
ρ	2.41829	2.40384	2.38683	2.36910	2.35157
ω	0.705073	0.712332	0.721069	0.730405	0.73988

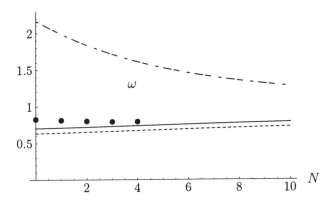

FIGURE 19.3   Two-loop (short-dashed) and three-loop (solid) critical exponent ω for different $O(N)$-symmetries. The ε-expansion (mixed-dashed) and the six and seven loops theoretical results to be listed in Tables 20.1 and 20.2 (dots) are displayed for comparison.

$$\nu = \frac{1}{2} - \frac{\varepsilon(N+2)\left[\varepsilon(N-3) - 2(N+8)\right]\rho(\rho+1)(2\rho-1)}{12\left[4(N+8)^2 - 2\varepsilon(2N^2 + N - 90) + \varepsilon^2(N^2 + 9N + 95)\right]}$$
$$+ \frac{\varepsilon(N+2)\left[\varepsilon(N-3) - 2(N+8)\right]^3 (2\rho-1)^3}{27\left[4(N+8)^2 - 2\varepsilon(2N^2 + N - 90) + \varepsilon^2(N^2 + 9N + 95)\right]^2}. \tag{19.100}$$

Figures 19.4 and 19.5 illustrate the accuracy of the two- and three-loop critical exponents γ and ν as a function of N.

If we keep ε in (19.95)–(19.98), we obtain

$$\rho = 2 + \frac{3(3N+14)}{(N+8)^2}\varepsilon - \frac{96\zeta(3)(5N+22)(N+8) + 33N^3 + 214N^2 + 1264N + 2512}{4(N+8)^4}\varepsilon^2. \tag{19.101}$$

Reexpanding the combination $\omega = \varepsilon/(\rho-1)$ in powers of ε up to ε^3, we rederive the corresponding terms in the ε-expansion (17.15). Inserting this expansion into ρ, we would have recovered the three-loop parts of the ε-expansions for γ and η of Eqs. (17.13) and (17.14) [recalling $\gamma = \nu(2 - \eta)$]. The critical exponent η is obtained using $2 - \gamma/\nu$, with γ and ν from (19.99) and (19.100), respectively. From this we can recover the ε-expansion (17.13).

Let us also calculate directly the strong-coupling limit of η from the three-loop extension of (19.80):

$$\eta(\bar{g}_B) = \frac{N+2}{18}\bar{g}_B^2 - \frac{(N+2)(N+8)}{216}\left(1 + \frac{8}{\varepsilon}\right)\bar{g}_B^3 + \cdots . \tag{19.102}$$

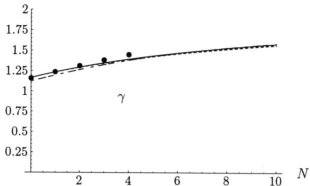

FIGURE 19.4 Two-loop (short-dashed) and three-loop (solid) critical exponent γ. For comparison, we also show the ε-expansion (mixed-dashed) and the theoretical values of Tables 20.1 and 20.2 (dots).

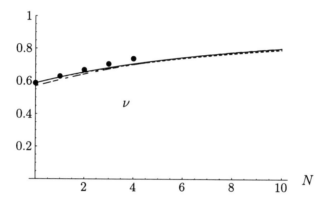

FIGURE 19.5 Two-loop (short-dashed) and three-loop (solid) critical exponent ν. For comparison, we also show the ε-expansion (mixed-dashed) and the theoretical values of Tables 20.1 and 20.2 (dots).

Since the linear term is absent, the optimum of variational perturbation theory is given by an extremum in which the root r in (19.70) is equal to -1, the optimum lying at $\bar{g}_B = -2\bar{f}_2/(3f_3)$, so that

$$\eta = \frac{4}{27}\frac{\bar{f}_2^3}{f_3^2} = \frac{32}{27}\frac{(2\rho-1)^3(N+2)}{(N+8)^2(8+\varepsilon)^2}\varepsilon^2. \tag{19.103}$$

With the three-loop ε-expansion (19.101) for ρ, this leads again to the correct terms in ε-expansion (17.13) for η. The difference between $\eta = 2 - \gamma/\nu$ and (19.103) at $\varepsilon = 1$ is illustrated in Figure 19.6 which also shows a direct plot of the ε-expansion for η as well as the theoretical values quoted in Tables 20.1 and 20.2 . In this case, the ε-expansion happens to yield the best critical exponent η, followed by the strong-coupling limit of the direct series (19.103). Comparing the different results, we see that they differ by about 30%. This is due to the smallness of η. Compared to unity, the error is small.

Let us finally calculate the critical exponent η_m up to three loops. The extension of the perturbation series (19.79) up to three loops reads

$$\eta_m(\bar{g}_B) = \frac{N+2}{3}\bar{g}_B - \frac{N+2}{18}\left(5 + 2\frac{N+8}{\varepsilon}\right)\bar{g}_B^2 + \frac{N+2}{108}\left[3\left(5N+37\right)\right.$$

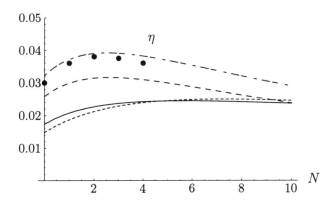

FIGURE 19.6 Two-loop (short-dashed) and three-loop (solid) critical exponent η from the definition $2 - \gamma/\nu$. For comparison, we also show the ε-expansion (short- and long-dashed), η from the strong-coupling limit of the direct (medium-dashed) series (19.103) and the theoretical values of Tables 20.1 and 20.2 (dots).

$$+ \left. \frac{2(19N + 122)}{\varepsilon} + \frac{4(N+8)^2}{\varepsilon^2} \right] \bar{g}_B^3 + \cdots . \tag{19.104}$$

From this we obtain the strong-coupling limit

$$\eta_m = \frac{\varepsilon(N+2)(2N + 16 + 5\varepsilon)\rho(\rho+1)(2\rho-1)}{3\left[4(N+8)^2 + \varepsilon(38N + 244) + 3\varepsilon^2(5N + 37)\right]} \\ - \frac{4\varepsilon(N+2)(2N + 16 + 5\varepsilon)^3(2\rho-1)^3}{27\left[4(N+8)^2 + \varepsilon(38N + 244) + 3\varepsilon^2(5N + 37)\right]^2} . \tag{19.105}$$

This result is analytically different from that obtained from (19.100) via the scaling relation $\eta_m = 2 - \nu^{-1}$. Numerically, however, it is close to it, as can be seen in Figure 19.7. For comparison, the figure also shows plots of the ε-expansion (19.106), and the theoretical values of Tables 20.1 and 20.2 .

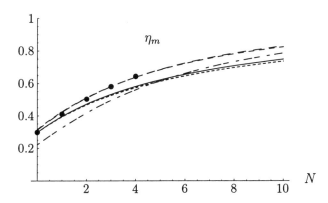

FIGURE 19.7 Two-loop (short-dashed) and three-loop (solid) critical exponent η_m from the definition $2 - \nu^{-1}$. For comparison, we also show the ε-expansion (short- and long-dashed), η_m from the strong-coupling limit of the direct two-loop (medium-dashed) and three-loop (long-dashed) series and the theoretical values of Tables 20.1 and 20.2 (dots).

Inserting into (19.105) the ε-expansion (19.101) for ρ, we find

$$
\begin{aligned}
\eta_m =\ & \frac{N+2}{N+8}\varepsilon + \frac{(N+2)(13N+44)}{2(N+8)^3}\varepsilon^2 \\
& + \frac{(N+2)\left[5312 + 2672N + 452N^2 - 3N^3 - 96(N+8)(5N+22)\zeta(3)\right]}{8(N+8)^5}\varepsilon^3.
\end{aligned}
\tag{19.106}
$$

For completeness, we list some numeric values of the critical exponents up to three loops in Table 19.3 .

TABLE 19.3   Critical exponents ν, η_m, η from three-loop strong-coupling theory in $4 - \varepsilon$ dimensions.

N	0	1	2	3	4
γ	1.16455	1.2338	1.29426	1.34697	1.39307
ν	0.587376	0.623381	0.654552	0.681561	0.705071
η_m	0.311607	0.421796	0.509799	0.580684	0.638337
η	0.0258218	0.029917	0.031452	0.0315846	0.03096

19.7   Five-Loop Resummation

The calculations in the last section have illustrated the power of the variational procedure, but do not yet give good results for the critical exponents. In order to achieve higher accuracy, we have to go to five loops in a straightforward extension of the two-loop treatment[13].

19.7.1   Critical Exponent ω

As in the two-loop calculations of the last section, we start by resumming the logarithmic derivative of $\bar{g}(\bar{g}_B)$ to determine the critical exponent ω of the approach to scaling.

As in Subsection 19.5.3, the exponent ω is determined by identifying the perturbation expansions of $\bar{g}(\bar{g}_B)$, truncated as in (19.53), with the function W_L in Eq. (19.54), and requiring that the theory scales as observed experimentally. This implies that $\bar{g}(\bar{g}_B)$ approaches a constant in the strong-coupling limit. The ratio of powers p/q on the left-hand side of (19.54) must therefore be zero. The expansion of the logarithmic derivative of $\bar{g}(\bar{g}_B)$ is calculated as indicated in Eq. (19.73) for a function $F(\bar{g}_B) = \bar{g}(\bar{g}_B)$. By Eq. (19.54) [or (19.61)], the resulting function $f(\bar{g}_B)$ has to be equal to $p/q = s$ in the strong-coupling limit, i.e., equal to zero.

Instead of the two-loop expansion for $\bar{g}(\bar{g}_B)$ in (19.76), we use the five-loop expansion in (15.18). Thus we obtain for $\rho = 1 + q/2 = 1 + \varepsilon/\omega$ higher L-loop approximations ρ_L. Altogether, we find four approximations ω_L to the approach parameter $\omega = 2/q = \varepsilon/(\rho - 1)$, with $L = 2, 3, 4, 5$. As a check we perform their ε-expansions and find that they agree with the initial ε-expansion (17.15) derived, in the renormalization group approach, from the derivative of the β-function at the fixed point.

Our variational expressions are evaluated without an ε-expansion at $\varepsilon = 1$. In Fig. 19.8, the resulting L-loop approximations ω_L are plotted as functions of L. The ω_L-values are obviously not yet asymptotic, calling for an extrapolation to infinite L. This is possible with the help of the theoretical knowledge on the L-dependence of the strong-coupling limit as determined in

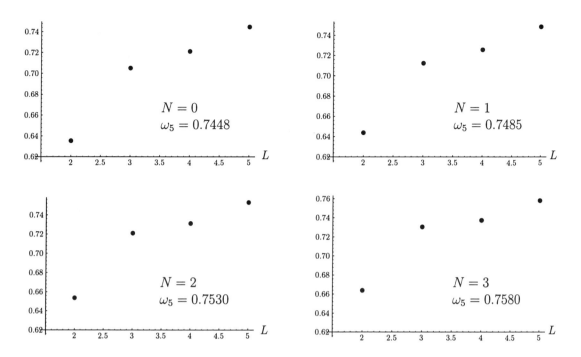

FIGURE 19.8 Critical exponent of approach ω calculated from $f_L^* = s_L^* = 0$, plotted against the order of approximation L.

Eq. (19.51). For an arbitrary function $f(\bar{g}_B)$ with a constant strong-coupling limit approachad as in Eq. (19.66), the L-dependence is of the form

$$f_L^* \propto f^* + \text{const} \times e^{-cL^{1-\omega}}. \tag{19.107}$$

The extrapolation to infinite order L may be carried out in two ways. We may choose an ω-value near the suspected correct exponent, and calculate the strong-coupling limits $f_L^* = s_L^*$ of the logarithmic derivative of $\bar{g}(\bar{g}_B)$. Through these we fit the theoretical L-behavior (19.107). Then we vary ω and the fit parameter c until the fit approaches as smoothly as possible the limit $s_L^* = 0$ for infinite L.

Alternatively, and apparently more reliably, we plot for a certain ω and c the strong-coupling limits $s_L^* = f_L^*$ against the variable $x_L = e^{-cL^{1-\omega}}$ instead of L. The parameter c is chosen such that the points lie as close as possible on a straight line. Since the points are obtained alternatively from minima and turning points for even and odd approximations, respectively, they are expected to lie on slightly different straight lines. We determine c by requiring these lines to cross at $x_L = 0$. With this condition, the two lines yield the same value for $s^* = s_\infty$. The trial ω is then varied until s^* is equal to zero. The final plots are shown in Fig. 19.9. For comparison, we also show the corresponding direct plots against the order L of the approximation in Fig. 19.10.

Table 19.4 lists in the first column the calculated values of ω for infinite L and $N = 0, 1, 2, 3$. The number in parentheses is the value ω_5 determined from the condition $s_5^* = f_5^* = 0$. It shows the importance of the extrapolation.

The extrapolated values for ω are now used to construct the strong-coupling limits for the exponents ν, γ and η.

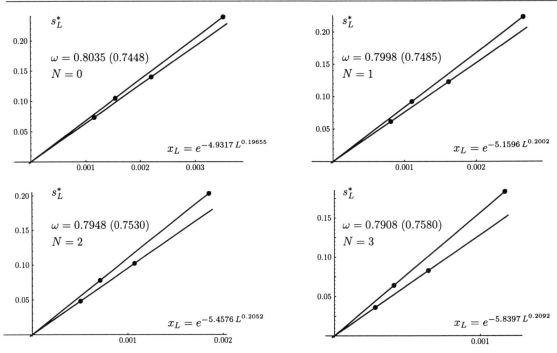

FIGURE 19.9 Approach to zero of the logarithmic derivative $(p/q)_L \equiv s_L^* = f_L^*$ of $\bar{g}(\bar{g}_B)$ as a function of the variable x_L. The critical exponent $\omega = 2/q$ is determined by requiring the straight lines to intersect at $x_L = 0$. Its value is given in the equation. The number in parentheses is the highest approximation ω_6 showing the exptrapolation distance to the final value. Both numbers are listed in Table 19.4 .

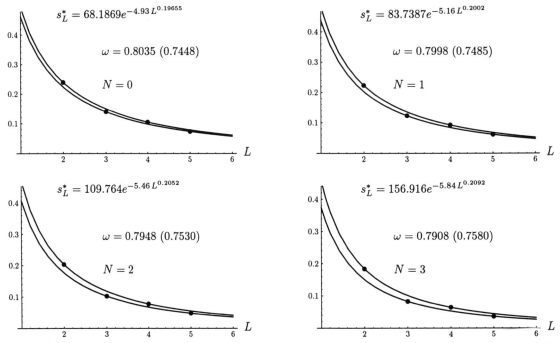

FIGURE 19.10 Same strong-coupling limits of logarithmic derivative of $\bar{g}(\bar{g}_B)$ as in Fig. 19.9, but plotted against the order L of the approximation. The extrapolating functions are written on top of each figure. Ihe extrapolated ω is given in the equation. The number in parentheses is again the highest approximation ω_6 showing the exptrapolation distance to the final value. Both numbers are listed in Table 19.4 .

ω	Var.-Pert.	Borel-Res. (GZ)	Chapter 17
	$\omega\,(\omega_5)$		
$N = 0$	0.8035(0.7448)	0.828 ± 0.023	0.817 ± 0.021
$N = 1$	0.7998(0.7485)	0.814 ± 0.018	0.806 ± 0.013
$N = 2$	0.7948(0.7530)	0.802 ± 0.018	0.800 ± 0.013
$N = 3$	0.7908(0.7580)	0.794 ± 0.018	0.796 ± 0.011

ν	Var.-Pert.		Borel-Res. (GZ)	Chapter 17
	$\nu\,(\nu_5)$ (I)	$\nu\,(\nu_5)$ (II)		
$N = 0$	0.5874(0.5809)	0.5878(0.5832)	0.5875 ± 0.0018	0.5865 ± 0.0013
$N = 1$	0.6292(0.6171)	0.6294(0.6222)	0.6293 ± 0.0026	0.6268 ± 0.0022
$N = 2$	0.6697(0.6509)	0.6692(0.6597)	0.6685 ± 0.0040	0.6642 ± 0.0111
$N = 3$	0.7081(0.6821)	0.7063(0.6951)	0.7050 ± 0.0055	0.6987 ± 0.0051

η	Var.-Pert.		Borel-Res. (GZ)	Chapter 17
	$\eta\,(\eta_5)$ (I)	$\eta\,(\eta_5)$ (II)		
$N = 0$	0.0316(0.0234)	0.0305(0.0234)	0.0300 ± 0.0060	0.0344 ± 0.0042
$N = 1$	0.0373(0.0308)	0.0367(0.0308)	0.0360 ± 0.0060	0.0395 ± 0.0043
$N = 2$	0.0396(0.0365)	0.0396(0.0365)	0.0385 ± 0.0065	0.0412 ± 0.0041
$N = 3$	0.0367(0.0409)	0.0402(0.0409)	0.0380 ± 0.0060	0.0366 ± 0.0020

γ	Var.-Pert.	Borel-Res. (GZ)
	$\gamma\,(\gamma_5)$	
$N = 0$	1.1576(1.1503)	1.1575 ± 0.0050
$N = 1$	1.2349(1.2194)	1.2360 ± 0.0040
$N = 2$	1.3105(1.2846)	1.3120 ± 0.0085
$N = 3$	1.3830(1.3452)	1.3830 ± 0.0135

TABLE 19.4 Extrapolated critical exponents from five-loop ε-expansions, compared with results of Chapter 17 and of Guida and Zinn-Justin (GZ), which agree with our present numbers. The difference between the versions (I) and (II) of the exponents ν and η are explained in the text. The numbers in parentheses are the six-loop approximations showing the exptrapolation distance to the final values.

19.7.2   Critical Exponent ν

For ν, we can proceed as in Eq. (19.91) of the last section, calculating the strong-coupling limit of $\eta_m(\bar{g}_B)$. Its perturbation expansion $\eta_m(\bar{g}_B) = 2\gamma_m(\bar{g}_B)$ is found by inserting (15.18) into (17.7). The critical exponents ν are obtained from the strong-coupling limits of η_m via the scaling relation $\nu = 1/(2 - \eta_m)$. The resulting exponents are shown in Table 19.4 as $\nu(\text{II})$.

Alternatively, we apply the procedure to the series $\nu(\bar{g}_B)$, and find directly the strong-coupling limits of the exponent ν as shown in Table 19.4 as $\nu(\text{I})$.

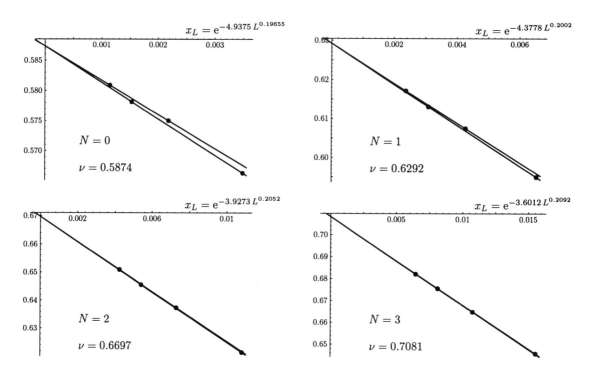

FIGURE 19.11 Critical exponent $\nu(\text{I})$ from variational perturbation theory, plotted as a function of x_L. Requiring the lines to cross at $x_L = 0$ determines the parameter c in x_L. See the description of the fitting procedure in the text and the final values in Table 19.4 .

In both cases, the extrapolation to infinite order proceeds as before. We use again a function of the form (19.107) to extrapolate the results for $L = 2, 3, 4, 5$ to infinite L. The results are shown in Figs. 19.11 and 19.12 for the direct calculation of ν. The horizontal lines indicate the limit $L \to \infty$. The extrapolating functions are displayed on top of each figure. The results of these calculations are listed in Table 19.4 . The value $\nu = 0.6697$ for $N = 2$ corresponds to a critical exponent

$$\alpha = 2 - 3\nu \approx -0.0091, \tag{19.108}$$

which is in satisfactory agreement with the experimental space shuttle value -0.01056 ± 0.0004 in Eq. (1.23)

All results depend on the value of ω used for the resummation. Let us estimate the error resulting from errors in ω. This is done by calculating ν for two adjacent ω-values, and we find

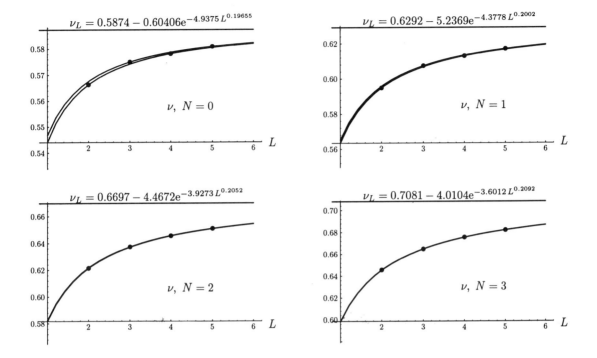

FIGURE 19.12 Critical exponent $\nu(\mathrm{I})$ from variational perturbation theory, as a function of the order L of approximation. The full fit function is written on top of the figure. The limiting values are listed in Table 19.4 .

$$\Delta\nu = \left\{ \begin{array}{l} -0.0900 \times (\omega - 0.8035) \\ -0.1375 \times (\omega - 0.7998) \\ -0.1853 \times (\omega - 0.7948) \\ -0.2271 \times (\omega - 0.7908) \end{array} \right\} \quad \text{for} \quad \left\{ \begin{array}{l} N = 0 \\ N = 1 \\ N = 2 \\ N = 3 \end{array} \right\}. \tag{19.109}$$

19.7.3   Critical Exponent η and γ

The calculation of the exponent η is always difficult, since its perturbation expansion starts out with $\bar{g}_B^2$. Thus we find only three approximants η_3, η_4, η_5, and these are not sufficient to carry out a reliable extrapolation procedure. In order to circumvent this, we study the strong-coupling limit of $\eta_m(\bar{g}_B) + \eta(\bar{g}_B)$, and subtract from this the strong-coupling limit of $\eta_m(\bar{g}_B)$. This is derived from $\eta_m = 2 - \nu^{-1}$ with either the values $\nu(\mathrm{I})$ or $\nu(\mathrm{II})$ of Table 19.4 . The results are called $\eta(\mathrm{I})$ and $\eta(\mathrm{II})$, respectively, and shown in Figs. 19.13 and 19.14. The horizontal line marks the limit $L \to \infty$. The extrapolating function is plotted on top of the figure.

We may calculate η also from another independent strong-coupling limit using the perturbation expansion for the critical exponent $\gamma = \nu(2 - \eta)$. The extrapolation for this exponent is shown in Figs. 19.15 and in 19.16. The resulting values for γ are also listed in Table 19.4 . The dependence on the value of ω is in the same order of magnitude as in the case of ν:

$$\Delta\gamma = \left\{ \begin{array}{l} -0.1500 \times (\omega - 0.8035) \\ -0.2237 \times (\omega - 0.7998) \\ -0.3147 \times (\omega - 0.7948) \\ -0.4014 \times (\omega - 0.7908) \end{array} \right\} \quad \text{for} \quad \left\{ \begin{array}{l} N = 0 \\ N = 1 \\ N = 2 \\ N = 3 \end{array} \right\}. \tag{19.110}$$

Table 19.4 contains the results of the extrapolation to infinite order followed in parentheses by the 5th-order approximation. For comparison with the results of other methods, see

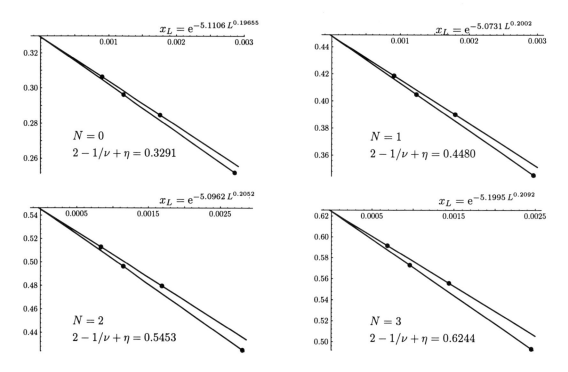

FIGURE 19.13 Determination of critical exponent η from the strong-coupling limit of $\eta_m + \eta$, plotted as a function of x_L. Requiring the lines to cross at $x_L = 0$ determines the parameter c in x_L. See the description of the fitting procedure in the text and the list of limiting values in Table 19.4 .

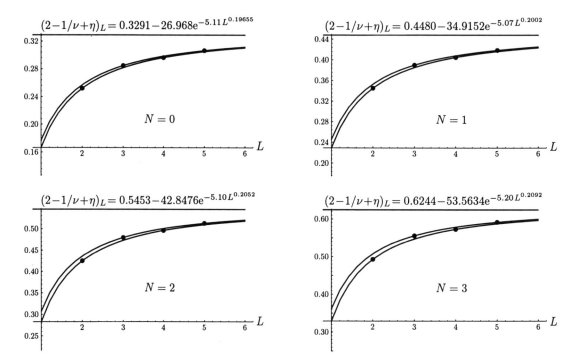

FIGURE 19.14 Same as above, plotted against the order of approximation L. The full fit function is written on top of the figure. The limiting values are listed in Table 19.4 .

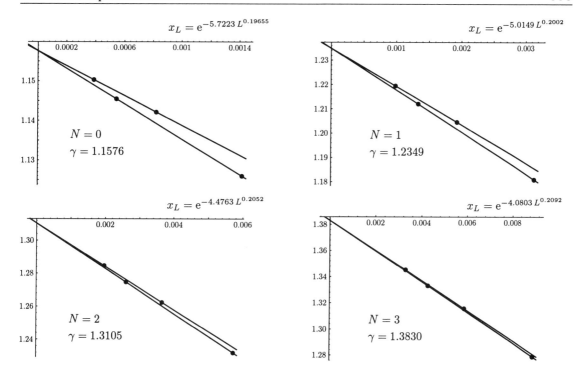

FIGURE 19.15 Critical exponent γ from variational perturbation theory, plotted as a function of x_L. Requiring the lines to cross at $x_L = 0$ determines the parameter c in x_L. See the description of the fitting procedure in the text. The limiting values are listed in Table 19.4 .

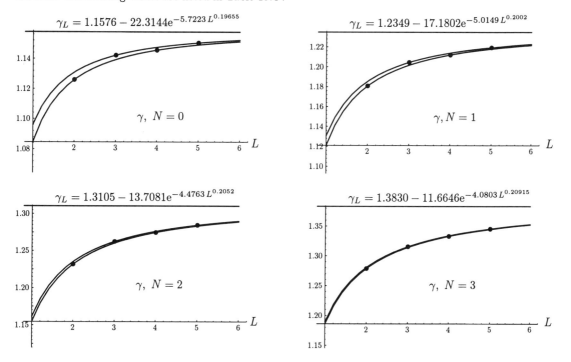

FIGURE 19.16 Critical exponent γ from variational perturbation theory, plotted as a function of the order of approximation L. The full fit function is written on top of the figure. The limiting values are listed in Table 19.4 .

Table 20.1 in the next chapter. Here, we only append the earlier results of the resummation of the ε-expansion in Chapter 17, and those of Guida and Zinn-Justin (GZ) [14]. The present numbers agree well with either of them. The differences between ν(I) and ν(II), and between η(I) and η(II), are much smaller than the errors given in Chapter 17, or in GZ.

It is interesting to note that the results of the strong-coupling theory agree very well with those derived from Borel-resummations, which make use of the known large-order behavior of the perturbation series. So far, the strong-coupling theory does not take advantage of this knowledge of the large-order behavior. It compensates for this by incorporating, in an essential way, the approach (19.66) to the strong-coupling limit. Further improvement comes from the theoretical knowledge of the large-L behavior (19.107) of the approximations.

The incorporation of the large-order behavior into the strong-coupling theory will be described in the Chapter 20.

19.8   Interpolating Critical Exponents between Two and Four Dimensions

Knowing the ε-expansions for the critical exponents, the question arises whether these expansions can also be resummed at $\varepsilon = 2$, i.e., in D=2 dimensions. However, with the difficulties encountered in going to $\varepsilon = 1$, we see little hope in obtaining reliable numbers this way.

Fortunately, there exists a completely independent access to the critical exponents ν and η from another field theory, the O(N)-symmetric *nonlinear σ-model*. The *universality hypothesis* of critical phenomena states that all systems with equal Goldstone bosons should have the same critical exponents. Thus renormalization group studies of O(N)-symmetric nonlinear σ-models in $D = 2 + \epsilon$ dimensions at $\epsilon = 1$, should yield the same critical exponents as O(N)-symmetric ϕ^4-theories as long as the second-order character of the transition is not destroyed by unexpected fluctuation effects. These conditions restrict the comparison to $N > 2$.

For $N = 1$ (Ising case), there are no Goldstone bosons, and for $N = 2$ (XY-model), the transition is of infinite order. In this case, the divergence of the correlation length with temperature cannot be parametrized like $\xi \propto |T - T_c|^{-\nu}$, as shown by Kosterlitz and Thouless [15].

The nonlinear σ-model with the same Goldstone bosons as the ϕ^4 theory for $N > 2$ can be studied in $D = 2 + \epsilon$ dimensions and yields expansions for the critical exponents in powers of ϵ. Note the distinct notation of ε in $D = 4 - \varepsilon$ and ϵ in $D = 2 + \epsilon$. Up to now, the latter expansions have remained rather useless for any practical calculation owing to their non-Borel character [16]. This has led some authors to doubt the use of such expansions around the *lower critical dimension* altogether [17]. This would be quite unfortunate, since it would jeopardize other interesting theories such as Anderson's theory of localization [18]. The basis of these doubts is the increasing relevance of higher powers of the derivative term [19]. However, hope for the utility of these expansions is not completely lost, since the argument requires interchanging two limits: one is the analytic continuation in ϵ, the other is the increasing to infinity of the number of derivatives [20]. The purpose of this section is to sustain this hope by combining the expansions in $2+\epsilon$ diemnsions with the previous expansions in $4-\varepsilon$ dimensions to obtain precise critical exponents for all dimensions D between four and two. This will be done explicitly [21] only for the critical exponent ν and the universality classes $N = 3, 4, 5$.

So far, the ϵ-expansions of ν^{-1} and the anomalous dimension η have been calculated up to the powers ϵ^4 [22]:

$$\nu^{-1} = \epsilon + \frac{\epsilon^2}{N-2} + \frac{\epsilon^3}{2(N-2)} - \left[30 - 14N + N^2 + (54 - 18N)\zeta(3)\right]\frac{\epsilon^4}{4(N-2)^3} + \ldots \quad (19.111)$$

$$\eta = \frac{\epsilon}{N-2} - \frac{(N-1)\,\epsilon^2}{(N-2)^2} + \frac{N\,(N-1)\,\epsilon^3}{2\,(N-2)^3}$$

$$-(N-1)\left[-6 + 2\,N + N^2 + (-12 + N + N^2)\,\zeta(3)\right]\frac{\epsilon^4}{4\,(N-2)^4} + \ldots \quad . \quad (19.112)$$

When evaluated at $\epsilon = 1$, the first series yields for the three-dimensional O(3)-model the diverging successive values $\nu^{-1} = (1,\,2,\,2.5,\,3.25)$. Padé approximations do not help; the best of them, the [1,2]-approximation, gives the too large value $\nu = 2$. So far, the only result which is not far from the true value has been obtained via the Padé-Borel transform

$$P^{[1,2]}(\epsilon, t) = \frac{\epsilon t}{1 - \epsilon t/2 + \epsilon^2 t^2/6}, \quad (19.113)$$

from which we obtain the ϵ-dependent critical exponent

$$\nu^{-1}(\epsilon) = \int_0^\infty dt\, e^{-t} P^{[1,2]}(\epsilon, t), \quad (19.114)$$

whose value at $\epsilon = 1$ is ≈ 1.252, corresponding to $\nu \approx 0.799$ still too large to be useful. The other Padé-Borel approximants are singular and thus of no use, as shown in Fig. 19.17.

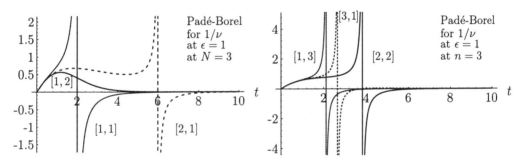

FIGURE 19.17 Integrands of the Padé-Borel transform (19.114) for the Padé approximants [1,1], [2,1], [1,2] and for [1,3], [3,1], [2,2] at $\epsilon = 1$. Only the last is integrable, yielding $\nu^{-1} \approx 1.25183 \approx 1/.79883$.

The full ϵ-dependence of the power series (19.111) and the Padé-Borel-approximation (19.114) can be seen in Fig. 19.18 on page 358.

A direct evaluation of the series for the anomalous dimension η yields the even worse values $(2, -2, 4, -5)$. Here the nonsingular Borel-Padé approximations [2, 1], [1, 2], and [1, 1] yield 0.147, 0.150, and 0.139, rather than the correct value 0.032. The full ϵ-dependence of the power series (19.112) and the Padé[2,1]-Borel-approximation are also shown in Fig. 19.18.

The remedy for these problems comes from the strong-coupling theory of Section 19.2. It allows us to find an extended variational perturbation expansion if we know not only L weak-coupling coefficients but also M strong-coupling expansion coefficients. We merely add to the set of coefficients $f_1, \ldots, f_L$ a set of M unknown ones, $f_{L+1}, \ldots, f_{L+M}$, and determine the latter via a fit of the resulting strong-coupling coefficients $b_0, \ldots, b_{M-1}$ of Eq. (19.14) to the known ones.

19.8.1   Critical Exponent ν.

This procedure will now be applied to the perturbation expansion (19.111) for ν^{-1} in $2 + \epsilon$ dimensions, considering it as the *strong-coupling expansion* of a series in the variable $\tilde{\varepsilon} = 2(4 - D)/(D - 2) = 4(1 - \epsilon/2)/\epsilon = \varepsilon/(1 - \varepsilon/2)$:

$$\nu^{-1} = 4\tilde{\varepsilon}^{-1} - 8\frac{N-4}{N-2}\tilde{\varepsilon}^{-2} + 16\frac{N-4}{N-2}\tilde{\varepsilon}^{-3} \tag{19.115}$$

$$-32\left[52 - 16N - 4N^2 + N^3 + 36(3-N)\zeta(3)\right]\frac{\tilde{\varepsilon}^{-4}}{(N-2)^3} + \cdots .$$

The weak-coupling expansion of this series is obtained from Eq. (17.14) by a change of variables from ε to $\tilde{\varepsilon}$, and a re-expansion:

$$\nu^{-1} = 2 - \frac{N+2}{N+8}\tilde{\varepsilon} + \left(20 + 3N + N^2\right)\frac{(2+N)\tilde{\varepsilon}^2}{2(8+N)^3} + \left[-2240 - 624N - 212N^2 - 9N^3 - 2N^4 + 8\cdot 12(8+N)(22+5N)\zeta(3)\right]\frac{(2+N)\tilde{\varepsilon}^3}{8(8+N)^5}$$

$$+ \Big[5\left(568576 + 382144N + 103920N^2 + 9532N^3 + 1142N^4 + 21N^5 + 4N^6\right)$$

$$+80\left(-249600 - 148960N - 42912N^2 - 6516N^3 - 350N^4 + 3N^5\right)\zeta(3) + 80\cdot 18(8+N)^3(22+5N)\zeta(4)$$

$$-80\cdot 80(8+N)^2\left(186 + 55N + 2N^2\right)\zeta(5)\Big]\frac{(2+N)}{160(8+N)^7}\tilde{\varepsilon}^4$$

$$+ \Big[945\left(-105091072 - 106771456N - 47635968N^2 - 11768576N^3 - 1835504N^4 - 122812N^5 - 6270N^6 - 45N^7 - 8N^8\right)$$

$$+15120\left(57911296 + 46323968N + 17913728N^2 + 3869024N^3 + 514592N^4 + 46900N^5 + 1902N^6 - 37N^7\right)\zeta(3)$$

$$+15120\left(-47874048 - 40615936N - 11928064N^2 - 1525888N^3 - 89408N^4 - 3200N^5 - 128N^6\right)\zeta^2(3)$$

$$+945\left(101376 + 61056N + 13392N^2 + 1278N^3 + 45N^4\right)\zeta(4)$$

$$+256\cdot 945(8+N)^2\left(345552 + 193822N + 48749N^2 + 6506N^3 + 235N^4\right)\zeta(5)$$

$$+945\cdot 56448(8+N)^3\left(526 + 189N + 14N^2\right)\zeta(7)\Big]\frac{(N+2)\tilde{\varepsilon}^5}{120960(8+N)^9} + \cdots ,$$

whose numerical forms are for $N = 3, 4, 5$

$$N = 3: \quad \nu^{-1} = 2 - 0.45455\,\tilde{\varepsilon} + 0.071375\,\tilde{\varepsilon}^2 + 0.15733\,\tilde{\varepsilon}^3 - 0.52631\,\tilde{\varepsilon}^4 + 1.5993\,\tilde{\varepsilon}^5 \tag{19.116}$$

$$N = 4: \quad \nu^{-1} = 2 - 0.5\,\tilde{\varepsilon} + 0.0833333\,\tilde{\varepsilon}^2 + 0.147522\,\tilde{\varepsilon}^3 - 0.499944\,\tilde{\varepsilon}^4 + 1.47036\,\tilde{\varepsilon}^5, \tag{19.117}$$

$$N = 5: \quad \nu^{-1} = 2 - 0.538462\,\tilde{\varepsilon} + 0.0955849\,\tilde{\varepsilon}^2 + 0.135442\,\tilde{\varepsilon}^3 - 0.469842\,\tilde{\varepsilon}^4 + 1.34491\,\tilde{\varepsilon}^5, \tag{19.118}$$

$$N = 1: \quad \nu^{-1} = 2 - 0.333333\,\tilde{\varepsilon} + 0.0493827\,\tilde{\varepsilon}^2 + 0.158478\,\tilde{\varepsilon}^3 - 0.539937\,\tilde{\varepsilon}^4 + 1.78954\,\tilde{\varepsilon}^5. \tag{19.119}$$

Extending these series by four more terms $f_6\,\tilde{\varepsilon}^6 + f_7\,\tilde{\varepsilon}^7 + f_8\,\tilde{\varepsilon}^8 + f_9\,\tilde{\varepsilon}^9$, we calculate the functions $b_n(\hat{g}_B)$ of Eq. (19.11), and from these the strong-coupling coefficients $\bar{b}_0, \bar{b}_1, \bar{b}_2, \bar{b}_3$, in (19.14) as described there, [after having identified g_B with $\tilde{\varepsilon}$, and the parameters (p, q) with $(-2, 2)$]. The coefficients f_6, f_7, f_8, f_9 are determined to make (19.14) agree with the expansion coefficient (19.115) [23]. The results are shown in Tables 19.5–19.8 .

TABLE 19.5   Coefficients of successive extensions of expansion coefficients in Eq. (19.116) of ν^{-1} with $N = 3$ determined from the strong-coupling coefficients $(4, 8, -16, 160)$ of Eq. (19.115) for $M = 1, 2, 3, 4$.

M	f_6	f_7	f_8	f_9
1	-203.827			
2	-5.67653	17.6165		
3	-4.25622	9.04109	-15.7331	
4	-3.80331	6.87304	-10.0012	12.3552

TABLE 19.6   Same as above for the expansion coefficients in Eq. (19.117) of ν^{-1} with $N = 4$. They are determined from the strong-coupling coefficients (4, 0, 0, 221.096) of Eq. (19.115) for $M = 1, 2, 3, 4$.

M	f_6	f_7	f_8	f_9
1	-147.508			
2	-7.91064	37.1745		
3	-4.59388	12.3044	-27.0837	
4	-3.72613	7.47851	-12.2129	16.9547

TABLE 19.7   Same as above the expansion coefficients in Eq. (19.118) of ν^{-1} with $N = 5$. They are determined from the strong-coupling coefficients (8, $-8/3$, 16/3, 106.131) of Eq. (19.115) for $M = 1, 2, 3, 4$.

M	f_6	f_7	f_8	f_9
1	-108.648			
2	-10.1408	60.7217		
3	-4.75598	15.1045	-38.9689	
4	-3.57909	7.84272	-14.1142	21.6045

TABLE 19.8   Coefficients of successive extensions of expansion coefficients in Eq. (19.119) of ν^{-1} with $N = 1$, determined from $M = 1, 2, 3, 4$ strong-coupling coefficients (4, -24, 48, 3825.54) of Eq. (19.115).

M	f_6	f_7	f_8	f_9
1	-413.921			
2	-5.25285	12.1104		
3	-442759	12450066	-196950675	
4	-5.7343	13.7134	-25.226	38.0976

In order to see how the result improves with the number M of additional strong-coupling coefficients, we go through this procedure successively for $M = 1, 2, 3, 4$. The four resulting curves for $\nu^{-1}(\varepsilon)$ of the $O(N)$ universality classes with $N = 3, 4, 5$ are shown in Figs. 19.18–19.20. In each figure, we have also plotted successive critical exponents ν at $\varepsilon = 1$ as a function of the variable $x = M^{-2}$, which makes them lie approximately on a straight line intercepting the ν-axis at the limiting value ν_∞. This extrapolated value is in excellent agreement with the seven-loop determination of critical exponents to be described in the next chapter. See in particular Table 20.2 .

Finally, we plot our highest ($M = 4$) approximations for $N = 3, 4, 5$ together with the large-N approximations for $N = \infty, 20, 10, 6$ in Fig. 19.21 to see the trend for increasing N, which shows that the latter for $N = 6$ is still far from the exact curve [24].

The relation between the ϵ- and ε-expansions is expected to be restricted to $N > 2$ for physical reasons. It is instructive to see that the variational interpolation method reflects this problem in two places. First, the expansion coefficients in Table 19.8 show a large irregularity for $N = 1$. Second, the successive approximations for ν^{-1} in Fig. 19.22 display no tendency of convergence with increasing order M of approximation.

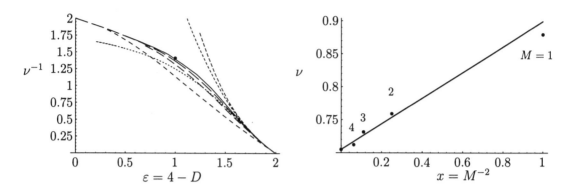

FIGURE 19.18 Inverse of critical exponent ν for classical Heisenberg model in O(3)-universality class, plotted as a function of $\varepsilon = 4-D$. Solid curve is obtained from variational interpolation between five-loop ε-expansion of ϕ^4-theory near $D = 4$ dimensions, and four-loop ϵ-expansion (19.111) of nonlinear σ-model around $D = 2$ dimensions. Long-dashed curves are successive approximations using strong-coupling expansion (19.115) up to the first, second, and third order. Short-dashed curves display the first three and four terms of ϵ-expansion and the associated Padé [1,2]-Borel approximations (lowest curve). The dot shows seven-loop results in $D = 3$ dimensions, $\nu = 0.705$. Values from our successive interpolations are $(\nu_1, \nu_2, \nu_3, \nu_4) = (0.87917, 0.75899, 0.731431, 0.712152)$. On the right, these values are extrapolated to infinite order by plotting them against $x = M^{-2}$ which makes them lie on a straight line with the intercept $\nu_\infty = 0.705$, in excellent agreement with the seven-loop result.

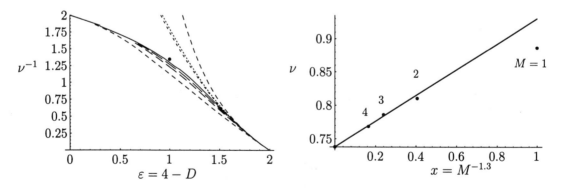

FIGURE 19.19 Analog of Fig. 19.18 ν^{-1} in for O(4)-universality class. The dot shows the six-loop result in $D = 3$ dimensions, $\nu = 0.737$. Values from our successive interpolation are $(\nu_1, \nu_2, \nu_3, \nu_4) = (0.88635, 0.810441, 0.786099, 0.768565)$. On the right, these values are extrapolated to infinite order by plotting them against $x = M^{-1.3}$, which makes them lie roughly on a straight line with intercept $\nu_\infty = 0.738$, in reasonable agreement with the six-loop result $\nu = 0.737$ (slightly worse than for $N = 3, 4$).

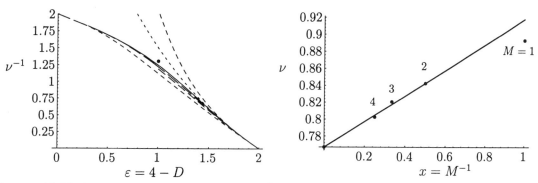

FIGURE 19.20 Analog of previous two figures for ν^{-1} in O(5)-universality class. Values from our successive interpolation are $(\nu_1, \nu_2, \nu_3, \nu_4) = (0.89278, 0.842391, 0.820491, 0.802416)$. On the right, these values are extrapolated to infinite order by plotting them against $x = M^{-1}$, which makes them lie roughly on a straight line with intercept $\nu_\infty = 0.767$, in excellent agreement with the six-loop result $\nu = 0.737$.

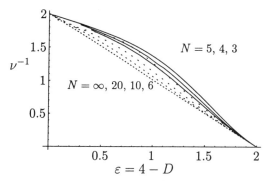

FIGURE 19.21 Highest approximations ($M = 4$) for ν^{-1} in O(N) universality class with $N = 3, 4, 5$ (counting from the top), and the $1/N$-expansions to order $1/N^2$ for $N = \infty, 20, 10, 6$ (counting from the bottom). The sixth-order result is still far from the exact one.

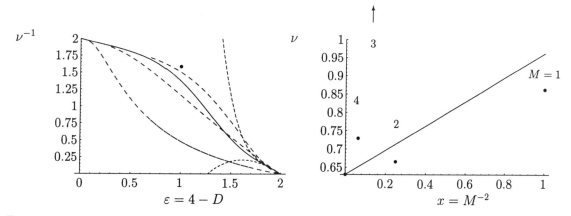

FIGURE 19.22 Same plot as in Fig. 19.18, but for ν^{-1} in O(1)-universality class (of Ising model). Again there is no Padé-Borel approximation. The dot represents seven-loop result in $D = 3$ dimensions $\nu = 0.6305$. The four interpolations give $(\nu_1, \nu_2, \nu_3, \nu_4) = (0.862357, 0.665451, 2.08686, 0.729231)$. They show no tendency of convergence towards the seven-loop exponent.

19.8.2   Critical Exponent η.

For the critical exponent η, the series (19.112) reads in the variable $\tilde{\varepsilon}$:

$$\eta = 4\frac{\tilde{\varepsilon}^{-1}}{N-2} + 8(4-3N)\frac{\tilde{\varepsilon}^{-2}}{(N-2)^2} + 16(12-18N+7N^2)\frac{\tilde{\varepsilon}^{-3}}{(N-2)^3}$$

$$+ 32\left[\left(20-56N+52N^2-15N^3\right)+2(N-1)\left(12-N-N^2\right)\zeta(3)\right]\frac{\tilde{\varepsilon}^{-4}}{(N-2)^4} + \cdots , \quad (19.120)$$

whereas the weak-coupling expansion is [25]

$$\frac{\eta}{\epsilon^2} = \frac{2+N}{2(N+8)^2} + \left(272+56\,N-N^2\right)\frac{(2+N)\,\epsilon}{8(N+8)^4} \quad\quad\quad\quad (19.121)$$

$$+ \left\{46144+17920\,N+1124\,N^2-230\,N^3-5\,N^4-32\cdot12\,(22+5\,N)\,\zeta(3)\right\}\frac{(2+N)\,\epsilon^2}{32(N+8)^6}$$

$$+ \left\{5655552+2912768\,N+262528\,N^2-121472\,N^3-27620\,N^4-946\,N^5-13\,N^6\right.$$

$$+16(N+8)\left(-171264-68672\,N-1136\,N^2+1220\,N^3+10\,N^4+N^5\right)\zeta(3)$$

$$\left.+128\cdot9\,(N+8)^3\,(22+5N)\,\zeta(4)+28\cdot40\,(N+8)^2\left(186+55\,N+2\,N^2\right)\,\zeta(5)\right\}\frac{(2+N)\,\epsilon^3}{128(N+8)^8} + \cdots .$$

whose numerical forms are, for $N = 3, 4, 5$:

$$N = 3: \quad \eta/\epsilon^2 = 5/242 + 0.0183987\,\tilde{\epsilon} - 0.0166488\,\tilde{\epsilon}^2 + 0.032432\,\tilde{\epsilon}^3 + \cdots , \quad (19.122)$$

$$N = 4: \quad \eta/\epsilon^2 = 1/48 + 0.0173611\,\tilde{\epsilon} - 0.0157657\,\tilde{\epsilon}^2 + 0.029057\,\tilde{\epsilon}^3 + \cdots , \quad (19.123)$$

$$N = 5: \quad \eta/\epsilon^2 = 7/338 + 0.0161453\,\tilde{\epsilon} - 0.0148734\,\tilde{\epsilon}^2 + 0.0259628\,\tilde{\epsilon}^3 + \cdots , \quad (19.124)$$

$$N = 1: \quad \eta/\epsilon^2 = 1/54 + 0.01869\,\tilde{\epsilon} - 0.0176738\,\tilde{\epsilon}^2 + 0.0386577\,\tilde{\epsilon}^3 + \cdots . \quad (19.125)$$

These series can again be extended by four more terms $f_4\,\tilde{\varepsilon}^4 + f_5\,\tilde{\varepsilon}^5 + f_6\,\tilde{\varepsilon}^6 + f_7\,\tilde{\varepsilon}^7$, that make the strong-coupling coefficients $\bar{b}_0, \bar{b}_1, \bar{b}_2, \bar{b}_3$, in (19.14) agree with those in Eq. (19.120), we obtain the additional weak-coupling coefficients f_4, f_5, $f_6\,f_7$ as described above. However, here we encounter problems: the η-values from the interpolation come out too large by about a factor 2. Also γ does not interpolate well. The reason for this failure seems to be that the strong-coupling expansions (19.115) and (19.120) are divergent, whereas the interpolation yields in general *convergent* strong-coupling expansions.

Notes and References

A simple exposition of variational perturbation theory in the context of quantum mechanics can be found in the textbook in Ref. [2], which also contains citations to the numerous publications preparing the grounds for the recent progress. A certainly incomplete list is given below.
There were two main predecessors coming from different directions. From the mathematical side, the seminal paper was Ref. [3]. From the physical side, the development was triggered by Feynman and Kleinert in Ref. [1], and its systematic extension to arbitrary order in

H. Kleinert, Phys. Lett. A **173**, 332 (1993) (quant-ph/9511020).

The main steps of variation perturbation theory from quantum mechanics to quantum field theory was made in Refs. [5,6,10]. First, still in quantum mechanics, by exploiting previously unused even approximants which do not have an extremum, as explained in Chapter 5 of the above textbook. Second, the determination of the exponent ω of approach to scaling from the assumption of scaling behavior in the strong-coupling limit, and third, a reliable extrapolation procedure to infinite order on the basis of the theoretically known analytic order dependence (the latter being inspired by the pioneering work of Seznec and Zinn-Justin in Ref. [3]). These three improvements were essential in obtaining extremely accurate critical exponents rivaling the powerful combination of renormalization group and Borel-type resummation methods.

Here is a selected list of contributions at the level of quantum mechanics:
T. Barnes and G.I. Ghandour, Phys. Rev. D **22**, 924 (1980);
B.S. Shaverdyan and A.G. Usherveridze, Phys. Lett. B **123**, 316 (1983);
P.M. Stevenson, Phys. Rev. D **30**, 1712 (1985);
D **32**, 1389 (1985);
P.M. Stevenson and R. Tarrach, Phys. Lett. B **176**, 436 (1986);
A. Okopinska, Phys. Rev. D **35**, 1835 (1987);
D **36**, 2415 (1987);
W. Namgung, P.M. Stevenson, and J.F. Reed, Z. Phys. C **45**, 47 (1989);
U. Ritschel, Phys. Lett. B **227**, 44 (1989);
Z. Phys. C **51**, 469 (1991);
M.H. Thoma, Z. Phys. C **44**, 343 (1991);
I. Stancu and P.M. Stevenson, Phys. Rev. D **42**, 2710 (1991);
R. Tarrach, Phys. Lett. B **262**, 294 (1991);
H. Haugerud and F. Raunda, Phys. Rev. D **43**, 2736 (1991);
A.N. Sissakian, I.L. Solivtosv, and O.Y. Sheychenko, Phys. Lett. B **313**, 367 (1993);
A. Duncan and H.F. Jones, Phys. Rev. D **47**, 2560 (1993);

For the anharmonic oscillator, the highest accuracy in the strong-coupling limit was reached with exponentially fast convergence in
W. Janke and H. Kleinert, Phys. Rev. Lett. **75**, 2787 (1995) (quant-ph/9502019).
That paper contains references to earlier less accurate calculations of strong-coupling expansion coefficients from weak-coupling perturbation theory, in particular
F.M. Fernández and R. Guardiola, J. Phys. A **26**, 7169 (1993);
F.M. Fernández, Phys. Lett. A **166**, 173 (1992);
R. Guardiola, M.A. Sol´is, and J. Ros, Nuovo Cimento B **107**, 713 (1992).
A.V. Turbiner and A.G. Ushveridze, J. Math. Phys. **29**, 2053 (1988);
B. Bonnier, M. Hontebeyrie, and E.H. Ticembal, J. Math. Phys. **26**, 3048 (1985);
These works were yet unable to extract the exponential law of convergence from their data.

This was shown to be related to the convergence radius of the strong-coupling expansion in Ref. [4]. Predecessors of these works which did not yet explain the exponentially fast convergence in the strong-couplings limit were
I.R.C. Buckley, A. Duncan, and H.F. Jones, Phys. Rev. D **47**, 2554 (1993);
C.M. Bender, A. Duncan, and H.F. Jones, Phys. Rev. D **49**, 4219 (1994);
A. Duncan and H.F. Jones, Phys. Rev. D **47**, 2560 (1993);
C. Arvanitis, H.F. Jones, and C.S. Parker, Phys. Rev. D **52**, 3704 (1995) (hep-th/9502386);
R. Guida, K. Konishi, and H. Suzuki, Annals Phys. **241**, 152 (1995) (hep-th/9407027).

The individual citations in the text refer to:

[1] R.P. Feynman, H. Kleinert, Phys. Rev. A **34**, 5080 (1986) (http://www.physik.fu-berlin.de/~kleinert/159).

[2] See Chapters 5 and 17 in the textbook
H. Kleinert, *Path Integrals in Quantum Mechanics, Statistics and Polymer Physics*, World Scientific Publishing Co., Singapore 1995 (www.physik.fu-berlin.de/~kleinert/b5).

[3] R. Seznec and J. Zinn-Justin, J. Math. Phys. **20**, 1398 (1979).

[4] H. Kleinert and W. Janke, Phys. Lett. A **206**, 283 (1995);
R. Guida K. Konishi, and H. Suzuki, Ann. Phys. **249**, 106 (1996).

[5] H. Kleinert, Phys. Lett. A **207**, 133 (1995) (quant-ph/9507005).

[6] H. Kleinert, Phys. Rev. D **57**, 2264 (1998) (www.physik.fu-berlin.de/~kleinert/re3.html #257); addendum (cond-mat/9803268); Phys. Rev. D **60**, 85001 (1999) (hep-th/9812197).

[7] The general discussion is contained in the forthcoming 3rd edition of the textbook [2] available on the internet.

[8] W. Janke and H. Kleinert, Phys. Rev. Lett. **75**, 2787 (1995) (quant-ph/9502019).

[9] For the anharmonic oscillator where $p = 1/3$ and $q = 3$, the proof is found in Ref. [4]. The generalization to any p, q was given in Ref. [6].

[10] H. Kleinert, Phys. Lett. B **434**, 74 (1998) (cond-mat/9801167); ibid. B **463**, 69 (1999) (cond-mat/9906359).

[11] For a discussion of such contributions see
B.G. Nickel, Physica A *177*, 189 (1991);
A. Pelissetto and E. Vicari, Nucl. Phys. B *519*, 626 (1998);
522, 605 (1998); Nucl. Phys. Proc. Suppl. *73*, 775 (1999). See also the review article by these authors in cond-mat/0012164.

[12] H. Kleinert and B. Van den Bossche, (cond-mat/0011329)

[13] H. Kleinert and V. Schulte-Frohlinde, J. Phys. A **34** 1037 (2001) (cond-mat/9907214).

[14] R. Guida, J. Zinn-Justin, J. Phys. A **31**, 8130 (1998) (cond-mat/9803240).

[15] J. Kosterlitz and D. Thouless, J. Phys. C **7**, 1046 (1973).

[16] S. Hikami and E. Brézin, J. Phys. A **11**, 1141 (1978).

[17] G.E. Castilly and S. Chakravarty, Nucl. Phys. B **485**, 613 (1997) (cond-mat/9605088).

[18] P.W. Anderson, Phys. Rev. **109**, 1429 (1958).

[19] F.J. Wegner, Z. Phys. B **78**, 33 (1990).

[20] E. Brézin and S. Hikami, Phys. Rev. B **55**, R10169 (1997) (cond-mat/9612016).

[21] H. Kleinert, Phys. Lett. A **264**, 357 (2000) (hep-th/9808145).

[22] S. Hikami and E. Brézin, J. Phys. A **11**, 1141 (1978);
W. Bernreuther and F.J. Wegner, Phys. Rev. Lett. **57**, 1383 (1986);
I. Jack, D.R.T. Jones, and N. Mohammedi, Phys. Lett. B **220**, 171 (1989), Nucl. Phys. B **322**, 431 (1989);
N.A. Kivel, A.A. Stepanenko, and A.N. Vasil'ev, Nucl.Phys. B **424**, 619 (1994) (hep-th/9308073).

[23] This technique is described in detail in Ref. [5].

[24] This can also be seen in Fig. 3 of the first paper in Ref. [6].

[25] H. Kleinert, J. Neu, V. Schulte-Frohlinde, K.G. Chetyrkin, S.A. Larin, Phys. Lett. B **272**, 39 (1991) (hep-th/9503230); Erratum ibid. **319**, 545 (1993).
See also
H. Kleinert and V. Schulte-Frohlinde, Phys. Lett. B **342**, 284 (1995) (cond-mat/9503038).

20

Critical Exponents from Other Expansions

It is useful to compare the results obtained so far with other approaches to the critical exponents. One is a similar field-theoretic approach based on perturbation expansions of ϕ^4-theories. But instead of working in $D = 4 - \varepsilon$ dimensions and continuing the results to $\varepsilon = 1$ to obtain critical exponents in the physical dimension $D = 3$, perturbation expansions can be derived directly in three dimensions, and studied in the regime of small masses. Fundamentally, this approach is less satisfactory than the previous one since there is no small parameter analogous to ε which permits us to prove the existence of a scaling limit at least for small ε. Another disadvantage is that it is impossible to calculate the expansion coefficients analytically, except for one- and two loop-diagrams. In one respect the three-dimensional approach has, however, an advantage: the power series have long been available up to six loops for ω, η, and ν, and there exist recent results up to seven loops for the critical exponents η and ν. Although numerical expansions are not as esthetic as the exact expansions derived in this book, their increased order leads to a higher accuracy in the critical exponents, as we shall see below in Section 20.4.

Another approach is based on *high-temperature expansions* for lattice models of the O(N)-symmetric classical Heisenberg model with the energy of Eq. (1.52). The critical exponents will be derived from such expansions in Section 20.8.

Finally we take advantage of the fact that the O(N)-symmetric classical Heisenberg model is exactly solvable in the limit of large N, where it reduces to the spherical model [recall the remark after Eq. (1.69)]. It is therefore possible to expand the ϕ^4-theory around this limit in powers of $1/N$. The results will be compared with those from the perturbation expansions in Section 20.2.

20.1 Sixth-Order Expansions in Three Dimensions

The three-dimensional ϕ^4-theory is defined by the bare energy functional

$$E[\phi_B] = \int d^D x \left\{ \frac{1}{2} \left[\partial \phi_B(\mathbf{x}) \right]^2 + \frac{1}{2} m_B^2 \phi_B^2(\mathbf{x}) + \frac{48\pi}{N+8} \frac{g_B}{4!} \left[\phi_B^2(\mathbf{x}) \right]^2 \right\}, \qquad (20.1)$$

where the coupling constant is normalized to obtain the most convenient perturbation expansions. The field $\phi_B(\mathbf{x})$ is an N-dimensional vector, and the action is O(N)-symmetric in this vector space.

By calculating the Feynman diagrams up to six loops, and imposing the normalization conditions (9.23), (9.24), and (9.33) we find renormalized values of mass, coupling constant, and field related to the bare input quantities by renormalization constants Z_ϕ, Z_{m^2}, Z_g:

$$m_B^2 = m^2 \, Z_{m^2} Z_\phi^{-1}, \qquad g_B = g \, Z_g Z_\phi^{-2}, \qquad \phi_B = \phi \, Z_\phi^{1/2}. \qquad (20.2)$$

The divergences are removed by analytic regularization [1]. In the literature, we find expansions for m_B^2/m^2, g_B/g, and ϕ_B/ϕ in powers of g up to g^6 (six loops), the latter two having recently

364

been extended to the power g^7 (seven loops) [2, 3]. We shall first discuss in detail the variational resummation of the six-loop expansions. The small improvements of the critical exponents brought about by the seven-loop terms will be calculated separately in Section 20.4.

Introducing the reduced dimensionless coupling constants $\bar{g} \equiv g/m$ and $\bar{g}_B \equiv g_B/m$, these determine the renormalization group functions by the following equations:

$$\omega(\bar{g}) = \frac{d\beta(\bar{g})}{d\bar{g}}, \quad \beta(\bar{g}) = -\left\{\frac{d}{d\bar{g}}\log\frac{\bar{g}Z_g}{Z_\phi^2}\right\}^{-1} = -\bar{g}_B\left(\frac{d\bar{g}_B}{d\bar{g}}\right)^{-1}, \tag{20.3}$$

$$\eta(\bar{g}) = \beta(\bar{g})\frac{d}{d\bar{g}}\log Z_\phi = \beta(\bar{g})\frac{d}{d\bar{g}}\log\frac{\phi_B^2}{\phi^2}, \tag{20.4}$$

$$\eta_m(\bar{g}) = -\beta(\bar{g})\frac{d}{d\bar{g}}\log\frac{Z_{m^2}}{Z_\phi} = -\beta(\bar{g})\frac{d}{d\bar{g}}\log\frac{m_B^2}{m^2}. \tag{20.5}$$

For our purpose, we must study all quantities as functions of the reduced bare coupling constant $\bar{g}_B$. In terms of this variable, the beta function $\beta(\bar{g})$ is obtained from the logarithmic derivative

$$\beta(\bar{g}_B) = -\frac{d\bar{g}(\bar{g}_B)}{d\log\bar{g}_B} = -\bar{g}_B\bar{g}'(\bar{g}_B). \tag{20.6}$$

The function $\omega(\bar{g}_B)$ may then be obtained from the function $\bar{g} = \bar{g}(\bar{g}_B)$ by the logarithmic derivative

$$\omega(\bar{g}_B) = -\frac{d\log\beta(\bar{g}_B)}{d\log\bar{g}_B} = -\frac{d\log[\bar{g}_B\bar{g}'(\bar{g}_B)]}{d\log\bar{g}_B} = -1 - \bar{g}_B\frac{\bar{g}''(\bar{g}_B)}{\bar{g}'(\bar{g}_B)}. \tag{20.7}$$

Comparison with Eq. (20.21) shows that if $\bar{g}(\bar{g}_B)$ goes to a constant $\bar{g}^*$ in the strong-coupling limit $\bar{g}_B \to \infty$, the limiting value $\omega \equiv \omega(\infty)$ plays the role of the approach parameter $2/q$ of the strong-coupling expansion of the function $\bar{g}(\bar{g}_B)$.

Similarly we convert Eqs. (20.4) and (20.5) into functions of the bare coupling constant g_B:

$$\eta_m(\bar{g}_B) = -\frac{d}{d\log\bar{g}_B}\log\frac{m^2}{m_B^2}, \qquad \eta(\bar{g}_B) = \frac{d}{d\log\bar{g}_B}\log\frac{\phi^2}{\phi_B^2}, \tag{20.8}$$

which correspond to the relations (19.67) in $4 - \varepsilon$ dimensions. If $\eta_m(\bar{g}_B)$ and $\eta(\bar{g}_B)$ have finite strong-coupling limits $\eta_m = \eta_m(\infty)$ and $\eta = \eta(\infty)$, these equations imply the strong-coupling behaviors

$$\frac{m^2}{m_B^2} \propto \bar{g}_B^{-\eta_m}, \qquad \frac{\phi^2}{\phi_B^2} \propto \bar{g}_B^\eta, \tag{20.9}$$

corresponding to (19.68) in $4 - \varepsilon$ dimensions. By replacing the reduced coupling constant $\bar{g}_B$ by g_B/m, this implies the small-mass behavior at a fixed bare coupling g_B:

$$\frac{m^2}{m_B^2} = \text{const} \times m^{\eta_m}, \qquad \frac{\phi^2}{\phi_B^2} = \text{const} \times m^{-\eta}. \tag{20.10}$$

The six-loop expansions are the following:

$$\omega(\bar{g}) = -1 + 2\bar{g}(8 + N) - 3\bar{g}^2(760/27 + 164N/27) + 4\bar{g}^3(199.640417 + 54.94037698N + 1.34894276N^2)$$
$$+ 5\bar{g}^4(-1832.206732 - 602.5212305N - 35.82020378N^2 + 0.15564589N^3)$$
$$+ 6\bar{g}^5(20770.17697 + 7819.564764N + 668.5543368N^2 + 3.2378762N^3 + 0.05123618N^4)$$

$$+7\bar{g}^6 \left(-271300.0372 - 114181.4357N - 12669.22119N^2 \right.$$
$$\left. -265.8357032N^3 + 1.07179839N^4 + 0.02342417N^5\right) \tag{20.11}$$

$$\eta(\bar{g}) = \bar{g}^2 \left(16/27 + 8N/27\right) + \bar{g}^3 \left(0.3949440224 + 0.246840014N + 0.0246840014N^2\right)$$
$$+\bar{g}^4 \left(6.512109933 + 4.609221057N + 0.6679859202N^2 - 0.0042985626N^3\right)$$
$$+\bar{g}^5 \left(-21.64720643 - 15.1880934N - 1.891139282N^2 + 0.1324510614N^3 - 0.0065509222N^4\right)$$
$$+\bar{g}^6 \left(369.7130739 + 300.7208933N + 64.07744656N^2 \right.$$
$$\left. +3.054030987N^3 - 0.0203994485N^4 - 0.0055489202N^5\right) \tag{20.12}$$

$$\eta_m(\bar{g}) = \bar{g}\left(2+N\right) - \bar{g}^2 \left(92/27 + 46N/27\right) + \bar{g}^3 \left(18.707787762 + 12.625201157N + 1.6356536385N^2\right)$$
$$+\bar{g}^4 \left(-134.28726152 - 98.33833174N - 15.117303198N^2 + 0.2400236453N^3\right)$$
$$+\bar{g}^5 \left(1318.4281763 + 1046.8184247N + 209.71327323N^2 + 8.143135609N^3 + 0.0937915707N^4\right)$$
$$+\bar{g}^6 \left(-15281.544489 - 12918.644832N - 2980.2279474N^2 \right.$$
$$\left. -164.6575873N^3 + 3.0931477063N^4 + 0.0495801299N^5\right). \tag{20.13}$$

Here and in the subsequent set of perturbative expansions, we save space by omitting in each term $\bar{g}^n$ a denominator $(N+8)^n$.

By integrating (20.3), we see that $\omega(\bar{g})$ implies the relation between bare and renormalized coupling constant:

$$\bar{g}_B = \bar{g}\left[1 + \bar{g}\left(8+N\right) + \bar{g}^2 \left(1348/27 + 350N/27 + N^2\right)\right.$$
$$+\bar{g}^3 \left(315.8307562667 + 120.7825947383N + 17.3632278267N^2 + N^3\right)$$
$$+\bar{g}^4 \left(1813.1642655362 + 949.9400421368N + 203.4347168377N^2 + 21.5210551192N^3 + N^4\right)$$
$$+\bar{g}^5 \left(11664.58684418 + 7259.6266136476N + 1965.0940131759N^2 \right.$$
$$\left. +298.9857773851N^3 + 25.5436671032N^4 + N^5\right)$$
$$+\bar{g}^6 \left(57253.8939657167 + 47753.8060061961N + 16981.2530394653N^2 + 3357.7450242179N^3 \right.$$
$$\left.\left. +407.679442164N^4 + 29.4800395765N^5 + N^6\right)\right]. \tag{20.14}$$

This can be inverted to

$$\bar{g} = \bar{g}_B\left[1 - \bar{g}_B\left(8+N\right) + \bar{g}_B^2 \left(2108/27 + 514N/27 + N^2\right)\right.$$
$$+\bar{g}_B^3 \left(-878.7937193 - 312.63444671N - 32.54841303N^2 - N^3\right)$$
$$+\bar{g}_B^4 \left(11068.06183 + 5100.403285N + 786.3665699N^2 + 48.21386744N^3 + N^4\right)$$
$$+\bar{g}_B^5 \left(-153102.85023 - 85611.91996N - 17317.702545N^2 - 1585.1141894N^3 - 65.82036203N^4 - N^5\right)$$
$$+\bar{g}_B^6 \left(2297647.148 + 1495703.313\,N + 371103.0896N^2 + 44914.04818N^3 \right.$$
$$\left.\left. +2797.291579N^4 + 85.21310501N^5 + N^6\right)\right], \tag{20.15}$$

where we suppress a denominator $(N+8)^n$ in each term $\bar{g}_B^n$. Inserting this into the functions (20.11), (20.12), and (20.13), they become

$$\omega(\bar{g}_B) = -1 + 2\bar{g}_B\left(8+N\right) - \bar{g}_B^2 \left(1912/9 + 452N/9 + 2N^2\right)$$
$$+ \bar{g}_B^3 \left(3398.857964 + 1140.946693N + 95.9142896N^2 + 2N^3\right)$$
$$+ \bar{g}_B^4 \left(-60977.50127 - 26020.14956N - 3352.610678N^2 - 151.1725764N^3 - 2N^4\right)$$
$$+ \bar{g}_B^5 \left(1189133.101 + 607809.998N + 104619.0281N^2 + 7450.143951N^3 + 214.8857494N^4 + 2N^5\right)$$
$$+ \bar{g}_B^6 \left(-24790569.76 - 14625241.87N - 3119527.967N^2 \right.$$
$$\left. -304229.0255N^3 - 14062.53135N^4 - 286.3003674N^5 - 2N^6\right), \tag{20.16}$$

$$\eta(\bar{g}_B) = \bar{g}_B^2 \left(16/27 + 8N/27\right) + \bar{g}_B^3 \left(-9.086537459 - 5.679085912N - 0.5679085912N^2\right)$$
$$+ \bar{g}_B^4 \left(127.4916153 + 94.77320534N + 17.1347755N^2 + 0.8105383221N^3\right)$$
$$+ \bar{g}_B^5 \left(-1843.49199 - 1576.46676N - 395.2678358N^2 - 36.00660242N^3 - 1.026437849N^4\right)$$
$$+ \bar{g}_B^6 \left(28108.60398 + 26995.87962N + 8461.481806N^2 + 1116.246863N^3 \right.$$

$$+ 62.8879068 N^4 + 1.218861532 N^5 \big) \tag{20.17}$$

$$
\begin{aligned}
\eta_m(\bar{g}_B) = {}& \bar{g}_B \left(2 + N\right) - \bar{g}_B^2 \left(-524/27 - 316N/27 - N^2\right) \\
& + \bar{g}_B^3 \left(229.3744544 + 162.8474234 N + 26.08009809 N^2 + N^3\right) \\
& + \bar{g}_B^4 \left(-3090.996037 - 2520.848751 N - 572.3282893 N^2 - 44.32646141 N^3 - N^4\right) \\
& + \bar{g}_B^5 \left(45970.71839 + 42170.32707 N + 12152.70675 N^2 + 1408.064008 N^3 + 65.97630108 N^4 + N^5\right) \\
& + \bar{g}_B^6 \left(-740843.1985 - 751333.064 N - 258945.0037 N^2 - 39575.57037 N^3 \right. \\
& \qquad \left. - 2842.8966 N^4 - 90.7145582 N^5 - N^6\right).
\end{aligned}
\tag{20.18}
$$

Let us also write down a power series expansion for the function $\gamma(g_B) = [2 - \eta(g_B)]/[2 - \eta_m(g_B)]$ which tends to the critical exponent γ of susceptibility. In resummation procedures applied to functions of the renormalized coupling constant g, this series has always been favored over that for $\eta(g)$ since, in contrast to $\eta(g)$, its expansion coefficients of $\gamma(g)$ have alternating signs permitting application of Padé-Borel resummation techniques [4]. Note, however, that the expansion (20.17) in powers of $\bar{g}_B$ does have alternating signs. For $\gamma(g_B)$, the series reads

$$
\begin{aligned}
\gamma(\bar{g}_B) = {}& 1 + \bar{g}_B \left(2 + N\right)/2 + \bar{g}_B^2 \left(-9 - 5N - N^2/4\right) \\
& + \bar{g}_B^3 \left(100.5267922 + 64.05955095 N + 7.148077413 N^2 + 0.125 N^3\right) \\
& + \bar{g}_B^4 \left(-1306.696473 - 953.5355208 N - 165.6165894 N^2 - 7.886473674 N^3 - 0.0625 N^4\right) \\
& + \bar{g}_B^5 \left(19047.24345 + 15717.20743 N + 3667.58258 N^2 + 300.9668324 N^3 + 7.848484825 N^4 + 0.03125 N^5\right) \\
& + \bar{g}_B^6 \left(-304324.882 - 279842.9929 N - 81107.12259 N^2 - 9519.124419 N^3 - 457.7147389 N^4 \right. \\
& \qquad \left. - 7.463312096 N^5 - 0.015625 N^6\right).
\end{aligned}
\tag{20.19}
$$

20.2 Critical Exponents up to Six Loops

We are now ready to apply our strong-coupling theory to these expansions. First we study the $\bar{g}_B \to \infty$ -limit of the series (20.15) for the renormalized reduced coupling constant $\bar{g}$. We expect the theory to reproduce experimentally observed scaling laws which means that $\bar{g}(\bar{g}_B)$ should tend to some constant value: $\bar{g}(\bar{g}_B) \to \bar{g}^*$ for $\bar{g}_B \to \infty$. The leading power parameter $p/q = s$ in $\bar{g}_B$ must be set equal to zero, implying for the power series of $\bar{g}(\bar{g}_B)$ the vanishing of the logarithmic derivative in the strong-coupling limit, as before in Eq (19.64):

$$
0 = \bar{g}_B \left. \frac{\bar{g}'(\bar{g}_B)}{\bar{g}(\bar{g}_B)} \right|_{\bar{g}_B = \infty}.
\tag{20.20}
$$

This equation is used to determine the critical exponent $\omega = 2/q$.

Alternatively, we may determine ω from the strong-coupling limit of the self-consistency condition Eq. (19.65) for ω, which reads here

$$
\omega = -1 - \bar{g}_B \left. \frac{\bar{g}''(\bar{g}_B)}{\bar{g}'(\bar{g}_B)} \right|_{\bar{g}_B = \infty}.
\tag{20.21}
$$

We now insert the expansion (20.15) for $\bar{g}(\bar{g}_B)$ into Eq. (20.20), re-expand up to order $\bar{g}_B^7$, and resum the series by variational perturbation theory for increasing orders L. The zeros determine the successive approximations ω_L for $L = 3, 4, 5, 6$. This is done for $O(N)$-symmetric theories with $N = 0, 1, 2, 3, \ldots$. The values of the highest available approximation ω_6 are found in the parentheses of the last column of Table 20.1 on page 373.

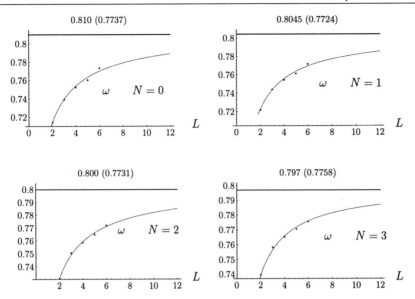

FIGURE 20.1 Behavior of strong-coupling values $\omega = \omega(\infty)$ with increasing order L of the approximation, for O(N)-symmetric theories with $N = 0, 1, 2, 3, \ldots$. These are the critical exponents observable in the approach to scaling of the second-order phase transitions in dilute polymer solutions, Ising magnets, superfluid helium, and the classical Heisenberg model. The points are fitted by $\omega - b\,e^{-cL^{1-\omega}}$ (dashed line), determining by extrapolation the limiting values indicated by the horizontal lines, which are plotted in the first plot of Fig. 20.2 and listed in Table 20.1 on page 373. The highest approximation ω_6 is indicated on top of the line in parentheses.

Since L is not very large, the results require extrapolation to infinite L. The functional form of the L-dependence was determined in (19.51), predicting the large-L behavior

$$\omega_L \approx \omega - b\,e^{-cL^{1-\omega}}. \tag{20.22}$$

In the upper left-hand plot of Fig. 20.1 we illustrate how the successive approximations ω_2—ω_6 are fitted by this asymptotic curve for O(N)-theories with $N = 0$ (dilute polymer solutions), $N = 1$, (Ising), $N = 2$ (superfluid helium), $N = 3$ (classical Heisenberg model). The extrapolated values obtained by such fits for infinite L are shown in the last column of Table 20.1 on page 373. They are plotted in the first plot of Fig. 20.2, together with the sixth-order approximation to show the significance of the $L \to \infty$ -extrapolation.

Our numbers merge smoothly with the $1/N$-expansion curve which has been calculated [5] to order $1/N^2$:

$$\omega = 1 - 8\frac{8}{3\pi^2}\frac{1}{N} + 2\left(\frac{104}{3} - \frac{9}{2}\pi^2\right)\left(\frac{8}{3\pi^2}\frac{1}{N}\right)^2 + \mathcal{O}(N^{-3}). \tag{20.23}$$

To judge the internal consistency of our procedure, we calculate ω once more from Eq. (20.21) for $\omega = 2/q$. Since ω appears also on the right-hand side of the series via the parameter q, this represents a self-consistency relation which can be iterated until input and output values for ω coincide for each L. The results are shown in Fig. 20.3 for increasing orders $L = 2, 3, 4, 5, 6$, in the Ising case $N = 1$. The data points are again fitted with the functional behavior (20.22), imposing the same extrapolated $\omega(\infty)$-values as in the previous fits.

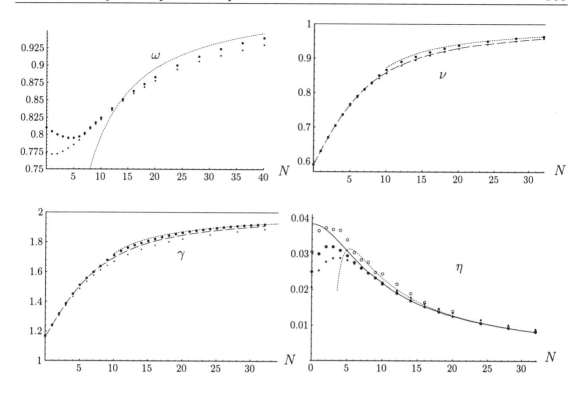

FIGURE 20.2 Sixth-order approximations to $\omega, \nu, \gamma, \eta$ (thin dots), and their $L \to \infty$ -limit (fat dots). The dashed lines in the second and third figures are an interpolation to the Padé-Borel resummations of S.A. Antonenko and A.I. Sokolov, Phys. Rev. E *51*, 1894 (1995) (where ω was not calculated). Their data for η scatter too much to be represented in this way—they are indicated by small circles. The dotted curves show $1/N$ -expansions for all four quantities. Note that our results lie closer to these than those of S.A. Antonenko and A.I. Sokolov. The solid η-curve is explained after Eq. (20.27).

The critical exponent ω can also be calculated from the expansions for γ and ν in this self-consistent way, as shown in Fig. 20.4.

Note that the agreement between the self-consistent ω-values with the previous ones determined from the $p = 0$ -condition can be considered as a confirmation of the hypothesis that the theory does indeed have a definite strong-coupling limit in which $\bar{g}(\bar{g}_B)$ tends to a constant $\bar{g}^*$ (an *infrared-stable fixed point* in the language of the renormalization group). It also implies all other scaling properties to be derived below.

Proceeding to other critical exponents, we now take the function $\nu^{-1}(\bar{g}_B) = 2 - \eta_m(\bar{g}_B)$ to the strong-coupling limit, to determine $\nu = \nu(\infty)$. The extrapolations to large L are done with the help of the approximations $\nu_2, \nu_3, \ldots, \nu_6$, as illustrated in Fig. 20.5. The resulting critical exponents are plotted against N in the second plot of Fig. 20.2, and listed in Table 20.1 on page 373. The points are seen to merge well with those of the $1/N$ -expansion [5] for ν:

$$\nu = 1 - 4\frac{8}{3\pi^2}\frac{1}{N} + \left(\frac{56}{3} - \frac{9}{2}\pi^2\right)\left(\frac{8}{3\pi^2}\frac{1}{N}\right)^2 + \mathcal{O}(N^{-3}). \tag{20.24}$$

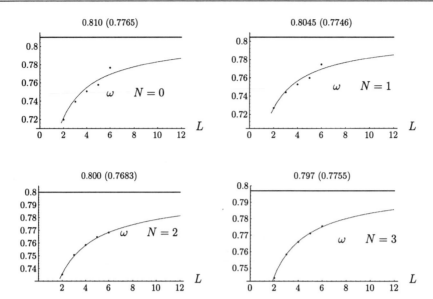

FIGURE 20.3 Behavior of self-consistent strong-coupling values $\omega = \omega(\infty)$ from Eq. (20.21) for $2/q = \omega$ with increasing order L of the approximation, for $O(N)$-symmetric theories with $N = 0,\ 1,\ 2,\ 3,\ \ldots$.

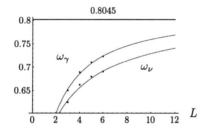

FIGURE 20.4 The ω-exponents obtained from the expansions of γ and ν for $N = 1$. Their convergence is much slower.

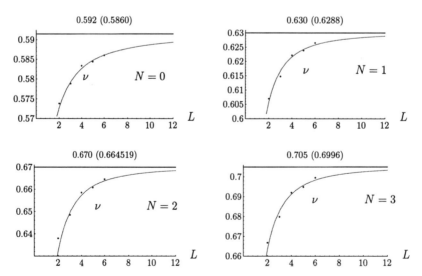

FIGURE 20.5 Plot analogous to Fig. 20.1 of critical exponent $\nu = \nu(\infty)$, illustrating the extrapolation procedure to $L \to \infty$ for $N = 0,\ 1,\ 2,\ 3$. The results of the extrapolation are plotted against L in the second plot of Fig. 20.2 and listed in Table 20.1 on page 373.

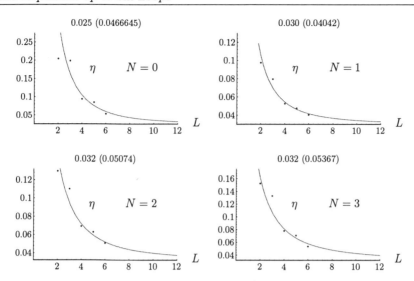

FIGURE 20.6 Plot analogous to Fig. 20.1 of critical exponent $\eta = \eta(\infty)$, illustrating the extrapolation procedure to $L \to \infty$ for $N = 0, 1, 2, 3$. The results of the extrapolation are plotted against N in the fourth plot of Fig. 20.2 and listed in Table 20.1 on page 373.

The function $\gamma(\bar{g}_B)$ is treated somewhat differently. Since γ serves to determine the critical exponent η via the scaling relation $\eta = 2 - \gamma/\nu$, and since this combination is very sensitive to small errors in γ (and in ν), we proceed as follows. After applying our method to the γ-series and calculating the approximations $\gamma_2, \gamma_3, \ldots, \gamma_6$, we go over to $\eta_2, \eta_3, \ldots, \eta_6$ via the scaling relation $\eta_L = 2 - \gamma_L/\nu$ and perform the extrapolation $L \to \infty$ on these η-values, as illustrated in Fig. 20.6. In this way, we obtain the smooth η-curves shown in the fourth plot of Fig. 20.2 and listed in Table 20.1 on page 373. The values of the highest approximation η_6 in the parentheses of Table 20.1 are obtained from a direct variational treatment of the perturbation expansion for η. These are closer to the final η-values than the approximations of $\eta_6 = 2 - \gamma_6/\nu$ used for the extrapolation, which are indicated in the parentheses on top of the plots of Fig. 20.6. Still, we have used the latter for extrapolation since there are five of them to be fitted with the asymptotic curve (20.22), while the η-expansion which has no linear term in $\hat{g}_B$ provides us only with four values, making a fit less reliable.

The extrapolated η-values are inserted into the scaling relation $\gamma = \nu(2 - \eta)$ to derive the extrapolated γ-values plotted in the third plot of Fig. 20.2 and listed in Table 20.1 on page 373. A direct extrapolation $L \to \infty$ of the variational approximations γ_L to the γ-expansion turns out to be fully compatible with the previous ones, as illustrated in Fig. 20.7.

For large N, our critical exponents η are in excellent agreement with the $1/N$-expansion, which is known [5, 6] up to order $1/N^3$:

$$\eta = \frac{8}{3\pi^2}\frac{1}{N} - \frac{28}{3}\left(\frac{8}{3\pi^2}\frac{1}{N}\right)^2 - \left[\frac{653}{18} - \left(27\log 2 + \frac{47}{4}\right)\zeta(2) + \frac{189}{4}\zeta(3)\right]\left(\frac{8}{3\pi^2}\frac{1}{N}\right)^3 + \mathcal{O}(N^{-4}),$$
(20.25)

where $\zeta(x)$ is Riemann's zeta function. Note that for η the finite-N corrections are very small. In fact, from the η_6-values near $N = 1000$, we can extract numerically the $1/N$-expansion

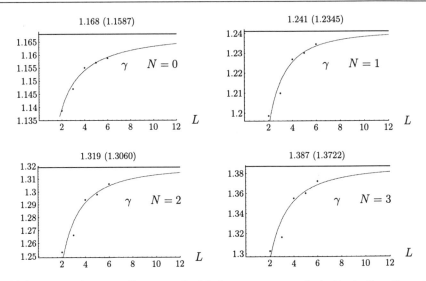

FIGURE 20.7 Plot analogous to Fig. 20.1 of critical exponent $\gamma = \gamma(\infty)$, illustrating the extrapolation procedure to $L \to \infty$ for $N = 0, 1, 2, 3$. The results of the extrapolation are plotted against N in the third plot of Fig. 20.2 and listed in Table 20.1 on page 373.

$$\eta \approx 0.303 \frac{1}{N} - 0.104 \frac{1}{N^2}, \tag{20.26}$$

which agrees reasonably well with the exact expansion

$$\eta \approx 0.270 \frac{1}{N} - 0.195 \frac{1}{N^2}. \tag{20.27}$$

It is worth pointing out that we may also apply our strong-coupling theory to $1/N$ - expansions for η, to find expressions valid for all N down to $N = 0$ by treating $1/N$ just like the variable g_B in Section 19.1. Taking the $1/N$-expansion as an example, we derive a smooth fit from large to rather small N by adding another term $-104/N^4$ to (20.25) before going through the resummation. The extra term improves the fit, such that it is probably a good prediction for the yet unknown next term in the $1/N$-expansion.

Since η starts out linearly at the "strong-coupling" value $1/N = \infty$, the parameters p and q are 0 and 2, respectively. The resulting curve is shown in the fourth plot of Fig. 20.2. It fits all data except for very small N. The figure shows also successive approximations provided by the $1/N$-expansion.

The exponents γ also merge well with their $1/N$-expansion [7]

$$\gamma = 2 - 9 \frac{8}{3\pi^2} \frac{1}{N} + \left(44 - 9\pi^2\right) \left(\frac{8}{3\pi^2} \frac{1}{N}\right)^2 + \mathcal{O}(N^{-3}), \tag{20.28}$$

as seen in the third plot of Fig. 20.2.

Finally, we exhibit the full power of our theory by plotting for the Ising case $N = 1$ the functions $\bar{g}(\bar{g}_B), \omega(\bar{g}_B), \nu(\bar{g}_B), \eta(\bar{g}_B)$ for all coupling strengths in Figs. 20.8–20.10, together with the diverging perturbative approximations as well as the convergent strong-coupling expansion. We do this successively for each increasing order L of the approximations. On a logarithmic plot, the quality of these very different approximations looks surprisingly similar.

Although the functions $\bar{g}(\bar{g}_B), \omega(\bar{g}_B), \nu(\bar{g}_B), \eta(\bar{g}_B)$ in Figs. 20.8–20.10 are derived by a numerical variational procedure, it is possible to write the results down in the form of new

TABLE 20.1   Critical exponents of Ref. [8] in comparison with those obtained by Padé-Borel resummation in Ref. [3], as well as earlier results (all cited in Notes and References). They refer to six-loop expansions in $D = 3$ dimensions ([2, 9]), or to five-loop expansions in $\epsilon = 4 - D$ ([10, 11]). For each of our results we give the $L = 6$ -approximation in parentheses to display the importance of the extrapolation procedure $L \to \infty$.

N	g_c	$\gamma(\gamma_6)$	$\eta(\eta_6)$	$\nu(\nu_6)$	α	β	$\omega\ (\omega_6)$	
0		1.168(1.159)	0.025(0.0206)	0.592(0.586)			0.810(0.7737)	
	1.402	1.160	0.034	0.589	0.231	0.305		[3]
	1.421±0.004	1.161±0.003	0.026±0.026	0.588±0.001	0.236±0.004	0.302±0.004	0.794 ± 0.06	[2]
	1.421 ± 0.008	1.1615 ± 0.002	0.027 ± 0.004	0.5880±0.0015		0.3020±0.0015	0.80 ± 0.04	[9]
		1.160 ± 0.004	0.031 ± 0.003	0.5885±0.0025		0.3025±0.0025	0.82 ± 0.04	[11]
1		1.241(1.235)	0.030(0.0254)	0.630(0.627)			0.805(0.7724)	
	1.419	1.239	0.038	0.631	0.107	0.327	0.781	[3]
	1.416±0.0015	1.241±0.004	0.031±0.011	0.630±0.002	0.110±0.008	0.324 ± 0.06	0.788±0.003	[2]
	1.416±0.004	1.2410±0.0020	0.031±0.004	0.6300±0.0015		0.3250±0.0015	0.79±0.03	[9]
			0.035 ± 0.002	0.628±0.001			0.80 ± 0.02	[10]
		1.1239 ± 0.004	0.037 ± 0.003	0.6305±0.0025		0.3265±0.0025	0.81 ± 0.04	[11]
2		1.318(1.306)	0.032(0.0278)	0.670(0.665)			0.800(0.7731)	
	1.408	1.315	0.039	0.670	-0.010	0.348	0.780	[3]
	1.406±0.005	1.316±0.009	0.032±0.015	0.669±0.003	-0.007±0.009	0.346±0.009	0.78 ± 0.01	[2]
	1.406±0.004	1.3160±0.0025	0.033±0.004	0.6690±0.0020		0.3455±0.002	0.78±0.025	[9]
			0.037 ± 0.002	0.665±0.001			0.79 ± 0.02	[10]
		1.315 ± 0.007	0.040 ± 0.003	0.671±0.005		0.3485±0.0035	0.80 ± 0.04	[11]
3		1.387(1.372)	0.032(0.0288)	0.705(0.700)			0.797(0.7758)	
	1.392	1.386	0.038	0.706	-0.117	0.366	0.780	[3]
	1.392±0.009	1.390±0.01	0.031±0.022	0.705±0.005	-0.115±0.015	0.362	0.78 ± 0.02	[2]
	1.391 ±0.004	1.386±0.004	0.033±0.004	0.705±0.003		0.3645±0.0025	0.78 ± 0.02	[9]
			0.037 ± 0.002	0.79±0.02			0.79 ± 0.02	[10]
		1.390 ± 0.010	0.040 ± 0.003	0.710±0.007		0.368±0.004	0.79 ± 0.04	[11]
4		1.451(1.433)	0.031(0.0289)	0.737(0.732)			0.795(0.780)	
	1.375	1.449	0.036	0.738	-0.213	0.382	0.783	
5		1.511(1.487)	0.0295(0.0283)	0.767(0.760)			0.795(0.785)	
	1.357	1.506	0.034	0.766	-0.297	0.396	0.788	[3]
6		1.558(1.535)	0.0276(0.0273)	0.790(0.785)			0.797(0.792)	
	1.339	1.556	0.031	0.790	-0.370	0.407	0.793	[3]
7		1.599(1.577)	0.0262(0.0260)	0.810(0.807)			0.802(0.800)	
	1.321	1.599	0.029	0.811	-0.434	0.417	0.800	[3]
8		1.638(1.612)	0.0247(0.0246)	0.829(0.825)			0.810(0.808)	
	1.305	1.637	0.027	0.830	-0.489	0.426	0.808	[3]
9		1.680(1.643)	0.0233(0.0233)	0.850(0.841)			0.817(0.815)	
	1.289	1.669	0.025	0.845	-0.536	0.433	0.815	[3]
10		1.713(1.670)	0.0216(0.0220)	0.866(0.854)			0.824(0.822)	
	1.275	1.697	0.024	0.859	-0.576	0.440	0.822	[3]
12		1.763(1.716)	0.0190(0.0198)	0.890(0.877)			0.838(0.835)	
	1.249	1.743	0.021	0.881	-0.643	0.450	0.836	[3]
14		1.795(1.750)	0.0169(0.0178)	0.905(0.894)			0.851(0.849)	
	1.227	1.779	0.019	0.898	-0.693	0.457	0.849	[3]
16		1.822(1.779)	0.0152(0.0161)	0.918(0.907)			0.862(0.860)	
	1.208	1.807	0.017	0.911	-0.732	0.463	0.861	[3]
18		1.845(1.803)	0.0148(0.0137)	0.929(0.918)			0.873(0.869)	
	1.191	1.829	0.015	0.921	-0.764	0.468	0.871	[3]
20		1.864(1.822)	0.0125(0.0135)	0.938(0.927)			0.883(0.878)	
	1.177	1.847	0.014	0.930	-0.789	0.471	0.880	[3]
24		1.890(1.850)	0.0106(0.0116)	0.950(0.939)			0.900(0.894)	
	1.154	1.874	0.012	0.942	-0.827	0.477	0.896	[3]
28		1.909(1.871)	0.009232(0.01010)	0.959(0.949)			0.913(0.906)	
	1.136	1.893	0.010	0.951	-0.854	0.481	0.909	[3]
32		1.920(1.887)	0.00814(0.00895)	0.964(0.955)			0.924(0.915)	
	1.122	1.908	0.009	0.958	-0.875	0.483	0.919	[3]

strong-coupling expansions which converge for *all* coupling strengths. We shall demonstrate this only for the Ising case $N = 1$, where $\omega = 0.805$. Consider first the function $\bar{g}(\bar{g}_B)$ with the sixth-order strong-coupling expansion (19.10) (whose $\bar{g}_B$-dependence is displayed in Fig. 20.8):

$$
\begin{aligned}
\bar{g} \;=\; & 1.400036164909792 - 2.015076019427151/\bar{g}_B^{\omega} + 2.512390732560552/\bar{g}_B^{2\omega} \\
& - 2.903034628806387/\bar{g}_B^{3\omega} + 3.123423917471507/\bar{g}_B^{4\omega} - 3.108796470872297/\bar{g}_B^{5\omega} \\
& + 2.844130229268904/\bar{g}_B^{6\omega} - 2.38207097645026/\bar{g}_B^{7\omega}. \tag{20.29}
\end{aligned}
$$

For $\bar{g}_B \to \infty$, this converges to the fixed point $\bar{g}^* = 1.400036164909792$, the critical coupling constant. The important observation is that by changing variables to $x = x(\bar{g}_B)$ defined by

$$
\bar{g}_B \equiv (1 - x)/x^{1/\omega}, \tag{20.30}
$$

and re-expanding up to the order x^6, we obtain a new modified strong-coupling expansion

$$
\begin{aligned}
\bar{g}(x) \;=\; & 1.400036164909792 - 2.015076019427151x + 0.890254536921696x^2 \\
& - 0.322063465947966x^3 + 0.02243448556302718x^4 + 0.02753503840016558x^5 \\
& + 0.004440133592881424x^6 - 0.0050524666607713413x^7. \tag{20.31}
\end{aligned}
$$

Numerically, this happens to converge for *all* $\bar{g}_B \in (0, \infty)$ where $x \in (1, 0)$. Indeed, when plotting this function in Fig. 20.8, it falls right on top of the previously calculated curve representing the full variational expression (19.6). To verify the convergence for small couplings,

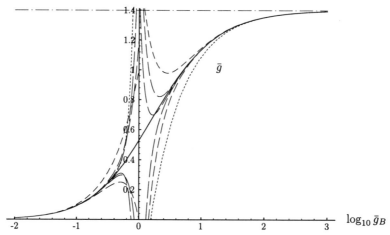

FIGURE 20.8 Logarithmic plot of variational perturbation result for expansion (20.15) of renormalized coupling constant $\bar{g}(\bar{g}_B)$ with $N = 1$ as a function of the bare coupling constant $\bar{g}_B$. The small-$\bar{g}_B$ regime shows truncated divergent perturbation expansions; the large-$\bar{g}_B$ regime shows the truncated strong-coupling expansions. Dash lengths increase with orders. The convergent re-expanded strong-coupling series (20.31) falls on top of the full curve.

we insert $x = 1$ corresponding to $\bar{g}_B = 0$ and obtain 0.0025, which misses only slightly the free-field value 0.

Similarly we obtain a strong-coupling expansion of the critical exponent $\nu(\bar{g}_B)$. In accordance with the above determination of ν from a resummation of the series for the inverse

$\nu^{-1}(\bar{g}_B) = 2 - \eta_m(\bar{g}_B)$, we first derive the strong-coupling expansion for the inverse, and invert the resulting power series. The result is

$$
\begin{aligned}
\nu(\bar{g}_B) &= 0.6264612502473953 - 0.2094930499887895/\bar{g}_B^{\omega} + 0.3174116223980956/\bar{g}_B^{2\omega} \\
&\quad - 0.4674929704457459/\bar{g}_B^{3\omega} + 0.6755577434122434/\bar{g}_B^{4\omega} - 0.961768075384486/\bar{g}_B^{5\omega} \\
&\quad + 1.35507538545696/\bar{g}_B^{6\omega}.
\end{aligned}
\tag{20.32}
$$

Changing again to the variable $x = x(\bar{g}_B)$, and re-expanding up to the order x^6, we obtain the convergent series

$$
\begin{aligned}
\nu(x) &= 0.6264612502473953 - 0.2094930499887895x + 0.1487697171571201x^2 \\
&\quad - 0.1086595778647923x^3 + 0.07115354531150436x^4 - 0.04710070728711805x^5 \\
&\quad + 0.0316053426101743x^6.
\end{aligned}
\tag{20.33}
$$

A plot of this expansion for ν in Fig. 20.9 is again indistinguishable from the full variational perturbation curve for all $\bar{g}_B$. At $x = 1$ corresponding to $\bar{g}_B = 0$, this series gives now 0.5127, which is only about 2% larger than the free-field value $1/2$.

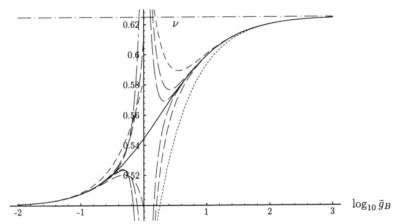

FIGURE 20.9 Logarithmic plot of variational perturbation result for exponent $\nu(\bar{g}_B) = 1/[2-\eta_m(\bar{g}_B)]$ obtained from expansion (20.18) with $N = 1$, and of the convergent re-expanded strong-coupling series (20.35), as a function of the bare coupling constant $\bar{g}_B$. The various dashed curves are explained in Fig. 20.8.

By inverting the series (20.31), we find x as a function of the deviation $\Delta\bar{g} \equiv \bar{g} - \bar{g}^*$ of the renormalized coupling from its strong-coupling value $\bar{g}^*$:

$$
\begin{aligned}
x(\bar{g}) &= -\,0.4962591933798516\Delta\bar{g} + 0.1088027548080915\Delta\bar{g}^2 - 0.02817579418977608\Delta\bar{g} \\
&\quad + 0.005412315522686147\Delta\bar{g}^4 + 0.00005845448187202598\Delta\bar{g}^5 \\
&\quad - 0.0006265703321966366\Delta\bar{g}^6 + 0.0002698406382563225\Delta\bar{g}^7.
\end{aligned}
\tag{20.34}
$$

After inserting this into (20.33), we obtain the critical exponent ν as a power series in $\Delta\bar{g}$:

$$
\begin{aligned}
\nu(\bar{g}) &= 0.6264612502473953 + 0.1039628520061216\Delta\bar{g} + 0.01384457142352891\Delta\bar{g}^2 \\
&\quad + 0.003117046054051534\Delta\bar{g}^3 + 0.0003684746290205897\Delta\bar{g}^4 \\
&\quad + 8.64736629778199 \times 10^{-5}\Delta\bar{g}^5 - 7.632599338747235 \times 10^{-6}\Delta\bar{g}^6.
\end{aligned}
\tag{20.35}
$$

In the weak-coupling limit, this is equal to 0.50039 rather than the exact $1/2$.

We now turn to $\gamma(\bar{g}_B)$, for which the initial strong-coupling expansion (19.10) reads

$$
\begin{aligned}
\gamma(\bar{g}_B) = \; & 1.400036164909792 - 2.015076019427151/\bar{g}_B^\omega + 2.512390732560552/\bar{g}_B^{2\omega} \\
& - 2.903034628806387/\bar{g}_B^{3\omega} + 3.123423917471507/\bar{g}_B^{4\omega} - 3.108796470872297/\bar{g}_B^{5\omega} \\
& + 2.844130229268904/\bar{g}_B^{6\omega} - 2.38207097645026/\bar{g}_B^{5\omega},
\end{aligned} \tag{20.36}
$$

which goes over into the following convergent new strong-coupling expansions in the variables x and $\Delta\bar{g}$:

$$
\begin{aligned}
\gamma(x) = \; & 1.234453309456454 - 0.3495068067938822x + 0.1754449364006056x^2 \\
& - 0.07496457184890283x^3 + 0.01553288625746019x^4 \\
& - 0.0007559094604789874x^5 - 0.00005490359805215839x^6, \tag{20.37} \\
\gamma(\bar{g}) = \; & 1.234453309456454 + 0.1734459660202996\Delta\bar{g} + 0.005180080229495593\Delta\bar{g}^2 \\
& + 0.00006337529084873493\Delta\bar{g}^3 + 7.604216767658758 \cdot 10^{-6}\Delta\bar{g}^4 \\
& + 0.00003970285111914907\Delta\bar{g}^5 - 0.00006597501217820878\Delta\bar{g}^6. \tag{20.38}
\end{aligned}
$$

Combining these expansions with (20.32), (20.33), and (20.35), we derive from the scaling relation $\eta = 2 - \gamma/\nu$ the corresponding expansions for η:

$$
\begin{aligned}
\eta(\bar{g}_B) = \; & 0.02948177725444778 - 0.101048653404458/\bar{g}_B^\omega + 0.2354470919307273/\bar{g}_B^{2\omega} \\
& - 0.452249390223013/\bar{g}_B^{3\omega} + 0.7620101849621524/\bar{g}_B^{4\omega} \\
& - 1.160617256113553/\bar{g}_B^{5\omega} + 1.627651002530155/\bar{g}_B^{6\omega}, \tag{20.39} \\
\eta(x) = \; & 0.02948177725444778 - 0.101048653404458x + 0.1541029259401388x^2 \\
& - 0.1465926820210476x^3 + 0.0958727624055483x^4 \\
& - 0.04186840682453081x^5 + 0.01220597704767396x^6, \tag{20.40} \\
\eta(\bar{g}) = \; & 0.02948177725444778 + 0.05014632323061653\Delta\bar{g} + 0.02695704683941055\Delta\bar{g}^2 \\
& + 0.004121619548050133\Delta\bar{g}^3 - 0.00038235521822759\Delta\bar{g}^4 \\
& + 0.00001736700551847262\Delta\bar{g}^5 + 0.0000435880986255714\Delta\bar{g}^6. \tag{20.41}
\end{aligned}
$$

The weak-coupling limits of these expansions are $\gamma(x = 1) = 1.00015$, $\gamma(\bar{g}_B = 0) = 1.00092$ rather than the exact 1; and $\eta(x = 1) = 0.0025$, $\eta(\bar{g}_B = 0) = 0.00043$ rather than the exact 0. The expansions for $\eta(x)$ converge rather slowly, so it is preferable to do calculations involving η by replacing it by $\eta = 2 - \gamma/\nu$ and using the expansions for γ and ν without re-expanding the ratio γ/ν. Then the plots of the convergent expansion for $\gamma(x)$ and of $\eta(x)$ are found once more to be very close to the plots of the corresponding full sixth-order approximations, as shown in Fig. 20.10.

It is now easy to give convergent expansions for the full m_B dependence of the renormalization factors. From (20.8) we see that

$$
\frac{m^2}{m_B^2} = \exp\left[-\int_0^{\bar{g}_B} \frac{d\bar{g}_B'}{\bar{g}_B'}\eta_m(\bar{g}_B')\right] = \frac{m^2}{g_B^2}\exp\left[-\int_x^1 \frac{dx'}{x'}\frac{\eta_m(x')}{f(x')}\right], \tag{20.42}
$$

$$
\frac{\phi^2}{\phi_B^2} = \exp\left[\int_0^{\bar{g}_B} \frac{d\bar{g}_B'}{\bar{g}_B'}\eta(\bar{g}_B')\right] = \exp\left[\int_x^1 \frac{dx'}{x'}\frac{\eta(x')}{f(x')}\right], \tag{20.43}
$$

where we have introduced the function

$$
f(x) \equiv -\frac{d\log x}{d\log\bar{g}_B} = \omega\frac{1 - x}{1 - (1 - \omega)x}. \tag{20.44}
$$

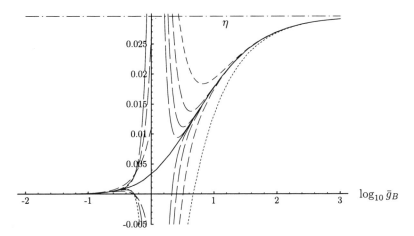

FIGURE 20.10 Logarithmic plot of variational perturbation result for expansion (20.17) of exponent $\eta(\bar{g}_B)$ with $N = 1$, and of the convergent re-expanded strong-coupling series (20.40) as a function of the bare coupling constant $\bar{g}_B$. The various dashed curves are explained in Fig. 20.8.

After isolating the logarithmic divergence of the integrals at $x = 0$, these can be rewritten as

$$\frac{m^2}{m_B^2} = x^{\eta_m/\omega} e^{-I_m(x)}, \qquad \frac{\phi^2}{\phi_B^2} = x^{-\eta/\omega} e^{I(x)}, \tag{20.45}$$

where $I(x)$ and $I_m(x)$ are the subtracted finite integrals

$$I_m(x) = \int_x^1 \frac{dx'}{x'} \left[\frac{\eta_m(x')}{f(x')} - \frac{\eta_m}{\omega} \right], \qquad I(x) = \int_x^1 \frac{dx'}{x'} \left[\frac{\eta(x')}{f(x')} - \frac{\eta}{\omega} \right]. \tag{20.46}$$

The integral $I_m(x)$ can readily be performed using a power series for $\eta_m(x)$ obtained from (20.33) via $\eta_m(x) = 2 - \nu^{-1}(x)$. The result is

$$
\begin{aligned}
I_m(x) = {} & 0.3023858220717581 - 0.3374211153180052x + 0.05257791758557147x^2 \\
& - 0.02006073290035280x^3 + 0.002720917209250741x^4 - 0.0001454953425150762x^5 \\
& - 0.00005731330570724153x^6. \tag{20.47}
\end{aligned}
$$

By combining the second equation in (20.45) with (20.30), rewritten as

$$\frac{m^2}{g_B^2} = x^{2/\omega}(1-x)^{-2}, \tag{20.48}$$

we find

$$\frac{m_B^2}{g_B^2} = x^{1/\nu\omega}(1-x)^{-2} e^{I_m(x)}. \tag{20.49}$$

In the weak-coupling limit $x \to 1$, the integral vanishes and the renormalized mass approaches the bare mass. In the strong-coupling limit $x \to 0$, on the other hand, $I_m(x)$ becomes a constant and we obtain once more the scaling relation $m \propto m_B^{2\nu}$. To study the crossover from the weak to the strong-coupling behavior exhibiting the critical behavior, we define a temperature interval ΔT_F within which fluctuations are important, and set

$$\frac{m_B^2}{g_B^2} \equiv \frac{T - T_c}{\Delta T_F} \equiv \tau_{\text{red}}. \tag{20.50}$$

A doubly logarithmic plot of the inverse square coherence length $\xi^{-2} \propto m^2/g_B$ in Fig. 20.11 shows how the slope changes from the free-field value 1 at large temperatures to 2ν near the critical temperature T_c.

Let us compare the fluctuation interval ΔT_F with the characteristic Ginzburg temperature interval $\Delta T_G = T_c^{\mathrm{MF}} \tau_G$ determined by Eq. (1.106). Recalling the normalization of the coupling strength in Eq. (20.1), we identify the parameter K in Eq. (1.105) as

$$K = 2^{D/2-1} \frac{48\pi}{N+8} \frac{g_B}{6}, \qquad (20.51)$$

having inserted $A_1 = 1$ and $a_2 = 1$, corresponding to $A_2 = m_B^2 = \tau = (T/T_c^{\mathrm{MF}} - 1)$ in (1.83). Using the cell size parameter $l = 1$ in (1.104), we insert the Yukawa potential at the orign $v_1^3(\mathbf{0}) \approx 0.1710$ from Table 1.1 , and obtain

$$\tau_G = \left[K v_1^3(\mathbf{0}) \right]^2 g_B^2 \approx 0.456\, g_B^2. \qquad (20.52)$$

Thus we have

$$\Delta T_F \approx 2.2 \Delta T_G, \qquad (20.53)$$

implying that the deviations from the mean-field behavior in Fig. 20.11 are in good agreement with Ginzburg's criterion.

For higher $O(N)$ symmetries, the same type of agreement would be found for the onset of directional fluctuations by applying Kleinert's criterion of Subsection 1.4.3. Indeed, the denominator $(N + 8)$ in Eq. (20.51) is canceled roughly by the factor N in formula (1.115) for Kleinert's temperature interval ΔT_K, such that the fluctuation temperature interval ΔT_F is of the same order of magnitude as $\Delta T_K = T_c^{\mathrm{MF}} \tau_K$.

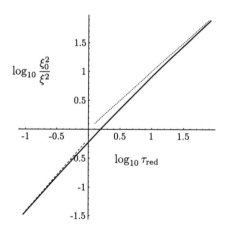

FIGURE 20.11 Doubly logarithmic plot of inverse square coherence length ξ^{-2} in arbitrary units against the reduced temperature $\tau_{\mathrm{red}} \equiv (T - T_c)/\Delta T_F$, where ΔT_F is the fluctuation temperature interval (20.50), closely related to Ginzburg's temperature by Eq. (20.53), where the theory crosses over from free field to critical behavior. The dotted line shows the free-field limit with unit slope, the dashed line the strong-coupling limit with slope $2\nu \approx 1.252$.

Finally, we use the series (20.31) for $\bar{g}(x)$ to calculate the β-function which plays an important role in the renormalization group approach to scaling (in contrast to our explicit theory). By definition, we obtain it as a function of x from the logarithmic derivative of $\bar{g}(\bar{g}_B)$ [compare (20.3)]

$$\beta(x) = -\bar{g}_B \frac{d\bar{g}}{d\bar{g}_B} = f(x)\, \bar{g}'(x), \qquad (20.54)$$

with $f(x)$ from (20.44), the result being

$$
\begin{aligned}
\beta(x) &= -1.622136195638856x + 2.73912944193321x^2 - 1.676962833531292x^3 \\
&\quad + 0.5230145612386834x^4 + 0.1405773254892622x^5 - 0.06197010583664305x^6 \\
&\quad - 0.06200066522622774x^7.
\end{aligned} \tag{20.55}
$$

The convergence of this strong-coupling expansion is seen by going to the weak-coupling limit $x = 1$ corresponding to $\bar{g}_B = 0$ where we find $\beta(1) = -0.0046$ rather than the exact value 0. Expressing x as a function of $\Delta \bar{g}$ via (20.34), we obtain

$$
\begin{aligned}
\beta(\bar{g}) &= 0.805\Delta \bar{g} + 0.4980812505494033\Delta \bar{g}^2 - 0.04513957559397346\Delta \bar{g}^3 \\
&\quad - 0.002836436593862963\Delta \bar{g}^4 + 0.000811067947065738\Delta \bar{g}^5 \\
&\quad + 0.002150487674009535\Delta \bar{g}^6 - 0.002024061592617085\Delta \bar{g}^7,
\end{aligned} \tag{20.56}
$$

which is plotted in Fig. 20.12.

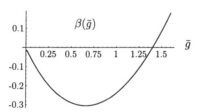

FIGURE 20.12 Plot of convergent strong-coupling expansion (20.56) for the beta function $\beta(\bar{g})$. The slope at the zero is the critical exponent $\omega = 0.805$. Note that the function converges well also at weak couplings. The curve misses the coordinate origin only by a very small amount.

From this we can derive the function $\omega(\bar{g})$ by a simple derivative with respect to $\bar{g}$:

$$
\begin{aligned}
\omega(\bar{g}) = \beta'(\bar{g}) &= 0.805 + 0.996162501098807\Delta \bar{g} - 0.1354187267819204\Delta \bar{g}^2 \\
&\quad - 0.01134574637545185\Delta \bar{g}^3 + 0.004055339735328691\Delta \bar{g}^4 \\
&\quad + 0.01290292604405721\Delta \bar{g}^5 - 0.01416843114831959\Delta \bar{g}^6.
\end{aligned} \tag{20.57}
$$

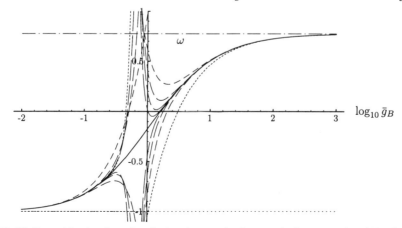

FIGURE 20.13 Logarithmic plot of variational perturbation result for expansion (20.16) with $N = 1$, and of the convergent re-expanded strong-coupling series (20.57) for exponent $\omega(\bar{g}_B)$, as a function of the bare coupling constant $\bar{g}_B$. The various dashed curves are explained in Fig. 20.8.

At $\bar{g} = \bar{g}^*$, the function $\omega(\bar{g}_B)$ has the value 0.805 as it should. In the weak-coupling limit $\bar{g}_B = 0$, it is equal to $-.984$, very close to the exact value -1. The full plot is shown in Fig. 20.13.

As a check for the consistent accuracy of our expansion procedures, we calculate ω once more from

$$\omega = -\frac{\bar{g}_B}{\beta(\bar{g})}\frac{d\beta(\bar{g})}{d\bar{g}_B} = f(x)\frac{x}{\beta(x)}\beta'(x). \tag{20.58}$$

After expressing x in terms of $\Delta\bar{g}$ we obtain a series with coefficients very close to those in (20.57), with only a slightly worse weak-coupling limit -1.08 rather than -1.

20.3   Improving the Graphical Extrapolation of Critical Exponents

In the last section, the critical exponents were obtained by extrapolating the approximations of order $2, 3, 4, 5$, and 6 to order $L \to \infty$ using the theoretically calculated large-L behavior const$+e^{-cL^{1-\omega}}$. The plots showed how the exponents approach their limiting values as functions of N. Exploiting this knowledge of the analytic form of the approach to infinite L, we plot the approximate exponents once more against the variable $x(L) = e^{-cL^{1-\omega}}$ rather than L [20], as done in Section 19.7. The parameter c is determined by varying it until the points merge approximately into a straight line for large L. Its intercept with the vertical axis yields the desired extrapolated critical exponent. The plots displayed in Figs. 20.14–20.17 show that variational strong-coupling theory provides us with a powerful tool for deriving the correct strong-coupling behavior and critical exponents of ϕ^4-theories in three dimensions.

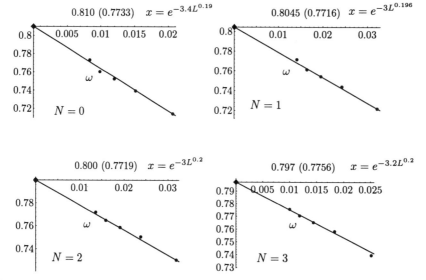

FIGURE 20.14 Behavior of strong-coupling values of the critical exponent of approach to scaling ω with increasing orders $L = 2, 3, 4, 5, 6$ in variational perturbation theory for $O(N)$-symmetric theories with $N = 0, 1, 2, 3, \dots$. The plot is against $x(L) = e^{-cL^{1-\omega}}$ with c chosen such that the points merge into a straight line. The numbers N correspond to different universality classes ($N = 0, 1, 2, 3$ for dilute polymer solutions, Ising magnets, superfluid helium, and the classical Heisenberg model). The numbers on top display the limiting value at the intercept, as well as the last calculated approximation ω_6 in parentheses (see Table 20.1 on page 373).

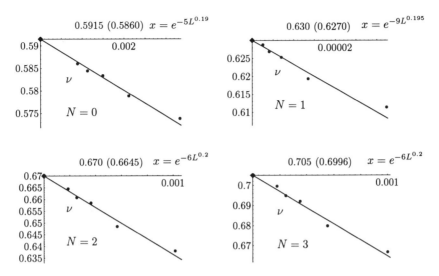

FIGURE 20.15 Plot analogous to Fig. 20.14 of critical exponents ν, for increasing orders $L = 2, 3, 4, 5, 6$ in variational perturbation theory, illustrating the extrapolation procedure to $N \to \infty$ at the intercept with the vertical axis, for $N = 0$, 1, 2, 3. The numbers on top display the limiting value at the intercept, as well as the last calculated approximation ν_6 in parentheses (see Table 20.1 on page 373).

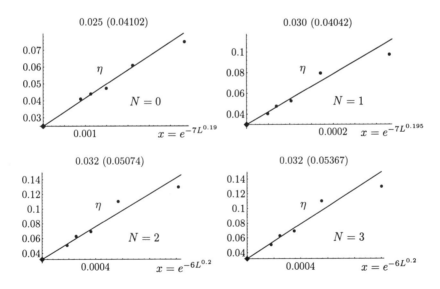

FIGURE 20.16 Plot analogous to Figs. 20.14 and 20.15 of critical exponents η, for increasing orders $L = 2, 3, 4, 5, 6$ in variational perturbation theory, illustrating the extrapolation procedure to $N \to \infty$ at the intercept with the vertical axis, for $N = 0$, 1, 2, 3. The approximations η_L are obtained from γ_L via the scaling relation $\eta_L = 2 - \gamma_L / \nu$. The numbers on top display the limiting value at the intercept, as well as the last calculated approximation η_6 in parentheses (see Table 20.1 on page 373).

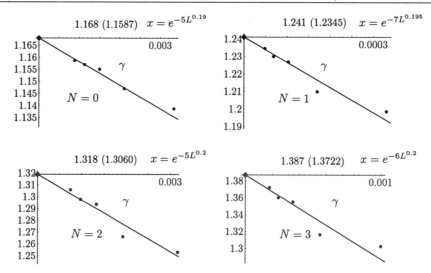

FIGURE 20.17 Plot analogous to Figs. 20.14, 20.15, and 20.16 of critical exponents γ, for increasing $L = 2, 3, 4, 5, 6$ in variational perturbation theory, illustrating the extrapolation procedure to $N \to \infty$ at the intercept with the vertical axis, for $N = 0$, 1, 2, 3. The numbers on top display the limiting value at the intercept, as well as the last calculated approximation γ_6 in parentheses (see Table 20.1 on page 373).

20.4   Seven-Loop Results for $N = 0, 1, 2,$ and 3

For $O(N)$-symmetric theories with $N = 0, 1, 2, 3$, the power series expansions (20.12) and (20.13) have recently been calculated up to seven loops [12]. The results are [13]

$$f_7^\eta = \left\{ \begin{matrix} 0.001901867 \\ 0.001697694 \\ 0.001395129 \\ 0.001111499 \end{matrix} \right\} \bar{g}^7, \quad f_7^{\eta m} = \left\{ \begin{matrix} 0.097383003 \\ 0.091551786 \\ 0.079018231 \\ 0.065974801 \end{matrix} \right\} \bar{g}^7 \quad \text{for} \quad \left\{ \begin{matrix} N = 0 \\ N = 1 \\ N = 2 \\ N = 3 \end{matrix} \right\} . \quad (20.59)$$

In the expansions (20.17), (20.18), and (20.19) of η, η_m, and γ in powers of the bare couplings $\bar{g}_B$, these add the terms

$$f_7^\eta = \left\{ \begin{matrix} -0.216423937 \\ -0.239546791 \\ -0.241424764 \\ -0.233364541 \end{matrix} \right\} \bar{g}_B^7, \quad f_7^{\eta m} = \left\{ \begin{matrix} 6.099829565 \\ 7.048219834 \\ 7.378080984 \\ 7.380848508 \end{matrix} \right\} \bar{g}_B^7, \quad f_7^\gamma = \left\{ \begin{matrix} 2.504064047 \\ 2.650615568 \\ 2.570336644 \\ 2.401546939 \end{matrix} \right\} \bar{g}_B^7, \quad (20.60)$$

respectively.

We now calculate the strong-coupling limit of the seven-loop power series expansions (20.18) for η_m, extended by $f_7^{\eta m}$. Using formula (19.16) with $q = 2/\omega$ and ω from Table 20.1 on page 373, we obtain the limiting values for $\nu = 1/(2 - \eta_m)$ shown in Fig. 20.18. They lead to the ν-values $\nu_7 = \{0.5886, 0.6311, 0.6713, 0.7072\}$, the entries in this vector referring to $N = 0, 1, 2, 3$. These results are derived using the six-loop ω-values listed in the last column of Table 20.1 on page 373:

$$\omega_6 = \left\{ \begin{matrix} 0.810 \\ 0.805 \\ 0.797 \\ 0.790 \end{matrix} \right\} \quad \text{for} \quad \left\{ \begin{matrix} n = 0 \\ n = 1 \\ n = 2 \\ n = 3 \end{matrix} \right\} . \quad (20.61)$$

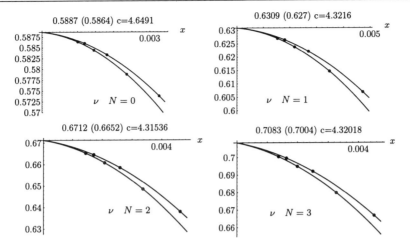

FIGURE 20.18 Strong-coupling values for the critical exponent ν^{-1} obtained from expansion (20.18) via formula (19.16), for increasing orders $L = 2, 3, \ldots, 7$ of the approximation. The exponents are plotted against the variable $x(L) = e^{-cL^{1-\omega}}$ and should lie on a straight line in the limit of large L. Here at finite L, even and odd approximants may be connected by slightly curved parabolas whose common intersection determines the critical exponents for $L = \infty$. More details on the determination of the constant c are given in the text.

It is useful to study the dependence of the extrapolation on ω. The result is

$$
\nu_7 = \left\{ \begin{array}{l} 0.5883 + 0.0417 \times (\omega - 0.810) \\ 0.6305 + 0.0400 \times (\omega - 0.805) \\ 0.6710 + 0.0553 \times (\omega - 0.800) \\ 0.7075 + 0.1891 \times (\omega - 0.797) \end{array} \right\} \quad \text{for} \quad \left\{ \begin{array}{l} N = 0 \\ N = 1 \\ N = 2 \\ N = 3 \end{array} \right\}. \tag{20.62}
$$

We extrapolate our results to $L = \infty$ by plotting the data against the variables $x(L) = e^{-cL^{1-\omega}}$, as in Section 20.3. However, since we are now in the possession of an even number of approximants $\nu_2, \ldots, \nu_7$, we may account for the fact that even and odd approximants are obtained differently, the former from extrema, the latter from turning points of the variational expression (19.16). We therefore plot even and odd approximants separately against $x(L)$, determining the unknown constants c by fitting to each set of points a slightly curved parabola and making them intersect the vertical axis at the same point. This yields the extrapolated critical exponent listed on top of each figure (together with the seventh-order value in parentheses, and the optimal parameter c).

For the critical exponent η we cannot use the same extrapolation procedure since the expansion (20.17) starts out with $\bar{g}_0^2$, so that there exists only an odd number of approximants η_L. We therefore use two alternative extrapolation procedures. In the first we connect the even approximants η_2 and η_4 by a straight line and the odd ones η_3, η_5, η_7 by a slightly curved parabola, allowing for smooth approach to the asymptotic behavior of the three available odd approximations. Then we vary c until there is an intersection at $x = 0$. This yields the critical exponents η shown in Fig. 20.19.

Allowing for the inaccurate knowledge of ω, the results may be stated as

$$
\eta_7 = \left\{ \begin{array}{l} 0.03215 + 0.1327 \times (\omega - 0.810) \\ 0.03572 + 0.0864 \times (\omega - 0.805) \\ 0.03642 + 0.0655 \times (\omega - 0.800) \\ 0.03549 + 0.0320 \times (\omega - 0.797) \end{array} \right\} \quad \text{for} \quad \left\{ \begin{array}{l} N = 0 \\ N = 1 \\ N = 2 \\ N = 3 \end{array} \right\}, \tag{20.63}
$$

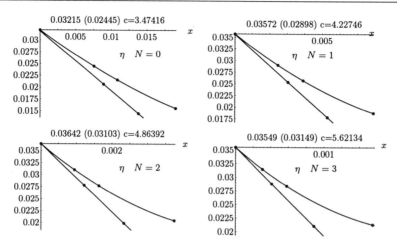

FIGURE 20.19 Strong-coupling values for the critical exponent η obtained from expansion (20.17), extended by f_7^η of Eq. (20.60), via formula (19.16), for increasing orders $L = 3, \ldots, 7$ of the approximation. The exponents are plotted against $x(L) = e^{-cL^{1-\omega}}$. Even approximants are connected by straight line and odd approximants by slightly curved parabolas, whose common intersection determines the critical exponents expected for $L = \infty$.

Alternatively, we connect the last odd approximants η_5 and η_7 also by a straight line and choose c to make the lines intersect at $x = 0$. This yields the exponents

$$\eta_7 = \left\{ \begin{array}{l} 0.03010 + 0.08760 \times (\omega - 0.810) \\ 0.03370 + 0.03816 \times (\omega - 0.805) \\ 0.03480 + 0.01560 \times (\omega - 0.800) \\ 0.03447 + 0.00588 \times (\omega - 0.797) \end{array} \right\} \quad \text{for} \quad \left\{ \begin{array}{l} N = 0 \\ N = 1 \\ N = 2 \\ N = 3 \end{array} \right\}, \tag{20.64}$$

as shown in Fig. 20.20, the ω dependences being somewhat weaker than in (20.63).

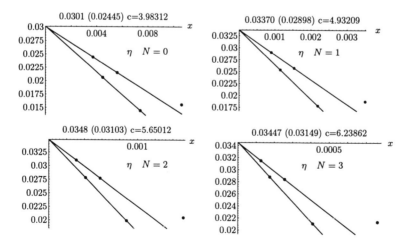

FIGURE 20.20 Plot analogous to Fig. 20.19, but with an extrapolation found from the intersection of the straight lines connecting the last two even and odd approximants. The resulting critical exponents differ only little from those obtained in Fig. 20.19, the differences giving an estimate for the systematic error of our results.

Combining the two results and using the difference to estimate the systematic error of the extrapolation procedure, we obtain for η the values

$$\eta_7 = \left\{ \begin{array}{l} 0.0311 \pm 0.001 \\ 0.0347 \pm 0.001 \\ 0.0356 \pm 0.001 \\ 0.0350 \pm 0.001 \end{array} \right\} \quad \text{for} \quad \left\{ \begin{array}{l} N = 0 \\ N = 1 \\ N = 2 \\ N = 3 \end{array} \right\}, \tag{20.65}$$

whose ω-dependence is the average of that in (20.63) and (20.64).

For our extrapolation procedure, the power series for the critical exponent $\gamma = \nu(2 - \eta)$ are actually better suited than those for η, since they possess three even and three odd approximants, just as ν^{-1}. Advantages of this expansion have been observed before in Ref. [4]. The associated plots are shown in Fig. 20.21.

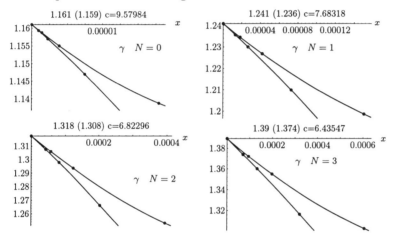

FIGURE 20.21 Strong-coupling values for the critical exponent $\gamma = \nu(2 - \eta) = (2 - \eta)/(2 - \eta_m)$ obtained from a combination of the expansions (20.17) and (20.18), optimizing the associated expression (19.16), for increasing orders $N = 2, 3, \ldots, 7$ of the approximation. The exponents are plotted against the variable $x(L) = e^{-cL^{1-\omega}}$ and should lie on a straight line in the limit of large L. Even and odd approximants are connected by slightly curved parabolas whose common intersection determines the critical exponents expected for $L = \infty$. The determination of the constant c is described in the text.

The extrapolated exponents are, including the ω-dependence:

$$\gamma_7 = \left\{ \begin{array}{l} 1.161 - 0.049 \times (\omega - 0.810) \\ 1.241 - 0.063 \times (\omega - 0.805) \\ 1.318 - 0.044 \times (\omega - 0.800) \\ 1.390 - 0.120 \times (\omega - 0.797) \end{array} \right\} \quad \text{for} \quad \left\{ \begin{array}{l} N = 0 \\ N = 1 \\ N = 2 \\ N = 3 \end{array} \right\}. \tag{20.66}$$

Unfortunately, the exponent $\gamma = \nu(2 - \eta)$ is not very sensitive to η since this is small compared to 2, so that the extrapolation results (20.63) are more reliable than those obtained from γ via the scaling relation $\eta = 2 - \gamma/\nu$. By combining (20.62) and (20.63), we find from $\gamma = \nu(2 - \eta)$:

$$\gamma_7 = \left\{ \begin{array}{l} 1.1589 \\ 1.2403 \\ 1.3187 \\ 1.3932 \end{array} \right\} \quad \text{for} \quad \left\{ \begin{array}{l} N = 0 \\ N = 1 \\ N = 2 \\ N = 3 \end{array} \right\}, \tag{20.67}$$

the difference with respect to (20.66) showing the typical small errors of our approximation.

As mentioned in the beginning, the knowledge of the large-order behavior does not help to improve significantly the accuracy of the approximation. In our theory, the most important information exploited by the resummation procedure is the knowledge of the exponentially fast convergence which leads to a linear behavior of the resummation results of order L in a plot against $x(L) = e^{-cL^{1-\omega}}$. This knowledge, which allows us to extrapolate our approximations for $L = 2, 3, 4, 5, 6, 7$ quite well to infinite order L, seems to be more powerful than the knowledge of the large-order behavior exploited by other authors (quoted in Table 20.1 on page 373).

The complete updated list of exponents is shown in Table 20.2 , which also contains values for the other critical exponents $\alpha = 2 - D\nu$ and $\beta = \nu(D - 2 + \eta)/2$.

TABLE 20.2   Seven-loop results for critical exponents $N = 0, 1, 2, 3$ from strong-coupling theory, improving the six-loop results of Table 20.1 on page 373. The ω-values in the last column are six-loops results, the references [14] and [12] seven-loop results. To facilitate the comparison with the earlier results in Table 20.1 we have repeated the former six-loop entries.

N	g_c	$\gamma(\gamma_6)$	$\eta(\eta_6)$	$\nu(\nu_6)$	α	β	ω (ω_6)	
0		1.161(1.159)	0.0311±0.001	0.5886(0.5864)	0.234		0.810(0.773)	
	1.413±0.006	1.160±0.002	0.0284±0.0025	0.5882±0.0011	0.235±0.003	0.3025±0.0008	0.812±0.016	[14]
	1.39	1.1569±0.0004	0.0297±0.0009	0.5872±0.0004	0.2384±0.0012			[12]
	1.402	1.160	0.034	0.589	0.231	0.305		[3]
	1.421±0.004	1.161±0.003	0.026±0.026	0.588±0.001	0.236±0.004	0.302±0.004	0.794 ± 0.06	[2]
	1.421 ± 0.008	1.1615 ± 0.002	0.027 ± 0.004	0.5880±0.0015		0.3020±0.0015	0.80 ± 0.04	[9]
		1.160 ± 0.004	0.031 ± 0.003	0.5885±0.0025		0.3025±0.0025	0.82 ± 0.04	[11]
1		1.241(1.236)	0.0347±0.001	0.6310(0.6270)	0.107		0.805(0.772)	
	1.411±0.004	1.240±0.001	0.0335±0.0025	0.6304±0.0013	0.109±0.004	0.3258±0.0014	0.799±0.011	[14]
	1.40	1.2378±0.0006	0.0355±0.0009	0.6301±0.0005	0.1097±0.0012			[12]
	1.419	1.239	0.038	0.631	0.107	0.327	0.781	[3]
	1.416±0.0015	1.241±0.004	0.031±0.011	0.630±0.002	0.110 ± 0.008	0.324 ± 0.06	0.788 ± 0.003	[2]
	1.416±0.004	1.2410±0.0020	0.031±0.004	0.6300±0.0015		0.3250±0.0015	0.79±0.03	[9]
			0.035 ± 0.002	0.628±0.001			0.80 ± 0.02	[10]
		1.1239 ± 0.004	0.037 ± 0.003	0.6305±0.0025		0.3265±0.0025	0.81 ± 0.04	[11]
2		1.318 ± 0.001	0.0356 ± 0.001	0.6713(0.6652)	-0.0129		0.800(0.772)	
	1.403±0.003	1.317±0.002	0.0354±0.0025	0.6703±0.0013	-0.011±0.004	0.3470±0.0014	0.789±0.011	[14]
	1.40	1.3178±0.001	0.0377±0.0006	0.6715±0.0007	-0.0145±0.0021			[12]
	1.408	1.315	0.039	0.670	-0.010	0.348	0.780	[3]
	1.406±0.005	1.316±0.009	0.032±0.015	0.669±0.003	-0.007±0.009	0.346±0.009	0.78 ± 0.01	[2]
	1.406±0.004	1.3160±0.0025	0.033±0.004	0.6690±0.0020		0.3455±0.002	0.78±0.025	[9]
			0.037 ± 0.002	0.665±0.001			0.79 ± 0.02	[10]
		1.315 ± 0.007	0.040 ± 0.003	0.671±0.005		0.3485±0.0035	0.80 ± 0.04	[11]
3		1.390(1.374)	0.0350 ± 0.0005	0.7072(0.7004)	-0.122		0.797(0.776)	
	1.391±0.004	1.390±0.005	0.0355±0.0025	0.7073±0.0030	-0.122±0.009	0.3662±0.0025	0.782±0.0013	[14]
	1.39	1.3926±0.001	0.0374±0.0004	0.7096±0.0008	-0.1288±0.0024			[12]
	1.392	1.386	0.038	0.706	-0.117	0.366	0.780	[3]
	1.392±0.009	1.390±0.01	0.031±0.022	0.705±0.005	-0.115±0.015	0.362	0.78 ± 0.02	[2]
	1.391 ±0.004	1.386±0.004	0.033±0.004	0.705±0.003		0.3645±0.0025	0.78 ± 0.02	[9]
			0.037 ± 0.002	0.79±0.02			0.79 ± 0.02	[10]
		1.390 ± 0.010	0.040 ± 0.003	0.710±0.007		0.368±0.004	0.79 ± 0.04	[11]

20.5   Large-Order Behavior

The new coefficients follow quite closely their theoretically expected large-order limiting behavior derived from instanton calculations, according to which the expansion coefficients with respect to the renormalized coupling $\bar{g}$ should grow for large order k as follows [recall Eq. (16.12)]:

$$f_k^\omega = \gamma_\omega(-a)^k k! k \Gamma(k + \beta_\omega) \left(1 + \frac{c_\beta^{(1)}}{k} + \frac{c_\beta^{(2)}}{k^2} + \dots \right), \tag{20.68}$$

$$f_k^\eta = \gamma_\eta(-\alpha)^k k! k \Gamma(k+\beta_\eta)\left(1+\frac{c_\eta^{(1)}}{k}+\frac{c_\eta^{(2)}}{k^2}+\dots\right), \tag{20.69}$$

$$f_k^{\bar\eta} = \gamma_{\bar\eta}(-\alpha)^k k! k \Gamma(k+\beta_{\bar\eta})\left(1+\frac{c_{\bar\eta}^{(1)}}{k}+\frac{c_{\bar\eta}^{(2)}}{k^2}+\dots\right), \tag{20.70}$$

where $\bar\eta \equiv \eta + \nu^{-1} - 2$. The growth parameter α is proportional to the inverse euclidean action of the classical *instanton* solution $\varphi_c(\mathbf{x})$ to the field equations, and an accurate numerical evaluation yields [15]

$$\alpha = (D-1)\frac{16\pi}{I_4}\frac{1}{N+8} \stackrel{D=3}{=} 0.14777423\frac{9}{N+8}. \tag{20.71}$$

The quantity I_4 denotes the integral

$$I_4 = \int d^D x [\varphi_c(\mathbf{x})]^4. \tag{20.72}$$

Its numerical values in two and three dimensions D are listed in Table 20.3 .

TABLE 20.3   Fluctuation determinants and integrals over extremal field solution.

D	D_L	D_T	I_1	I_4	I_6	H_3
3	10.544 ± 0.004	$1.4571\pm .0001$	31.691522	75.589005	659.868352	13.563312
2	135.3 ± 0.1	1.465 ± 0.001	15.10965	23.40179	71.08023	9.99118

The growth parameters $\beta_\omega, \beta_\eta, \beta_{\bar\eta}$, are related to the number of zero-modes in the fluctuation determinant around the instanton, and are

$$\beta_\omega \equiv \beta_\beta + 1 = \frac{1}{2}(D+5+N), \quad \beta_\eta = \frac{1}{2}(D+1+N), \quad \beta_{\bar\eta} = \frac{1}{2}(D+3+N). \tag{20.73}$$

The prefactors $\gamma_\beta, \gamma_\eta, \gamma_{\bar\eta}$ in (20.68)–(20.70) require the calculation of the full fluctuation determinants. This yields for γ_β the somewhat lengthy expression

$$\gamma_\beta \equiv \frac{(n+8)2^{(n+D-5)/2}3^{-3(D-2)/2}}{\pi^{3+D/2}\Gamma\left(2+\frac{1}{2}n\right)}\left(\frac{I_1^2}{I_4}\right)^2\left(\frac{I_6}{I_4}-1\right)^{D/2}D_L^{-1/2}D_T^{-(n-1)/2}e^{-1/a}. \tag{20.74}$$

The constants I_1, I_2, I_6 are generalizations of the above integral I_4:

$$I_p \equiv \int d^D x [\varphi_c(\mathbf{x})]^p, \tag{20.75}$$

and D_L and D_T are found from the longitudinal and transverse parts of the fluctuation determinants. Their numerical values are given in Table 20.3 . The constant γ_β is the prefactor of growth in the expansion coefficients of the β-function

$$f_k^\beta \approx \gamma_\beta(-\alpha)^k k! \Gamma(k+\beta_\beta+1). \tag{20.76}$$

The prefactors in $\gamma_\omega, \gamma_\eta$, and $\gamma_{\bar\eta}$ in (20.68)–(20.70) are related to γ_β by

$$\gamma_\omega = -\alpha\gamma_\beta, \quad \gamma_\eta = \gamma_{\bar\eta}\frac{2H_3}{I_1 D(4-D)}, \quad \gamma_{\bar\eta} = \gamma_\beta\frac{n+2}{n+8}(D-1)4\pi\frac{I_2}{I_1^2}, \tag{20.77}$$

TABLE 20.4 Growth parameter of $D = 3$ perturbation expansions of $\beta(\bar{g})$, $\eta(\bar{g})$ and $\bar{\eta} = \eta + \nu^{-1} - 2$.

	$N = 0$	$N = 1$	$N = 2$	$N = 3$
a	0.1662460	0.14777422	0.1329968	0.12090618
β_ω	4	9/2	9	11/2
$\beta_{\bar{\eta}}$	3	7/2	4	9/2
β_η	2	5/2	3	7/2
$10^2 \times \gamma$	8.5489(16)	3.9962(6)	1.6302(3)	0.59609(10)
$10^3 \times \gamma_{\bar{\eta}}$	10.107	6.2991	3.0836	1.2813
$10^3 \times \gamma_\eta$	2.8836	1.7972	0.8798	0.3656

where $I_2 = (1 - D/4)I_4$ and H_3 are listed in Table 20.3 . The numerical values of all growth parameters are listed in Table 20.4 . In Fig. 20.22 we show a comparison between the exact coefficients and their asymptotic forms (20.70).

It is possible to use the information coming from the theoretical large-order behavior (20.68)–(20.70) to *predict* approximately the values of the expansion coefficients beyond the seven-loop order. For this we choose the coefficients $c^{(i)}$ in the asymptotic formulas (20.68)–(20.70) to fit exactly the six known expansion coefficients of $\omega(\bar{g})$ and the seven of $\bar{\eta}(\bar{g})$ and $\eta(\bar{g})$. The coefficients are listed in Table 20.8 . The quality of the fits can be seen in Fig. 20.22.

Note that even and odd coefficients f_k^η lie on two smooth separate curves, so that we fit the two sets separately. These fits permit us now to extend the list of presently available coefficients as shown in Table 20.5 up to g^{20}. The errors in these predictions are expected to be smallest for f_k^ω, as illustrated in Fig. 20.23.

At this place we observe an interesting phenomenon. According to Table 20.5 , the expansion coefficients f_k^ω of $\omega(\bar{g})$ have alternating signs and grow rapidly, reaching precociously their asymptotic form (20.68), as visible in Fig. 20.22. Now, from $\omega(\bar{g})$ we can derive the so called β-function, $\beta(\bar{g}) \equiv \int d\bar{g}\, \omega(\bar{g})$, and from this the expansion for the bare coupling constant $\bar{g}_B(\bar{g}) = - \int d\bar{g}/\beta(\bar{g})$, with coefficients $f_k^{\bar{g}_0}$ listed in Table 20.6 . From the standard instanton analysis we know that the function $\bar{g}_0(\bar{g})$ has the same left-hand cut in the complex $\bar{g}$-plane as the functions $\omega(\bar{g}), \bar{\eta}(\bar{g}), \eta(\bar{g})$, with the same discontinuity proportional to $e^{-\text{const}/g}$ at the tip of the cut.

Hence the coefficients $f_k^{\bar{g}_0}$ must have asymptotically similar alternating signs and a factorial growth. Surprisingly, this expectation is not borne out by the explicit seven-loop coefficients $\bar{g}_0^{(k)}$ following from (20.68) in Table 20.6 . However, if we look at the higher-order coefficients derived from the extrapolated f_k^ω sequence which are also listed in that table, we see that sign change and factorial growth do eventually set in at the rather high order 11. Before this order, the coefficients $f_k^{\bar{g}_0}$ look like those of a convergent series. If we were to make a plot analogous to those in Fig. 20.22 for $f_k^{\bar{g}_0}$, we would observe huge deviations from the asymptotic form up to an order much larger than 10. In contrast, the inverse series $\bar{g}(\bar{g}_0)$ has expansion coefficients $f_k^{\bar{g}}$ which do approach rapidly their asymptotic form, as seen in Table 20.7 . This is the reason why our resummation of the critical exponents $\omega, \bar{\eta}, \eta$ as power series in $\bar{g}_B$ already yields good results at the available low order seven.

TABLE 20.5 Coefficients of extended perturbation expansions obtained from large-order expansions (20.68)–(20.70) for ω, $\bar{\eta} \equiv \nu^{-1} + \eta - 2$, η up to g^{12}.

	k	$N = 0$	$N = 1$	$N = 2$	$N = 3$
f_k^{ω}	0	-1	-1	-1	-1
	1	2	2	2	2
	2	-95/72	-308/243	-272/225	-1252/1089
	3	1.559690758	1.404278391	1.259667768	1.131786725
	4	-2.236580484	-1.882634142	-1.589642400	-1.351666500
	5	3.803133000	2.973285060	2.346615000	1.875335400
	6	-7.244496000	-5.247823000	-3.867143000	-2.904027000
	7	15.0706772	10.0938530	6.9384728	4.8954471
	8	-33.8354460	-20.9045761	-13.3833570	-8.8630280
	9	81.4263429	46.2983010	27.5543342	17.1018561
	10	-209.0371337	-109.1428445	-60.2679848	-34.9985085
	11	570.2558985	272.8574773	139.5403648	75.6925030
	12	−1647.63898	−721.159283	−340.986931	−172.506443
	13	5027.12671	2009.473994	877.142753	413.269514
	14	−16154.2792	−5888.53514	−2369.63316	−1038.433113
	15	54539.7867	18105.83253	6708.76515	2731.28823
	16	−193034.402	−58292.0930	−19865.5739	−7505.78230
	17	714771.195	196130.5369	61414.0151	21513.8526
	18	2.7637289×10^6	−688418.829	−197883.530	−64215.5872
	19	1.1139530×10^7	2.5166119×10^6	663509.086	199303.824
	20	-4.6728706×10^7	-9.5668866×10^6	-2.3117713×10^6	−642301.398
$f_k^{\bar{\eta}}$	1	-1/4	-1/3	-2/5	-5/11
	2	1/16	2/27	2/25	10/121
	3	-0.0357672729	-0.0443102531	-0.0495134446	-0.0525519564
	4	0.0343748465	0.0395195688	0.0407881055	0.0399640005
	5	-0.0408958349	-0.0444003474	-0.0437619509	-0.0413219917
	6	0.0597050472	0.0603634414	0.0555575703	0.0490929344
	7	-0.09928487	-0.09324948	-0.08041336	-0.06708630
	8	0.18143353	0.15857090	0.12955711	0.10413882
	9	-0.35946458	-0.29269274	-0.22839265	-0.17925852
	10	0.76759881	0.58218392	0.43525523	0.33488318
	11	-1.75999735	-1.24181846	-0.88911482	-0.66904757
	12	4.31887516	2.82935836	1.93487570	1.41644564
	13	−11.3068155	−6.86145603	−4.46485563	−3.15991301
	14	31.4831400	17.65348358	10.8846651	7.40110473
	15	−92.9568675	−48.04185493	−27.9476939	−18.1528875
	16	290.205144	137.9015950	75.3808299	46.5326521
	17	−955.369710	−416.4425396	−213.088140	−124.454143
	18	3308.08653	1319.8954890	630.008039	346.784997
	19	−12019.6749	−4380.9238169	−1944.51060	−1005.36571
	20	45726.095	15196.764595	6254.75115	3028.67211
f_k^{η}	1	0	0	0	0
	2	1/108	8/729	8/675	40/3267
	3	0.0007713749	0.0009142223	0.0009873600	0.0010200000
	4	0.0015898706	0.0017962229	0.0018368107	0.0017919257
	5	-0.0006606149	-0.0006536980	-0.0005863264	-0.0005040977
	6	0.0014103421	0.0013878101	0.0012513930	0.0010883237
	7	-0.001901867	-0.0016976941	-0.001395129	-0.001111499
	8	0.003178395	0.0026439888	0.002043629	0.001544149
	9	-0.006456700	-0.0049783320	-0.003585593	-0.002532983
	10	0.012015200	0.0084255120	0.005570210	0.003647578
	11	-0.029656348	-0.0194143738	-0.012066168	-0.007451622
	12	0.064239639	0.0378738590	0.021403479	0.012148673
	13	−0.180415293	−0.0992734993	−0.0527914785	−0.0282931664
	14	0.4519047994	0.22304200134	0.10748443321263	0.0528085190
	15	−1.4092869972	−0.6472476781	−0.2928360472879924	−0.135567321
	16	4.0214900375	1.65386975	0.677414388712502	0.287414739
	17	−13.7588144054	−5.24609037	−2.01071514	−0.801301742
	18	44.0902845294	15.0426293	5.21919799	1.907241838
	19	−164.205876	−51.7544723	−16.7458885	−5.728643910
	20	583.728411	164.571258	48.2146655	15.13540671

TABLE 20.6   Coefficients $f_k^{\bar{g}_B}$ of expansion $\bar{g}_B(\bar{g}) = \sum_{k=1}^{20} f_k^{\bar{g}_B} \bar{g}^k$, defining extended perturbation expansions deduced from large-order expansions (20.68)–(20.70) for $\omega(\bar{g})$.

k	$N = 0$	$N = 1$	$N = 2$	$N = 3$
1	1	1	1	1
2	+1	+1	+1	+1
3	+337/432	+575/729	+539/675	+2641/3267
4	+0.61685694588	+0.62411053351	+0.63484885720	+0.64721832545
5	+0.44266705709	+0.45557995443	+0.47149516705	+0.48876206059
6	+0.35597494073	+0.35927512536	+0.36876801981	+0.38195333853
7	+0.21840619207	+0.23668638696	+0.25507866294	+0.27372501773
8	+0.23516444398	+0.22010271935	+0.21833333377	+0.22423492600
9	+0.02522653990	+0.07797541233	+0.11146939079	+0.13619054953
10	+0.32466738893	+0.21722566733	+0.17071122132	+0.15281461436
11	−0.46084539160	−0.17781419227	−0.04796874299	+0.01851106465
12	+1.36111296151	+0.62177013621	+0.32371445346	+0.19688967179
13	−3.42004319798	−1.33935153089	−0.55625249070	−0.23297770291
14	+9.68597708110	+3.55457753745	+1.44715002648	+0.65263956302
15	−28.5286709455	−9.51594412468	−3.51833733708	−1.41477489238
16	+88.9376821020	+27.1477264424	+9.31404148366	+3.53850316476
17	−291.235785543	−81.0609653416	−25.6008150903	−9.00262320492
18	+1000.66241399	+253.799830529	+73.8458792207	+24.1544067361
19	−3599.15484483	−830.784519325	−222.359395181	−67.4743858406
20	+13526.5566605	+2838.71379781	+698.348588943	+196.518945901

TABLE 20.7   Coefficients $f_k^{\bar{g}}$ of expansion $\bar{g}(\bar{g}_B) = \sum_{k=1}^{20} f_k^{\bar{g}} \bar{g}_B^k$, obtained by inverting the extended series in Table 20.6 .

k	$N = 0$	$N = 1$	$N = 2$	$N = 3$
1	1	1	1	1
2	−1	−1	−1	−1
3	+527/432	+883/729	+811/675	+3893/3267
4	−1.7163939829	−1.680351960126292	−1.642256264617284	−1.60528382897736
5	+2.7021635328	+2.591685040643859	+2.481604560563785	+2.378891143794822
6	−4.6723281932	−4.363908063002809	−4.073635397816119	−3.813515390028028
7	+8.7648283753	+7.926093595753771	+7.180326093595318	+6.539645290718699
8	−17.684135663	−15.39841276963578	−13.47981441366666	−11.90293506879397
9	+38.129348202	+31.80063328573243	+26.79259688548747	+22.86325133485651
10	−87.419391225	−69.48420478282783	−56.1279351033013	−46.14596304145893
11	+212.28789113	+160.0400066477353	+123.4985362910675	+97.5437851896555
12	−544.33806227	−387.4479410496121	−284.6297746951519	−215.3826650602743
13	+1470.2445538	+983.719405302971	+685.668309006505	+495.7770927688912
14	−4175.1804881	−2614.933427024693	−1723.672999416843	−1187.794187410145
15	+12447.739474	+7268.064649337187	+4516.120357408118	+2958.336103932099
16	−38915.141370	−21101.49568383381	−12320.85534817637	−7652.516371929849
17	+127440.33105	+63943.24392789235	+34975.98186824855	+20545.02631707489
18	− 436738.21140	−202094.1329180427	−103252.1798678474	−57215.98372843337
19	+1564637.2472	+665710.523944826	+316810.7604431689	+165210.8008728902
20	−5853354.4104	−2283830.09806744	−1009811.938755735	−494409.476944406

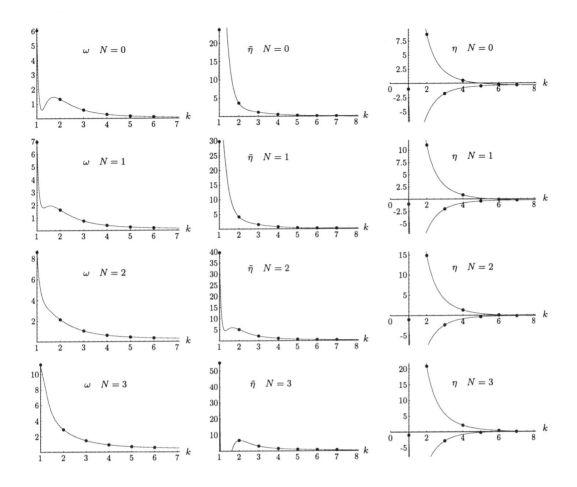

FIGURE 20.22 Early onset (precocity) of large-order behavior of coefficients of the expansions of the critical exponents ω, η, and $\bar{\eta} \equiv \nu^{-1} + \eta - 2$ in powers of the renormalized coupling constant. The dots show the relative deviations (exact value)/(asymptotic value) $- 1$. The curves are plots of the asymptotic expressions in Eqs. (20.68)–(20.70) listed in Table 20.8 . The curve for ω is the smoothest, promising the best extrapolation to the next orders, with consequences to be discussed in Section 20.6.

20.6   Influence of Large-Order Information

The critical exponents obtained so far have quite a high accuracy. The question arises whether this can be increased further by making use of the extended set of expansion coefficients in Table 20.5 derived from the theoretically known large-order behavior (20.68)–(20.70). We shall now demonstrate that the gain of accuracy is very modest. Within strong-coupling theory, the knowledge of the large-order behavior has little influence on the results except for ω whose values are slightly lowered (by less than $\sim 0.2\%$). The reason for this seems to be that in the present approach the critical exponents are obtained from evaluations of expansions at infinite bare couplings. The information on the large-order behavior, on the other hand, specifies the discontinuity at the tip of the left-hand cut at the origin of the complex-coupling constant plane, which is too far from the infinite-coupling limit to be of relevance. The use of ω-information in our extrapolation is crucial to obtaining high accuracy when resumming the series in the bare coupling constant. This information is more useful than the large-order information in previous

TABLE 20.8 Coefficients of the large-order expansions (20.68)–(20.70) fitting known expansion coefficients of ω, η, $\bar{\eta}$. The coefficients f_k^η possess two separate expansions for even and odd k.

	N	$c^{(1)}$	$c^{(2)}$	$c^{(3)}$	$c^{(4)}$	$c^{(5)}$	$c^{(6)}$
ω	0	0.26301475112	3.4408182822	-31.76733359043	209.94304685908	-387.9820769504	212.1565739538
	1	1.63535099051	-8.7629408561	32.52987246310	49.56939798556	-198.5501186375	130.5393597659
	2	4.19032409932	-32.5218822010	159.23160834532	-271.56782378290	185.5214629862	-36.2203620935
	3	8.06590542355	-69.1380037623	356.19870179271	-773.40843073419	787.4105682986	-297.8631178069
$\bar{\eta}$	0	15.47452873233	-263.105249597	1695.8521799417	-4797.254788814	6198.211268910	-2825.378774427
	1	10.94704206385	-169.697930580	1074.8269224230	-2886.578089415	3577.486553055	-1577.198376659
	2	1.24814548715	60.9324565140	-409.5953535647	1526.620407734	-2300.494640749	1163.425537324
	3	-25.80328671245	508.523659337	-3253.9391201198	9876.171576908	-13307.48621904	6257.590654494
η_o	0	-6.363429671227	54.7969857339	-209.2126943952	159.77913833249		
	1	-5.860815634115	58.1732922278	-237.1581744239	183.84569783020		
	2	-5.108698105700	64.4651051506	-285.1161542307	224.75974718583		
	3	-4.203986342723	76.2691471289	-364.4529959457	291.38783515954		
η_e	0	-5.6929922203758	15.551243915764	61.12469347544379			
	1	-5.3245881267711	14.110708087849	81.2312043328075			
	2	-4.5203425601138	9.799960635959	117.4131477198922			
	3	-3.1970976073075	1.705210978430	176.4615812743069			

resummation schemes in which the critical exponents are determined at a finite renormalized coupling constant g^* of order unity, which lies at a finite distance from the left-hand cut in the complex g-plane. Although this determination is sensitive to the discontinuity at the tip of the cut, it must be realized that the influence of the cut is very small due to the smallness of the *fugacity* of the leading instanton, which is equal to a Boltzmann factor $e^{-\text{const}/g}$.

Let us see the results obtained by using the extended list of expansion coefficients in Table 20.5 . The extrapolations are shown in Fig. 20.24, for an extension of the known six-loops coefficients of $\omega(\bar{g}_B)$ and $\eta(\bar{g}_B)$ by one extrapolated coefficient. This produces an even number of approximants which can be most easily extrapolated to infinite order. For $\bar{\eta}(\bar{g}_B)$ we use two more coefficients for the same reason. The resulting ω_8-values are lowered somewhat with respect to ω_6 from (20.61) to

$$\omega_8 = \begin{Bmatrix} 0.7935 \\ 0.7916 \\ 0.7900 \\ 0.7880 \end{Bmatrix} \quad \text{for} \quad \begin{Bmatrix} N = 0 \\ N = 1 \\ N = 2 \\ N = 3 \end{Bmatrix}. \tag{20.78}$$

The new η values are

$$\eta_8 = \begin{Bmatrix} 0.02829 - 0.01675 \times (\omega - 0.7935) \\ 0.03319 - 0.01523 \times (\omega - 0.7916) \\ 0.03503 - 0.02428 \times (\omega - 0.7900) \\ 0.03537 - 0.01490 \times (\omega - 0.7880) \end{Bmatrix} \quad \text{for} \quad \begin{Bmatrix} N = 0 \\ N = 1 \\ N = 2 \\ N = 3 \end{Bmatrix}, \tag{20.79}$$

lying reasonably close to the previous seven-loop results (20.63), (20.64) for the smaller ω-values (20.78). The first set yields $\eta_8 = \{0.0300, 0.0356, 0.0360, 0.0354\}$; the second $\eta_8 = \{0.0315, 0.0342, 0.0349, 0.0345\}$.

For $\bar{\eta}$ we find the results

$$\bar{\eta}_9 = \begin{Bmatrix} -0.2711 + 0.0400 \times (\omega - 0.810) \\ -0.3803 + 0.0974 \times (\omega - 0.805) \\ -0.4735 + 0.1240 \times (\omega - 0.800) \\ -0.5506 + 0.4761 \times (\omega - 0.797) \end{Bmatrix} \quad \text{for} \quad \begin{Bmatrix} N = 0 \\ N = 1 \\ N = 2 \\ N = 3 \end{Bmatrix}. \tag{20.80}$$

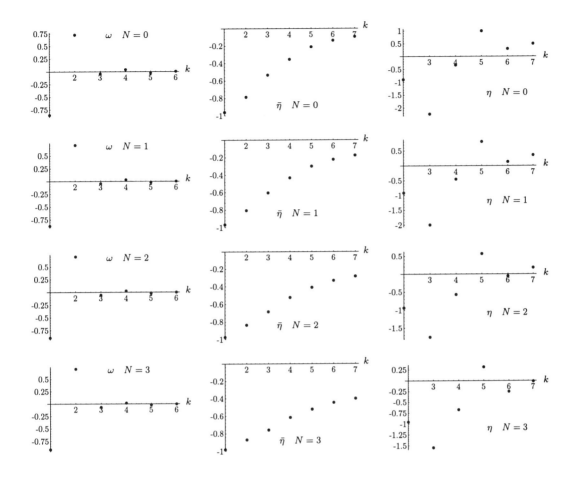

FIGURE 20.23 Relative errors $f_k/f_k^{\text{large order}} - 1$ in predicting the kth expansion coefficient by fitting the large-order expressions (20.68)–(20.70) for ω, , η, and $\bar{\eta} \equiv \nu^{-1} + \eta - 2$ to the first $k - 1$ expansion coefficients.

It is interesting to observe how the resummed values $\omega_L, \bar{\eta}_L, \eta_L$ obtained from the extrapolated expansion coefficients in Table 20.5 continue to higher orders in N. This is shown in Fig. 20.25. The dots converge to some specific values which, however, are different from the extrapolation results in Fig. 20.24 based on the theoretical convergence behavior error $\approx e^{-cN^{1-\omega}}$. We shall argue below that these results are worse than the properly extrapolated values.

All the above numbers agree reasonably well with each other and with other estimates in the literature listed in Table 20.1 on page 373. The only comparison with experiment which is sensitive enough to judge the accuracy of the results and the reliability of the resummation procedure is provided by the measurement of the critical exponent $\alpha = 2 - 3\nu$ in the space shuttle experiment of Lipà et al. [16] whose data were plotted in Fig. 1.2. By going into a vicinity of the critical temperature with $\Delta T \approx 10^{-8}$ K, their fit to the experimental singularity $C \propto |1 - T/T_c|^{-\alpha}$ in the specific heat at the λ-point of superfluid helium, yields the highly accurate value [recall Eq. (1.23)]

$$\alpha = -0.01056 \pm 0.00038. \tag{20.81}$$

Since ν is of the order $2/3$, this measurement is extremely sensitive to ν, which is

$$\nu = 0.6702 \pm 0.0001. \tag{20.82}$$

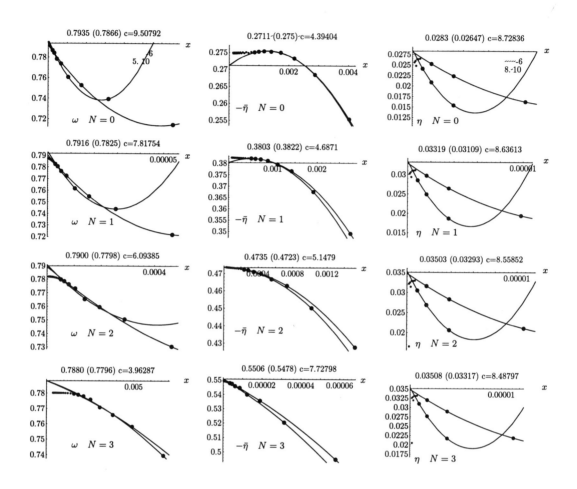

FIGURE 20.24 Extrapolation of resummed $\omega, \bar{\eta}, \eta$-values if one (ω, η) or two $(\bar{\eta})$ more expansion coefficients of Table 20.5 are taken into account. The fat dots show the resummed values used for extrapolation, the small dots higher resummed values not used for the extrapolation. The numbers on top specify the extrapolated values and the values of the last approximation, corresponding to the leftmost fat dot.

It is therefore useful to perform resummation and extrapolation at $N = 2$ directly for the approximate α-values $\alpha_L = 2 - 3\nu_L$, once for the six-loop ω-value $\omega = 0.8$, and once for a neighboring value $\omega = 0.790$, to see the ω-dependence. The results are shown in Fig. 20.26. The extrapolated values for our $\omega = 0.8$ in Table 20.1 on page 373 yield

$$\alpha = -0.01294 \pm 0.00060, \tag{20.83}$$

in very good agreement with experiment.

Note that the results get worse if we use more than eleven orders of the extended series. This shows that the higher extrapolated expansion coefficients in Table 20.5 do not really carry more information on the critical exponent ν. Although these coefficients lie closer and closer to the true ones as expected from the decreasing errors in the plots in Fig. 20.23, this does not increase their usefulness. The errors are samll only in relation to the huge factorially growing expansion coefficients. The resummation procedure removes the factorial growth and becomes extremely sensitive to very small deviations from these huge coefficients. This is the numerical consequence of the fact discussed earlier that the information residing in the exponentially

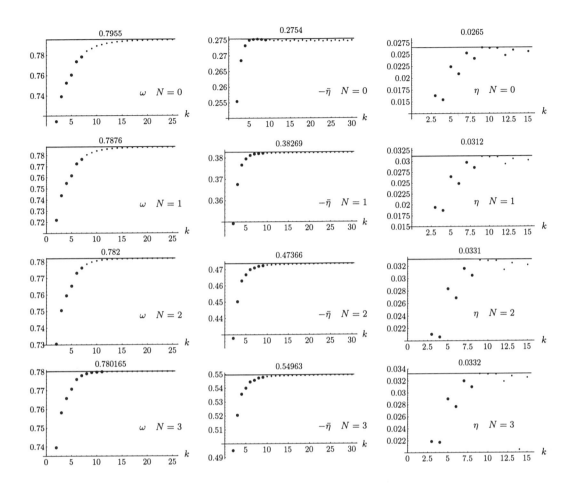

FIGURE 20.25 Direct plots of the resummed $\omega, \bar{\eta}, \eta$-values for all resummed values from all extrapolated expansion coefficients of Table 20.5 . The line is fitted to the maximum of all dots at the place specified by the number on top. Fat and small dots distinguish the resummed exponents used in the previous extrapolations from the unused ones.

small imaginary part of all critical exponents near the tip of the left-hand cut in the complex $\bar{g}_B$-plane has practically no effect upon the strong-coupling results at infinite $\bar{g}_B$.

Therefore, the knowledge of the large-order behavior is only of limited help in improving the accuracy of the approximation. In our theory, an additional important information is contained in the theoretically known exponential convergence behavior which predicts a linear behavior of the approximations when plotted against $x(L) = e^{-cL^{1-\omega}}$. This knowledge, which allows us to extrapolate our approximations for $L = 2, 3, 4, 5, 6, 7$ well to infinite order L, seems to be more powerful than that of the large-order behavior.

The complete updated list of exponents is shown in Table 20.2 , which also contains values for the other critical exponents $\alpha = 2 - D\nu$ and $\beta = \nu(D - 2 + \eta)/2$.

A further improvement of the above results should be possible by taking into account the existence of further terms $1/g_B^{\omega'}$, $(1/g_B^{2\omega'}),\ldots$ in the strong-coupling expansion (19.52), corresponding to *confluent singularities* [17] in the renormalized coupling constant $\bar{g}$. This can be done following the strategy developed for simple quantum-mechanical systems [18].

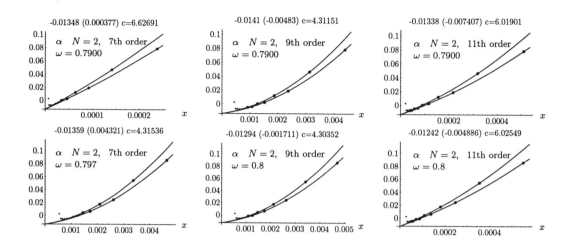

FIGURE 20.26 Extrapolation of resummed α-values if two more expansion coefficients are used from Table 20.5. The small dots near the origins show how the resummed results start deviating from the smooth exptrapolation curve if more than two coefficients are taken into account.

20.7   Another Variational Resummation Method

With the experimental value of α offering a sensitive method for the quality of various resummation schemes, let us also try another such scheme based on variational perturbation theory [19]. Consider the seven-loop power series expansion for $\nu^{-1}(\bar{g}_B) = 2 - \eta_m(\bar{g}_B)$ in powers of the unrenormalized coupling constant of O(2)-invariant ϕ^4-theory [recall Eqs. (20.18) and (20.60)]:

$$
\begin{aligned}
\nu^{-1} = {} & 2 - 0.4\,\bar{g}_B + 0.4681481481482289\,\bar{g}_B^2 - 0.66739\,\bar{g}_B^3 + 1.079261838589703\,\bar{g}_B^4 \\
& - 1.91274\,\bar{g}_B^5 + 3.644347291527398\,\bar{g}_B^6 - 7.37808\,\bar{g}_B^7 + \ldots .
\end{aligned}
\tag{20.84}
$$

Using the fits in Tables 20.8 to the theoretical large-order behavior we extend this series to higher orders as follows (see Table 20.5) [20]:

$$
\begin{aligned}
\Delta\nu^{-1} = {} & 15.75313406543747\,\bar{g}_B^8 - 35.2944\,\bar{g}_B^9 + 82.6900901520064\,\bar{g}_B^{10} \\
& - 202.094\,\bar{g}_B^{11} + 514.3394395526179\,\bar{g}_B^{12} - 1361.42\,\bar{g}_B^{13} \\
& + 3744.242656157152\,\bar{g}_B^{14} - 10691.7\,\bar{g}_B^{15} + \ldots .
\end{aligned}
\tag{20.85}
$$

We now calculate the expansion for the logarithmic derivative $s = \bar{g}_B \bar{g}'(\bar{g}_B)/\bar{g}(\bar{g}_B)$ [recall (19.61)] of the renormalized coupling constant in powers of $\bar{g}_B$ from (20.15):

$$
\begin{aligned}
s = {} & 1 - \bar{g}_B + \frac{947\,\bar{g}_B^2}{675} - 2.322324349407407\,\bar{g}_B^3 + 4.276203609026057\,\bar{g}_B^4 \\
& - 8.51611440473227\,\bar{g}_B^5 + 18.05897631325589\,\bar{g}_B^6 + \ldots .
\end{aligned}
\tag{20.86}
$$

A fit with the theoretical large-order behavior extends this series by [see Table (20.7)]

$$
\begin{aligned}
\Delta s = {} & 40.386572287301\,\bar{g}_B^7 + 94.645339912347\,\bar{g}_B^8 - 231.39224421625\,\bar{g}_B^9 \\
& + 588.32061725791\,\bar{g}_B^{10} - 1552.1163584042\,\bar{g}_B^{11} + 4242.3726850801\,\bar{g}_B^{12} \\
& - 12001.188664918\,\bar{g}_B^{13} + 35115.230066461\,\bar{g}_B^{14} - 106234.46430864\,\bar{g}_B^{15} + \ldots . \;(20.87)
\end{aligned}
$$

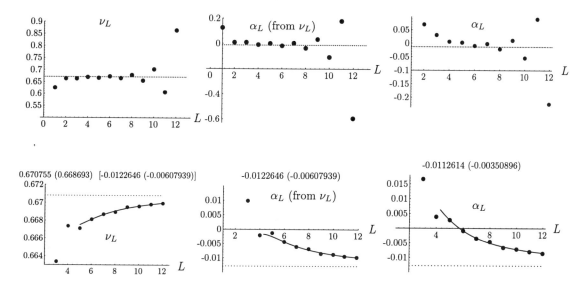

FIGURE 20.27 Upper plots: Results of partial sums of series (20.88) for $\nu^{-1}(h = 1)$ up to order L, once plotted as $\nu_L = 1/\nu_L^{-1}$, and once as $\alpha_L = 2 - 3\nu_L$. The right-hand plot shows the corresponding partial sums of the series (20.89) for $\alpha(h = 1)$. The dotted lines indicate the experimental value $\alpha^{ss} = -0.01285 \pm 0.00038$ of the space shuttle data in Fig. 1.2.   Lower plots: The corresponding resummed values and a fit of them by $c_0 + c_1/L^2 + c_2/L^4$. The constant c_0 is written on top, together with the seventh-order approximation (in parentheses). The square brackets on top of the left-hand plot for ν shows the corresponding α-values.

The scaling properties of the theory near T_c imply that $g(\bar{g}_B)$ becomes a constant for $\bar{g}_B \to \infty$, so that $s(\bar{g}_B)$ goes to zero in this limit. By inverting the expansion for $s(\bar{g}_B) + \Delta s(\bar{g}_B)$, we obtain an expansion for $\bar{g}_B$ in powers of $h \equiv 1 - s$, and subsequently from (20.84), (20.85) an expansion for ν^{-1} in powers of $h \equiv 1 - s$ as follows:

$$
\begin{aligned}
\nu^{-1}(h) &= 2 - 0.4\,h - 0.093037\,h^2 + 0.000485012\,h^3 - 0.0139286\,h^4 + 0.007349\,h^5 - 0.0140478\,h^6 \\
&\quad + 0.0159545\,h^7 - 0.029175\,h^8 + 0.0521537\,h^9 - 0.102226\,h^{10} + 0.224026\,h^{11} - 0.491045\,h^{12} \\
&\quad + 1.22506\,h^{13} - 3.00608\,h^{14} + 8.29528\,h^{15} + \dots\,.
\end{aligned} \tag{20.88}
$$

The critical exponent ν^{-1} is obtained by evaluating this series at $h = 1$ via variational perturbation theory. From the result we obtain the critical exponent $\alpha = 2 - 3\nu$.

For estimating the systematic errors of our resummation, we also re-expand (20.88) to find the corresponding power series for $\alpha = 2 - 3\nu$:

$$
\begin{aligned}
\alpha(h) &= 0.5 - 0.3\,h - 0.129778\,h^2 - 0.0395474\,h^3 - 0.0243203\,h^4 - 0.0032498\,h^5 - 0.0121091\,h^6 \\
&\quad + 0.00749308\,h^7 - 0.0194876\,h^8 + 0.0320172\,h^9 - 0.0651726\,h^{10} + 0.14422\,h^{11} - 0.315055\,h^{12} \\
&\quad + 0.802395\,h^{13} - 1.95455\,h^{14} + 5.49143\,h^{15} + \dots\,,
\end{aligned} \tag{20.89}
$$

whose value at $h = 1$ gives directly the critical exponent α, from which we obtain another approximation for $\nu = (2 - \alpha)/3$.

In order to obtain a first idea of the behavior of the expansions (20.88) and (20.89), we plot their partial sums at $h = 1$ in the upper row of Fig. 20.27. After an initial apparent convergence, the partial sums show the typical divergence of asymptotic series which call for

TABLE 20.9   Results of Padé approximations $P_{MN}(h)$ at $h = 1$ to power series $\nu^{-1}(h)$ and $\alpha(h)$ in Eq. (20.88) and Eq. (20.89), respectively. The parentheses show the associated values of α and ν.

MN	ν	(α)	(ν)	α
4 4	0.678793	(-0.0363802)	(0.678793)	-0.0363802
5 4	0.671104	(-0.0133107)	(0.670965)	-0.0128940
4 5	0.670965	(-0.0128940)	(0.670901)	-0.0127031
5 5	0.670756	(-0.0122678)	(0.670756)	-0.0122678

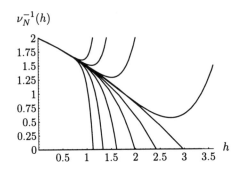

FIGURE 20.28   Successive truncated expansions of $\nu^{-1}(h)$ of orders $N = 2, \dots, 12$.

a resummation. A rough resummation is possible using Padé approximants. The results are shown in Table 20.9 . The highest Padé approximants yield

$$\alpha^{\mathrm{P}} = -0.0123 \pm 0.0050. \tag{20.90}$$

The error is estimated by the distance from the next lower approximation.

We now resum the expansions (20.88) and (20.89) by variational perturbation theory. Since we now want to know the functions at a finite argument h, we cannot use (19.16) but must first adapt our function to the general form to which we can apply our resummation procedure. Plotting the successive truncated power series for $\nu^{-1}(h)$ against h in Fig. 20.28, we see that this function curve seems to have a zero somewhere above $h = h_0 = 3$. We therefore go over to the variable x defined by $h = h(x) \equiv h_0 x/(h_0 - 1 + x)$, in terms of which $f(x) = \nu^{-1}(h(x))$. It has a power series expansion

$$f(x) = \sum_{n=0}^{\infty} f_n x^n, \tag{20.91}$$

which behaves like (20.9) with $p = 0$ and $q = 2$, and has to be evaluated at $x = 1$ corresponding to $\bar{g}_B = 0$. This large-x behavior is obtained in the resummation of (20.91) by introducing an auxiliary scale parameter κ and forming the truncated functions

$$f_L(x) \equiv \kappa^p \sum_{n=0}^{L} f_n \left(\frac{x}{\kappa^q} \right)^n. \tag{20.92}$$

The parameter κ will be set equal to 1 at the end. Then we introduce a variational parameter K by the replacement

$$\kappa \to \sqrt{K^2 + \kappa^2 - K^2}. \tag{20.93}$$

The functions $f_L(x)$ are so far independent of K. This is changed by expanding the square root in (20.93) in powers of $\kappa^2 - K^2$, thereby treating this quantity as being of order h. The terms $\kappa^p x^n / \kappa^{qn}$ in (20.92) are then expanded as

$$\kappa^p \frac{x^n}{\kappa^{qn}} \rightarrow K^p \frac{x^n}{K^{qn}} \left[1 + \binom{(p-qn)/2}{1} r + \binom{(p-qn)/2}{2} r^2 + \ldots + \binom{(p-qn)/2}{N-n} r^{N-n} \right],$$
(20.94)

where $r = (\kappa^2 - K^2)/K^2$. Setting $\kappa = 1$ and replacing the variational parameter K by v defined by $K^2 \equiv v^{-1}x$, we obtain from (20.92) at $x = 1$ the variational expansions

$$W_L(1, K) = \sum_{n=0}^{L} f_n v^{qn-p/2} [1 - (1-v)]_{L-n}^{(p-qn)/2},$$
(20.95)

where the symbol $[1 - A]_{N-n}^{(p-qn)/2}$ was defined in Eq. (19.17) by the binomial expansion of $(1 - A)^{(p-qn)/2}$ in powers of A up to the order A^{N-n}. This is of course the expansion (19.6) in which the role of g_B is played by x. At $g_B \equiv x = 1$, the paramter $\sigma(K)$ in Eq. (19.5) is equal to $\sigma = (1 - 1/K^2)K^q$ at $\kappa = 1$, and $K^2 = v^{-1}$.

The appropriate modification of formula (19.16) for the evaluation at unit argument rather than the strong-coupling limit is

$$f_L^*(1) = \operatorname*{opt}_{v} \left[\sum_{l=0}^{L} f_l v^{l-p/2} [1 - (1-v)]_{L-l}^{(p-ql)/2} \right].$$
(20.96)

The variational expansions are listed in Table 20.10 . They are optimized in v by minima for odd and by turning points for even L, as shown in Fig. 20.29. The extrema are plotted as

TABLE 20.10   Variational expansions $\nu_L^{-1}(x)$ and $\alpha_L(x)$ for $L = 2, \ldots, 9$. They depend on the variational parameter K via $v = x/K^2$, and are plotted for $x = 1$ in Fig. 20.29. Their minima or turning points are extrapolated to $L = \infty$ in the lower plots of Fig. 20.27. The table goes only up to $L = 9$, to save space, the plots go to $L = 12$.

$$\nu_2^{-1} = 2 - 1.2v + 0.69067v^2$$
$$\nu_3^{-1} = 2 - 1.8v + 2.07200v^2 - 0.72036v^3$$
$$\nu_4^{-1} = 2 - 2.4v + 4.14400v^2 - 2.88145v^3 + 0.53412v^4$$
$$\nu_5^{-1} = 2 - 3.0v + 6.90667v^2 - 7.20363v^3 + 2.67060v^4 + 0.28949v^5$$
$$\nu_6^{-1} = 2 - 3.6v + 10.3600v^2 - 14.4073v^3 + 8.01180v^4 + 1.73692v^5 - 2.96286v^6$$
$$\nu_7^{-1} = 2 - 4.2v + 14.5040v^2 - 25.2127v^3 + 18.6942v^4 + 6.07922v^5 - 20.7401v^6 + 11.1835v^7$$
$$\nu_8^{-1} = 2 - 4.8v + 19.3387v^2 - 40.3403v^3 + 37.3884v^4 + 16.2113v^5 - 82.9602v^6 + 89.4683v^7 - 36.9575v^8$$
$$\nu_9^{-1} = 2 - 5.4v + 24.8640v^2 - 60.5105v^3 + 67.2992v^4 + 36.4753v^5 - 248.881v^6 + 402.607v^7 - 332.617v^8 + 121.914v^9$$

$$\alpha_2 = 0.5 - 0.90v + 0.3830v^2$$
$$\alpha_3 = 0.5 - 1.35v + 1.1490v^2 - 0.26997v^3$$
$$\alpha_4 = 0.5 - 1.80v + 2.2980v^2 - 1.07989v^3 + 0.025254v^4$$
$$\alpha_5 = 0.5 - 2.25v + 3.8300v^2 - 2.69972v^3 + 0.126271v^4 + 0.57604v^5$$
$$\alpha_6 = 0.5 - 2.70v + 5.7450v^2 - 5.39945v^3 + 0.378812v^4 + 3.45629v^5 - 2.19244v^6$$
$$\alpha_7 = 0.5 - 3.15v + 8.0430v^2 - 9.44903v^3 + 0.883895v^4 + 12.0970v^5 - 15.3471v^6 + 6.89011v^7$$
$$\alpha_8 = 0.5 - 3.60v + 10.724v^2 - 15.1184v^3 + 1.767790v^4 + 32.2587v^5 - 61.3884v^6 + 55.1208v^7 - 21.5704v^8$$
$$\alpha_9 = 0.5 - 4.05v + 13.788v^2 - 22.6777v^3 + 3.182020v^4 + 72.5821v^5 - 184.165v^6 + 248.044v^7 - 194.134v^8 + 70.781v^9$$

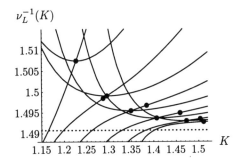
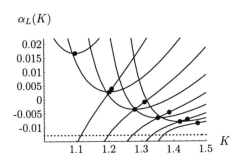

FIGURE 20.29 Successive variational functions $\nu_L^{-1}(h)$ and $\alpha_L(h)$ of Table 20.10 plotted for $h = x = 1$ against the variational parameter $K = \sqrt{x/v}$, together with their minima for odd L, or turning points for even L. These points are plotted against L in the lower row of Fig. 20.27, where they are extrapolated to $L \to \infty$, yielding the critical exponents.

a function of the order L in the lower row of Fig. 20.27. The left-hand plot shows directly the extremal values of $\nu_L^{-1}(v)$, the middle plot shows the α-values $\alpha_L = 2 - 3\nu_L$ corresponding to these. The right-hand plot, finally, shows the extremal values of $\alpha_L(v)$. All three sequences of approximations are fitted very well by a large L expansion $c_0 + c_1/L^2 + c_2/L^4$, if we omit the lowest five data points which are not yet very regular. The inverse powers of L were determined by making a general ansatz $c_0 + c_1/L^{p_1} + c_2/L^{p_2}$ and varying p_1, p_2 until the sum of the square deviations of the fit from the points is minimal. The highest data point is chosen to be the one with $L = 12$ since, up to this order, the successive asymptotic values c_0 change monotonously by decreasing amounts. Starting with $L = 13$, the changes increase and reverse direction. In addition, the mean square deviations of the fits increase drastically, indicating a decreasing usefulness of the extrapolated expansion coefficients in (20.85) and (20.87) for the extrapolation $N \to \infty$. From the parameter c_0 of the best fit for α which is indicated on top of the lower right-hand plot in Fig. 20.27, we find the critical exponent

$$\alpha = -0.01126 \pm 0.0010, \tag{20.97}$$

where the error estimate takes into account the basic systematic errors indicated by the difference between the resummation of $\alpha = 2 - 3\nu$, and of ν^{-1}, which by the lower middle plot in Fig. 20.27 yields $\alpha = -0.01226$. It also accommodates our earlier results from the previous seven-loop resummation in Ref. [20] for α in Eq. (20.83).

The dependence on the choice of h_0 is negligible as long as the resummed series $\nu^{-1}(x)$ and $\alpha(x)$ do not change their Borel character. Thus $h_0 = 2.2$ leads to results well within the error limits in (20.97). The critical exponent (20.97) is in excellent agreement with the experimental space shuttle value (20.81). A graphical comparison of our result with experiment and the results of numerous other authors will be given in the next chapter, in Fig. 21.7.

As before, a further improvement of all resummation results should be possible by taking into account the existence of further terms $1/g_B^{\omega'}$, $(1/g_B^{\omega'})^2, \ldots$ in the strong-coupling expansion corresponding to confluent singularities [recall the remarks in Section 19.4] [17] in the renormalized coupling constant $\bar{g}$.

20.8   High-Temperature Expansions of Lattice Models

It is also possible to extract the critical value β_c directly from the following combination of ratios:

$$\beta_n = [nR_n - (n-1)R_{n-1}]^{-1}. \tag{20.98}$$

Since this combination contains no corrections of order $1/n$, we plot the sequence against $1/n^2$, with the results shown in Fig. 20.30. The critical exponent γ can be obtained most directly from the $n \to \infty$ -limit of the sequence

$$\gamma_n = 1 - n(n-1)\frac{R_n - R_{n-1}}{nR_n - (n-1)R_{n-1}}. \tag{20.99}$$

An alternative and more ancient access to critical exponents of the $O(N)$ universality classes is possible via the classical Heisenberg model, whose energy has been given in Eq. (1.52). For this model, the literature has by now offered extensive high-temperature expansions. The basics of such models are explained in many textbooks [21]. The classical Heisenberg model has the partition function

$$Z(\beta) = \prod_i \int d\mathbf{S}(\mathbf{x}_i)\exp\left[\frac{\beta}{2}\sum_{\{i,j\}} \mathbf{S}(\mathbf{x}_i)\mathbf{S}(\mathbf{x}_j)\right], \tag{20.100}$$

where $\beta \equiv J/k_BT$ is the inverse temperature in natural units, $\mathbf{S}_i$ are unit vectors on the lattice sites $\mathbf{x}_i$ covered by the sum over i. For each i, the sum over j runs over all nearest neighbors of the site $\mathbf{x}_i$. The integrals $d\mathbf{S}(\mathbf{x}_i)$ cover the surface of an N-dimensional sphere.

For this model, several quantities have been calculated [23] on square, simple-cubic (sc), and body-centered cubic (bcc) lattices as a power series in β up to the order β^{22}. In particular, we know the susceptibility

$$\chi(\beta) = \sum_i \langle \mathbf{S}(\mathbf{x}_i)\mathbf{S}(\mathbf{x}_j)\rangle = 1 + \sum_{k=1}^{L} a_k\beta^k \tag{20.101}$$

up to order β^{22}. For $N = 2$ on a bcc lattice, $\chi(\beta)$ starts out like

$$\chi(\beta) = 1 + 8\frac{\beta}{2} + 56\left(\frac{\beta}{2}\right)^2 + 388\left(\frac{\beta}{2}\right)^3 + \dots . \tag{20.102}$$

Note that since the susceptibility diverges near β_c like $(\beta_c - \beta)^{-\gamma}$, and since the correlation function $\langle \mathbf{S}(\mathbf{x}_i)\mathbf{S}(\mathbf{x}_j)\rangle$ falls off exponentially with the correlation length like

$$\langle \mathbf{S}(\mathbf{x}_i)\mathbf{S}(\mathbf{x}_j)\rangle \approx g(\beta)\, e^{-|\mathbf{x}_i-\mathbf{x}_j|/\xi(\beta)}, \tag{20.103}$$

the prefactor diverges like

$$g(\beta) \approx (\beta_c - \beta)^{-\gamma+3\nu}. \tag{20.104}$$

Another calculated lattice sum is

$$\mu_2(\beta) = \sum_i \mathbf{x}_i^2\langle \mathbf{S}(\mathbf{x}_i)\mathbf{S}(\mathbf{x}_j)\rangle = \sum_{k=1}^{L} h_k\beta^k, \tag{20.105}$$

which for $N = 2$ on a bcc lattice, starts out like

$$\mu_2(\beta) = 8\frac{\beta}{2} + 128\cdot 56\left(\frac{\beta}{2}\right)^2 + 1412\left(\frac{\beta}{2}\right)^3 + \dots . \tag{20.106}$$

This diverges near the critical point like

$$\xi^5(\beta) g(\beta) \approx (\beta_c - \beta)^{-2\nu + \gamma}. \tag{20.107}$$

If we form the combination

$$X^2(\beta) \equiv \frac{\mu_2(\beta)}{6\chi(\beta)}, \tag{20.108}$$

this quantity is proportional to the square of the coherence length, and diverges like

$$X^2(\beta) \approx (\beta_c - \beta)^{-2\nu}. \tag{20.109}$$

This function gives direct access to the critical exponent ν.

In a finite system, all lattice functions such as $\chi(\beta)$ and $X^2(\beta)$ are polynomials in β. These have complex-conjugate pairs of zeros in the complex plane corresponding to the degree of the polynomials. For a growing number of lattice sites, these zeros approach the real β in the form of two sequences from above and below. In the thermodynamic limit of an infinite system, the points move arbitrarily close to each other and form two continuous lines whose end points pinch the β-axis at β_c [22]. When taking the logarithm of Z and forming $-\beta f'$, the lines of zeros become lines of singularities, giving rise to a critical power behavior, such as

$$\chi(\beta) \propto (\beta_c - \beta)^{-\gamma}. \tag{20.110}$$

The critical exponent γ can be extracted from the power series in the following way. Expanding (20.110) in powers of β, we obtain a series

$$\chi(\beta) \propto \sum_n \binom{-\gamma}{n} \left(\frac{\beta}{\beta_c}\right)^n. \tag{20.111}$$

For the ratios R_n of two successive large order coefficients, this implies the following limit:

$$
\begin{aligned}
R_n &= \frac{1}{\beta_c} \binom{-\gamma}{n} \bigg/ \binom{-\gamma}{n-1} = \frac{1}{\beta_c} \frac{\gamma(\gamma+1)\cdots(\gamma+n-1)}{n!} \bigg/ \frac{\gamma(\gamma+1)\cdots(\gamma+n-2)}{(n-1)!} \\
&= \frac{1}{\beta_c}\left(1 + \frac{\gamma-1}{n}\right). \tag{20.112}
\end{aligned}
$$

If an expansion is known up to a sufficiently high order in β, this can be used to estimate the critical β_c as well as the critical exponent γ. We simply plot $1/R_n$ against $1/n$ and read off β_c from the extrapolated intercept. This procedure is called the *ratio test*. The slope at the origin is $-(\gamma - 1)\beta_c$, and yields the critical index γ. The plots for $N = 0, 1, 2$, and 3 are shown in Fig. 20.31. The extrapolation of the average of even and odd sequences is shown in Fig. 20.32.

Similar procedures can be applied to the power series for $X^2(\beta)$ and the critical exponent 2ν. In Fig. 20.33 we determined the critical exponent ν by plotting the ratios R_n against $1/n$, fitting the last four points of even and odd sequences by parabolas, and deducing from the average slope the critical exponent ν.

Butera and Comi have performed much more sophisticated analyses which are summarized in Tables 20.11 and 20.12 at the end of this section. For an explanation of their procedures see the original paper in Ref. [23]. Their critical exponents agree reasonably well with those in Tables 20.1 on page 373 and 20.2 on page 386.

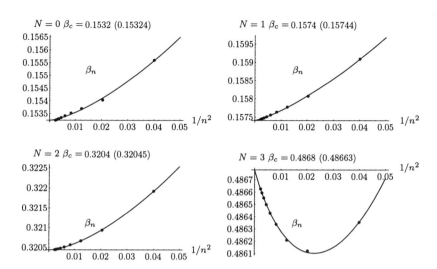

FIGURE 20.30 Critical inverse temperature β_c from a parabolic fit to the average of the even and odd sequences (20.98), plotted against $1/n^2$. The value at $1/n^2 = 0$ yields the critical β_c. The numbers in parentheses show the highest approximation to β_c before the extrapolation.

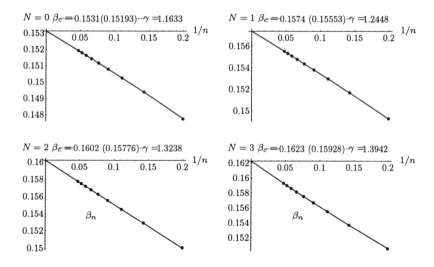

FIGURE 20.31 Critical inverse temperature β_c from the last four terms of the average of the even and odd sequences (20.98), plotted against $1/n$. An almost straight parabola is fitted through the points whose value at $1/n = 0$ yields the critical reduced temperature $t_c \equiv 1/\beta_c$. The slope at the origin is an approximation for the critical exponent γ. The numbers in parentheses show the highest approximations to β_c before the extrapolation.

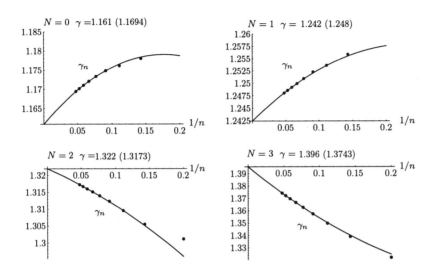

FIGURE 20.32 Critical Exponents γ from the average of the sequences (20.99) for even and odd n, plotted against $1/n$. The extrapolation curve at $1/n = 0$ yields the critical exponent γ. Only the last four averages are extrapolated parabolically.

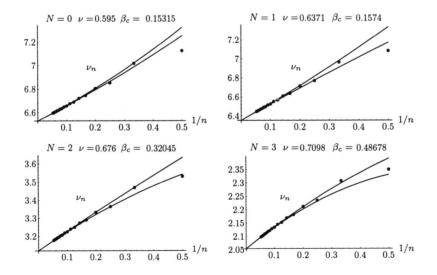

FIGURE 20.33 Critical Exponents ν from the average of the slopes of the even and odd ratios.

TABLE 20.11 Summary of lattice estimates of critical exponents for $0 \leq N \leq 3$.

N	Method and Ref.	β_c	γ	ν
0	HTE sc unbiased [24]	0.2134987(10)	1.16193(10)	
	HTE sc unbiased	0.213497(6)	1.161(2)	0.592(2)
	HTE sc θ-biased	0.213493(3)	1.1594(8)	0.5878(6)
	MonteCarlo sc [25]	0.2134969(10)		0.5877(6)
	MonteCarlo sc [26]	0.213492(1)	1.1575(6)	
	HTE bcc unbiased	0.153131(2)	1.1612(8)	0.591(2)
	HTE bcc θ-biased	0.153128(3)	1.1582(8)	0.5879(6)
	R.G. FD perturb. [12]		1.1569(8)	0.5872(8)
	R.G. ϵ−expansion [27]		1.157(3)	0.5880(15)
1	HTE sc unbiased	0.221663(9)	1.244(3)	0.634(2)
	HTE sc θ-biased	(0.2216544(3))	1.2388(10)	0.6315(8)
	MonteCarlo sc [28]	0.2216595(26)		0.6289(8)
	MonteCarlo sc [29]	0.2216544(3)	1.237(2)	0.6301(8)
	HTE bcc [30]		1.2395(4)	0.632(1)
	HTE bcc [31]		1.237(2)	0.6300(15)
	HTE bcc unbiased	0.157379(2)	1.243(2)	0.634(2)
	HTE bcc θ-biased	(0.157373(2))	1.2384(6)	0.6308(5)
	R.G. FD perturb. [12]		1.2378(12)	0.6301(10)
	R.G. ϵ−expansion [27]		1.2390(25)	0.6310(15)
2	Experiment [16]			0.6702(1)
	HTE sc unbiased	0.45419(3)	1.327(4)	0.677(3)
	HTE sc θ-biased	0.45419(3)	1.325(3)	0.675(2)
	MonteCarlo sc [32]	0.45420(2)	1.308(16)	0.662(7)
	MonteCarlo sc [33]	0.4542(1)	1.316(5)	0.670(7)
	MonteCarlo sc [34]	0.454165(4)	1.319(2)	0.672(1)
	HTE bcc unbiased	0.320428(3)	1.322(3)	0.674(2)
	HTE bcc θ-biased	0.320427(3)	1.322(3)	0.674(2)
	HTE fcc [35]	0.2075(1)	1.323(15)	0.670(7)
	R.G. FD perturb. [12]		1.318(2)	0.6715(15)
	R.G. ϵ−expansion [27]		1.315(7)	0.671(5)
3	HTE sc unbiased	0.69303(3)	1.404(4)	0.715(3)
	HTE sc θ-biased	0.69305(4)	1.406(3)	0.716(2)
	MonteCarlo sc [36]	0.693035(37)	1.3896(70)	0.7036(23)
	MonteCarlo sc [34]	0.693002(12)	1.399(2)	0.7128(14)
	HTE bcc unbiased	0.486805(4)	1.396(3)	0.711(2)
	HTE bcc θ-biased	0.486820(4)	1.402(3)	0.714(2)
	MonteCarlo bcc [36]	0.486798(12)	1.385(10)	0.7059(37)
	HTE fcc [37]	0.3149(6)	1.40(3)	0.72(1)
	R.G. FD perturb. [12]		1.3926(26)	0.7096(16)
	R.G. ϵ−expansion [27]		1.39(1)	0.710(7)

TABLE 20.12   Summary of lattice estimates of critical parameters for $4 \leq N \leq 12$.

N	Method and Ref.	β_c	γ	ν
4	HTE sc unbiased	0.93589(6)	1.474(4)	0.750(3)
	HTE sc θ-biased	0.93600(4)	1.491(4)	0.759(3)
	MonteCarlo sc [38]	0.9360(1)	1.477(18)	0.7479(90)
	MonteCarlo sc [34]	0.935861(8)	1.478(2)	0.7525(10)
	HTE bcc unbiased	0.65531(6)	1.461(4)	0.744(3)
	HTE bcc θ-biased	0.65542(3)	1.484(4)	0.756(3)
	R.G. FD perturb. [3]		1.449	0.738
6	HTE sc unbiased	1.42859(6)	1.582(5)	0.804(3)
	HTE sc θ-biased	1.42895(6)	1.614(5)	0.821(3)
	HTE bcc unbiased	0.99613(6)	1.566(4)	0.796(3)
	HTE bcc θ-biased	0.99644(4)	1.608(4)	0.819(3)
	R.G. FD perturb. [3]		1.556	0.790
8	HTE sc unbiased	1.9263(2)	1.656(5)	0.840(3)
	HTE sc θ-biased	1.92705(7)	1.701(4)	0.864(3)
	HTE bcc unbiased	1.33984(7)	1.644(5)	0.833(3)
	HTE bcc θ-biased	1.34040(6)	1.696(4)	0.862(3)
	R.G. FD perturb. [3]		1.637	0.830
	1/N expansion [7]		1.6449	0.8355
10	HTE sc unbiased	2.4267(2)	1.712(6)	0.867(4)
	HTE sc θ-biased	2.42792(8)	1.763(4)	0.894(4)
	HTE bcc unbiased	1.68509(8)	1.699(5)	0.860(4)
	HTE bcc θ-biased	1.68586(7)	1.761(4)	0.893(3)
	R.G. FD perturb. [3]		1.697	0.859
	1/N expansion [7]		1.7241	0.8731
12	HTE sc unbiased	2.9291(3)	1.759(6)	0.889(4)
	HTE sc θ-biased	2.9304(1)	1.812(5)	0.916(4)
	HTE bcc unbiased	2.03130(8)	1.741(6)	0.881(4)
	HTE bcc θ-biased	2.03230(8)	1.808(5)	0.914(3)
	R.G. FD perturb. [3]		1.743	0.881
	1/N expansion [7]		1.7746	0.8969

Notes and References

The individual citations in the text refer to the papers below.

[1] G. Parisi, J. Stat. Phys. **23**, 49 (1980).

[2] Expansions up to six loops for all O(N) were calculated in
G.A. Baker, Jr., B.G. Nickel, and D.I. Meiron, Phys. Rev.B **17**, 1365 (1978).
More details are containeed in an unpublished report
B.G. Nickel, D.I. Meiron, and G.A. Baker, Jr., University of Guelph report, 1977 .
A copy of this report can be downloaded in PDF format from our Berlin server starting
at www.physik.fu-berlin.de/~kleinert/re.html#b8.

[3] S.A. Antonenko and A.I. Sokolov, Phys. Rev. E **51**, 1894 (1995); Fiz. Tverd. Tela
(Leningrad) **40**, 1284 (1998) [Sov. Phys. Sol. State **40**, 1169 (1998)].

[4] See the remark at the end of
G.A. Baker, Jr., B.G. Nickel, M.S. Green, and D.I. Meiron, Phys. Rev. Lett. **36**, 1351 (1976).

[5] J.A. Gracey, *Progress with Large-Nf β-Functions*, Talk presented at 5th International Workshop on Software Engineering, Neural Nets, Genetic Algorithms, Expert Systems, Symbolic Algebra and Automatic Calculations, Lausanne, Switzerland, 2-6th September, 1996, publ. in Nucl. Instrum. Meth. A **389**, 361 (1997). (hep-th/9609409).
See also
D.J. Broadhurst, J.A. Gracey, and D. Kreimer, Z. Phys. C **75**, 559 (1997) (hep-th/9607174).
S.E. Derkachov, J.A. Gracey, and A.N. Manashov, Eur. Phys. J. C **2**, 569 (1998) (hep-ph/9705268).

[6] A.N. Vasil'ev, Yu.M. Pis'mak, and J.R. Honkonen, Theor. Math. Phys. **50**, 127 (1982).

[7] Y. Okabe and M. Oku, Prog. Theor. Phys. **60**, 1287 (1978).

[8] H. Kleinert, Phys. Rev. D **57**, 2264 (1998) (E-Print aps1997jun25_001); addendum (cond-mat/9803268); See also H. Kleinert, Phys. Lett. B **434**, 74 (1998) (cond-mat/9801167).

[9] J.C. Le Guillou and J. Zinn-Justin, Phys. Rev. B **21**, 3976 (1980); Phys. Rev. Lett. **39**, 95 (1977).

J.-B. Zuber, Nucl. Phys. B **220**, 415 (1983).

[10] S.G. Gorishny, S.A. Larin, and F.V. Tkachov, Phys. Lett. A **101**, 120 (1984).
The critical exponents obtained there are based on the ε-expansions calculated in
A.A. Vladimirov, D.I. Kazakov, and O.V. Tarasov, Zh. Eksp. Teor. Fiz. **77**, 1035 (1979) [Sov. Phys. JETP **50**, 521 (1979)]; K.G. Chetyrkin, S.G. Gorishny, S.A. Larin, and F.V. Tkachov, Phys. Lett. B **132**, 351 (1983); K.G. Chetyrkin, A.L. Kataev, and F.V. Tkachov, Phys. Lett. B **99**, 147 (1981); **101**, 457 (1981); D.I. Kazakov, *ibid.* **133**, 406 (1983). The recent corrections of their five-loop expansion coefficients in H. Kleinert, J. Neu, V. Schulte-Frohlinde, K.G. Chetyrkin, S.A. Larin, Phys. Lett. B **272**, 39 (1991) (hep-th/9503230); Erratum ibid. **319**, 545 (1993) have only little effects on these results.

[11] J.C. Le Guillou and J. Zinn-Justin, J. de Phys. Lett **46**, L137 (1985). Their expansion coefficients are the same as in [10].

[12] D.B. Murray and B.G. Nickel, *Revised estimates for critical exponents for the continuum N-vector model in 3 dimensions*, Unpublished Guelph University report (1991).

[13] In contrast to Eqs. (20.12) and (20.13), the numbers on the right are now the true coefficients, i.e., we have *not* suppressed a factor $(N+8)^7$ accompanying the power $\bar{g}^7$, to save space.

[14] R. Guida, J. Zinn-Justin, J. Phys. **A 31**, 8103 (1998) (cond-mat/9803240).

[15] The instanton action and the numbers in Tables 20.3 and 20.4 were first calculated in Ref. [1].

[16] J.A. Lipa, D.R. Swanson, J. Nissen, T.C.P. Chui, and U.E. Israelson, Phys. Rev. Lett. **76**, 944 (1996).
The initially published fit to the data was erroneous and corrected in a footnote (number [15]) of a subsequent paper
J.A. Lipa, D.R. Swanson, J. Nissen, Z.K. Geng, P.R. Williamson, D.A. Stricker, T.C.P. Chui, U.E. Israelson, and M. Larson, Phys. Rev. Lett. **84**, 4894 (2000).

[17] For a discussion of such contributions see
B.G. Nickel, Physica A **177**, 189 (1991);
A. Pelissetto and E. Vicari, Nucl. Phys. B **519**, 626 (1998);
522, 605 (1998); Nucl. Phys. Proc. Suppl. **73**, 775 (1999). See also the review article by these authors in cond-mat/0012164.

[18] H. Kleinert, Phys. Lett. B **360**, 65 (1995) (quant-ph/9507009).

[19] H. Kleinert, Phys. Lett. A **277**, 205 (2000) (cond-mat/9906107).

[20] H. Kleinert, Phys. Rev. D **60**, 085001 (1999) (hep-th/9812197).

[21] See the book by
H.E. Stanley, *Introduction to Phase Transitions and Critical Phenomena*, Oxford Science Publications, Oxford, 1971;
or
H. Kleinert, *Gauge Fields in Condensed Matter*, World Scientific, Singapore, 1989, Vol. I, *Superflow and Vortex Lines, Disorder Fields and Phase Transitions*, Sections 4.5 and 9.5 (www.physik.fu-berlin.de/~kleinert/re.html#b1).

[22] C. Itzykson, R.B. Pearson

[23] P. Butera and M. Comi, Phys. Rev. B **56**, 8212 (1997) (hep-lat/973018).

[24] D. MacDonald, D.L. Hunter, K. Kelly and N. Jan, J. Phys. A **25**, 1429 (1992).

[25] B. Li, N. Madras and A.D. Sokal,, J. Stat. Phys. **80**, 661 (1995).

[26] S. Caracciolo, M.S. Causo and A. Pelissetto, cond-mat/9703250.

[27] J. Zinn-Justin, *Quantum field theory and critical phenomena* (Clarendon, Oxford, 1989, third edition 1996).

[28] A.M. Ferrenberg and D.P. Landau, Phys. Rev. B **44**, 5081 (1991).

[29] H.W.J. Blöte, E. Luijten and J.R. Heringa, J. Phys. A **28**, 6289 (1995); A.L. Talapov and H.W.J. Blöte, J. Phys. A **29**, 5727 (1996).

[30] J.H. Chen, M.E. Fisher and B.G. Nickel, Phys. Rev. Lett. **48**, 630 (1982); M.E. Fisher and J.H. Chen, J. Physique **46**, 1645 (1985).

[31] B.G. Nickel and J.J. Rehr, J. Stat. Phys. **61**, 1 (1990).

[32] M. Hasenbusch and S. Meyer, Phys. Lett. B **241**, 238 (1990);
A.P. Gottlob and M. Hasenbusch , Physica A **201**, 593 (1993).

[33] W. Janke, Phys. Lett. A **143**, 306 (1990).

[34] H.G. Ballesteros, L.A. Fernandez, V. Martin-Mayor and A. Munoz Sudupe, Phys. Lett. B **387**, 125 (1996).

[35] M. Ferer, M.A. Moore and M. Wortis, Phys. Rev. B **8**, 5205 (1973).

[36] K. Chen, A.M. Ferrenberg and D.P. Landau, Phys. Rev. B **48**, 3249 (1993); J. Appl. Phys., **73**, 5488 (1993);
C. Holm and W. Janke, Phys. Lett. A **173**, 8 (1993) and Phys. Rev. B **48**, 936 (1993).

[37] D.S. Ritchie and M.E. Fisher, Phys. Rev. B **5**, 2668 (1972);
S. McKenzie, C. Domb and D.L. Hunter, J. Phys. A **15**, 3899 (1982);
M. Ferer and A. Hamid-Aidinejad, Phys. Rev. B **34**, 6481 (1986).

[38] K. Kanaya and S. Kaya, Phys. Rev. D **51**, 2404 (1995).

Several interesting papers appeared after this chapter was written:
M. Campostrini, A. Pelissetto, P. Rossi, E. Vicari, Phys. Rev. E **60**, 3526 (1999); Phys. Rev. B **61**, 5905 (2000),
the latter producing a number for the critical exponent α of the specific heat in superfluid helium very close to our result from strong-coupling theory in the paper cited in Ref. [19], whose results are contained in Section 20.7.
Recent Monte Carlo studies are
M. Weigel and W. Janke, *High-precision Monte Carlo study of universal correlation lengths scaling in three-dimensional O(n) spin models*, Leipzig preprint March 2000 (cond-mat/0003124);
H.W.J. Blöte, J.R. Heringa, M.M. Tsypin, *Three-dimensional Ising model in the fixed-magnetization ensemble: a Monte Carlo study* (cond-mat/9910145).
There are also interesting new results in
P. Butera and M. Comi, Phys. Rev. B **60**, 6749 (1999) (hep-lat/9903010); Phys. Rev. B **50**, 3052 (1994) (cond-mat/9902326); Phys. Rev. B **58**, 11552 (1998).

Recently, accurate five-loop studies appeared for two-dimensional ϕ^4 theories. Extending the perturbation expansions for a single real field in Ref. [2] to any O(N), the critical exponents were calculated by
E.V. Orlov and A.I. Sokolov, *Critical thermodynamics of the two-dimensional systems in five-loop renormalization-group approximation*, Fiz. Tverdogo Tela, **42** (2000), November issue (hep-th/0003140).

21

New Resummation Algorithm

A rapidly convergent resummation algorithm has recently been developed by Jasch and Kleinert [1]. It is applicable to many types of divergent power series, and yields critical exponents with high accuracy. It possesses similar virtues as variational perturbation theory: the approach to scaling at large bare couplings $\bar{g}_B$ is of the form $c + c'/g_B^\omega$, and there are variational parameters to optimize the results. In addition, it exploits the knowledge on the large-order behavior of the expansion coefficients, or equivalently, on the discontinuity across the tip of the left-hand cut in the complex coupling constant plane. Moreover, it is formulated in a simple algebraic way similar to the Janke-Kleinert algorithm in Section 16.5.

The combination of these properties leads to a considerable increase in the speed of convergence and a high accuracy of the results.

In this chapter we explain this algorithm and illustrate its power by calculating the critical exponents of $O(N)$-symmetric ϕ^4-theories from the six- and seven-loop perturbation expansions.

21.1 Hyper-Borel Transformation

The basis for this algorithm is a generalization of the Borel transformation to a *hyper-Borel transformation*. Like the ordinary Borel or Borel-Leroy transforms in Section 16.3 [recall (16.25) and (16.43)] it is constructed by removing from the expansion coefficients f_n in Eq. (19.1) [recall (17.23)]

$$f_n = \gamma(-\alpha)^n n^\beta n![1 + O(1/n)] \tag{21.1}$$

the factorial growth which produces a convergent series denoted by $\tilde{B}(t)$. From this the original function $f(g_B)$ can be recovered by some integral over $\tilde{B}(t)$ along the positive t-axis. Since $\tilde{B}(t)$ is initially obtained from $f(g_B)$ as a power series with a finite convergence radius, an analytic continuation is necessary to achieve convergence. This is done as before by a transformation like (16.45) from the complex t- into a complex w-plane such that the image of the positive t-axis lies inside the maximal circle of convergence. As before, a suitable modification of the expansion for $\tilde{B}(t)$ of the type (16.55) allows us to account for the leading strong-coupling behavior g_B^s.

The important additional property of the new transformation is that the nonleading powers of the strong-coupling expansion (19.2) are also matched. All previous approaches based on a Borel or Borel-Leroy transformation do not allow this. In order to understand the difficulty, let us insert the strong-coupling expansion (19.2) in the form

$$f(g_B) = g_B^s \sum_{m=0}^{\infty} b_m \, g_B^{-\omega m} \tag{21.2}$$

into Eq. (16.28) for the Borel transform. We have set the leading power p/q equal to s, as in Eq. (16.54), and identified the powers $(g_B^{-2/q})^m$ directly with $(g_B^{-\omega})^m$, where ω is the exponent of approach to scaling. Performing the contour integral, we find

$$B(t) = t^s \sum_{m=0}^{M} \frac{\sin \pi (\omega m - s)}{\pi} \Gamma(\omega m - s) b_m \, t^{-\omega m}, \qquad (21.3)$$

For dimensional reasons, the large-t expansion of the Borel transform $B(t)$ contains the same powers of t as the original function $f(g_B)$ does in g_B. However, while the series (19.2) has a finite radius of convergence, the large-t expansion of $B(t)$ is a divergent asymptotic one, due to the factor $\Gamma(m\omega - s)$ in each expansion coefficient. Thus, if we want to account for the power structure of the strong-coupling expansion (19.2) within the ordinary Borel framework, we have to find a way of re-expanding the Borel series which will guarantee the large-t property (21.3). This is an interesting open problem.

Note that, in general, an expansion in the Borel-plane with a power sequence in t as in (21.3) is not sufficient to ensure an expansion in the same powers in the g_B-plane as in (19.2), because of the appearance of extra integer powers in g_B. This is illustrated by the simple function $B(t) = (1+t)^s$, which possesses a strong-coupling expansion in the powers t^{s-k}. If s is non-integer the expansion of the corresponding function $f(g_B)$ reads

$$f(g_B) = \int_0^\infty dt \, e^{-t}(1 + g_B t)^s = e^{1/g_B} \Gamma(s+1) g_B^s + e^{1/g_B} \sum_{k=0}^{\infty} \frac{(-1)^k}{(k+s+1)k!} g_B^{-k-1}. \qquad (21.4)$$

Expanding the exponential we see that the sum contains integer powers which are not contained in the strong-coupling expansion of $B(t)$.

A match of the strong-coupling power structure can be achieved with the help of the following hyper-Borel transform defined by

$$\tilde{B}(y) = \sum_{k=0}^{\infty} \tilde{B}_k \, y^k, \qquad (21.5)$$

with coefficients

$$\tilde{B}_k \equiv \omega \frac{\Gamma\left(k(1/\omega - 1) + \beta_0\right)}{\Gamma\left(k/\omega - s/\omega\right)\Gamma(\beta_0)} f_k. \qquad (21.6)$$

The original function $f(g_B)$ is recovered from $\tilde{B}(y)$ by the integral

$$f(g_B) = \frac{\Gamma(\beta_0)}{2\pi i} \oint_C dt \, e^t \, t^{-\beta_0} \int_0^\infty \frac{dy}{y} \left[\frac{g_B}{y \, t^{(1-\omega)/\omega}}\right]^s \exp\left[\frac{y \, t^{(1-\omega)/\omega}}{g_B}\right]^\omega \tilde{B}(y). \qquad (21.7)$$

The proof of this is straightforward with the help of the integral representation of the inverse Gamma function

$$\frac{1}{\Gamma(z)} = \frac{1}{2\pi i} \int_C dt \, e^t \, t^{-z}. \qquad (21.8)$$

The transformation possesses a free parameter β_0 which can be used to optimize the approximation $f_L(g_B)$ for each order L. The power ω of the strong-coupling expansion is assumed to lie in the interval $0 < \omega < 1$.

The hyper-Borel transformation has the desired property of allowing for a resummation of $f_L(g_B)$ with the correct powers of g_B of the strong-coupling expansion (19.2). In order to show this we first observe that, as in the ordinary Borel transform (16.25), the large-argument behavior of the Gamma function [recall Sterling's formula (16.11)] removes the factorial growth (21.1) from the expansion coefficients f_k, and leads to a simple power behavior of the coefficients $\tilde{B}_k$:

$$\tilde{B}_k \overset{k\to\infty}{=} \text{const.} \times \left[\alpha\omega(1-\omega)^{1/\omega-1}\right]^k k^{\beta+\beta_0+1/2+s/\omega}[1+\mathcal{O}(1/k)]. \tag{21.9}$$

Thus our new transform $\tilde{B}(y)$ shares with the ordinary Borel transform $B(t)$ the property of being analytic at the origin. Its radius of convergence is determined by the singularity on the negative real axis at

$$y_s = -\frac{1}{\sigma} \equiv -\frac{1}{\alpha}\frac{1}{\omega(1-\omega)^{1/\omega-1}}. \tag{21.10}$$

A resummation procedure can now be set up on the basis of the transform $\tilde{B}(y)$ as before. The inverse transformation (21.7) contains an integral over the entire positive axis, requiring again an analytic continuation of the Taylor expansion of $\tilde{B}(y)$ beyond the convergence radius. In the previous resummation scheme using the ordinary Borel of Borel-Leroy transforms, this is achieved in two ways: either by a conformal mapping such as (16.45), or by a re-expansion of $B(g)$ in terms of a complete set of basis functions (16.75), (16.76). The relation between the two methods was explained in Section 16.5.3.

Here we shall follow the second method and re-expand the hyper-Borel transform $\tilde{B}(y)$ in a complete set of basis functions $\tilde{B}_{I_n}(y)$, which all possess a singularity at $y = -1/\sigma$. As before, we choose $\tilde{B}_{I_n}(y)$ to have Taylor series starting with y^n, which will lead to a triangular matrix to be inverted when re-expanding $\tilde{B}(y)$ in terms of $\tilde{B}_{I_n}(y)$.

The reason for introducing the new transform $\tilde{B}(y)$ is to allow us to reproduce the complete power structure of the strong-coupling expansion (19.2), with a leading power g^s and a sub-leading string of powers $g^{s-k\omega}$, $k = 1, 2, 3, \ldots$. This is achieved by the following basis functions to span the space of transforms $\tilde{B}(y)$:

$$\tilde{B}_{I_n}(y) = e^{-\rho\sigma y}\frac{(\sigma y)^n}{(1+\sigma y)^{n+\delta}}. \tag{21.11}$$

By analogy with the re-expansion in the Borel-Leroy case (16.55), this may be written as

$$\tilde{B}(y) \equiv \sum_{k=0}^{\infty} \tilde{B}_k\, y^k = e^{-\rho\sigma y}[1-w(y)]^\delta \sum_{k=0}^{\infty} W_k[w(y)]^k, \tag{21.12}$$

with

$$w(y) = \frac{\sigma y}{1+\sigma y}. \tag{21.13}$$

The inverse transform of $\tilde{B}_{I_n}(y)$ yields the basis functions

$$I_n(g_B) = \frac{\Gamma(\beta_0)}{2\pi i}\oint_C dt\, e^t\, t^{-\beta_0}\int_0^\infty \frac{dy}{y}\left[\frac{g_B}{y\,t^{1/\omega-1}}\right]^s \exp\left[-\frac{y\,t^{1/\omega-1}}{g_B}\right]^\omega \tilde{B}_{I_n}(y), \tag{21.14}$$

which span the space of functions $f(g_B)$. The functions $I_n(g_B)$ are used as basis functions to re-expand the truncated expansions $f_L(g_B)$ in the form

$$f_L(g_B) = \sum_{n=0}^{L} h_n I_n(g_B). \tag{21.15}$$

The complete list of parameters of the functions $I_n(g_B)$ reads as follows:

$$I_n(g_B) = I_n(g_B, \omega, s, \rho, \sigma, \delta, \beta_0) = I_n(\sigma g_B, \omega, s, \rho, 1, \delta, \beta_0), \tag{21.16}$$

but in the following we shall mostly use the shorter notation $I_n(g_B)$.

The integral representation of $I_n(g_B)$ breaks down at $s = n$, requiring an analytical continuation. For our applications it will be sufficient to perform this continuation in the convergent strong-coupling expansion of $I_n(g_B)$. This is obtained by performing a Taylor expansion of the exponential function in (21.14), which is an expansion in powers of $1/g_B^\omega$. After integrating over t and y using (21.8), we obtain the expansion

$$I_n(g_B) = g_B^p \sum_{k=0}^{\infty} b_k^{(n)} g_B^{-k\omega}, \qquad (21.17)$$

which agrees with the strong-coupling behavior (19.2). The expansion coefficients are

$$b_k^{(n)} = \frac{(-1)^k}{k!} \frac{\sigma^{s-k\omega}\Gamma(\beta_0)}{\Gamma[(\omega-1)k + \beta_0 + (1/\omega-1)s]} i_k^{(n)}, \qquad (21.18)$$

where $i_k^{(n)}$ denotes the integral

$$i_k^{(n)} = \int_0^\infty dy e^{-\rho y}(1+y)^{-\delta-n} y^{k\omega+n-s-1}. \qquad (21.19)$$

This integral is seen to coincide with a Kummer function

$$U(\alpha, \gamma, z) \equiv \frac{1}{\Gamma(\alpha)} \int_0^\infty dy e^{-zy} y^{\alpha-1}(1+y)^{\gamma-\alpha-1}, \qquad (21.20)$$

so that we can write

$$i_k^{(n)} = \Gamma(k\omega + n - s) U(k\omega + n - s, k\omega - s - \delta + 1, \rho). \qquad (21.21)$$

The latter expression is useful since in some applications the integral (21.19) may diverge requiring an analytic continuation of the contour of integration. Such deformations are automatically supplied by other representations of the Kummer function, for instance

$$U(\alpha, \gamma, z) = \frac{\pi}{\sin \pi\gamma}\left[\frac{M(\alpha, \gamma, z)}{\Gamma(1+\alpha-\gamma)\Gamma(\gamma)} - z^{1-\gamma}\frac{M(1+\alpha-\gamma, 2-\gamma, z)}{\Gamma(\alpha)\Gamma(2-\gamma)}\right], \qquad (21.22)$$

where $M(\alpha, \gamma, z)$ is the confluent hypergeometric function which has the Taylor expansion

$$M(\alpha, \gamma, z) = 1 + \frac{\alpha}{\gamma}\frac{z}{1!} + \frac{\alpha(\alpha-1)}{\gamma(\gamma-1)}\frac{z^2}{2!} + \cdots . \qquad (21.23)$$

The alternative expression (21.21) for $i_k^{(n)}$ with (21.22) and (21.23) is useful for resumming various asymptotic expansions, for example, that of the ground state energy of the anharmonic oscillator, in which case the leading strong-coupling power s has the value $1/3$. There the integral representation (21.19) will have to be evaluated for values $n = 0$, $k = 0$, where the integral does not exist, whereas formula (21.21) with (21.22), (21.23) is well-defined.

For large k, the integral on the right-hand side of (21.19) can be estimated with the help of the saddle point approximation. The saddle point lies at

$$y_s \approx \frac{k\omega}{\rho}, \qquad (21.24)$$

leading to the asymptotic estimate

$$i_k^{(n)} \overset{k\to\infty}{=} \left(\frac{k\omega}{\rho}\right)^{-\delta-n} \int_0^\infty dy\, e^{-\rho y} y^{\omega k+n-s-1} [1+\mathcal{O}(1/k)]$$

$$= \left(\frac{\omega k}{\rho}\right)^{-\delta-n} \rho^{-k\omega-n+s} \Gamma(k\omega+n-s) [1+\mathcal{O}(1/k)]. \qquad (21.25)$$

The behavior of the strong-coupling coefficients $b_k^{(n)}$ for large k is obtained with the help of the identity

$$\Gamma(z)\Gamma(1-z) = \frac{\pi}{\sin \pi z} \qquad (21.26)$$

and Stirling's formula (16.11), yielding

$$b_k^{(n)} \overset{k\to\infty}{=} \gamma \sin \pi[k(\omega-1)+\beta_0+(1/\omega-1)s] \left[-\frac{(1-\omega)^{(1-\omega)}}{(\sigma\rho)^\omega}\right]^k k^{\gamma_1}[1+\mathcal{O}(1/k)]. \qquad (21.27)$$

The real constants γ, γ_1 will not be needed for the upcoming discussions and are therefore not calculated explicitly. Equation (21.27) shows that the strong-coupling expansion (19.2) has a convergence radius

$$|g_B| > \frac{(\rho\sigma)^\omega}{(1-\omega)^{1-\omega}}, \qquad (21.28)$$

which means that the basis functions $I_n(g_B)$, and certainly also $f(g_B)$ itself, possess additional singularities beside $g_B = 0$. The parameter ρ will be optimally adjusted to match the positions of these singularities.

For re-expanding $f_L(g_B)$ in terms of the basis functions $I_n(g_B)$, we must know the Taylor series of the basis functions $I_n(g_B)$. This is obtained by substituting into (21.14) the variable y with $g_B y'$, and expanding the integrand of (21.14) in powers of g_B. After performing the integrals over y' and t, we obtain the expansion

$$I_n(g_B) = \sum_{k=n}^\infty f_k^{(n)} g_B^k, \qquad (21.29)$$

with the coefficients

$$f_k^{(n)} = \sum_{k=n}^\infty \frac{1}{\omega} \frac{\Gamma(\beta_0)\Gamma(k/\omega-s/\omega)}{\Gamma(k(1/\omega-1)+\beta_0)} \sum_{j=0}^{k-n} \binom{-\delta-n}{j} \frac{(-\rho)^{k-n-j}}{(k-n-j)!} \sigma^k. \qquad (21.30)$$

The coefficients in the second sum come from the t integral:

$$\sum_{j=0}^{k-n} \binom{-n-\delta}{j} \frac{(-\rho)^{k-n-j}}{(k-n-j)!} = \frac{(-1)^{k-n}}{\Gamma(k-n+1)\Gamma(n+\delta)} \int_0^\infty dt\, e^{-t} t^{\delta+n-1} (\rho+t)^{k-n}. \qquad (21.31)$$

For large k, we may evaluate the integral with the help of the saddle-point approximation. Using also Stirling's formula (16.11), we find

$$\sum_{j=0}^{k-n} \binom{-n-\delta}{j} \frac{(-\rho)^{k-n-j}}{(k-n-j)!} \overset{k\to\infty}{=} \frac{(-1)^{k-n} e^\rho}{\Gamma(\delta+n)} k^{\delta+n-1} [1+\mathcal{O}(1/k)]. \qquad (21.32)$$

Inserted into (21.30) and using once more Stirling's formula we obtain for the expansion coefficients $f_k^{(n)}$ the following factorial growth

$$f_k^{(n)} \stackrel{k\to\infty}{=} \frac{(-1)^n e^\rho \Gamma(\beta_0)}{\sqrt{2\pi}\Gamma(\delta+n)}(1-w)^{1/2-\beta_0}w^{\beta_0-1+s/w}k^{\delta-\beta_0+n-3/2-s/w}\left[-\frac{\sigma}{w(1-w)^{1/w-1}}\right]^k k!$$
$$\times[1+\mathcal{O}(1/k)]. \qquad (21.33)$$

For an optimal re-expansion (21.15), we shall choose the free parameters of the basis functions $I_n(g_B, w, s, \rho, \sigma, \delta, \beta_0)$ to match the large-order behavior of the coefficients f_k in (21.1).

For an understanding of the convergence properties of the resummed $f_L(g_B)$ to the correct strong-coupling expansions (19.2) we shall also need the large-n behavior of the expansion coefficients $b_k^{(n)}$ in the strong-coupling expansion (21.17) of $I_n(g_B)$. This is determined by the saddle point approximation to the integral $i_k^{(n)}$ in Eq. (21.19), which we rewrite as

$$i_k^{(n)} = \int_0^\infty dy e^{-\rho y - n\ln(1+1/y)}(1+y)^{-\delta}y^{kw-s-1}. \qquad (21.34)$$

The saddle point lies at

$$y_s = \sqrt{\frac{n}{\rho}}\left[1+\mathcal{O}(1/\sqrt{n})\right]. \qquad (21.35)$$

At this point, the total exponent in the integral is

$$-\rho y_s - n\ln\left(1+\frac{1}{y_s}\right) = -2\sqrt{\rho n}\left[1+\mathcal{O}(1/\sqrt{n})\right], \qquad (21.36)$$

implying the large-n behavior

$$b_k^{(n)} \stackrel{k\to\infty}{=} \text{const.} \times n^{kw-s-1-\delta}e^{-2\sqrt{\rho n}}\left[1+\mathcal{O}(1/\sqrt{n})\right]. \qquad (21.37)$$

21.2 Convergence Properties

Let us denote the strong-coupling coefficients b_k of the truncated function $f_L(g)$ by b_k^L. They are linear combinations of the coefficients $b_k^{(n)}$ of the basis functions $I_n(g_B)$:

$$b_k^L = \sum_{n=0}^L b_k^{(n)}h_n. \qquad (21.38)$$

Let us estimate the expected speed of convergence with which the coefficients b_k^L converge against the strong-coupling coefficients b_k of $f(g_B)$ as the number L goes to infinity. It is governed by the growth with n of the re-expansion coefficients h_n, and of the coefficients $b_k^{(n)}$ in Eq. (21.37). For the series to be resummed, the re-expansion coefficients h_n will have to grow at most like some power n^r, implying that the approximations b_k^L approach their $L\to\infty$-limit b_k with an error proportional to

$$b_k^L - b_k \sim L^{r+kw-s-\delta-1/2} \times e^{-2\sqrt{\rho L}}. \qquad (21.39)$$

The leading exponential falloff of the error $e^{-2\sqrt{\rho L}}$ is independent of the other parameters in the basis functions $I_n(g_B, w, p, \rho, \sigma, \delta, \beta_0)$ to be adjusted below. This is the important advantage of the present resummation method with respect to variational perturbation theory where the error decreases like $e^{-L^{1-w}}$ with $1-w$ close to $1/4$ [recall (19.51) as well as Figs. 19.10–19.16 in $4-\varepsilon$ and Figs. 20.14–20.17 in $D=3$ dimensions].

21.2.1 Parameters s and ω

The perturbation expansions for the critical exponents are power series in the bare coupling constant g_B whose strong-coupling limit is a constant. The same is true for the series expressing the renormalized coupling constant g in powers of the bare coupling constant. This implies that the growth parameter s for the basis functions $I_n(g_B)$ should be set equal to zero. The constant asymptotic values are approached with a subleading power behavior $\propto 1/g_B^{k\omega-s}$, where ω is the universal experimentally measurable critical exponent governing the approach to scaling.

21.2.2 Parameter σ

In the ordinary Borel-transformation, the parameter α in the large-order behavior of the expansion coefficients f_k in Eq. (21.1) specifies also the position of the singularity on the negative y-axis in $B(y)$. It is determined directly by the inverse reduced action of the classical solution to the field equations. In our new transform $\tilde{B}(y)$, the growth parameter and the inverse reduced action α is no longer directly given by the nearest singularity in $\tilde{B}(y)$, which now lies at [see Eq. (21.10)]

$$\sigma = \alpha\omega(1-\omega)^{1/\omega-1}. \tag{21.40}$$

This value of σ ensures that the expansion coefficients f_k^n of the basis functions $I_n(g_B)$ in Eq. (21.33) grow for large k with the same factor $(-\alpha)^k$ as the expansion coefficients for $f(g_B)$ in Eq. (21.1).

The conformal mapping (21.13) maps the singularity at $y = -1/\sigma$ to $w = \infty$, and converts the cut along the negative y-axis into a cut in the w-plane from 1 to ∞. The growth of the re-expansion coefficients h_n in (21.15) with n is therefore determined by the nature of the singularity of $\tilde{B}(y)$ at ∞.

In the upcoming applications to critical exponents both, the value (21.40) following from the inverse action of the solution to the classical field equations and ω, will not yield the fastest convergence of the approximations $f^L(g_B)$ towards $f(g_B)$. A slightly smaller value will turn out to give better results. This seems to be due to the fact that the classical solution gives only the nearest singularity in the hyper-Borel transform $\tilde{B}(y)$ of $f(g_B)$. In reality, there are many additional cuts from other fluctuating field configurations which determine the size of the expansion coefficients f_k at asymptotic orders k. Since the few known f_ks are in praxis quite far from the asymptotic estimates, they are best accounted for by an effective shift of the position of the singularity in the direction of the additional cuts at larger negative y, corresponding to a smaller σ.

21.2.3 Parameter ρ

According to Eq. (21.28), the parameter ρ determines the radius of convergence of the strong-coupling expansion of the basis functions $I_n(g_B)$. It should therefore be adjusted to fit optimally the corresponding radius of the original function $f(g_B)$. Since we do not know this radius, this adjustment will be done phenomenologically by varying ρ to optimize the speed of convergence. Specifically, we shall search at each order L for a vanishing highest re-expansion coefficient h_L or, if it does not vanish anywhere, for a vanishing derivative with respect to ρ:

$$h_L(\rho) = 0, \quad \text{or} \quad \frac{dh_L(\rho)}{d\rho} = 0. \tag{21.41}$$

21.2.4 Parameter δ

From Eq. (21.33) we see that the parameter δ influences the power k^β in the large-order behavior (21.1). By comparing the two equations, we identify the growth parameter β of $I_n(g_B)$ as being

$$\beta = \delta - \beta_0 - 3/2 - s/\omega + n. \tag{21.42}$$

At first it would appear impossible to give *all* basis functions $I_n(g_B)$ the same growth power β in (21.33) by simply letting δ depend on the order n as required by (21.42). If we were to do this, we will have to assign to δ the value

$$\delta = \delta_n \equiv \beta + \beta_0 + 3/2 + s/\omega - n, \tag{21.43}$$

which depends on the index n of the function $I_n(g_B)$. This means that we perform an analytical continuation of the power series expansion of $\tilde{B}(y)$ by re-expanding it as follows:

$$\tilde{B}(y) = \sum_{k=0}^{\infty} \tilde{B}_k\, y^k = e^{-\rho\sigma y}(1 + \sigma y)^{-\delta} \sum_{k=0}^{\infty} h_k(\sigma y)^k. \tag{21.44}$$

But the series in this formula which is obtained from the series of $\tilde{B}(y)$ by removing a simple factor still has the same finite radius of convergence and cannot be used to estimate $\tilde{B}(y)$ for large values of y needed to perform the back-transform (21.7). It is, however, possible to sidetrack this problem by letting ρ grow linearly with the order L. Then the exponential factor of (21.44) suppresses the integrals over y for large y sufficiently to make the divergence of the re-expanded series (21.44) at large y irrelevant. If we determine ρ from the condition (21.41), the growth of ρ with L will come about by itself.

21.2.5 Parameter β_0

The parameter β_0 has two effects. From Eq. (21.18) we see that for

$$k > k_c \equiv \frac{\beta_0 + (1/\omega - 1)s}{1 - \omega} \tag{21.45}$$

the signs of the strong-coupling expansion coefficients start to alternate irregularly. This irregularity weakens the convergence of the higher strong-coupling coefficients b_k^L with $k > k_c$ against b_k. The convergence can therefore be improved by choosing a β_0 which grows proportionally to the order L of the approximation.

In addition, β_0 appears in the power of k in (21.33) because it determines the nature of the cut in $\tilde{B}(y)$ in the complex y-plane starting at $y = -1/\sigma$ [see Eq. (21.12)].

If we expand both sides of (21.12) in powers of $w = \sigma y/(1 + \sigma y)$ and compare the coefficients of powers of w, it is easy to write down an explicit formula for the re-expansion coefficients h_n in terms of the coefficients $\tilde{B}_j$ of $\tilde{B}(y)$:

$$h_n = \sum_{k=0}^{n}\sum_{j=0}^{k} \frac{\tilde{B}_j \sigma^{-j}\rho^{k-j}}{(k-j)!}\binom{\delta + n - 1}{n - k}, \tag{21.46}$$

where $\tilde{B}_j$ are obtained from the original expansion coefficients f_k of $f(g_B)$ by the relation (21.6).

Before beginning with the resummation of the perturbation expansions for the critical exponents of ϕ^4-field theories, it will be useful to have a feel for the quality of the new resummation procedure and for the significance of the parameters to the speed of convergence by resumming the asymptotic expansion for the ground state energy of the anharmonic oscillator.

21.3   Resummation of Ground State Energy of Anharmonic Oscillator

Consider the one-dimensional anharmonic oscillator with the Hamiltonian

$$H = \frac{1}{2}p^2 + \frac{m^2}{2}x^2 + g_B x^4. \tag{21.47}$$

The ground state energy has a perturbation expansion

$$E^{(0)}(g_B) = \sum_{k}^{\infty} f_k g_B^k, \tag{21.48}$$

whose coefficients can be calculated by a recursion relation of Bender and Wu [2] to arbitrarily high orders, with a large-order behavior

$$f_k = -\sqrt{\frac{6}{\pi^3}} k! (-3)^k k^{-1/2} [1 + \mathcal{O}(1/k)]. \tag{21.49}$$

By comparison with (21.1) we identify the growth parameters

$$\alpha = 3, \quad \beta = -1/2. \tag{21.50}$$

By performing a scale transformation $x \to g^{1/6}x$ on the Hamiltonian (21.47), one finds the scaling property

$$E(m^2, g_B) = g_B^{1/3} E(g_B^{-2/3} m^2, 1) \tag{21.51}$$

for the energy considered as a function of g_B and m^2. Combining this with the knowledge [3] that $E(m^2, 1)$ is an analytic function at $m^2 = 0$, we see that $E(1, g)$ possesses an expansion of the form (19.2) with the powers $s = 1/3$, $\omega = 2/3$. Inserting the latter number together with α from Eq. (21.50) into (21.40), we identify

$$\sigma = \frac{2}{\sqrt{3}}. \tag{21.52}$$

The ground state energy $E^{(0)}(g)$ obeys a once-subtracted dispersion relation

$$E^{(0)}(g) = \frac{1}{2} + \frac{g}{\pi} \int_0^\infty \frac{dg'}{g'} \frac{\mathrm{Im}\, E^{(0)}(-g')}{g' + g}. \tag{21.53}$$

The perturbative expansion (21.48) is obtained from this by expanding the integrand in powers of g, and performing the integral term by term. This shows explicitly that the large-order behavior (21.49) corresponds to an imaginary part

$$\mathrm{Im}\, E^{(0)}(-|g_B|) = \sqrt{\frac{6}{\pi}} \sqrt{\frac{1}{3|g_B|}} e^{-1/3|g_B|} [1 + \mathcal{O}(|g_B|)] \tag{21.54}$$

near the tip of the left-hand cut in the complex g_B-plane, in agreement with the general form (16.7) associated with the large-order behavior (21.1) [compare (16.12)]. Equation (21.33) shows us that if we want to match the imaginary part of $E^{(0)}(g_B)$ at small negative g_B, we have to set $\delta = \beta_0 + 3/2 - n$, in accordance with the general relation (21.42). Thus we resum the perturbative expansion by a re-expansion in terms of the basis functions

$$I_n(g, 2/3, 1/3, \rho, 2/\sqrt{3}, \beta_0 + 3/2 - n, \beta_0), \tag{21.55}$$

where the parameters ρ and β_0 are still undetermined.

Note that by (21.12), the resulting re-expansion corresponds to an analytic continuation of our transform $\tilde{B}(y)$ written as follows:

$$\tilde{B}(y) = \sum_{k=0}^{\infty} \tilde{B}_k\, y^k = e^{-2\rho y/\sqrt{3}}(1 + 2y/\sqrt{3})^{-\beta_0+3/2} \sum_{k=0}^{\infty} W_k\,(2y/\sqrt{3})^k. \tag{21.56}$$

The parameter ρ will be determined by an order-dependent optimization of the approximations. Specifically, we shall search in each order L for a vanishing highest re-expansion coefficient h_L with a vanishing derivative with respect to ρ:

$$h_L(\rho) = 0, \qquad \frac{dh_L(\rho)}{d\rho} = 0. \tag{21.57}$$

From Eq. (21.30) for the expansion coefficients $f_k^{(n)}$ of the basis functions $I_n(g_B)$, we see that a large parameter β_0 shifts the large-k-regime, where the large-order formula Eq. (21.33) begins dominating the growth, to increasing values of k. To match these growth properties with those of the expansion coefficients f_k of the ground state energy, we assign to β_0 the value $\beta_0 = 5$.

Let us test the convergence of our algorithm at small negative coupling constants g_B, i.e., near the tip of the left-hand cut in the complex g_B-plane by calculating the prefactor γ in the large-order behavior (21.1). With the large-order behavior (21.33) of the basis functions $I_n(g_B)$, we find the resummed functions $f_L(g_B)$ of Lth order $\sum_{n=0}^{L} h_n I_n(g_B)$ having a large-order behavior (21.1) with a prefactor

$$\gamma_L = \frac{e^\rho \Gamma(\beta_0)}{\sqrt{2\pi}\Gamma(\delta)} \sum_{k=0}^{L} (-1)^k w_k. \tag{21.58}$$

The values of these sums for increasing L are shown in Fig. 21.1. They converge exponentially

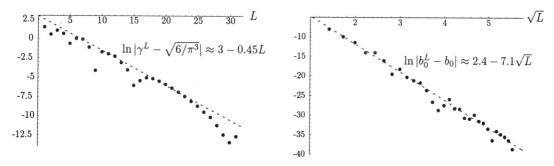

FIGURE 21.1 Logarithmic plot of the convergence behavior of the constant γ^L which normalizes the asymptotic behavior of the perturbative expansion, and of the leading strong coupling coefficient b_0^L.

fast against the exact limiting value

$$\gamma = -\sqrt{\frac{6}{\pi^3}}, \tag{21.59}$$

with superimposed oscillations. The oscillations are of the same kind as those observed in variational perturbation theory for the convergence of the approximations to the strong-coupling coefficients b_k (see Figs. 5.19 and 5.20 in Ref. [5]).

Here also, the strong-coupling coefficients b_k^L converge exponentially fast towards b_k, but with a larger power of L in the exponent of the last term $\approx e^{-\mathrm{const}\sqrt{L}}$ [see Eq. (21.39)], rather

n	b_n
0	0.667 986 259 155 777 108 270 962 02
1	0.143 668 783 380 864 910 020 319
2	−0.008 627 565 680 802 279 127 963
3	0.000 818 208 905 756 349 542 41
4	−0.000 082 429 217 130 077 219 91
5	0.000 008 069 494 235 040 964 75
6	−0.000 000 727 977 005 945 772 63
7	0.000 000 056 145 997 222 351 17
8	−0.000 000 002 949 562 732 709 36
9	−0.000 000 000 064 215 331 956 97
10	0.000 000 000 048 214 263 789 07

TABLE 21.1 Strong-coupling coefficients b_n of the 70-th order approximants $E_{70}^0(g) = \sum_{n=0}^{70} h_n I_n(g)$ to ground state energy $E^0(g)$ of the anharmonic oscillator. They have the same accuracy as the variational perturbation theoretic calculations up to order 251 in Ref. [4].

than $\approx e^{-\mathrm{const}L^{1/3}}$ for variational perturbation theory as in Eq. (19.51). This is seen on the right of Fig. 21.1.

We apply the resummation method to the first ten strong-coupling coefficients using the expansion coefficients f_k up to order 70. The results are shown in Table 21.3. Comparison with a similar table in Refs. [4, 5] shows that the new resummation method yields in 70th order the same accuracy as variational perturbation theory did in 251st order. In all cases the optimal parameter ρ turns out to be a slowly growing function with L.

In the strong-coupling regime, the convergence is fastest by choosing for β_0 an L-dependent value

$$\beta_0 = L. \tag{21.60}$$

Note that this choice of β_0 ruins the convergence to the imaginary part for small negative g_B which was resummed best with $\beta_0 = 5$.

21.4 Resummation for Critical Exponents

Having convinced ourselves of the fast convergence of our new resummation method, let us now apply our technique to the perturbation expansions for $\bar{g}(\bar{g}_B)$, and the critical exponents $\eta(\bar{g}_B)$, $\eta_m(\bar{g}_B)$, and $\gamma(\bar{g}_B)$ in Eqs. (20.15)–(20.19) of the $O(N)$-symmetric ϕ^4-theory in $D = 3$ dimensions, with the latter extended to seven loops in Eq. (20.60).

Now the power s in Eq. (19.2) is equal to zero for the critical exponents to approach a constant strong-coupling value. Similar to variational perturbation theory, we now use Eqs. (19.64) and (20.20), to determine the critical exponent of approach to scaling ω. In slight contrast to the earlier approach, we proceed here by resumming the power series of

$$-\bar{g}(\bar{g}_B)s(\bar{g}_B) = -\bar{g}_B \frac{\bar{g}'(\bar{g}_B)}{\bar{g}(\bar{g}_B)}\bigg|_{\bar{g}_B=\infty}, \tag{21.61}$$

which is simply the β-function:

$$\beta(\bar{g}_B) = -\bar{g}_B \frac{d\bar{g}(\bar{g}_B)}{d\bar{g}_B} \tag{21.62}$$

[recall (19.63)]. We do this for various values of ω and find that ω-value for which the approximations β_L^* extrapolate best to zero for $L \to \infty$. Thus we re-expand $\beta(\bar{g}_B)$ in terms of the basis functions $I_n(\bar{g}_B, \omega)$ up to the order L:

$$\beta_L(\bar{g}_B) = \sum_{n=0}^{L} h_n I_n(\bar{g}_B, \omega), \qquad (21.63)$$

and plot β_L^* for various values of ω as shown in Fig. 21.2.

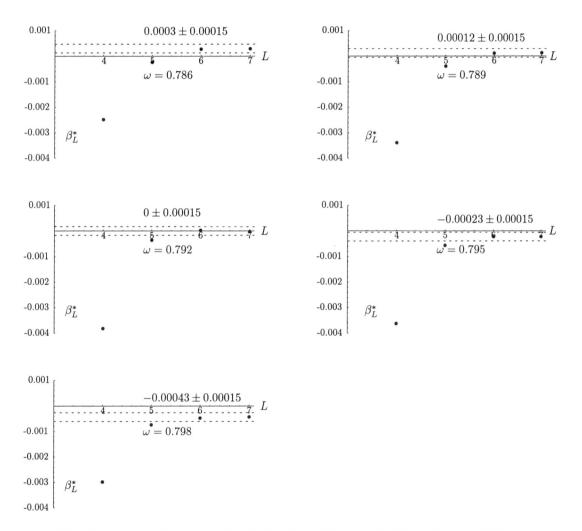

FIGURE 21.2 Convergence of strong-coupling limits of the β-function (21.62) for $N = 1$ and different values of ω. The upper and lower dashed lines denote the range of the $L \to \infty$ limit of β_L^* from which the value of ω is determined in Fig. 21.3 to be equal to $\omega = 0.792 \pm 0.003$.

We must explain how we have chosen the parameters in the basis functions $I_n(g_B) = I_n(g_B, \omega, s, \rho, \sigma, \delta, \beta_0)$. After trying out a few choices, we have given the parameters β and ρ the fixed values 1 and 10, respectively, to accelerate the convergence. To determine the remaining parameters, we recall the growth parameters of the large-order behavior of the perturbative series for the critical exponents listed in Table 20.4 . Having omitted a factor $1/(N+8)$ in the

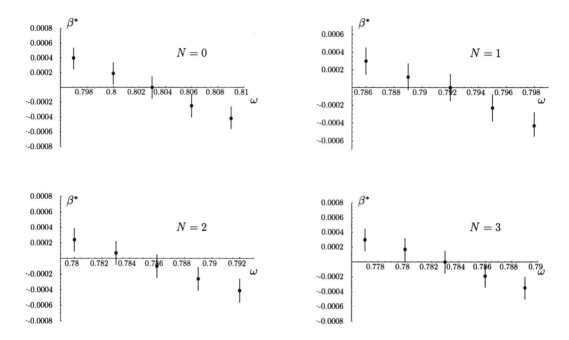

FIGURE 21.3 Plot of resummed values of β^* against ω. The true value of ω is deduced from the condition $\beta^* = 0$ and the errors are determined from the range of ω where the error bars from the resummation of β^* intersect with the x-axis.

bare coupling constants $\bar{g}_B$ on the right-hand sides in expansions (20.15)–(20.19), the α-value of these expansions is from Eq. (20.71):

$$\alpha = (D - 1)\frac{16\pi}{I_4} \stackrel{D=3}{=} 0.14777423 \cdot \frac{9}{N + 8}, \qquad (21.64)$$

while the parameter β_β is from Eq. (20.73)

$$\beta_\beta = 4 + N/2. \qquad (21.65)$$

The prefactor γ_β in the asymptotic growth equation is listed for $N = 0, 1, 2, 3$ in the fourth row of Table 20.4 .

Before starting with the resummation it is important to make the following observation: For the expansion coefficients of the critical exponents in powers of the bare coupling constant $\bar{g}_B$, large-order estimates of the type (20.68)–(20.70) become reliable only at much larger orders L than those in powers of the renormalized coupling constant, where the convergence was quite fast. The lack of precocity is illustrated for the β-function in Table 21.2 , which gives the β_k divided by their leading asymptotic estimates $f_k^{\beta \, \text{as}}$:

$$f_k^\beta / f_k^{\beta \, \text{as}} \equiv f_k^\beta / k!(-\alpha)^k k^\beta. \qquad (21.66)$$

The first six approximations approach their large-order limits quite slowly. For this reason we prefer to adapt σ not from α by using Eq. (21.40), but by an optimization of the convergence. Since the re-expanded series converges for fixed values of δ and σ, it is reasonable to determine

N	0	1	2	3
k	$f_k^\beta / f_k^{\beta\,\mathrm{as}}$	$f_k^\beta / f_k^{\beta\,\mathrm{as}}$	$f_k^\beta / f_k^{\beta\,\mathrm{as}}$	$f_k^\beta / f_k^{\beta\,\mathrm{as}}$
2	0.57	0.45	0.35	0.27
3	0.61	0.45	0.32	0.22
4	0.73	0.51	0.34	0.22
5	0.89	0.61	0.40	0.25
6	1.07	0.73	0.47	0.29
7	1.26	0.88	0.56	0.34
$\vdots$	$\vdots$	$\vdots$	$\vdots$	$\vdots$
γ_β	110.0	97.0	75.5	53.2

TABLE 21.2 First six perturbative coefficients in expansions of β-function in powers of bare coupling constant g_B, divided by their asymptotic large-order estimates $(-\alpha)^k k! k^{\beta_\beta}$. The ratios $f_k^\beta / f_k^{\beta\,\mathrm{as}}$ increase quite slowly towards the theoretically predicted normalization constant γ_β in the lowest row.

these parameters by searching for a point of least dependence in the largest available order L. This is done by imposing the conditions

$$\frac{d\kappa_L}{d\sigma} = 0 \quad \text{and} \quad \frac{d^2\kappa_L}{d\sigma^2} = 0 \qquad (21.67)$$

to determine *both* parameters δ, σ, where κ_L denotes the Lth approximation to any of the exponents γ, ν or η. In accordance with the discussions in Subsection 21.2.2, this procedure provides a value of σ which is smaller than that given by (21.40).

The results for the critical exponents of all O(N)-symmetries are shown in Figs. 21.2–21.6 and Table 21.3 .

N	γ	η	ν	ω
0	1.1604[8] (4) {0.075}	0.0285[6] (4) {0.037}	0.5881[8] (4) {0.075}	0.803[3] {1}
1	1.2403[8] (4) {0.110}	0.0335[6] (3) {0.043}	0.6303[8] (4) {0.065}	0.792[3] {1}
2	1.3164[8] (5) {0.033}	0.0349[8] (5) {0.042}	0.6704[7] (4) {0.098}	0.784[3] {1}
3	1.3882[10] (7) {0.210}	0.0350[8] (5) {0.043}	0.7062[7] (4) {0.110}	0.783[3] {1}

TABLE 21.3 Critical exponents of O(N)-symmetric ϕ^4-theory from new resummation method. The numbers in square brackets indicate the total errors. They arise form the error of the resummation at fixed values of ω indicated in parentheses, and the errors coming from the inaccurate knowledge of ω. The former are estimated from the scattering of the approximants around the graphically determined large-L limit, the latter follow from the errors in ω and the derivatives of the critical exponents with respect to changes of ω indicated in the curly brackets.

The total errors are indicated in the square brackets. They are deduced from the errors of resummation of the critical exponents at a fixed value of ω indicated in the parentheses, and from the error $\Delta\omega$ of ω, using the derivative of the exponent with respect to ω given in curly brackets. Symbolically, the relation between these errors is

$$[\ldots] = (\ldots) + \Delta\omega\{\ldots\}. \qquad (21.68)$$

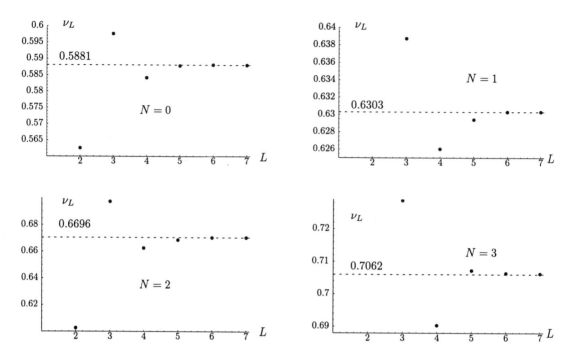

FIGURE 21.4 Convergence of critical exponent ν for different values of N.

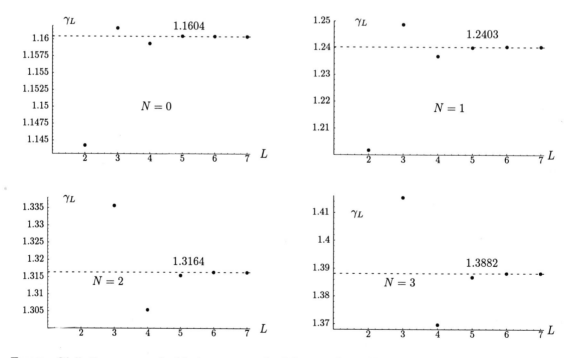

FIGURE 21.5 Convergence of critical exponent γ for different values of N.

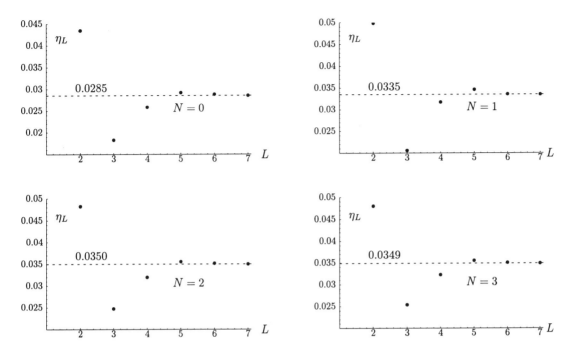

FIGURE 21.6 Convergence of critical exponent η for different values of N.

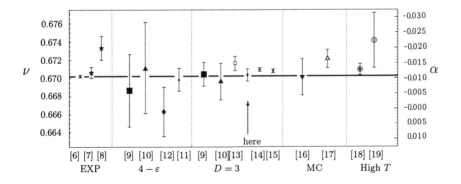

FIGURE 21.7 Comparison of our results for critical exponents α of superfluid helium with experiments and results from other theories. The value marked by an arrow points to the result (21.71) of the resummation in this chapter.

The accuracy of our results can be judged by comparison with the most accurately measured critical exponent α found in the space shuttle experiment in Ref. [6] [see Eq. (1.23)]:

$$\alpha = -0.01056 \pm 0.00038. \tag{21.69}$$

Our value for ν in Table 21.3 is

$$\nu = 0.6704 \pm 0.007 \tag{21.70}$$

and yields via the scaling relation $\alpha = 2 - 3\nu$:

$$\alpha = -0.0112 \pm 0.0021, \tag{21.71}$$

in best agreement with the experimental number (21.69).

A comparison between these numbers and the results of other authors in Fig. 21.7 shows that our results are among the more accurate ones.

Notes and References

The hyper-Borel transformation was first introduced in Ref. [1]. In the mathematical literature, this transformation has never been investigated, although it is contained in a class of quite general mathematical transformations introduced in the textbook of Hardy, *Divergent Series* (Oxford University Press, Oxford 1949 in the context of *moment constant methods*. These comprise transformation $B(y) = \sum f_k y^k / \mu_k$, where the μ_k are given by a Stieltjes integral $\mu_k = \int_0^\infty x^k d\chi(x)$ and χ is a bounded and increasing function of x guaranteeing the convergence of the Stieltjes integral. This definition includes our transformation for the somewhat complicated choice $d\chi(x) = \Gamma(\beta_0) x^{-s-1} \oint_C dt \, e^{t + x^\omega t^{1-\omega}} t^{s(1-1/\omega) - \beta_0} \, dx/2\pi i$.

The references quoted in the text and Fig. 21.7 are

[1] F. Jasch and H. Kleinert, J. Math. Phys. 2000 (in press) (cond-mat/9906246).

[2] C.M. Bender and T.T. Wu, Phys. Rev. **184**, 1231 (1969).

[3] B. Simon, Ann. Phys. **58**, 76 (1970)

[4] W. Janke and H. Kleinert, Phys. Rev. Lett. **75**, 2787 (1995) (quant-ph/9502019).

[5] H. Kleinert, *Path Integrals in Quantum Mechanics Statistics and Polymer Physics*, World Scientific Singapore 1995.

[6] J.A. Lipa, D.R. Swanson, J. Nissen, T.C.P. Chui and U.E. Israelson, Phys. Rev. Lett. **76**, 944 (1996).
    As mentioned in Ref. [16] at the end of Chapter 20, the initially published fit to the data was erroneous and corrected in a footnote (number [15]) of a subsequent paper J.A. Lipa, D.R. Swanson, J. Nissen, Z.K. Geng, P.R. Williamson, D.A. Stricker, T.C.P. Chui, U.E. Israelson, and M. Larson, Phys. Rev. Lett. **84**, 4894 (2000).

[7] L.S. Goldner, N. Mulders, and G. Ahlers, J. Low Temp. Phys. **93**, 131 (1992).

[8] G. Ahlers, Phys. Rev. A **3**, 696 (1971);
    K.H. Mueller, G. Ahlers, F. Pobell, Phys. Rev. B **14**, 2096 (1976);

[9] R. Guida and J. Zinn-Justin, J. Phys. A **31**, 8103 (1998)

[10] J.C. Le Guillou and J. Zinn-Justin, Phys. Rev. Lett. **39**, 95 (1977); Phys. Rev. B **21**, 3976 (1980); J. de Phys. Lett. **46**, L137 (1985).

[11] Result of Section 19.7, first calculated in
H. Kleinert and V. Schulte-Frohlinde, J. Phys. A **34** 1037 (2001) (cond-mat/9907214).

[12] A. Pelissetto and E. Vicari, Nucl. Phys. B *519*, 626 (1998);
522, 605 (1998); Nucl. Phys. Proc. Suppl. *73*, 775 (1999). See also the review article by these authors in cond-mat/0012164.

[13] D.B. Murray and B.G. Nickel, unpublished.

[14] Result of Section 20.4, first calculated in
H. Kleinert, *Seven Loop Critical Exponents from Strong-Coupling ϕ^4-Theory in Three Dimensions*, to appear in Phys. Rev. D (hep-th/9812197).

[15] Result of Section 20.7, first published in
H. Kleinert, Phys. Lett. A **277**, 205 (2000) (cond-mat/9906107).

[16] W. Janke, Phys. Lett. A **148**, 306 (1990).

[17] H.G. Ballesteros, L.A. Fernandez, V. Martin-Mayor, and A. Munoz-Sudupe, Phys. Lett. B **387**, 125 (1996).

[18] M. Ferer, M.A. Moore, and M. Wortis, Phys. Rev. B **65**, 2668 (1972).
See Section 20.8 for the associated theory.

[19] P. Butera and M. Comi, Phys. Rev. B **56**, 8212 (1997) (hep-lat/9703018).
See Section 20.8 for the associated theory.

Conclusion

In this book we have explained the presently available calculational techniques to calculate the critical exponents of ϕ^4-theories in $D = 4 - \varepsilon$ dimensions. The calculations were organized with the help of Feynman diagrams, which were considered up to five loops. The analytic effort going into the evaluation of the associated Feynman integrals was considerable, and would have been unsurmountable without the use of computer-algebraic programs such as Reduce and Mathematica. The calculations yielded perturbation expansions for the critical exponents, which led to power series in ε to the order ε^5. These power series were calculated for interactions with $O(N)$ symmetry and a combination of $O(N)$ and cubic symmetry.

An important tool for deriving these series was the technique dimensional regularization with minimal subtraction developed by t'Hooft and Veltman.[1] This leads to ε-expansions for all observable critical quantities in less than $D = 4$ dimensions of maximal simplicity.

The calculation of the renormalization constants was done most efficiently with the help of the R-operation developed by Bogoliubov and Shirkov.[2] This operation must be performed tediously diagram by diagram.

The most important simplification brought about by the minimal subtraction is the independence of the renormalization constants on the mass and the external momenta of the Feynman integrals. In this way, only massless integrals need to be calculated, with at most one external momentum. The external vertices of this momentum may be chosen quite freely, restricted only by the requirement that the infrared behavior of the integral remains unchanged (method of IR-rearrangement). This allowed us to use existing algorithms[3] for the reduction of the massless integrals to nested one-loop integrals, which can all be expressed in terms of Gamma-functions. Unfortunately, however, these algorithms are applicable to only some generic two- and three-loop diagrams which contain a subdiagram in triangle form. So far, no generalization has been found if a subdiagram has a square form.

A further class of diagrams becomes accessible when transforming the massless, dimensionally regularized integrals into their dual form by Fourier transformation. In **x**-space, the integrals are solvable or may be reduced to solvable integrals by applying the reduction algorithms in **x**-space (method of ideal index constellations).[4]

Most of the diagrams can be brought to one of the calculable forms by IR-rearrangement. In some cases, the IR-rearrangement produces artificial IR-divergences. In dimensional regularization such IR-divergences appear also in the form of poles in ε and must be removed by a procedure analogous to the R-operation, the R^*-operation.

In this way, all integrals up to five loops have been calculated algebraically, with only six exceptions, which require special methods based on partial integration, clever differentiation, and a final application of the R-operation. All calculations were done with the help of computer-

[1]G.'t Hooft and M. Veltman, Nuclear Physics B *44*, 189-213 (1972); G.'t Hooft, Nucl. Phys. B *61*, 455 (1973).

[2]N.N. Bogoliubov and D.V. Shirkov, *Introduction to the Theory of Quantized Fields*, Wiley Interscience, New York, 1958.

[3]K.G. Chetyrkin, A.L. Kataev, and F.V. Tkachov, Nucl. Phys. B *174*, 345-377 (1980).

[4]D.I. Kazakov, Phys. Lett. B *133*, 406-410 (1983); Theor. Math. Phys. *61*, 84-89 (1985).

algebraic programs to avoid trivial errors. The implementation of the subtraction procedures $\bar{R}$ and $\bar{R}^*$ on the computer has an obstacle in the ambiguities of IR-rearrangement. The choice of the external vertices selects different integrals to be calculated. Since it was not clear at the beginning which integrals would arise and which were calculable, the IR-rearrangement had to be done tediously by hand. For many diagrams there is only one arrangement of the external vertices which allowed us to calculate the corresponding integral.

At the six-loop level, new generic types of diagrams are encountered. A subdiagram of the square type cannot be avoided by IR-rearrangement. The above-described methods are therefore insufficient to calculate all six-loop diagrams, and new methods had to be developed.

In the case of fields with several indices and tensorial interactions, each Feynman diagram characterizes not only a momentum integral, but also an index sum. This sum leads to a symmetry factor for each diagram. In this work, a combination of $O(N)$ and cubic symmetry was considered in the interaction. For $N = 2, 3$, the $O(N)$ and the mixed $O(N)$-cubic symmetry cover all symmetries which lead to only one length scale. For $N = 4$, there are several other symmetries with this property. In a two-loop calculation,[5] however, only two other stable fixed points were found in addition to the isotropic and the cubic one. In this text, we have extended these results to the five-loop level.

The five-loop expansions calculated in this text lead to critical exponents as power series ε up to ε^5 for critical phenomena of systems with a combination of $O(N)$ and cubic symmetry. For the comparison with critical exponents measured by experiments in $D = 3$ dimensions, these series have to be evaluated at $\varepsilon = 1$. This is not directly possible since the series are divergent and require special techniques for their resummation.

The most simple technique, the Padé approximation does not yield satisfactory results since it does not include other available information on the series. This comes from a knowledge of the behavior of all series at large orders. This knowledge can be exploited by re-expanding all series into functions with the same large-order behavior. This method has been applied to systems with $O(N)$ and with a mixture of $O(N)$ and cubic symmetry.

Two other resummation schemes have been developed. One is based on the fact that critical phenomena correspond to a strong-coupling limit of the divergent series when expanded in powers of the *bare* coupling constant.[6]

In $D = 3$ dimensions, six-loop calculations have been done a long time ago.[7] For the critical exponents η and η_m, even seven-loop expansions have recently become available.[8] Although the additional terms have large coefficients, their effect upon the critical exponents after resummation is very small.[9] These alternative expansions have been described and resummed in this text for comparison.

Other series for critical exponents have been derived from lattice models, the most extensive ones by Butera and Comi.[10] Also these have been discussed here in some detail.

An extension of the perturbation expansions in powers of ε in $4 - \varepsilon$ dimensions to more than five loops seems prohibitively difficult at the present time. New theoretical ideas will

---

[5]J.-C. Toledano, L. Michel, P. Toledano, E. Brézin, Phys. Rev. B *31*, 7171 (1985).

[6]H. Kleinert, Phys. Rev. D **57**, 2264 (1998) (E-Print aps1997jun25_001); addendum (cond-mat/9803268); Phys. Rev. D (in press) (1999) (hep-th/9812197).

[7]B.G. Nickel, D.I. Meiron, and G.A. Baker, Jr., University of Guelph report, 1977 (unpublished).

[8]D.B. Murray and B.G. Nickel, unpublished.

[9]See R. Guida and J. Zinn-Justin, Saclay preprint 1098, cond-mat/9803240 and H. Kleinert, Phys. Rev. D (in press) (1999) (hep-th/9812197).

[10]See the references in the footnotes of Section 20.8.

probably be needed for such extensions. Recent knot-theoretic work[11] may be a first step in this direction.

[11]D.J. Broadhurst, Z. Phys. C *32*, 249 (1986); D. Kreimer, Phys. Lett. B *273*, 177 (1991); D.J. Broadhurst and D. Kreimer, Int. J. Mod. Physics C *6* 519 (1995); UTAS-HHYS-96-44 (hep-th/9609128).

Appendix A

Diagrammatic R-Operation up to Five Loops

A.1 Tadpole-Free Logarithmically Divergent Diagrams for Z_g and Z_{m^2}

$$\underline{1-\ \text{Loop}\ \sim g}$$

No. 1: $\mathcal{K}\bar{R}(\,\text{⊸⊙⊸}\,) = \mathcal{K}\left[\ \text{⊸⊙⊸}\ \right]$

$$\underline{2-\ \text{Loop}\ \sim g^2}$$

No. 1 : $\mathcal{K}\bar{R}(\text{⊸◯◯⊸}) = -\mathcal{K}\left[\ \text{⊸⊙⊸}\ \right]\mathcal{K}\left[\ \text{⊸⊙⊸}\ \right]$

No. 2 : $\mathcal{K}\bar{R}(\text{⊸⊘⊸}) = \mathcal{K}\left[\ \text{⊸⊘⊸}\ -\ \mathcal{K}(\text{⊸⊙⊸})\text{⊸⊙⊸}\ \right]$

$$\underline{3-\ \text{Loop}\ \sim g^3}$$

No. 1 : $\mathcal{K}\bar{R}(\text{⊸◯◯◯⊸}) = \mathcal{K}\left[\ \text{⊸⊙⊸}\ \right]\mathcal{K}\left[\ \text{⊸⊙⊸}\ \right]\mathcal{K}\left[\ \text{⊸⊙⊸}\ \right]$

No. 2 : $\mathcal{K}\bar{R}(\text{⊸◯◯⊘⊸}) = -\mathcal{K}\left[\ \text{⊸⊙⊸}\ \right]\mathcal{K}\bar{R}\left[\ \text{⊸⊘⊸}\ \right]$

No. 3 : $\mathcal{K}\bar{R}(\text{⊸⊘⊸}) = -\mathcal{K}\left[\ \text{⊸⊘⊸}\ -\ \partial_{\mathbf{k}^2}\mathcal{K}(\text{⊸⊖⊸})\ \text{⊸⊙⊸}\ \right]$

No. 4 : $\mathcal{K}\bar{R}(\text{⊸⧄⊸}) = \mathcal{K}\left[\ \text{⊸⧄⊸}\ \right]$

No. 5 : $\mathcal{K}\bar{R}(\text{⊸⊖⊸}) = \mathcal{K}\left[\ \text{⊸⊖⊸} - 2\mathcal{K}(\text{⊸⊙⊸})\text{⊸⊖⊸} - \mathcal{K}\bar{R}(\text{⊸◯◯⊸})\text{⊸⊙⊸}\ \right]$

No. 6 : $\mathcal{K}\bar{R}(\text{⊸⊗⊸}) = \mathcal{K}\left[\ \text{⊸⊗⊸} - 2\mathcal{K}(\text{⊸⊙⊸})\text{⊸⊘⊸} + \mathcal{K}(\text{⊸⊙⊸})\mathcal{K}(\text{⊸⊙⊸})\text{⊸⊙⊸}\ \right]$

No. 7 : $\mathcal{K}\bar{R}(\text{⊸⧄⊸}) = \mathcal{K}\left[\ \text{⊸⧄⊸} - \mathcal{K}(\text{⊸⊙⊸})\text{⊸⊘⊸} - \mathcal{K}\bar{R}(\text{⊸⊘⊸})\text{⊸⊙⊸}\ \right]$

No. 8 : $\mathcal{K}\bar{R}(\text{⊸⧉⊸}) = \mathcal{K}\left[\ \text{⊸⧉⊸} - \mathcal{K}(\text{⊸⊙⊸})\text{⊸◯◯⊸} - 2\mathcal{K}\bar{R}(\text{⊸⊘⊸})\text{⊸⊙⊸}\ \right]$

$$\underline{4-\ \text{Loop}\ \sim g^4}$$

No. 1 : $\mathcal{K}\bar{R}(\text{⊸◯◯◯◯⊸}) = -\mathcal{K}\left[\ \text{⊸⊙⊸}\ \right]\mathcal{K}\left[\ \text{⊸⊙⊸}\ \right]\mathcal{K}\left[\ \text{⊸⊙⊸}\ \right]\mathcal{K}\left[\ \text{⊸⊙⊸}\ \right]$

No. 2 : $K\bar{R}(\text{⊙⊙⟨⟩}) = K\left[\ \text{⊖}\ \right]K\left[\ \text{⊖}\ \right]K\bar{R}\left[\ \text{⟨⟩}\ \right]$

No. 3 : $K\bar{R}(\text{⟨⟩⟨⟩}) = -K\bar{R}\left[\ \text{⟨⟩}\ \right]K\bar{R}\left[\ \text{⟨⟩}\ \right]$

No. 4 : $K\bar{R}(\text{⊙⟨⟩}) = -K\left[\ \text{⊖}\ \right]K\bar{R}\left[\ \text{⟨⟩}\ \right]$

No. 5 : $K\bar{R}(\text{⊙⟨⟩}) = -K\left[\ \text{⊖}\ \right]K\bar{R}\left[\ \text{⟨⟩}\ \right]$

No. 6 : $K\bar{R}(\text{⟨▷⊙}) = -K\left[\ \text{⊖}\ \right]K\bar{R}\left[\ \text{⟨▷}\ \right]$

No. 7 : $K\bar{R}(\text{⟨◊▷⊙}) = -K\left[\ \text{⊖}\ \right]K\bar{R}\left[\ \text{⟨◊▷}\ \right]$

No. 8 : $K\bar{R}(\text{⟨▽▷}) = K\left[\ \text{⟨▽▷} - K(\text{⊖})\,\text{⟨◊▷} - K\bar{R}(\text{⟨⊖})\,\text{⊙⊙}\right.$

$$\left. -2K\bar{R}(\text{⟨▷})\,\text{⊖}\ \right]$$

No. 9 : $K\bar{R}(\text{⟨◊▷}) = K\left[\ \text{⟨◊▷} - K(\text{⊖})\,\text{⟨◊▷} - K(\text{⊖})\,\text{⊙⟨▷}\right.$

$$-K\bar{R}(\text{⟨▷})\,\text{⟨▷} + K\bar{R}(\text{⟨▷})\,K(\text{⊖})\,\text{⊖}$$

$$\left. -K\bar{R}(\text{⟨◊▷})\,\text{⊖}\ \right]$$

No. 10 : $K\bar{R}(\text{⟨◈▷}) = K\left[\ \text{⟨◈▷}\ \right]$

No. 11 : $K\bar{R}(\text{⟨◈▷}) = K\left[\ \text{⟨◈▷} - K(\text{⊖})\,\text{⟨◈▷}\ \right]$

No. 12 : $K\bar{R}(\text{⟨◊▷}) = K\left[\ \text{⟨◊▷} - K(\text{⊖})\,\text{⟨◊▷} - K(\text{⊖})\,\text{⟨◊▷} + K(\text{⊖})^{2}\,\text{⟨▷}\right.$

$$\left. -K\bar{R}(\text{⟨◊})\,\text{⊖} + K\bar{R}(\text{⟨◊})\,K(\text{⊖})\,\text{⊖}\ \right]$$

No. 13 : $K\bar{R}(\text{⟨◊▷}) = K\left[\ \text{⟨◊▷} - K(\text{⊖})\,\text{⟨◊▷} - 2K(\text{⊖})\,\text{⟨◊▷}\right.$

$$+2K(\text{⊖})K(\text{⊖})\,\text{⟨▷} - K\bar{R}(\text{⊙⊙})\,\text{⟨◊▷}$$

$$\left. +K(\text{⊖})K\bar{R}(\text{⊙⊙})\,\text{⊖}\ \right]$$

No. 14 : $K\bar{R}(\text{⟨◊▷}) = K\left[\ \text{⟨◊▷} - 2K(\text{⊖})\,\text{⟨◊▷} - K\bar{R}(\text{⊙⊙})\,\text{⊙⊙}\right.$

$$\left. -2K\bar{R}(\text{⟨◊})\,\text{⊖}\ \right]$$

No. 15 : $\mathcal{K}\bar{R}(\ \ ) = \mathcal{K}\left[\ \ - 2\mathcal{K}(\ \ )\ \ - \partial_{k^2}\mathcal{K}\bar{R}(\ \ )\ \ \right]$

No. 16 : $\mathcal{K}\bar{R}(\ \ ) = \mathcal{K}\left[\ \ - \partial_{k^2}\mathcal{K}(\ \ )\ \ - \mathcal{K}\bar{R}(\ \ )\ \ \right]$

No. 17 : $\mathcal{K}\bar{R}(\ \ ) = \mathcal{K}\left[\ \ - \mathcal{K}(\ \ )\ \ - 2\mathcal{K}\bar{R}(\ \ )\ \right.$
$\left. - \mathcal{K}\bar{R}(\ \ )\ \ \right]$

No. 18 : $\mathcal{K}\bar{R}(\ \ ) = \mathcal{K}\left[\ \ - 3\mathcal{K}(\ \ )\ \ + \mathcal{K}(\ \ )\mathcal{K}(\ \ )\ \right.$
$\left. - 2\mathcal{K}\bar{R}(\ \ )\ \ - \mathcal{K}\bar{R}(\ \ )\ \ \right]$

No. 19 : $\mathcal{K}\bar{R}(\ \ ) = \mathcal{K}\left[\ \ - \partial_{k^2}\mathcal{K}(\ \ )\ \ - \mathcal{K}(\ \ )\ \right.$
$\left. + \mathcal{K}(\ \ )\ \partial_{k^2}\mathcal{K}(\ \ )\ \ \right]$

No. 20 : $\mathcal{K}\bar{R}(\ \ ) = \mathcal{K}\left[\ \ - \mathcal{K}(\ \ )\ \ - \mathcal{K}(\ \ )\ \right.$
$\left. + \mathcal{K}(\ \ )\mathcal{K}(\ \ )\ \ - \mathcal{K}\bar{R}(\ \ )\ \ \right]$

No. 21 : $\mathcal{K}\bar{R}(\ \ ) = \mathcal{K}\left[\ \ - \mathcal{K}(\ \ )\ \ \right]$

No. 22 : $\mathcal{K}\bar{R}(\ \ ) = \mathcal{K}\left[\ \ - 2\mathcal{K}(\ \ )\ \ + \mathcal{K}(\ \ )\mathcal{K}(\ \ )\ \right.$
$\left. - \mathcal{K}\bar{R}(\ \ )\ \ \right]$

No. 23 : $\mathcal{K}\bar{R}(\ \ ) = \mathcal{K}\left[\ \ - \mathcal{K}(\ \ )\ \ - \mathcal{K}(\ \ )\ \right.$
$+ \mathcal{K}(\ \ )\mathcal{K}(\ \ )\ $
$\left. - \mathcal{K}\bar{R}(\ \ )\ \ - \mathcal{K}\bar{R}(\ \ )\ \ \right]$

No. 24 : $\mathcal{K}\bar{R}(\ \ ) = \mathcal{K}\left[\ \ - 2\mathcal{K}(\ \ )\ \ - \mathcal{K}\bar{R}(\ \ )\ \ \right.$
$\left. - \mathcal{K}\bar{R}(\ \ )\ \ \right]$

No. 25 : $\mathcal{K}\bar{R}(\ \ ) = \mathcal{K}\left[\ \ - \mathcal{K}(\ \ )\ \ - \mathcal{K}\bar{R}(\ \ )\ \ \right.$
$\left. - \mathcal{K}\bar{R}(\ \ )\ \ \right]$

No. 26 : $\quad \mathcal{K}\bar{R}(\ \text{⬡}\ ) = \mathcal{K}\big[\ \text{⬡} - \mathcal{K}(\text{⊙})\ \text{⬡} - \mathcal{K}\bar{R}(\text{⬡}\,)\ \text{⬡}$

$\qquad\qquad\qquad\qquad -\mathcal{K}\bar{R}(\text{⬡})\,\text{⊙}\ \big]$

$$\underline{5-\ \text{Loop}\ \sim\ g^5}$$

No. 1 : $\quad \mathcal{K}\bar{R}(\ \text{⊕}\ ) = \mathcal{K}\big[\ \text{⊕}\ \big]$

No. 2 : $\quad \mathcal{K}\bar{R}(\ \text{⊕}\ ) = \mathcal{K}\big[\ \text{⊕}\ \big]$

No. 3 : $\quad \mathcal{K}\bar{R}(\ \text{⬡}\ ) = \mathcal{K}\big[\ \text{⬡}\ \big]$

No. 4 : $\quad \mathcal{K}\bar{R}(\ \text{⬡}\ ) = \mathcal{K}\big[\ \text{⬡} - \mathcal{K}(\text{⊙})\ \text{⬡} - \partial_{k^2}\mathcal{K}(\text{⊖})\ \text{⬡}$

$\qquad\qquad\qquad\qquad + \mathcal{K}(\text{⊖})\ \partial_{k^2}\mathcal{K}(\text{⊖})\ \text{⬡} - \mathcal{K}\bar{R}(\text{⬡})\ \text{⊙}\ \big]$

No. 5 : $\quad \mathcal{K}\bar{R}(\ \text{⬡}\ ) = \mathcal{K}\big[\ \text{⬡} - \mathcal{K}(\text{⊙})\ \text{⬡} - \partial_{k^2}\mathcal{K}(\text{⊖})\ \text{⬡}$

$\qquad\qquad\qquad\qquad + \mathcal{K}(\text{⊙})\ \partial_{k^2}\mathcal{K}(\text{⊖})\ \text{⬡} - \mathcal{K}\bar{R}(\text{⬡})\ \text{⊙}\ \big]$

No. 6 : $\quad \mathcal{K}\bar{R}(\ \text{⬡}\ ) = \mathcal{K}\big[\ \text{⬡} - \mathcal{K}(\text{⊙})\ \text{⬡} + \mathcal{K}(\text{⊙})\ \partial_{k^2}\mathcal{K}(\text{⊖})\ \text{⬡}$

$\qquad\qquad\qquad\qquad - \mathcal{K}\bar{R}(\text{⬡})\ \text{⬡} + \mathcal{K}\bar{R}(\text{⬡})\ \partial_{k^2}\mathcal{K}(\text{⊖})\ \text{⊙}$

$\qquad\qquad\qquad\qquad - \partial_{k^2}\mathcal{K}(\text{⊖})\ \text{⬡}\ \big]$

No. 7 : $\quad \mathcal{K}\bar{R}(\ \text{⬡}\ ) = \mathcal{K}\big[\ \text{⊙}\ \big]\ \mathcal{K}\big[\ \text{⊙}\ \big]\ \mathcal{K}\bar{R}\big[\ \text{⬡}\ \big]$

No. 8 : $\quad \mathcal{K}\bar{R}(\ \text{⬡}\ ) = \mathcal{K}\big[\ \text{⊙}\ \big]\ \mathcal{K}\big[\ \text{⊙}\ \big]\ \mathcal{K}\bar{R}\big[\ \text{⬡}\ \big]$

No. 9 : $\quad \mathcal{K}\bar{R}(\ \text{⬡}\ ) = -\mathcal{K}\bar{R}\big[\ \text{⬡}\ \big]\mathcal{K}\bar{R}\big[\ \text{⬡}\ \big]$

No. 10 : $\quad \mathcal{K}\bar{R}(\ \text{⬡}\ ) = \mathcal{K}\big[\ \text{⬡} - \mathcal{K}(\text{⊙})\ \text{⬡} - \partial_{k^2}\mathcal{K}(\text{⊖})\ \text{⬡}$

$\qquad\qquad\qquad\qquad + \mathcal{K}(\text{⊙})\ \partial_{k^2}\mathcal{K}(\text{⊖})\ \text{⬡} - \mathcal{K}\bar{R}(\text{⬡})\ \text{⊙}$

$\qquad\qquad\qquad\qquad + \mathcal{K}\bar{R}(\text{⬡})\mathcal{K}(\text{⊙})\ \text{⊙}\ \big]$

No. 11 : $\quad \mathcal{K}\bar{R}(\ \text{⬡}\ ) = \mathcal{K}\big[\ \text{⬡} - 2\mathcal{K}(\text{⊙})\ \text{⬡} + \mathcal{K}(\text{⊙})\mathcal{K}(\text{⊙})\ \text{⬡}$

$$-\partial_{k^2}\mathcal{K}(\ominus)\; \bigcirc\!\!\!\!\bigcirc + 2\mathcal{K}(\circleddash)\partial_{k^2}\mathcal{K}(\ominus)\; \bigcirc\!\!\!\!\bigcirc$$

$$-\mathcal{K}(\circleddash)\mathcal{K}(\circleddash)\partial_{k^2}\mathcal{K}(\ominus)\circleddash\;\Big]$$

No. 12 : $\;\mathcal{K}\bar{R}(\; \diagup\!\!\!\!\bigcirc\;) \;=\; \mathcal{K}\Big[\; \diagup\!\!\!\!\bigcirc \;-\; \mathcal{K}(\circleddash)\; \diagup\!\!\!\!\bigcirc \;-\; \mathcal{K}\bar{R}(\diagup\!\!\!\!\bigcirc)\; \bigcirc$

$$-\partial_{k^2}\mathcal{K}(\ominus)\; \diagup\!\!\!\!\bigcirc \;+\; \mathcal{K}(\circleddash)\,\partial_{k^2}\mathcal{K}(\ominus)\circleddash$$

$$+\partial_{k^2}\mathcal{K}(\ominus)\mathcal{K}\bar{R}(\diagup\!\!\!\!\bigcirc)\;\circleddash\;\Big]$$

No. 13 : $\;\mathcal{K}\bar{R}(\oplus) \;=\; \mathcal{K}\Big[\; \oplus \;-\; \partial_{k^2}\mathcal{K}(\ominus)\; \oplus \;\Big]$

No. 14 : $\;\mathcal{K}\bar{R}(\bigcirc) \;=\; \mathcal{K}\Big[\; \bigcirc \;-\; 2\,\partial_{k^2}\mathcal{K}(\ominus)\; \bigcirc \;+\; \partial_{k^2}\mathcal{K}(\ominus)\partial_{k^2}\mathcal{K}(\ominus)\circleddash$

No. 15 : $\;\mathcal{K}\bar{R}(\ominus\!\!\!\!\ominus) \;=\; \mathcal{K}\Big[\; \ominus\!\!\!\!\ominus \;-\; \partial_{k^2}\mathcal{K}(\ominus)\; \bigcup \;-\; \mathcal{K}\bar{R}(\bigcup)\;\circleddash\;\Big]$

No. 16 : $\;\mathcal{K}\bar{R}(\; \diagdown\!\!\!\!\bigcirc\;) \;=\; \mathcal{K}\Big[\; \diagdown\!\!\!\!\bigcirc \;-\; \partial_{k^2}\mathcal{K}(\ominus)\; \diagdown\!\!\!\!\bigcirc \;-\; \mathcal{K}\bar{R}(\bigcup)\; \diagdown\!\!\!\!\bigcirc$

$$-\mathcal{K}\bar{R}(\diagdown\!\!\!\!\bigcirc)\;\circleddash\;\Big]$$

No. 17 : $\;\mathcal{K}\bar{R}(\diagdown\!\!\!\!\bigcirc\!\!\!\!\diagup) \;=\; \mathcal{K}\Big[\; \diagdown\!\!\!\!\bigcirc\!\!\!\!\diagup \;-\; \mathcal{K}(\circleddash)\; \diagdown\!\!\!\!\bigcirc \;-\; \partial_{k^2}\mathcal{K}(\ominus)\; \diagdown\!\!\!\!\bigcirc$

$$+\mathcal{K}(\circleddash)\,\partial_{k^2}\mathcal{K}(\ominus)\; \diagdown\!\!\!\!\bigcirc \;-\; \mathcal{K}\bar{R}(\bigcup)\; \diagdown\!\!\!\!\bigcirc$$

$$+\mathcal{K}\bar{R}(\bigcup)\mathcal{K}(\circleddash)\;\circleddash\;\Big]$$

No. 18 : $\;\mathcal{K}\bar{R}(\circ\!\!\!\!\diagup\!\!\!\!\bigcirc) \;=\; -\mathcal{K}\Big[\;\circleddash\;\Big]\mathcal{K}\bar{R}\Big[\;\diagup\!\!\!\!\bigcirc\;\Big]$

No. 19 : $\;\mathcal{K}\bar{R}(\circledcirc) \;=\; \mathcal{K}\Big[\; \circledcirc \;-\; \partial_{k^2}\mathcal{K}(\ominus)\; \oplus \;-\; \mathcal{K}\bar{R}(\bigcup)\;\circ\!\!\!\circ$

$$-2\,\mathcal{K}\bar{R}(\diagdown\!\!\!\!\bigcirc)\;\circleddash\;\Big]$$

No. 20 : $\;\mathcal{K}\bar{R}(\bigcirc) \;=\; \mathcal{K}\Big[\; \bigcirc \;-\; 2\mathcal{K}(\circleddash)\; \bigcirc \;-\; \mathcal{K}\bar{R}(\circ\!\!\!\circ)\; \bigcup$

$$-\,\partial_{k^2}\mathcal{K}(\ominus)\; \bigcirc \;+\; 2\mathcal{K}(\circleddash)\,\partial_{k^2}\mathcal{K}(\ominus)\circleddash$$

$$+\,\partial_{k^2}\mathcal{K}(\ominus)\mathcal{K}(\circ\!\!\!\circ)\;\circleddash\;\Big]$$

No. 21: $\;\mathcal{K}\bar{R}^{\star}(\diagup\!\!\!\!\bigcirc) \;=\; \mathcal{K}\Big[\; \diagup\!\!\!\!\bigcirc \;-\; \mathcal{K}\bar{R}(\diagdown\!\!\!\!\bigcirc)\; \bigcup \;-\; \partial_{k^2}\mathcal{K}(\ominus)\; \diagup\!\!\!\!\bigcirc$

$$+\,\partial_{k^2}\mathcal{K}(\ominus)\mathcal{K}\bar{R}(\diagdown\!\!\!\!\bigcirc)\;\circleddash$$

$$+ (\,[\text{diagram}]\,)_{IR} \Big\{ [\text{diagram}] - \mathcal{K}([\text{diagram}])\,[\text{diagram}] - \partial_{k^2}\mathcal{K}([\text{diagram}])\,[\text{diagram}]$$

$$+ \mathcal{K}([\text{diagram}])\,\partial_{k^2}\mathcal{K}([\text{diagram}])\,[\text{diagram}] - \mathcal{K}\bar{R}([\text{diagram}]) \Big\} \Big]$$

No. 22 : $\mathcal{K}\bar{R}([\text{diagram}]) = -\mathcal{K}\big[\,[\text{diagram}]\,\big]\,\mathcal{K}\bar{R}\big[\,[\text{diagram}]\,\big]$

No. 23 : $\mathcal{K}\bar{R}([\text{diagram}]) = \mathcal{K}\big[\,[\text{diagram}] - 2\,\partial_{k^2}\mathcal{K}([\text{diagram}])\,[\text{diagram}] + \partial_{k^2}\mathcal{K}([\text{diagram}])\,\partial_{k^2}\mathcal{K}([\text{diagram}])\,[\text{diagram}]\,\big]$

No. 24 : $\mathcal{K}\bar{R}([\text{diagram}]) = \mathcal{K}\big[\,[\text{diagram}] - \mathcal{K}([\text{diagram}])\,[\text{diagram}]\,\big]$

No. 25 : $\mathcal{K}\bar{R}^{*}([\text{diagram}]) = \mathcal{K}\big[\,[\text{diagram}] + (\,[\text{diagram}]\,)_{IR}\big\{[\text{diagram}] - \mathcal{K}([\text{diagram}])\big\}\big]$

No. 26 : $\mathcal{K}\bar{R}([\text{diagram}]) = \mathcal{K}\big[\,[\text{diagram}] - \mathcal{K}([\text{diagram}])\,[\text{diagram}]\,\big]$

No. 27 : $\mathcal{K}\bar{R}([\text{diagram}]) = \mathcal{K}\big[\,[\text{diagram}] - \mathcal{K}([\text{diagram}])\,[\text{diagram}]\,\big]$

No. 28 : $\mathcal{K}\bar{R}([\text{diagram}]) = \mathcal{K}\big[\,[\text{diagram}]\,\big]$

No. 29 : $\mathcal{K}\bar{R}([\text{diagram}]) = \mathcal{K}\big[\,[\text{diagram}] - \mathcal{K}([\text{diagram}])\,[\text{diagram}]\,\big]$

No. 30 : $\mathcal{K}\bar{R}([\text{diagram}]) = \mathcal{K}\big[\,[\text{diagram}] - 2\mathcal{K}([\text{diagram}])\,[\text{diagram}] - \mathcal{K}\bar{R}([\text{diagram}])\,[\text{diagram}]\,\big]$

No. 31 : $\mathcal{K}\bar{R}([\text{diagram}]) = -\mathcal{K}\big[\,[\text{diagram}]\,\big]\,\mathcal{K}\bar{R}\big[\,[\text{diagram}]\,\big]$

No. 32 : $\mathcal{K}\bar{R}^{*}([\text{diagram}]) = \mathcal{K}\big[\,[\text{diagram}] - \mathcal{K}\bar{R}([\text{diagram}])\,[\text{diagram}]$

$$+ (\,[\text{diagram}]\,)_{IR}\big\{[\text{diagram}] - \mathcal{K}\bar{R}([\text{diagram}])\,[\text{diagram}]$$

$$- \mathcal{K}\bar{R}([\text{diagram}])\big\}\big]$$

No. 33 : $\mathcal{K}\bar{R}([\text{diagram}]) = \mathcal{K}\big[\,[\text{diagram}] - \mathcal{K}([\text{diagram}])\,[\text{diagram}] - \mathcal{K}\bar{R}([\text{diagram}])\,[\text{diagram}]\,\big]$

No. 34 : $\mathcal{K}\bar{R}^{*}([\text{diagram}]) = \mathcal{K}\big[\,[\text{diagram}] + (\,[\text{diagram}]\,)_{IR}\big\{[\text{diagram}] - \mathcal{K}\bar{R}([\text{diagram}])\big\}$

$$+ (\,[\text{diagram}]\,)_{IR}\big\{[\text{diagram}] - \mathcal{K}([\text{diagram}])\big\}\big]$$

No. 35 : $\mathcal{K}\bar{R}([\text{diagram}]) = \mathcal{K}\big[\,[\text{diagram}] - \mathcal{K}([\text{diagram}])\,[\text{diagram}] - \mathcal{K}([\text{diagram}])\,[\text{diagram}]$

$$+\mathcal{K}(\,\oplus\,)\mathcal{K}(\,\multimap\,)\,\multimap\,\Big]$$

No. 36 : $\mathcal{K}\bar{R}^{\star}(\,\oplus\,)$ $=$ $\mathcal{K}\Big[\;\oplus\; +(\,\big\vert\,)_{IR}\big\{\,\oplus\; -\mathcal{K}\bar{R}(\,\multimap\,)\big\}$

$$+(\,\sqcap\,)_{IR}\big\{\,\oplus\; -\mathcal{K}(\,\oplus\,)\big\}\Big]$$

No. 37 : $\mathcal{K}\bar{R}^{\star}(\,\multimap\,)$ $=$ $\mathcal{K}\Big[\;\multimap\; +(\,\big\vert\,)_{IR}\big\{\,\multimap\; -\mathcal{K}\bar{R}(\,\multimap\,)\big\}$

$$+(\,\triangle\,)_{IR}\big\{\,\multimap\; -\mathcal{K}(\,\multimap\,)\big\}\Big]$$

No. 38 : $\mathcal{K}\bar{R}(\,\multimap\,)$ $=$ $\mathcal{K}\Big[\;\multimap\; -\mathcal{K}(\,\multimap\,)\,\multimap\; -\mathcal{K}\bar{R}(\,\multimap\,)\,\multimap\;\Big]$

No. 39 : $\mathcal{K}\bar{R}(\,\multimap\,)$ $=$ $\mathcal{K}\Big[\;\multimap\; -\mathcal{K}(\,\multimap\,)\,\multimap\; -\mathcal{K}(\,\multimap\,)\,\multimap\;$

$$+\mathcal{K}(\,\multimap\,)\mathcal{K}(\,\multimap\,)\,\multimap\;\Big]$$

No. 40 : $\mathcal{K}\bar{R}(\,\multimap\,)$ $=$ $\mathcal{K}\Big[\;\multimap\; -\mathcal{K}(\,\multimap\,)\,\multimap\; -\mathcal{K}(\,\multimap\,)\,\multimap\;$

$$+\mathcal{K}(\,\multimap\,)\mathcal{K}(\,\multimap\,)\,\multimap\;\Big]$$

No. 41 : $\mathcal{K}\bar{R}(\,\multimap\,)$ $=$ $\mathcal{K}\Big[\;\multimap\; -2\mathcal{K}(\,\multimap\,)\,\multimap\; +\mathcal{K}(\,\multimap\,)\mathcal{K}(\,\multimap\,)\,\multimap\;\Big]$

No. 42 : $\mathcal{K}\bar{R}(\,\multimap\,)$ $=$ $\mathcal{K}\Big[\;\multimap\; -\mathcal{K}(\,\multimap\,)\,\multimap\; -\mathcal{K}\bar{R}(\,\multimap\,)\,\multimap\;\Big]$

No. 43 : $\mathcal{K}\bar{R}(\,\multimap\,)$ $=$ $\mathcal{K}\Big[\;\multimap\; -\mathcal{K}(\,\multimap\,)\,\multimap\; -\mathcal{K}\bar{R}(\,\multimap\,)\,\multimap\;\Big]$

No. 44 : $\mathcal{K}\bar{R}(\,\multimap\,)$ $=$ $\mathcal{K}\Big[\;\multimap\; -2\mathcal{K}(\,\multimap\,)\,\multimap\; +\mathcal{K}(\,\multimap\,)\mathcal{K}(\,\multimap\,)\,\multimap\;\Big]$

No. 45 : $\mathcal{K}\bar{R}(\,\multimap\,)$ $=$ $\mathcal{K}\Big[\;\multimap\; -2\mathcal{K}(\,\multimap\,)\,\multimap\;$

$$-\mathcal{K}\bar{R}(\,\multimap\,)\,\multimap\; -\mathcal{K}(\,\multimap\,)\,\multimap\;$$

$$+2\mathcal{K}(\,\multimap\,)\mathcal{K}(\,\multimap\,)\,\multimap\; +\mathcal{K}\bar{R}(\,\multimap\,)\mathcal{K}(\,\multimap\,)\,\multimap\;$$

$$-\mathcal{K}\bar{R}(\,\multimap\,)\,\multimap\; +2\mathcal{K}(\,\multimap\,)\mathcal{K}\bar{R}(\,\multimap\,)\,\multimap\;$$

$$+\mathcal{K}\bar{R}(\,\multimap\,)\mathcal{K}\bar{R}(\,\multimap\,)\,\multimap\;\Big]$$

No. 46 : $\mathcal{K}\bar{R}(\,\multimap\,)$ $=$ $-\mathcal{K}\Big[\;\multimap\;\Big]\mathcal{K}\bar{R}\Big[\;\multimap\;\Big]$

No. 47 : $K\bar{R}^*(\,\diagram\,) = K\big[\,\diagram\, - K(\diagram)\,\diagram$

$\qquad\qquad\qquad -K\bar{R}(\,\diagram\,)\,\diagram\, - K\bar{R}(\,\diagram\,)\,\diagram$

$\qquad\qquad\qquad +(\,\diagram\,)_{IR}\big\{\,\diagram\, - K(\diagram)\,\diagram\, - K\bar{R}(\,\diagram\,)\,\diagram$

$\qquad\qquad\qquad\qquad -K\bar{R}(\,\diagram\,)\,\diagram\, - K\bar{R}(\,\diagram\,)\big\}\big]$

No. 48 : $K\bar{R}(\,\diagram\,) = -K\big[\,\diagram\,\big]K\bar{R}\big[\,\diagram\,\big]$

No. 49 : $K\bar{R}(\,\diagram\,) = K\big[\,\diagram\, - 2K(\diagram)\,\diagram\, - K\bar{R}(\,\diagram\,)\,\diagram$

$\qquad\qquad\qquad -K\bar{R}(\,\diagram\,)\,\diagram\, - K(\diagram)\,\diagram$

$\qquad\qquad\qquad +2K(\diagram)K(\diagram)\,\diagram\, + K(\diagram)K\bar{R}(\,\diagram\,)\,\diagram$

$\qquad\qquad\qquad +K(\diagram)K\bar{R}(\,\diagram\,)\,\diagram\,\big]$

No. 50 : $K\bar{R}(\,\diagram\,) = K\big[\,\diagram\, - 2K(\diagram)\,\diagram\, - K\bar{R}(\,\diagram\,)\,\diagram$

$\qquad\qquad\qquad -K\bar{R}(\,\diagram\,)\,\diagram\, - K\bar{R}(\,\diagram\,)\,\diagram\,\big]$

No. 51 : $K\bar{R}(\,\diagram\,) = K\big[\,\diagram\, - 2K(\diagram)\,\diagram\, - K\bar{R}(\,\diagram\,)\,\diagram$

$\qquad\qquad\qquad -K\bar{R}(\,\diagram\,)\,\diagram\, - K\bar{R}(\,\diagram\,)\,\diagram\,\big]$

No. 52 : $K\bar{R}(\,\diagram\,) = K\big[\,\diagram\, - 2K(\diagram)\,\diagram\, + K(\diagram)K(\diagram)\,\diagram$

$\qquad\qquad\qquad -2K\bar{R}(\,\diagram\,)\,\diagram\, + 2K\bar{R}(\,\diagram\,)K(\diagram)\,\diagram$

$\qquad\qquad\qquad +K\bar{R}(\,\diagram\,)K\bar{R}(\,\diagram\,)\,\diagram\,\big]$

No. 53 : $K\bar{R}^*(\,\diagram\,) = K\big[\,\diagram\, + (\,\diagram\,)_{IR}\big\{\,\diagram\, - K\bar{R}(\,\diagram\,)\big\}$

$\qquad\qquad\qquad +(\,\diagram\,)_{IR}\big\{\,\diagram\, - K\bar{R}(\,\diagram\,)\big\}$

$\qquad\qquad\qquad +(\,\diagram\,)_{IR}\big\{\,\diagram\, - K\bar{R}(\,\diagram\,)\big\}$

$\qquad\qquad\qquad +(\,\diagram\,)_{IR}\big\{\,\diagram\, - K\bar{R}(\,\diagram\,)\big\}\big]$

No. 54 : $K\bar{R}(\,\diagram\,) = K\big[\,\diagram\, - K(\diagram)\,\diagram\, - K\bar{R}(\,\diagram\,)\,\diagram$

$$-\mathcal{K}\bar{R}(\ \cdot\ )\ \cdot\ -\mathcal{K}\bar{R}(\ \cdot\ )\ \cdot\ \Big]$$

No. 55 : $\mathcal{K}\bar{R}(\ \cdot\ ) = \mathcal{K}\Big[\ \cdot\ -\mathcal{K}(\ \cdot\ )\ \cdot\ -\mathcal{K}(\ \cdot\ )\ \cdot$

$$+\mathcal{K}(\ \cdot\ )\mathcal{K}(\ \cdot\ )\ \cdot\ -\mathcal{K}\bar{R}(\ \cdot\ )\ \cdot$$

$$+\mathcal{K}\bar{R}(\ \cdot\ )\mathcal{K}(\ \cdot\ )\ \cdot\ -\mathcal{K}\bar{R}(\ \cdot\ )\ \cdot$$

$$+\mathcal{K}\bar{R}(\ \cdot\ )\mathcal{K}(\ \cdot\ )\ \cdot\ \Big]$$

No. 56 : $\mathcal{K}\bar{R}(\ \cdot\ ) = \mathcal{K}\Big[\ \cdot\ -\mathcal{K}(\ \cdot\ )\ \cdot\ -\mathcal{K}(\ \cdot\ )\ \cdot$

$$+\mathcal{K}(\ \cdot\ )\mathcal{K}(\ \cdot\ )\ \cdot$$

$$-\mathcal{K}\bar{R}(\ \cdot\ )\ \cdot\ +\mathcal{K}\bar{R}(\ \cdot\ )\mathcal{K}(\ \cdot\ )\ \cdot$$

$$-\mathcal{K}\bar{R}(\ \cdot\ )\ \cdot\ +\mathcal{K}\bar{R}(\ \cdot\ )\mathcal{K}(\ \cdot\ )\ \cdot\ \Big]$$

No. 57 : $\mathcal{K}\bar{R}(\ \cdot\ ) = \mathcal{K}\Big[\ \cdot\ -\mathcal{K}(\ \cdot\ )\ \cdot\ -\mathcal{K}(\ \cdot\ )\ \cdot$

$$+\mathcal{K}(\ \cdot\ )\mathcal{K}(\ \cdot\ )\ \cdot\ -\mathcal{K}\bar{R}(\ \cdot\ )\ \cdot$$

$$+\mathcal{K}\bar{R}(\ \cdot\ )\mathcal{K}(\ \cdot\ )\ \cdot\ -\mathcal{K}\bar{R}(\ \cdot\ )\ \cdot\ \Big]$$

No. 58 : $\mathcal{K}\bar{R}^{\star}(\ \cdot\ ) = \mathcal{K}\Big[\ \cdot\ -\mathcal{K}(\ \cdot\ )\ \cdot\ -\mathcal{K}(\ \cdot\ )\ \cdot$

$$+\mathcal{K}(\ \cdot\ )\mathcal{K}(\ \cdot\ )\ \cdot$$

$$-\mathcal{K}\bar{R}(\ \cdot\ )\ \cdot\ +\mathcal{K}\bar{R}(\ \cdot\ )\mathcal{K}(\ \cdot\ )\ \cdot$$

$$(\ \cdot\ )_{IR}\Big\{\ \cdot\ -\mathcal{K}(\ \cdot\ )\ \cdot\ -\mathcal{K}(\ \cdot\ )\ \cdot$$

$$+\mathcal{K}(\ \cdot\ )\mathcal{K}(\ \cdot\ )\ \cdot\ -\mathcal{K}\bar{R}(\ \cdot\ )\ \cdot$$

$$+\mathcal{K}\bar{R}(\ \cdot\ )\mathcal{K}(\ \cdot\ )\ \cdot\ -\mathcal{K}\bar{R}(\ \cdot\ )\Big\}\Big]$$

No. 59 : $\mathcal{K}\bar{R}(\ \cdot\ ) = \mathcal{K}\Big[\ \cdot\ -\mathcal{K}(\ \cdot\ )\ \cdot\ -\mathcal{K}(\ \cdot\ )\ \cdot$

$$+\mathcal{K}(\ \cdot\ )\mathcal{K}(\ \cdot\ )\ \cdot$$

$$-\mathcal{K}\bar{R}(\ \cdot\ )\ \cdot\ +\mathcal{K}\bar{R}(\ \cdot\ )\mathcal{K}(\ \cdot\ )\ \cdot$$

$$-\mathcal{K}\bar{R}(\ \cdot\ )\ \cdot\ -\mathcal{K}\bar{R}(\ \cdot\ )\ \cdot\ \Big]$$

No. 60 : $\mathcal{K}\bar{R}(\text{⬡}) = \mathcal{K}\Big[\text{⬡} -2\mathcal{K}(\text{o})\text{⬡} -\mathcal{K}(\text{o})\text{⬡}$

$+\mathcal{K}(\text{o})\mathcal{K}(\text{o})\text{⬡} +2\mathcal{K}(\text{o})\mathcal{K}(\text{o})\text{⬡}$

$-\mathcal{K}(\text{o})\mathcal{K}(\text{o})\mathcal{K}(\text{o})\text{⬡} -\mathcal{K}\bar{R}(\text{⬡})\text{⬡}$

$+\mathcal{K}\bar{R}(\text{⬡})\mathcal{K}(\text{o})\text{⬡} -\mathcal{K}\bar{R}(\text{⬡})\text{⬡}$

$+\mathcal{K}\bar{R}(\text{⬡})\mathcal{K}(\text{o})\text{o} \quad\Big]$

No. 61 : $\mathcal{K}\bar{R}^{\star}(\text{⬡}) = \mathcal{K}\Big[\text{⬡} -\mathcal{K}(\text{o})\text{⬡} -\mathcal{K}(\text{o})\text{⬡}$

$+\mathcal{K}(\text{o})\mathcal{K}(\text{o})\text{⬡}$

$-\mathcal{K}\bar{R}(\text{⬡})\text{⬡} +\mathcal{K}\bar{R}(\text{⬡})\mathcal{K}(\text{o})\text{o}$

$+(\text{⦙})_{IR}\Big\{\text{⬡} -\mathcal{K}(\text{o})\text{⬡} -\mathcal{K}(\text{o})\text{⬡}$

$+\mathcal{K}(\text{o})\mathcal{K}(\text{o})\text{⬡} -\mathcal{K}\bar{R}(\text{⬡})\text{⬡}$

$+\mathcal{K}\bar{R}(\text{⬡})\mathcal{K}(\text{o})\text{o} -\mathcal{K}\bar{R}(\text{⬡})\Big\}\Big]$

No. 62 : $\mathcal{K}\bar{R}^{\star}(\text{⬡}) = \mathcal{K}\Big[\text{⬡} -\mathcal{K}(\text{o})\text{⬡} -\mathcal{K}\bar{R}(\text{⬡})\text{⬡}$

$-\mathcal{K}\bar{R}(\text{⬡})\text{⬡} +\mathcal{K}(\text{o})\mathcal{K}\bar{R}(\text{⬡})\text{⬡}$

$+\mathcal{K}\bar{R}(\text{⬡})\mathcal{K}\bar{R}(\text{⬡})\text{o}$

$+(\text{⦙})_{IR}\Big\{\text{⬡} -\mathcal{K}(\text{o})\text{⬡}$

$-\mathcal{K}(\text{o})\text{⬡} +\mathcal{K}(\text{o})\mathcal{K}(\text{o})\text{⬡}$

$-\mathcal{K}\bar{R}(\text{⬡})\text{⬡} +\mathcal{K}\bar{R}(\text{⬡})\mathcal{K}(\text{o})\text{o}$

$-\mathcal{K}\bar{R}(\text{⬡})\Big\}\Big]$

No. 63 : $\mathcal{K}\bar{R}(\text{⬡}) = \mathcal{K}\Big[\text{⬡} -\mathcal{K}(\text{o})\text{⬡} -\mathcal{K}(\text{o})\text{⬡}$

$+\mathcal{K}(\text{o})\mathcal{K}(\text{o})\text{⬡} -\mathcal{K}\bar{R}(\text{⬡})\text{⬡}$

$+\mathcal{K}\bar{R}(\text{⬡})\mathcal{K}(\text{o})\text{⬡} -\mathcal{K}\bar{R}(\text{⬡})\text{o} \quad\Big]$

No. 64 : $\mathcal{K}\bar{R}(\text{⬡}) = \mathcal{K}\Big[\text{⬡} -\mathcal{K}(\text{o})\text{⬡} -\mathcal{K}(\text{o})\text{⬡}$

$$+ \mathcal{K}(\multimap\!\!\!-\!\!\!\circ)\mathcal{K}(\multimap\!\!\!-\!\!\!\circ) \, \langle\!\!\!\bigcirc\!\!\!\rangle$$

$$- \mathcal{K}\bar{R}(\langle\!\!\!\bigcirc\!\!\!\rangle)\!\!\!-\!\!\!\circ \quad - \mathcal{K}\bar{R}(\langle\!\!\!\bigcirc\!\!\!\rangle) \, \circ\!\!\!-\!\!\!\circ \quad]$$

No. 65 : $\mathcal{K}\bar{R}^{\star}(\langle\!\!\!\bigcirc\!\!\!\rangle) = \mathcal{K}\Big[\langle\!\!\!\bigcirc\!\!\!\rangle - \mathcal{K}(\multimap\!\!\!-\!\!\!\circ)\langle\!\!\!\bigcirc\!\!\!\rangle - \mathcal{K}(\multimap\!\!\!-\!\!\!\circ)\langle\!\!\!\bigcirc\!\!\!\rangle$

$$+ \mathcal{K}(\multimap\!\!\!-\!\!\!\circ)\mathcal{K}(\multimap\!\!\!-\!\!\!\circ)\!\!-\!\!\langle\!\!\!\bigcirc\!\!\!\rangle$$

$$- \mathcal{K}\bar{R}(\langle\!\!\!\bigcirc\!\!\!\rangle)\!\!-\!\!\langle\!\!\!\bigcirc\!\!\!\rangle + \mathcal{K}\bar{R}(\langle\!\!\!\bigcirc\!\!\!\rangle)\mathcal{K}(\multimap\!\!\!-\!\!\!\circ)\langle\!\!\!\bigcirc\!\!\!\rangle$$

$$+ \big(\,\substack{\textstyle|\\\textstyle\bullet}\,\big)_{IR}\Big\{ \langle\!\!\!\bigcirc\!\!\!\rangle - \mathcal{K}(\multimap\!\!\!\circ\!\!\!-\!\!\!)\langle\!\!\!\bigcirc\!\!\!\rangle - \mathcal{K}(\multimap\!\!\!-\!\!\!\circ)\,\langle\!\!\!\bigcirc\!\!\!\rangle$$

$$+ \mathcal{K}(\multimap\!\!\!-\!\!\!\circ)\mathcal{K}(\multimap\!\!\!-\!\!\!\circ)\langle\!\!\!\bigcirc\!\!\!\rangle - \mathcal{K}\bar{R}(\langle\!\!\!\bigcirc\!\!\!\rangle)\!\!-\!\!\langle\!\!\!\bigcirc\!\!\!\rangle$$

$$+ \mathcal{K}\bar{R}(\langle\!\!\!\bigcirc\!\!\!\rangle)\mathcal{K}(\multimap\!\!\!-\!\!\!\circ)\,\circ\!\!\!-\!\!\!\circ \quad - \mathcal{K}\bar{R}(\langle\!\!\!\bigcirc\!\!\!\rangle)\Big\}\Big]$$

No. 66 : $\mathcal{K}\bar{R}(\langle\!\!\!\bigcirc\!\!\!\rangle) = \mathcal{K}\Big[\langle\!\!\!\bigcirc\!\!\!\rangle - \mathcal{K}(\multimap\!\!\!-\!\!\!\circ)\langle\!\!\!\bigcirc\!\!\!\rangle$

$$- 2\mathcal{K}(\multimap\!\!\!-\!\!\!\circ)\langle\!\!\!\bigcirc\!\!\!\rangle + 2\mathcal{K}(\multimap\!\!\!-\!\!\!\circ)\mathcal{K}(\multimap\!\!\!-\!\!\!\circ)\langle\!\!\!\bigcirc\!\!\!\rangle$$

$$- \mathcal{K}(\multimap\!\!\!-\!\!\!\circ)\mathcal{K}(\multimap\!\!\!-\!\!\!\circ)\mathcal{K}(\multimap\!\!\!-\!\!\!\circ)\langle\!\!\!\bigcirc\!\!\!\rangle + \mathcal{K}(\multimap\!\!\!-\!\!\!\circ)\mathcal{K}(\multimap\!\!\!-\!\!\!\circ)\langle\!\!\!\bigcirc\!\!\!\rangle$$

$$- \mathcal{K}\bar{R}(\langle\!\!\!\bigcirc\!\!\!\rangle)\!\!-\!\!\langle\!\!\!\bigcirc\!\!\!\rangle + \mathcal{K}\bar{R}(\langle\!\!\!\bigcirc\!\!\!\rangle)\mathcal{K}(\multimap\!\!\!-\!\!\!\circ)\,\circ\!\!\!-\!\!\!\circ \quad]$$

No. 67 : $\mathcal{K}\bar{R}^{\star}(\langle\!\!\!\oplus\!\!\!\rangle) = \mathcal{K}\Big[\langle\!\!\!\oplus\!\!\!\rangle - \mathcal{K}(\multimap\!\!\!-\!\!\!\circ)\langle\!\!\!\oplus\!\!\!\rangle - \mathcal{K}(\multimap\!\!\!-\!\!\!\circ)\langle\!\!\!\oplus\!\!\!\rangle$

$$+ \mathcal{K}(\multimap\!\!\!-\!\!\!\circ)\mathcal{K}(\multimap\!\!\!-\!\!\!\circ)\langle\!\!\!\oplus\!\!\!\rangle$$

$$+ \big(\,\substack{\textstyle|\\\textstyle\circ}\,\big)_{IR}\Big\{ \langle\!\!\!\oplus\!\!\!\rangle - 2\mathcal{K}(\multimap\!\!\!-\!\!\!\circ)\langle\!\!\!\bigcirc\!\!\!\rangle$$

$$+ \mathcal{K}(\multimap\!\!\!-\!\!\!\circ)\mathcal{K}(\multimap\!\!\!-\!\!\!\circ)\langle\!\!\!\bigcirc\!\!\!\rangle - \mathcal{K}\bar{R}(\langle\!\!\!\bigcirc\!\!\!\rangle)\Big\}$$

$$+ \big(\langle\!\!\!\bigcirc\!\!\!\rangle\big)_{IR}\Big\{ \langle\!\!\!\bigcirc\!\!\!\rangle - 2\mathcal{K}(\multimap\!\!\!-\!\!\!\circ)\langle\!\!\!\bigcirc\!\!\!\rangle$$

$$+ \mathcal{K}(\multimap\!\!\!-\!\!\!\circ)\mathcal{K}(\multimap\!\!\!-\!\!\!\circ)\,\circ\!\!\!-\!\!\!\circ \quad - \mathcal{K}\bar{R}(\langle\!\!\!\bigcirc\!\!\!\rangle)\Big\}\Big]$$

No. 68 : $\mathcal{K}\bar{R}(\langle\!\!\!\oplus\!\!\!\rangle) = \mathcal{K}\Big[\langle\!\!\!\bigcirc\!\!\!\rangle - \mathcal{K}(\multimap\!\!\!-\!\!\!\circ)\langle\!\!\!\bigcirc\!\!\!\rangle - 2\mathcal{K}(\multimap\!\!\!-\!\!\!\circ)\langle\!\!\!\bigcirc\!\!\!\rangle$

$$+ 2\mathcal{K}(\multimap\!\!\!-\!\!\!\circ)\mathcal{K}(\multimap\!\!\!-\!\!\!\circ)\langle\!\!\!\bigcirc\!\!\!\rangle + \mathcal{K}(\multimap\!\!\!-\!\!\!\circ)\mathcal{K}(\multimap\!\!\!-\!\!\!\circ)\langle\!\!\!\bigcirc\!\!\!\rangle$$

$$- \mathcal{K}(\multimap\!\!\!-\!\!\!\circ)\mathcal{K}(\multimap\!\!\!-\!\!\!\circ)\mathcal{K}(\multimap\!\!\!-\!\!\!\circ)\langle\!\!\!\bigcirc\!\!\!\rangle - \mathcal{K}\bar{R}(\langle\!\!\!\bigcirc\!\!\!\rangle)\,\langle\!\!\!\bigcirc\!\!\!\rangle$$

$$+ \mathcal{K}\bar{R}(\langle\!\!\!\bigcirc\!\!\!\rangle)\mathcal{K}(\multimap\!\!\!-\!\!\!\circ)\,\circ\!\!\!-\!\!\!\circ \quad]$$

No. 69 : $\mathcal{K}\bar{R}(\!\!\bigcirc\!\!) = \mathcal{K}\Big[\, \bigcirc - \mathcal{K}(\!-\!\circ\!) \bigcirc - \mathcal{K}(\!-\!\circ\!-) \bigcirc$

$\qquad\qquad + \mathcal{K}(\!-\!\circ\!)\mathcal{K}(\!\circ\!-) \bigcirc$

$\qquad\qquad - \mathcal{K}\bar{R}(\!\bigcirc\!) \bigcirc + \mathcal{K}(\!-\!\circ\!)\mathcal{K}\bar{R}(\!\bigcirc\!) \bigcirc$

$\qquad\qquad - \mathcal{K}\bar{R}(\!\bigcirc\!) \circ \,\Big]$

No. 70 : $\mathcal{K}\bar{R}(\!\!\bigcirc\!\!) = \mathcal{K}\Big[\, \bigcirc - \mathcal{K}(\!-\!\circ\!) \bigcirc - \mathcal{K}(\!-\!\circ\!-) \bigcirc$

$\qquad\qquad + \mathcal{K}(\!-\!\circ\!)\mathcal{K}(\!-\!\circ\!) \bigcirc - \mathcal{K}\bar{R}(\!\bigcirc\!) \bigcirc$

$\qquad\qquad + \mathcal{K}(\!-\!\circ\!)\mathcal{K}\bar{R}(\!\bigcirc\!) \bigcirc - \mathcal{K}\bar{R}(\!\bigcirc\!) \circ \,\Big]$

No. 71 : $\mathcal{K}\bar{R}(\!\!\bigcirc\!\!) = \mathcal{K}\Big[\, \bigcirc - 2\mathcal{K}(\!-\!\circ\!) \bigcirc - \mathcal{K}\bar{R}(\!-\!\circ\!\circ\!-) \bigcirc$

$\qquad\qquad - \mathcal{K}(\!-\!\circ\!-) \bigcirc + 2\mathcal{K}(\!-\!\circ\!)\mathcal{K}(\!\circ\!-) \bigcirc$

$\qquad\qquad + \mathcal{K}(\!-\!\circ\!-)\mathcal{K}\bar{R}(\!-\!\circ\!\circ\!-) \bigcirc - \mathcal{K}\bar{R}(\!\bigcirc\!) \circ \,\Big]$

No. 72 : $\mathcal{K}\bar{R}(\!\circ\!\!\bigcirc\!\!) = -\mathcal{K}\Big[\, \circ\!-\, \Big]\mathcal{K}\bar{R}\Big[\!\bigcirc\!\Big]$

No. 73 : $\mathcal{K}\bar{R}^{\star}(\!\!\bigcirc\!\!) = \mathcal{K}\Big[\, \bigcirc - \mathcal{K}(\!\circ\!) \bigcirc$

$\qquad\qquad + 2\big(\big\downarrow\big)_{IR}\Big\{ \bigcirc - \mathcal{K}(\!-\!\circ\!-) \bigcirc - \mathcal{K}\bar{R}(\!\bigcirc\!)\Big\}$

$\qquad\qquad + \big(\big\downarrow\big)_{IR}\big(\big\downarrow\big)_{IR}\Big\{ \bigcirc - \mathcal{K}(\!\circ\!) \bigcirc - \mathcal{K}\bar{R}(\!\bigcirc\!)\Big\}$

$\qquad\qquad + (\!\circ\!\!\bigcirc\!\!)_{IR}\Big\{ -\mathcal{K}(\!-\!\circ\!-) \circ + \mathcal{K}(\!-\!\circ\!)\mathcal{K}(\!-\!\circ\!)\Big\}$

$\qquad\qquad + (\!\bigcirc\!\!)_{IR}\Big\{ \circ\!- - \mathcal{K}(\!-\!\circ\!)\Big\}\Big]$

No. 74 : $\mathcal{K}\bar{R}(\!\!\bigcirc\!\!) = \mathcal{K}\Big[\, \bigcirc - \mathcal{K}(\!-\!\circ\!) \bigcirc - \mathcal{K}\bar{R}(\!\bigcirc\!) \bigcirc$

$\qquad\qquad - \mathcal{K}\bar{R}(\!\bigcirc\!) \bigcirc - \mathcal{K}\bar{R}(\!\bigcirc\!) \circ \,\Big]$

No. 75 : $\mathcal{K}\bar{R}^{\star}(\!\!\bigcirc\!\!) = \mathcal{K}\Big[\, \bigcirc + \big(\big\downarrow\big)_{IR}\Big\{ \bigcirc - \mathcal{K}\bar{R}(\!\bigcirc\!)\Big\}$

$\qquad\qquad\qquad + \big(\big\downarrow\big)_{IR}\Big\{ \bigcirc - \mathcal{K}\bar{R}(\!\bigcirc\!)\Big\}$

$$+(\;)_{IR}\left\{\; -\mathcal{K}\bar{R}(\;)\right\}$$

$$+(\;)_{IR}\left\{\; -\mathcal{K}\bar{R}(\;)\right\}\Big]$$

No. 76 : $\mathcal{K}\bar{R}(\;) = \mathcal{K}\Big[\; -\mathcal{K}(\;)\; -\mathcal{K}(\;)\;$

$$+\mathcal{K}(\;)\mathcal{K}(\;)\; -\mathcal{K}\bar{R}(\;)\;$$

$$+\mathcal{K}\bar{R}(\;)\mathcal{K}(\;)\; -\mathcal{K}\bar{R}(\;)\;$$

$$+\mathcal{K}(\;)\mathcal{K}\bar{R}(\;)\; \Big]$$

No. 77 : $\mathcal{K}\bar{R}(\;) = \mathcal{K}\Big[\; -2\mathcal{K}(\;)\; -2\mathcal{K}\bar{R}(\;)\;$

$$+2\mathcal{K}\bar{R}(\;)\mathcal{K}(\;)\; +\mathcal{K}(\;)\mathcal{K}(\;)\;$$

$$+\mathcal{K}\bar{R}(\;)\mathcal{K}\bar{R}(\;)\; \Big]$$

No. 78 : $\mathcal{K}\bar{R}(\;) = -\mathcal{K}\Big[\;\Big]\mathcal{K}\bar{R}\Big[\;\Big]$

No. 79 : $\mathcal{K}\bar{R}^\star(\;) = \mathcal{K}\Big[\; -3\mathcal{K}(\;)\;$

$$+\mathcal{K}(\;)\mathcal{K}(\;)\; -2\mathcal{K}\bar{R}(\;)\;$$

$$+2(\;)_{IR}\left\{\; -3\mathcal{K}(\;)\;\right.$$

$$+\mathcal{K}(\;)\mathcal{K}(\;)\;$$

$$\left. -2\mathcal{K}\bar{R}(\;)\; -\mathcal{K}\bar{R}(\;)\right\}$$

$$+(\;)_{IR}(\;)_{IR}\left\{\; -3\mathcal{K}(\;)\;\right.$$

$$-2\mathcal{K}\bar{R}(\;)\;$$

$$\left.+\mathcal{K}(\;)\mathcal{K}(\;)\; -\mathcal{K}\bar{R}(\;)\right\}\Big]$$

No. 80 : $\mathcal{K}\bar{R}(\;) = \mathcal{K}\Big[\;\Big]\mathcal{K}\Big[\;\Big]\mathcal{K}\bar{R}\Big[\;\Big]$

No. 81 : $\mathcal{K}\bar{R}(\;) = -\mathcal{K}\Big[\;\Big]\mathcal{K}\bar{R}\Big[\;\Big]$

No. 82 : $\mathcal{K}\bar{R}(\;) = \mathcal{K}\Big[\; -3\mathcal{K}(\;)\; +\mathcal{K}(\;)\mathcal{K}(\;)\;$

$$-2\mathcal{K}\bar{R}(\text{⬭}) \cdot \text{⬭} - \mathcal{K}\bar{R}(\text{⬭}) \cdot \text{⬭}$$

$$-\mathcal{K}\bar{R}(\text{⬭}) \cdot \text{⬭} \;]$$

No. 83 : $\quad \mathcal{K}\bar{R}(\text{⬭}) = \mathcal{K}\left[\; \text{⬭} \;\right]\mathcal{K}\left[\; \text{⬭} \;\right]\mathcal{K}\bar{R}\left[\; \text{⬭} \;\right]$

No. 84 : $\quad \mathcal{K}\bar{R}(\text{⬭}) = -\mathcal{K}\left[\; \text{⬭} \;\right]\mathcal{K}\bar{R}\left[\; \text{⬭} \;\right]$

No. 85 : $\quad \mathcal{K}\bar{R}(\text{⬭}) = \mathcal{K}\left[\; \text{⬭} - 4\mathcal{K}(\text{⬭}) \; \text{⬭} + 3\mathcal{K}(\text{⬭})\mathcal{K}(\text{⬭}) \; \text{⬭}\right.$

$$-3\mathcal{K}\bar{R}(\text{⬭}) \cdot \text{⬭} + 2\mathcal{K}(\text{⬭})\mathcal{K}\bar{R}(\text{⬭}) \cdot \text{⬭}$$

$$\left. -2\mathcal{K}\bar{R}(\text{⬭}) \cdot \text{⬭} - \mathcal{K}\bar{R}(\text{⬭}) \cdot \text{⬭} \;\right]$$

No. 86 : $\quad \mathcal{K}\bar{R}(\text{⬭}) = \mathcal{K}\left[\; \text{⬭} - 3\mathcal{K}(\text{⬭}) \cdot \text{⬭} + \mathcal{K}(\text{⬭})\mathcal{K}(\text{⬭}) \; \text{⬭}\right.$

$$-2\mathcal{K}\bar{R}(\text{⬭}) \cdot \text{⬭} - \mathcal{K}\bar{R}(\text{⬭}) \cdot \text{⬭}$$

$$\left. -\mathcal{K}\bar{R}(\text{⬭}) \cdot \text{⬭} \;\right]$$

No. 87 : $\quad \mathcal{K}\bar{R}(\text{⬭}) = -\mathcal{K}\left[\; \text{⬭} \;\right]\mathcal{K}\bar{R}\left[\; \text{⬭} \;\right]$

No. 88 : $\quad \mathcal{K}\bar{R}^*(\text{⬭}) = \mathcal{K}\left[\; \text{⬭} - 2\mathcal{K}(\text{⬭}) \cdot \text{⬭} + \mathcal{K}(\text{⬭})\mathcal{K}(\text{⬭}) \; \text{⬭}\right.$

$$-\mathcal{K}\bar{R}(\text{⬭}) \cdot \text{⬭}$$

$$+\left(\; \vdots \;\right)_{IR}\left\{\; \text{⬭} - 2\mathcal{K}(\text{⬭}) \; \text{⬭}\right.$$

$$+\mathcal{K}(\text{⬭})\mathcal{K}(\text{⬭}) \; \text{⬭}$$

$$\left.\left. -\mathcal{K}\bar{R}(\text{⬭}) \cdot \text{⬭} - \mathcal{K}\bar{R}(\text{⬭})\right\}\right]$$

No. 89 : $\quad \mathcal{K}\bar{R}(\text{⬭}) = \mathcal{K}\left[\; \text{⬭} - 2\mathcal{K}(\text{⬭}) \cdot \text{⬭} - \mathcal{K}\bar{R}(\text{⬭}) \cdot \text{⬭}\right.$

$$-\mathcal{K}\bar{R}(\text{⬭}) \cdot \text{⬭} - \mathcal{K}(\text{⬭}) \cdot \text{⬭}$$

$$+2\mathcal{K}(\text{⬭})\mathcal{K}(\text{⬭}) \cdot \text{⬭} + \mathcal{K}(\text{⬭})\mathcal{K}\bar{R}(\text{⬭}) \cdot \text{⬭}$$

$$\left. +\mathcal{K}(\text{⬭})\mathcal{K}\bar{R}(\text{⬭}) \cdot \text{⬭} \;\right]$$

No. 90 : $\quad \mathcal{K}\bar{R}(\text{⬭}) = -\mathcal{K}\left[\; \text{⬭} \;\right]\mathcal{K}\bar{R}\left[\; \text{⬭} \;\right]$

No. 91 : $\mathcal{K}\bar{R}^\star(\text{⬡}) = \mathcal{K}\Big[\,\text{⬡} - 2\mathcal{K}(\text{⊸})\,\text{⬡}$

$+2\mathcal{K}(\text{⊸})\mathcal{K}(\text{⊸})\,\text{⬡} - \mathcal{K}\bar{R}(\text{⬡})\,\text{⬡}$

$+\mathcal{K}\bar{R}(\text{⬡})\mathcal{K}(\text{⊸})\,\text{⊸} \quad - \mathcal{K}(\text{⊸})\,\text{⬡}$

$+(\,\vdots\,)_{IR}\Big\{\text{⬡} - 2\mathcal{K}(\text{⊸})\,\text{⬡}$

$-\mathcal{K}\bar{R}(\text{⊸⊸})\,\text{⬡} \quad - \mathcal{K}(\text{⊸})\,\text{⬡}$

$+2\mathcal{K}(\text{⊸})\mathcal{K}(\text{⊸})\,\text{⬡}$

$+\mathcal{K}(\text{⊸})\mathcal{K}\bar{R}(\text{⊸⊸})\,\text{⊸} \quad - \mathcal{K}\bar{R}(\text{⬡})\Big\}\Big]$

No. 92 : $\mathcal{K}\bar{R}^\star(\text{⬡}) = \mathcal{K}\Big[\,\text{⬡} - 2\mathcal{K}(\text{⊸})\,\text{⬡} - \mathcal{K}\bar{R}(\text{⊸⊸})\,\text{⬡}$

$-\mathcal{K}\bar{R}(\text{⬡})\,\text{⬡} + 2\mathcal{K}(\text{⊸})\mathcal{K}\bar{R}(\text{⬡})\,\text{⊸}$

$+\mathcal{K}\bar{R}(\text{⊸⊸})\mathcal{K}\bar{R}(\text{⬡})\,\text{⊸}$

$+(\,\vdots\,)_{IR}\Big\{\text{⬡} - 2\mathcal{K}(\text{⊸})\,\text{⬡}$

$+2\mathcal{K}(\text{⊸})\mathcal{K}(\text{⊸})\,\text{⬡} - \mathcal{K}\bar{R}(\text{⊸⊸})\,\text{⬡}$

$+\mathcal{K}(\text{⊸})\mathcal{K}\bar{R}(\text{⊸⊸})\,\text{⊸} - \mathcal{K}(\text{⊸})\,\text{⬡}$

$-\mathcal{K}\bar{R}(\text{⬡})\Big\}\Big]$

No. 93 : $\mathcal{K}\bar{R}^\star(\text{⬡}) = \mathcal{K}\Big[\,\text{⬡} - \mathcal{K}(\text{⊸})\,\text{⬡} - \mathcal{K}\bar{R}(\text{⬡})\,\text{⬡}$

$+2\,(\,\vdots\,)_{IR}\Big\{\text{⬡} \quad - \mathcal{K}(\text{⊸})\,\text{⬡}$

$-\mathcal{K}\bar{R}(\text{⬡}\,)\,\text{⬡} - \mathcal{K}\bar{R}(\text{⬡})\Big\}$

$+(\,\vdots\,)_{IR}(\,\vdots\,)_{IR}\Big\{-\mathcal{K}\bar{R}(\text{⬡}\,)\,\text{⊸}$

$+\mathcal{K}\bar{R}(\text{⬡}\,)\mathcal{K}(\text{⊸})\Big\}$

$+(\text{⊸⊸})_{IR}\Big\{-\mathcal{K}(\text{⊸})\,\text{⊸} \quad +\mathcal{K}(\text{⊸})\mathcal{K}(\text{⊸})\Big\}$

$+(\text{⬡})_{IR}\Big\{\text{⊸} \quad - \mathcal{K}(\text{⊸})\Big\}\Big]$

No. 94 : $\mathcal{K}\bar{R}(\,\blacksquare\,) = -\mathcal{K}\big[\,\blacksquare\,\big]\,\mathcal{K}\bar{R}\big[\,\blacksquare\,\big]$

No. 95 : $\mathcal{K}\bar{R}^{\star}(\,\blacksquare\,) = \mathcal{K}\Big[\,\blacksquare\, - 2\mathcal{K}(\,\blacksquare\,)\,\blacksquare\, - \mathcal{K}\bar{R}(\,\blacksquare\,)\,\blacksquare\,$

$\qquad\qquad\qquad\quad -\mathcal{K}\bar{R}(\,\blacksquare\,)\,\blacksquare\,$

$\qquad\qquad\qquad\quad +\big(\,\blacksquare\,\big)_{IR}\big\{\,\blacksquare\, - 2\mathcal{K}(\,\blacksquare\,)\,\blacksquare\,$

$\qquad\qquad\qquad\qquad\quad -\mathcal{K}\bar{R}(\,\blacksquare\,)\,\blacksquare\,$

$\qquad\qquad\qquad\qquad\quad -\mathcal{K}\bar{R}(\,\blacksquare\,)\,\blacksquare\, - \mathcal{K}\bar{R}(\,\blacksquare\,)\big\}\Big]$

No. 96 : $\mathcal{K}\bar{R}(\,\blacksquare\,) = \mathcal{K}\Big[\,\blacksquare\, - 4\mathcal{K}(\,\blacksquare\,)\,\blacksquare\, + 4\mathcal{K}(\,\blacksquare\,)\mathcal{K}(\,\blacksquare\,)\,\blacksquare\,$

$\qquad\qquad\qquad\quad -2\mathcal{K}\bar{R}(\,\blacksquare\,)\,\blacksquare\, + 4\mathcal{K}\bar{R}(\,\blacksquare\,)\mathcal{K}(\,\blacksquare\,)\,\blacksquare\,$

$\qquad\qquad\qquad\quad +\mathcal{K}\bar{R}(\,\blacksquare\,)\mathcal{K}\bar{R}(\,\blacksquare\,)\,\blacksquare\,\Big]$

No. 97 : $\mathcal{K}\bar{R}(\,\blacksquare\,) = \mathcal{K}\Big[\,\blacksquare\, - 2\mathcal{K}(\,\blacksquare\,)\,\blacksquare\, - \partial_{k^2}\mathcal{K}\bar{R}(\,\blacksquare\,)\,\blacksquare\,$

$\qquad\qquad\qquad\quad -\mathcal{K}\bar{R}(\,\blacksquare\,)\,\blacksquare\,\Big]$

No. 98 : $\mathcal{K}\bar{R}(\,\blacksquare\,) = \mathcal{K}\Big[\,\blacksquare\, - \mathcal{K}(\,\blacksquare\,)\,\blacksquare\, - 2\mathcal{K}\bar{R}(\,\blacksquare\,)\,\blacksquare\,$

$\qquad\qquad\qquad\quad - \partial_{k^2}\mathcal{K}\bar{R}(\,\blacksquare\,)\,\blacksquare\,\Big]$

No. 99 : $\mathcal{K}\bar{R}(\,\blacksquare\,) = -\mathcal{K}\big[\,\blacksquare\,\big]\,\mathcal{K}\bar{R}\big[\,\blacksquare\,\big]$

No. 100 : $\mathcal{K}\bar{R}(\,\blacksquare\,) = \mathcal{K}\Big[\,\blacksquare\, - 3\mathcal{K}(\,\blacksquare\,)\,\blacksquare\, + \mathcal{K}(\,\blacksquare\,)\mathcal{K}(\,\blacksquare\,)\,\blacksquare\,$

$\qquad\qquad\qquad\quad -2\mathcal{K}\bar{R}(\,\blacksquare\,)\,\blacksquare\, - \partial_{k^2}\mathcal{K}\bar{R}(\,\blacksquare\,)\,\blacksquare\,\Big]$

No. 101 : $\mathcal{K}\bar{R}^{\star}(\,\blacksquare\,) = \mathcal{K}\Big[\,\blacksquare\, - \mathcal{K}(\,\blacksquare\,)\,\blacksquare\, - \mathcal{K}(\,\blacksquare\,)\,\blacksquare\,$

$\qquad\qquad\qquad\quad +\mathcal{K}(\,\blacksquare\,)\mathcal{K}(\,\blacksquare\,)\,\blacksquare\,$

$\qquad\qquad\qquad\quad -\mathcal{K}\bar{R}(\,\blacksquare\,)\,\blacksquare\, + \mathcal{K}(\,\blacksquare\,)\mathcal{K}\bar{R}(\,\blacksquare\,)\,\blacksquare\,$

$\qquad\qquad\qquad\quad +\big(\,\blacksquare\,\big)_{IR}\big\{\,\blacksquare\, - \mathcal{K}(\,\blacksquare\,)\,\blacksquare\, - \mathcal{K}\bar{R}(\,\blacksquare\,)\,\blacksquare\,$

$\qquad\qquad\qquad\qquad\quad +\mathcal{K}(\,\blacksquare\,)\mathcal{K}\bar{R}(\,\blacksquare\,)\,\blacksquare\, - \mathcal{K}\bar{R}(\,\blacksquare\,)\big\}$

$$+(\,⬡\,)_{IR}\Big\{ \,⬡\, - 2\mathcal{K}(\,⬡\,)\,⬡\,$$

$$+\mathcal{K}(\,⬡\,)\mathcal{K}(\,⬡\,)\,⬡\, - \mathcal{K}\bar{R}(\,⬡\,)\Big\}\Big]$$

No. 102 : $\mathcal{K}\bar{R}(\,⬡\,) = \mathcal{K}\Big[\,⬡\, - 2\mathcal{K}(\,⬡\,)\,⬡\, - 2\mathcal{K}\bar{R}(\,⬡\,)\,⬡\,$

$$+2\mathcal{K}\bar{R}(\,⬡\,)\mathcal{K}(\,⬡\,)\,⬡\,$$

$$+\mathcal{K}(\,⬡\,)\mathcal{K}(\,⬡\,)\,⬡\, - \mathcal{K}\bar{R}(\,⬡\,)\,⬡\,\Big]$$

No. 103 : $\mathcal{K}\bar{R}^{\star}(\,⬡\,) = \mathcal{K}\Big[\,⬡\, - 2\mathcal{K}(\,⬡\,)\,⬡\, + \mathcal{K}(\,⬡\,)\mathcal{K}(\,⬡\,)\,⬡\,$

$$+2(\,⬡\,)_{IR}\Big\{ \,⬡\, - \mathcal{K}(\,⬡\,)\,⬡\, - \mathcal{K}\bar{R}(\,⬡\,)\Big\}$$

$$+(\,⬡\,)_{IR}(\,⬡\,)_{IR}\Big\{ \,⬡\, - \mathcal{K}(\,⬡\,)\Big\}\Big]$$

No. 104 : $\mathcal{K}\bar{R}(\,⬡\,) = \mathcal{K}\Big[\,⬡\, - \mathcal{K}(\,⬡\,)\,⬡\, - 2\mathcal{K}(\,⬡\,)\,⬡\,$

$$+2\mathcal{K}(\,⬡\,)\mathcal{K}(\,⬡\,)\,⬡\, - \mathcal{K}\bar{R}(\,⬡\,)\,⬡\,$$

$$+\mathcal{K}\bar{R}(\,⬡\,)\mathcal{K}(\,⬡\,)\,⬡\, - \mathcal{K}\bar{R}(\,⬡\,)\,⬡\,$$

$$+2\mathcal{K}(\,⬡\,)\mathcal{K}\bar{R}(\,⬡\,)\,⬡\, - \mathcal{K}\bar{R}(\,⬡\,)\,⬡\,\Big]$$

No. 105 : $\mathcal{K}\bar{R}(\,⬡\,) = -\mathcal{K}\Big[\,⬡\,\Big]\mathcal{K}\bar{R}\Big[\,⬡\,\Big]$

No. 106 : $\mathcal{K}\bar{R}^{\star}(\,⬡\,) = \mathcal{K}\Big[\,⬡\, - 2\mathcal{K}\bar{R}(\,⬡\,)\,⬡\,$

$$+\mathcal{K}\bar{R}(\,⬡\,)\mathcal{K}\bar{R}(\,⬡\,)\,⬡\,$$

$$+2(\,⬡\,)_{IR}\Big\{ \,⬡\, - \mathcal{K}(\,⬡\,)\,⬡\, - \mathcal{K}\bar{R}(\,⬡\,)\,⬡\,$$

$$+\mathcal{K}\bar{R}(\,⬡\,)\mathcal{K}(\,⬡\,)\,⬡\, - \mathcal{K}\bar{R}(\,⬡\,)\Big\}$$

$$+(\,⬡\,)_{IR}(\,⬡\,)_{IR}\Big\{ \,⬡\, - 2\mathcal{K}(\,⬡\,)\,⬡\,$$

$$+\mathcal{K}(\,⬡\,)\mathcal{K}(\,⬡\,)\,⬡\, - \mathcal{K}\bar{R}(\,⬡\,)\Big\}\Big]$$

No. 107 : $\mathcal{K}\bar{R}^{\star}(\,⬡\,) = \mathcal{K}\Big[\,⬡\, - 2\mathcal{K}(\,⬡\,)\,⬡\,$

$$+ \left(\, \vert \, \right)_{IR} \left\{ \ominus\!\!\!\!\ominus - 2\mathcal{K}(\multimap) \ominus\!\!\!\bullet - \mathcal{K}\bar{R}(\ominus\!\!\!\!\ominus) \right\}$$

$$+ 2 \left(\mathfrak{M} \right)_{IR} \left\{ \triangle\!\!\!\!\!- - 2\mathcal{K}(\multimap) \triangle\!\!\!\!\!- - \mathcal{K}\bar{R}(\text{-}\!\circleddash) \right\}$$

$$+ \left(\textcircled{\triangle} \right)_{IR} \left\{ \multimap\!\!\!\circ - 2\mathcal{K}(\multimap) \multimap - \mathcal{K}\bar{R}(\multimap\!\!\!\circ\text{-}) \right\} \Big]$$

No. 108 : $\quad \mathcal{K}\bar{R}(\multimap\!\!\circledcirc) \;=\; -\mathcal{K}\left[\multimap \right] \mathcal{K}\bar{R}\left[\circledcirc\text{-} \right]$

No. 109 : $\quad \mathcal{K}\bar{R}^{\star}(\circledcirc\text{-}) \;=\; \mathcal{K}\Big[\circledcirc\text{-} - \mathcal{K}(\multimap)\circledcirc\text{-} - \mathcal{K}\bar{R}(\circledcirc\text{-})\multimap$

$$- 2\mathcal{K}\bar{R}(\circleddash\text{-})\circleddash$$

$$+ \left(\, \vert \, \right)_{IR} \left\{ \circledcirc\text{-} - \mathcal{K}(\multimap)\circleddash - 2\mathcal{K}\bar{R}(\circleddash)\circleddash \right.$$

$$\left. - \mathcal{K}\bar{R}(\text{-}\!\circleddash\text{-})\multimap - \mathcal{K}\bar{R}(\circledcirc\text{-}) \right\} \Big]$$

No. 110 : $\quad \mathcal{K}\bar{R}(\mathbb{\infty}\!\!\circledcirc) \;=\; -\mathcal{K}\bar{R}\left[\circleddash \right] \mathcal{K}\bar{R}\left[\circledcirc \right]$

No. 111: $\quad \mathcal{K}\bar{R}^{\star}(\textcircled{\triangledown}) \;=\; \mathcal{K}\Big[\textcircled{\triangledown} + \left(\, \vert \, \right)_{IR} \left\{ \circledcirc\text{-} - \mathcal{K}\bar{R}(\text{-}\!\textcircled{\triangledown}\text{-}) \right\}$

$$+ 2 \left(\mathfrak{M} \right)_{IR} \left\{ \circleddash - \mathcal{K}\bar{R}(\text{-}\!\textcircled{\triangleleft}\text{-}) \right\}$$

$$+ \left(\mathfrak{M} \right)_{IR} \left\{ \circleddash - \mathcal{K}\bar{R}(\text{-}\!\circleddash) \right\}$$

$$+ \left(\textcircled{\triangle} \right)_{IR} \left\{ \multimap - \mathcal{K}\bar{R}(\multimap) \right\} \Big]$$

No. 112: $\quad \mathcal{K}\bar{R}(\text{-}\!\circleddash\!\circ) \;=\; \mathcal{K}\Big[\circleddash\!\circ - \mathcal{K}(\multimap) \circleddash - \mathcal{K}(\multimap) \circleddash$

$$+ \mathcal{K}(\multimap)\mathcal{K}(\multimap) \circleddash$$

$$- 2\mathcal{K}\bar{R}(\circleddash) \circleddash + 2\mathcal{K}(\multimap)\mathcal{K}\bar{R}(\circleddash)\multimap$$

$$- \mathcal{K}\bar{R}(\text{-}\!\circleddash\text{-}) \circleddash + \mathcal{K}(\multimap)\mathcal{K}\bar{R}(\text{-}\!\circleddash\text{-})\multimap \Big]$$

No. 113: $\quad \mathcal{K}\bar{R}(\textcircled{\circleddash}) \;=\; \mathcal{K}\Big[\circleddash - \mathcal{K}(\multimap) \circleddash - 2\mathcal{K}\bar{R}(\circleddash) \circleddash$

$$- \mathcal{K}\bar{R}(\text{-}\!\circleddash\text{-}) \circleddash - \mathcal{K}\bar{R}(\textcircled{\circleddash}) \multimap \Big]$$

No. 114: $\quad \mathcal{K}\bar{R}(\multimap\!\!\circledcirc) \;=\; -\mathcal{K}\bar{R}\left[\circleddash \right] \mathcal{K}\bar{R}\left[\circleddash \right]$

No. 115 : $\mathcal{K}\bar{R}(\;\diagdown\;) = \mathcal{K}\Big[\;\diagdown\; - \mathcal{K}(\multimap)\;\diagdown\; - 2\mathcal{K}(\multimap)\;\diagdown\;$

$$+ 2\mathcal{K}(\multimap)\mathcal{K}(\multimap)\;\diagdown\;$$

$$-\,\partial_{k^2}\mathcal{K}\bar{R}(\multimap)\multimap + \mathcal{K}(\multimap)\,\partial_{k^2}\mathcal{K}\bar{R}(\multimap)\multimap\Big]$$

No. 116 : $\mathcal{K}\bar{R}^\star(\multimap) = \mathcal{K}\Big[\;\diagdown\; - \mathcal{K}(\multimap)\;\diagdown\; - \mathcal{K}\bar{R}(\multimap)\;\diagdown\;$

$$+\,(\;\vdots\;)_{IR}\Big\{2\varepsilon\,\diagdown\; + 2\varepsilon\,\diagdown\; + 8\,\diagdown\;$$

$$-\,2\varepsilon\,\mathcal{K}(\multimap)\,\diagdown\; - 2\varepsilon\,\mathcal{K}\bar{R}(\multimap)\,\diagdown\;$$

$$-\,2D\,\partial_{k^2}\mathcal{K}\bar{R}(\multimap)\Big\}$$

$$+\,(\;\diagdown\;)_{IR}\Big\{\diagdown\; - \mathcal{K}\bar{R}(\multimap)\Big\}$$

$$+\,(\;\diagdown\;)_{IR}\Big\{-\mathcal{K}(\multimap)\,\multimap + \mathcal{K}(\multimap)\mathcal{K}(\multimap)\Big\}$$

$$+\,(\;\diagdown\;)_{IR}\Big\{\multimap - \mathcal{K}(\multimap)\Big\}\Big]$$

No. 117 : $\mathcal{K}\bar{R}(\multimap\!\multimap\!\multimap) = \mathcal{K}\big[\multimap\big]\mathcal{K}\big[\multimap\big]\mathcal{K}\big[\multimap\big]\mathcal{K}\big[\multimap\big]\mathcal{K}\big[\multimap\big]$

No. 118 : $\mathcal{K}\bar{R}(\multimap) = \mathcal{K}\Big[\multimap - 4\mathcal{K}(\multimap)\,\multimap + 3\mathcal{K}(\multimap)\mathcal{K}(\multimap)\,\multimap$

$$-\,3\mathcal{K}\bar{R}(\multimap)\,\multimap + 2\mathcal{K}\bar{R}(\multimap)\mathcal{K}(\multimap)\,\multimap$$

$$-\,2\mathcal{K}\bar{R}(\multimap)\,\multimap - \mathcal{K}\bar{R}(\multimap)\,\multimap\;\Big]$$

No. 119 : $\mathcal{K}\bar{R}(\multimap) = -\mathcal{K}\big[\multimap\big]\mathcal{K}\big[\multimap\big]\mathcal{K}\big[\multimap\big]\mathcal{K}\bar{R}\big[\multimap\big]$

No. 120 : $\mathcal{K}\bar{R}(\multimap) = \mathcal{K}\big[\multimap\big]\mathcal{K}\big[\multimap\big]\mathcal{K}\bar{R}\big[\multimap\big]$

No. 121 : $\mathcal{K}\bar{R}(\multimap) = \mathcal{K}\big[\multimap\big]\mathcal{K}\big[\multimap\big]\mathcal{K}\bar{R}\big[\multimap\big]$

No. 122 : $\mathcal{K}\bar{R}(\multimap) = \mathcal{K}\big[\multimap\big]\mathcal{K}\bar{R}\big[\multimap\big]\mathcal{K}\bar{R}\big[\multimap\big]$

No. 123 : $\mathcal{K}\bar{R}(\multimap) = \mathcal{K}\Big[\multimap - \mathcal{K}(\multimap)\,\multimap - \mathcal{K}(\multimap)\,\multimap$

$$+\,\mathcal{K}(\multimap)\mathcal{K}(\multimap)\,\multimap - 2\mathcal{K}\bar{R}(\multimap)\,\multimap$$

$$+\,\mathcal{K}(\multimap)\mathcal{K}\bar{R}(\multimap)\,\multimap - \mathcal{K}\bar{R}(\multimap)\,\multimap$$

$$-\,\mathcal{K}\bar{R}(\multimap)\,\multimap\;\Big]$$

No. 124 : $\mathcal{K}\bar{R}(\multimap) = -\mathcal{K}\bar{R}\big[\multimap\big]\mathcal{K}\bar{R}\big[\multimap\big]$

A.2 Tadpole-Free Quadratically Divergent Diagrams for Z_ϕ

$$\underline{2-\ \text{Loop}\ \sim g^2}$$

No. 1: $\quad \mathcal{K}\bar{R}(\ominus) \;=\; \mathcal{K}\big[\ \ominus\ \big]$

$$\underline{3-\ \text{Loop}\ \sim g^3}$$

No. 1: $\quad \mathcal{K}\bar{R}(\ominus\!\!\ominus) \;=\; \mathcal{K}\big[\ \ominus\!\!\ominus\ -\ 2\mathcal{K}(\ominus)\,\ominus\ \big]$

$$\underline{4-\ \text{Loop}\ \sim g^4}$$

No. 1: $\quad \mathcal{K}\bar{R}(\ominus\!\!\ominus\!\!\ominus) \;=\; \mathcal{K}\big[\ \ominus\!\!\ominus\!\!\ominus\ -\ 3\mathcal{K}(\ominus)\,\ominus\!\!\ominus\ +\ \mathcal{K}(\ominus)\mathcal{K}(\ominus)\,\ominus$
$$-\ 2\mathcal{K}\bar{R}(\ominus\!\!\ominus)\,\ominus\ \big]$$

No. 2: $\quad \mathcal{K}\bar{R}(\ominus) \;=\; \mathcal{K}\big[\ \ominus\ -\ \partial_{k^2}\mathcal{K}(\ominus)\,\ominus\ \big]$

No. 3: $\quad \mathcal{K}\bar{R}(\ominus) \;=\; \mathcal{K}\big[\ \ominus\ -\ \mathcal{K}(\ominus)\,\ominus\ -\ 2\mathcal{K}\bar{R}(\ominus)\,\ominus\ \big]$

No. 4: $\quad \mathcal{K}\bar{R}(\ominus) \;=\; \mathcal{K}\big[\ \ominus\ -\ 2\mathcal{K}(\ominus)\,\ominus\ +\ \mathcal{K}(\ominus)\mathcal{K}(\ominus)\,\ominus$
$$-\ 2\mathcal{K}\bar{R}(\ominus)\,\ominus\ \big]$$

$$\underline{5-\ \text{Loop}\ \sim g^5}$$

No. 1: $\quad \mathcal{K}\bar{R}(\ominus\!\!\ominus\!\!\ominus\!\!\ominus) \;=\; \mathcal{K}\big[\ \ominus\!\!\ominus\!\!\ominus\!\!\ominus\ -\ 4\mathcal{K}(\ominus)\,\ominus\!\!\ominus\!\!\ominus\ +\ 3\mathcal{K}(\ominus)\mathcal{K}(\ominus)\,\ominus\!\!\ominus$
$$-\ 3\mathcal{K}\bar{R}(\ominus\!\!\ominus)\,\ominus\!\!\ominus\ +\ 2\mathcal{K}(\ominus)\mathcal{K}\bar{R}(\ominus\!\!\ominus)\,\ominus$$
$$-\ 2\mathcal{K}\bar{R}(\ominus\!\!\ominus\!\!\ominus)\,\ominus\ \big]$$

No. 2: $\quad \mathcal{K}\bar{R}(\ominus) \;=\; \mathcal{K}\big[\ \ominus\ -\ \mathcal{K}(\ominus)\,\ominus\ -\ \partial_{k^2}\mathcal{K}(\ominus)\,\ominus\!\!\ominus$
$$+\ \mathcal{K}(\ominus)\,\partial_{k^2}\mathcal{K}(\ominus)\,\ominus\ -\ \mathcal{K}\bar{R}(\ominus)\,\ominus\ \big]$$

No. 3: $\quad \mathcal{K}\bar{R}(\ominus) \;=\; \mathcal{K}\big[\ \ominus\ -\ 2\mathcal{K}(\ominus)\,\ominus\ -\ \partial_{k^2}\mathcal{K}(\ominus)\,\ominus$
$$+\ 2\mathcal{K}(\ominus)\,\partial_{k^2}\mathcal{K}(\ominus)\,\ominus\ \big]$$

No. 4 : $\mathcal{K}\bar{R}(\ominus) = \mathcal{K}\Big[\ominus - 2\mathcal{K}(\multimap)\,\ominus - \partial_{k^2}\mathcal{K}\bar{R}(\multimap)\,\ominus\Big]$

No. 5 : $\mathcal{K}\bar{R}(\ominus) = \mathcal{K}\Big[\ominus - \mathcal{K}(\multimap)\ominus - \mathcal{K}(\multimap)\ominus$
$\qquad\qquad\qquad + \mathcal{K}(\multimap)\mathcal{K}(\multimap)\ominus - 2\mathcal{K}\bar{R}(\ominus)\,\ominus$
$\qquad\qquad\qquad + 2\mathcal{K}(\multimap)\mathcal{K}\bar{R}(\ominus)\,\ominus - \mathcal{K}\bar{R}(\ominus)\,\ominus\Big]$

No. 6 : $\partial_{k^2}\mathcal{K}\bar{R}(\ominus) = \mathcal{K}\Big[\dfrac{1}{4-\varepsilon}\big\{-\varepsilon\mathcal{K}\bar{R}(\ominus) - \varepsilon\mathcal{K}\bar{R}(\ominus) + 4\mathcal{K}\bar{R}(\ominus)\big\}\Big]$
$\qquad\qquad\qquad = \mathcal{K}\Big[\dfrac{1}{4-\varepsilon}\big\{-\varepsilon\mathcal{K}\bar{R}(\text{No. }60) - \varepsilon\mathcal{K}\bar{R}(\text{No. }50)$
$\qquad\qquad\qquad\qquad + 4[\ominus - 2\mathcal{K}(\multimap)\,\ominus$
$\qquad\qquad\qquad\qquad - \mathcal{K}\bar{R}(\multimap)\,\ominus]\big\}\Big]$

No. 7 : $\partial_{k^2}\mathcal{K}\bar{R}(\ominus) = \mathcal{K}\Big[\dfrac{-\varepsilon}{4-\varepsilon}\mathcal{K}\bar{R}^\star(\ominus)\Big] = \mathcal{K}\Big[\dfrac{-\varepsilon}{4-\varepsilon}\mathcal{K}\bar{R}^\star(\text{No. }107)\Big]$

No. 8 : $\partial_{k^2}\mathcal{K}\bar{R}(\ominus) = \mathcal{K}\Big[\dfrac{-\varepsilon}{4-\varepsilon}\mathcal{K}\bar{R}^\star(\ominus)\Big] = \mathcal{K}\Big[\dfrac{-\varepsilon}{4-\varepsilon}\mathcal{K}\bar{R}(\text{No. }111)\Big]$

No. 9 : $\partial_{k^2}\mathcal{K}\bar{R}(\ominus) = \mathcal{K}\Big[\dfrac{1}{4-\varepsilon}\big\{-2\varepsilon\mathcal{K}\bar{R}(\ominus) + 4\mathcal{K}\bar{R}(\ominus)\big\}\Big]$
$\qquad\qquad\qquad = \mathcal{K}\Big[\dfrac{1}{4-\varepsilon}\big\{-2\varepsilon\mathcal{K}\bar{R}(\text{No. }56)$
$\qquad\qquad\qquad\qquad + 4[\ominus - 2\mathcal{K}(\multimap)\,\ominus$
$\qquad\qquad\qquad\qquad + \mathcal{K}(\multimap)\mathcal{K}(\multimap)\,\ominus]\big\}\Big]$

No. 10 : $\partial_{k^2}\mathcal{K}\bar{R}(\ominus) = \mathcal{K}\Big[\dfrac{1}{4-\varepsilon}\big\{-\varepsilon\mathcal{K}\bar{R}(\ominus) - \varepsilon\mathcal{K}\bar{R}(\ominus) + 4\mathcal{K}\bar{R}(\ominus)\big\}\Big]$
$\qquad\qquad\qquad = \mathcal{K}\Big[\dfrac{1}{4-\varepsilon}\big\{-\varepsilon\mathcal{K}\bar{R}(\text{No. }59) - \varepsilon\mathcal{K}\bar{R}(\text{No. }46)$
$\qquad\qquad\qquad\qquad + 4[\ominus - \mathcal{K}(\multimap)\,\ominus - \mathcal{K}(\multimap)\,\ominus$
$\qquad\qquad\qquad\qquad + \mathcal{K}(\multimap)\mathcal{K}(\multimap)\,\ominus]\big\}\Big]$

No. 11 : $\partial_{k^2}\mathcal{K}\bar{R}(\oplus) = \mathcal{K}\Big[\dfrac{2}{4-\varepsilon}\big\{-\varepsilon\mathcal{K}\bar{R}^\star(\oplus) + 2\mathcal{K}(\oplus)\big\}\Big]$
$\qquad\qquad\qquad = \mathcal{K}\Big[\dfrac{2}{4-\varepsilon}\big\{-\varepsilon\mathcal{K}\bar{R}^\star(\text{No. }36) + 2\mathcal{K}(\oplus)\big\}\Big]$

A.3   Calculation of IR-Counterterms

The counterterms are calculated with the R^*-operation shown diagrammatically in the first subsection. The results follow in the next subsection.

A.3.1   Determination of IR-Counterterms by R^*-Operation

$$\mathrm{IR}_1 : (\,\vcenter{\hbox{⦶}}\,)_{IR} \;=\; -k^2\mathcal{K}\Big[\;\bigcirc\!\!-\;\Big]$$

$$\mathrm{IR}_2 : (\,\vcenter{\hbox{⧲}}\,)_{IR} \;=\; -k^2\mathcal{K}\Big[\;\bigcirc\!\!\bigcirc\; -\;\mathcal{K}(\,\oslash\,)\,\bigcirc\!\!-\;\Big]$$

$$\mathrm{IR}_3 : (\,\vcenter{\hbox{⦶}}\,)_{IR} \;=\; -\frac{k^6}{8}\mathcal{K}\Big[\;\bigcirc\!\!\bigcirc\; +\,(\,\vcenter{\hbox{⦶}}\,)_{IR}\;\frown\;\Big]$$

$$\mathrm{IR}_4 : (\,\vcenter{\hbox{⧩}}\,)_{IR} \;=\; -\frac{k^4}{2}\mathcal{K}\Big[\;\oplus\; +\,(\,\vcenter{\hbox{⦶}}\,)_{IR}\,\bigcirc\!\!-\;\Big]$$

$$\mathrm{IR}_5 : (\,\vcenter{\hbox{⊕}}\,)_{IR} \;=\; -k^2\mathcal{K}\Big[\;\oplus\; +\,(\,\vcenter{\hbox{⦶}}\,)_{IR}\oplus\; +\,2(\,\vcenter{\hbox{⧩}}\,)_{IR}\;\leftrightsquigarrow\;\Big]$$

$$\mathrm{IR}_{6a} : (\,\vcenter{\hbox{⫝̸}}\,)_{IR} \;=\; -k^4\mathcal{K}\Big[\;\boxtimes\; +\,(\,\vcenter{\hbox{⦶}}\,)_{IR}\,\boxtimes\; +\,2(\,\vcenter{\hbox{⧲}}\,)_{IR}\;\curvearrowright\;\Big]$$

$$\mathrm{IR}_{6b} : (\,\vcenter{\hbox{⫝̸}}\,)_{IR} \;=\; -k^4\mathcal{K}\Big[\;\boxtimes\; +\,(\,\vcenter{\hbox{⦶}}\,)_{IR}\,\boxtimes\; +\,(\,\vcenter{\hbox{⧩}}\,)_{IR}\,\curvearrowright\;\Big]$$

$$\mathrm{IR}_{6c} : (\,\vcenter{\hbox{⫝̸}}\,)_{IR} \;=\; -k^4\mathcal{K}\Big[\;\boxtimes\; +\,(\,\vcenter{\hbox{⦶}}\,)_{IR}\,\boxtimes\; +\,(\,\vcenter{\hbox{⧲}}\,)_{IR}\,\curvearrowright\;\Big]$$

$$\mathrm{IR}_{7a} : (\,\vcenter{\hbox{⟁}}\,)_{IR} \;=\; -k^2\mathcal{K}\Big[\;\boxtimes\; +\,(\,\vcenter{\hbox{⦶}}\,)_{IR}\,\boxtimes\; +\,2(\,\vcenter{\hbox{⧩}}\,)_{IR}\;\curvearrowright$$
$$+\,(\,\vcenter{\hbox{⫝̸}}\,)_{IR}\;\leftrightsquigarrow\;\Big]$$

$$\mathrm{IR}_{7b} : (\,\vcenter{\hbox{⟁}}\,)_{IR} \;=\; -k^2\mathcal{K}\Big[\;\boxtimes\; +\,(\,\vcenter{\hbox{⦶}}\,)_{IR}\,\boxtimes\; +\,(\,\vcenter{\hbox{⧩}}\,)_{IR}\,\curvearrowright$$
$$+\,(\,\vcenter{\hbox{⫝̸}}\,)_{IR}\;\leftrightsquigarrow\;\Big]$$

$$\mathrm{IR}_{7c} : (\,\vcenter{\hbox{⟁}}\,)_{IR} \;=\; -k^2\mathcal{K}\Big[\;\boxtimes\; +\,(\,\vcenter{\hbox{⦶}}\,)_{IR}\,\boxtimes\; +\,(\,\vcenter{\hbox{⧲}}\,)_{IR}\,\curvearrowright$$
$$+\,(\,\vcenter{\hbox{⫝̸}}\,)_{IR}\;\leftrightsquigarrow\;\Big]$$

$$\mathrm{IR}_{7d} : (\,\vcenter{\hbox{⟁}}\,)_{IR} \;=\; -k^2\mathcal{K}\Big[\;\boxtimes\; +\,(\,\vcenter{\hbox{⦶}}\,)_{IR}\,\curvearrowright\;\Big]$$

$$
\mathrm{IR_8} \ : \ (\!\langle\!\mid\!\mid\!\rangle\!)_{IR} \ = \ -k^2\mathcal{K}\left[\ \langle\!\mid\!\mid\!\rangle\!-\ \right]
$$

$$
\mathrm{IR_9} \ : \ (\!\langle\!\hat{0}\!\rangle\!)_{IR} \ = \ -k^4\mathcal{K}\Big[\ \bigotimes\ -\ \mathcal{K}(\!\!-\!\!\bigcirc\!\!-\!)\ \langle\!\langle\!\bigcirc\!\!-\ \ -\mathcal{K}\bar{R}(\!\!-\!\!\bigcirc\!\!-\!)\ \cdots
$$

$$
+\left(\ \substack{\circ\\\circ}\ \right)_{IR}\Big\{2\varepsilon\big[\ \cdots\!\bigcirc\ -\mathcal{K}(\!\!-\!\!\bigcirc\!\!-\!)\ \bullet\!\!-\!\!\bullet\!\!-\!\!\bigcirc
$$

$$
-\mathcal{K}\bar{R}(\!\!\bigcirc\!\!-\!)\ \bullet\!\!-\!\!\bullet\!\!-\!\!\bullet\ \big]
$$

$$
+2\varepsilon\ \cdot\!\bigcirc\ +8\ \cdot\!\!-\!\!\bigcirc\!\!\cdot\ \Big\}\Big]
$$

$$
\mathrm{IR_{10a}}: \ (\!\langle\!\bigcirc\!\bigcirc\!\rangle\!)_{IR} \ = \ -k^2\mathcal{K}\Big[\!-\!\bigcirc\!\bigcirc\!-\ -\mathcal{K}(\!\!-\!\!\bigcirc\!\!-\!)\!-\!\bigcirc\!-
$$

$$
+2\left(\ \substack{\circ\\\circ}\ \right)_{IR}\Big\{\!-\!\bigcirc\!\!-\!\!\bullet\ -\mathcal{K}(\!\!-\!\!\bigcirc\!\!-\!)\ \bullet\!\!-\!\!\smile\ \Big\}
$$

$$
+\left(\ \substack{\circ\\\circ}\ \right)_{IR}\left(\ \substack{\circ\\\circ}\ \right)_{IR}\Big\{-\mathcal{K}(\!\!-\!\!\bigcirc\!\!-\!)\ \smile\ \Big\}\Big]
$$

$$
\mathrm{IR_{10b}}: \ (\!\langle\!\bigcirc\!\bigcirc\!\rangle\!)_{IR} \ = \ -k^2\mathcal{K}\Big[\!-\!\bigcirc\!\bigcirc\!-\ +2\left(\ \substack{\circ\\\circ}\ \right)_{IR}\!-\!\bigcirc\!-
$$

$$
+\left(\ \substack{\circ\\\circ}\ \right)_{IR}\left(\ \substack{\circ\\\circ}\ \right)_{IR}\ \smile\ \Big]
$$

$$
\mathrm{IR_{10c}}: \ (\!\langle\!\bigcirc\!\bigcirc\!\rangle\!)_{IR} \ = \ -k^2\mathcal{K}\Big[\!-\!\bigcirc\!\bigcirc\!-\ -\mathcal{K}(\!\!-\!\!\bigcirc\!\!-\!)\!-\!\bigcirc\!-
$$

$$
+\left(\ \substack{\circ\\\circ}\ \right)_{IR}\Big\{2\varepsilon\ -\!\bigcirc\!-\ -2\varepsilon\,\mathcal{K}(\!\!-\!\!\bigcirc\!\!-\!)\ \smile\ \Big\}\Big]
$$

$$
\mathrm{IR_{11a}}: \ (\!\langle\!\hat{\bigcirc}\!\rangle\!)_{IR} \ = \ -k^2\mathcal{K}\Big[\ \bigotimes\ -\mathcal{K}(\!\!-\!\!\bigcirc\!\!-\!)\ -\!\bigcirc\!-\ -\mathcal{K}\bar{R}(\!\!-\!\!\bigcirc\!\!-\!)\ -\!\bigcirc\!-
$$

$$
+2\left(\ \substack{\circ\\\circ}\ \right)_{IR}\Big\{\!-\!\bigcirc\!-\ -\mathcal{K}(\!\!-\!\!\bigcirc\!\!-\!)\ -\!\bigcirc\!-\ -\mathcal{K}\bar{R}(\!\!-\!\!\bigcirc\!\!-\!)\ -\!\smile\ \Big\}
$$

$$
-\left(\ \substack{\circ\\\circ}\ \right)_{IR}\left(\ \substack{\circ\\\circ}\ \right)_{IR}\mathcal{K}\bar{R}(\!\!-\!\!\bigcirc\!\!-\!)\ \smile
$$

$$
-(\!\langle\!\bigcirc\!\bigcirc\!\rangle\!)_{IR}\,\mathcal{K}(\!\!-\!\!\bigcirc\!\!-\!)\ \smile\ \Big]
$$

$$
\mathrm{IR_{11b}}: \ (\!\langle\!\hat{0}\!\rangle\!)_{IR} \ = \ -k^2\mathcal{K}\Big[\ \bigotimes\ -\mathcal{K}(\!\!-\!\!\bigcirc\!\!-\!)\ -\!\bigcirc\!-
$$

$$
+2\left(\ \substack{\circ\\\circ}\ \right)_{IR}\Big\{\!-\!\bigcirc\!-\ -\mathcal{K}(\!\!-\!\!\bigcirc\!\!-\!)\ -\!\bigcirc\!-\ \Big\}
$$

$$
+\left(\ \substack{\circ\\\circ}\ \right)_{IR}\left(\ \substack{\circ\\\circ}\ \right)_{IR}\Big\{\!-\!\bigcirc\!-\ -\mathcal{K}(\!\!-\!\!\bigcirc\!\!-\!)\ -\!\bigcirc\!-\ \Big\}
$$

$$- (\,\text{⊗⊃}\,)_{IR}\, \mathcal{K}(\text{─O─})\; \text{⌣} \;\Big]$$

$$\mathrm{IR}_{11c}: (\,\text{⟨⟩}\,)_{IR} \;=\; -\mathrm{k}^2 \mathcal{K}\Big[\; \text{⊕} \;-\; \mathcal{K}(\text{─O─})\; \text{⊕} \;-\; \mathcal{K}\bar{R}(\text{⟨⟩})\, \text{⊕}$$

$$+(\;\text{┊}\;)_{IR} \Big\{ 2\varepsilon\,[\, \text{⊕} \;-\mathcal{K}(\text{─O─})\; \text{⊕}$$

$$-\mathcal{K}\bar{R}(\text{⟨⟩})\; \text{⟶} \;]$$

$$+2\varepsilon\; \text{⊕} \;+8\; \text{⊕} \Big\}$$

$$+(\,\text{⊗O}\,)_{IR} \Big\{ -\mathcal{K}(\text{─O─})\; \text{⌣} \Big\}$$

$$+(\,\text{⟨⟩}\,)_{IR}\; \text{⟶} \;\Big]$$

$$\mathrm{IR}_{12}: (\,\text{⟨⟩}\,)_{IR} \;=\; -\mathrm{k}^4 \mathcal{K}\Big[\, \text{⟨⟩} \,\Big]$$

$$\mathrm{IR}_{13}: (\,\text{⟨⟩}\,)_{IR} \;=\; -\mathrm{k}^2 \mathcal{K}\Big[\; \text{⊕} \;+(\,\text{⟨⟩}\,)_{IR}\; \text{⟶} \;\Big]$$

A.3.2 Pole Terms of IR-Counterterms

The results of the calculation of the IR counterterms with the R^*-operation in the last subsection are listed in the following table.

IR_1	$2\varepsilon^{-1}$
IR_2	$\varepsilon^{-1} - 2\varepsilon^{-2}$
IR_3	$\frac{1}{4}\varepsilon^{-1}$
IR_4	$\varepsilon^{-1} + 2\varepsilon^{-2}$
IR_5	$-\frac{2}{3}\varepsilon^{-1} + \frac{8}{3}\varepsilon^{-2} + \frac{8}{3}\varepsilon^{-3}$
IR_{6a}	$-\frac{2}{3}\varepsilon^{-1} + \frac{8}{3}\varepsilon^{-2} + \frac{8}{3}\varepsilon^{-3}$
IR_{6b}	$\frac{4}{3}\varepsilon^{-1} + 2\varepsilon^{-2} + \frac{4}{3}\varepsilon^{-3}$
IR_{6c}	$\frac{4}{3}\varepsilon^{-1} + 2\varepsilon^{-2} + \frac{4}{3}\varepsilon^{-3}$
IR_{7a}	$[2\zeta(3) - 1]\varepsilon^{-1} + \frac{1}{3}\varepsilon^{-2} + \frac{8}{3}\varepsilon^{-3} + \frac{4}{3}\varepsilon^{-4}$
IR_{7b}	$-[2\zeta(3) - \frac{5}{2}]\varepsilon^{-1} + \frac{19}{6}\varepsilon^{-2} + 2\varepsilon^{-3} + \frac{2}{3}\varepsilon^{-4}$
IR_{7c}	$-[2\zeta(3) - \frac{5}{2}]\varepsilon^{-1} + \frac{19}{6}\varepsilon^{-2} + 2\varepsilon^{-3} + \frac{2}{3}\varepsilon^{-4}$
IR_{7d}	$[\frac{3}{2}\zeta(4) + 3\zeta(3)]\varepsilon^{-1} + 2\zeta(3)\varepsilon^{-2}$
IR_8	$10\zeta(5)\varepsilon^{-1}$
IR_9	$-\frac{1}{2}\varepsilon^{-1} - \frac{1}{3}\varepsilon^{-2}$
IR_{10a}	$\frac{2}{3}\varepsilon^{-1} + \frac{4}{3}\varepsilon^{-2} + \frac{8}{3}\varepsilon^{-3}$
IR_{10b}	$\frac{2}{3}\varepsilon^{-1} + \frac{4}{3}\varepsilon^{-2} + \frac{8}{3}\varepsilon^{-3}$
IR_{10c}	$-\frac{1}{2}\varepsilon^{-1} - \frac{1}{3}\varepsilon^{-2}$
IR_{11a}	$-[\zeta(3) - \frac{11}{6}]\varepsilon^{-1} + \frac{7}{3}\varepsilon^{-2} + 2\varepsilon^{-3} + \frac{4}{3}\varepsilon^{-4}$
IR_{11b}	$-[\zeta(3) - \frac{11}{6}]\varepsilon^{-1} + \frac{7}{3}\varepsilon^{-2} + 2\varepsilon^{-3} + \frac{4}{3}\varepsilon^{-4}$
IR_{11c}	$-\frac{1}{2}\varepsilon^{-1} - \frac{5}{3}\varepsilon^{-2} - \frac{2}{3}\varepsilon^{-3}$
IR_{12}	$4\zeta(3)\varepsilon^{-1}$
IR_{13}	$-[\frac{3}{2}\zeta(4) - 3\zeta(3)]\varepsilon^{-1} + 6\zeta(3)\varepsilon^{-2}$

Appendix B

Contributions to Renormalization-Constants

The following tables collect the pole terms $\mathcal{K}\bar{R}G$, the weight factors W, and symmetry factors S needed for the calculation of the RG-constants up to five loops. Their use was explained in detail in Chapter 15. The first table contains the four-point diagrams needed for the mass and the coupling constant renormalization. The two-point diagrams needed for the field renormalization are listed in the second table.

In the first column, the numbers of the diagrams n_L appear which run for each L from 1 to N_L, and the second column shows the result of the $\mathcal{K}\bar{R}$-operation which was illustrated in A.

As indicated on page 89, the tadpole-free two- and four-point diagrams contributing to the tables are generated out of vacuum diagrams in Fig. 14.3. The numbers of the generating vacuum diagram, n_{L+3}^{Vac} for the four-point diagrams and n_{L+1}^{Vac} for the two-point diagrams appear in the last column of the tables.

On page 88, a relationship was established between the symmetry factors of the vacuum diagrams and the symmetry factors of the diagrams generated from them. A vacuum diagram n_{L+1}^{Vac} generates different two-point diagrams, which will all have the same symmetry factor $S_2(n_{L+1}^{\text{Vac}})$. The same holds for the four-point diagrams in O(N) symmetry, but not in cubic symmetry, where some more symmetry factors arise due to the different types of vertices. Anyhow, there are much less different symmetry factors than two- or four-point diagrams. The symmetry factors S_2 and S_4 are therefore listed separately in B.3 and B.4 labeled by serial numbers i_2, $i_4^{\text{O}(N)}$ and i_4^{cub}. Instead of the numbers n_{L+1}^{Vac} and n_{L+3}^{Vac} of the vacuum diagrams we used the numbers i_2 and $i_4^{\text{O}(N)}$. In fact, the numbers $i_4^{\text{O}(N)}$ equal the numbers n_{L+3}^{Vac}. But since Fig. 14.3 does not list the vacuum diagrams with tadpoles, not all two-point diagrams for the mass renormalization can be generated from them, and thus not all the symmetry factors for the mass renormalization. Remember, that, for the mass renormalization, we need the symmetry factor of two-point diagram, even though we calculate the pole term of the four point diagram. Thus, only the symmetry factors for the field renormalization are labeled by n_{Vac}.

In general, the numbers i_4 differ in O(N) and in cubic symmetry. The symmetry factors themselves are always different in O(N) and in cubic symmetry. They will therefore carry a superscript $S^{\text{O}(N)}$ or S^{cub}.

Finally, W_2 and W_4 are the weight factors for the two- and four-point diagrams, respectively. The contribution to Z_{m^2} is calculated with the pole terms of the four-point diagrams. The factors F which are generated by the differentiation of the lines of the two-point diagrams may be included in the weight factor such that the weight factor for the mass renormalization is not W_2 but $W_{m^2} = W_2 \cdot F$.

B.1 Contributions to Z_g and Z_{m^2}

For the calculation of Z_g and Z_{m^2} we need the pole term $\mathcal{K}\bar{R}G$, the weight factor, and the symmetry factor for each tadpole-free 1PI four-point diagram, called G_4. The calculation of

$\mathcal{K}\bar{R}G_4$ is shown diagrammatically in Appendix A, where the diagrams are listed in the same order as here. The columns labeled $i_4^{O(N)}$ and i_2 give the serial numbers of the symmetry factors, while n_{L+3}^{Vac} specifies the number of the vacuum diagram with $L+3$ loops in Fig. 14.3, from which the corresponding L-loop four-point diagram is generated.

n_L	$I_4 = \mathcal{K}\bar{R}G_4$	W_4	$i_4^{O(N)}$	i_4^{cub}	W_{m^2}	i_2	n_{L+3}^{Vac}
n_1	$1-\text{loop}$						
1	$2\varepsilon^{-1}$	$\frac{3}{2}$	1	1	$\frac{1}{2}$	1	1
n_2	$2-\text{loop}$						
1	$-4\varepsilon^{-2}$	$\frac{3}{4}$	1	1	$\frac{1}{4}$	1	1
2	$\varepsilon^{-1}-2\varepsilon^{-2}$	3	2	2	$\frac{1}{2}$	2	2
n_3	$3-\text{loop}$						
1	$8\varepsilon^{-3}$	$\frac{3}{8}$	1	1	$\frac{1}{8}$	1	1
2	$-2\varepsilon^{-2}+4\varepsilon^{-3}$	$\frac{3}{2}$	2	2	$\frac{1}{4}$	2	2
3	$-\frac{3}{4}\varepsilon^{-1}+\frac{2}{3}\varepsilon^{-2}$	$\frac{1}{2}$	3	3	$\frac{1}{6}$	2	3
4	$4\zeta(3)\varepsilon^{-1}$	1	4	4			4
5	$-\frac{2}{3}\varepsilon^{-1}-\frac{4}{3}\varepsilon^{-2}+\frac{8}{3}\varepsilon^{-3}$	$\frac{3}{2}$	2	5	$\frac{1}{4}$	3	2
6	$-\frac{2}{3}\varepsilon^{-1}-\frac{4}{3}\varepsilon^{-2}+\frac{8}{3}\varepsilon^{-3}$	$\frac{3}{2}$	5	6			5
7	$\frac{4}{3}\varepsilon^{-1}-2\varepsilon^{-2}+\frac{4}{3}\varepsilon^{-3}$	6	5	7	1	3	5
8	$\frac{2}{3}\varepsilon^{-1}-\frac{8}{3}\varepsilon^{-2}+\frac{8}{3}\varepsilon^{-3}$	$\frac{3}{4}$	2	2	$\frac{1}{4}$	2	2

n_L	$I_4 = \mathcal{K}\bar{R}G_4$	W_4	$i_4^{O(N)}$	i_4^{cub}	W_{m^2}	i_2	n_{L+3}^{Vac}
n_4	$4-\text{loop}$						
1	$-16\varepsilon^{-4}$	$\frac{3}{16}$	1	1	$\frac{1}{16}$	1	1
2	$4\varepsilon^{-3}-8\varepsilon^{-4}$	$\frac{3}{4}$	2	2	$\frac{1}{8}$	2	2
3	$-\varepsilon^{-2}+4\varepsilon^{-3}-4\varepsilon^{-4}$	$\frac{3}{4}$	3	3			3
4	$\frac{4}{3}\varepsilon^{-2}+\frac{8}{3}\varepsilon^{-3}-\frac{16}{3}\varepsilon^{-4}$	$\frac{3}{4}$	4	4	$\frac{1}{8}$	3	4
5	$\frac{3}{2}\varepsilon^{-2}-\frac{4}{3}\varepsilon^{-3}$	$\frac{1}{2}$	5	5	$\frac{1}{6}$	2	5
6	$-\frac{8}{3}\varepsilon^{-2}+4\varepsilon^{-3}-\frac{8}{3}\varepsilon^{-4}$	3	6	6	$\frac{1}{2}$	3	6
7	$-\frac{4}{3}\varepsilon^{-2}+\frac{16}{3}\varepsilon^{-3}-\frac{16}{3}\varepsilon^{-4}$	$\frac{3}{4}$	2	2	$\frac{1}{4}$	2	2
8	$[\frac{11}{6}-\zeta(3)]\varepsilon^{-1}-\frac{13}{3}\varepsilon^{-2}+\frac{10}{3}\varepsilon^{-3}-\frac{4}{3}\varepsilon^{-4}$	$\frac{3}{2}$	6	6	$\frac{1}{2}$	3	6
9	$-\frac{1}{2}\varepsilon^{-1}+\frac{1}{6}\varepsilon^{-2}+\frac{10}{3}\varepsilon^{-3}-\frac{10}{3}\varepsilon^{-4}$	$\frac{3}{2}$	7	3	$\frac{1}{4}$	4	7
10	$10\zeta(5)\varepsilon^{-1}$	3	8	7			8
11	$[3\zeta(3)-\frac{3}{2}\zeta(4)]\varepsilon^{-1}-2\zeta(3)\varepsilon^{-2}$	6	9	8			9
12	$-\frac{2}{3}\varepsilon^{-1}-\frac{5}{6}\varepsilon^{-2}+\frac{8}{3}\varepsilon^{-3}-2\varepsilon^{-4}$	6	10	9			10
13	$[\frac{1}{2}-\zeta(3)]\varepsilon^{-1}+\varepsilon^{-2}+2\varepsilon^{-3}-4\varepsilon^{-4}$	$\frac{3}{2}$	6	10			6
14	$-[2-2\zeta(3)]\varepsilon^{-1}+\frac{4}{3}\varepsilon^{-2}+\frac{8}{3}\varepsilon^{-3}-\frac{8}{3}\varepsilon^{-4}$	$\frac{3}{8}$	4	4	$\frac{1}{8}$	3	4
15	$-\frac{7}{12}\varepsilon^{-1}+\varepsilon^{-2}-\frac{2}{3}\varepsilon^{-3}$	$\frac{3}{4}$	11	11	$\frac{1}{4}$	3	11
16	$-\frac{121}{96}\varepsilon^{-1}+\frac{11}{8}\varepsilon^{-2}-\frac{1}{2}\varepsilon^{-3}$	1	12	12	$\frac{1}{6}$	4	12
17	$-[1-2\zeta(3)]\varepsilon^{-1}-\frac{5}{3}\varepsilon^{-2}+\frac{8}{3}\varepsilon^{-3}-\frac{4}{3}\varepsilon^{-4}$	$\frac{3}{2}$	3	13	$\frac{1}{4}$	5	3
18	$[\frac{1}{2}-\zeta(3)]\varepsilon^{-1}+\varepsilon^{-2}+2\varepsilon^{-3}-4\varepsilon^{-4}$	$\frac{3}{4}$	2	14	$\frac{1}{8}$	6	2
19	$\frac{37}{96}\varepsilon^{-1}+\frac{5}{8}\varepsilon^{-2}-\frac{5}{6}\varepsilon^{-3}$	1	13	12	$\frac{1}{6}$	4	13
20	$-[\frac{5}{6}-\zeta(3)]\varepsilon^{-1}-\frac{1}{3}\varepsilon^{-2}+2\varepsilon^{-3}-\frac{4}{3}\varepsilon^{-4}$	$\frac{3}{2}$	14	15	$\frac{1}{4}$	6	14
21	$[3\zeta(3)+\frac{3}{2}\zeta(4)]\varepsilon^{-1}-6\zeta(3)\varepsilon^{-2}$	3	9	16	$\frac{1}{2}$	5	9
22	$-\frac{5}{6}\varepsilon^{-1}-\frac{1}{3}\varepsilon^{-2}+2\varepsilon^{-3}-\frac{4}{3}\varepsilon^{-4}$	$\frac{3}{2}$	15	17	$\frac{1}{4}$	5	15
23	$-\frac{2}{3}\varepsilon^{-1}-\frac{5}{6}\varepsilon^{-2}+\frac{8}{3}\varepsilon^{-3}-2\varepsilon^{-4}$	3	7	13	$\frac{1}{2}$	5	7
24	$-[\frac{5}{6}-\zeta(3)]\varepsilon^{-1}-\frac{1}{3}\varepsilon^{-2}+2\varepsilon^{-3}-\frac{4}{3}\varepsilon^{-4}$	3	6	18	$\frac{1}{2}$	6	6
25	$[\frac{5}{2}-2\zeta(3)]\varepsilon^{-1}-\frac{19}{6}\varepsilon^{-2}+2\varepsilon^{-3}-\frac{2}{3}\varepsilon^{-4}$	6	10	19	1	5	10
26	$\frac{5}{2}\varepsilon^{-1}-\frac{19}{6}\varepsilon^{-2}+2\varepsilon^{-3}-\frac{2}{3}\varepsilon^{-4}$	6	10	19	1	5	10

n_L	$I_4 = \mathcal{K}\bar{R}G_4$	W_4	$i_4^{O(N)}$	i_4^{cub}	W_{m^2}	i_2	n_{L+3}^{Vac}
n_5	5$-$loop						
1	$\frac{72}{5}\zeta(3)^2\varepsilon^{-1}$	2	1	1			1
2	$\frac{72}{5}\zeta(3)^2\varepsilon^{-1}$	$\frac{3}{2}$	1	2			1
3	$\frac{441}{20}\zeta(7)\varepsilon^{-1}$	12	2	3			2
4	$-[\frac{2}{5}\zeta(3)-\frac{103}{160}]\varepsilon^{-1}+\frac{37}{240}\varepsilon^{-2}-\frac{19}{20}\varepsilon^{-3}+\frac{7}{15}\varepsilon^{-4}$	1	3	4	$\frac{1}{6}$	1	3
5	$\frac{103}{160}\varepsilon^{-1}+\frac{37}{240}\varepsilon^{-2}-\frac{19}{20}\varepsilon^{-3}+\frac{7}{15}\varepsilon^{-4}$	1	3	4	$\frac{1}{6}$	1	3
6	$-\frac{11}{192}\varepsilon^{-1}+\frac{33}{80}\varepsilon^{-2}-\frac{11}{12}\varepsilon^{-3}+\frac{3}{5}\varepsilon^{-4}$	1	3	4	$\frac{1}{6}$	1	3
7	$-3\varepsilon^{-3}+\frac{8}{3}\varepsilon^{-4}$	$\frac{1}{4}$	4	5	$\frac{1}{12}$	2	4
8	$-3\varepsilon^{-3}+\frac{8}{3}\varepsilon^{-4}$	$\frac{1}{8}$	4	5	$\frac{1}{24}$	2	4
9	$\frac{3}{4}\varepsilon^{-2}-\frac{13}{6}\varepsilon^{-3}+\frac{4}{3}\varepsilon^{-4}$	$\frac{1}{2}$	5	6	$\frac{1}{12}$	3	5
10	$\frac{151}{192}\varepsilon^{-1}+\frac{197}{240}\varepsilon^{-2}-\frac{103}{60}\varepsilon^{-3}+\frac{11}{15}\varepsilon^{-4}$	1	5	7	$\frac{1}{6}$	1	5
11	$[\frac{2}{5}\zeta(3)-\frac{151}{480}]\varepsilon^{-1}-\frac{53}{120}\varepsilon^{-2}-\frac{23}{30}\varepsilon^{-3}+\frac{6}{5}\varepsilon^{-4}$	$\frac{1}{2}$	6	8			6
12	$-\frac{11}{192}\varepsilon^{-1}+\frac{33}{80}\varepsilon^{-2}-\frac{11}{12}\varepsilon^{-3}+\frac{3}{5}\varepsilon^{-4}$	1	6	4	$\frac{1}{6}$	1	6
13	$[\frac{3}{5}\zeta(4)-\frac{27}{10}\zeta(3)]\varepsilon^{-1}+\frac{4}{5}\varepsilon^{-2}\zeta(3)$	1	7	9			7
14	$-\frac{5}{96}\varepsilon^{-1}-\frac{3}{10}\varepsilon^{-2}+\frac{4}{15}\varepsilon^{-3}$	$\frac{1}{24}$	8	10	$\frac{1}{72}$	3	8
15	$\frac{857}{960}\varepsilon^{-1}-\frac{13}{20}\varepsilon^{-2}+\frac{2}{15}\varepsilon^{-3}$	$\frac{1}{4}$	9	10	$\frac{1}{12}$	3	9
16	$-\frac{2387}{960}\varepsilon^{-1}+\frac{41}{15}\varepsilon^{-2}-\frac{13}{10}\varepsilon^{-3}+\frac{4}{15}\varepsilon^{-4}$	2	10	4	$\frac{1}{3}$	1	10
17	$\frac{151}{192}\varepsilon^{-1}+\frac{197}{240}\varepsilon^{-2}-\frac{103}{60}\varepsilon^{-3}+\frac{11}{15}\varepsilon^{-4}$	1	10	8			10
18	$\frac{121}{48}\varepsilon^{-2}-\frac{11}{4}\varepsilon^{-3}+\varepsilon^{-4}$	$\frac{1}{2}$	11	6	$\frac{1}{12}$	3	11
19	$-\frac{215}{96}\varepsilon^{-1}+\frac{25}{6}\varepsilon^{-2}-\frac{7}{3}\varepsilon^{-3}+\frac{8}{15}\varepsilon^{-4}$	$\frac{1}{4}$	11	6	$\frac{1}{12}$	3	11
20	$[\frac{2}{5}\zeta(3)-\frac{151}{480}]\varepsilon^{-1}-\frac{53}{120}\varepsilon^{-2}-\frac{23}{30}\varepsilon^{-3}+\frac{6}{5}\varepsilon^{-4}$	$\frac{1}{2}$	12	7	$\frac{1}{12}$	1	12
21	$\frac{157}{320}\varepsilon^{-1}-\frac{11}{60}\varepsilon^{-2}-\frac{7}{5}\varepsilon^{-3}+\frac{16}{15}\varepsilon^{-4}$	$\frac{1}{2}$	12	6	$\frac{1}{6}$	3	12
22	$-\frac{37}{48}\varepsilon^{-2}-\frac{5}{4}\varepsilon^{-3}+\frac{5}{3}\varepsilon^{-4}$	$\frac{1}{2}$	12	6	$\frac{1}{12}$	3	12
23	$-\frac{5}{96}\varepsilon^{-1}-\frac{3}{10}\varepsilon^{-2}+\frac{4}{15}\varepsilon^{-3}$	$\frac{1}{12}$	13	10	$\frac{1}{36}$	3	13
24	$-[5\zeta(6)-10\zeta(5)-\frac{2}{5}\zeta(3)^2]\varepsilon^{-1}-4\zeta(5)\varepsilon^{-2}$	12	14	11			14
25	$-[5\zeta(6)-10\zeta(5)+\frac{22}{5}\zeta(3)^2]\varepsilon^{-1}-4\varepsilon^{-2}\zeta(5)$	3	14	12			14
26	$[5\zeta(6)+6\zeta(5)-\frac{2}{5}\zeta(3)^2]\varepsilon^{-1}-16\zeta(5)\varepsilon^{-2}$	6	14	13	1	4	14
27	$-[5\zeta(6)-10\zeta(5)+2\zeta(3)^2]\varepsilon^{-1}-4\zeta(5)\varepsilon^{-2}$	6	15	14			15

n_5	$I_4 = \mathcal{K}\bar{R}G_4$	W_4	$i_4^{O(N)}$	i_4^{cub}	W_{m^2}	i_2	n_{L+3}^{Vac}
28	$-[5\zeta(6)-10\zeta(5)+\frac{34}{5}\zeta(3)^2]\varepsilon^{-1}-4\zeta(5)\varepsilon^{-2}$	$\frac{3}{2}$	15	15			15
29	$[5\zeta(6)+6\zeta(5)-\frac{14}{5}\zeta(3)^2]\varepsilon^{-1}-16\zeta(5)\varepsilon^{-2}$	3	15	16	$\frac{1}{2}$	5	15
30	$[\frac{14}{5}\zeta(5)-\frac{9}{5}\zeta(4)+\frac{2}{5}\zeta(3)]\varepsilon^{-1}+[\frac{6}{5}\zeta(4)-\frac{12}{5}\zeta(3)]\varepsilon^{-2}+\frac{8}{5}\zeta(3)\varepsilon^{-3}$	3	16	17			16
31	$-[3\zeta(4)+6\zeta(3)]\varepsilon^{-2}+12\zeta(3)\varepsilon^{-3}$	$\frac{3}{2}$	16	18	$\frac{1}{4}$	6	16
32	$-[\frac{16}{5}\zeta(5)-\frac{9}{5}\zeta(4)-\frac{14}{5}\zeta(3)]\varepsilon^{-1}-[\frac{24}{5}\zeta(4)+\frac{48}{5}\zeta(3)]\varepsilon^{-2}+\frac{48}{5}\zeta(3)\varepsilon^{-3}$	$\frac{3}{4}$	16	18	$\frac{1}{4}$	6	16
33	$[\frac{7}{5}\zeta(5)-\frac{21}{10}\zeta(4)+\frac{31}{5}\zeta(3)]\varepsilon^{-1}+[\frac{3}{5}\zeta(4)-\frac{14}{5}\zeta(3)]\varepsilon^{-2}+\frac{4}{5}\zeta(3)\varepsilon^{-3}$	6	17	19			17
34	$-[\frac{23}{5}\zeta(5)+\frac{21}{10}\zeta(4)-\frac{31}{5}\zeta(3)]\varepsilon^{-1}+[\frac{3}{5}\zeta(4)-\frac{14}{5}\zeta(3)]\varepsilon^{-2}+\frac{4}{5}\zeta(3)\varepsilon^{-3}$	6	17	19			17
35	$-[\frac{7}{5}\zeta(5)-\frac{3}{10}\zeta(4)+\frac{13}{5}\zeta(3)]\varepsilon^{-1}-[\frac{3}{5}\zeta(4)+\frac{18}{5}\zeta(3)]\varepsilon^{-2}+\frac{36}{5}\zeta(3)\varepsilon^{-3}$	3	17	20			17
36	$-[\frac{23}{5}\zeta(5)-\frac{21}{10}\zeta(4)-5\zeta(3)]\varepsilon^{-1}-[\frac{12}{5}\zeta(4)+\frac{36}{5}\zeta(3)]\varepsilon^{-2}$ $+\frac{24}{5}\zeta(3)\varepsilon^{-3}$	6	17	21	1	5	17
37	$-[\frac{24}{5}\zeta(5)-\frac{16}{5}\zeta(3)]\varepsilon^{-1}+[\frac{9}{5}\zeta(4)-6\zeta(3)]\varepsilon^{-2}+\frac{12}{5}\zeta(3)\varepsilon^{-3}$	6	18	22	1	4	18
38	$[\frac{6}{5}\zeta(5)+\frac{16}{5}\zeta(3)]\varepsilon^{-1}+[\frac{9}{5}\zeta(4)-6\zeta(3)]\varepsilon^{-2}+\frac{12}{5}\zeta(3)\varepsilon^{-3}$	6	18	22	1	4	18
39	$-[\frac{6}{5}\zeta(5)+\frac{9}{5}\zeta(4)-\frac{2}{5}\zeta(3)]\varepsilon^{-1}+[\frac{6}{5}\zeta(4)-\frac{12}{5}\zeta(3)]\varepsilon^{-2}+\frac{8}{5}\zeta(3)\varepsilon^{-3}$	6	18	23			18
40	$-[\frac{16}{5}\zeta(5)+\frac{9}{5}\zeta(4)-\frac{2}{5}\zeta(3)]\varepsilon^{-1}+[\frac{6}{5}\zeta(4)-\frac{12}{5}\zeta(3)]\varepsilon^{-2}+\frac{8}{5}\zeta(3)\varepsilon^{-3}$	3	18	24			18
41	$[\frac{14}{5}\zeta(5)-\frac{9}{5}\zeta(4)+\frac{2}{5}\zeta(3)]\varepsilon^{-1}+[\frac{6}{5}\zeta(4)-\frac{12}{5}\zeta(3)]\varepsilon^{-2}+\frac{8}{5}\zeta(3)\varepsilon^{-3}$	$\frac{3}{2}$	19	25			19
42	$[\frac{16}{5}\zeta(5)+\frac{16}{5}\zeta(3)]\varepsilon^{-1}+[\frac{9}{5}\zeta(4)-6\zeta(3)]\varepsilon^{-2}+\frac{12}{5}\zeta(3)\varepsilon^{-3}$	3	19	26	$\frac{1}{2}$	7	19
43	$-[\frac{14}{5}\zeta(5)-\frac{16}{5}\zeta(3)]\varepsilon^{-1}+[\frac{9}{5}\zeta(4)-6\zeta(3)]\varepsilon^{-2}+\frac{12}{5}\zeta(3)\varepsilon^{-3}$	3	19	26	$\frac{1}{2}$	7	19
44	$-[\frac{6}{5}\zeta(5)+\frac{9}{5}\zeta(4)-\frac{2}{5}\zeta(3)]\varepsilon^{-1}+[\frac{6}{5}\zeta(4)-\frac{12}{5}\zeta(3)]\varepsilon^{-2}+\frac{8}{5}\zeta(3)\varepsilon^{-3}$	3	19	27			19
45	$[\frac{3}{10}\zeta(4)-\zeta(3)+\frac{3}{5}]\varepsilon^{-1}+[\frac{2}{5}\zeta(3)+\frac{3}{5}]\varepsilon^{-2}+\frac{4}{5}\varepsilon^{-3}-4\varepsilon^{-4}+\frac{16}{5}\varepsilon^{-5}$	3	20	24			20
46	$-5\varepsilon^{-2}+\frac{19}{3}\varepsilon^{-3}-4\varepsilon^{-4}+\frac{4}{3}\varepsilon^{-5}$	3	20	29	$\frac{1}{2}$	6	20
47	$-[\frac{3}{10}\zeta(4)+3\zeta(3)-\frac{73}{15}]\varepsilon^{-1}+[\frac{18}{5}\zeta(3)-9]\varepsilon^{-2}$ $+6\varepsilon^{-3}-\frac{12}{5}\varepsilon^{-4}+\frac{8}{15}\varepsilon^{-5}$	3	20	29	1	6	20
48	$[4\zeta(3)-5]\varepsilon^{-2}+\frac{19}{3}\varepsilon^{-3}-4\varepsilon^{-4}+\frac{4}{3}\varepsilon^{-5}$	3	20	29	$\frac{1}{2}$	6	20
49	$-[\frac{3}{10}\zeta(4)+\frac{1}{5}\zeta(3)]\varepsilon^{-1}-[\frac{2}{5}\zeta(3)-\frac{19}{15}]\varepsilon^{-2}-\frac{44}{15}\varepsilon^{-4}+\frac{32}{15}\varepsilon^{-5}$	3	20	30			20
50	$[\frac{3}{10}\zeta(4)+\frac{9}{5}\zeta(3)-\frac{2}{5}]\varepsilon^{-1}-[\frac{8}{5}\zeta(3)+\frac{4}{15}]\varepsilon^{-1}+\frac{26}{15}\varepsilon^{-3}-\frac{8}{5}\varepsilon^{-4}+\frac{8}{15}\varepsilon^{-5}$	3	20	31	$\frac{1}{2}$	7	20
51	$[\frac{3}{2}\zeta(4)-\frac{3}{5}\zeta(3)-\frac{2}{5}]\varepsilon^{-1}-\frac{4}{15}\varepsilon^{-2}+\frac{26}{15}\varepsilon^{-3}-\frac{8}{5}\varepsilon^{-4}+\frac{8}{15}\varepsilon^{-5}$	3	20	31	$\frac{1}{2}$	7	20
52	$\frac{8}{15}\varepsilon^{-1}-\frac{8}{15}\varepsilon^{-2}+\frac{34}{15}\varepsilon^{-3}-\frac{16}{5}\varepsilon^{-4}+\frac{8}{5}\varepsilon^{-5}$	3	21	32			21
53	$[\frac{3}{5}\zeta(4)-\frac{26}{5}\zeta(3)+\frac{28}{5}]\varepsilon^{-1}+[\frac{4}{5}\zeta(3)-\frac{19}{3}]\varepsilon^{-2}+\frac{11}{3}\varepsilon^{-3}-\frac{4}{3}\varepsilon^{-4}+\frac{4}{15}\varepsilon^{-5}$	6	21	33	1	4	21

n_5	$I_4 = \mathcal{K}\bar{R}G_4$	W_4	$i_4^{O(N)}$	i_4^{cub}	W_{m^2}	i_2	n_{L+3}^{Vac}
54	$\frac{28}{5}\varepsilon^{-1}-\frac{19}{3}\varepsilon^{-2}+\frac{11}{3}\varepsilon^{-3}-\frac{4}{3}\varepsilon^{-4}+\frac{4}{15}\varepsilon^{-5}$	6	21	33	1	4	21
55	$[\frac{3}{5}\zeta(4)-\frac{2}{5}\zeta(3)-\frac{28}{15}]\varepsilon^{-1}+[\frac{4}{5}\zeta(3)-\frac{19}{15}]\varepsilon^{-2}+\frac{56}{15}\varepsilon^{-3}-\frac{44}{15}\varepsilon^{-4}+\frac{16}{15}\varepsilon^{-5}$	6	21	34			21
56	$-\frac{28}{15}\varepsilon^{-1}-\frac{19}{15}\varepsilon^{-2}+\frac{56}{15}\varepsilon^{-3}-\frac{44}{15}\varepsilon^{-4}+\frac{16}{15}\varepsilon^{-5}$	3	22	35	$\frac{1}{2}$	4	22
57	$-[\frac{3}{10}\zeta(4)-\frac{7}{5}\zeta(3)+\frac{4}{3}]\varepsilon^{-1}-[\frac{2}{5}\zeta(3)+\frac{2}{15}]\varepsilon^{-2}+\frac{11}{5}\varepsilon^{-3}-\frac{32}{15}\varepsilon^{-4}+\frac{4}{5}\varepsilon^{-5}$	3	22	36	$\frac{1}{2}$	7	22
58	$[\frac{3}{10}\zeta(4)-\frac{3}{5}\zeta(3)-\frac{4}{3}]\varepsilon^{-1}+[\frac{2}{5}\zeta(3)-\frac{2}{15}]\varepsilon^{-2}+\frac{11}{5}\varepsilon^{-3}-\frac{32}{15}\varepsilon^{-4}+\frac{4}{5}\varepsilon^{-5}$	3	22	36	$\frac{1}{2}$	7	22
59	$[\frac{3}{5}\zeta(4)-\frac{2}{5}\zeta(3)-\frac{28}{15}]\varepsilon^{-1}+[\frac{4}{5}\zeta(3)-\frac{19}{15}]\varepsilon^{-2}+\frac{56}{15}\varepsilon^{-3}-\frac{44}{15}\varepsilon^{-4}+\frac{16}{15}\varepsilon^{-5}$	3	22	35	$\frac{1}{2}$	4	22
60	$[\frac{3}{10}\zeta(4)-\zeta(3)+\frac{3}{5}]\varepsilon^{-1}+[\frac{2}{5}\zeta(3)+\frac{3}{5}]\varepsilon^{-2}+\frac{4}{5}\varepsilon^{-3}-4\varepsilon^{-4}+\frac{16}{5}\varepsilon^{-5}$	3	22	37			22
61	$[\frac{3}{10}\zeta(4)-\frac{1}{5}\zeta(3)-\frac{11}{15}]\varepsilon^{-1}+[\frac{2}{5}\zeta(3)-\frac{4}{15}]\varepsilon^{-2}+\frac{61}{15}\varepsilon^{-3}-\frac{64}{15}\varepsilon^{-4}+\frac{28}{15}\varepsilon^{-5}$	3	22	38	$\frac{1}{2}$	1	22
62	$-\frac{16}{15}\varepsilon^{-1}+\frac{7}{15}\varepsilon^{-2}+\frac{43}{15}\varepsilon^{-3}-\frac{68}{15}\varepsilon^{-4}+\frac{12}{5}\varepsilon^{-5}$	3	22	38	$\frac{1}{2}$	1	22
63	$-[\frac{3}{10}\zeta(4)-\frac{7}{5}\zeta(3)+\frac{4}{3}]\varepsilon^{-1}-[\frac{2}{5}\zeta(3)+\frac{2}{15}]\varepsilon^{-2}+\frac{11}{5}\varepsilon^{-3}-\frac{32}{15}\varepsilon^{-4}+\frac{4}{5}\varepsilon^{-5}$	3	23	39	$\frac{1}{2}$	7	23
64	$[\frac{9}{10}\zeta(4)-\frac{9}{5}\zeta(3)-\frac{2}{15}]\varepsilon^{-1}+[-\frac{4}{5}\zeta(3)-\frac{4}{15}]\varepsilon^{-2}+\frac{26}{15}\varepsilon^{-3}-\frac{8}{5}\varepsilon^{-4}+\frac{8}{15}\varepsilon^{-5}$	3	23	40	$\frac{1}{2}$	4	23
65	$[\frac{3}{10}\zeta(4)-\frac{3}{5}\zeta(3)-\frac{4}{3}]\varepsilon^{-1}+[\frac{2}{5}\zeta(3)-\frac{2}{15}]\varepsilon^{-2}+\frac{11}{5}\varepsilon^{-3}-\frac{32}{15}\varepsilon^{-4}+\frac{4}{5}\varepsilon^{-5}$	3	23	39	$\frac{1}{2}$	7	23
66	$-[\frac{3}{10}\zeta(4)+\frac{1}{5}\zeta(3)]\varepsilon^{-1}-[\frac{2}{5}\zeta(3)-\frac{19}{15}]\varepsilon^{-2}-\frac{44}{15}\varepsilon^{-4}+\frac{32}{15}\varepsilon^{-5}$	$\frac{3}{2}$	23	41			23
67	$[\frac{3}{5}\zeta(4)-\frac{6}{5}\zeta(3)-\frac{2}{5}]\varepsilon^{-1}+[\frac{4}{5}\zeta(3)-\frac{4}{15}]\varepsilon^{-2}+\frac{26}{15}\varepsilon^{-3}-\frac{8}{5}\varepsilon^{-4}+\frac{8}{15}\varepsilon^{-5}$	3	24	42	$\frac{1}{2}$	4	24
68	$\frac{19}{15}\varepsilon^{-2}-\frac{44}{15}\varepsilon^{-4}+\frac{32}{15}\varepsilon^{-5}$	$\frac{3}{2}$	24	43			24
69	$[\frac{3}{5}\zeta(4)+\frac{2}{5}\zeta(3)-\frac{4}{3}]\varepsilon^{-1}+[\frac{4}{5}\zeta(3)-\frac{2}{15}]\varepsilon^{-2}+\frac{11}{5}\varepsilon^{-3}-\frac{32}{15}\varepsilon^{-4}+\frac{4}{5}\varepsilon^{-5}$	3	24	44	$\frac{1}{2}$	4	24
70	$-\frac{4}{3}\varepsilon^{-1}-\frac{2}{15}\varepsilon^{-2}+\frac{11}{5}\varepsilon^{-3}-\frac{32}{15}\varepsilon^{-4}+\frac{4}{5}\varepsilon^{-5}$	3	24	44	$\frac{1}{2}$	4	24
71	$-[\frac{3}{5}\zeta(4)+\frac{4}{5}\zeta(3)-\frac{3}{5}]\varepsilon^{-1}+[\frac{6}{5}\zeta(3)+\frac{1}{5}]\varepsilon^{-2}+\frac{2}{5}\varepsilon^{-3}-\frac{12}{5}\varepsilon^{-4}+\frac{8}{5}\varepsilon^{-5}$	$\frac{3}{2}$	25	45	$\frac{1}{4}$	7	25
72	$\frac{5}{3}\varepsilon^{-2}+\frac{2}{3}\varepsilon^{-3}-4\varepsilon^{-4}+\frac{8}{3}\varepsilon^{-5}$	$\frac{3}{4}$	25	46	$\frac{1}{8}$	6	25
73	$[\frac{3}{5}\zeta(4)+\frac{2}{5}\zeta(3)-\frac{16}{15}]\varepsilon^{-1}+[\frac{4}{5}\zeta(3)+\frac{6}{5}]\varepsilon^{-2}+\frac{28}{15}\varepsilon^{-3}-\frac{8}{3}\varepsilon^{-4}+\frac{16}{15}\varepsilon^{-5}$	$\frac{3}{8}$	25	46	$\frac{1}{8}$	6	25
74	$-[\frac{3}{5}\zeta(4)+\frac{14}{5}\zeta(3)-\frac{28}{5}]\varepsilon^{-1}+[\frac{16}{5}\zeta(3)-\frac{19}{3}]\varepsilon^{-2}$ $+\frac{11}{3}\varepsilon^{-3}-\frac{4}{3}\varepsilon^{-4}+\frac{4}{15}\varepsilon^{-5}$	6	26	33	1	4	26
75	$-[\frac{16}{5}\zeta(3)-\frac{28}{5}]\varepsilon^{-1}-[4\zeta(3)+\frac{19}{3}]\varepsilon^{-1}+\frac{11}{3}\varepsilon^{-3}-\frac{4}{3}\varepsilon^{-4}+\frac{4}{15}\varepsilon^{-5}$	6	26	33	1	4	26
76	$-\frac{28}{15}\varepsilon^{-1}-\frac{19}{15}\varepsilon^{-2}+\frac{56}{15}\varepsilon^{-3}-\frac{44}{15}\varepsilon^{-4}+\frac{16}{15}\varepsilon^{-5}$	6	26	34			26
77	$\frac{8}{15}\varepsilon^{-1}-\frac{8}{15}\varepsilon^{-2}+\frac{34}{15}\varepsilon^{-3}-\frac{16}{5}\varepsilon^{-4}+\frac{8}{5}\varepsilon^{-5}$	3	26	32			26
78	$[2\zeta(3)-1]\varepsilon^{-2}-2\varepsilon^{-3}-4\varepsilon^{-4}+8\varepsilon^{-5}$	$\frac{3}{8}$	27	47	$\frac{1}{16}$	8	27
79	$-[\frac{12}{5}\zeta(4)-\frac{6}{5}\zeta(3)-\frac{7}{5}]\varepsilon^{-1}+[\frac{4}{5}\zeta(3)+\frac{4}{5}]\varepsilon^{-2}-\frac{8}{5}\varepsilon^{-3}-\frac{16}{5}\varepsilon^{-4}+\frac{16}{5}\varepsilon^{-5}$	$\frac{3}{16}$	27	47	$\frac{1}{16}$	8	27

n_5	$I_4 = \mathcal{K}\bar{R}G_4$	W_4	$i_4^{O(N)}$	i_4^{cub}	W_{m^2}	i_2	n_{L+3}^{Vac}
80	$-\frac{8}{3}\varepsilon^{-3}-\frac{16}{3}\varepsilon^{-4}+\frac{32}{3}\varepsilon^{-5}$	$\frac{3}{8}$	27	48	$\frac{1}{16}$	9	27
81	$-[4\zeta(3)-4]\varepsilon^{-2}-\frac{8}{3}\varepsilon^{-3}-\frac{16}{3}\varepsilon^{-4}+\frac{16}{3}\varepsilon^{-5}$	$\frac{3}{8}$	27	48	$\frac{1}{8}$	9	27
82	$-[\frac{6}{5}\zeta(4)-\frac{2}{5}\zeta(3)-\frac{3}{5}]\varepsilon^{-1}+[\frac{2}{5}\zeta(3)+\frac{1}{5}]\varepsilon^{-2}+\frac{2}{5}\varepsilon^{-3}-\frac{12}{5}\varepsilon^{-4}+\frac{8}{5}\varepsilon^{-5}$	$\frac{3}{2}$	28	49	$\frac{1}{4}$	10	28
83	$\frac{16}{3}\varepsilon^{-3}-8\varepsilon^{-4}+\frac{16}{3}\varepsilon^{-5}$	$\frac{3}{2}$	28	50	$\frac{1}{4}$	9	28
84	$[2\zeta(3)-\frac{11}{3}]\varepsilon^{-2}+\frac{26}{3}\varepsilon^{-3}-\frac{20}{3}\varepsilon^{-4}+\frac{8}{3}\varepsilon^{-5}$	$\frac{3}{2}$	28	50	$\frac{1}{2}$	9	28
85	$[\frac{6}{5}\zeta(4)-\frac{4}{5}\zeta(3)-\frac{2}{5}]\varepsilon^{-1}+[\frac{8}{5}\zeta(3)-\frac{4}{5}]\varepsilon^{-2}-\frac{8}{5}\varepsilon^{-3}-\frac{16}{5}\varepsilon^{-4}+\frac{32}{5}\varepsilon^{-5}$	$\frac{3}{4}$	28	51			28
86	$-[\frac{6}{5}\zeta(4)-\frac{2}{5}\zeta(3)-\frac{3}{5}]\varepsilon^{-1}+[\frac{2}{5}\zeta(3)+\frac{1}{5}]\varepsilon^{-2}$ $+\frac{2}{5}\varepsilon^{-3}-\frac{12}{5}\varepsilon^{-4}+\frac{8}{5}\varepsilon^{-5}$	$\frac{3}{2}$	29	52	$\frac{1}{4}$	10	29
87	$-[2\zeta(3)-\frac{5}{3}]\varepsilon^{-2}+\frac{2}{3}\varepsilon^{-3}-4\varepsilon^{-4}+\frac{8}{3}\varepsilon^{-5}$	$\frac{3}{4}$	29	53	$\frac{1}{8}$	8	29
88	$[\frac{6}{5}\zeta(4)-\frac{16}{15}]\varepsilon^{-1}-[\frac{12}{5}\zeta(3)-\frac{6}{5}]\varepsilon^{-2}+\frac{28}{15}\varepsilon^{-3}-\frac{8}{3}\varepsilon^{-4}+\frac{16}{15}\varepsilon^{-5}$	$\frac{3}{8}$	29	53	$\frac{1}{8}$	8	29
89	$-[\frac{3}{10}\zeta(4)+\frac{1}{5}\zeta(3)]\varepsilon^{-1}-[\frac{2}{5}\zeta(3)-\frac{19}{15}]\varepsilon^{-2}-\frac{44}{15}\varepsilon^{-4}+\frac{32}{15}\varepsilon^{-5}$	$\frac{3}{2}$	30	54	$\frac{1}{4}$	7	30
90	$\frac{4}{3}\varepsilon^{-2}+\frac{5}{3}\varepsilon^{-3}-\frac{16}{3}\varepsilon^{-4}+4\varepsilon^{-5}$	$\frac{3}{2}$	30	55	$\frac{1}{4}$	6	30
91	$-[\frac{9}{10}\zeta(4)-\frac{1}{5}\zeta(3)-\frac{22}{15}]\varepsilon^{-1}+[\frac{4}{5}\zeta(3)+\frac{4}{5}]\varepsilon^{-2}$ $-\frac{22}{15}\varepsilon^{-3}-\frac{56}{15}\varepsilon^{-4}+\frac{56}{15}\varepsilon^{-5}$	$\frac{3}{4}$	30	56	$\frac{1}{8}$	1	30
92	$-[\frac{3}{10}\zeta(4)+\frac{3}{5}\zeta(3)-\frac{1}{3}]\varepsilon^{-1}+[\frac{8}{5}\zeta(3)-\frac{2}{15}]\varepsilon^{-2}-\frac{2}{15}\varepsilon^{-3}$ $-\frac{24}{5}\varepsilon^{-4}+\frac{24}{5}\varepsilon^{-5}$	$\frac{3}{4}$	30	56	$\frac{1}{8}$	1	30
93	$-[\frac{3}{10}\zeta(4)-\frac{11}{5}\zeta(3)+\frac{14}{5}]\varepsilon^{-1}-[\frac{2}{5}\zeta(3)-\frac{29}{15}]\varepsilon^{-2}$ $+\frac{34}{15}\varepsilon^{-3}-\frac{52}{15}\varepsilon^{-4}+\frac{8}{5}\varepsilon^{-5}$	$\frac{3}{4}$	30	55	$\frac{1}{4}$	6	30
94	$-[2\zeta(3)-\frac{5}{3}]\varepsilon^{-2}+\frac{2}{3}\varepsilon^{-3}-4\varepsilon^{-4}+\frac{8}{3}\varepsilon^{-5}$	$\frac{3}{2}$	31	57	$\frac{1}{4}$	8	31
95	$[\frac{6}{5}\zeta(4)-\frac{16}{15}]\varepsilon^{-1}-[\frac{12}{5}\zeta(3)-\frac{6}{5}]\varepsilon^{-2}+\frac{28}{15}\varepsilon^{-3}-\frac{8}{3}\varepsilon^{-4}+\frac{16}{15}\varepsilon^{-5}$	$\frac{3}{4}$	31	57	$\frac{1}{4}$	8	31
96	$[\frac{6}{5}\zeta(4)-\frac{4}{5}\zeta(3)-\frac{2}{5}]\varepsilon^{-1}+[\frac{8}{5}\zeta(3)-\frac{4}{5}]\varepsilon^{-2}-\frac{8}{5}\varepsilon^{-3}-\frac{16}{5}\varepsilon^{-4}+\frac{32}{5}\varepsilon^{-5}$	$\frac{3}{8}$	31	57			31
97	$-[\frac{1}{5}\zeta(3)+\frac{841}{480}]\varepsilon^{-1}+\frac{49}{24}\varepsilon^{-2}-\frac{13}{10}\varepsilon^{-3}+\frac{2}{5}\varepsilon^{-4}$	$\frac{3}{2}$	32	59	$\frac{1}{4}$	1	32
98	$-[\frac{4}{5}\zeta(3)+\frac{193}{480}]\varepsilon^{-1}+\frac{19}{15}\varepsilon^{-2}-\frac{14}{15}\varepsilon^{-3}+\frac{4}{15}\varepsilon^{-4}$	$\frac{3}{4}$	32	60	$\frac{1}{4}$	6	32
99	$\frac{7}{6}\varepsilon^{-2}-2\varepsilon^{-3}+\frac{4}{3}\varepsilon^{-4}$	$\frac{3}{4}$	33	61	$\frac{1}{4}$	9	33
100	$-[\frac{3}{10}\zeta(3)-\frac{81}{160}]\varepsilon^{-1}+\frac{1}{5}\varepsilon^{-2}-\frac{6}{5}\varepsilon^{-3}+\frac{4}{5}\varepsilon^{-4}$	$\frac{3}{8}$	33	62	$\frac{1}{8}$	8	33
101	$-[\frac{3}{5}\zeta(4)-\frac{8}{5}\zeta(3)+\frac{2}{3}]\varepsilon^{-1}-[\frac{4}{5}\zeta(3)-\frac{4}{3}]\varepsilon^{-2}+\varepsilon^{-3}-\frac{8}{3}\varepsilon^{-4}+\frac{4}{3}\varepsilon^{-5}$	3	34	63	$\frac{1}{2}$	4	34
102	$\frac{8}{15}\varepsilon^{-1}-\frac{8}{15}\varepsilon^{-2}+\frac{34}{15}\varepsilon^{-3}-\frac{16}{5}\varepsilon^{-4}+\frac{8}{5}\varepsilon^{-5}$	$\frac{3}{2}$	34	63	$\frac{1}{4}$	4	34

n_5	$I_4 = \mathcal{K}\bar{R}G_4$	W_4	$i_4^{O(N)}$	i_4^{cub}	W_{m^2}	i_2	n_{L+3}^{Vac}
103	$\frac{8}{15}\varepsilon^{-1}+\frac{6}{5}\varepsilon^{-2}-\frac{4}{3}\varepsilon^{-3}-\frac{64}{15}\varepsilon^{-4}+\frac{64}{15}\varepsilon^{-5}$	$\frac{3}{4}$	34	64			34
104	$[\frac{3}{10}\zeta(4)-\zeta(3)+\frac{3}{5}]\varepsilon^{-1}+[\frac{2}{5}\zeta(3)+\frac{3}{5}]\varepsilon^{-2}+\frac{4}{5}\varepsilon^{-3}-4\varepsilon^{-4}+\frac{16}{5}\varepsilon^{-5}$	$\frac{3}{2}$	35	65	$\frac{1}{4}$	7	35
105	$\varepsilon^{-2}-\frac{1}{3}\varepsilon^{-3}-\frac{20}{3}\varepsilon^{-4}+\frac{20}{3}\varepsilon^{-5}$	$\frac{3}{4}$	35	66	$\frac{1}{8}$	3	35
106	$-\frac{2}{15}\varepsilon^{-1}+\frac{16}{15}\varepsilon^{-2}+\frac{8}{5}\varepsilon^{-3}-\frac{32}{5}\varepsilon^{-4}+\frac{64}{15}\varepsilon^{-5}$	$\frac{3}{8}$	35	66	$\frac{1}{8}$	3	35
107	$-[\frac{3}{5}\zeta(4)-\frac{28}{5}\zeta(3)+\frac{17}{3}]\varepsilon^{-1}-[\frac{24}{5}\zeta(3)-\frac{24}{5}]\varepsilon^{-2}$ $-\frac{4}{15}\varepsilon^{-3}-\frac{32}{15}\varepsilon^{-4}+\frac{16}{15}\varepsilon^{-5}$	$\frac{3}{4}$	36	54	$\frac{1}{8}$	7	36
108	$-[4\zeta(3)-2]\varepsilon^{-2}+\frac{10}{3}\varepsilon^{-3}-\frac{16}{3}\varepsilon^{-4}+\frac{8}{3}\varepsilon^{-5}$	$\frac{3}{4}$	36	55	$\frac{1}{8}$	6	36
109	$[\frac{3}{5}\zeta(4)+\frac{28}{5}\zeta(3)-\frac{101}{15}]\varepsilon^{-1}-[\frac{36}{5}\zeta(3)-\frac{52}{15}]\varepsilon^{-2}$ $+\frac{16}{5}\varepsilon^{-3}-\frac{16}{5}\varepsilon^{-4}+\frac{16}{15}\varepsilon^{-5}$	$\frac{3}{8}$	36	55	$\frac{1}{8}$	6	36
110	$\frac{2}{3}\varepsilon^{-2}-\frac{16}{3}\varepsilon^{-4}+\frac{16}{3}\varepsilon^{-5}$	$\frac{3}{4}$	36	56			36
111	$-[\frac{9}{10}\zeta(4)-\frac{9}{5}\zeta(3)+\frac{13}{15}]\varepsilon^{-1}+[\frac{4}{5}\zeta(3)-\frac{58}{15}]\varepsilon^{-2}$ $+\frac{22}{5}\varepsilon^{-3}-\frac{32}{15}\varepsilon^{-4}+\frac{8}{15}\varepsilon^{-5}$	3	37	35	$\frac{1}{2}$	4	37
112	$-[\frac{3}{10}\zeta(4)+\zeta(3)-\frac{29}{15}]\varepsilon^{-1}-[\frac{2}{5}\zeta(3)-\frac{1}{3}]\varepsilon^{-2}$ $+\frac{32}{15}\varepsilon^{-3}-4\varepsilon^{-4}+\frac{32}{15}\varepsilon^{-5}$	$\frac{3}{2}$	37	37			37
113	$[\frac{3}{10}\zeta(4)+\frac{23}{5}\zeta(3)-\frac{18}{5}]\varepsilon^{-1}-[\frac{18}{5}\zeta(3)-\frac{7}{15}]\varepsilon^{-2}$ $+\frac{38}{15}\varepsilon^{-3}-\frac{28}{15}\varepsilon^{-4}+\frac{8}{15}\varepsilon^{-5}$	3	37	36	$\frac{1}{2}$	7	37
114	$-\frac{4}{3}\varepsilon^{-2}+\frac{14}{3}\varepsilon^{-3}-\frac{16}{3}\varepsilon^{-4}+\frac{8}{3}\varepsilon^{-5}$	3	37	38			37
115	$[\frac{1}{5}\zeta(3)+\frac{293}{480}]\varepsilon^{-1}+\frac{1}{40}\varepsilon^{-2}-\frac{11}{10}\varepsilon^{-3}+\frac{14}{15}\varepsilon^{-4}$	$\frac{3}{2}$	38	59	$\frac{1}{4}$	1	37
116	$[\frac{4}{5}\zeta(3)-\frac{25}{12}]\varepsilon^{-1}+\frac{32}{15}\varepsilon^{-2}-\frac{4}{3}\varepsilon^{-3}+\frac{8}{15}\varepsilon^{-4}$	$\frac{3}{4}$	38	60	$\frac{1}{4}$	6	38
117	$32\,\varepsilon^{-5}$	$\frac{3}{32}$	39	67	$\frac{1}{32}$	11	39
118	$[\frac{6}{5}\zeta(4)-\frac{4}{5}\zeta(3)-\frac{2}{5}]\varepsilon^{-1}+[\frac{8}{5}\zeta(3)-\frac{4}{5}]\varepsilon^{-2}-\frac{8}{5}\varepsilon^{-3}-\frac{16}{5}\varepsilon^{-4}+\frac{32}{5}\varepsilon^{-5}$	$\frac{3}{8}$	40	68	$\frac{1}{16}$	10	40
119	$-8\varepsilon^{-4}+16\varepsilon^{-5}$	$\frac{3}{8}$	40	69	$\frac{1}{16}$	2	40
120	$\frac{8}{3}\varepsilon^{-3}-\frac{32}{3}\varepsilon^{-4}+\frac{32}{3}\varepsilon^{-5}$	$\frac{3}{8}$	40	69	$\frac{1}{8}$	2	40
121	$\frac{8}{3}\varepsilon^{-3}-\frac{32}{3}\varepsilon^{-4}+\frac{32}{3}\varepsilon^{-5}$	$\frac{3}{16}$	40	69	$\frac{1}{16}$	2	40
122	$2\varepsilon^{-3}-8\varepsilon^{-4}+8\varepsilon^{-5}$	$\frac{3}{8}$	41	66			41
123	$-[\frac{3}{10}\zeta(4)+\zeta(3)-\frac{29}{15}]\varepsilon^{-1}-[\frac{2}{5}\zeta(3)-\frac{1}{3}]\varepsilon^{-2}+\frac{32}{15}\varepsilon^{-3}-4\varepsilon^{-4}+\frac{32}{15}\varepsilon^{-5}$	$\frac{3}{2}$	41	65	$\frac{1}{4}$	7	41
124	$-\frac{2}{3}\varepsilon^{-2}+4\varepsilon^{-3}-8\varepsilon^{-4}+\frac{16}{3}\varepsilon^{-5}$	$\frac{3}{4}$	41	66	$\frac{1}{8}$	3	41

B.2   Contribution to Z_ϕ

For the calculation of Z_ϕ we need the pole term $\mathcal{K}\bar{R}G$, the weight factor, and the symmetry factor for each tadpole-free 1PI two-point diagram, listed in the subsequent tables. The column labeled i_2 gives the serial numbers of the symmetry factors. The entry n_{L+1}^{Vac} specifies the number of the vacuum diagram with $L+1$ loops in Fig. 14.3, from which the corresponding L-loop two-point diagram is generated. The calculation of $\mathcal{K}\bar{R}G_2$ is shown diagrammatically in Appendix A, where the diagrams are listed in the same order as here.

n_L	$I_2 = \partial_{\mathbf{k}^2}\mathcal{K}\bar{R}G_2$	W_2	i_2	n_{L+1}^{Vac}
n_2	2 − loop			
1	$-\frac{1}{2}\varepsilon^{-1}$	$\frac{1}{6}$	2	1
n_3	3 − loop			
1	$-\frac{1}{6}\varepsilon^{-1} + \frac{2}{3}\varepsilon^{-2}$	$\frac{1}{4}$	3	1
n_4	4 − loop			
1	$\frac{5}{16}\varepsilon^{-1} + \frac{1}{4}\varepsilon^{-2} - \varepsilon^{-3}$	$\frac{1}{8}$	6	1
2	$\frac{5}{16}\varepsilon^{-1} - \frac{1}{8}\varepsilon^{-2}$	$\frac{1}{12}$	4	3
3	$-\frac{13}{48}\varepsilon^{-1} + \frac{7}{12}\varepsilon^{-2} - \frac{1}{3}\varepsilon^{-3}$	$\frac{1}{4}$	5	2
4	$-\frac{2}{3}\varepsilon^{-1} + \frac{2}{3}\varepsilon^{-2} - \frac{2}{3}\varepsilon^{-3}$	$\frac{1}{4}$	5	2
n_5	5 − loop			
1	$[\frac{2}{5}\zeta(3) - \frac{13}{40}]\varepsilon^{-1} - \frac{1}{2}\varepsilon^{-2} - \frac{2}{5}\varepsilon^{-3} + \frac{8}{5}\varepsilon^{-4}$	$\frac{1}{16}$	10	1
2	$\frac{7}{64}\varepsilon^{-1} - \frac{23}{60}\varepsilon^{-2} + \frac{11}{60}\varepsilon^{-3}$	$\frac{1}{6}$	1	3
3	$-\frac{149}{480}\varepsilon^{-1} - \frac{17}{60}\varepsilon^{-2} + \frac{7}{30}\varepsilon^{-3}$	$\frac{1}{24}$	1	3
4	$\frac{209}{480}\varepsilon^{-1} - \frac{3}{10}\varepsilon^{-2} + \frac{1}{10}\varepsilon^{-3}$	$\frac{1}{8}$	1	3
5	$-[\frac{1}{10}\zeta(3) - \frac{13}{80}]\varepsilon^{-1} + \frac{19}{60}\varepsilon^{-2} - \frac{13}{15}\varepsilon^{-3} + \frac{8}{15}\varepsilon^{-4}$	$\frac{1}{4}$	7	2
6	$-[\frac{3}{10}\zeta(3) - \frac{137}{480}]\varepsilon^{-1} + \frac{11}{40}\varepsilon^{-2} - \frac{9}{10}\varepsilon^{-3} + \frac{14}{15}\varepsilon^{-4}$	$\frac{1}{4}$	7	2
7	$-[\frac{6}{5}\zeta(3) - \frac{277}{240}]\varepsilon^{-1} - \frac{11}{60}\varepsilon^{-2} - \frac{7}{15}\varepsilon^{-3} + \frac{4}{15}\varepsilon^{-4}$	$\frac{1}{8}$	7	2
8	$[\frac{1}{5}\zeta(3) - \frac{347}{480}]\varepsilon^{-1} + \frac{39}{40}\varepsilon^{-2} - \frac{1}{2}\varepsilon^{-3} + \frac{2}{15}\varepsilon^{-4}$	$\frac{1}{2}$	4	5
9	$-\frac{14}{15}\varepsilon^{-1} + \frac{22}{15}\varepsilon^{-2} - \frac{16}{15}\varepsilon^{-3} + \frac{8}{15}\varepsilon^{-4}$	$\frac{1}{4}$	4	5
10	$[\frac{3}{5}\zeta(3) - \frac{331}{480}]\varepsilon^{-1} + \frac{77}{120}\varepsilon^{-2} - \frac{23}{30}\varepsilon^{-3} + \frac{2}{5}\varepsilon^{-4}$	$\frac{1}{2}$	4	5
11	$-[\frac{6}{5}\zeta(4) + \frac{3}{5}\zeta(3)]\varepsilon^{-1} + \frac{12}{5}\varepsilon^{-2}\zeta(3)$	$\frac{1}{6}$	5	4

B.3 Symmetry Factors in O(N) Symmetric Theory

In this section, we list all symmetry factors for the O(N)-symmetric theory.

B.3.1 Symmetry Factors $S_2^{O(N)}$ of Two-Point Diagrams

The symmetry factors S_2 of the two-point diagrams contribute to the calculation of Z_{m^2} and of Z_ϕ.

i_2	$S_2^{O(N)}(i_2)$
1-loop	
1	$(2+N)/3$
2-loop	
1	$(4+4N+N^2)/9$
2	$(2+N)/3$
3-loop	
1	$(8+12N+6N^2+N^3)/27$
2	$(12+12N+3N^2)/27$
3	$(16+10N+N^2)/27$
4-loop	
1	$(16+32N+24N^2+8N^3+N^4)/81$
2	$(24+36N+18N^2+3N^3)/81$
3	$(32+36N+12N^2+N^3)/81$
4	$(36+36N+9N^2)/81$
5	$(44+32N+5N^2)/81$
6	$(40+32N+8N^2+N^3)/81$
5-loop	
1	$(96+108N+36N^2+3N^3)/243$
2	$(48+96N+72N^2+24N^3+3N^4)/243$
3	$(72+108N+54N^2+9N^3)/243$
4	$(120+100N+22N^2+N^3)/243$
5	$(132+96N+15N^2)/243$
6	$(88+108N+42N^2+5N^3)/243$
7	$(112+100N+28N^2+3N^3)/243$
8	$(80+104N+48N^2+10N^3+N^4)/243$
9	$(64+104N+60N^2+14N^3+N^4)/243$
10	$(96+96N+40N^2+10N^3+N^4)/243$
11	$(32+80N+80N^2+40N^3+10N^4+N^5)/243$

B.3.2   Symmetry Factors $S_4^{O(N)}$ of Four-Point Diagrams

The symmetry factors S_4 of the four-point diagrams contribute to the calculation of Z_g. Some symmetry factors are referred to by more than one number since different vacuum diagrams may result in the same symmetry factor.

$i_4^{O(N)}$	$S_4^{O(N)}(i_4^{O(N)})$
1-loop	
1	$(8+N)/9$
2-loop	
1	$(20+6N+N^2)/27$
2	$(22+5N)/27$
3-loop	
1	$(48+24N+8N^2+N^3)/81$
2	$(56+22N+3N^2)/81$
3	$(48+30N+3N^2)/81$
4	$(66+15N)/81$
5	$(60+20N+N^2)/81$
4-loop	
1	$(112+80N+40N^2+10N^3+N^4)/243$
2	$(136+80N+24N^2+3N^3)/243$
3, 7	$(156+76N+11N^2)/243$
4	$(144+80N+18N^2+N^3)/243$
5	$(120+96N+24N^2+3N^3)/243$
6	$(152+76N+14N^2+N^3)/243$
8	$(186+55N+2N^2)/243$
9, 15	$(176+62N+5N^2)/243$
10	$(164+72N+7N^2)/243$
11	$(128+96N+18N^2+N^3)/243$
12, 13	$(132+96N+15N^2)/243$
14	$(160+72N+10N^2+N^3)/243$

$i_4^{O(N)}$	$S_4^{O(N)}(i_4^{O(N)})$
5-loop	
1	$(528+186N+15N^2)/729$
2	$(526+189N+14N^2)/729$
3, 6, 10	$(360+300N+66N^2+3N^3)/729$
4	$(288+288N+120N^2+30N^3+3N^4)/729$
5, 11, 12	$(336+300N+84N^2+9N^3)/729$
7, 13	$(396+288N+45N^2)/729$
8, 9	$(288+324N+108N^2+9N^3)/729$
14	$(492+210N+26N^2+N^3)/729$
15	$(504+206N+19N^2)/729$
16	$(440+232N+52N^2+5N^3)/729$
17	$(484+220N+25N^2)/729$
18	$(472+224N+32N^2+N^3)/729$
19	$(464+224N+38N^2+3N^3)/729$
20, 25	$(416+252N+56N^2+5N^3)/729$
21, 26	$(448+244N+36N^2+N^3)/729$
22, 37	$(424+252N+50N^2+3N^3)/729$
23	$(440+244N+42N^2+3N^3)/729$
27	$(352+264N+96N^2+16N^3+N^4)/729$
28	$(368+256N+88N^2+16N^3+N^4)/729$
29	$(400+248N+68N^2+12N^3+N^4)/729$
30, 36	$(400+260N+64N^2+5N^3)/729$
31	$(384+256N+76N^2+12N^3+N^4)/729$
32, 38	$(352+300N+72N^2+5N^3)/729$
33	$(320+296N+96N^2+16N^3+N^4)/729$
34	$(432+252N+44N^2+N^3)/729$
35, 41	$(384+260N+76N^2+9N^3)/729$
39	$(256+240N+160N^2+60N^3+12N^4+N^5)/729$
40	$(320+256N+120N^2+30N^3+3N^4)/729$

B.4   The Symmetry Factors for Cubic Symmetry

In this section, we list all symmetry factors for the theory with $O(N)$-cubic symmetry. Some relations for the $O(N)$-cubic symmetry factors were discussed in Section 6.4.2 and on page 89.

B.4.1   Symmetry Factors $S^{\mathrm{cub}}_{2;(L-k\,,k)}$ of Two-Point Diagrams

The symmetry factors of the two-point diagrams appear in sums of the form $\sum_{k=0}^{L}(\bar{g}_1)^{L-k}(\bar{g}_2)^k S^{\mathrm{cub}}_{2;(L-k,k)}$, where $L = V$. The S_2 contribute to the calculation of Z_ϕ and of Z_{m^2}. The coefficient of the power $(g_2)^L$ is always equal to unity, $S^{\mathrm{cub}}_{2;(0,L)}(i_{m^2}) = 1$, and therefore not listed in the table. Note that $S^{\mathrm{cub}}_{2;(L,0)}(i_2)$ equals $S_2^{O(N)}(i_2)$.

1-loop	
i_2	$S^{\mathrm{cub}}_{2;(1,0)}(i_2)$
1	$(2+N)/3$

2-loop		
i_2	$S^{\mathrm{cub}}_{2;(2,0)}(i_2)$	$S^{\mathrm{cub}}_{2,(1,1)}(i_2)$
1	$(4+4N+N^2)/9$	$(4+2N)/3$
2	$(6+3N)/9$	$6/3$

3-loop			
i_2	$S^{\mathrm{cub}}_{2;(3,0)}(i_2)$	$S^{\mathrm{cub}}_{2;(2,1)}(i_2)$	$S^{\mathrm{cub}}_{2;(1,2)}(i_2)$
1	$(8+12N+6N^2+N^3)/27$	$(12+12N+3N^2)/9$	$(6+3N)/3$
2	$(12+12N+3N^2)/27$	$(18+9N)/9$	$(8+N)/3$
3	$(16+10N+N^2)/27$	$(24+3N)/9$	$9/3$

4-loop				
i_2	$S^{\mathrm{cub}}_{2;(4,0)}(i_2)$	$S^{\mathrm{cub}}_{2;(3,1)}(i_2)$	$S^{\mathrm{cub}}_{2;(2,2)}(i_2)$	$S^{\mathrm{cub}}_{2;(1,3)}(i_2)$
1	$(16+32N+24N^2+8N^3+N^4)/81$	$(32+48N+24N^2+4N^3)/27$	$(24+24N+6N^2)/9$	$(8+4N)/3$
2	$(24+36N+18N^2+3N^3)/81$	$(48+48N+12N^2)/27$	$(34+19N+N^2)/9$	$(10+2N)/2$
3	$(32+36N+12N^2+N^3)/81$	$(64+40N+4N^2)/27$	$(42+12N)/9$	$(11+N)/3$
4	$(36+36N+9N^2)/81$	$(72+36N)/27$	$(48+6N)/9$	$12/3$
5	$(44+32N+5N^2)/81$	$(88+20N)/27$	$(52+2N)/9$	$12/3$
6	$(40+32N+8N^2+N^3)/81$	$(80+24N+4N^2)/27$	$(48+6N)/9$	$12/3$

5-loop

i_2	$S^{\text{cub}}_{2;(5,0)}(i_2)$	$S^{\text{cub}}_{2;(4,1)}(i_2)$	$S^{\text{cub}}_{2;(3,2)}(i_2)$	$S^{\text{cub}}_{2;(2,3)}(i_2)$	$S^{\text{cub}}_{2;(1,4)}(i_2)$
1	$(96+108N+36N^2+3N^3)/243$	$(240+150N+15N^2)/81$	$(214+55N+N^2)/27$	$(84+6N)/9$	$15/3$
2	$(48+96N+72N^2+24N^3+3N^4)/243$	$(120+180N+90N^2+15N^3)/81$	$(116+120N+33N^2+N^3)/27$	$(54+33N+3N^2)/9$	$(12+3N)/3$
3	$(72+108N+54N^2+9N^3)/243$	$(180+180N+45N^2)/81$	$(168+96N+6N^2)/27$	$(72+18N)/9$	$(14+N)/3$
4	$(120+100N+22N^2+N^3)/243$	$(300+100N+5N^2)/81$	$(244+26N)/27$	$(88+2N)/9$	$15/3$
5	$(132+96N+15N^2)/243$	$(330+75N)/81$	$(260+10N)/27$	$90/9$	$15/3$
6	$(88+108N+42N^2+5N^3)/243$	$(220+160N+25N^2)/81$	$(192+76N+2N^2)/27$	$(76+14N)/9$	$(14+N)/3$
7	$(112+100N+28N^2+3N^3)/243$	$(280+110N+15N^2)/81$	$(230+39N+N^2)/27$	$(86+4N)/9$	$15/3$
8	$(80+104N+48N^2+10N^3+N^4)/243$	$(200+160N+40N^2+5N^3)/81$	$(176+84N+10N^2)/27$	$(72+18N)/9$	$(14+N)/3$
9	$(64+104N+60N^2+14N^3+N^4)/243$	$(160+180N+60N^2+5N^3)/81$	$(148+106N+16N^2)/27$	$(64+25N+N^2)/9$	$(13+2N)/3$
10	$(96+96N+40N^2+10N^3+N^4)/243$	$(240+120N+40N^2+5N^3)/81$	$(200+60N+10N^2)/27$	$(80+10N)/9$	$15/3$
11	$(32+80N+80N^2+40N^3+10N^4+N^5)/243$	$(80+160N+120N^2+40N^3+5N^4)/81$	$(80+120N+60N^2+10N^3)/27$	$(40+40N+10N^2)/9$	$(10+5N)/3$

B.4.2 Symmetry Factors $S_{4;\,(L+1-k,k)}^{\text{cub}}$ of Four-Point Diagrams

As shown in Section 6.4.2, the symmetry factors of the four-point diagrams appear in sums of the form $\sum_{k=0}^{L+1}(\bar{g}_1)^{L+1-k}(\bar{g}_2)^k\left[S_{4_1;(L+1-k,k)}^{\text{cub}}T_{\alpha\beta\gamma\delta}^{(1)}+S_{4_2;(L+1-k,k)}^{\text{cub}}T_{\alpha\beta\gamma\delta}^{(2)}\right]$, where $L+1=V$. The symmetry factors S_{4_1} contribute to Z_{g_1}, whereas S_{4_2} contributes to Z_{g_2}. The relation of the symmetry factors of four-point diagrams to those of vacuum diagrams is explained on page 89.

Contribution to Z_{g_1}

Note that the symmetry factor $S_{4_1;(L+1,0)}^{\text{cub}}(i_4^{\text{cub}})$ equals the $O(N)$ symmetry factor $S_4^{O(N)}(i_4)$. This was pointed out in Section 6.4.2.

1-loop		
i_4^{cub}	$S_{4_1;(2,0)}^{\text{cub}}(i_4^{\text{cub}})$	$S_{4_1;(1,1)}^{\text{cub}}(i_4^{\text{cub}})$
1	$(8+N)/9$	$2/3$

2-loop			
i_4^{cub}	$S_{4_1;(3,0)}^{\text{cub}}(i_4^{\text{cub}})$	$S_{4_1;(2,1)}^{\text{cub}}(i_4^{\text{cub}})$	$S_{4_1;(1,2)}^{\text{cub}}(i_4^{\text{cub}})$
1	$(20+6N+N^2)/27$	$(12+3N)/9$	1
2	$(22+5N)/27$	$12/9$	$1/3$

3-loop				
i_4^{cub}	$S_{4_1;(4,0)}^{\text{cub}}(i_4^{\text{cub}})$	$S_{4_1;(3,1)}^{\text{cub}}(i_4^{\text{cub}})$	$S_{4_1;(2,2)}^{\text{cub}}(i_4^{\text{cub}})$	$S_{4_1;(1,3)}^{\text{cub}}(i_4^{\text{cub}})$
1	$(48+24N+8N^2+N^3)/81$	$(48+24N+4N^2)/27$	$(24+6N)/9$	$4/3$
2	$(56+22N+3N^2)/81$	$(52+12N)/27$	$(18+N)/9$	$2/3$
3	$(48+30N+3N^2)/81$	$(60+12N)/27$	$(20+N)/9$	$2/3$
4	$(66+15N)/81$	$48/27$	$6/9$	0
5	$(56+22N+3N^2)/81$	$(52+9N)/27$	$15/9$	$1/3$
6	$(60+20N+N^2)/81$	$(48+4N)/27$	$8/9$	0
7	$(60+20N+N^2)/81$	$(54+4N)/27$	$13/9$	$1/3$

4-loop				
i_4^{cub}	$S_{4_1;(5,0)}^{\text{cub}}(i_4^{\text{cub}})$	$S_{4_1;(4,1)}^{\text{cub}}(i_4^{\text{cub}})$	$S_{4_1;(3,2)}^{\text{cub}}(i_4^{\text{cub}})$	$S_{4_1;(2,3)}^{\text{cub}}(i_4^{\text{cub}})$	$S_{4_1;(1,4)}^{\text{cub}}(i_4^{\text{cub}})$
1	$(112+80N+40N^2+10N^3+N^4)/243$	$(160+120N+40N^2+5N^3)/81$	$(120+60N+10N^2)/27$	$(40+10N)/9$	$5/3$
2	$(136+80N+24N^2+3N^3)/243$	$(184+90N+15N^2)/81$	$(114+33N+N^2)/27$	$(32+3N)/9$	$3/3$
3	$(156+76N+11N^2)/243$	$(192+52N)/81$	$(88+6N)/27$	$16/9$	$1/3$
4	$(144+80N+18N^2+N^3)/243$	$(192+68N+5N^2)/81$	$(100+15N)/27$	$(22+N)/9$	$2/3$
5	$(120+96N+24N^2+3N^3)/243$	$(192+90N+15N^2)/81$	$(110+33N+N^2)/27$	$(30+3N)/9$	$3/3$
6	$(152+76N+14N^2+N^3)/243$	$(200+60N+5N^2)/81$	$(102+16N)/27$	$(24+N)/9$	$2/3$
7	$(186+55N+2N^2)/243$	$(192+10N)/81$	$50/27$	$4/9$	0
8	$(176+62N+5N^2)/243$	$(188+22N)/81$	$(56+N)/27$	$5/9$	0
9	$(164+72N+7N^2)/243$	$(188+32N)/81$	$(62+2N)/27$	$6/9$	0
10	$(152+76N+14N^2+N^3)/243$	$(176+48N+5N^2)/81$	$(64+9N)/27$	$8/9$	0
11	$(128+96N+18N^2+N^3)/243$	$(224+68N+5N^2)/81$	$(120+15N)/27$	$(26+N)/9$	$2/3$
12	$(132+96N+15N^2)/243$	$(204+66N)/81$	$(100+8N)/27$	$18/9$	$1/3$
13	$(156+76N+11N^2)/243$	$(204+46N)/81$	$(92+4N)/27$	$16/9$	$1/3$
14	$(136+80N+24N^2+3N^3)/243$	$(184+72N+12N^2)/81$	$(96+18N)/27$	$20/9$	$1/3$
15	$(160+72N+10N^2+N^3)/243$	$(196+40N+5N^2)/81$	$(80+10N)/27$	$16/9$	$1/3$
16	$(176+62N+5N^2)/243$	$(206+25N)/81$	$(80+2N)/27$	$14/9$	$1/3$
17	$(164+72N+7N^2)/243$	$(200+32N)/81$	$(76+2N)/27$	$12/9$	$1/3$
18	$(152+76N+14N^2+N^3)/243$	$(200+48N+5N^2)/81$	$(90+9N)/27$	$17/9$	$1/3$
19	$(164+72N+7N^2)/243$	$(212+32N)/81$	$(88+2N)/27$	$15/9$	$1/3$

| i_4^{cub} | $S^{cub}_{41;(6,0)}(i_4^{cub})$ | $S^{cub}_{41;(5,1)}(i_4^{cub})$ | 5-loop | | | |
			$S^{cub}_{41;(4,2)}(i_4^{cub})$	$S^{cub}_{41;(3,3)}(i_4^{cub})$	$S^{cub}_{41;(2,4)}(i_4^{cub})$	$S^{cub}_{41;(1,5)}(i_4^{cub})$
1	$(528+186N+15N^2)/729$	$(702+81N)/243$	$(288+6N)/81$	$48/27$	$3/9$	0
2	$(528+186N+15N^2)/729$	$(696+84N)/243$	$(286+7N)/81$	$48/27$	$3/9$	0
3	$(526+189N+14N^2)/729$	$(708+78N)/243$	$(290+5N)/81$	$48/27$	$3/9$	0
4	$(360+300N+66N^2+3N^3)/729$	$(684+306N+18N^2)/243$	$(462+83N+N^2)/81$	$(138+7N)/27$	$19/9$	$1/3$
5	$(288+288N+120N^2+30N^3+3N^4)/729$	$(576+432N+144N^2+18N^3)/243$	$(480+276N+50N^2+N^3)/81$	$(216+72N+4N^2)/27$	$(48+6N)/9$	$4/3$
6	$(336+300N+84N^2+9N^3)/729$	$(648+360N+54N^2)/243$	$(476+154N+6N^2)/81$	$(172+24N)/27$	$(30+N)/9$	$2/3$
7	$(336+300N+84N^2+9N^3)/729$	$(648+342N+45N^2)/243$	$(458+121N+3N^2)/81$	$(148+12N)/27$	$21/9$	$1/3$
8	$(360+300N+66N^2+3N^3)/729$	$(648+288N+18N^2)/243$	$(396+68N+N^2)/81$	$(96+4N)/27$	$8/9$	0
9	$(396+288N+45N^2)/729$	$(684+234N)/243$	$(390+33N)/81$	$84/27$	$6/9$	0
10	$(288+324N+108N^2+9N^3)/729$	$(648+432N+54N^2)/243$	$(528+168N+6N^2)/81$	$(192+24N)/27$	$(32+N)/9$	$2/3$
11	$(492+210N+26N^2+N^3)/729$	$(684+126N+6N^2)/243$	$(306+23N)/81$	$(58+N)/27$	$4/9$	0
12	$(492+210N+26N^2+N^3)/729$	$(696+120N+6N^2)/243$	$(316+18N)/81$	$60/27$	$4/9$	0
13	$(492+210N+26N^2+N^3)/729$	$(750+141N+6N^2)/243$	$(418+31N)/81$	$(118+2N)/27$	$17/9$	$1/3$
14	$(504+206N+19N^2)/729$	$(684+102N)/243$	$(280+8N)/81$	$42/27$	$2/9$	0
15	$(504+206N+19N^2)/729$	$(708+102N)/243$	$(312+8N)/81$	$56/27$	$4/9$	0
16	$(504+206N+19N^2)/729$	$(768+108N)/243$	$(404+10N)/81$	$104/27$	$15/9$	$1/3$
17	$(440+232N+52N^2+5N^3)/729$	$(648+198N+27N^2)/243$	$(338+57N+N^2)/81$	$(78+3N)/27$	$6/9$	0
18	$(440+232N+52N^2+5N^3)/729$	$(720+252N+30N^2)/243$	$(494+105N+2N^2)/81$	$(176+16N)/27$	$(30+N)/9$	$2/3$
19	$(484+220N+25N^2)/729$	$(696+132N)/243$	$(326+13N)/81$	$60/27$	$4/9$	0
20	$(484+220N+25N^2)/729$	$(684+138N)/243$	$(322+15N)/81$	$60/27$	$4/9$	0
21	$(484+220N+25N^2)/729$	$(786+135N)/243$	$(452+14N)/81$	$122/27$	$17/9$	$1/3$
22	$(472+224N+32N^2+3N^3)/729$	$(732+162N+6N^2)/243$	$(412+32N)/81$	$(112+2N)/27$	$16/9$	$1/3$
23	$(472+224N+32N^2+3N^3)/729$	$(660+150N+6N^2)/243$	$(298+25N)/81$	$(52+N)/27$	$3/9$	0
24	$(472+224N+32N^2+3N^3)/729$	$(672+156N+6N^2)/243$	$(320+30N)/81$	$(64+2N)/27$	$5/9$	0
25	$(464+224N+38N^2+3N^3)/729$	$(648+168N+18N^2)/243$	$(304+48N+N^2)/81$	$(64+4N)/27$	$5/9$	0
26	$(464+224N+38N^2+3N^3)/729$	$(720+180N+18N^2)/243$	$(418+55N+N^2)/81$	$(124+5N)/27$	$18/9$	$1/3$
27	$(464+224N+38N^2+3N^3)/729$	$(660+168N+15N^2)/243$	$(320+42N)/81$	$(68+2N)/27$	$5/9$	0
28	$(416+252N+56N^2+5N^3)/729$	$(624+222N+27N^2)/243$	$(322+63N+N^2)/81$	$(70+4N)/27$	$5/9$	0
29	$(416+252N+56N^2+5N^3)/729$	$(720+264N+30N^2)/243$	$(492+104N+2N^2)/81$	$(170+16N)/27$	$(29+N)/9$	$2/3$
30	$(416+252N+56N^2+5N^3)/729$	$(648+222N+27N^2)/243$	$(354+63N+N^2)/81$	$(84+4N)/27$	$7/9$	0
31	$(416+252N+56N^2+5N^3)/729$	$(720+222N+27N^2)/243$	$(450+63N+N^2)/81$	$(132+4N)/27$	$18/9$	$1/3$
32	$(448+244N+36N^2+N^3)/729$	$(672+180N+6N^2)/243$	$(336+34N)/81$	$(68+2N)/27$	$5/9$	0
33	$(448+244N+36N^2+N^3)/729$	$(768+180N+6N^2)/243$	$(464+34N)/81$	$(130+2N)/27$	$18/9$	$1/3$
34	$(448+244N+36N^2+N^3)/729$	$(684+174N+6N^2)/243$	$(346+29N)/81$	$(70+N)/27$	$5/9$	0
35	$(424+252N+50N^2+3N^3)/729$	$(732+228N+15N^2)/243$	$(468+58N)/81$	$(138+4N)/27$	$19/9$	$1/3$

5-loop

i_4^{cub}	$S_{4i;(6,0)}^{cub}(i_4^{cub})$	$S_{4i;(5,1)}^{cub}(i_4^{cub})$	$S_{9i;(4,2)}^{cub}(i_4^{cub})$	$S_{4i;(3,3)}^{cub}(i_4^{cub})$	$S_{4i;(2,4)}^{cub}(i_4^{cub})$	$S_{4i;(1,5)}^{cub}(i_4^{cub})$
36	$(424+252N+50N^2+3N^3)/729$	$(732+222N+18N^2)/243$	$(462+59N+N^2)/81$	$(136+5N)/27$	$19/9$	$1/3$
37	$(424+252N+50N^2+3N^3)/729$	$(648+216N+18N^2)/243$	$(344+54N+N^2)/81$	$(76+4N)/27$	$6/9$	0
38	$(424+252N+50N^2+3N^3)/729$	$(684+246N+18N^2)/243$	$(434+73N+N^2)/81$	$(132+7N)/27$	$19/9$	$1/3$
39	$(440+244N+42N^2+3N^3)/729$	$(708+198N+18N^2)/243$	$(410+57N+N^2)/81$	$(118+5N)/27$	$17/9$	$1/3$
40	$(440+244N+42N^2+3N^3)/729$	$(732+192N+15N^2)/243$	$(436+46N)/81$	$(124+2N)/27$	$17/9$	$1/3$
41	$(440+244N+42N^2+3N^3)/729$	$(648+192N+18N^2)/243$	$(324+52N+N^2)/81$	$(72+4N)/27$	$6/9$	0
42	$(448+244N+36N^2+N^3)/729$	$(744+180N+6N^2)/243$	$(436+34N)/81$	$(120+2N)/27$	$17/9$	$1/3$
43	$(448+244N+36N^2+N^3)/729$	$(648+168N+6N^2)/243$	$(296+24N)/81$	$48/27$	$2/9$	0
44	$(448+244N+36N^2+N^3)/729$	$(720+180N+6N^2)/243$	$(404+34N)/81$	$(106+2N)/27$	$15/9$	$1/3$
45	$(416+252N+56N^2+5N^3)/729$	$(672+222N+27N^2)/243$	$(390+63N+N^2)/81$	$(108+4N)/27$	$15/9$	$1/3$
46	$(416+252N+56N^2+5N^3)/729$	$(696+264N+30N^2)/243$	$(464+104N+2N^2)/81$	$(160+16N)/27$	$(28+N)/9$	$2/3$
47	$(352+264N+96N^2+16N^3+N^4)/729$	$(624+336N+72N^2+6N^3)/243$	$(456+144N+14N^2)/81$	$(160+24N)/27$	$(28+N)/9$	$2/3$
48	$(352+264N+96N^2+16N^3+N^4)/729$	$(624+372N+84N^2+6N^3)/243$	$(492+192N+21N^2)/81$	$(196+42N+N^2)/27$	$(39+3N)/9$	$3/3$
49	$(368+256N+88N^2+16N^3+N^4)/729$	$(648+288N+72N^2+6N^3)/243$	$(440+120N+14N^2)/81$	$(144+16N)/27$	$21/9$	$1/3$
50	$(368+256N+88N^2+16N^3+N^4)/729$	$(648+360N+84N^2+6N^3)/243$	$(512+198N+21N^2)/81$	$(210+43N+N^2)/27$	$(41+3N)/9$	$3/3$
51	$(368+256N+88N^2+16N^3+N^4)/729$	$(576+288N+72N^2+6N^3)/243$	$(344+120N+14N^2)/81$	$(96+16N)/27$	$10/9$	0
52	$(400+248N+68N^2+12N^3+N^4)/729$	$(648+252N+60N^2+6N^3)/243$	$(400+114N+15N^2)/81$	$(132+19N)/27$	$21/9$	$1/3$
53	$(400+248N+68N^2+12N^3+N^4)/729$	$(672+288N+60N^2+6N^3)/243$	$(468+144N+15N^2)/81$	$(176+28N)/27$	$(32+N)/9$	$2/3$
54	$(400+260N+64N^2+5N^3)/729$	$(696+258N+27N^2)/243$	$(454+75N+N^2)/81$	$(136+6N)/27$	$19/9$	$1/3$
55	$(400+260N+64N^2+5N^3)/729$	$(696+288N+30N^2)/243$	$(484+106N+2N^2)/81$	$(164+16N)/27$	$(28+N)/9$	$2/3$
56	$(400+260N+64N^2+5N^3)/729$	$(648+270N+27N^2)/243$	$(414+79N+N^2)/81$	$(122+6N)/27$	$17/9$	$1/3$
57	$(384+256N+76N^2+12N^3+N^4)/729$	$(672+300N+60N^2+6N^3)/243$	$(476+138N+15N^2)/81$	$(172+27N)/27$	$(31+N)/9$	$2/3$
58	$(384+256N+76N^2+12N^3+N^4)/729$	$(576+264N+60N^2+6N^3)/243$	$(312+108N+15N^2)/81$	$(80+18N)/27$	$9/9$	0
59	$(352+300N+72N^2+5N^3)/729$	$(720+306N+27N^2)/243$	$(502+91N+N^2)/81$	$(154+8N)/27$	$21/9$	$1/3$
60	$(352+300N+72N^2+5N^3)/729$	$(792+312N+30N^2)/243$	$(592+108N+2N^2)/81$	$(200+16N)/27$	$(32+N)/9$	$2/3$
61	$(320+296N+96N^2+16N^3+N^4)/729$	$(672+372N+84N^2+6N^3)/243$	$(516+192N+21N^2)/81$	$(200+42N+N^2)/27$	$(39+3N)/9$	$3/3$
62	$(320+296N+96N^2+16N^3+N^4)/729$	$(720+336N+72N^2+6N^3)/243$	$(544+144N+14N^2)/81$	$(192+24N)/27$	$(32+N)/9$	$2/3$
63	$(432+252N+44N^2+N^3)/729$	$(720+216N+6N^2)/243$	$(444+42N)/81$	$(124+2N)/27$	$17/9$	$1/3$
64	$(432+252N+44N^2+N^3)/729$	$(648+216N+6N^2)/243$	$(360+36N)/81$	$80/27$	$6/9$	0
65	$(384+260N+76N^2+9N^3)/729$	$(672+294N+45N^2)/243$	$(466+105N+3N^2)/81$	$(150+10N)/27$	$21/9$	$1/3$
66	$(384+260N+76N^2+9N^3)/729$	$(648+336N+54N^2)/243$	$(488+152N+6N^2)/81$	$(184+24N)/27$	$(32+N)/9$	$2/3$
67	$(256+240N+160N^2+60N^3+12N^4+N^5)/729$	$(480+480N+240N^2+60N^3+6N^4)/243$	$(480+360N+120N^2+15N^3)/81$	$(240+120N+20N^2)/27$	$(60+15N)/9$	$6/3$
68	$(320+256N+120N^2+30N^3+3N^4)/729$	$(576+360N+120N^2+15N^3)/243$	$(440+180N+30N^2)/81$	$(160+30N)/27$	$25/9$	$1/3$
69	$(320+256N+120N^2+30N^3+3N^4)/729$	$(576+432N+144N^2+18N^3)/243$	$(512+276N+50N^2+N^3)/81$	$(232+72N+4N^2)/27$	$(50+6N)/9$	$4/3$

Contribution to Z_{g_2}

The coefficient of the power $(g_2)^{L+1}$ is always equal to unity, $S^{\mathrm{cub}}_{42;(0,L+1)}(i^{\mathrm{cub}}_4) = 1$, as explained in Section 6.4.2, and is therefore not listed.

	1-loop
i^{cub}_4	$S^{\mathrm{cub}}_{42;(1,1)}(i^{\mathrm{cub}}_4)$
1	4/3

	2-loop	
i^{cub}_4	$S^{\mathrm{cub}}_{42;(2,1)}(i^{\mathrm{cub}}_4)$	$S^{\mathrm{cub}}_{42;(1,2)}(i^{\mathrm{cub}}_4)$
1	12/9	6/3
2	$(14+N)/9$	8/3

	3-loop		
i^{cub}_4	$S^{\mathrm{cub}}_{42;(3,1)}(i^{\mathrm{cub}}_4)$	$S^{\mathrm{cub}}_{42;(2,2)}(i^{\mathrm{cub}}_4)$	$S^{\mathrm{cub}}_{42;(1,3)}(i^{\mathrm{cub}}_4)$
1	32/27	24/9	8/3
2	$(40+4N)/27$	$(34+N)/9$	10/3
3	$(24+12N)/27$	$(30+3N)/9$	10/3
4	$(56+4N)/27$	48/9	12/3
5	$(40+6N+N^2)/27$	$(36+3N)/9$	11/3
6	$(48+8N)/27$	$(44+2N)/9$	12/3
7	$(44+6N)/27$	$(40+N)/9$	11/3

	4-loop			
i^{cub}_4	$S^{\mathrm{cub}}_{42;(4,1)}(i^{\mathrm{cub}}_4)$	$S^{\mathrm{cub}}_{42;(3,2)}(i^{\mathrm{cub}}_4)$	$S^{\mathrm{cub}}_{42;(2,3)}(i^{\mathrm{cub}}_4)$	$S^{\mathrm{cub}}_{42;(1,4)}(i^{\mathrm{cub}}_4)$
1	80/81	80/27	40/9	10/3
2	$(104+12N)/81$	$(116+6N)/27$	$(54+N)/9$	12/3
3	$(132+28N+N^2)/81$	$(160+16N)/27$	$(72+2N)/9$	14/3
4	$(112+24N+4N^2)/81$	$(136+18N+N^2)/27$	$(64+3N)/9$	13/3
5	$(72+36N)/81$	$(108+18N)/27$	$(54+3N)/9$	12/3
6	$(120+20N)/81$	$(144+8N)/27$	$(64+N)/9$	13/3
7	$(182+21N)/81$	$(216+4N)/27$	86/9	15/3
8	$(168+26N+N^2)/81$	$(202+11N)/27$	$(84+N)/9$	15/3
9	$(148+36N+N^2)/81$	$(188+18N)/27$	$(82+2N)/9$	15/3
10	$(136+36N+4N^2)/81$	$(172+24N+N^2)/27$	$(78+4N)/9$	15/3
11	$(64+40N+4N^2)/81$	$(112+22N+N^2)/27$	$(60+3N)/9$	13/3
12	$(84+48N+3N^2)/81$	$(132+30N)/27$	$(68+4N)/9$	14/3
13	$(124+28N+3N^2)/81$	$(156+18N)/27$	$(72+2N)/9$	14/3
14	$(104+24N+8N^2+N^3)/81$	$(128+24N+4N^2)/27$	$(64+6N)/9$	14/3
15	$(136+28N)/81$	$(168+12N)/27$	$(72+2N)/9$	14/3
16	$(156+18N)/81$	$(184+4N)/27$	76/9	14/3
17	$(140+32N+N^2)/81$	$(176+16N)/27$	$(76+2N)/9$	14/3
18	$(120+28N+4N^2)/81$	$(152+18N+N^2)/27$	$(70+3N)/9$	14/3
19	$(132+28N+N^2)/81$	$(168+12N)/27$	$(74+N)/9$	14/3

	5-loop				
i_4^{cub}	$S_{42;(5,1)}^{cub}(i_4^{cub})$	$S_{42;(4,2)}^{cub}(i_4^{cub})$	$S_{42;(3,3)}^{cub}(i_4^{cub})$	$S_{42;(2,4)}^{cub}(i_4^{cub})$	$S_{42;(1,5)}^{cub}(i_4^{cub})$
1	$(588+84N+3N^2)/243$	$(882+39N)/81$	$(488+4N)/27$	$132/9$	$18/3$
2	$(592+84N+2N^2)/243$	$(884+38N)/81$	$(488+4N)/27$	$132/9$	$18/3$
3	$(580+90N+2N^2)/243$	$(880+40N)/81$	$(488+4N)/27$	$132/9$	$18/3$
4	$(264+168N+18N^2)/243$	$(504+162N+3N^2)/81$	$(350+45N)/27$	$(112+4N)/9$	$17/3$
5	$(192+96N)/243$	$(336+72N)/81$	$(224+24N)/27$	$(78+3N)/9$	$14/3$
6	$(240+144N+12N^2)/243$	$(444+132N+3N^2)/81$	$(304+40N)/27$	$(100+4N)/9$	$16/3$
7	$(240+156N+24N^2+3N^3)/243$	$(456+162N+15N^2)/81$	$(322+57N+N^2)/27$	$(108+6N)/9$	$17/3$
8	$(288+192N+24N^2)/243$	$(552+192N+6N^2)/81$	$(384+56N)/27$	$(122+5N)/9$	$18/3$
9	$(336+192N+12N^2)/243$	$(624+168N)/81$	$(416+40N)/27$	$(126+3N)/9$	$18/3$
10	$(144+144N+36N^2)/243$	$(324+180N+9N^2)/81$	$(264+60N)/27$	$(96+6N)/9$	$16/3$
11	$(528+108N+6N^2)/243$	$(814+71N+N^2)/81$	$(466+15N)/27$	$(130+N)/9$	$18/3$
12	$(520+108N+8N^2)/243$	$(806+73N+2N^2)/81$	$(464+16N)/27$	$(130+N)/9$	$18/3$
13	$(484+76N+N^2)/243$	$(732+34N)/81$	$(416+4N)/27$	$118/9$	$17/3$
14	$(552+116N+4N^2)/243$	$(854+73N)/81$	$(484+14N)/27$	$(132+N)/9$	$18/3$
15	$(536+108N+4N^2)/243$	$(830+65N)/81$	$(472+12N)/27$	$(130+N)/9$	$18/3$
16	$(496+84N+2N^2)/243$	$(762+39N)/81$	$(432+4N)/27$	$120/9$	$17/3$
17	$(448+116N+20N^2+N^3)/243$	$(700+108N+11N^2)/81$	$(424+34N+N^2)$	$(126+3N)/27$	$18/9$
18/3	$(400+56N)/243$	$(588+26N)/81$	$(344+4N)/27$	$104/9$	$16/3$
19	$(504+120N+6N^2)/243$	$(792+84N)/81$	$(464+16N)/27$	$(130+N)/9$	$18/3$
20	$(512+120N+4N^2)/243$	$(796+82N)/81$	$(464+16N)/27$	$(130+N)/9$	$18/3$
21	$(444+88N+5N^2)/243$	$(698+51N)/81$	$(412+6N)/27$	$118/9$	$17/3$
22	$(456+96N+6N^2)/243$	$(704+66N+N^2)/81$	$(412+14N)/27$	$(118+N)/9$	$17/3$
23	$(504+128N+10N^2)/243$	$(788+102N+2N^2)/81$	$(462+25N)/27$	$(130+2N)/9$	$18/3$
24	$(496+120N+8N^2)/243$	$(772+92N+N^2)/81$	$(452+22N)/27$	$(128+2N)/9$	$18/3$
25	$(496+120N+8N^2)/243$	$(768+92N+2N^2)/81$	$(448+24N)/27$	$(128+2N)/9$	$18/3$
(26	$(448+88N+4N^2)/243$	$(684+56N+N^2)/81$	$(398+13N)/27$	$(116+N)/9$	$17/3$
27	$(488+116N+10N^2+N^3)/243$	$(756+92N+5N^2)/81$	$(444+26N)/27$	$(128+2N)/9$	$18/3$
28	$(416+148N+20N^2+N^3)/243$	$(680+138N+11N^2)/81$	$(422+43N+N^2)/27$	$(126+4N)/9$	$18/3$
29	$(352+88N+4N^2)/243$	$(556+60N+N^2)/81$	$(340+14N)/27$	$(104+N)/9$	$16/3$
30	$(400+140N+20N^2+N^3)/243$	$(656+130N+11N^2)/81$	$(410+41N+N^2)/27$	$(124+4N)/9$	$18/3$
31	$(352+116N+20N^2+N^3)/243$	$(584+106N+11N^2)/81$	$(370+33N+N^2)/27$	$(114+3N)/9$	$17/3$
32	$(448+144N+8N^2)/243$	$(732+112N+N^2)/81$	$(444+26N)/27$	$(128+2N)/9$	$18/3$
33	$(384+112N+8N^2)/243$	$(636+80N+N^2)/81$	$(392+16N)/27$	$(116+N)/9$	$17/3$
34	$(440+144N+10N^2)/243$	$(724+114N+2N^2)/81$	$(442+27N)/27$	$(128+2N)/9$	$18/3$
35	$(360+108N+14N^2+N^3)/243$	$(588+96N+5N^2)/81$	$(372+26N)/27$	$(114+2N)/9$	$17/3$
36	$(360+112N+14N^2)/243$	$(592+98N+3N^2)/81$	$(374+25N)/27$	$(114+2N)/9$	$17/3$
37	$(416+144N+16N^2)/243$	$(680+132N+4N^2)/81$	$(424+36N)/27$	$(126+3N)/9$	$18/3$
38	$(392+112N+6N^2)/243$	$(616+90N+N^2)/81$	$(378+23N)/27$	$(114+2N)/9$	$17/3$
39	$(408+120N+6N^2)/243$	$(656+90N+N^2)/81$	$(394+23N)/27$	$(116+2N)/9$	$17/3$
40	$(392+116N+10N^2+N^3)/243$	$(636+92N+5N^2)/81$	$(388+26N)/27$	$(116+2N)/9$	$17/3$
41	$(448+144N+8N^2)/243$	$(720+116N+2N^2)/81$	$(432+32N)/27$	$(126+3N)/9$	$18/3$
42	$(400+120N+8N^2)/243$	$(652+92N+N^2)/81$	$(396+22N)/27$	$(116+2N)/9$	$17/3$
43	$(464+160N+12N^2)/243$	$(756+136N+3N^2)/81$	$(456+36N)/27$	$(130+3N)/9$	$18/3$
44	$(416+128N+8N^2)/243$	$(676+100N+N^2)/81$	$(408+24N)/27$	$(118+2N)/9$	$17/3$
45	$(384+132N+20N^2+N^3)/243$	$(624+126N+11N^2)/81$	$(386+41N+N^2)/27$	$(116+4N)/9$	$17/3$
46	$(368+96N+4N^2)/243$	$(572+72N+N^2)/81$	$(344+20N)/27$	$(104+2N)/9$	$16/3$
47	$(288+96N+32N^2+4N^3)/243$	$(456+120N+24N^2+N^3)/81$	$(304+48N+4N^2)/27$	$(100+6N)/9$	$16/3$
48	$(288+72N+12N^2)/243$	$(432+72N+6N^2)/81$	$(276+24N+N^2)/27$	$(90+3N)/9$	$15/3$
49	$(304+104N+32N^2+4N^3)/243$	$(496+120N+24N^2+N^3)/81$	$(328+48N+4N^2)/27$	$(108+6N)/9$	$17/3$
50	$(304+56N)/243$	$(448+36N)/81$	$(276+10N)/27$	$(90+N)/9$	$15/3$

	5-loop				
i_4^{cub}	$S_{4_2;(5,1)}^{cub}(i_4^{cub})$	$S_{4_2;(4,2)}^{cub}(i_4^{cub})$	$S_{4_2;(3,3)}^{cub}(i_4^{cub})$	$S_{4_2;(2,4)}^{cub}(i_4^{cub})$	$S_{4_2;(1,5)}^{cub}(i_4^{cub})$
51	$(352+128N+32N^2+4N^3)/243$	$(568+144N+24N^2+N^3)/81$	$(368+56N+4N^2)/27$	$(118+7N)/9$	$18/3$
52	$(368+112N+12N^2)/243$	$(584+96N+6N^2)/81$	$(356+32N+N^2)/27$	$(110+4N)/9$	$17/3$
53	$(352+80N)/243$	$(536+52N)/81$	$(320+16N)/27$	$(100+2N)/9$	$16/3$
54	$(336+116N+24N^2+N^3)/243$	$(552+122N+11N^2)/81$	$(358+39N+N^2)/27$	$(112+4N)/9$	$17/3$
55	$(336+96N+12N^2)/243$	$(532+88N+3N^2)/81$	$(336+24N)/27$	$(104+2N)/9$	$16/3$
56	$(368+124N+20N^2+N^3)/243$	$(584+126N+11N^2)/81$	$(370+41N+N^2)/27$	$(114+4N)/9$	$17/3$
57	$(320+88N+12N^2)/243$	$(504+76N+6N^2)/81$	$(316+24N+N^2)/27$	$(100+3N)/9$	$16/3$
(58	$(384+144N+24N^2)/243$	$(624+144N+12N^2)/81$	$(392+48N+2N^2)/27$	$(120+6N)/9$	$18/3$
59	$(224+156N+24N^2+N^3)/243$	$(464+146N+11N^2)/81$	$(334+43N+N^2)/27$	$(110+4N)/9$	$17/3$
60	$(176+128N+20N^2)/243$	$(396+112N+5N^2)/81$	$(296+28N)/27$	$(100+2N)/9$	$16/3$
61	$(192+120N+12N^2)/243$	$(384+96N+6N^2)/81$	$(268+28N+N^2)/27$	$(90+3N)/9$	$15/3$
62	$(160+128N+32N^2+4N^3)/243$	$(360+128N+24N^2+N^3)/81$	$(272+48N+4N^2)/27$	$(96+6N)/9$	$16/3$
63	$(384+120N+12N^2)/243$	$(620+108N+N^2)/81$	$(388+26N)/27$	$(116+2N)/9$	$17/3$
64	$(432+144N+12N^2)/243$	$(684+132N+3N^2)/81$	$(424+36N)/27$	$(126+3N)/9$	$18/3$
65	$(320+100N+24N^2+3N^3)/243$	$(512+114N+15N^2)/81$	$(338+41N+N^2)/27$	$(110+4N)/9$	$17/3$
66	$(336+80N+4N^2)/243$	$(500+68N+N^2)/81$	$(312+20N)/27$	$(100+2N)/9$	$16/3$
67	$192/243$	$240/81$	$160/27$	$60/9$	$12/3$
68	$(256+80N+40N^2+10N^3+N^4)/243$	$(400+120N+40N^2+5N^3)/81$	$(280+60N+10N^2)/27$	$(100+10N)/9$	$17/3$
69	$(256+32N)/243$	$(352+24N)/81$	$(224+8N)/27$	$(78+N)/9$	$14/3$

Index